Volume 6

Ceramic Transactions

MULLITE AND MULLITE MATRIX COMPOSITES

Edited by Shigeyuki Sōmiya, Nishi Tokyo University
Robert F. Davis, North Carolina State University
Joseph A. Pask, University of California, Berkeley

The American Ceramic Society, Inc.
Westerville, Ohio

Proceedings of the International Conference on Mullite held In Tokyo, Japan, November 9–10, 1987.

ISBN 0-944904-05-x

Printed in the United States of America

1 2 3 4 5—94 93 92 91 90

Foreword

Mullite is now considered to be a strong candidate material for structural applications at high temperatures because of its intrinsic strength, low thermal expansion, and good chemical stability at such temperatures. Mullite is the only compound in the Al_2O_3–SiO_2 system under standard conditions. Most traditional ceramic products have mullite as part of their final phase composition, since they usually contain some clay and silicon as starting materials. The microstructures of the materials also contain a relatively low melting glass. Thus, the significance of the inherently favorable properties of mullite was not recognized until about fifteen years ago when the mechanical behavior in compression at high temperature (1500 °C) of mullite specimens free or nearly free of a glassy phase was first studied.

However, one characteristic in which mullite materials presently are considered to be deficient is fracture toughness. It was and is evident that microstructure and composite development studies were necessary. As a result, during this fifteen-year period much research in many countries, most extensively in Japan, has been conducted on the processing of mullite. This work has also been closely coupled with the determination of corresponding physical and chemical properties. It was further evident that, although points of agreement regarding the physical and chemical characteristics of mullite and the Al_2O_3–SiO_2 existed, points of disagreement definitely were present.

Thus, it was appropriate to bring the active participants together to discuss these problems. As a consequence, the First International Conference on Mullite was organized and held in Tokyo, Japan, on November 9–10, 1987. Participants came from many parts of the world. Most of the presentations are included in this volume. The presentations and discussions dealt with the following topics:

- Phase equilibria and immiscibility in the Al_2O_3–SiO_2 system
- Crystal growth, crystalline and amorphous structures and impurity distribution
- Reactions and mechanisms of mullitization
- Starting powder preparation and characterization
- Forming and sintering of mullite and composites containing mullite
- Mullite matrix composites
- Mechanical properties and behavior
- Applications and process characterization

The Conference was well attended and the discussions were penetrating and informative. The number of studies reported was greater than expected. It was worthwhile to bring these studies together. It is expected that this volume will become the basis for continuing studies which, in turn, will be the basis for the Second Conference on Mullite.

On behalf of the participants and the mullite community, we gratefully acknowledge the indispensable financial support provided by the Nikkan Kogyo Shimbun. We also express our appreciation for the sponsorship of this conference by the Japan Society for the Promotion of Science, Japan Fine Ceramics Association, The Ceramic Society of Japan, The Chemical Society of Japan, The Crystallographic Society of Japan, The Material Science Society of Japan, The Mineralogical Society of Japan, The Society of Materials Science and The Japan Technical Association of Refractories.

We also acknowledge the considerable assistance in the planning and conduct of this conference by the International Advisory, Domestic Advisory, Organizing, Executive, General Affairs and Program Committees. We extend our thanks to the speakers and authors. This volume would not have materialized without their participation. Finally, it is appropriate to acknowledge with real affection the patience, tolerance, and tangible and moral support we have been accorded by our colleagues, our families and our friends through those extended periods of time we have had to commit to planning, organizing and editing.

Shigeyuki Sōmiya
Robert F. Davis
Joseph A. Pask

Contents

Section I. Phase Equilibria and Immisibility

Section II. Crystal Growth, Crystalline and Amorphous Structures and Impurity Distribution

Section III. Powder Preparation, Densification Behavior and Mechanical Properties

Section IV. Additional Methods of Mullite Preparation

Section V. Mullite Matrix Composites

Section VI. Applications and Process Characterization

__Ceramic Transactions__ is a new proceedings series designed to meet two needs: high quality content and rapid publication. Volumes in the series come from meetings, symposia, and forums. Each paper is reviewed by two peers, and final manuscripts are prepared by authors in a ''camera-ready'' format. The volumes in this series would not be possible without the hard work, dedication, and cooperation of editors, reviewers, and authors, who all deserve a great deal of thanks.

This volume in the __Ceramic Transactions__ series was generated electronically from disks prepared by authors. Text was converted to files readable by Word Perfect, Version 5.0, and was output in a consistent format on a Hewlett-Packard Laser Jet Series II printer.

Your comments, questions, and suggestions for future __Ceramic Transactions__ volumes are welcomed and should be addressed to the Director of Publications, The American Ceramic Society, Inc.

Critical Review of Phase Equilibria in the Al_2O_3-SiO_2 System

J. A. Pask

Materials and Chemical Sciences Division
Lawrence Berkeley Laboratory and Department of Materials Science
and Mineral Engineering
University of California
Berkeley, California 94720

Abstract

Many phase equilibria studies have been made of the Al_2O_3-SiO_2 binary system. Two controversial areas have emerged: nature of the melting and the nature of the solid solution range of mullite. Mullite has been reported to melt incongruently and also congruently. It is shown that this difference is associated with the presence or absence of αAl_2O_3 nuclei which assist in determining the position of the αAl_2O_3 liquidus. The stable solid solution range is obtained when processing conditions result in homogenization of the starting materials on an atomic or molecular level; for example a 3:2 mullite composition forms a single phase mullite microstructure. When the mixture is prepared under conditions resulting in non-homogeneity, a two-phase microstructure consisting of mullite with a higher Al_2O_3 content and a liquid phase is formed.

Introduction

Because of potentially favorable properties, mullite (nominally $3Al_2O_3 \cdot 2SiO_2$) is a candidate as an advanced ceramic material. It occurs extensively in ceramic products made from alumino-silicate materials, e.g. clay. Such products are practically always associated with a glassy phase due to presence of impurity oxides and an excess of silica. Thus, materials have not been made extensively from mullite as the only phase. Furthermore, when a single phase composition is successfully obtained, the achievement of theoretical density is difficult. Consequently, properties of mullite materials with controlled microstructure have not been fully investigated and studied. Hence, the current interest in a potentially attractive material that has not been completely explored.

It is the only compound in the αAl_2O_3-SiO_2 system at one atmosphere pressure. This system is one of the most important binary systems in understanding the phase compositions constituting microstructure development. Determination of phase equilibrium diagrams is normally accomplished experimentally and based on identification of the phase composition at a constant temperature for a given chemical composition with the tacit assumption that the approach being used leads to stable phase equilibria.

The experimental procedure generally followed is referred to as the static method of quenching. As an example, a given chemical composition is heat-treated above the liquidus temperature to achieve homogenization, lowered to the test temperature of interest, annealed to obtain equilibrium phases, and quenched. Another experimental approach commonly used is to carry on a solid state reaction of a selected mixture of powders at a given temperature to reach a steady state phase composition, and quenching. In both approaches, the resulting specimens are characterized to determine the phase composition and microstructure and utilized for the developing phase diagram.

Although the Al_2O_3-SiO_2 system appears to be straightforward, it has actually become controversial. Results have been reported alternately indicating that mullite melts incongruently with other publications indicating congruent melting. Variations in the mullite solid solution range have also been reported. It would be illogical to consider that only one set of experimental data is correct. It is more logical to consider that both sets of data are correct experimentally but that only one set must be stable thermodynamically and that the others are metastable. Thus, the objective of this review paper is to provide a status report, or an overview, of the current understanding of this system and critically discuss the principles associated with the controversies in terms of identifying the stable thermodynamic equilibria.

Historical Summary

A number of reviews on the Al_2O_3-SiO_2 system have been published.[1–4] An overview of a selected number of critical studies of the Al_2O_3-SiO_2 system will be presented here chronologically. The objective of the abbreviated selections is to provide a background for the theme of our discussion and to indicate the trends with the years.

1. In 1865, Sainte-Claire Deville and Caron[5] identified the existence of a compound with a composition of ≈ 70.5 wt.% Al_2O_3.
2. In 1865 Vernadsky[6] identified a similar compound with ≈ 70.3 wt.% Al_2O_3 and proposed the formula $11Al_2O_3 \cdot 8SiO_2$.
3. In 1909, Shepherd, Rankin and Wright[7] published the first phase equilibrium diagram in this system. They believed that sillimanite ($Al_2O_3 \cdot SiO_2$) was the binary compound and that it melted congruently.
4. In 1924, Bowen and Greig[8] published the first phase equilibrium diagram showing mullite ($3Al_2O_3 \cdot SiO_2$) as the only compound. It was shown to melt incongruently and with no solid solution range, Fig. 1. (Sillimanite is a high pressure mineral. Mullite was named after a similar mineral occurring naturally on the island of Mull in Scotland).
5. In 1950 and 1951, Bauer et al.[9] grew a single crystal of mullite with 83 wt.% Al_2O_3 by the flame fusion process. The high Al_2O_3 content was presumably due to loss of SiO_2 by volatilization because of the manner of achieving the high fusion temperature. Doubts as to the incongruent melting of mullite were raised by this experiment.
6. In 1951, Toropov and Galakhov[10] published a diagram showing mullite with a congruent melting point.
7. In 1954, Shears and Archibald[11] published a diagram showing mullite with a solid solution range from 3:2 to 2:1 of Al_2O_3:SiO_2 below 1600°C and a decreasing range richer in Al_2O_3 up to an indicated incongruent melting point of ≈ 1810°C.
8. In 1960, Welch[12] supported the proposed solid solution range and reported observations of incongruent melting with a high temperature microscope.

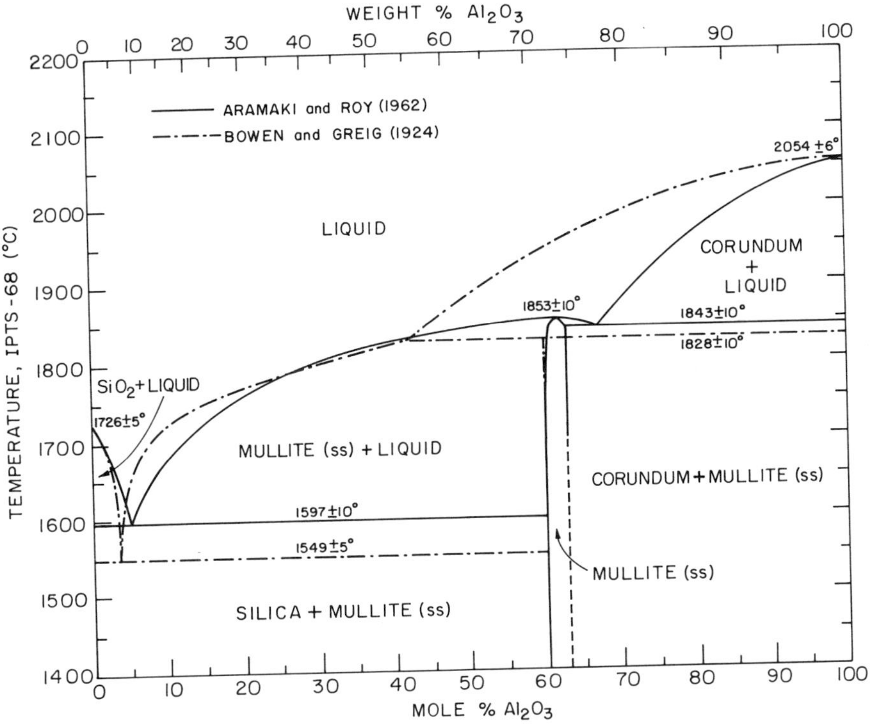

Fig. 1. SiO_2-Al_2O_3 phase equilibrium diagrams as determined by Bowen and Greig[8] and by Aramaki and Roy.[14]

9. In 1958 and again in 1964, Toropov and Galakhov[13] again showed mullite with a congruent melting point and a solid solution range.
10. In 1962, Aramaki and Roy[14] showed mullite with a congruent melting point and a solid solution range of 71.8 to 74.3 wt.% Al_2O_3 which was extended to 77.3 wt.% due to disordering under metastable conditions, Fig. 1.
11. In 1972, Davis and Pask[15] in studies using semi-infinite diffusion couples of sapphire and fused SiO_2 up to 1800°C determined the solid solution range of mullite to be 71.0 to 74.0 wt.% Al_2O_3. The range was determined by microprobe analysis of mullite layers grown at the interfaces of the couples.
12. In 1974, Guse, and Guse and Mateika[16] showed that Czochralski grown mullite single crystals were of the 2:1 type with a solid solution range of 75 to 77.3 wt.% Al_2O_3. A 3:2 mullite (71.8 wt.% Al_2O_3) was not observed.
13. In 1975, Aksay and Pask[17] reported an extension of the diffusion couple studies above 1800°C which showed that the αAl_2O_3 liquidus profile essentially matched that of Bowen and Greig[8] in support of mullite having an incongruent melting point, Fig. 2. The peritectic reaction of αAl_2O_3 plus liquid to form mullite on cooling was also observed.

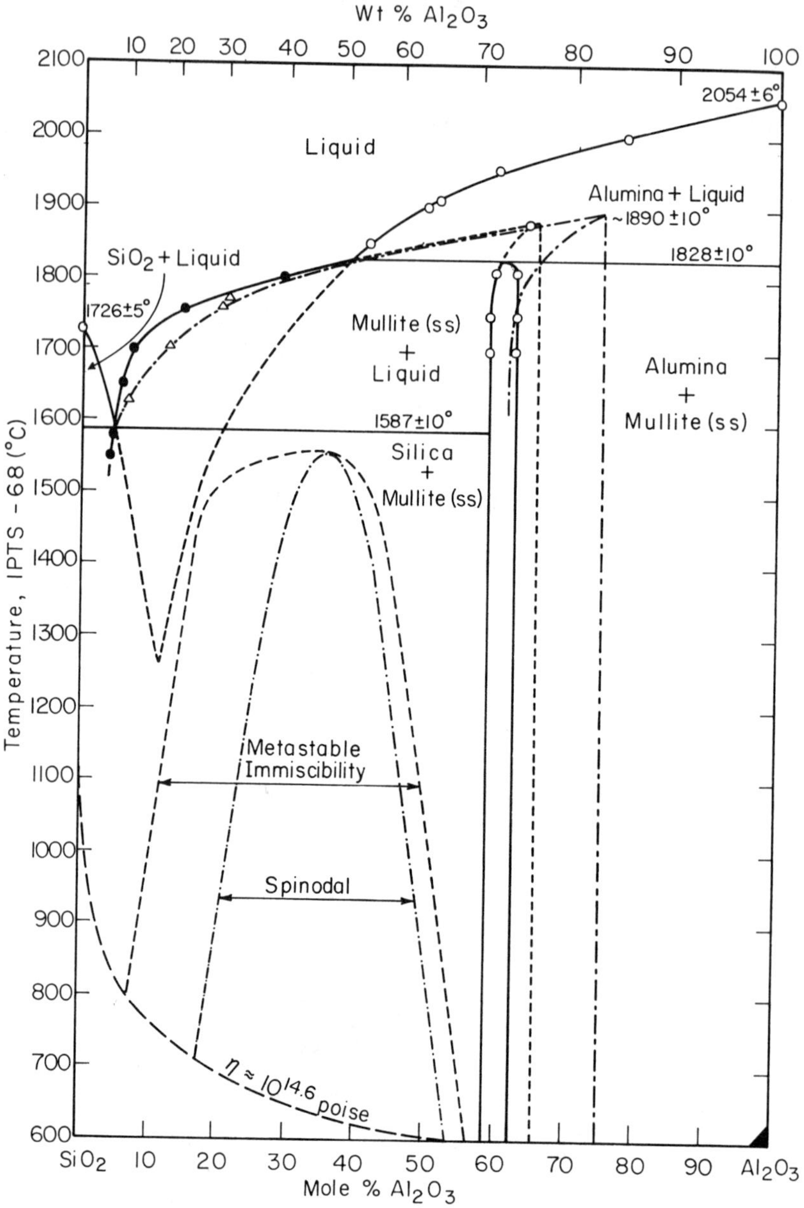

Fig. 2. Stable and metastable equilibrium phase diagrams within the SiO_2-Al_2O_3 system.[15,17,22,23]

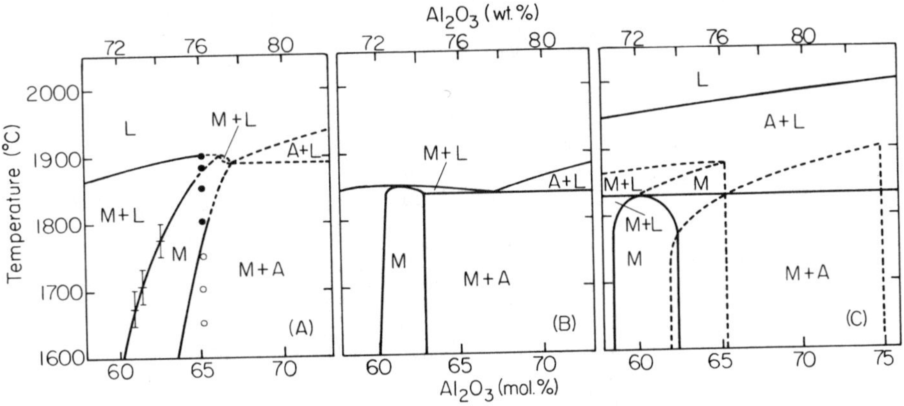

Fig. 3. Solid solution range of mullite: (A) Ref. [20], (B) Ref. [14], and (C) Ref. [17].

14. In 1980, Shindo[18] completed a DSc thesis in which he grew single crystals of mullite by the Slow Cooling Float Zone Method. The mullite in all cases was of the 2:1 type and was determined to melt incongruently.
15. In 1983, Prochazka and Klug[19] proposed a congruent melting point and a solid solution range that shifted to higher Al_2O_3 contents in its entirety with increase in temperature above 1600°C, Fig. 3.
16. In 1986, Klug, Prochazka and Doremus[20] reported on a modification of the position of the αAl_2O_3 liquidus indicating that mullite had an incongruent melting point.
17. In 1987, Pask, Zhang, Tomsia and Yoldas[21] prepared specimens with 72 wt.% Al_2O_3 from powders prepared by sol-gel colloidal and sol-gel polymeric techniques and fired at 1700°C in air. The microstructure of the former specimens showed the presence of a glassy phase with the mullite; the latter specimens showed no glassy phase, Fig. 4. This result indicates that the atomic homogeneity as well as disordering within the mullite grains has to be considered.

In analyzing these reports critically, it is evident, as previously stated, that the major disagreements can be combined into two categories. The first category deals with the question of whether mullite melts congruently or incongruently. Associated factors are nucleation and growth phenomena of αAl_2O_3 and the position of the αAl_2O_3 liquidus. The second category deals with the range and nature of the mullite solid solution. Associated factors are the degree of disorder and homogeneity of the mullite atomic structure and their effect on the solid solution behavior. These areas will be discussed with an objective of identifying the thermodynamically stable and metastable phase relationships.

Nucleation and Growth Phenomena

When mullite melts congruently at some temperature, it forms a complete liquid. The same overall composition then remains in the liquid phase field with increasing temperature. On the other hand, when mullite melts incongruently, it should dissociate to form an αAl_2O_3 crystalline phase and a liquid that is saturated with

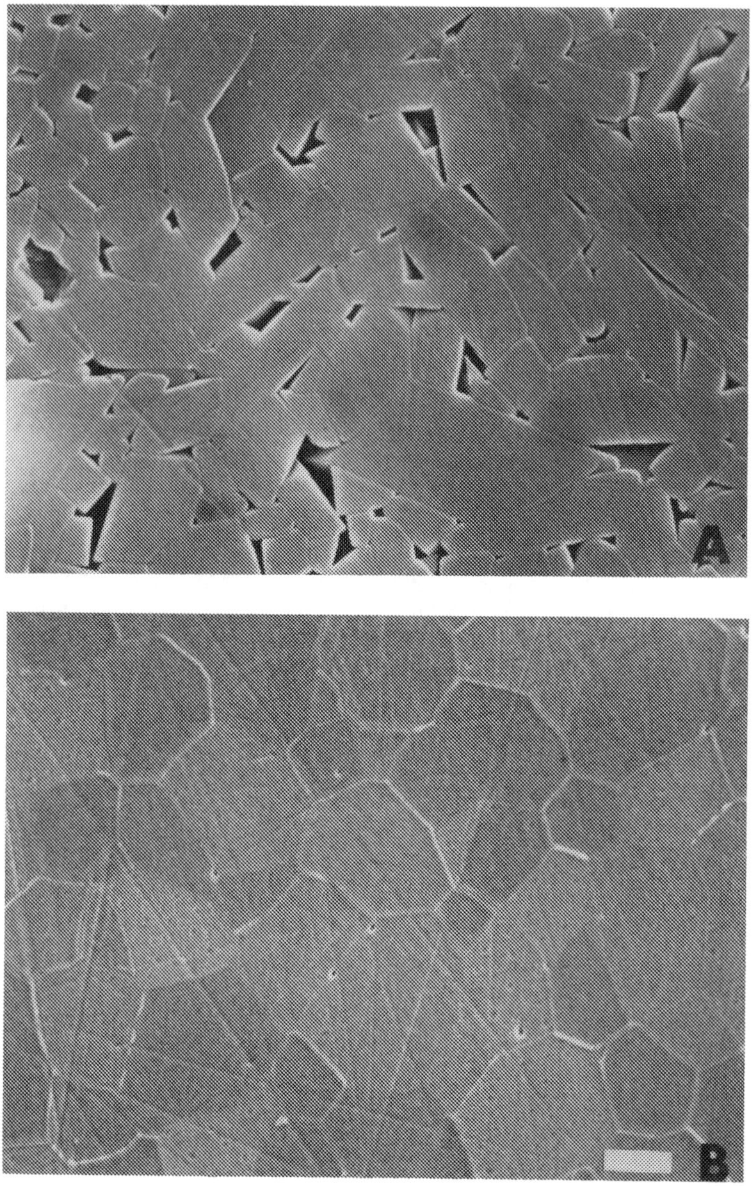

Fig. 4. SEM micrographs[21] of (A) a specimen prepared by sol-gel colloidal method, and (B) a specimen prepared by sol-gel polymeric method. Both were sintered at 1700°C for 6 hrs in air. (Bar = 2 μmm).

Al_2O_3 relative to αAl_2O_3 as the solute. With increasing temperature the αAl_2O_3 is continuously dissolved since the solubility limit of the liquid for Al_2O_3 molecules is increased and not reached, and the driving force for solution of αAl_2O_3 by the liquid solvent is increased. The temperature at which the αAl_2O_3 is completely dissolved determines the point for the mullite composition on the αAl_2O_3 liquidus. Thus, in this case the αAl_2O_3 liquidus outlining the primary phase field of αAl_2O_3 lies above the melting point of mullite and extends to lower Al_2O_3 contents than that of mullite (Figs. 1-3). Furthermore, identification of such a field is tantamount to the fact that mullite melts incongruently. It should be further emphasized that in the case of congruent melting, i.e. melting without the crystallization of αAl_2O_3, the αAl_2O_3 primary phase field and the αAl_2O_3 liquidus are at higher Al_2O_3 contents (Figs. 1 and 3). This limit should now correspond to the saturation of the liquid structure with Al_2O_3 molecules. The determination of the location of the αAl_2O_3 liquidus is thus crucial to the point of this discussion.

Let us now consider the experimental evidence supporting the αAl_2O_3 liquidus position indicated by Bowen and Greig.[8] They prepared their mullite specimens by crushing and grinding mixtures of corundum (αAl_2O_3) and quartz, heating to a high but below the melting temperature, and repeating the crushing and grinding cycle five times to obtain maximum crystallization of mullite. Considering the high activation energy of chemical interdiffusivity of Si^{4+} and Al^{3+},[22] it can be postulated that the reaction was not complete and that unreacted αAl_2O_3 particles were still present as nuclei. However, these particles would have gone into solution at temperatures above the melting point of mullite if the specimen was in a liquid phase field, but the αAl_2O_3 crystals did not go into solution. Aksay and Pask[23] conducted a similar experiment by exposing a mixture of powders equivalent to 3:2 mullite composition in a sealed Mo crucible at the temperature of the αAl_2O_3 liquidus (1953°C) for eight hours. A peritectic reaction was observed with slow cooling to and soaking at 1753°C for 29.4 days which would not have happened if αAl_2O_3 nuclei were not present after homogenizing at 1953°C. Even more significant results[15,17] were based on semi-infinite diffusion couples of sapphire and fused SiO_2 because the solution reaction was with the surface of a large αAl_2O_3 crystal which acted as a nucleation surface. The amount of αAl_2O_3 dissolved (i.e., the composition of the liquid in contact with the crystal) at temperatures above the melting point of mullite determined the αAl_2O_3 liquidus which had a lower Al_2O_3 content than mullite up to about 1955°C (Fig. 2). Energy-wise these events are based on an equilibrium between the bonding forces of αAl_2O_3 which have to be broken in order to release an Al_2O_3 molecule, and the driving force of the liquid to incorporate additional amounts of Al_2O_3 into its structure thereby lowering its internal free energy.

From the viewpoint of melting determinations of single crystal or polycrystalline mullite similar relationships were obtained. On heating single phase mullite it did not dissociate at the peritectic temperature of ≈ 1830°C but melted congruently at ≈ 1890°C. On the other hand, if the polycrystalline mullite was not completely reacted and some αAl_2O_3 remained, the αAl_2O_3 nuclei did not melt above the melting point and became the basis for the observed peritectic melting reaction of mullite.

Further significant information is that the single crystal grown by the Czochralski method had about 77 wt.% Al_2O_3 which corresponds to 2:1 mullite.[16] This composition even formed from a series of melts containing approximately 70 to 80 wt.% Al_2O_3, including 72 wt.% corresponding to 3:2 mullite. In all of these cases, the liquid was homogenized above the liquidus temperature and thus free of αAl_2O_3 nuclei. The question at this point arises as to whether a continuous series exists between 3:2 and 2:1 mullites, or whether two types of mullite [3:2 (grown by solid state reaction) and 2:1 (grown from the melt)] actually exist.

Another significant study that contributes to an understanding of this question is one made by Greskovich et al.[24] who were developing a high temperature Al_2O_3-SiO_2 material for use as a shell mold for direct solidification of eutectic superalloys. They were working with a composition of 88 wt.% Al_2O_3 and 12 wt.% SiO_2. Shell molds prepared with ≈3.5 μmm Al_2O_3 were completely reacted to form αAl_2O_3 and mullite solid solution between 72.0 and 74.5 wt.% Al_2O_3 after firing at 1725°C for 1 h in Ar. This value corresponded to that indicated by the phase equilibrium diagram of Fig. 2. A reaction occurred after firing at 1850°C, however, and the mullite composition was found to range between 77 and 77.9 wt.% Al_2O_3 which was retained on cooling. This reaction to form the 2:1 mullite is significant because it occurred in the absence of a liquid (i.e., it was not grown from a melt) above the peritectic temperature of ≈1830°C. The authors then observed that at ≈ 1890°C this phase underwent incongruent melting as indicated by the presence of αAl_2O_3. This peritectic temperature is expected to be thermodynamically metastable because no mullite formed at the interface of fused SiO_2-sapphire diffusion couples at 1850°C. This observation, however, should be investigated further.

Still another significant related series of experiments should be mentioned.[25] The series consisted of binary mixtures with 50 to 85 wt.% Al_2O_3. After homogenizing above the liquidus temperatures so that no αAl_2O_3 nuclei would be present, the mixtures were slowly cooled. No αAl_2O_3 crystallized from liquids containing up to ≈ 83 wt.% Al_2O_3 but at lower temperatures mullite with a higher Al_2O_3 content crystallized. At about 83 wt.% Al_2O_3 and higher amounts, αAl_2O_3 crystallized spontaneously above the mullite melting temperature. These points would be expected to approximate the αAl_2O_3 liquidus shown in the diagram of Aramaki and Roy[14] (Fig. 1).

It can thus be postulated that at a constant temperature (e.g. 1850°C) with increasing amounts of αAl_2O_3 the liquid silicate reaches a limit of about 55 wt.% Al_2O_3 in the sense that it can not obtain more Al_2O_3 molecules from αAl_2O_3. That is, a balance of the energy released by the solution on gaining a molecule of Al_2O_3 and the energy required to release a molecule of Al_2O_3 from αAl_2O_3 crystal is attained. The structure of the liquid, however, does not become saturated with Al_2O_3 molecules until about 83 wt.% is present. That is, in the range of about 55 to 83 wt.% Al_2O_3 and above the melting temperature in the absence of αAl_2O_3 nuclei, the liquid is supersaturated relative to αAl_2O_3, but the liquid's structure can still absorb additional Al_2O_3 molecules. Consequently, the experimental difficulties and disagreements are the result of the need of αAl_2O_3 nuclei to obtain stable phase equilibria in mixtures up to 83 wt.% Al_2O_3. Thus, whenever such mixtures were heated so that complete homogeneity of the liquid phase with an absence of αAl_2O_3 nuclei was obtained, αAl_2O_3 did not appear as a primary phase on cooling, as noted above, before the precipitation of mullite became energetically favorable.

In summary, it is evident that experimentally mullite melts congruently since αAl_2O_3 nuclei do not form. But with αAl_2O_3 crystals present that remain when mullite melts, melting is inherently incongruent. Also, taking into account the results of the semi-infinite diffusion couples, it should be emphasized that the binary phase equilibrium diagram for the SiO_2-Al_2O_3 system showing mullite melting incongruently must be the thermodynamically stable diagram. This concept, discussed in detail, can be supported thermodynamically by postulating that a liquid phase separation must precede crystallization and could be shown on the basis of free energy curves as a function of composition at a number of temperatures.[26]

Order-Disorder in Mullite

Another area of disagreement among many research reports is the width, shape and position of the solid solution range for mullite primarily at the higher temperatures as surmised from the published phase equilibrium diagrams shown in Fig. 3 and others. These variations presumably are the result of the fact that mullite shows strong disordering with a variable Al/Si ratio which result in these variations. Studies devoted to determining the ranges of compositions and their causes are limited. This discussion, therefore, is primarily based on logical and deductive reasoning which, in turn, is based on available experimental evidence. It is thus subject to verification or modification by continued crystallographic studies.

Available high resolution transmission electron microscope photos indicate an atomic structure with a regular but random distribution of cations. The XRD patterns of mullite with compositions within the solid solution range indicate the same crystal structure with changes in the lattice parameters. This variability indicates that Al and Si ions can be interchanged at least at some lattice sites with an appropriate adjustment of the oxygen anions in the structure to achieve a charge balance. There are no recognized phase transformations. The variability in the mullite composition can then crystallographically be attributed to disordering. It should also be visualized that any irregularity in the distribution can also lead to non-homogeneity.

The most convincing evidence that variations in the structure and composition of mullite grains exist is provided by two specimens with the same overall chemical composition (71.3 wt.% Al_2O_3) made from powders prepared by two sol-gel processing methods and fired together.[21] One powder was labeled sol-gel colloidal and the other, sol-gel polymeric. The former formed a two-phase and the latter, a single phase microstructure (Fig. 4). As can be seen the two-phase microstructure consists of prismatic grains and a liquid. In order to maintain a mass balance in a two phase microstructure, the mullite grains must be higher in Al_2O_3 content than the starting mixture with the excess SiO_2 being the base for the liquid phase. Verification of the composition of the grains was achieved with a microprobe analysis. In the sol-gel colloidal procedure, it is postulated that kernels of Al_2O_3 several hundred nanometers in size are present which are not homogeneously dispersed with SiO_2 on a molecular level. It appears that such a kernel or clump of Al_2O_3 can be assimilated into the mullite structure and correspond to a region where the cation sites are predominantly Al. The energy barrier that has to be surmounted to separate the atoms on a regular basis can be too high. The result is a disordered mullite grain with a higher Al_2O_3 content. Furthermore, the question also arises as to whether there might be a composition gradient within the grain since it would be expected that the surface composition would be in equilibrium with the liquid phase. Another possibility is that the liquid would adjust its composition to be in equilibrium with the higher Al_2O_3-containing mullite grains. The sol-gel polymeric procedure resulted in a homogeneous atomic or molecular mixture probably forming disordered mullite grains that are equivalent in composition to the starting mixture. In both cases, the basic mullite structure is the same.

The usual procedure of using mixtures of Al_2O_3 and SiO_2 starting powders produces similar results to the sol-gel[27] colloidal procedure. With 3:2 compositions two-phase microstructures are formed. In this case, however, firing has to be done at higher temperatures and longer times because the particles of the starting powders are larger than the colloidal particles and diffusion distances are greater.

The semi-infinite diffusion couples of sapphire and fused SiO_2, previously described, conveniently indicate the solid solution range of mullite whose maximum Al_2O_3 content is controlled by equilibrium with αAl_2O_3 and minimum Al_2O_3 content by equilibrium with the liquid phase. With this technique no clumps or kernels of Al_2O_3 are possible. In the sol-gel colloidal procedure the kernels of Al_2O_3 are readily assimilated into the mullite structure since they are less stable than crystalline αAl_2O_3. Complete dispersion, however, is not attained resulting in non-homogeneous disordered mullite grains. In the case of powder mixtures, mullite forms by chemical interdiffusion while the αAl_2O_3 particles are completely surrounded by SiO_2. It can be postulated that when the αAl_2O_3 particles become small enough, they begin to lose their crystalline stability. At this point, a similar situation to that of the sol-gel colloidal mixtures exists.

Single crystals of mullite grown by Czochralski technique at a constant temperature with no αAl_2O_3 particles present have a 2:1 composition ($2Al_2O_3 \cdot SiO_2$). Metastable superheated polycrystalline 3:2 mullite reacts with excess αAl_2O_3 which also is more reactive because of the small particles, to form 2:1 mullite above the stable peritectic temperature of 1829°C.[24] This data suggest that 2:1 mullite has the lowest free energy.

Lastly, mullite with Al_2O_3 contents extending beyond 77 wt.% Al_2O_3 (2:1) form by crystallizing from quenched melts with Al_2O_3 contents up to 83 wt.%. The Al_2O_3 contents of these mullites were larger than those of the melts except that the difference decreased with increasing Al_2O_3 content of the melt and dropped to zero at $\approx$ 83 wt.% Al_2O_3.[28] The question arises as to whether this series can be considered to be a different type of disordered mullite with some introduced ordering. It is of interest to note that 83 wt.% Al_2O_3:17 wt.% SiO_2 approximates a composition of $3Al_2O_3 \cdot SiO_2$.

Significant contributions towards unraveling some of the indicated crystallographic questions have been made by H. Saalfeld and co-workers,[29] Nakajima and Ribbe,[30] Kriven and Pask,[28] and Risbud and Pask.[27] Nevertheless a considerable amount of work and study is still necessary in this area.

Summary and Conclusions

The general conclusion is that practically all phase equilibrium diagrams that have been published are probably experimentally correct and can be justified on the basis of relative thermodynamic stability. The exercise becomes one of identifying stable and metastable features.

The nature of the starting materials and the processing conditions are critical factors since they affect and control the appearance of αAl_2O_3 nuclei. The importance of this occurrence arises from the fact that as the liquid dissolves αAl_2O_3 it reaches an internal energy value that balances the bond energies of αAl_2O_3 and further solution does not occur. This occurs at about 55 wt.% αAl_2O_3 at about 1830°C. This balance is attained before the liquid structure is saturated with Al_2O_3 molecules which occurs at about 83 wt.% Al_2O_3 at 1830°C. It thus means that the αAl_2O_3 liquidus at the lower Al_2O_3 contents is stable thermodynamically in the presence of αAl_2O_3 nuclei and crystals, and the Al_2O_3 liquidus at the higher Al_2O_3 contents is metastable thermodynamically. It also means that it is difficult to spontaneously nucleate αAl_2O_3 in this inbetween range from liquids cooled from high temperatures. On the other hand, αAl_2O_3 crystallizes readily from liquids with Al_2O_3 contents higher than 83 wt.%.

Consequently, the binary system showing incongruent melting should then be referred to as the SiO_2-αAl_2O_3 stable phase equilibrium system in the presence of αAl_2O_3. And, the binary system showing congruent melting should be referred to as the SiO_2-Al_2O_3 metastable phase equilibrium system in the absence of αAl_2O_3 nuclei. Therefore, the processing conditions are extremely critical in terms of their effect on either retaining or eliminating αAl_2O_3 nuclei. For example, homogenizing liquids with 55 to 83 wt.% Al_2O_3 at temperatures above the liquidus dissolves all αAl_2O_3 nuclei which do not reform on cooling through the αAl_2O_3 primary stable phase field.

It now appears that there may be still another metastable phase equilibrium diagram showing a 2:1 type mullite with an incongruent melting point at ≈ 1890°C.[20] This arises from the fact that polycrystalline 3:2 mullite in a metastable state, i.e. above the peritectic temperature of ≈ 1830°C, reacts with excess αAl_2O_3 to form the 2:1 composition mullite. This product apparently melts at ≈ 1890°C. Its melting is peritectic in nature because the αAl_2O_3 that is present does not melt indicating the presence of an αAl_2O_3 primary phase field above the mullite composition range. This possibility requires further study.

The nature of the starting materials and the processing conditions are critical since they affect the structure and composition of the mullite grains, i.e. the structural disorder and the molecular homogeneity characteristics. Such structures can lead to the growth of mullite with higher Al_2O_3 contents resulting in the formation of some liquid phase with the mullite in the case of an overall 3:2 composition. It appears evident that possibly the only way to obtain mullite homogeneous and essentially free from a nonuniform disordered structure is to follow processing techniques that result in atomic or molecular mixing of the composition prior to high temperature heating for crystallization, e.g. powders formed by the sol-gel polymeric technique.

In this report an effort has been made to analyze published phase equilibria and interpret them on the basis of processing variations and thermodynamic stability. Studies on metastable liquid immiscibilities have not been covered. One extensive study has been made.[31] Risbud[32] will discuss this subject later in this workshop conference.

References

[1]Janos Grofcsik, "Mullite, Its Structure, Formation and Significance," Akademiai Klado, Publishing House of the Hungarian Academy of Sciences, Budapest, 163 p. (1961).

[2]R. F. Davis and J. A. Pask, "Mullite," Chapt. 3 in High Temperature Oxides—Part IV, Refractory Glasses, Glass-Ceramics and Ceramics, edited by Allen M. Alper, Academic Press, pp. 37–76 (1971).

[3]J. A. Pask, "Stable and Metastable Phase Equilibria and Reactions in the SiO_2-Alpha Al_2O_3 System," *Ceramics International*, **9** [4] 107–113 (1983).

[4]S. P. Chaudhuri, "Melting/Decomposition of Mullite: Incongruent or Congruent? I. Phase Equilibria of the System Al_2O_3•SiO_2," *Ceramics International*, **13** [3] 167–176 (1987) and S. P. Chaudhuri, "Melting/Decomposition of Mullite: Incongruent or Congruent? II. Responsible Factors for Dual Nature of Mullite," ibid., 177–182 (1987).

[5]Sainte-Claire Deville and A. Caron, *Ann. Chim. Phys.*, **5**, 114 (1865).

[6]W. I. Vernadsky, *Bull. Soc. Franc. Miner.*, **13**, 270 (1890).

[7]E. S. Shepherd, G. A. Rankin and W. Wright, "The Binary Systems of Alumina and Silica, Lime and Magnesia," *Am. J. Sci.*, **28**, 301 (1909).

[8]N. L. Bowen and J. W. Greig, "The System Al_2O_3-SiO_2," *J. Am. Ceram. Soc.*, 7, p. 238 (1924); ibid., p. 410.

[9]a)W. H. Bauer, I. Gordon and C. H. Moore, "Flame Fusion Synthesis of Mullite Single Crystals," *J. Am. Ceram. Soc.*, **33**, p. 140 (1950).

b)W. H. Bauer and I. Gordon, "Flame Fusion Synthesis of Several Types of Silicate Structures," *J. Am. Ceram. Soc.*, **34**, p. 250 (1951).

[10]N. A. Toropov and F. Y. Galakhov, "New Data for the System Al_2O_3-SiO_2," *Dokl. Akad. Nauk SSSR*, **78**, p. 299 (1951).

[11]E. C. Shears and W. A. Archibald, "Aluminosilicate Refractories," *Iron Steel (London)*, **27**, p. 26 (1954).

[12]J. H. Welch, "New Interpretation of the Mullite Problem," *Nature (London)*, **186**, [4724] pp. 545–546 (1960).

[13]a)N. A. Toropov and F. Y. Galakhov, "Solid Solutions in the System Al_2O_3-SiO_2," *Izv. Akad. Nauk SSSR Otd. Khim. Nauk*, **1958**, p. 8 (1958).

b)N. A. Toropov and F. Y. Galakhov, "Al_2O_3-SiO_2 System," *Eksp. Tekh. Mineral. Petrogr. Mater. Sovetch 7 th L.*, **1966**, p. 3 (1964).

[14]S. Aramaki and R. Roy, "Revised Phase Diagram for the System Al_2O_3-SiO_2," *J. Am. Ceram. Soc.*, **45**, pp. 229–242 (1962).

[15]R. F. Davis and J. A. Pask, "Diffusion and Reaction Studies in the System Al_2O_3-SiO_2," *J. Am. Ceram. Soc.*, **55**, pp. 525–531 (1972).

[16]a)W. Guse and D. Mateika, "Growth of Mullite Single Crystals ($2Al_2O_3{\cdot}SiO_2$) by the Czochralski Method," *J. Crystal Growth*, **22**, pp. 237–240 (1974).

b)W. Guse, "Compositional Analysis of Czochralski Grown Mullite Single Crystals," *J. Crystal Growth*, **26**, pp. 151–152 (1974).

[17]I. A. Aksay and J. A. Pask, "Stable and Metastable Equilibria in the System SiO_2-Al_2O_3," *J. Am. Ceram. Soc.*, **58**, pp. 507–512 (1975).

[18]I. Shindo, "Applications of the Floating Zone Technique in Phase Equilibria Study and in Single Crystal Growth," D.Sc. Thesis, Tohoku Univ., 1980.

[19]S. Prochazka and F. J. Klug, "Infrared Transparent Mullite Ceramic," *J. Am. Ceram. Soc.*, **66** [12] 874–880 (1983).

[20]F. J. Klug, S. Prochazka and R. H. Doremus, "Al_2O_3-SiO_2 Phase Diagram in the Mullite Region," *J. Am. Ceram. Soc.*, **70** [10] 750–759 (1987).

[21]J. A. Pask, X. W. Zhang, A. P. Tomsia and B. E. Yoldas, "Effect of Sol-Gel Mixing on Mullite Microstructure and Phase Equilibria in the αAl_2O_3-SiO_2 System," *J. Am. Ceram. Soc.*, **70** [10] 704–707 (1987).

[22]I. A. Aksay, "Diffusion and Phase Relationship Studies in the Alumina-Silica System," (Ph.D. Thesis), University of California, Berkeley, April 1973.

[23]I. A. Aksay and J. A. Pask, "The Silica-Alumina System: Stable and Metastable Equilibria at 1.0 Atmosphere," *Science*, **183**, No. 4120, pp. 69–71 (1974).

[24]C. D. Greskovich, M. F. X. Gigliotti and P. Svec, "Preparation and Microstructural Developments of Al_2O_3-SiO_2 Shell Molds," *Trans. and J. Brit. Ceram. Soc.*, 77 [3] 98–103 (1978).

[25]V. F. Draper, "Mullite Phase Equilibria in the System: CaO-Al_2O_3-SiO_2," M.S. Thesis, University of California, Berkeley, October 1976.

[26]J. W. Cahn, "The Metastable Liquidus and Its Effect on the Crystallization of Glass," *J. Am. Ceram. Soc.*, **52** [3] 118–121 (1961).

[27]M. D. Sacks and J. A. Pask, "Sintering of Mullite Containing Materials: I. Effect of Composition," *J. Am. Ceram. Soc.*, **65**, pp. 65–70 (1982).

[28]W. M. Kriven and J. A. Pask, "Solid Solution Range and Microstructures of Melt-Grown Mullite," *J. Am. Ceram. Soc.*, **66**, pp. 649–654 (1983).

[29]a)H. Saalfeld, "The Domain Structure of 2:1 Mullite ($2Al_2O_3{\cdot}SiO_2$)," *N. Jb. Miner. Abh.*, **134**, pp. 305–316 (1979).

b)H. Saalfeld and W. Guse, "Structure Refinement of 3:2 Mullite ($3Al_2O_3{\cdot}2SiO_2$), *N. Jb. Miner. Mh.*, **1981**, H4, pp. 145–150 (1981).

[30]Y. Nakajima and P. H. Ribbe, "Twinning and Superstructure of Al-Rich Mullite," *Am. Mineral.*, **66**, pp. 142–147 (1981).

[31]S. H. Risbud and J. A. Pask, "Calculated Thermodynamic Data and Metastable Immiscibility in the System SiO_2-Al_2O_3," *J. Am. Ceram. Soc.*, **60**, pp. 418–424 (1977).

[32]S. H. Risbud, "Revisiting Liquid Immiscibility and Structure in SiO_2-Al_2O_3 Glasses," this volume.

Alumina-Silica Phase Diagram in the Mullite Region

F. J. KLUG AND S. PROCHAZKA

General Electric Corporate Research & Development
Ceramics Branch
Materials Laboratory
Schenectady, New York 12301

R. H. DOREMUS

Materials Engineering Department
Rensselaer Polytechnic Institute
Troy, New York 12181

Abstract

The alumina-silica phase relations were studied in samples annealed in oxygen and quenched, using optical microscopy, image analysis, X-ray diffraction, and the electron microprobe. The solid solution boundaries of mullite changed with increasing temperature and joined at 1890°C and a composition of 77.15 wt % alumina. The melting point of mullite was 1890°C with a peritectic between 76.5 and 77.0% alumina.

Introduction

Alumina and silica are components of many commercial ceramics and important minerals, so their phase diagram is one of the most important of all oxide pairs. Nevertheless there is substantial disagreement about the diagram in the region of the compound mullite ($3Al_2O_3$-$2SiO_2$).[1–3,11–20] Experimental difficulties because of the high temperatures, volatility of silica, slow reaction rates in solids, and temperature measurements have led to different experimental results and interpretations. Even, recent and carefully done investigations have produced very different results, as shown in Figs. 1 and 2.[1,2]

Aramaki and Roy[1] studied the mullite region with samples of mixed alumina and silica powders heated to the desired temperature and quenched to room temperature (see Fig. 1). Their samples were sealed in Pt-Rh containers to prevent loss of silica. Phases were identified by optical microscopy.

Aksay and Pask[2] deduced a quite different equilibrium diagram from an electron microprobe study of reaction zones in diffusion couples between sapphire, and aluminosilicate glass (Fig. 2). Concentration profiles in quenched couples were measured with the microprobe, and the concentration of alumina at phase interfaces was measured. Aksay and Pask also encapsulated samples to prevent silica loss.

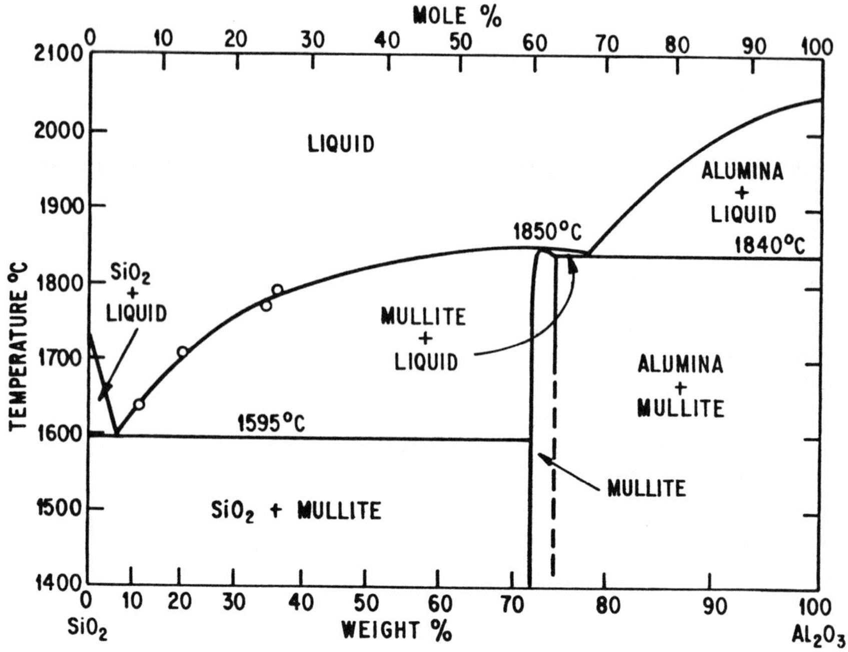

Fig. 1. Alumina-silica phase diagram derived by Aramaki and Roy.[1]

Aksay and Pask found that at temperatures of 1853°C and above, mullite would not grow on the sapphire interface, but that below 1800°C a coherent layer of mullite grew. At temperature of 1853°C and above, mullite crystals were found close to the interface; Aksay and Pask claimed these crystals grew from the melt on cooling, because the composition and morphology were the same as for crystals grown a distance from the interface. From the morphology and the composition of the mullite growing at the interface, and the position of the liquidus curves, Aksay and Pask concluded that mullite melts incongruently at 1828°C ± 10°C with a peritectic at 52.3 weight % alumina. The equilibrium solid solution limits were found to be almost identical with those of Aramaki and Roy[1], ranging from 70.5% to 74.0% alumina. The liquidus lines in the phase diagram shown in Fig. 2 are almost the same as those of Bowen and Greig.[3]

Aksay and Pask also studied the effect of cooling rate. The precipitation of alumina was cooling rate dependent. Alumina-fused silica diffusion couples cooled rapidly formed mullite and glass, and when slowly cooled formed alumina and glass. Intermediate cooling rates formed all three phases. The results were attributed to a metastable Al_2O_3-SiO_2 phase diagram[4] concomitant with supercooling and nucleation phenomena. Similar results were observed with 78 to 80 weight % alumina containing melt. The mullite formed from the high alumina silicate melts formed metastable mullite with alumina contents as high as 83.2%

Aksay and Pask found the melting point of hot-pressed mullite (71.8% alumina) was 1880°C, in conflict with the 1828°C peritectic temperature. They assumed this resulted from superheating of the mullite. The dashed lines in Fig. 2 show this and

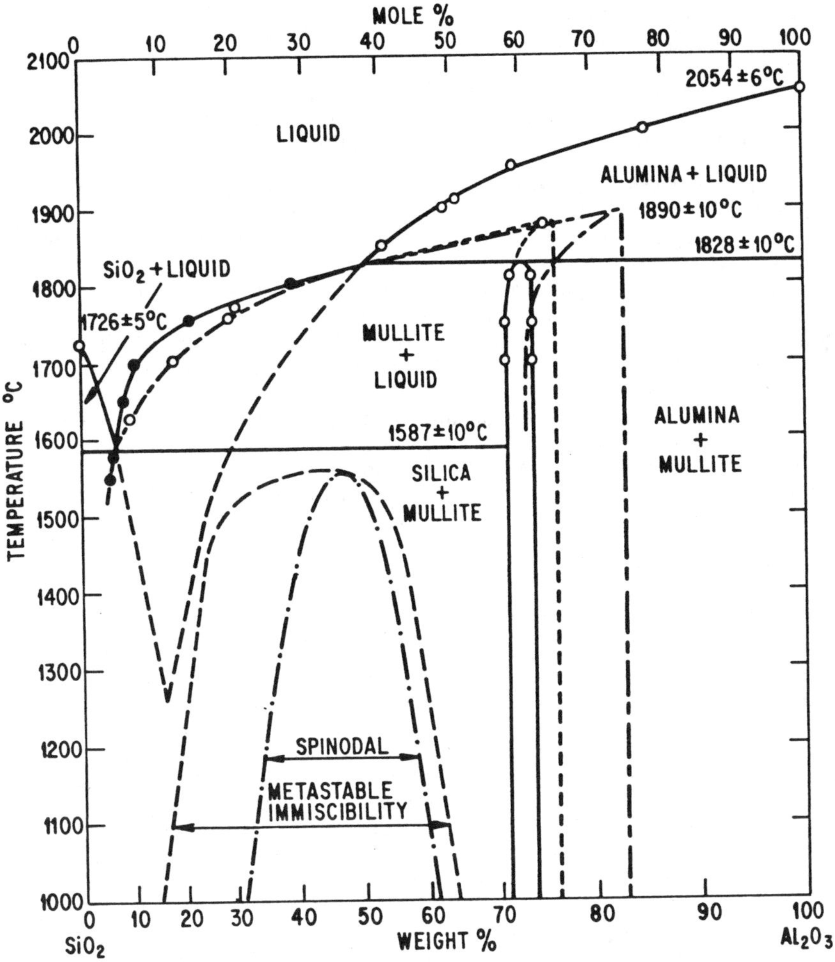

Fig. 2. Alumina-silica phase diagram derived by Aksay and Pask,[2] Risbud and Pask.[6]

other metastable regions in the phase diagram. See also more recent papers by Pask and coworkers.[4–10]

Russian workers[11,12] have found a diagram similar to that of Aramaki and Roy, with congruent melting of mullite. Rehfield and Schwiete[13] also agreed with most of the diagram proposed by Aramaki and Roy, as did Staronka et al.,[14] although the latter authors found high (about 1930°C) melting temperatures for mullite.

A diagram more like that in Fig. 2, with incongruent melting of mullite, was found by Troemel et al.[15] Muller[16] also found incongruent melting of mullite, as did Welch,[17] although the peritectic composition of the latter author was higher than that of Aksay and Pask. Horibe and Kuwabara[18] found a peritectic decomposition temperature of 1890°C and a peritectic composition of 76.7% alumina.

The purpose of this work was to reinvestigate the high-temperature phase equilibria associated with mullite and to answer the following questions:

1. Does mullite melt congruently or incongruently?
2. At what temperature does mullite melt or decompose?
3. What are the solid solution limits of mullite?

Other portions of the alumina-silica diagram, such as the eutectic between silica and mullite, are well established and were not reexamined.

High-purity polycrystalline samples of various compositions were prepared and the phase changes that occurred during heating and cooling were examined. Prochazka and Klug[19] prepared amorphous alumino-silicates by a sol-gel process and translucent single-phase mullite by solid state reaction, and these methods were repeated in the present study. The low pore fraction and consequent translucency of these samples allowed for optical identification and measurement of small amounts of phases.

X-ray diffraction, the electron microprobe, and image analysis were used to determine the solid solubility limits in mullite.

This paper is based on the Ph.D. thesis of F. J. Klug.[20] The thesis includes a thorough review of references and previous work on the Al_2O_3-SiO_2 phase diagram and extensive experimental details of the work summarized here.

Experimental Methods

Powder and Sample Preparation

A sol-gel produced homogeneous aluminosilicate powders of different compositions, which were converted to mullite or phase mixtures by heating above 1300°C. The starting materials were AlO(OH)* and tetraethylorthoslicate† (TEOS). The TEOS was added to an aqueous colloidal dispersion of AlO(OH) and intimately mixed for many hours, after which the TEOS was completely hydrolyzed. The mixture was then cooled to 4–7°C by adding an equal volume of ice made from distilled water. A gel was formed in 600 ml of the cooled mixture in a blender running at high speed as 10 ml of concentrated $NH_4(OH)$ was added. The gel was then frozen, which broke it into a solid-liquid mixture that was filtered through ordinary filter paper in a Buchner funnel. The filter cake was washed with anhydrous methanol at least five times to remove water and prevent aggregation, and then dried under a heat lamp. The dried powder was calcined by heating at 100°C/h to 800°C where it was held 12 h to remove residual hydrocarbons.

Samples were prepared from weighed amounts of chemically analyzed powders. Discs 16 mm in diameter and 0.4g by weight were die pressed at 350 MPa. The discs were heated at 100°C/h to 1325°C and held there for 12 h in air to burn out any residual carbon.

Furnace and Temperature Measurements

An electrically-heated alumina tube furnace was especially made for this work. Construction details are given in the thesis.[20] The furnace atmosphere was flowing oxygen, in which all anneals were carried out, and could operate to 1900°C.

***CATAPAL (TM) SB alumina Conoco Chemicals Division Continental Oil Co., 5 Greenway Plaza E., P. O. 2197, Houston, TX 77001.**

†Reagent Grade, Fisher Scientific Co., Fairlawn, New York 07410.

The furnace was controlled with a W5%Re-W26%Re thermocouple sheathed in a closed-end sapphire tube placed directly in the hot zone. A flowing H_2 atmosphere prevented the thermocouple from oxidation.

Sample temperatures were measured with a Pt6%Rh-Pt30%Rh (type B) thermocouple and an optical pyrometer. Variations of only ± 2°C were observed on the sample thermocouple after 1 h at 1750°C and on an adjacent type B thermocouple. A Leeds and Northrup optical pyrometer was frequently calibrated with an NBS standard source and checked against the melting point of platinum while sighting inside a closed-end alumina tube that served as a black body. The black body was inserted into the furnace before the samples, removed, and inserted again and after they were quenched. Temperatures were measured by sighting directly on the specimens with the pyrometer, which was calibrated against the black body. Below 1800°C the temperature was periodically checked with a type B thermocouple next to the sample. Agreement was always better than ± 5°C. The optical pyrometer had an accuracy of better than ± 10°C. All temperatures reported are on the International Practical Temperature Scale.[21]

The furnace had a hot zone of at least 5 cm with a less than 2°C longitudinal variation measured by the type B thermocouple at less than 1800°C. At temperatures greater than 1800°C no temperature differences could be detected over 5 cm within the measuring limit of the optical pyrometer.

Other Measurements

Starting materials and powders were analyzed for silica and alumina gravimetrically. Trace impurities in the powder were analyzed by plasma emission spectroscopy. Surface areas were determined with Micromeritic BET‡ apparatus. The morphologies of starting powders were examined in the SEM.

Sintering and Annealing of Samples

Sample discs were fired in one atm. of oxygen to reduce silica loss. For times at temperature greater than one hour the samples were tightly wrapped in 0.1 mm rhodium foil. Weight loss during sintering for more than one hour was measured.

For anneals of one hour or less the compacted discs were held in a rack of rhodium rods, so that three samples could be seen simultaneously. The samples were slowly pushed into the furnace with a rhodium rod until they reached 1200°C. After 5 min. the samples were rapidly pushed into the furnace until they rested against the end of the thermocouple protection tube. After one min. no difference in color temperature between the rhodium rack, the alumina thermocouple protection tube, and the mullite samples could be detected with the optical pyrometer. For melting experiments, the furnace was then ramped at 1°C/min from 1880°C until melting of all three samples occurred. The temperature of the W5%Re-W26%Re thermocouple was recorded as each sample melted. A record of the order of melting as a function of position in the rack and composition was also kept. After the desired time at temperature the samples were rapidly pulled out of the furnace and quenched to room temperature in air.

Analysis of Samples

Each sample was cut in half and polished, and phases identified in the optical microscope with reflected light (bright field) or Nomarski contrast. Grain boundaries were etched with boiling conc. NaOH for five min.

---

‡Micromeritics Instrument Corp., 5680 Goshen Springs Rd., Norcross, GA 30093.

Table 1. Diffraction peaks of mullite for measurement of lattice parameters

2	hkl
26.2	210
31.1	001
33.2	220
37.0	130
39.2	201
42.6	230
48.0	400
50.7	330
58.2	401
59.0	141
63.5	421
64.5	002
65.5	250
87.0	261
102.0	641
102.8	442
137.8	372

The volume fraction of alumina in selected samples was determined with a Zeiss IBAS image analysis system. A minimum of 30 fields of view (130 × 130 μm) were examined on each sample. Alumina particles below the plane of polish were detected because of the translucency of the samples, and were manually edited from the image before measurement. Holes of similar size and shape to alumina precipitates were measured as alumina particles; they were infrequent and resulted from polishing pullouts. A minimum of 30 areas of each sample were selected by the computer for measurement.

The alumina/silica ratio of mullite in a few fired samples was measured with the electron microprobe with a voltage of 15.0 kV. A sintered mullite disc with a starting composition of 76.5% was used as a standard; it contained less than 0.1% of second-phase alumina by image analysis and X-ray diffraction. Repeatability with the microprobe was + 3%. The alumina/silica ratio in the mullite phase was also determined from the starting compositions and the amount of second phase.

X-ray Diffraction

Half of each sample was prepared for X-ray diffraction analysis by crushing in a steel mortar and screening through 100 mesh nylon. NBS standard silicon #640 was added to the powder in an agate mortar and the mixture plus amyl acetate and Duco cement was lightly ground with an agate pestle. The resulting slurry was poured onto a glass microscope slide and dried with a heat lamp. A Siemens D500 diffractometer was operated at 50 kV and 20 mA to give copper Kα radiation. The diffractometer had two 3° divergence slits, a 3° scatter slit, and a .05° receiving slit.

Steps of .05° in 2θ for 20 s counting time were measured from a 2θ of 20° to 140°. A least-squares fit of the background was subtracted from the diffraction pattern by computer. The computer also eliminated Kα2 peaks.[22,23] The 2θ positions of six silicon peaks were used to correct 2θ positions for the samples, based on NBS positions for the silicon peaks. Lattice parameters of mullite were calculated from

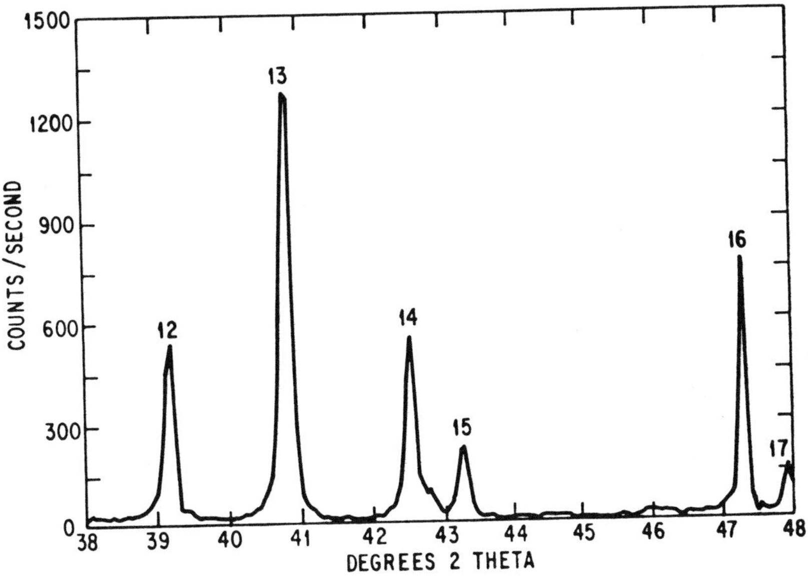

Fig. 3. X-ray diffraction spectrum of sample 9 used for determination of second phase alumina. Peak 15 is for α-alumina, (113), peak 16 is the silicon standard. The other peaks are mullite: 12 is 201, 13 is 121 and 211, and 14 is 230.

the diffraction peaks given in Table 1. These peaks were chosen because they did not overlap with alumina or silicon peaks and had enough intensity and definite hkl values. The lattice parameters were calculated by a computer program.[24]

The amount of second-phase alumina was determined from the ratio of peaks areas of an alumina peak to four mullite peaks, using the (113) alpha alumina peak and the sum of the areas of the 201, 121, 211 and 230 mullite peaks. A sample diffraction pattern is shown in Fig. 3. Standards made of mixtures of alumina and mullite powders gave unreliable results. Therefore, a calibration curve was first prepared from measurements of the amount of second-phase alumina by image analysis of optical micrographs. The samples chosen for this plot are shown by an asterisk in Table 2. A graph was then made of the ratio of the alumina and mullite peaks for the samples as a function of the percent alumina determined by image analysis, as shown in Fig. 4. The line the figure is a least-square representation of the data and was used to calculate the percent second-phase alumina in a sample from the X-ray diffraction pattern. The accuracy for the determination of second-phase alumina by X-ray diffraction was about ±0.1%, and the minimum detection limit with 90% confidence was 0.1% alumina.[25]

Experimental Results

Phase Analysis

In Table 2 the results of measurements of the amounts of second-phase alumina after heating at different temperatures are recorded. The standard deviations in the image analysis measurements ranged from about .1% at low amounts (<.3%) of second-phase alumina to about .7% at higher amounts (3-5%); at least 30 different

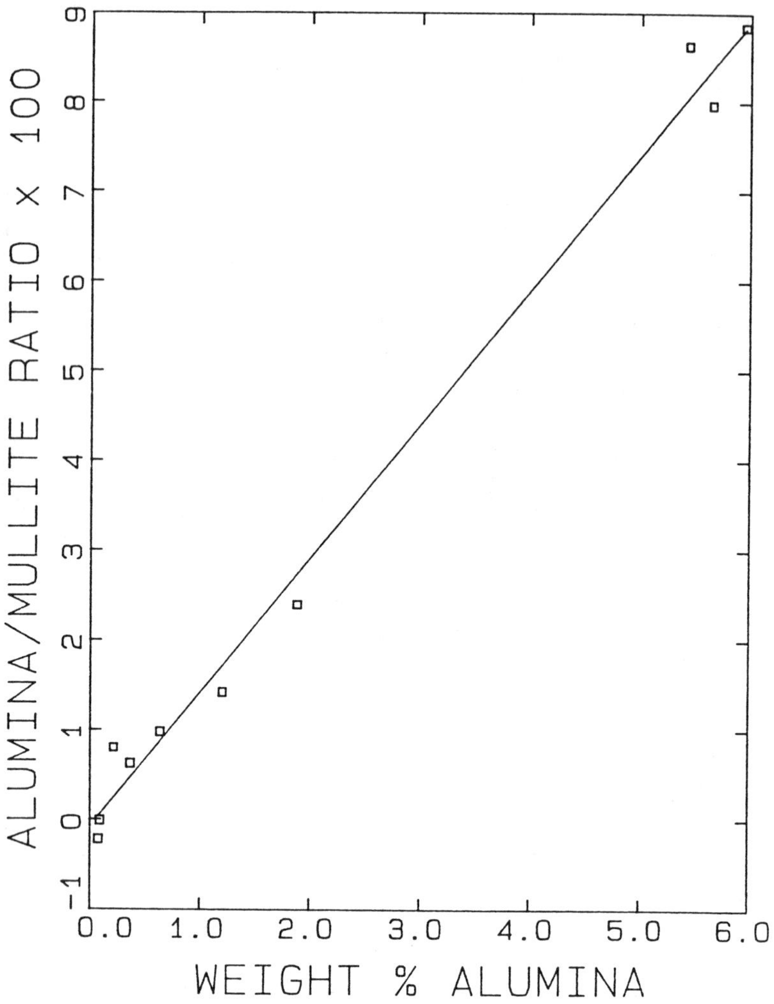

Fig. 4. Calibration curve for the determination of % second-phase alumina by X-ray diffraction. The ordinate is the ratio of the area under the (113) α-alumina peak to the sum of the areas of the (201), (121), (211), and (230) mullite peaks (see Fig. 3). The abscissa is the % alumina determined by image analysis.

Table 2. Amount of second-phase alumina and solid solubility limit of mullite samples held one hour at different temperatures

Sample	Annealing Temp. °C	Composition Wt. %	Measured Amount of Second-Phase Alumina Wt. %		Calculated Solubility Limit, Wt. %	
			Image Analysis	X-ray Diffraction	From X-ray Diffraction	From Image Analysis
1A	1880	78.0	4.60			76.94
45	1880	77.5		2.20	76.99	77.0#
* 5	1870	78.0	5.66	5.38	76.75	76.68
* 2	1870	77.0	.21	.58	76.87	76.95
2A	1860	78.0	5.80			76.65
4A	1857	77.0	.75			76.83
* 9	1850	78.0	5.96	5.97	76.60	76.61
44	1850	77.5		2.83	76.84	
* 7	1850	77.0	1.20	.99	76.77	76.72
*42	1850	77.0	.64	.70	76.84	76.85
3A	1837	78.0	5.87			76.76
5A	1837	77.0	1.02			76.76
* 3	1827	78.0	5.44	5.83	76.64	76.73
6A	1827	77.0	1.47			76.65
43	1825	77.5	1.98	2.89	76.83	77.05
*39	1825	76.5	.36	.46	76.39	76.41
* 1	1820	77.0	1.89	1.65	76.61	76.56
46	1800 (16h)	76.5		1.54	76.13	
62	1730 (48h)	76.5		4.18	75.47	
59	1680	76.5		5.04	75.25	

* These samples were used for the calibration curve for X-ray determination of % second phase alumina (Fig. 4).
\# Microprobe instead of Image Analysis.

areas were measured for each sample. The weight percent W_A of second-phase alumina was calculated from the average volume fraction V and the formula

$$W_A = \frac{3.97V}{3.976V+(1-V)3.16} \tag{1}$$

where 3.97 is the density of alumina and 3.16 that of mullite. The amount of second-phase alumina was also calculated from the X-ray diffraction peaks and the calibration curve of Fig. 4, as recorded in Table 2. Then the solid solubility limits in mullite were calculated from the lever rule, assuming no solid solubility in α-alumina. The weight percent of mullite at the solid solubility limit W_M is given by

$$W_M = (W_T - W_A)/(1 - W_A) \tag{2}$$

where W_T is the overall (total) weight percent alumina and W_A is the weight percent of second-phase alumina as recorded in Table 2.

In one atm. of oxygen the partial pressure of SiO_2 over mullite is about $10^{-5.5}$ atm at 1880°C, so the loss of silica under these conditions should be negligible.

All of the samples in Table 2 were examined in the optical microscopy at 1000x and found to contain two crystalline phases, mullite and alumina. The alumina was uniformly distributed throughout the cross-section. No increase in the amount of second-phase alumina could be defected near the sintered surfaces. No glass was observed in any of these samples.

For any particular composition the amount of second-phase alumina decreased with increasing temperature until the sample became single-phase or started to melt. At a particular temperature the amount of second-phase alumina decreased with decrease in total alumina concentration until the phase boundary was reached and single-phase mullite was formed.

In three samples no second-phase alumina was detected by X-ray diffraction. They had a total alumina concentration of 75.0%, 76.5% and 76.5% and were annealed at 1860°C, 1880°C and 1850°C, respectively. A small quantity of alpha alumina was always observed with optical microscopy below the melting point of mullite. Previous investigators[1,13] stated that a trace of alpha alumina can be discounted as non-equilibrium and the result of mixing in homogeneity. A similar practice will be followed here. Samples with less than 0.2% second-phase alumina and no glass were considered to be single phase.

Photomicrographs of etched and unetched surfaces of the samples in Table 2 are in the Klug thesis. Examples are in Figs. 5–8. All the micrographs were consistent with the phase analysis reported in Table 2 and 3. At temperatures below about 1800°C the grain size was smaller than 24 μm and the alumina particles were on grain boundaries; at higher temperatures the grains grew, trapping some alumina particles inside the grains. Above approximately 1850°C the grains became elongated, compared to equiaxed grains at lower temperatures; however, there was no evidence of liquid at these temperatures. After 1 hr. at 1870°C some mullite grains were as large as 150 μm long and 30 μm wide.

Single-phase mullite samples were translucent because of low pore volume, as shown in Fig. 9. Those with appreciable second-phase alumina or glass from melting were opaque.

There was no precipitated alumina at the sample surfaces; for example sample 38 after 1 hr at 1880°C, shows that the loss of silica from the surface was negligibly small.

Samples containing 81%, 83% and 85% alumina were held 1 hr at 1850°C. All microstructures had large amounts of second-phase alumina, mostly on the mullite grain boundaries; an example is in Fig. 8. It was not possible to get an accurate measure of the amount of second-phase alumina in these samples.

Melting Point of Mullite

Samples were observed three at a time with the optical pyrometer while heating at 1°C/min from 1880°C until they melted. A total of 34 samples with compositions of 75.0, 76.5, 77.0, 78.0 and 79.0% alumina were observed to melt at 1890°C. The furnace was then ramped at 1°C/min from 1880°C until melting of all three samples occurred. The temperature of the W5%Re-W26%Re thermocouple was recorded as each sample melted. A record of the order of melting as a function of position in the rack and composition was also kept.

The results of the melting experiments are in Table 6. Run #6 and run #9 show that small but reproducible temperature gradient of about 1–2°C was present across

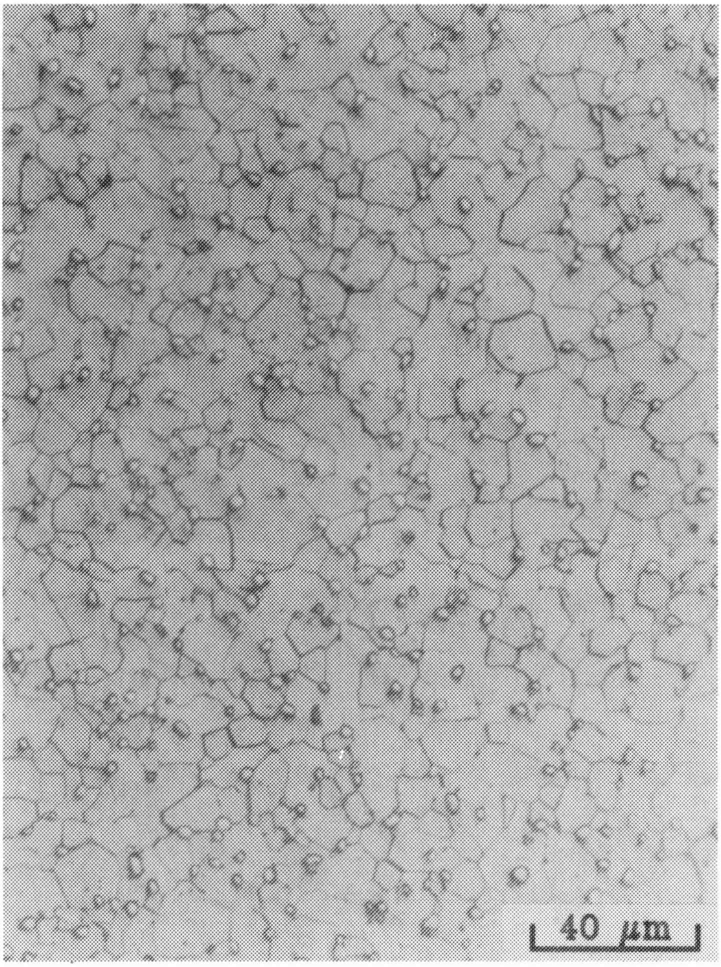

Fig. 5. Nomarski photomicrograph of etched mullite (76.5% alumina) after 48 h at 1730° (sample 62). Alumina particles are light grey in darker grey mullite matrix.

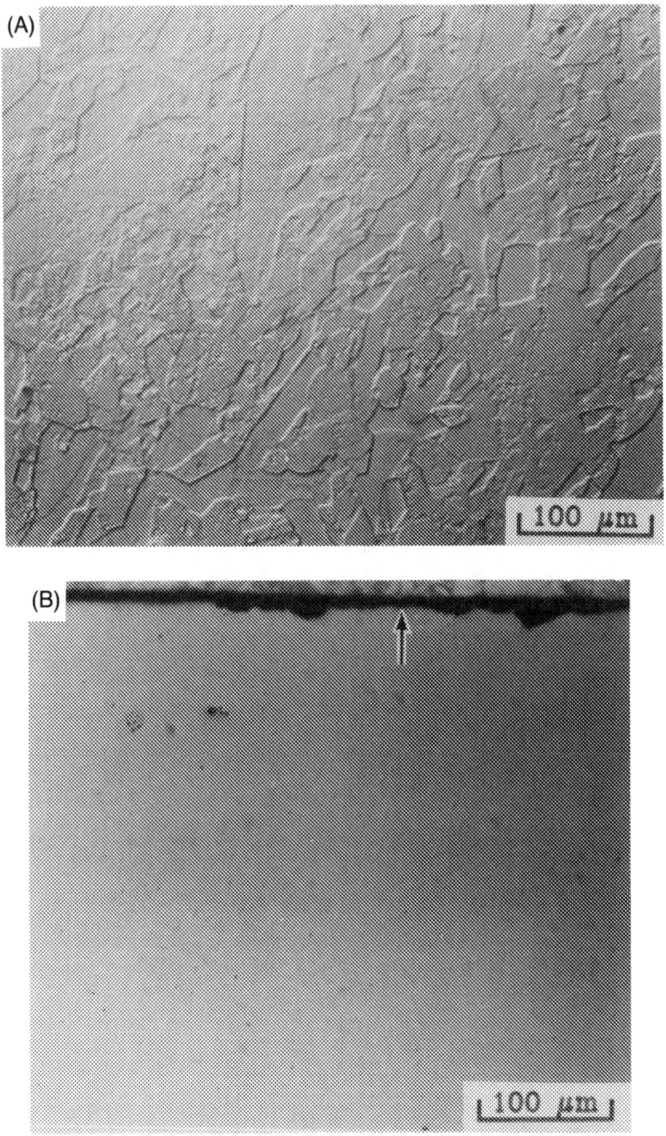

Fig. 6. Nomarski photomicrograph of etched mullite (76.5% alumina) after one h at 1880° (sample 38). The roughness in some grains is an artifact of etching. (A) etched in NaOH (B) unetched. The arrow points the surface of the sample.

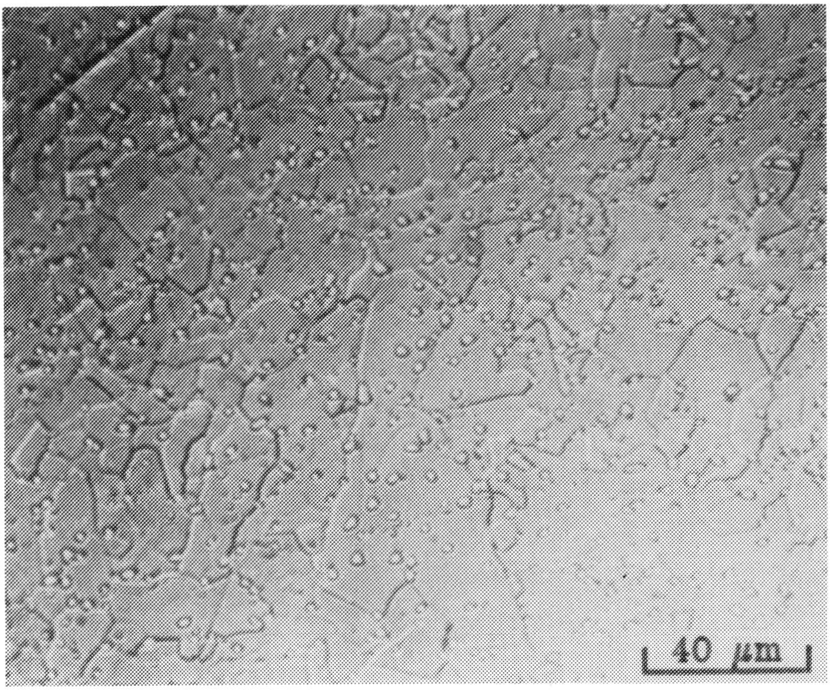

Fig. 7. Nomarski photomicrograph of etched mullite (78.0% alumina) after one h at 1860° (sample 2A). The light dots are alumina in the darker mullite matrix.

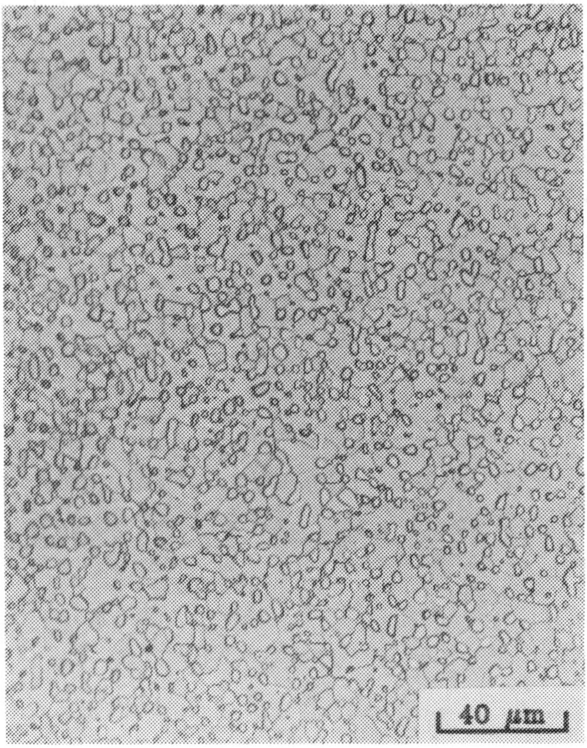

Fig. 8. Bright-field photomicrograph of etched mullite (81% alumina) after h at 1850°C. Rounded alumina particles are on mullite grain boundaries.

Fig. 9. Mullite discs (75% alumina) 1 mm thick after sintering at 1845°, 1860° and 1870°C for 1 h.

Table 3. Single-phase mullite samples (<0.2% second phase) held one hour at temperature, unless otherwise noted

Sample No.	Temp. °C	Mixture Comp. wt.% Al_2O_3
38	1880	76.5
79	1860	75.0
41	1850	76.0
P&K	1850	76.0
	1845	75.0
P&K	1800	76.0
	1790	75.0
P&K	1630	72.3

P&K - Prochazka and Klug[19]

Table 6. The melting point of mullite

The melting point as a function of composition is reported. The optical pyrometer temperature is on IPTS-68 (56) +/-10°C. The W-Re thermocouple compares the change in the melting point +/-2°C as a function of composition, but is not an absolute measure of the melting temperature (see text for details).

RUN NO.	PYROMETER TEMP°C	W5%-W26%Re TEMP°C	COMPOSITION WT% ALUMINA	POSITION IN FURNACE	MELTING ORDER
1	1890	------	76.5	LEFT	1
1	1890	------	77.5	MIDDLE	1
1	1890	------	77.0	RIGHT	3
2	1890	1867	76.5	LEFT	1
2	1890	1870	77.0	MIDDLE	3
2	1890	1869	77.5	RIGHT	2
3	1890	------	77.0	LEFT	3
3	1890	------	76.5	MIDDLE	1
3	1890	------	77.5	RIGHT	1
4	1890	1875	77.5	LEFT	1
4	1890	1875	77.0	MIDDLE	1
4	1890	1875	76.5	RIGHT	1
5	1890	------	76.5	LEFT	1
5	1890	------	77.5	MIDDLE	2
5	1890	------	77.0	RIGHT	3
6	1890	1870	76.5	LEFT	1
6	1890	1872	76.5	MIDDLE	3
6	1890	1871	76.5	RIGHT	2
7	1890	------	77.0	LEFT	1
7	1890	------	77.5	MIDDLE	2
7	1890	------	76.5	RIGHT	3
8	1890	1869	75.0	LEFT	1
8	1890	1872	78.0	MIDDLE	3
8	1890	1872	77.0	RIGHT	2
9	1890	1868-9	75.0	LEFT	1
9	1890	1870	75.0	MIDDLE	3
9	1890	1870	75.0	RIGHT	2
10	1890	1869	77.5	LEFT	1
10	1890	1872	78.0	MIDDLE	2
10	1890	1872	79.0	RIGHT	2
11	1890	1868	79.0	LEFT	
11	1890	1870	78.0	MIDDLE	3
11	1890	1869	77.5	RIGHT	2
12+	1888	1865	76.5	LEFT	PARTIALLY MELTED
12+	1888	1865	77.5	MIDDLE	CORNERS ROUNDED
12+	1888	1865	77.0	RIGHT	CORNERS ROUNDED
13	1890	----	76.5	LEFT	
13	1885	----	72.3	MIDDLE	
13	----	----	EMPTY	RIGHT	

\+ Samples in run 12 were left in the furnace for 1 h. Melting of the samples was observed after 10 minutes at temperature.

the furnace diameter as measured by the W-Re thermocouple. Based on the melting order and the W-Re thermocouple no differences between the 76.5, 77.0, 77.5, 78.0 and 79.0% alumina samples could be found. They all melted at 1890°C. (Run 4 lodged 10 mm from the end of the thermocouple protection tube accounting for the slightly higher thermocouple temperatures when compared to other runs.) The 75.0 weight % sample when placed in the lowest temperature position based on runs #6 and #9 was the first sample to show signs of melting. The 72.3 wt % sample melted at 1885° which was the lowest melting sample measured. One set of samples heated to 1888°C for 1 hr showed: 76.5% alumina, partially melted; 77.0% and 77.5%, corners rounded, alumina particles in the mullite matrix, as shown in Fig. 10. Grey areas similar to those in Fig. 10 are taken as evidence of melting and were seen in all samples that had completely or partially melted (Table 4). The microprobe analyzed these areas as high in silica. They were often featureless, probably glassy, but in some cases they contained some fine mullite crystals formed on cooling.

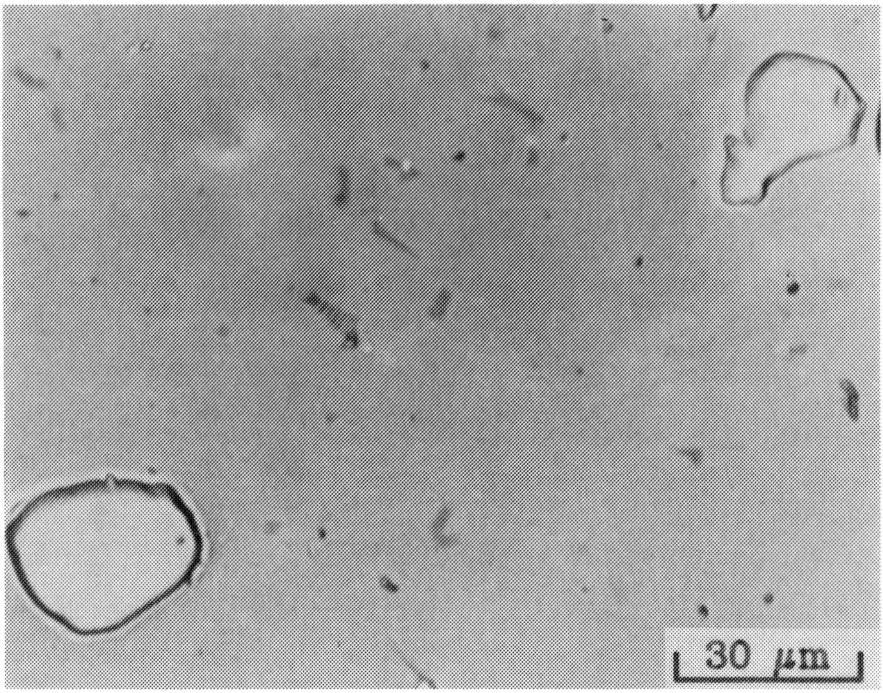

Fig. 10. Bright field photomicrograph of mullite (77.5% alumina) after 1 h at 1888°C. Angular grey areas in the mullite matrix result from some melting; large bright particles are alumina.

Table 4. Mullite samples showing evidence of melting held one hour at temperature unless otherwise noted

Sample No.	Temp.	Mixture Comp.	Comp. Analyzed by Microprobe wt.% Al_2O_3	Comp. Calculated from Glassy Phase	Notes
M12+	1888	77.5	77.0		
M12	1888	77.0	77.2		
M12	1888	76.5	77.0		
81	1880	76.0			
77	1730	72.3			
*80	1730	72.3		74.1	Hot Pressed
*70	1680	72.3		72.7	Hot Pressed

* Samples of Prochazka and Klug[19]
\+ Run #12 in Table 6

These results suggest that the melting temperature of mullite at these compositions is 1890°C. The pyrometer temperatures were consistent with the temperatures read from the W-WRe thermocouple, although this thermocouple read about 20°C below the pyrometer temperature because of the protective flow of hydrogen gas inside the thermocouple sheath and thermal conduction in the molybdenum shield.

A sample with 76.5% alumina was translucent after sintering at 1880°C and contained only a trace of second-phase alumina. This sample was heated at 1895°C where it melted completely and formed a drop on the rhodium foil. The polished surface of this sample showed only mullite and grey melted regions (glass), which were high in silica by microprobe analysis. Only a trace of second-phase alumina was found in the melted sample.

Samples containing 77.0%, 77.5%, 78.0% and 79% alumina were all melted similarly at 1895°C and formed drops on rhodium foil. A typical microstructure is shown in Fig. 11 and contains large alumina particles in a largely mullite matrix containing a small amount of amorphous material. As the overall alumina concentration increased, the amount of second-phase alumina increased. At the melting temperature the melt contained some solid alumina; on cooling most of the liquid crystallized to mullite, leaving a small amount of glassy material.

A sample containing 77.0% alumina was rapidly pushed into the furnace at 1920°C where it melted completely. After quenching to room temperature it contained only mullite and glass, with no second-phase alumina.

Precipitation of Alumina from Mullite

Several mullite samples of 76.5% alumina composition, containing < 0.1 w/o of second-phase alumina, were heated for long times at different temperatures, as shown in Table 5. Second-phase alumina formed in all these experiments, almost all on mullite grain boundaries. There was little grain growth in the mullite after the treatments listed in Table 5. An example of microstructure is shown in Fig. 12.

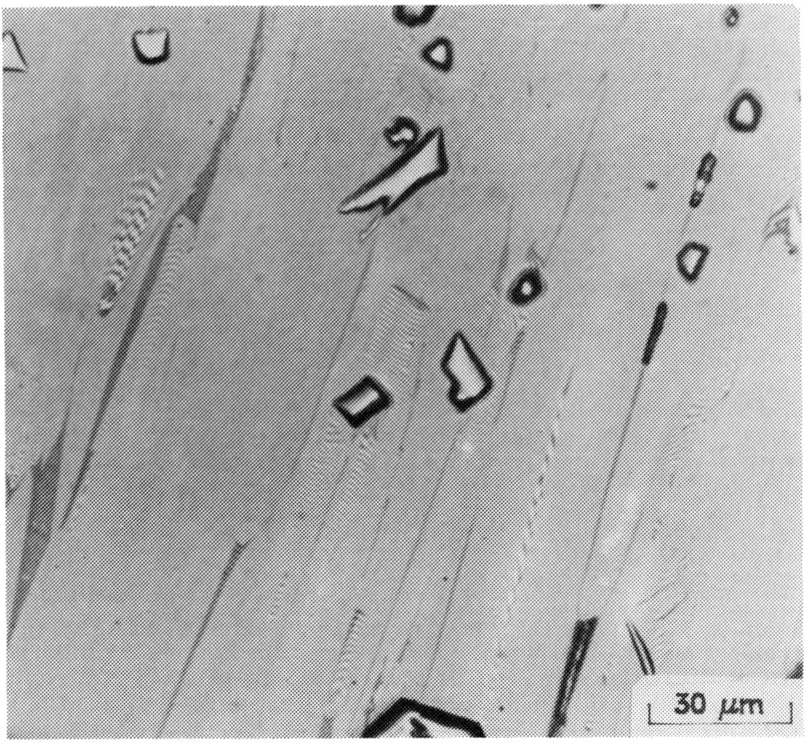

Fig. 11. Bright field photomicrograph of melted mullite (77.0% alumina). The blocky particles are alumina; dark grey areas are amorphous and high in silica in a mullite matrix.

Table 5. Precipitation of alumina from single-phase mullite containing 76.5% alumina

Sample No.	Temp. °C	Time, hrs.	Wt.% Alumina Precipitated	Calculated Mullite Comp.
33	1800	4	.64	76.35
34	1800	16	.75	76.32
60	1680	144	.45	76.37
61	1730	48	1.48	76.12

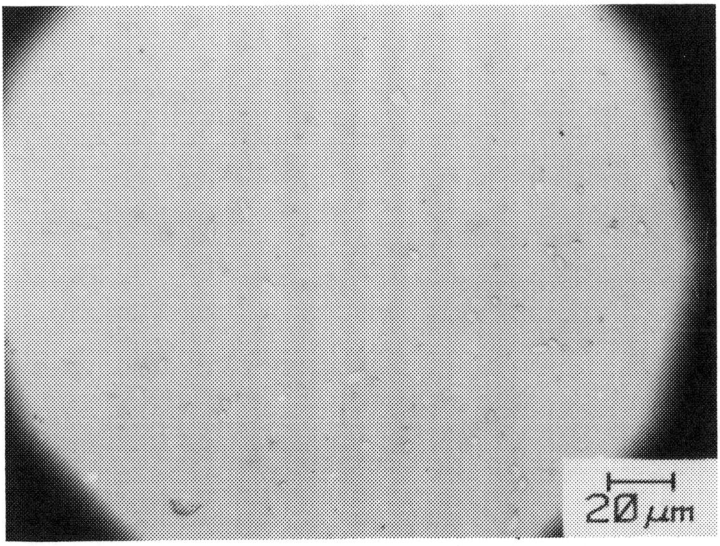

Fig. 12. Bright field photomicrograph of single-phase mullite (76.5% alumina) after holding 18 h at 1800°C, showing growth of alumina particles.

After further heating of this sample for 5 h at 1880°C, most of the second-phase alumina dissolved, as shown in Fig. 13. Thus, the precipitation and resolution of alumina from solid solution in mullite was reversible.

The mullite compositions (wt% alumina) in this sample were analyzed by electron microprobe to obtain initial, 76.5%, 76.2% after heating at 1800°C, and 76.5% after heating at 1880°C, with an error of ±0.3%. These measurements show that alumina precipitation was not caused by silica loss.

Lattice Parameters

The lattice parameters of a number of the mullite phases were measured as described in the experimental methods section and are given in Table 7. A plot of the a spacings are is given in Fig. 14 as a function of composition. They agree well with those measured by Cameron,[26] as shown in figure 14.

Discussion

Phase Diagram

The results have been used to plot the alumina-silica phase diagram in the mullite region, as shown in Fig. 15. The mullite and mullite + alumina boundary was determined from the two-phase samples listed in Table 2. A least-squares line was drawn through the data to the peritectic temperature of 1890°C; it has the formula

$$W = 58.20 + 0.01003T \tag{3}$$

where W is the weight percent alumina at temperature T. It is possible that this line is slightly curved, but any curvature was below the experimental error of the measurements.

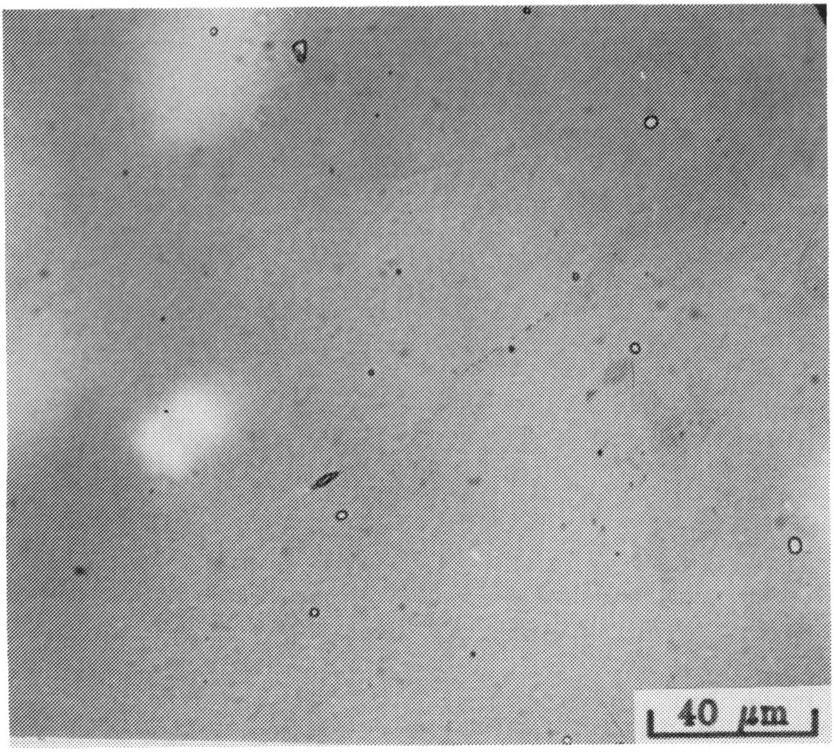

Fig. 13. Bright field photomicrograph of mullite (76.5% alumina) sample of Fig. 12 after further heating for 5 h at 1880°C. White circular particles are alumina.

Single-phase samples from Table 3 were plotted in the mullite solid solution field. The mullite and mullite + liquid boundary line was determined from the samples in Table 4.

The M12 samples in Table 4, analyzed by microprobe after annealing to have an average mullite composition of 77.1% alumina, were partially melted at 1888°C. This mullite composition was therefore taken as that with the highest decomposition temperature. The least-squares fit of the mullite and mullite + alumina boundary intersected the mullite melting temperature of 1890°C at 77.15% alumina. Thus both of the solid solution boundaries appear to intersect the melt (peritectic) line at 77.15% alumina, as drawn in Fig. 15.

The melting point of mullite was ± 2°C over the composition range from 75.0 to 79.0% alumina as measured by the W-Re thermocouple, the melting point of mullite was 1890°C + 10°C. At 72.3% alumina the melting point was slightly lower at 1885°C, as expected from the liquidus line in other diagrams. The 72.3% sample was completely molten before a companion 76.5% sample in the same boat showed any signs of melting, conclusively demonstrating that 72% alumina is not the highest melting mullite composition. There was no evidence of lower-melting compositions that might be expected to result from a eutectic such as that in Fig. 1. The following experimental precautions give additional confidence in this result: (1) The melting

Table 7. Lattice parameters of mullite as a function of composition

The lattice parameters of mullite as a function of composition and the annealing conditions are reported. All compositions were obtained by the lever rule from lever rule from quantitative XRD measurements of the amount of second-phase alumina.

Composition wt.% alumina	ID#	Time (h)	Temp.(^{o}C)	Lattice Parameter (nm) A	B	C
76.99	45	1	1880	0.7583(1)	0.7687(1)	0.28876(4)
76.97	23	1	1865	0.7583(1)	0.7685(1)	0.28874(4)
76.97	40	1	1880	0.7582(2)	0.7686(2)	0.28873(7)
76.87	2	1	1870	0.7581(1)	0.7685(1)	0.28869(4)
76.84	42	1	1850	0.7580(2)	0.7688(2)	0.28871(8)
76.84	43	1	1825	0.7580(1)	0.7686(1)	0.28868(6)
76.75	5	1	1870	0.7582(1)	0.7684(1)	0.28867(3)
76.64	3	1	1827	0.7576(1)	0.7686(1)	0.28863(5)
76.61	1	1	1820	0.7576(1)	0.7687(1)	0.28867(4)
76.60	9	1	1850	0.7580(1)	0.7686(1)	0.28871(3)
76.50	38	1	1880	0.7579(2)	0.7686(2)	0.28870(6)
76.50	41	1	1850	0.7578(2)	0.7687(2)	0.28869(8)
76.49	37	1	1850	0.7581(1)	0.7687(1)	0.28868(4)
76.39	39	1	1825	0.7579(1)	0.7688(1)	0.28871(5)
76.37	60	1+144	1850+1680	0.7580(2)	0.7687(2)	0.28871(5)
76.35	33	1+4	1850+1800	0.7578(2)	0.7687(2)	0.28864(4)
76.32	34	1+16	1850+1800	0.7578(2)	0.7687(2)	0.28867(6)
76.13	46	16	1800	0.7578(1)	0.7690(1)	0.28869(3)
76.12	61	1+48	1850+1730	0.7576(1)	0.7689(1)	0.28869(4)
75.47	62	48	1730	0.7569(2)	0.7689(2)	0.28853(7)
75.25	59	144	1680	0.7567(2)	0.7690(2)	0.28858(8)
74.95	58	144	1680	0.7567(2)	0.7690(2)	0.28856(7)
75.0$	77	1	1870	0.7571(2)	0.7690(2)	0.28861(7)
75.0++	79	1	1860	0.7571(1)	0.7690(1)	0.28858(4)
72.3*	75	HOT PRESS	1630	0.7539(3)	0.7696(3)	0.28841(13)
72.7*	70	HOT PRESS	1680	0.7543(2)	0.7696(2)	0.28844(6)

++ Starting composition reported. It was determined to have less than 0.1% second phase alumina by optical microscopy.
$ Starting composition reported. It has approximately 0.1% glass as second phase by optical microscopy.
* Samples and composition obtained from Prochazka and Klug (11).

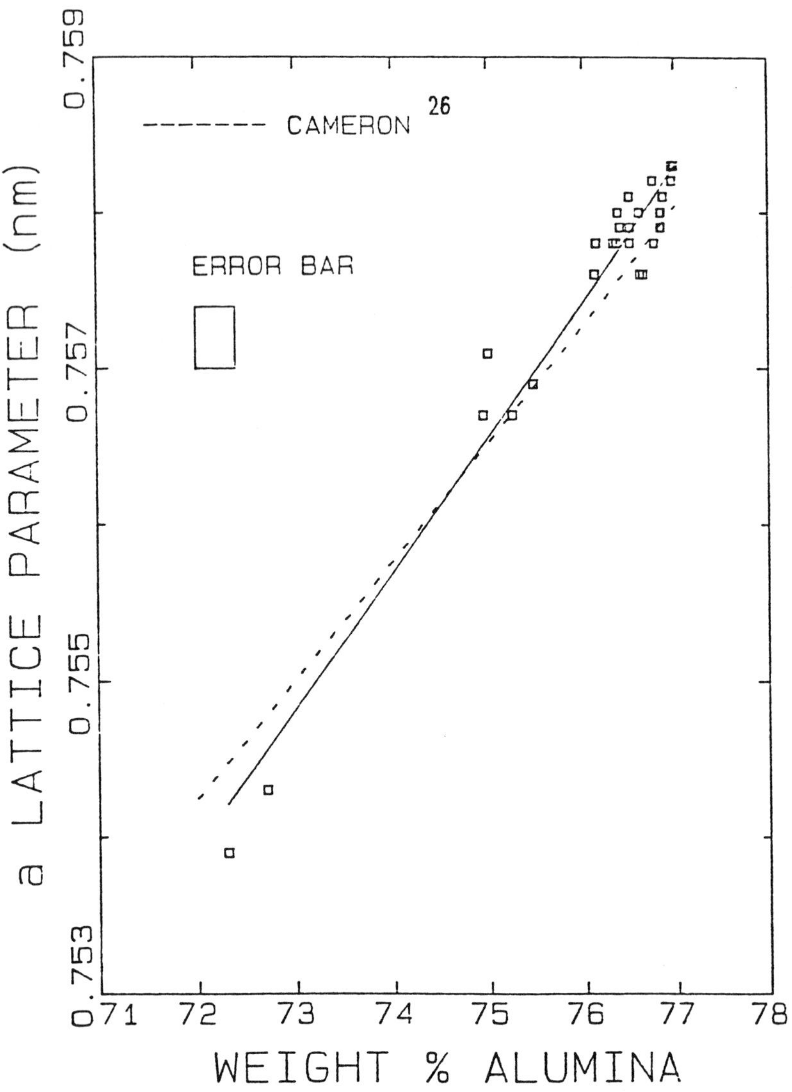

Fig. 14. The a lattice distance of mullite as a function of composition. The data of Cameron are shown as the dashed line.

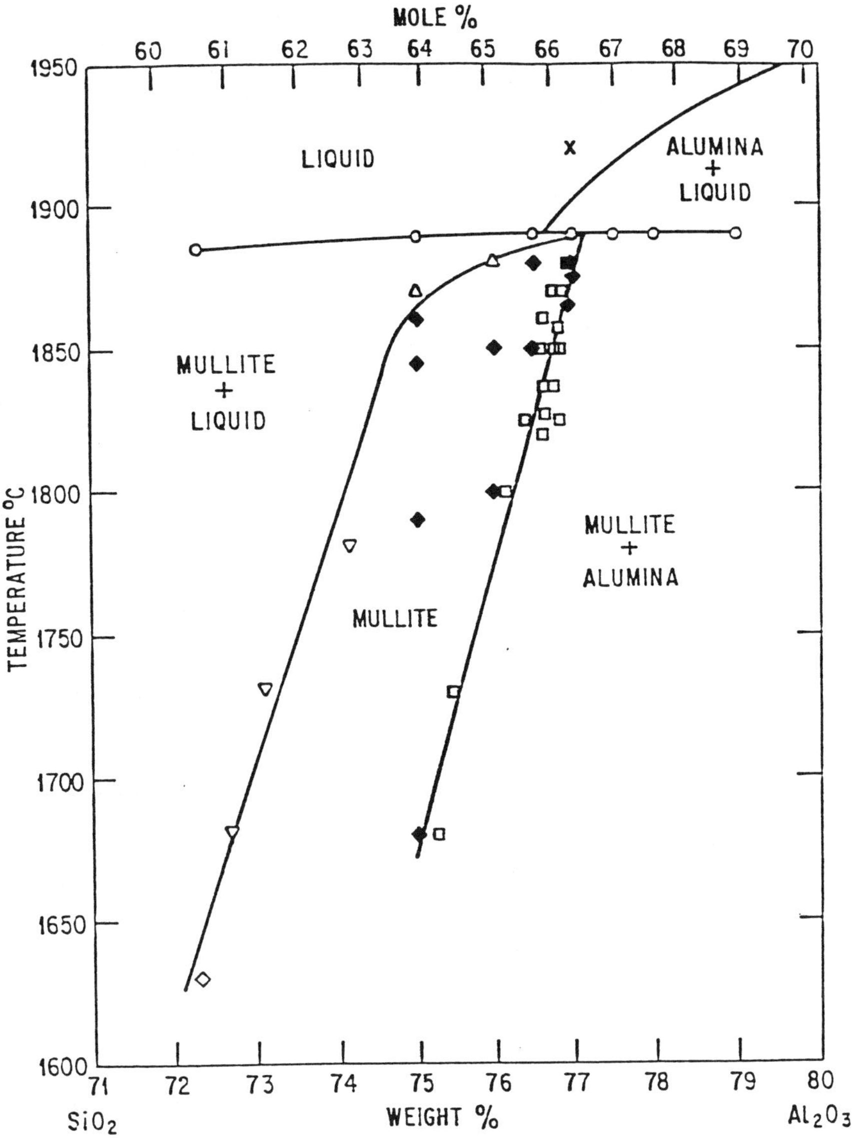

Fig. 15. The alumina-silica phase diagram. The □ mullite and mullite + alumina phase boundary calculated from the lever rule (Table 2); ◊ single-phase mullite (Table 3); Δ mullite + glass (Table 4), ∇ ref. 19, melting of mullite.

temperatures of three different specimens of the same composition placed across the diameter of the furnace were the same. (2) The sample boat was placed in exactly the same position in the furnace for each determination of the melting temperature, and (3) repeated measurements with samples in different positions in the same boat gave the same result.

The position of the liquidus line is defined by the melting experiments at 1895°C. Samples of 76.0% and 76.5% alumina show only mullite and glass phases after melting. Samples with more than 77% alumina had three phases, mullite, glass, and alumina. The 77.0% sample was melted in a two-phase region of alumina and liquid. Alumina particles nucleated and grew to large size (Fig. 11), and the remelting liquid phase had more silica than the original composition. During quenching to room temperature some mullite crystallized and the resulting silica-rich liquid was retained as glass.

The composite alumina-silica phase diagram is shown in Fig. 16.

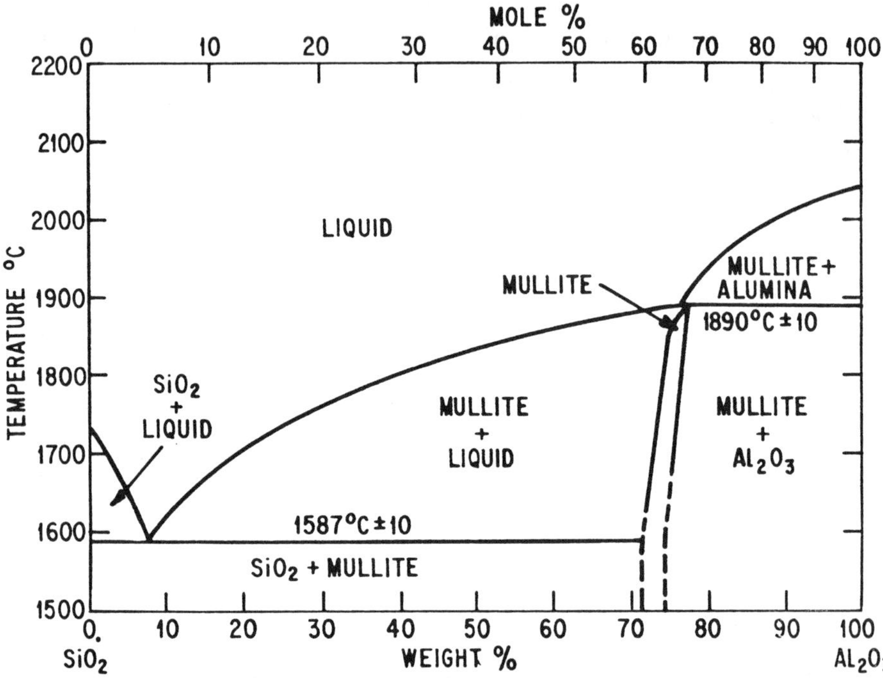

Fig. 16. Composite alumina-silica phase diagram incorporating Fig. 15. The diagram from 0 to 70% alumina is from Aramaki and Roy, except that the eutectic temperature of 1587°C is from Aksay and Pask (Reference 2). The mullite + alumina liquidus line is from Horibe and Kuwabara (Reference 18).

Comparison with Work of Others

Much of the diagram of Aramaki and Roy (Fig. 1) is consistent with the present work. The main difference is in the melting temperature of mullite and the details of the mullite melting region of the diagram. In much recent work[2,14,19,21,27,28] the melting temperature of mullite is between 1880°C and 1900°C, consistent with the present results. Staronka et al.[14] were the only workers to report considerably higher melting temperatures of mullite.

Single crystals of mullite grown by Welch[17,27] from slowly cooled melts had an analyzed composition of 77.2% alumina, which Welch concluded was the composition of mullite in equilibrium with alumina and liquid at the peritectic decomposition temperature. This composition is very close to the value of 77.15% found in the present work.

The alumina liquidus found by Horibe and Kuwabara[18] is consistent with the present work. They studied cooling curves and microstructure. They determined the composition of mullite in equilibrium with alumina in 83.6% starting composition after 1/2 and 2 h at 1880°C, and found 77.0% and 76.7%, in close agreement with Fig. 15. However, there was probably silica loss during their experiments, and if their measured lattice parameters are compared to those in Fig. 14, composition of 78% and 77.6% are calculated, suggesting that the system was not completely at equilibrium.

A melting temperature of 1880°C for a 71.8% alumina sample was found by Aksay and Pask.[2] This sample contained no second-phase alumina, and the mullite composition was 76.8%; these results are consistent with Fig. 15. However, Aksay and Pask claimed the mullite was superheated, because alumina was difficult to nucleate. They proposed a solid solution range of mullite from 70.5 to 74.0% alumina, and that all mullite compositions above 74% are metastable. Tromel et al.[15] also postulated that alumina was difficult to nucleate.

There was no difficulty in nucleating alumina in the present work. The volume fraction of alumina in samples of total composition 77% or greater increased rapidly above the melting temperature of 1890°C, as shown in Fig. 11. The 77% samples were single-phase mullite before melting. It is therefore doubtful that there is any barrier to the nucleation and growth of alumina.

It is unlikely that the absence of alumina influences superheating of mullite. The melting temperature of mullite was 1890°C whether or not alumina was present. In the precipitation experiments a sample was heated at 1880°C, where second-phase alumina dissolved into the mullite, leaving a few small particles after 5 h at 1880°C (Fig. 13). It is likely that mullite would not melt under these conditions if it were 50°C above its melting point.

Superheating is found when a melt is viscous, but not when the viscosity is as low as 4 poise, which is the viscosity of a 50% alumina-silica melt at 1850°C.[29] Addition of alumina to alumina-silica melts reduces the viscosity. An equation for melting rate is[30]:

$$u = \frac{\Delta H_f (T - T_m)}{2\pi n a^2 T_m N}$$

where ΔH_f is the heat of fusion per atom, T the supercooling, n the melt viscosity, T_m the melting temperature, and a the lattice spacing. Horibe and Kuwabara[18] found 63 kj/mole for ΔH_f, and a is ~3.8×10^{-10}m, or twice the cation- oxygen distance in mullite. If n = 4P for a mullite melt at 1850°C (conservative), then for superheating20 at 1850°C, the melting rate is about 1.1 cm/min. It is therefore highly unlikely that any superheating of mullite would be observed.

Ternary diagrams such as that of Roeder et al.[31] for Al_2O_3-(Cr_2O_3-SiO_2) are consistent with the diagram in Fig. 15. The boundary between the crystallization fields of mullite and alumina end near the peritectic at 76.7% alumina, which is much different from the peritectic of 54% of Aksay and Pask.

In conclusion, based on our experimental observations, we believe Fig. 15 and Fig. 16 represent the stable phase equilibria for the alumina-silica system.

References

[1]S. Aramaki and R. Roy, "Revised Equilibrium Diagram for the System Al_2O_3-SiO_2," *J. Am. Ceram. Soc.*, **45** (5) 229–42 (1962).

[2]I. A. Aksay and J. A. Pask, "Stable and Metastable Phase Equilibria in the System Al_2O_3," *J. Am. Ceram. Soc.*, **58** (11–12) 507–512 (1975).

[3]N. L. Bowen and Greig, "The System: Al_2O_3," *J. Am. Ceram. Soc.*, **7** (4) 238–54 (1924); ibid (5) 410.

[4]R. F. Davis and J. A. Pask, "Diffusion and Reaction Studies in the System Al_2O_3-SiO_2," *J. Am. Ceram. Soc.*, **55** (10) 525–31 (1972).

[5]S. H. Risbud and J. A. Pask, "Mullite Crystallization from SiO_2-Al_2O_3 Melts," *J. Am. Ceram. Soc.*, **61** (1–2) 63–7 (1978).

[6]S. H. Risbud and J. A. Pask, "SiO_2-Al_2O_3 Metastable Phase Equilibrium Diagram without Mullite," *J. Mater. Sci.*, **13** (11) 1449–54 (1978).

[7]S. H. Risbud, V. F. Draper, and J. A. Pask, "Dependence of Phase Composition on Nuclei Available in Al_2O_3-SiO_2 Mixtures," *J. Am. Ceram. Soc.*, **61** (9–10) 471–2 (1978).

[8]Y. Nurishi and J. A. Pask, "Sintering of $Al_2O_3$0-Amorphous Silica Compacts," *Ceram. Int.*, **8** (2) (1982).

[9]W. M. Kriven and J. A. Pask, "Solid Solution Range and Microstructures of Melt Grown Mullite," *J. Am. Ceram. Soc.*, **66** (9) 649–654 (1983).

[10]J. A. Pask, "Stable and Metastable Phase Equilibria and Reactions in the SiO_2-alpha Al_2O_3 System," *Ceram. Int.*, **9** (4) 107–13 (1983).

[11]F. Y. Galakhov, "Crystallization of Mullite in Ternary Silica-Alumina Systems," *Bull. Soc. Franc. Ceram.*, 1985 No. 38 pp. 11–16 (1958).

[12]F. Y. Galakhov, "The Character of Melting of Mullite," *Inorganic Materials*, **6** (2) 201–204 (1980).

[13]G. Rehfeld and H. E. Schwiete, "Recent Results on the Melting Behavior of Mullite," *Ber. Deut. Keram. Ges.*, **48** (6) 258–61 (1971).

[14]A. Staronka, H. Pham, and M. Rolin, "Study of the System Silica-Alumina by the Method of Freezing Curves," Rev. Int. Hautes. Temper. et. Refract. 5 pp. 111–15 (1968).

[15]G. Troemel, K. H. Obst, K. Konopicky, H. Bauer, and I. Patzak, "Investigations in the $SiO_2$0-Al_2O_3 System," *Ber. Deut. Keram. Ges.*, **34** (12) 397–402 (1957).

[16]H. R. Mueller, "Solidification Behavior of Mullite in Bulk," *Ber. Deut. Keram. Ges.*, **40** (5) 316–19 (1963).

[17]J. H. Welch, "New Interpretation of the Mullite Problem," *Nature (London)*, **186** (4724) 545–46 (1960).

[18]T. Horibe and S. Kuwabara, "Thermo-Analytical Investigation of the Phase Equilibria in the Al_2O_3-SiO_2 System," *Bull. Chem. Soc. Jpn.*, **40** (4) 972–82 (1967).

[19]S. Prochazka and F. J. Klug, "Infrared-Transparent Mullite Ceramic," *J. Am. Ceram. Soc.*, **66** (5) 874–80 (1983).

[20]F. J. Klug, "Alumina-Silica Phase Equilibria in the Mullite Region," Ph.D. Thesis, Rensselaer Polytechnic Institute, Troy, NY 12181. Obtainable from University Microfilms.

[21]F. D. Rossini, "International Practical Temperature Scale of 1968," *J. Chem. Thermodyn.*, **2**, 447–59 (1970).

[22]R. P. Goehner, "SPECPLOT-An Interactive Data Reduction and Display Program for Spectral Data," In *Advances in X-ray Analysis*, 23, 305–311 (1980) ed. J. R. Rhodes.

[23]W. A. Rachinger, "A Correction for the Alpha 1, Alpha-2 Doublet in the Measurement of Width of X-ray Diffraction Lines," *J. Sci. Instrum.*, **25**, 254–255 (1948).

[24]M. H. Mueller, L. Heaton, and K. T. Miller, Acta Cryst., Vol 13 p. 828,1960 A later improved version of this, B106 developed by L. Heaton, J. Gvildys and M. H. Mueller, "A Collection of Fortran Programs for Crystal Structure Analysis," NRL Report 5885, AD404 653, U.S. Department of Commerce, Office of Technical Services, Washington, D.C., 1963.

[25]Gilfrich and Birks, "Estimation of Detection Limits in X-ray Fluorescence Spectroscopy," *Anal. Chem.*, **56** (1) 77–9 (1984).

[26]W. E. Cameron, "Composition and Cell Dimensions of Mullite," *Am. Ceram. Soc. Bull.*, **56** (11) 1003–11 (1977).

[27]J. H. Welch, "Observations on the Composition and Melting Behavior of Mullite," in Transactions of the VIIth International Ceramic Congress, British Ceramic Society, London (1960).

[28]K. S. Mazdiyasni and L. M. Brown, "Synthesis and Mechanical Properties of Stoichiometric Aluminum Silicate (Mullite)," *J. Am. Ceram. Soc.*, **55** (11) 548–552 (1972).

[29]R. Rossin, J. Bersan, and G. Urbain, "The Study of the Viscosity of the Liquids Belonging to the Ternary System SiO_2-Al_2O_3-CaO," *Rev. Hautes. Temper. Refract.* **1** (2) 159–70 (1964).

[30]N. G. Ainsle, J. D. Mackensie and D. Turnbull, "The Melting Kinetics of Quartz and Cristobalite," *J. Phys. Chem.*, **65** (10) 1718–24, (1961).

[31]R. L. Roeder, F. P. Glasser, and E. F. Osborn, "The System Al_2O_3-Cr2O3-SiO_2," *J. Am. Ceram. Soc.*, **51** (10) 585–91 (1968).

The Al_2O_3-SiO_2 Phase Diagram: Metastability and Order-Disorder

RUSTUM ROY

Materials Research Laboratory
The Pennsylvania State University
University Park, PA 16802

Abstract

This paper discusses the questions which can still be raised regarding the Al_2O_3-SiO_2 phase diagram the broad outlines of which now seem to be getting more general agreement. The first concerns the importance of the ternary phase boundary location in indicating binary equilibria. The second deals with the likely generation of a metastable liquidus for the "Al_2O_3 + liquid" field by the particular experimental conditions of Aksay and Pask. The third concerns the question of the extremely complex nature of the order-disorder and crystalline solubility (stable and metastable) in the mullite phase. Very little attention has been paid by those doing the phase equilibrium work to the mineralogical-crystallographic work on this phenomenon and the impact it could have on melting relations. A summary of questions which remain open and some thought experiments which can be conducted to resolve them are presented.

Introduction

The interest in the Al_2O_3-SiO_2 system is intense, and appropriately so, since it is the key system for the ceramic community. The present author has been studying one or other aspect of the system continuously for forty years. In 1948, I began a systematic study of the system Al_2O_3-SiO_2-H_2O[1] which initiated my interest in the relations among sillimanite, andalusite, kyanite and mullite. It was this study which directly led to the discovery of the sol-gel method because of the total unreactivity of virtually *any* starting materials in this system in attempting to attain equilibrium. The use of organic precursors, silicon ethoxide and aluminum isopropoxide, and later inorganic sols, made reactions orders of magnitude more rapid in all ceramic systems including this system, Al_2O_3-SiO_2.[2] Several observations in this work suggested that the *anhydrous* equilibrium diagram of Bowen and Greig[3] which was universally accepted may not be correct in two respects. First, the extent of solid solution in mullite (which they showed as non-existent) and second the incongruent melting point of mullite. Hence in 1956 Aramaki and Roy started a careful examination of the phase equilibria in sealed noble metal containers using the quenching method. This led to a revised phase diagram which became widely accepted[4] (whether or not it is correct is another matter). This is the departure point for the present short comments which will respond to the following issues:

1. By reading the literature since Aramaki and Roy what questions can be raised about the different papers?
2. What questions remain unanswered?
3. What experiments could be done to add more evidence?

Questions Raised By Recent Work

Paucity of Low-Temperature Data on Crystalline Solutions in Mullite Phase

I take as my departure point the paper by Klug, Prochazka and Doremus,[5] the latest detailed and careful study of the system. As stated by these authors their data are essentially consistent with the Aramaki and Roy diagram although it is clearly more precise especially with respect to the melting temperature of mullite. However, in their Fig. 15 they have only five data points below 1750° and the runs are short (typically 1 hr.). Hence I would question whether the uncommon t-x *slope* of the two boundaries of the mullite crystalline solution field are sufficiently well established. Although KPD showed data supporting isothermal reversal the extrapolate to the lower temperatures (near 1000°C) of importance in the geosciences, would require peculiar changes in slope of the exsolution curve. Such changes may *not* in fact be impossible in such a complex system but we have not seen any validated so far.

Experimental Forcing of Metastability

Aksay and Pask[6] showed mullite as melting incongruently with a liquidus (strangely enough) nearly identical to the original Bowen and Greig diagram. Yet in private discussions with Dr. J. W. Greig, then at Penn State, in the seventies he did not believe that their experimental methods of fifty years earlier could justify the confidence generated by the single fact of reprinting the diagram. Especially at the high-Al_2O_3 end of the diagram he knew that they had not been able to avoid SiO_2 evaporation. Although a few quenching runs were carried out by Aksay and Pask their significant data derive from their diffusion couple experimental approach to equilibria in oxide systems. While such diffusion couple studies have been used in several metallurgical systems, there has been no example tested of their applicability to silicate and other oxide systems. The viscosities in such systems, the well known tendency to metastability not only of order-disorder say in Al-Si, but even in melting (e.g. in pure albite will tend to enhance the probability of being "trapped" in a metastable equilibrium potential well). The experimental arrangements clearly favor equilibrium between two phases—one of which is fixed—which may be either stable or metastable with each other. In the relevant case one of the two phases is *always* a "large" single crystal of sapphire, the other is a siliceous liquid with which it could be in equilibrium—either metastable or stable—only at the interface. Assuming these two phases have come to equilibrium at the surface, what Aksay and Pask failed to establish was that that assemblage was not the metastable set rather than the stable set. For example, at the 3:2 mullite composition the two optional sets of equilibrium phases at 1850° (Aramaki and Roy data) are *either* 3:2 mullite xl + 3:2 mullite liq. *or* cor. + liq. (about 50% Al_2O_3-50% SiO_2). Aksay and Pask made no attempt to establish the latter by melting mullite to nearly 50% cor. + liquid. The experimental conditions superimposed—i.e. the pre-existing nucleus of an 'infinite' supply of corundum—may be adduced to support the presumption of metastability of this assemblage. Under these conditions, I believe that the Aksay-Pask liquidus could be the metastable corundum-SiO_2 liquidus. Metastable liquidus lines have been observed in other silicate systems. They seem especially to occur whenever one adds an ion with a strong octahedral field preference (hence tending to "unmix" on a nanoscale). [See the system $CaO-TiO_2-SiO_2$ by Iwase and Fukusima.[7]]

Conclusive Evidence From Phase Boundary Location In Ternary Systems

One of the arguments which led Aramaki and Roy[4] to question the incongruent melting of mullite was the location of the phase boundary of the corundum field in a variety of ternary systems with other simple oxides. Toropov and Galakhov[9] had already used this argument in their 1951 paper questioning the location of this boundary in the ternary systems with CaO and MgO. Aramaki and Roy[10] re-determined the equilibria in both these systems and showed very strong evidence against the peritectic composition essential to the Bowen and Greig, Ref. 3 (and later Aksay and Pask, Ref. 6) diagram. Following the data in Ref. 10, it is inconceivable that any phase boundary could run from the proposed peritectic (of Refs. 3,6) around composition #9 in the CaO system and composition #5 in the MgO system and then turn around to the rather well-established ternary eutetics between mullite, corundum and respectively anorthite (CaO system) and indialite or cordierite (MgO system). Klug et al. in their closing paragraph agree with this line of argument and do in fact cite this same kind of evidence for the Cr_2-O_3-Al_2O_3-SiO_3 system by Roeder, Glasser and Osborn[11] but curiously fail to mention the second Aramaki and Roy[10] paper. Indeed a very large number of ternary Al_2O_3-SiO_2-X systems have been determined after 1950 by many different workers (see e.g. Diagrams #734, 556, 2586 in Ref. 12) and none show any support for locating the peritectic at the Bowen and Greig composition.

Order Disorder in Mullite Superimposed on Stable and Metastable Crystalline Solubility

One of the difficulties in studying equilibria in this system is the enormously complicated crystallographic order-disorder phenomena which occur in the mullite phase. Many workers have commented on this but have failed to note its connection with melting relations. Aksay and Pask[6] do in fact show two dotted extensions of a so-called liquidus for "disordered" and "ordered-mullites" but these in fact on their Fig. 1 do not refer to isoplethal structural disorder but to quite different *compositions* of mullite. Hence they are concerned with Al-Si and vacancy ordering. Aside from these speculations let us consider first the data.

a. There is incontrovertible evidence of stable crystalline solubility of at least roughly 3-4 mole % Al_2O_3 in the mullite phase. This range brackets the 61 mole % at low temperatures from 1700°C down[4] and moves to bracket 64.5% composition at about 1850°C according to Klug et al. In nature the most siliceous mullite (at much lower temperatures) is found to have only 57 mole % Al_2O_3 [Cameron[13]] and the highest Al_2O_3 natural mullite is only 60% Al_2O_3 corresponding roughly to the Aramaki and Roy diagram. Kriven and Pask[8] studied melt-derived mullites which they believe to be metastable which extended the metastable crystalline solubility to 75 mol%.

Cameron has shown a structural similarity between mullite and i-Al_2O_3, the tetragonal polymorph and *speculates* on possible (but not yet found) continuous crystalline solutions from pure iota-Al_2O_3 via various ordering schemes of Al^{3+}, Si^{4+} and vacancies, to nearly pure sillimanite and the orthorhombic mullite and its putative tetragonal polymorph of pure Al_2O_3 composition.

While Aksay and Pask refer almost exclusively to variations in the mullite structure carried by *compositional* changes. Aramaki and Roy showed data and claimed that "one of the most important new findings is that in addition to changes in the cell parameters caused 'by compositional changes. . .' *there are changes of nearly the same magnitude due to heat treatment of one mullite of fixed composition.*" These changes are quite slow (hours to days even in the 1600-1850 range). Hence the possibility, nay the probability, exists that various thermal reaction studies may in fact be dealing with mullite which is metastable under the new conditions. *How drastically this Al-Si-[ ]-O order disorder will impact melting behavior is difficult to predict.* From the kinetics observed for the disordering, as in the feldspars, it should be easy to superheat the ordered phases, and possibly metastably melt them. In other words this fact introduces yet one more uncertainty into the *details* of the phase diagram. Figures 1 and 2 summarize qualitatively some of the stable and metastable relations that exist, although we do not yet know the magnitude of the thermal effects which can be attributed to structural order-disorder.

Some Proposed Experiments to Shed Light on the Al_2O_3-SiO_2 System

1. The nature of the *isoplethal* structural disorder in *mullites* should be studied in detail, especially using solid state NMR techniques, and equilibrating as below.
2. Long-term runs dry at 1700-1850 for many hours but also hydrothermally for weeks at 800°C, and from various starting materials such as monophasic and disphasic xerogels [Komarneni and Roy[14]] should be carried out with the same fixed composition to see if there is a *clearcut equilibrium order-disorder transition* in isoplethal material.

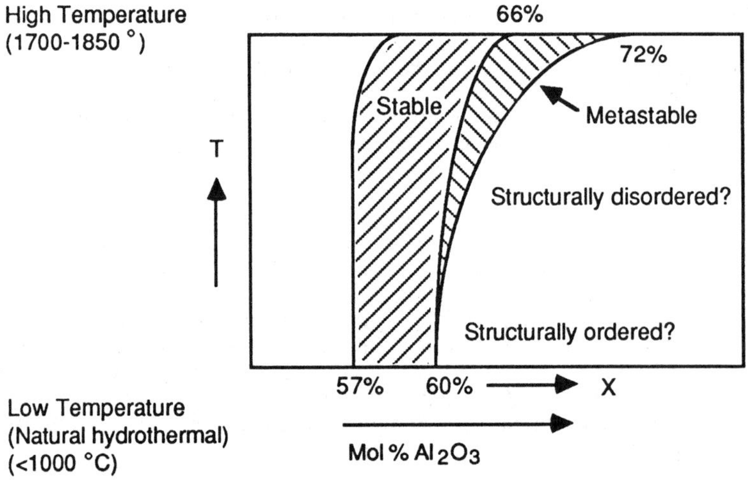

Fig. 1. Schematic representation of crystalline solubility in the mullite phase; showing the metastable extension which itself must be temperature dependent. Super-imposed on these compositional effects there are structural transitions (see Fig. 2) which themselves can be metastably overriden. Overlapping of structural and compositional effects provide enormous possibilities for confusion.

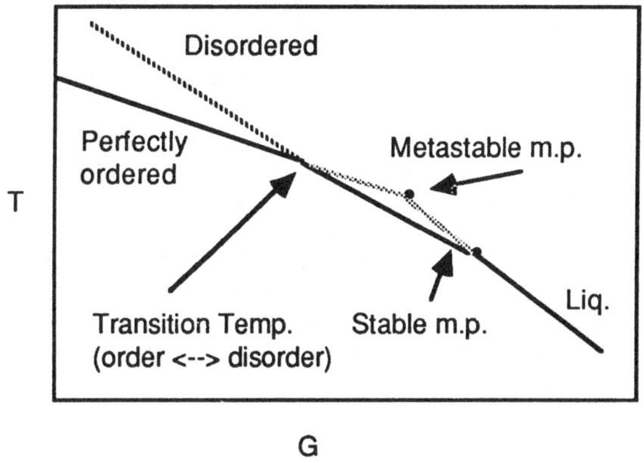

Fig. 2. Free-energy temperature diagram for mullites assuming a (unlikely) first order transition. This illustrates the fact that an ordered mullite will melt metastably below the equilibrium melting point. Melting points should therefore be determined only on samples thoroughly equilibrated just below the m.p.

3. Since there are certainly some equilibrium phase transitions in mullite (at any fixed composition) then the metastable melting point of the ordered low-temperature phases must be determined in a rapid rise time furnace, such as a strip furnace. Then assuming that there is a measurable difference, we can ascribe this term to the lower melting of the metastable phase (in this case the ordered one). These data *may* have relevance to some of the differences observed among experimental determinations of the phase diagram.
4. Another set of experiments which could resolve the issue would be to duplicate the Aksay and Pask method in other well-known *silicate* systems to see whether in fact the "crystallographically induced metastability" alluded to above manifests itself there also.

Acknowledgments

This research has been supported by National Science Foundation Grant #CBT-8619064.

References

[1]R. Roy and E. F. Osborn, "The System Al_2O_3-SiO_2-H_2O," *Am. Min.*, **39**, 853–85 (1954).

[2]R. Roy, "Aids in Hydrothermal Experimentation II. Methods for Making Mixtures," *J. Am. Cer. Soc.*, **39**, 145–146 (1956).

[3]N. L. Bowen and J. W. Greig, "The System Al_2O_3-SiO_2," *J. Am. Cer. Soc.*, **7**, 238–254 (1924).

[4]S. Aramaki and R. Roy, "Revised Phase Diagram for the System Al_2O_3-SiO_2," *J. Am. Cer. Soc.*, **45**, 229–242 (1962).

[5]F. J. Klug, S. Prochazka and R. H. Doremus, "Al_2O_3-SiO_2 Phase Diagram in the Mullite Region," Tech. Info. Series 86 CRD007, Gen. Elect. Co., Schenectady, NY (1986); now publ. in *J. Am. Cer. Soc.*, **70**, 750–759 (1987).

[6]I. A. Aksay and J. A. Pask, "Stable and Metastable Equilibria in the System Al_2O_3-SiO_2," *J. Am. Cer. Soc.*, **58**, 507–512 (1975).

[7]K. Iwase and M. Fukusima, "Metastable Crystallization of $CaTiO_3$, TiO_2, SiO_2," Science Repts. Tohoku Univ. Serie 1, Honda Vol. pp. 454–464 (1936).
[8]N. A. Toropov and F. Y. Galakhov, "The System Al_2O_3-SiO_2," *Dokl. Akad. Nauk, SSSR*, **78**, 299–302 (1951).
[9].S. Aramaki and R. Roy, "The Mullite-Corundum Boundary in CaO-Al_2O_3-SiO_2 and MgO-Al_2O_3-SiO_2," *J. Am. Cer. Soc.*, **42**, 644–645 (1959).
[10]R. L. Roeder, F. P. Glasser and E. F. Osborn, "The System Al_2O_3-Cr_2-O_3-SiO_2," *J. Am. Cer. Soc.*, **51**, 585–91 (1968).
[11]Phase Diagrams for Ceramists, 5 Volumes, Publ. by Am. Cer. Soc., 1964 ff.
[12]Cameron, "Mullite—A Substituted Alumina," *Am. Min.*, **62**, 747–755 (1977).
[13]S. Komarneni, Y. Suwa and R. Roy, "Application of Compositionally Diphasic Xerogels for Enhanced Densification: System Al_2O_3-SiO_2," *J. Am. Ceram. Soc.*, **69**, pp. C155–C156 (1986).

Phase Equilibrium and Metaequilibrium Calculations in the System Al_2O_3-SiO_2

CHU KUN KUO

Shanghai Institute of Ceramics
Chinese Academy of Sciences
865 Changning Road
Shanghai 200050, China

Abstract

Phase diagrams of the binary system Al_2O_3-SiO_2 representing various stable and metastable phase assemblages have been calculated, including crystobalite-mullite, mullite-corundum, crystobalite-sillimanite, crystobalite-kyanite, sillimanite-corundum, kyanite-corundum, crystobalite-corundum, mullite-liquid, sillimanite-liquid and kyanite-liquid systems. Simple, polymerized molecule and cation sublattice models are used to evaluate the mixing free energy terms for the liquid state. Calculated phase diagrams are constructed and compared with literature data.

Introduction

The phases in the system Al_2O_3-SiO_2 are constituents of many ceramic products. This system also forms an important basic binary system in a series of multi-component systems of interest in ceramic industries. Phase diagram studies of this system[1–14] indicate that mullite is the only stable intermediate phase. Moreover these studies show that mullite has limited solid solubility in Al_2O_3. However, several conflicting results on the melting and metaequilibrium behavior have been observed. These discrepancies in experiments have attracted great attention and led to several systematic and sophisticated investigations.

In addition to the experimental studies, Howald and Eliezer,[15] Kaufman[16] and Doerner et al.[17] have computed the Al_2O_3-SiO_2 phase diagram from first principals of thermodynamics. Their computation results show the mullite to melt either congruently or incongruently. On the other hand, Pask and coworkers[6,8] have reported a metastable phase diagram of Al_2O_3 and SiO_2 without the formation of mullite. The present investigation computes both the equilibrium and metaequilibrium diagrams for different stable and metastable phase assemblages. The mixing free energy is calculated from the structural model of a melt based on molecular or cation sublattice species.

Basic Equations For Equilibrium Calculation

The total free energy of binary system A-B is evaluated by the following equations:

$$\text{Solid A + Solid B: } G = n_A G_A^s + n_B G_B^s \quad (1)$$

$$\text{Solid A + Liquid: } G = n_A^s G_A^s + (n_A - n_A^s) G_A^l + n_B G_B^l +$$

$$RT\left(\frac{n_A - n_A^s}{n_A - n_A^s + n_B} \ln \frac{n_A - n_A^s}{n_A - n_A^s + n_B} + \frac{n_B}{n_A - n_A^s + n_B} \ln \frac{n_B}{n_A - n_A^s + n_B}\right) + {}^{ex}G^l \quad (2)$$

$$\text{Solid B + Liquid: } G = n_B^s G_B^s + n_A G_A^l + (n_B - n_B^s) G_B^l +$$

$$RT\left(\frac{n_A}{n_A + n_B - n_B^s} \ln \frac{n_A}{n_A + n_B - n_B^s} + \frac{n_B - n_B^s}{n_A + n_B - n_B^s} \ln \frac{n_B - n_B^s}{n_A + n_B - n_B^s}\right) + {}^{ex}G^l \quad (3)$$

$$\text{Liquid: } G + n_A G_A^l + n_B G_B^L + RT\left(\frac{n_A}{n_A + n_B} \ln \frac{n_A}{n_A + n_B} + \frac{n_B}{n_A + n_B} \ln \frac{n_B}{n_A + n_B}\right) + {}^{ex}G^l \quad (4)$$

where n_A and n_B represent the number of moles of the component A and B, G is the Gibbs free energy, ${}^{ex}G$ represents the excess free energy, superscripts s and l denote solid and liquid state respectively, and R and T have their usual meaning. In the calculation ${}^{ex}G^l$ is taken as zero for a quasi-ideal solution consisting of supposed structural species, and the solubilities in solids are ignored. The equilibria are computed by minimizing the free energies described in Equations (1)-(4).

Calculated Phase Diagrams

The free energy at various temperatures for individual species in the calculation is taken from JANAF Tables[18] and listed in Table I. The free energies of the liquid phases $3Al_2O_3{\cdot}2SiO_2$ and $Al_2O_3{\cdot}SiO_2$ have been estimated by summing the free energy of each of the oxide components and adjusting the resultant values using known melting point data. The following equilibrium and metaequilibrium diagrams are calculated:

Equilibrium diagrams: SiO_2-$3Al_2O_3{\cdot}2SiO_2$, $3Al_2O_3{\cdot}2SiO_2$-Al_2O_3.
Metaequilibrium diagrams: SiO_2-$Al_2O_3{\cdot}SiO_2$, $Al_2O_3{\cdot}SiO_2$-Al_2O_3,
SiO_2-Al_2O_3, $3Al_2O_3{\cdot}2SiO_2$-liquid, $Al_2O_3{\cdot}SiO_2$-liquid.

Table 1. Free energy of the phases in Al_2O_3-SiO_2 system

Phases	m.p. in K	G in J/mole
SiO_2 (s,crystobalite)	1995	-926020 + 187.99T/K
SiO_2 (l)		-918080 + 184.01T/K
$3Al_2O_3{\cdot}2SiO_2$ (s, mullite)	2183	-6879280+1321.89T/K
$3Al_2O_3{\cdot}2SiO_2$ (l)		-6567120+1178.90T/K
Al_2O_3 (s,corundum)	2315	-1676560+ 319.66T/K
Al_2O_3 (l)		-1560940+ 269.71T/K
$Al_2O_3{\cdot}SiO_2$ (s,sillimanite)	1845	-2612950+ 514.77T/K
$Al_2O_3{\cdot}SiO_2$ (s,kyanite)	1715	-2615720+ 521.92T/K
$Al_2O_3{\cdot}SiO_2$ (l)		-2496790+ 454.31T/K

Four structural models of melt are used to evaluate the free energy of liquid mixtures. The supposed species in the melt are (1) SiO_2, $Al_2O_3{\cdot}SiO_2$, $3Al_2O_3{\cdot}2SiO_2$ and Al_2O_3; (2) SiO_2 and Al_2O_3; (3) Al+Si cation sublattice; and (4) $(SiO_2)_n$ and $(3Al_2O_3{\cdot}2SiO_2)_m$. The model (4) is employed only to calculate the equilibria in the SiO_2-rich portion.

Considering a given initial molar composition of $Al_2O_3=n_A$ and $SiO_2=n_B$, according to the four models described above we have

Model 1: $SiO_2-3Al_2O_3{\cdot}SiO_2$: $3Al_2O_3{\cdot}2SiO_2=n_A/3$, $SiO_2=n_B-2n_A/3$.
$3Al_2O_3{\cdot}2SiO_2-Al_2O_3$: $3Al_2O_3{\cdot}2SiO_2=n_B/2$,
$Al_2O_3=n_A-3n_B/2$.
$SiO_2-Al_2O_3{\cdot}SiO_2$: $Al_2O_3{\cdot}SiO_2=n_A$, $SiO_2=n_B-n_A$.
$Al_2O_3{\cdot}SiO_2-Al_2O_3$: $Al_2O_3{\cdot}SiO_2=n_B$, $Al_2O_3=n_A-n_B$.
$SiO_2-Al_2O_3$: $Al_2O_3=n_A$, $SiO_2=n_B$.
Model 2: $Al_2O_3=n_A$, $SiO_2=n_B$.
Model 3: $Al=2n_A$, $Si=n_B$.
Model 4: $SiO_2-3Al_2O_3{\cdot}2SiO_2$: $(SiO_2)_n=(n_B-2n_A/3)/n$,
$(3Al_2O_3{\cdot}2SiO_2)_m=n_A/3m$.

The computational results are shown in Fig. 1. Table II gives the calculated invariant points.

Discussion

Melting Behavior of Mullite

The above computation shows that the thermal stability of mullite is different for different structural models of melt assumed in the equilibrium and metaequilibrium calculations. The model which consists of a $3Al_2O_3{\cdot}2SiO_2$ molecule leads to a congruent melting point of mullite, whereas the simple molecule Al_2O_3 + SiO_2 and cation sublattice Al+Si models produce a primary crystallization field of corundum at the mullite compositional point, e. i. the mullite melts incongruently. We are led to the conclusion that the discrepancy regarding the melting behavior of mullite may reflect a structural change in the existing liquid phase which arises from different experimental conditions. According to the investigations of Howald and Eliezer[15] and Cameron[13] a new silica-dissolving Al_2O_3 phase may be formed which changes the crystallization behavior of the binary melt. It seems logical to assume that the formation of the new stable or metastable crystalline phase might be associated with the species existing in the liquid where crystallization has occurred on cooling.

Metastable Phase Diagram

Three types of metastable diagrams are illustrated in Fig. 1. The first concerns the formation of the metastable phases sillimanite and kyanite. As shown in Figs. 1a-1c, the crystallization fields of the $Al_2O_3{\cdot}SiO_2$ phases always occur beneath the primary field of mullite or corundum. As a consequence, the separation of the $Al_2O_3{\cdot}SiO_2$ crystals must be prevented by crystallization of the more stable phases. The second type of metastable phase diagrams involve metastable coexistence of crystalline Al_2O_3 and SiO_2 phases. The liquidus curves of Al_2O_3 and SiO_2 extend into the metastable region when mullite fails to separate from melt. Phase diagrams of the third type arise simply from retarded crystallization. The undercooled liquid which persists after crystallization solidifies into a glassy substance during cooling. Frequently the melting point of a given composition decreases as a metastable phase appears.

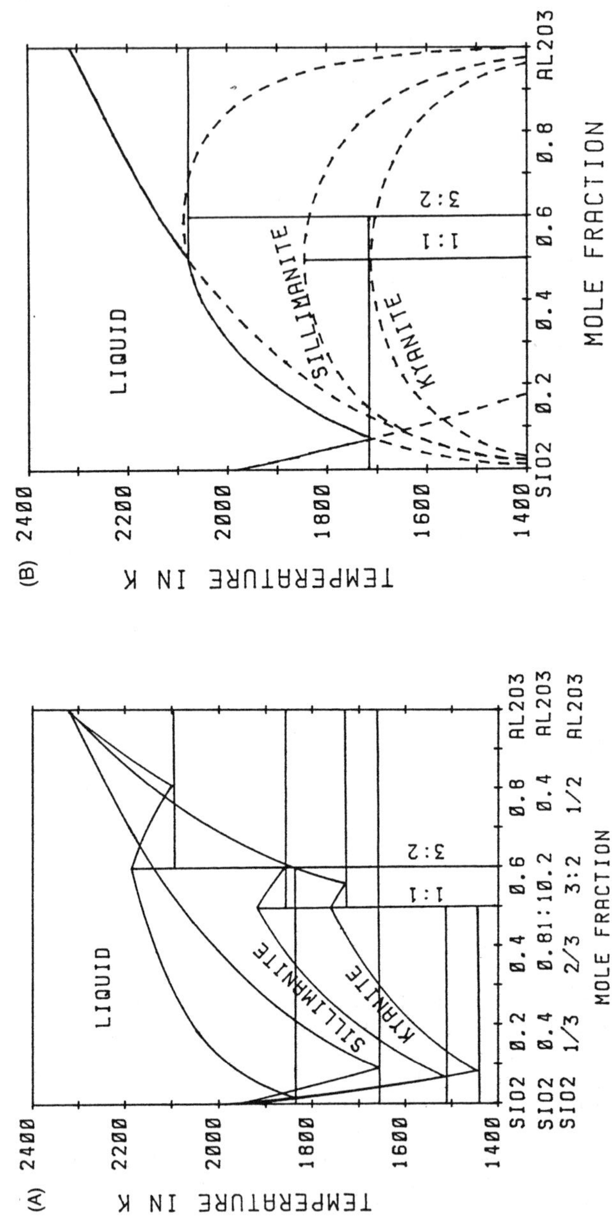
(A)
TEMPERATURE IN K
2400
2200
2000
1800
1600
1400
LIQUID
SILLIMANITE
KYANITE
1:1
3:2
SIO2 0.2 0.4 0.6 0.8 AL2O3
SIO2 0.4 0.81:10.2 0.4 AL2O3
SIO2 1/3 2/3 3:2 1/2 AL2O3
MOLE FRACTION
(B)
TEMPERATURE IN K
2400
2200
2000
1800
1600
1400
LIQUID
SILLIMANITE
KYANITE
1:1
3:2
SIO2 0.2 0.4 0.6 0.8 AL2O3
MOLE FRACTION

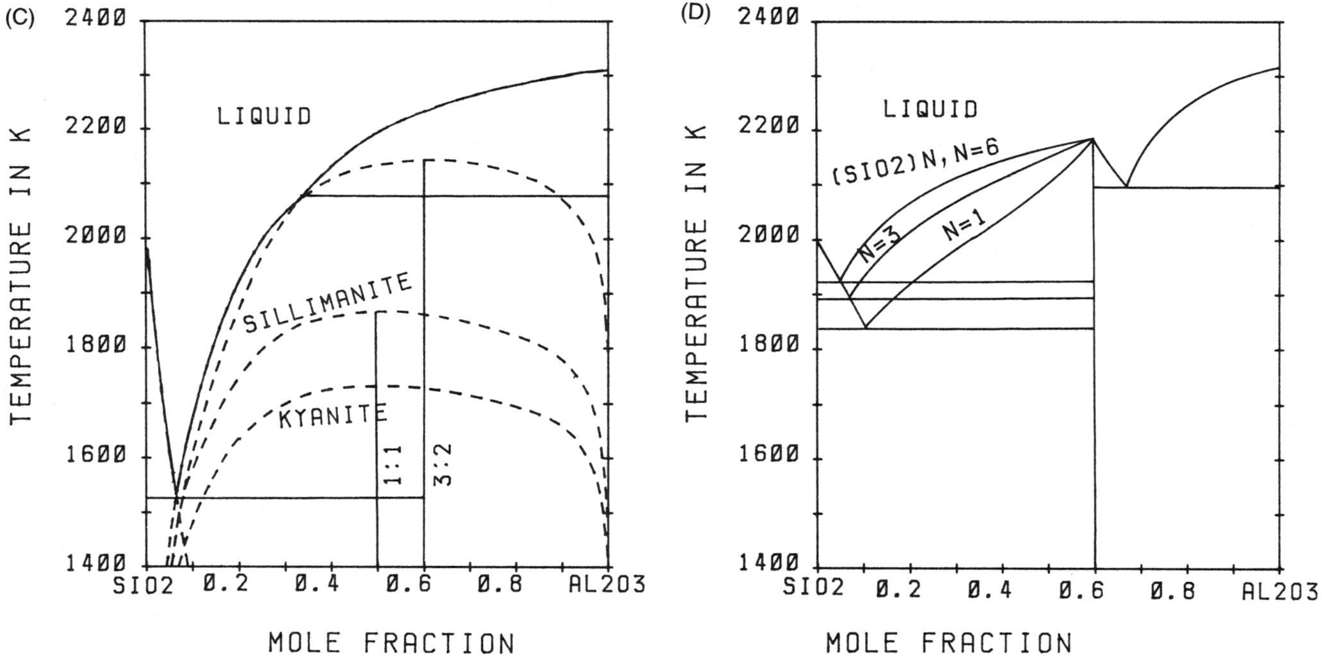

Fig. 1. Calculated phase diagrams of the Al_2O_3-SiO_2 system, (a) Model 1; (b) Model 2, the dashed lines indicate metastable liquidus curves; (c) Model 3, the dashed lines indicate metastable liquidus curves; (d) Model 4.

Table 2. Invariant points in Al_2O_3-SiO_2 system

T/K	mole% Al_2O_3	Invariant point	Phases in coexistence
Model 1			
1840	0.1032	Stable, eutectic	crystobalite-mullite-liquid
2090	0.6729	Stable, eutectic	mullite-corundum-liquid
1514	0.1408	metastable, eutectic	crystobalite-sillimanite-liquid
1854	0.5636	metastable, eutectic	sillimanite-corundum-liquid
1444	0.1657	metastable, eutectic	crystobalite-kyanite-liquid
1730	0.5356	metastable, eutectic	kyanite-corundum-liquid
1660	0.0927	metastable, eutectic	crystobalite-corundum-liquid
Model 2			
1712	0.0760	stable, eutectic	crystobalite-mullite-liquid
2078	0.5035	stable, peritectic	mullite-corundum-liquid
1648	0.0962	metastable, eutectic	crystobalite-sillimanite-liquid
1576	0.0598	metastable, peritectic	sillimanite-corundum-liquid
1573	0.1207	metastable, eutectic	crystobalite-kyanite-liquid
1658	0.0927	metastable, eutectic	crystobalite-corundum-liquid
Model 3			
1523	0.0664	stable, eutectic	crystobalite-mullite-liquid
2078	0.3354	stable, peritectic	mullite-corundum-liquid
1483	0.0738	metastable, eutectic	crystobalite-sillimanite-liquid
1474	0.0702	metastable, peritectic	sillimanite-corundum-liquid
1433	0.0838	metastable, eutectic	crystobalite-kyanite-liquid
1486	0.0732	metastable, eutectic	crystobalite-corundum-liquid
Model 4			
$(SiO_2)_{n,\ n=3}$			
1893	0.0751	stable, eutectic	crystobalite-mullite-liquid
$(SiO_2)_{n,\ n=6}$			
1923	0.0534	stable, eutectic	crystobalite-mullite-liquid

Comparisons Between the Calculated and Observed Phase Diagrams

Figure 2 illustrates the liquidus curves computed using melt models 3 and 4 together with the experimental data points taken from the literature. Except for the fusion data from Troemel et al.,[12] most of the experimental points are present around either of the calculated curves. It can be seen that the points in the high alumina region can be grouped into two separate sets of points that agree with models 3 and 4 respectively. Computation with these models results in either congruent or incongruent melting of mullite. On the other hand, the points within the compositions 0.1–0.6 Al_2O_3 are less dispersive. A comparison between the calculated and experimental liquidus temperatures suggests the applicability of model 4 with the degree of polymerization of SiO_2 equal to 3 or 6, this is sometimes considered to be the most stable configuration of the polymerized SiO_2 unit rings in a molten state in

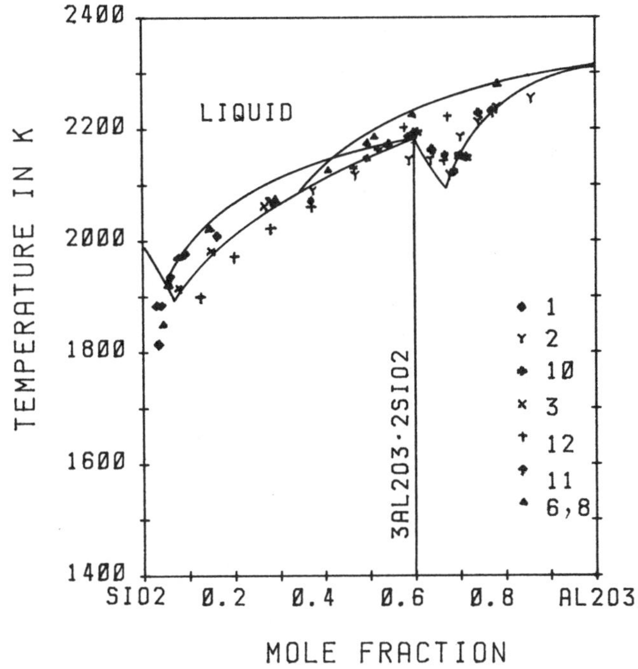

Fig. 2. A comparison of liquidus curves calculated by models 3 and 4 with literature data.

the case of the dissociation of mullite. Figure 3 gives the liquidus curves calculated from model 4 with different degrees of both the polymerization of silica and the dissociation of the mullite molecules.

Based on the above discussion, the following model of the melt structure can be tentatively suggested as an interpretation of the melting point depression in the Al_2O_3-SiO_2 binary system.

$$[6iAl + (2i+nj)Si]O_{13i+2nj,\ n=3\ \text{or}\ 6}$$

↑ Increasing temperature and probably decreasing SiO_2 concentration

$$i(3Al_2O_3{\bullet}2SiO_2) + j(SiO_2)_{n,\ n=3\ \text{or}\ 6}$$

↓ Increasing SiO_2 concentration

$$2i(3Al_2O_3{\bullet}2SiO_2)_{0.5} + j(SiO_2)_{n,\ n=3\ \text{or}\ 6}$$

where i and nj are determined by the initial molar composition and the degree of polymerization of SiO_2.

It is known that in a homogeneous phase in which the activity of each of the components varies with concentration, multiple species can coexist without changing the degree of freedom. The agreement between the calculated and experimental liquidus temperatures may likely suggest that the number of dominate species controlling the melting point depression could be small and the effect of the minor species, if present, might be ignored.

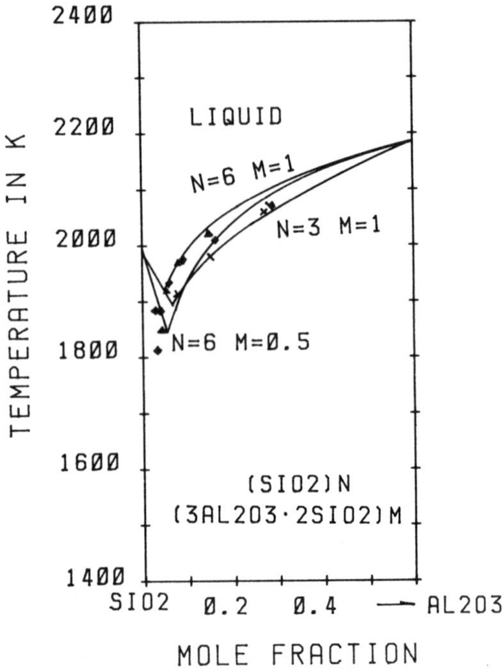

Fig. 3. Calculated SiO_2-rich liquidus curves obtained by using a $(3Al_2O_3\bullet 2SiO_2)_m$-$(SiO_2)_n$ structural model and their comparison with literature data.

Conclusions

1. Equilibrium and metaequilibrium phase diagrams of the system Al_2O_3-SiO_2 have been computed by minimizing the free energy of assumed stable or metastable phase assemblages.
2. The metaequilibrium diagram calculation involved in the presence of $Al_2O_3\bullet SiO_2$ phases. The results showed that the occurrence of the field of the metastable crystalline phases of either sillimanite or kyanite is hindered by the primary field of the more stable phases mullite and corundum.
3. Computational results suggest the melting behavior of mullite to be associated with the formation of mullite-like species in the liquid state. The melt model consisting of such species leads to congruent melting of mullite. By contrast, the melt model containing more simple molecules or a cation sublattice leads to incongruent melting.
4. The combination of a polymerized SiO_2 molecule and a cation sublattice model for the melt can provide reasonably interpretation for the liquidus temperatures in the Al_2O_3-SiO_2 binary system.

Acknowledgments

I deeply thank Ms. Yi Min Yan for her assistance in the computer work. This research program is partially supported by the Science Foundation of China.

References

[1]N. L. Bowen and J. W. Greig, "The System: Al_2O_3-SiO_2," *J. Am. Ceram. Soc.*, **7** [4] 238–54 (1924); Corrections, [5] 410.

[2]N. A. Toropov and F. Ya. Galakhov, "New Data on the System Al_2O_3-SiO_2," *Dokl. Akad. Nauk S.S.S.R.*, **78** [2] 299–302 (1951).

[3]S. Aramaki and R. Roy, "Revised Phase Diagram for the System Al_2O_3-SiO_2," *J. Am. Ceram. Soc.*, **45** [5] 229–42 (1962).

[4]W. H. Bauer, I. Gordon and C. H. Moore, "Flame-Fusion Synthesis of Mullite Single Crystals," *J. Am. Ceram. Soc.*, **33** [4] 140–43 (1950).

[5]K. Konopicky, "Equilibrium diagram Al_2O_3-SiO_2," *Ber. Deut. Keram. Ges.*, **40** [5] 286 (1963).

[6]I. A. Aksay and J. A. Pask, "Stable and Metastable Equilibria in the System SiO_2-Al_2O_3," *J. Am. Ceram. Soc.*, **58** [11–12] 507–12 (1975).

[7]S. H. Risbud and J. A. Pask, "Mullite Crystallization from SiO_2-Al_2O_3 melts," *J. Am. Ceram. Soc.*, **61** [1–2] 63–67 (1978).

[8]R. F. Davis and J. A. Pask, "Diffusion and Reaction Studies in the System Al_2O_3-SiO_2," *J. Am. Ceram. Soc.*, **55** [10] 525–31 (1972).

[9]H. Mueller-Hesse, "Development and Present State of the Investigation of the System Al_2O_3-SiO_2," *Ber. Deut. Keram. Ges.*, **40** [5] 281–85 (1963).

[10]N. A. Toropov and F. Ya. Galakhov, "Solid Solutions in the System Al_2O_3-SiO_2," *Izvest. Akad. Nauk S.S.S.R., Otdel. Khim. Nauk*, **1957**, No. 1, pp. 8–11.

[11]P. P. Budnikov, S. G. Tresvyatski and V. I. Kushakovski, "Making the Phase Diagram of the System Al_2O_3-SiO_2 More Precise," *Dokl. Akad. Nauk S.S.S.R.*, **93** [2] 281–83 (1953).

[12]G. Troemel, K. H. Obst, K. Konopicky, H. Bauer and I. Patzak, "Investigations in the System Al_2O_3-SiO_2," *Ber. Deut. Keram. Ges.*, **34** [12] 397–402 (1957).

[13]W. E. Cameron, "Mullite: A Substituted Alumina," *Am. Mineral.*, **62** [7–8] 747–55 (1977).

[14]F. J. Klug, S. Prochazka and R. H. Doremus, "Alumina-Silica Phase Diagram in the Mullite Region," *J. Am. Ceram. Soc.*, **70** [10] 750–59 (1987).

[15]R. A. Howald and I. Eliezer, "Thermodynamic Properties of Mullite," *J. Phys. Chem.*, **82** [20] 2199–204 (1978).

[16]L. Kaufman, "Calculation of Quasibinary and Quasiternary Oxide Systems—II," *CALPHAD*, **3** [1] 27–44 (1979).

[17]P. Doerner, L. J. Gauckler, H. Krieg, H. L. Lukas, G. Petrow and J. Weiss, "On the Calculation and Representation of Multicomponent Systems," *CALPHAD*, **3** [4] 241–57 (1979).

[18]D. R. Stull and H. Prophet, JANAF Thermochemical Tables, 2nd ed., NSRDS-NBS 37 (1971).

Revisiting Metastable Immiscibility and Structure in Silica-Alumina Glasses

SUBHASH H. RISBUD

Department of Materials Science and Engineering
University of Arizona
Tucson, AZ 85721

Abstract

The existence and location of metastable liquid-liquid immiscibility phase boundaries in the silica-mullite and mullite-alumina regions of the silica-alumina system are reviewed. A survey of the experimental and theoretical work in the last 25 years indicates that the composition boundaries of the miscibility gap in the silica-mullite system are known with reasonable degree of agreement but the consolute temperature of the gap is still not established. Little experimental data exists for the thermodynamically predicted metastable liquid immiscibility dome in the mullite-alumina region of the phase diagram. Microstructures resulting from the presence of metastable liquid immiscibility have been observed in rapidly quenched glasses by transmission electron microscopy (TEM). Local atomic coordination probed by magic angle sample spinning nuclear magnetic resonance spectroscopy (MASS-NMR) shows the presence of 4, 5, and 6 fold Al sites in rapidly quenched phase separated silica-alumina glasses.

Introduction

The history and current status of phase equilibria in the silica-alumina system dating back to the work of Bowen and Greig in 1924[1] have been thoroughly reviewed in this workshop. Although much new understanding has evolved,[2,3] many basic points of conflict related to stable equilibrium boundaries, melting behavior of mullite, and metastability still remain unresolved. The location of metastable liquid immiscibility in this system, specially the upper consolute temperature of the proposed miscibility gaps(s), is also an area of considerable disagreement. Attempting to outline the liquid immiscibility regions had at one time, in 1974, seemed to me to be a somewhat routine exercise.[4] The fact that miscibility gaps in this system are not known with precision even today and that some curiosity driven scientists still find this a fascinating subject is testimony to the enduring and astonishing complexity of the silica-alumina system.

In this overview, I wish to present my perspective of the efforts to locate the metastable liquid-liquid immiscibility boundaries in the silica-alumina system. Metastable liquid-liquid immiscibility in silica-alumina glasses has been investigated by microscopic observations, solution thermodynamic calculations, x-ray and neutron scattering, and scanning transmission electron microscopy (STEM). Some of our recent work has also used high resolution electron microscopy (HREM) and magic angle sample spinning nuclear magnetic resonance spectroscopy (MASS-NMR) to study the structure of rapidly quenched phase separated silica-alumina glasses.

Metastable Immiscibility in the Silica-Mullite Region

The first suggestion of possible immiscibility in this system dates back to the famous phase diagram study of Aramaki and Roy[2] in which they stated that, on the basis of Dietzel's field strength calculations,[5a] a wide (60 mole%) miscibility gap might be expected at any temperature below about 2000°C. This was based on Roy's earlier work[5b] on the connection between immiscibility, subsolidus nucleation, and the shape of liquidus curves. These studies led Aramaki and Roy to suggest that the flat mullite liquidus indicated a strong unmixing tendency. MacDowell[6] showed experimentally that the formation of mullite in aluminosilicate glass-ceramics was, in fact, preceded by immiscibility. Galakhov and Konovalova[7] postulated micro and macro immiscibility based on petrographic microscopy observations while Ganguli and Saha[8] noted that their observations on glasses of composition 3.5 to 13.5 mole% alumina were not consistent with an immiscibility dome region. Clear evidence for even an approximate location of the miscibility gap boundary was not obtained in any of the above studies although the existence of subliquidus metastable immiscibility was widely suspected.

In a classic paper published in 1969, MacDowell and Beall[9] proposed the compositional boundaries of immiscibility in the silica-mullite region of the silica-alumina system based on investigations of glasses from 5 to 60 mole% alumina. They prepared glasses by pouring melts on water cooled stainless steel rollers and studied the morphology of quenched and heat treated samples by replica electron microscopy. Textures observed in electron microscopy suggested that a miscibility gap extended from about 7 to 55 mole% alumina with the possibility of spinodal decomposition in glasses in the range 20 to 35 mole% alumina. Figure 1 shows the gap proposed by MacDowell and Beall[9] along with several metastable immiscibility domes proposed by subsequent research investigations.[10–12]

Takamori and Roy[13] studied the crystallization of splat cooled and flame sprayed silica-alumina glasses and reported that over a wide range of compositions crystallization occurred rapidly at near 1000°C. Takamori and Roy reasoned that since crystallization was rapid at 1000°C, MacDowell and Beall must have studied glasses that had been through recrystallization and the morphology of the crystallized phases may not reflect the original glass-in-glass microstructure. In any case, Takamori and Roy's data did not point to the location of the miscibility gap although their work did support the tendency toward unmixing.

Risbud and Pask[14] performed thermodynamic calculations based on the regular solution model, following the procedures described by Rey[15] and Charles,[16] and using the liquidus curve data from the phase diagram proposed by Aksay and Pask.[3] The activities of silica, alumina, and mullite species of various assumed compositions showed strong deviations from ideal mixing signalling impending immiscibility. While the data obtained were approximate due to the limitations of the solution model it was possible to bracket a most probable miscibility gap extending from 11 to 50 mole% alumina (at 1100°C) with an upper consolute temperature of 1540°C and a critical composition of 36 mole% alumina. The calculations also indicated another miscibility gap at the alumina end of the diagram (see later). TEM data also showed that immiscibility was realizable in a slowly cooled 15 mole% alumina glass sample. The gap proposed by Risbud and Pask was similar to that suggested by the experimental work of MacDowell and Beall and the composition limits of immiscibility seemed to be in reasonable agreement. Howald and Eliezer[17] studied the

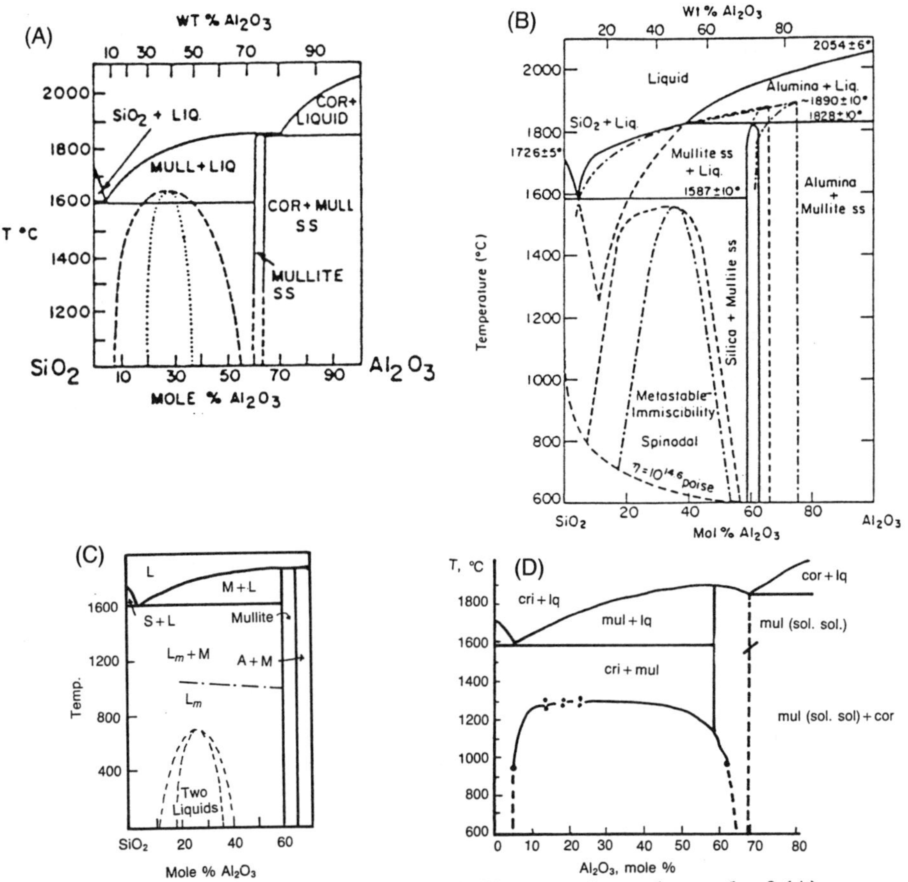

Fig. 1. Proposed metastable liquid miscibility gaps from the work of (A) MacDowell and Beall (ref. 9); (B) Risbud and Pask (ref. 14); (C) Jantzen and Herman (ref. 11); and (D) Aver'yanov et al. (ref. 12).

thermodynamic properties of mullite and liquids in the alumina-silica system. Redlich-Kister coefficients obtained for the liquid phase from the liquidus curve data of Davis and Pask[18] were in agreement with the existence of liquid immiscibility at temperatures below 1375°C. The difficulty of determining the temperature limits of immiscibility with adequate precision remains unsolved, no doubt complicated by the fact that homogeneous glasses cannot be obtained even by rapid quenching and crystallization intervenes at about 950 to 1000°C.[13,19,20]

Jantzen et al.[21] used neutron scattering experiments on rapidly quenched glasses (containing from 20 to 35 mole% alumina) to outline a spinodal boundary in the silica-mullite region. The spinodal was then used to calculate the boundary of the miscibility gap following the method described by Hilliard and Cook.[22] Jantzen et al. found that the interconnected quenched microstructure could be enhanced by heat treatment at 100 to 400°C, and the neutron scattering profiles suggested homogenization of the quenched glasses in to a single phase glass above 700°C.

Based on these data Jantzen et al. made the startling prediction of a miscibility gap with a consolute temperature of 650 to 700°C and a critical composition of 28 mole% alumina, with limits of 12 to 43 mole % alumina at ambient temperatures (see Fig. 1).

Risbud and Pask[23] attempted to explain the low temperature of this gap by suggesting that the rapidly cooled glasses studied by Jantzen et al. must have started to decompose at low temperatures because the mobility of species in the fast quenched frozen state is perhaps higher. Also, much of the previous data was strongly indicative of phase separation in glasses processed outside the 12 to 43 mole% alumina composition range. Yet, as Jantzen and Herman[21] pointed out phase separation observed in quenched glasses could occur during any part of the quench and it thus becomes difficult to estimate the temperature at which the undercooled melt crosses the miscibility gap boundary. Risbud and Pask's data[24] on mullite crystallization from homogenized melts (held at 2090°C for about 1 hour) indirectly suggests that for melts containing 47 mole% alumina phase separation might start (during cooling) at about 1325°C. Mehta and Risbud's[25] STEM microanalyses of a 15 mole % alumina glass cooled at normal rates showed that phase separated glassy droplets (40 nm) of composition greater than 51 mole% alumina were dispersed in a glass matrix containing less than 6 mole% alumina. Howitt and Jantzen[26] also studied the immiscibility in this system by analytical electron microscopy of samples aged at several temperatures.

Aver'yanov et al.[12] used a combination of small angle X-ray scattering (SAXS), and electron microscopy to study the phase decomposition of silica-alumina glasses. X-ray amorphous plates (0.1 to 0.2 mm thick) of glasses containing up to 50 mole% alumina were obtained by collapsing melted droplets between two polished metal plates. The estimated cooling rates were 10 K/sec in the 1300 to 1000°C interval. Aver'yanov et al. used the SAXS method which they state made it possible to record the SAXS intensity on the same volume of the specimen through its complete heat treatment cycle (samples were heat treated at temperatures from 100 to 800°C for 2 to 3 hours). Thus, the SAXS intensity did not change in samples isothermally heated between 100 and 800°C indicating that the structure had not changed by heat treatment. Aver'yanov et al. compared their SAXS experiments with the neutron scattering work (SANS) of Jantzen et al.[21] and felt that the changes in the SANS profiles observed by Jantzen et al. were due to random differences in the structure caused by deviations in specimen composition and cooling rates. Aver'yanov et al. proposed a miscibility gap with a consolute temperature of 1300°C and a critical composition of 30 mole% alumina with limits of 5 and 62 mole% alumina at ambient temperatures (see Fig. 1).

Galakhov et al.[27] made a deliberate attempt to determine the "upper temperature of phase separation" by melting a 20 mole% alumina composition at 1950°C, undercooling to a temperature of 1700 to 900°C and rapidly cooling to room temperature (this was called "heat treatment from above"). In a second series, "heat treatment from below," quenched samples were reheated to 1000 to 1850°C and returned to room temperature for analyses. The results of their experiments by the "heat treatment from above" indicate that phase separation commenced at some temperature between 1275 and 1300°C. They were not able to show phase separation using the "heat treatment from below" processing due to crystallization of the samples. The temperature of 1275 to 1300°C as the approximate miscibility gap boundary for the 20 mole% alumina melt appears to be consistent with the phase separation region determined by extrapolation of the ternary surface in the gallia-alumina-silica system.[28]

Other Metastable Liquid Miscibility Gaps

Thermodynamic calculations by Risbud and Pask[14] raised the intriguing possibility of another liquid-liquid miscibility gap between mullite and alumina. Calculated activities in this region show pronounced positive deviations from ideal mixing and the free energy of mixing versus composition curves point to subliquidus immiscibility at a critical composition of 78 mole% alumina with an upper consolute temperature of 1120°C. A calculation using mullite and alumina as chemical species in the solution model even predicted stable liquid immiscibility at temperatures up to and greater than 2100°C. Perhaps this large variation in the consolute temperature of the immiscibility region is merely an artifact of the molecular species chosen for the calculation. Nonetheless, in a system that is known to yield its share of surprises, demonstration of immiscibility in high alumina liquids in the mullite-alumina range can not be totally ruled out. Indeed, one result of regular solution thermodynamic activities of silica and alumina obtained by Risbud's[4] calculations suggest that a single broad miscibility gap may be a possibility in the silica-alumina system. The regular solution interaction parameter indicates that such a gap has a consolute temperature of 1350°C (± 50) with limits of about 22 and 75 mole% alumina at 600°C. Attempts to verify this gap experimentally may be fruitful in light of current efforts in low temperature processing of aluminosilicates.

Encouraging experimental evidence for the possible existence of a metastable gap in the mullite-alumina system was provided by McPherson[29] in rapid solidification investigations of plasma quenched silica-alumina powders of compositions greater than 70 wt% alumina. McPherson reasoned that his results (double DTA exotherms) on the 0.2 micron plasma prepared powders could be interpreted on the basis of a miscibility gap with a consolute temperature 100 to 200°C higher than that proposed by Risbud and Pask.[14] McPherson's work and similar research on novel processing of ceramics and glasses prove that awareness of immiscibility regions in unconventional compositions may be more than just an academic curiosity. Continued emphasis on new synthesis techniques and sophisticated characterization may one day give definitive answers as to the exact nature of metastable glass phase equilibria in the silica-alumina-mullite phase fields.

Local Structure of Silica-Alumina Glasses

The structure of silica-alumina glasses and melts is clearly of importance in connection with immiscibility. Concepts of structural groups in the liquid, non-uniformity of melt structure, and melt regions resembling the future precipitating phase have been mentioned in the literature associated with immiscibility and crystallization of silica-alumina melts and glasses. The tetrahedral coordination of Si is in little doubt but the coordination of Al in silica-alumina glasses has been harder to pinpoint. Previous work has suggested that Al may exist in 6 fold sites,[30] as interstitial Al,[31] or as triclusters.[32] Raman spectroscopy,[33–36] density data,[37] and radial distribution function x-ray scattering analysis[38] have been used to probe the structure of melts and glasses in the silica-alumina system. The general thinking appears to favor 4 fold Al sites at low alumina contents with a gradual change to 6 fold Al at higher alumina concentrations. Morikawa et al.,[38] for example obtained an average coordination number of 4.7 in alumina-rich glasses.

The use of magic angle sample spinning nuclear magnetic resonance spectroscopy (MASS-NMR) for probing the structure of glasses and ceramics[39] has recently provided interesting results on the structural units in rapidly quenched phase separated silica-alumina glasses. Figures 2 and 3 show the data obtained by Risbud

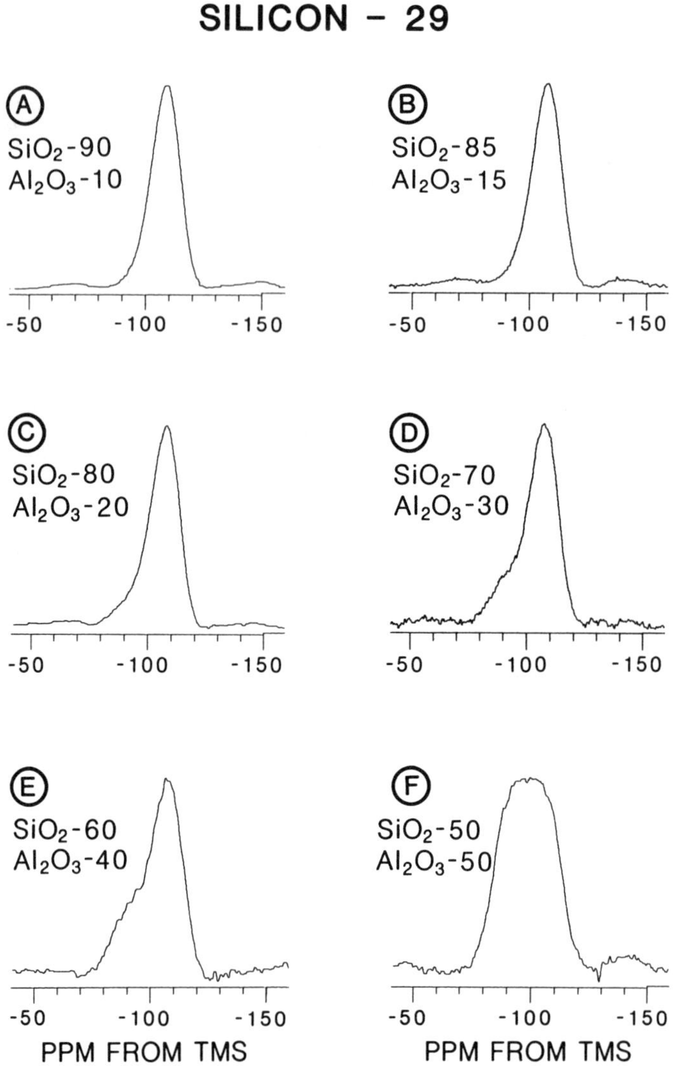

Fig. 2. Si-29 MASS-NMR spectra of silica-alumina glasses (compositions labelled in wt%) obtained at a field strength of 8.45T with a 30 degree spin-system excitation and recycle times from 90 to 600 seconds. Tetramethylsilane (TMS) was used as the standard.

ALUMINUM - 27

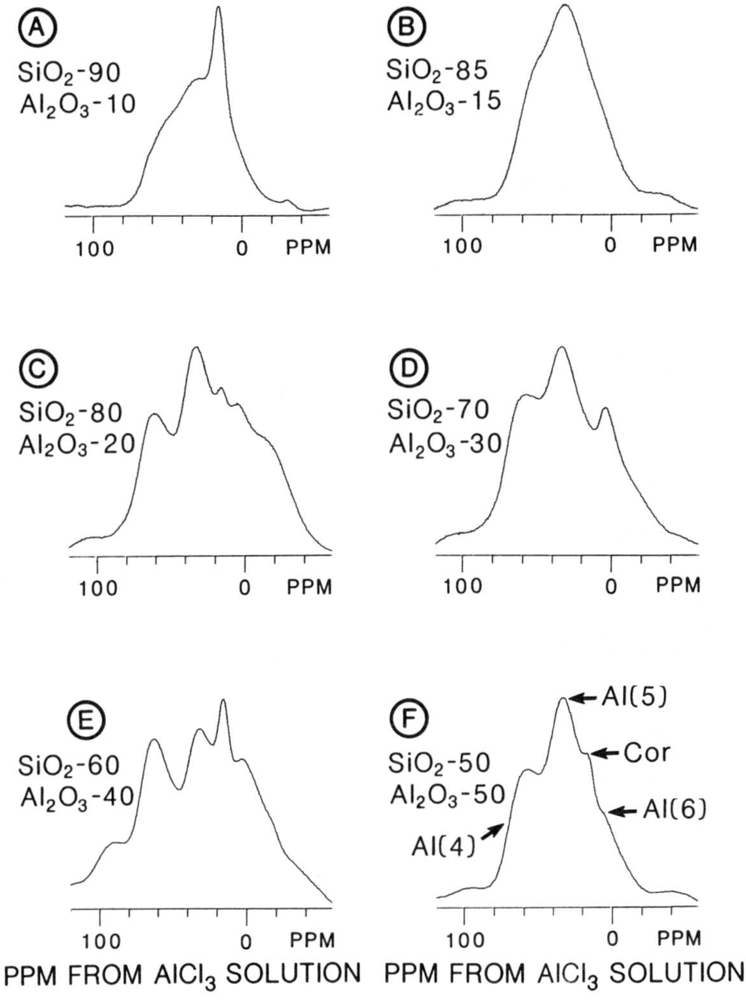

Fig. 3. Al-27 MASS-NMR obtained at a field strength of 11.7T with sample spinning at speeds greater than 7 kHz, a pulse width of 2 microseconds, and a recycle time of 1 second. All peaks shown between 0 and +70 ppm are due to true intensity (because of high spinning speeds) and not due to spinning sidebands.

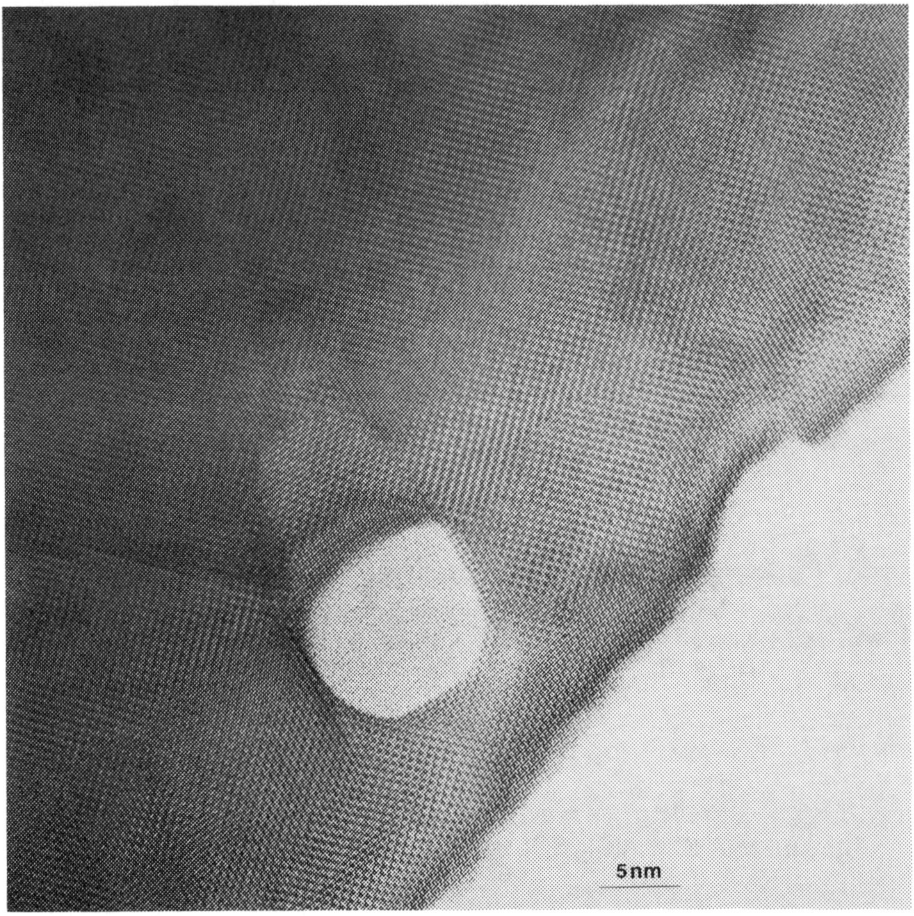

Fig. 4. HREM micrograph of a roller quenched 70 wt% alumina-30 wt% silica glass showing local structure of mullite phase(s) in the darker areas and white particle like amorphous structures. (point to point resolution is about 1.7 nm).

et al.[40] for Si-29 and Al-27 MASS-NMR on a range of glass compositions. The highlight of these data is that the Al-27 spectra reveal chemical shifts for 4 fold Al (at +60 ppm), 6 fold Al (at 0 ppm), and 5 fold Al (at +30 ppm). While Al is known to occupy 5 fold sites in, andalusite for example, the presence of 5 fold Al sites in the glass structure represents a new observation. Application of MASS-NMR techniques to glasses prepared under high pressures have been reported by Ohtani et al.[41] and Komarneni et. al. applied MASS-NMR to studies of silica-alumina gels.[42] We have recently begun to probe local structure in roller quenched glasses and crystallized glasses by high resolution electron microscopy (HREM). While this work is still in the early stages the HREM micrograph of a high alumina (70 wt%) glass shown in Fig. 4 gives indications of the atomistic structure of both the mullite areas and the amorphous regions (the white precipitate-like area, 9–10 nm in size). We are continuing to probe the local structure of glasses, electronic and optical materials by HREM, MASS-NMR, Raman spectroscopy, and molecular dynamic simulation techniques.[43]

Acknowledgments

I am grateful to Joseph Pask and Alan Searcy for their encouragement during my work on the silica-alumina system at Berkeley. I also wish to acknowledge productive collaborations with Waltraud Kriven and Jim Kirkpatrick at the University of Illinois in Urbana Champaign. It is a pleasure to thank the National Science Foundation for continuing support over the past 10 years, most recently under grant NSF-DMR 87-03721, Ceramics and Electronic Materials Program. The HREM experiments were conducted in the Facility for High Resolution Electron Microscopy in the Center for Solid State Science at Arizona State University supported by NSF grant DMR-86-11609.

References

[1]N. L. Bowen and J. W. Greig, "The System Alumina-Silica," *J. Amer. Ceram. Soc.*, **7**, 238–54 (1924).

[2]S. Aramaki and R. Roy, "Revised Phase Diagram for the System Alumina-Silica," ibid, **45**, 229–42 (1962).

[3]I. A. Aksay and J. A. Pask, "Stable and Metastable Equilibria in the System Silica-Alumina," ibid, **58**, 507–12 (1975).

[4]S. H. Risbud, "Metastability and Crystallization Studies in the Silica-Alumina System," Ph.D. thesis, Univ. of California, Lawrence Berkeley Laboratory Report LBL-5453, 1976.

[5a]F. P. Glasser, I. Warshaw, and R. Roy, "Liquid Immiscibility in Silicate Systems," *Physics and Chem. of Glasses*, **1**, 39–45 (1960).

[5b]R. Roy, "Metastable Liquid Immiscibility and Subsolidus Nucleation," *J. Amer. Ceram. Soc.*, **43**, 670 (1960).

[6]J. F. MacDowell, *Ind. Eng. Chem.*, **58**, 38 (1966).

[7]F. Y. Galakhov and S. F. Konovalova, "Liquation Phenomena in the System Alumina-Silica," Izv. Akad. Nauk. SSSR, Ser. Khim., No. 8, 1373–77 (1964).

[8]D. Ganguli and P. Saha, "Macroliquation in the System Alumina-Silica," *Mater. Res. Bulletin*, **2**, 25–36 (1967).

[9]J. F. MacDowell and G. H. Beall, "Immiscibility and Crystallization in Alumina-Silica Glasses," *J. Amer. Ceram. Soc.*, **52**, 17–25 (1969).

[10]S. H. Risbud, "STEM and EELS Analysis of Multiphase Microstructures in Oxide and Non-oxide Glasses," *J. Non-Cryst. Solids*, **49**, 241–51 (1982).

[11]C. M. Jantzen and H. Herman, "Phase Equilibria in the System Silica-Alumina," *J. Amer. Ceram. Soc.*, **62**, 212–14 (1979).
[12]V. I. Aver'yanov, M. P. Areshev, V. V. Golubkov, and L. A. Doronina, "Phase Decomposition in the Alumina-Silica System," *Sov. J. of Glass Phys. and Chem.*, **10**, 155–62 (1984).
[13]T. Takamori and R. Roy, "Rapid Crystallization of Silica-Alumina Glasses," *J. Amer. Ceram. Soc.*, **56**, 639–44 (1973).
[14]S. H. Risbud and J. A. Pask, "Calculated Thermodynamic Data and Metastable Immiscibility in the System Silica-Alumina," ibid, **60**, 418–24 (1977).
[15]M. Rey, "The Thermodynamic Activity of Silica and of Oxides in Silicate Melts," Disc. Faraday Soc., No. 4, 257 (1948).
[16]R. J. Charles, "Activities in Lithia-, Soda-, Potash-Silica Solutions," *J. Amer. Ceram. Soc.*, **50**, 631- (1967).
[17]R. A. Howald and I. Eliezer, "Thermodynamic Properties of Mullite," *J. of Physical Chemistry*, **82**, 2199 (1978).
[18]R. F. Davis and J. A. Pask, "Diffusion and Reaction Studies in the System Alumina-Silica," *J. Amer. Ceram. Soc.*, **55**, 525 (1972).
[19]N. A. Toropov and F. Y. Galakhov, "Questions of Petrography and Mineralogy" Izv. Akad. Nauk SSSR, Otd. Khim. Nauk., No.1, 8 (1958).
[20]A. P. Taglialavore, W. M. Kriven, and S. H. Risbud, "Microstructural Development of Rapidly Solidified, Phase-Separated Silica-Alumina Glasses," pp. 442–43, Proc. of the 44th Annual Meeting of EMSA, Edited by G. W. Bailey, 1986.
[21]C. M. Jantzen, D. Schwahn, J. Schelten, and H. Herman, "The Silica-Alumina System, Part I: Later Stage Spinodal Decomposition and Metastable Immiscibility," *Phys. Chem. Glasses*, **22**, 122 (1981).
[22]H. E. Cook and J. E. Hilliard, "A Simple Method of Estimating the Chemical Spinodal," *Trans. AIME*, **233**, 142 (1965).
[23]S. H. Risbud and J. A. Pask, "On the Location of Metastable Immiscibility in the System Silica-Alumina," *J. Amer. Ceram. Soc.*, **62**, 214–15 (1979).
[24]S. H. Risbud and J. A. Pask, "Mullite Crystallization from Silica-Alumina Melts," ibid, **61**, 63–67 (1978).
[25]S. Mehta and S. H. Risbud, "Characterization of a Phase Separated Aluminosilicate Glass by Analytical Electron Microscopy," ibid, **62**, 641–42 (1979).
[26]D. G. Howitt and C. M. Jantzen, "Location of the Immiscibility Gap in Aluminosilicate Glass," *Bull. Amer. Ceram. Soc.*, **62**, 420 (1983).
[27]F. Y. Galakhov, V. I. Aver'yanov, V. T. Vavilonova, and M. P. Areshev, *Sov. J. of Phys. and Chem*, **8**, 186 (1982).
[28]F. Y. Galakhov, V. I. Aver'yanov, V. T. Vavilonova, and M. P. Areshev, *Fiz. Khim. Stekla.*, **2**, 129 (1976).
[29]R. McPherson, "Evidence for Metastable Immiscibility between Mullite and Alumina," *J. Amer. Ceram. Soc.*, **63**, 110 (1980).
[30]H. Moore and P. W. McMillan, *J. Soc. Glass Tech.*, **40**, 66 (1956).
[31]D. E. Day and G. E. Rindone, *J. Amer. Ceram. Soc.*, **45**, 489 (1962).
[32]E. D. Lacy, *Phys. Chem. Glasses*, **4**, 234 (1963).
[33]D. Kato, *J. Appl. Phys.*, **47**, 2050 (1976).
[34]B. O. Mysen, D. Virgo, I. Kushiro, *Amer. Mineral.*, **66**, 678 (1981).
[35]B. O. Mysen, D. Virgo, and C. Scarfe, ibid, **65**, 690 (1980).
[36]P. McMillan and B. Piriou, *J. Non-Cryst. Solids*, **53**, 279 (1982).
[37]I. A. Aksay, J. A. Pask, and R. F. Davis, *J. Amer. Ceram. Soc.*, **62**, 332 (1979).
[38]H. Morikawa, S. Miwa, M. Miyake, F. Marumo, and T. Sata, ibid, **65**, 78 (1982).
[39]G. Turner, R. J. Kirkpatrick, S. H. Risbud, and E. Oldfield, *Bull. Amer. Ceram. Soc.*, **66**, 656 (1987).

[40]S. H. Risbud, R. J. Kirkpatrick, A. P. Taglialavore, and B. Montez, *J. Amer. Ceram. Soc.*, **70**, C-10 (1987).

[41]E. Ohtani, F. Taulelle, and C. A. Angell, *Nature*, **314**, 78 (1985).

[42]S. Komarneni, R. Roy, C. Fyfe, G. J. Kennedy, and H. Strobl, *J. Amer. Ceram. Soc.*, **69**, C-42 (1986).

[43]P. N. Kumta, P. A. Deymier, and S. H. Risbud, "Improved Rigid Ion Model of Molten Zinc Chloride," submitted to *Physica B*.

Mullite Single Crystal Growth and Characterization

H. SAALFELD AND W. GUSE

Mineralogisch-Petrographisches Institut der Universität Hamburg
Grindelallee 48, 2 Hamburg 13, W.-Germany

Abstract

In this paper a review on the single crystal growth of pure and impure mullite and the characterization of the resulting materials is given. It also presents some results, which have been unpublished so far. The first part starts with a report on the growth of 2:1-mullite, which has been the starting-point for many investigations on (1) chemical variations, (2) crystal structure and order/disorder phenomena, (3) X-ray topography, (4) optical data, (5) dielectric properties, (6) microhardness and (7) incorporation of foreign ions. The following results were obtained:

1. **The chemical variations in Czochralski-grown mullite show, that there is no solid solution between the mullites $3Al_2O_3{\cdot}2SiO_2$ and $2Al_2O_3{\cdot}1SiO_2$.**
2. **Mullite represents a defect structure and due to the vacancies there are diffuse reflections and satellites in reciprocal space. The results of the structure analysis revealed that**

   •only tetrahedra and vacancies with their distribution are responsible for the appearance of the satellites. The octahedral chains are not effected by the crystal disorder.
   •there are tetrahedral double chains ("mullite chains") different from those in sillimanite.
   •at a first approximation, supercells can be postulated containing both sillimanite double chains and mullite double chains.

   The existence of domains within the mullite structure has been confirmed.
3. **X-ray topographic studies showed, that the main defects in Czochralski-grown mullite are growth striations and dislocations. The observed loops are edge-type dislocations that can be explained by a strain field produced by a glassy phase localized at cracks parallel to (100) and (010).**
4. **The mullites with varying composition and nearly identical crystal structure show an almost linear decrease in the refractive indices with increasing Al_2O_3-content. New data measured by means of a spindle stage are presented.**
5. **The electronic structure of 2:1-mullite is anisotropic. There is an energy band gap of 7.7 ± 0.2 eV, which proves mullite to be a good insulating material similar to SiO_2(10.5 ± 0.2 eV) and Al_2O_3 (5.2 ± 0.2 eV).**

6. **Large mullite single crystals enabled the measurement of microhardness, which was found to be anisotropic e.g., 19.2% on (001); 24.4% on (100) and 24.9% on (010).**
7. **The incorporation of foreign ions obviously depends on the growth conditions of mullite. The experiments showed, that Czochralski-grown mullite only takes up Cr_2O_3 in solid solution. Fe^{3+} and Ti^{4+} could not be incorporated.**

In the second part of this paper the substitution of Si^{4+} by Ge^{4+} in mullite is described. The structure refinement of 3:2 (Al,Ge)-mullite and the incorporation of large ions like Nd^{3+} and Pb^{2+} in its lattice are reported. It is shown, that these heavy cations occupy the vacancies within the mullite lattice. The influence of the incorporated large ions on the optical behaviour of mullite has been determined. It is clearly seen, that they cause an increase of the refractive indices. Axial angle and birefringence remain nearly unchanged.

Electron diffraction studies of 3:2(Al,Ge)-mullite revealed a complicated arrangement of satellites (besides the sharp reflections) due to a possible domain structure. This will be the subject of another paper.

Introduction

In ceramic materials, mullite always occurs in the forms of small grains and fine, 1-5 μm needles. In general, these sinter products have a composition of $3Al_2O_3 \cdot 2SiO_2$. Fusing calcined diaspore or bauxite in an electric furnace yields larger crystals. Mullite needles can sometimes be found with a length of a few centimeters, especially in the voids of the cooled melts.

The first systematic experiments to grow mullite single crystals were successfully performed by means of the Verneuil method.[1,2] The boules produced had a composition of $2Al_2O_3 \cdot 1SiO_2$ with little stoichiometric variations.[2]

The flux-technique has been employed to produce mullite having special compositions, e.g., (Al,Ge)-mullite. In this technique, the flux medium strongly influences the chemical composition of the mullite single crystals (see chapter III in Ref. (2)).

The application of the Czochralski-method finally allowed the synthesis of large mullite single crystals under controlled conditions.[3] These crystals proved to be of high quality and thus were suitable for further investigation.

In the following, the Czochralski synthesis is briefly described, and the properties of the resulting mullite are reported.

The Czochralski-Growth of 2:1 Mullite

For the investigation of the mechanical, optical and other properties of mullite, large crystals of high quality are needed. Guse and Mateika[3] have succeeded in growing such crystals by the Czochralski-method. Figure 1 shows the pulling chamber used in this research. An iridium tube containing the seed crystal is mounted at the end of the pulling rod. An iridium crucible, covered to prevent heat radiation is also used. Mixtures of pure alumina and silica in different ratios are used as starting materials. Because of its lower melting point, the silica powder is introduced first and alumina, in the form of sintered cylinders, is subsequently placed on the upper side of the crucible. The iridium crucible is slowly heated in a N_2-atmosphere to the melting point of mullite (1850°C). After one hour, a seed crystal is dipped into and vertically pulled from the melt. Table I presents some processing parameters, and Fig. 2 shows a mullite single crystal.

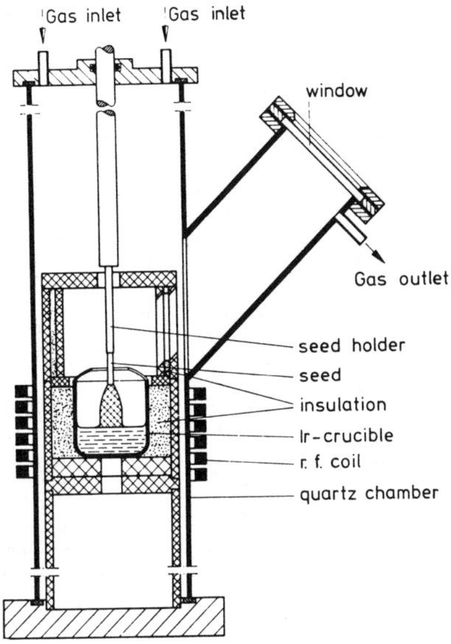

Fig. 1. Schematic of the pulling chamber.

Table I. Processing Parameters for Mullite Synthesis

Pulling rate	1.5 mm/h
Rotation	10 rpm
Composition with best result	77.3 wt% Al_2O_3 22.7 wt% SiO_2
Growth directions	[001], [100]

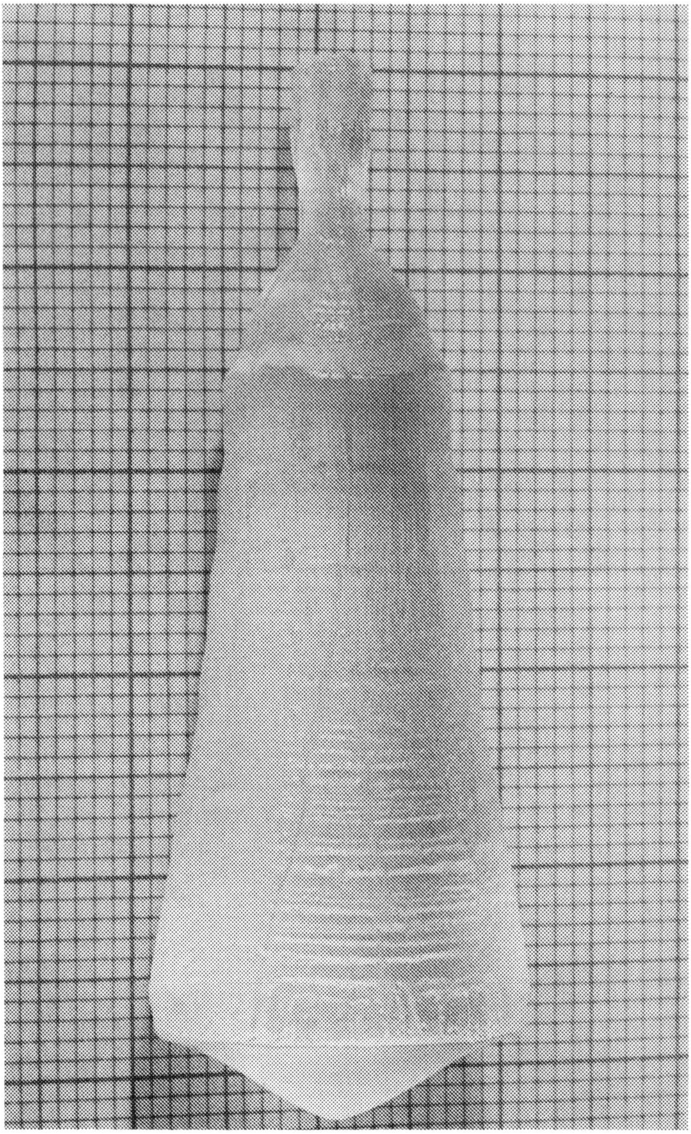

Fig. 2. 2:1(Al,Si)-mullite single crystal.

Fig. 3. Cleavage cracks of a Czochralski (Al,Si)-mullite crystal grown from a starting mixture having the 3:2-composition.

Cleavage cracks parallel to (100), (010) and (110) occurred with decreasing alumina concentration as shown in Fig. 3. These cracks were filled with a silica-rich glassy phase. Fast pulling rates led to an increase in the number of cracks and inclusions. The main pulling direction was [001] but the growth in [100] has also been successful.

Chemical Variation

Guse[4] has attempted to determine the solid solution range of Czochralski-grown mullites. Table II presents the results. The chemical analyses were performed by the electron microprobe technique (ARL-EMX-SM). In all cases the mullites had a composition of 2 Al_2O_3•1 SiO_2 (2:1-mullite) with a variation of at most 2 wt%. Under these experimental conditions there is no solid solution between the mullites 3 Al_2O_3•2 SiO_2 and 2 Al_2O_3•1 SiO_2. This is not in accordance with the results given by other authors[5,6] who postulated a solid solution in this composition range. Apparently the formation of a solid solution also depends on the method of crystal synthesis.

Crystal Structure and Order/Disorder Phenomena

Mullite represents a defect structure due to the substitution scheme $2Si^{4+} + O^{2-}$ <---> $2Al^{3+}$ + vacancy. Thus the general formula is $Al_{4+2x}Si_{2-2x}O_{10-x}$ where x is the number of possible oxygen atoms missing per unit cell.[6] For example x = 0.4 is calculated for the 2:1-mullite. The consequence of the appearance and distribution of vacancies are diffuse reflections and satellites in reciprocal space.

On the basis of the sharp reflections alone, the ordinary unit cell (subcell) results with its "average structure" only. This crystal structure[7–10] closely resembles that of sillimanite. However, there are distinct differences from the sillimanite structure. Both substances are built up of chains of AlO_6-octahedra running parallel to the c-axis. These octahedra are cross-linked by chains of (Si,Al)-tetrahedra. In

Table II. Chemical Variation of the Czochralski-Grown Mullite

Starting composition wt %		Microprobe analysis			
Al_2O_3	SiO_2	Al_2O_3	SiO_2	Total	No.
77.3	22.7	77.81	23.61	101.42	Ia
		77.61	23.35	100.96	Ib
		77.66	23.24	100.90	IIa
		77.35	23.11	100.46	IIb
75.0	25.0	76.47	24.70	101.18	Ia
		76.65	24.75	101.40	Ib
		75.57	25.55	101.12	IIa
		75.47	25.02	100.49	IIb
73.0	27.0	76.08	25.09	101.17	Ia
		75.57	25.14	100.71	Ib
		76.44	25.02	101.46	IIa
		76.22	25.43	101.65	IIb
71.8	28.2	75.85	25.23	101.08	Ia
		75.67	25.32	100.99	Ib
		75.24	25.53	100.77	IIa
		75.42	25.78	101.20	IIb

I: Slice, cut close to the neck of the crystal
II: Slice, cut close to the end of the crystal
a: Analysis made near the edge of the slice
b: Analysis made in the middle of the slice

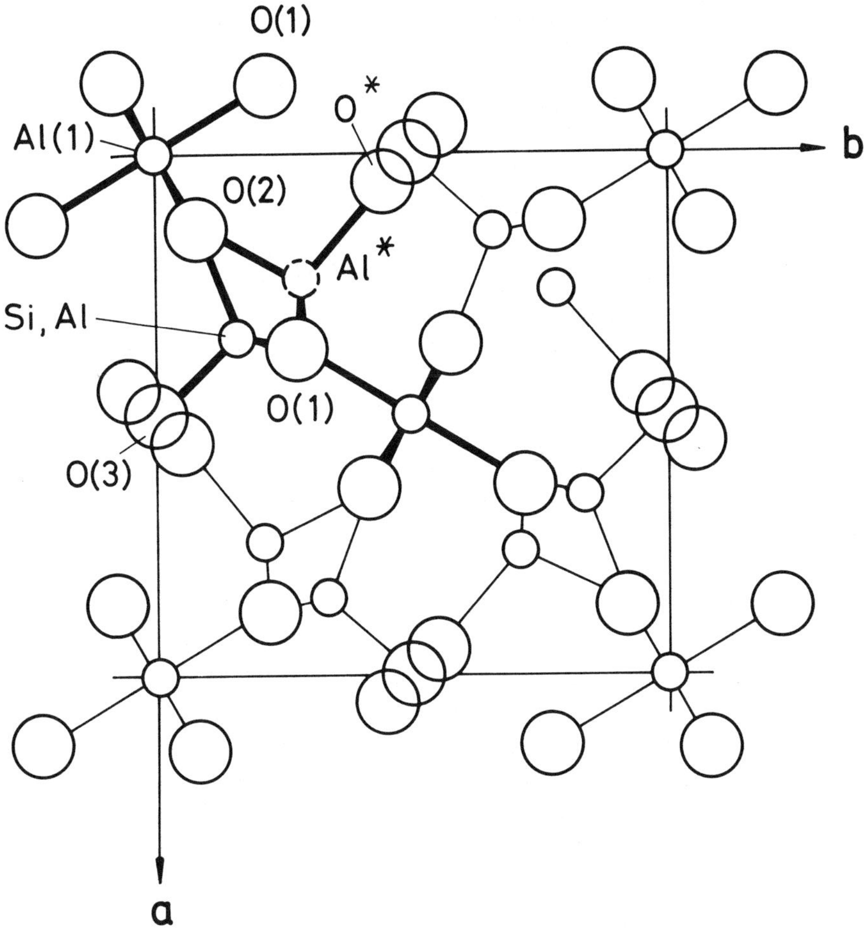

Fig. 4. Projection of the mullite structure.

sillimanite Si and Al are ordered giving rise to a c-axis of 5.8 Å. In mullite there is cation disordering in the (Si,Al)-positions, which reduces the c-axis to 2.9 Å. Moreover a new tetrahedral site (Al*) occurs with a site-occupancy of 0.19 (Fig. 4). Some crystallographic data for the 2:1-mullite are listed in Table III.

Many studies have been performed to describe and locate the diffuse reflections and satellites.[6,11–17] Figure 5 shows the electron diffraction pattern of the reciprocal h0l-plane with pairs of satellites connected by diffuse streaks. Here the reciprocal coordinates of the satellites can be expressed as follows:

$$h' = h \pm 0.30 \ (h = 2n+1)$$
$$k' = 0$$
$$l' = l + 1/2$$

Table III. Crystallographic Data of the "Average" Structure of 2:1-Mullite

Lattice constants	a=7.579(5) b=7.683(3) c=2.888(3) Å
Space group	Pbam
Z	1 x $\|Al_{4.8}Si_{1.2}O_{9.6}\|$
s-vector of satellites	0.30
D_x	3.12 $g \cdot cm^{-3}$

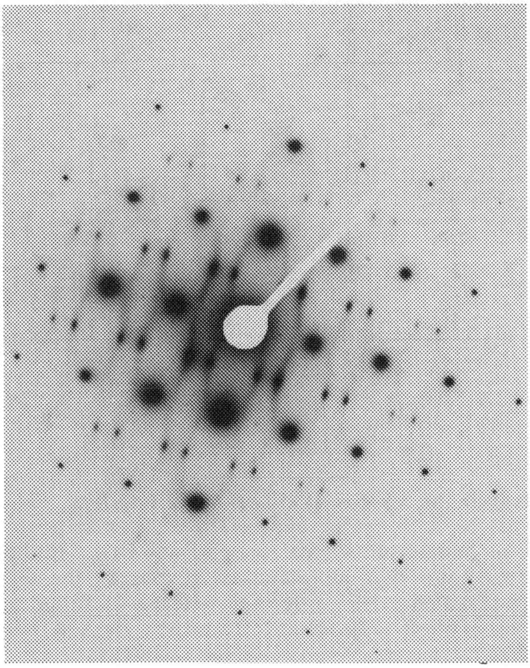

Fig. 5. Electron diffraction pattern of the h0l reciprocal lattice section of 2:1(Al,Si)-mullite. Satellite spots appear about the missing h=odd l=1/2 n reflections.

The satellites in the h0l-plane obey the extinctions of the a-glide plane, i.e. the h0l-plane contains only half of all possible satellites. The distance between the pair of satellites (s-vector) reveals a modulation wave in the a-direction (superstructure). Its variation is sensitive to the chemical composition.[6] An arc-fused mullite crystal with the chemical composition 1.99 $Al_2O_3{\cdot}SiO_2$ showed the satellites with relatively strong intensities.[18]

The intensities of more than 500 satellites were measured taking into account their non-integral positions on a single crystal X-ray diffractometer. On the basis of these 3-dimensional data a partial Patterson synthesis was calculated with the following results:

- Only tetrahedra and vacancies with their distribution are responsible for the appearance of the satellites. The octahedral chains are not affected by the crystal disorder.
- There are tetrahedral double chains ("mullite chains") different from those in sillimanite. They form sixmembered rings of tetrahedra with vertices pointing in alternate directions (Fig. 6).
- At a first approximation, supercells can be postulated containing both sillimanite double chains and mullite double chains.

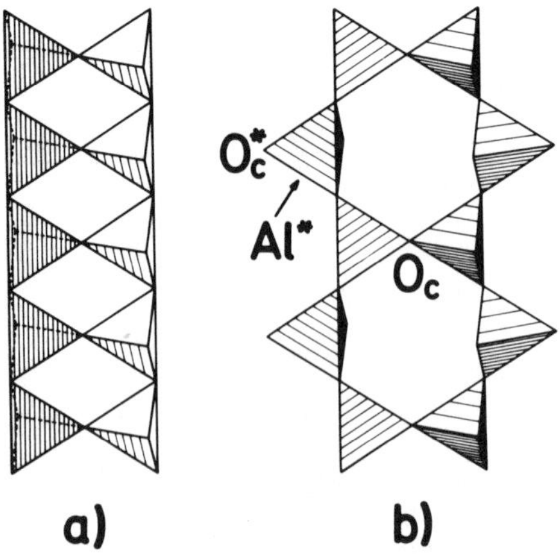

Fig. 6. Tetrahedral double chains a) sillimanite b) mullite.

A study of the incommensurate structure and stability of mullite[19] has confirmed the existence of two kinds of double chains. Based on the average structure, a symmetry analysis revealed Al/Si ordering patterns as found in sillimanite and mullite. The overlap of these patterns in the mullite structure increases the stability of the lattice.

Another attempt to solve the satellite problem was made by Tokonami et al.[14] The periodic unitary intensity distribution was deduced using information from the average structure and from the oxygen sublattice where the vacancies are located. They postulated domains of lower symmetry of about 40,000 atoms each. However, a direct high-magnification lattice image confirming this model has not yet been reported.

The existence of domains has been confirmed[20] by means of high-resolution-transmission-electron-microscopy (HRTEM). They exhibit antiphase relations parallel to the c-axis and antiphase boundaries parallel to (100). An antiphase-domain-structure with antiphase boundaries parallel to (601) is developed[15,17] with increasing Al_2O_3-content. Simultaneously twinning to (100) also occurs.

X-Ray Topography

Topographical methods are usually applied to investigate imperfections (defects) in a crystal. Such defects are dislocations, cracks, precipitations, strains etc. Stacy and Guse[21] analysed Czochralski-grown mullite (2 Al_2O_3•1 SiO_2) with X-ray topography. The topographs were taken both in transmission (Lang projection topography) and in reflection (double crystal configuration). Only optically perfect specimens were chosen (no cracks, inclusions or twinning). Figure 7 shows the typical imperfections revealed in a transmission X-ray topograph of a (001)-slice examined with Mok alpha(1) - radiation in a 210-reflection.

The main defects are growth striations. The strong ones are irregular with a periodicity of about 1mm; the weak ones have a periodicity of about 20 μm. The strain amplitudes of these striations were analysed by means of double crystal topographs taken at different diffraction angle settings and were found to be Delta d/d $\approx 10^{-4}$ and 10^{-6}, respectively.

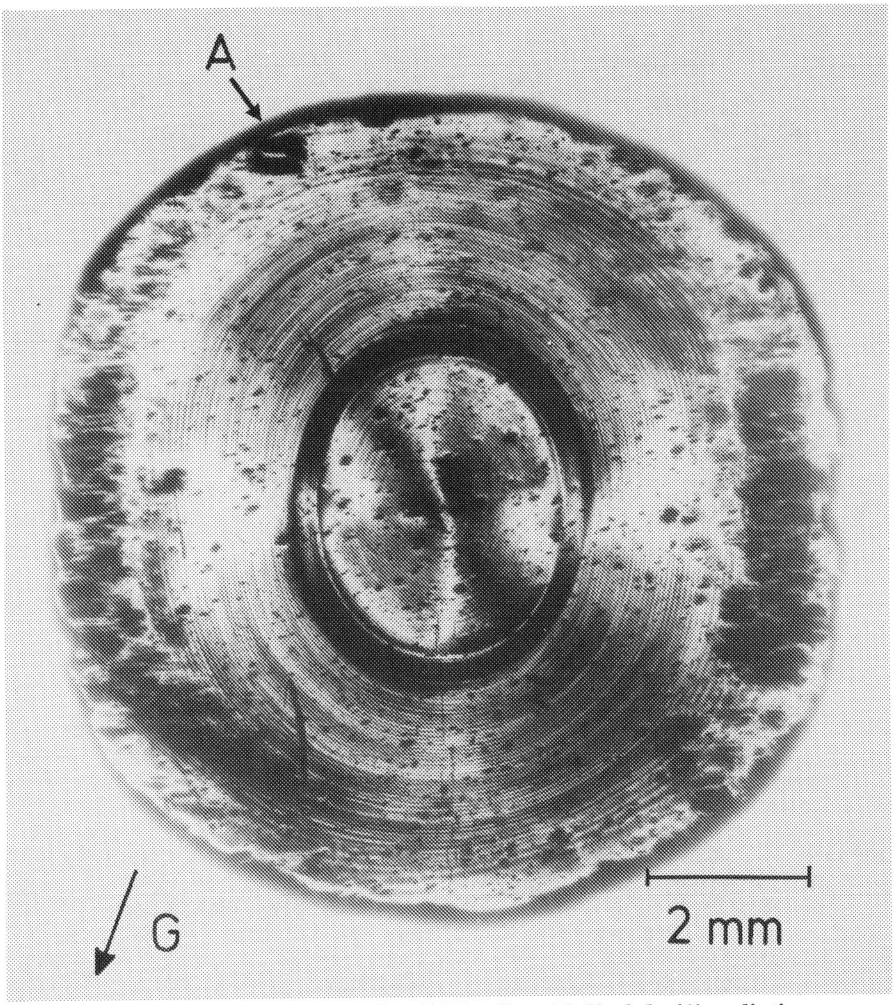

Fig. 7. Transmission topograph of a (001)-slice, MoK alpha(1) radiation, diffraction vector G = 210.

The origin of these striations is not yet completely understood. The weak striations may be related to the seed rotation, since the pulling rate was 5μm during 1 rotation and thereby a simple fraction of the observed spacing. The strong striations also did not correspond to a change in interface shape and therefore are not the "melt-back" striations common to other Czochralski-grown crystals. The strong striations could not be correlated with any known mechanical or electrical fluctuations which might have occurred during the growth process.

In reflection topographs many individual dislocation lines are resolved (Fig. 8). The dislocation loops are generally rectangular in shape with the loop axis parallel to [001]. There are no dislocation glide networks as in Czochralski-grown corundum

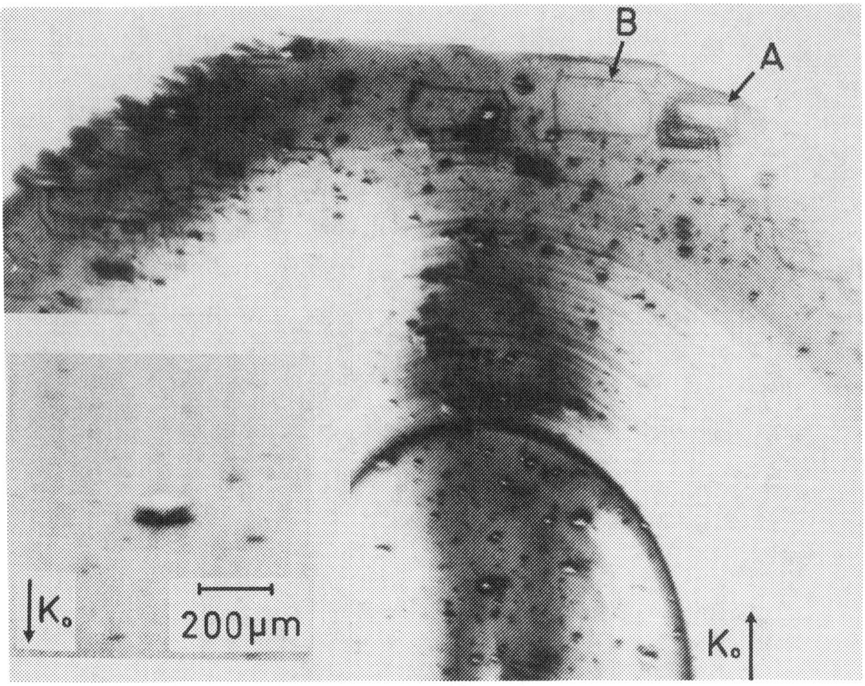

Fig. 8. Dislocation loops (A,B) as revealed by double crystal reflection topographs of the upper region in Fig. 7. K_0 indicates the direction of the incident X-ray beam projected in the specimen surface. Note that the image is mirrored with respect to Fig. 7.

(alpha-Al_2O_3). The rectangular edges are parallel to [100] and [010] with the latter edge always the longest. Sometimes [110]-segments were also observed. Loop diameters vary from 10 to 650μm. A typical loop density in this crystal was found to be 3 $\times 10^4 cm^{-3}$. For an average loop circumference of 200μm this corresponds to a line density of $6 \times 10^2\ cm^{-2}$. The observed loops are edge-type dislocations that can be explained by a strain field produced by submicroscopic precipitates. This could be the above mentioned glassy phase localized at cracks parallel to (100) and (010), characteristic of the loop segments.

Optical Data

Optical data primarily depend on chemical composition and crystal structure. The mullites with varying composition and nearly identical crystal structures show an almost linear decrease in the refractive indices with increasing Al_2O_3-content.[6,12,22] The variation of n_x(Na-light) extends from 1.642 (3:2-mullite) to 1.650 (2:1-mullite). The reported range for n_z varies from 1.653 (3:2-mullite) to 1.662 (2:1-mullite).

However, most of the optical values are incomplete. No spindle stage was used to collect and to determine all optical data. Therefore a crystal fragment of a Czochralski-grown 2:1-mullite was mounted on a spindle stage of a micro-refractometer to measure the refractive indices at different wave lengths. The refractive

Table IV. Optical Data of Czochralski-Grown 2:1-Mullite

Orientation	X=a Y=b Z=c
Refractive indices	n_x=1.6370 C, 1.6421 D, 1.6546 F n_y=1.6376 C, 1.6429 D, 1.6560 F n_z=1.6474 C, 1.6525 D, 1.6650 F C=653.3, D=589.3, F=486.1 nm
Axial angles *)	(+) $2V_C$ = 60.1(2)° (+) $2V_D$ = 59.5(2)° (+) $2V_F$ = 59.1(5)°
Chromatic dispersion	r > v, very weak
Birefringence	n_z-n_x = 0.0104 (for C, D, F)

*) Determined by the extinction curves using the program EXCALIBR (Bloss and Riess)

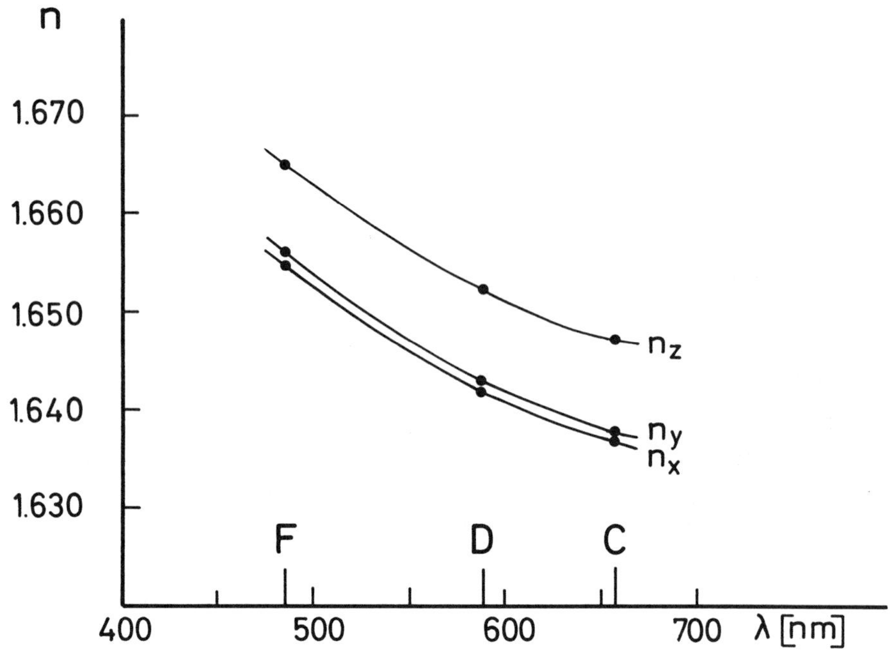

Fig. 9. Dispersion curves of 2:1(Al,Si)-mullite. C=653.3, D=589.3, F=486.1 nm.

indices were corrected by the empirical Hartmann-dispersion-formula. The orientation relation to the crystal axes were controlled by single crystal X-ray diagrams. The results are listed in Table IV. Figure 9 illustrates the dispersion curves of the refractive indices. The measured value (n_x only) of Cameron[6] approaches the spindle stage data quite well.

The Dielectric Properties

Mullite is an excellent refractory and dielectric material. Despite its importance only a few studies of the electrical properties of mullite have been reported. Bárta and Bartuska[23] measured the specific electrical resistance, and Kirillova[24] reported on the dielectric loss angle tan δ in mullite. These mullites however, all had the composition 3 Al_2O_3•2 SiO_2, typical of sintered mullite.

Büchner and Guse[25] investigated the electronic structure of Czochralski-grown 2:1-mullite. The electron energy loss measurements were carried out with 55 keV electrons in an electron spectrometer. Polished thin slices parallel to the planes (010) and (100) were used. The thin sections were thinned by sputtering to about 1000 Å and placed perpendicular to the electron beam of the spectrometer. The spectra were taken in a scattering angle of 10^{-3} rad. The results are given in Fig. 10 a-c. The data suggest that the electronic structure of 2:1-mullite is anisotropic. The crystallographic axes of the crystal coincide with the main axes of the dielectric tensor (i.a. a≙x, b≙y, c≙z). Above 18 eV this anisotropic behaviour is no longer observed. Below 7.7 eV there are no excitations of the mullite electrons. There is an energy band gap of 7.7 ± 0.2 eV. This result proves that mullite belongs to the same group of good insulating materials as SiO_2 (10.5 ± 0.2 eV) and corundum (5.2 ± 0.2 eV).

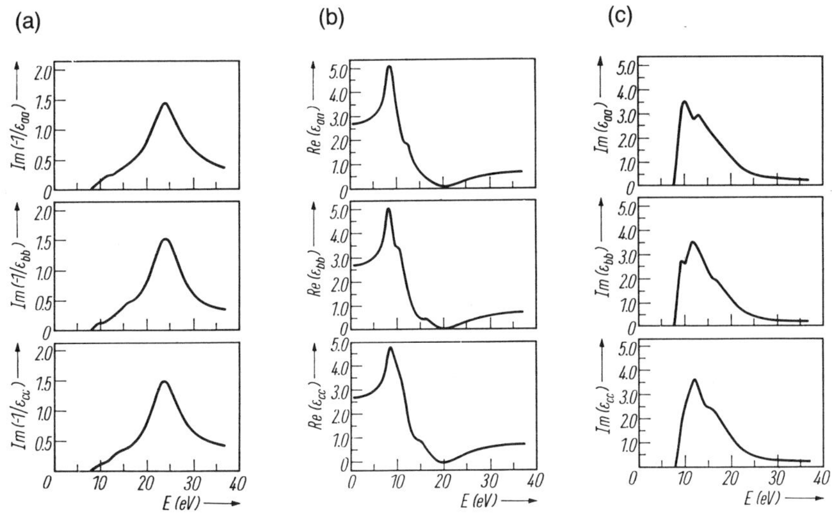

Fig. 10.a) The energy loss function Im $(-\epsilon_{ii})$ of the 2:1-mullite b) The dielectric tensor components Re(ϵ_{ii}) c) The dielectric tensor components Im(ϵ_{ii}).

An assignment of the Im(ϵ_{ii})-maxima to interband transitions at symmetry points of the Brillouin zone of mullite cannot be made as no band structure calculations have yet been carried out.

A sense of the anisotropic behaviour can be obtained from the atomic arrangement within the orthorhombic symmetry of the average-structure. Further work is necessary to determine if size and orientation of the domains will have an influence on the dielectric properties.

Microhardness

The successful growth of large 2:1-mullite single crystals offered the possibility to test mechanical properties such as hardness, which is dependent on crystal structure, chemical bond, valency, packing of the atoms etc. Hardness measurements were reported by Grofcsik and Tamas.[26] They found a Mohs-hardness of 7.5 that refers, however, to 3:2-mullite and does not imply any information about anisotropy. Scratch hardness measured on (001) of a 2:1-mullite revealed a Mohs-hardness of 8.5.[27] Mazdiyasni and Brown[28] reported on Knoop microhardness of 3:2-mullite (hot pressed specimens) without giving experimental details.

A systematic investigation of the microhardness of 2:1-mullite single crystals was first reported by Guse.[27] The Knoop pyramid of a micro-indentation hardness tester was pressed on the surface of a crystal plate and the major diagonal of the resulting indentation was measured by a microscope. Plates (thickness: 1mm, diameter: 10mm) were cut parallel to the (001), (010) and (100) planes and polished. The exact orientation of the crystal slices was controlled by Laue X-ray photographs. Realizable results were attained with a load of 30p, which was applied for 15 seconds. The measurements were carried out in steps of 10°.

The results are plotted in polar diagrams (Fig. 11 a-c). Figure 11a shows the hardness geometry in the (001) plane. Here the hardness in [100] was determined as 1419 kp/mm^2; that of [100] as 1508 kp/mm^2.

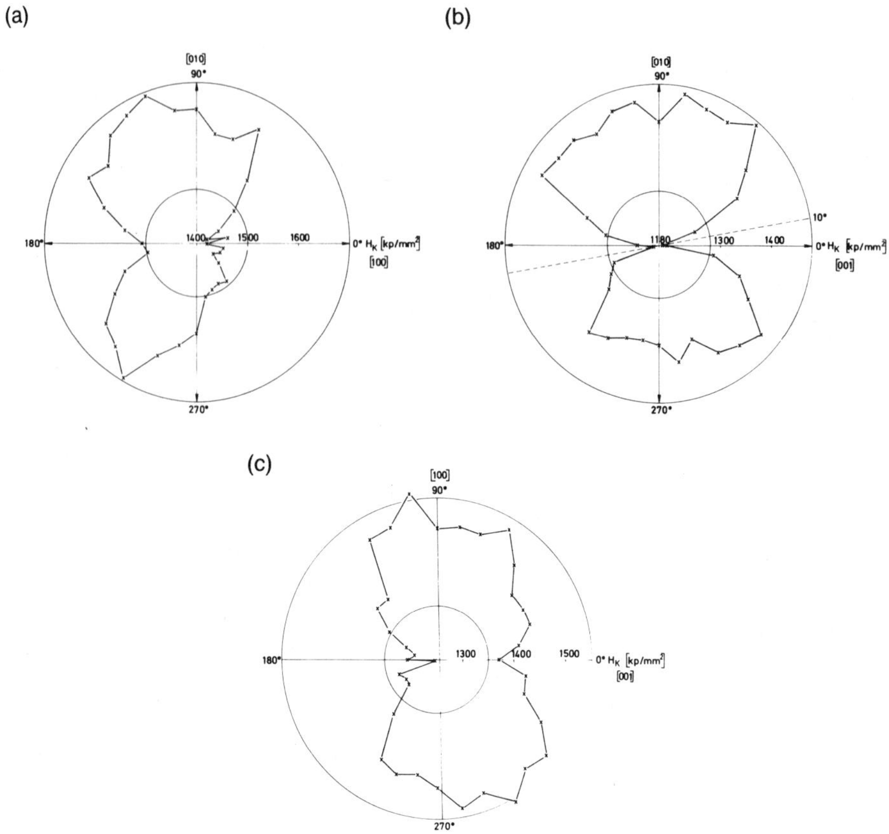

Fig. 11. Microhardness on a) (001) b) (100) c) (010) P = 30 p = constant.

Fig. 11b and 11c represent the slope of hardness on (100) and (010). The diagrams reveal the strong anisotropy of microhardness, which may be quantitatively expressed by the following equation[29]:

$$\frac{H_{max} - H_{min}}{H_{min}} = \frac{\delta}{H_{min}} \tag{1}$$

The result is given in % after multiplying the quotient by 100.
The corresponding values are: (001) 19.2%; (100) 24.4%; (010) 24.9%.

The dependence of load on diagonal length may be approximated to a linear fit.[30] The Knoop hardness is given by

$$H_K = K \cdot \frac{p}{d^2} \left[kp/mm^2 \right] \tag{2}$$

where p is the load in pond, d the major diagonal in μm and K is a constant. This equation is modified to

$$p^{1/2} = \left(\frac{H_K}{K}\right)^{1/2} \cdot d \qquad (3)$$

The evaluation of a plot of $p^{1/2}$ vs d should be a straight line with a slope, which gives the magnitude of hardness. As an example, the load dependence of microhardness on (001), (100) and (010) is shown in Fig. 12 a-c. Approximation to the measured values was achieved by the different lines denoted 1-3. It can be seen that the hardness is greater under a lighter load. The reason for this behaviour is seen as a process of gliding at low loadings and as initial cleavage at greater pressures.

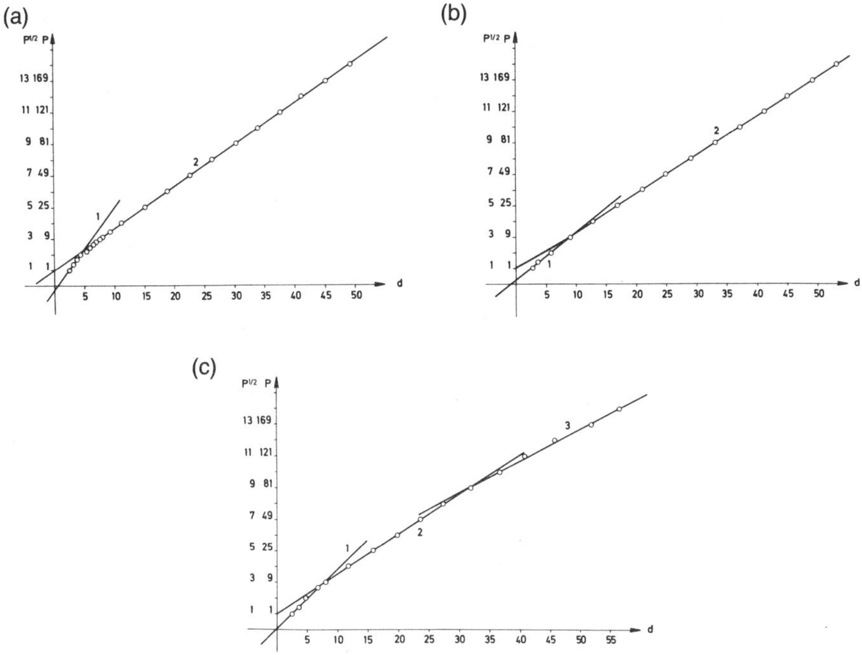

Fig. 12. Load dependence of microhardness: Values are plotted in [Kp/mm^2] as function of the diagonal length d in | μm | : a) measured on (001) | | [100] b) measured on (100) | | [001] c) measured on (010) | | [001].

A correlation between microhardness and the crystal structure of mullite seems plausible. Mullite has a chain structure with AlO_6-octahedra chains running parallel to [001] and that are interconnected by chains of $(Si,Al)O_4$– tetrahedra. The lowest hardness is found on (100) in [001]; the greatest on (001) in [010] due to the cohesion forces which attract the chains. The geometry of anisotropy fits well in this picture. The asymmetry, however, cannot yet be explained.

Incorporation of Foreign Ions

The incorporation of foreign ions (Fe^{2+}, Fe^{3+}, Cr^{3+}, Ti^{4+}) into the mullite lattice has been the subject of many studies.[31–36] In nearly all cases, powdered mixtures have been heated in special atmospheres to different temperatures to determine the solubility of the oxides. The solid solution mechanism is primarily based on substituting Al^{3+} in octahedral or tetrahedral sites. Thus Fe^{3+} can occupy both sites to a limited extent.[36] Even larger ions can be incorporated.[37] In this case the vacancies in the lattice fulfill the geometrical requirements as a possible site.

The results published so far refer almost exclusively to sinter-mullite, $3Al_2O_3 \cdot 2SiO_2$, and cannot be transferred to fused mullite or Czochralski-grown single crystals of 2:1-mullite. One reason for this is the dependence of the solubility of many additives on temperature. Small differences in crystal structure and disorder may also play a certain role. Schneider and Rager[38] presented such differences from the results of EPR-spectra. Incorporation of Fe^{3+} and Cr^{3+} in Verneuil-boules was reported by Neuhaus and Richartz.[2] However, no details were given.

In a series of experiments Guse investigated the solubility of Fe^{3+}, Cr^{3+} and Ti^{4+} in Czochralski-grown mullite. The unpublished results of this research are given in the following:

Fe_2O_3

Powders of pure alumina and silica containing 0.1 wt% Fe_2O_3 were used for the Czochralski synthesis. The grown crystals had a dim appearance and showed many cleavage cracks. Microscopic examination of polished slices led to the conclusion that iron-oxide is accumulated in the glass phase located in the cracks. This was confirmed by electron microprobe measurements (Table V). At a growth temperature of 1850°C the content of incorporated Fe_2O_3 is very low. Sinter-mullites, however, are able to incorporate up to 13 wt% Fe_2O_3.[32-35,38] The temperature dependence of Fe_2O_3 in mullite clearly shows that the iron content of mullite decreases with increasing temperature (1300°C: 10.5 wt%; 1670°C: 2.5 wt%).[37] Extrapolation to higher temperatures would reduce the solubility to zero above 1800°C. Therefore it becomes clear that the Czochralski-grown mullites do not take up Fe_2O_3 in solid solution.

Cr_2O_3

Several studies[31,34] on the solubility of Cr_2O_3 in mullite show that it is possible to incorporate up to 10 wt% into the lattice. Czochralski-grown 2:1-mullites also take up Cr_2O_3 in solid solution. Table VI presents the results of the chemical analyses (electron microprobe).

TiO_2

Incorporation of TiO_2 has been observed in natural and in sintered mullites[12,34,36,37,39] with different contents of TiO_2 (0.5–6.0 wt%).

For the Czochralski synthesis the starting mixture was 76.95 wt% Al_2O_3, 21.54 wt% SiO_2 and 1.5 wt% TiO_2. During the synthesis the formation of a boule was observed but an undisturbed growth was not achieved. After cooling to room temperature, the boule proved to be of poor quality having many cracks and imperfections.

Thin sections showed the inhomogenity of the boule. The microstructure consisted of single mullite blocks separated by a glass phase, similar to that found with the Fe_2O_3 additions. Microprobe analyses of the mullite blocks revealed a TiO_2 content of about 0.5 wt%. The content of TiO_2 in the glass phase varied from 15.6

Table V. Average Values of the Constituent Oxides in the Fe_2O_3-Containing Mullites (wt %). Limits of Variation within Parentheses.

	Al_2O_3	SiO_2	Fe_2O_3	Total
Lower part of the crystal (neck)	76.65 (76.58-76.72)	23.07 (22.92-23.22)	0.04 (0.04)	99.76
Upper part of the crystal	76.48 (76.41-76.54)	22.89 (22.87-22.91)	0.03 (0.03)	99.40
Glass phase	11.67 (11.23-12.11)	29.19 (28.24-30.15)	59.73 (58.21-61.25)	100.59

Table VI. Average Values of the Constituent Oxides in the Cr_2O_3-Containing Mullites. Limits of Variation are given in Parentheses.

Starting mixtures (wt %)			Mullite composition (wt %)			
Al_2O_3	SiO_2	Cr_2O_3	Al_2O_3	SiO_2	Cr_2O_3	Total
77.22	22.68	0.1	77.73 (77.65-77.88)	23.03 (22.98-23.09)	0.06 (0.05-0.07)	100.82
77.07	22.63	0.3	77.26 (77.01-77.43)	22.98 (22.74-23.21)	0.19 (0.17-0.20)	100.43
76.92	22.59	0.5	76.73 (76.54-76-81)	23.19 (22.84-23.50)	0.31 (0.28-0.32)	100.23
76.76	22.54	0.7	77.08 (76.89-77.26)	22.79 (22.53-23.06)	0.44 (0.41-0.46)	100.31
76.54	22.46	1.0	77.21 (76.99-77.43)	22.75 (22.51-22.91)	0.61 (0.58-0.63)	100.57

to 27.4 wt%. The solid solution of TiO_2 in mullite was found to be very low at high temperatures. It may increase at lower temperatures. At 1400°C 3:2-mullite incorporates 2–4 wt% TiO_2.[31,34,39] Quantitative data, however, are not yet available.

The remaining melt is an eutectic mixture of mullite and corundum. Thin sections reveal oriented mullite needles on the platy Al_2O_3-crystals. Weissenberg diagrams (Fig. 13) showed texture reflections of mullite (elongated streaks) on well crystallized corundum. Two orientation relations were found:

Fig. 13. Weissenberg photograph of the zone [100] of corundum with the elongated reflections hkO of mullite.

$[001]_m \parallel [100]_{cor}$ and $[010]_m \parallel [100]_{cor}$

In both cases $[100]_m$ and $[001]_{cor}$ coincided. It is not clear if geometrical or other reasons are responsible for the observed orientations.

Substitution of Si^{4+} by Ge^{4+} in Mullite

Chemical similarity of silicon and germanium and comparable sizes of the ionic radii (Si^{4+}:0.40 Å; Ge^{4+} :0.54 Å) favour the formation of isomorphic compounds. This substitution was observed in samples of mullite powder[31] with the composition $3Al_2O_3{\cdot}2GeO_2$ (germanium mullite). A solid solution from $3Al_2O_3{\cdot}2GeO_2$ to $3Al_2O_3{\cdot}2SiO_2$ and to $2Al_2O_3{\cdot}GeO_2$ was proposed. Other authors[40,41] have confirmed the existence of germanium-mullite. Single crystals were grown by reaction of corundum grains in a melt of GeO_2 at 1600°C.[42] The crystals attained a length of 1 mm and had a uniform composition of $1.54Al_2O_3{\cdot}GeO_2$. Structure analysis showed that germanium-mullite is completely isostructural with silicon-mullite. In Table VIII, the refined atomic parameters are listed together with those found by Neumann.[43]

Table VII. Crystallographic Data of 3:2(Al,Ge)-Mullite (Neumann[43]). The Values Published by Durovič and Fejdi[42] are given in Parentheses.

Lattice constants	a=7.644 (7.650) b=7.779 (7.779) c=2.925 (2.925) Å
Space group	Pbam
Z	$1 \times \|Al_{4.5}Ge_{1.5}O_{9.75}\|$
Crystal forms	{110} {001}
D_{exp}	3.6874 g cm^{-3}
D_X	3.6876 g cm^{-3}

Table VIII. Atomic Parameters and Site Occupancies of the Components in 3:2(Al,Ge)-Mullite (Neumann[43]). The Values Published by Durovič and Fejdi[42] are given in Parentheses.

	Occupancy	x	y	z
Al(1)	1.00 (1.00)	0 (0)	0 (0)	0 (0)
Al*	Al 0.1081 Ge 0.0169 (Al 0.125) (Ge 0.007)	0.2302 (0.2310)	0.2949 (0.2949)	0.5 (0.5)
(Al,Ge)	Al 0.5169 Ge 0.3581 (Al 0.507) (Ge 0.361)	0.3505 (0.3506)	0.1605 (0.1605)	0.5 (0.5)
O(1)	1.00 (1.00)	0.3708 (0.3709)	0.2869 (0.2868)	0 (0)
O(2)	1.00 (1.00	0.1369 (0.1372)	0.0795 (0.0795)	0.5 (0.5)
O(3)	0.625 (0.603)	0 (0)	0.5 (0.5)	0.5 (0.5)
O*	0.125 (0.132)	0.0467 (0.0474)	0.4510 (0.4504)	0.5 (0.5)

Structure Refinement of 3:2(Al,Ge)-Mullite

The successful Czochralski syntheses of mullites were sufficiently encouraging to try the growth of GeO_2-containing (Si,Al)-mullites or even that of pure germanium-mullite.[43] These experiments, however, failed because the GeO_2-containing melt was too viscous and the volatization of GeO_2 was too high. Therefore a molten mixture of 76 mol% GeO_2 and 24 mol% Al_2O_3 was produced at 1700°C in an iridium crucible under a N_2-atmosphere and slowly cooled (20°C/min). The reaction product proved to be a mixture of 3:2-germanium-mullite, GeO_2, GeO and Ge. The separation of mullite was achieved by treatment of the mixture with dilute hydrofluoric acid. The remaining mullite needles had lengths up to 4 mm and a thickness of about 0.2 mm. Microprobe analyses gave the composition $3Al_2O_3{\cdot}2GeO_2$. Some crystallographic data are listed in Table VII. The refinement on the basis of 1756 measured reflections with anisotropic temperature factors resulted in a reliability of R = 0.017. The atomic parameters are listed in Table VIII. The most important distances are given in Table IX.

The occupancies of the sites (Al,Ge) and Al* are of special interest:

(Al,Ge)	:	51.7 % Al (50.7)
		35.8 % Ge (36.1)
Al*	:	10.8 % Al (12.5)
		1.7 % Ge (0.7)

The parentheses contain the values published by Durović and Fejdi[42] for the 1.54:1(Al,Ge)-mullite. The Al* position is occupied predominantly by Al^{3+}. However, a minor amount of Ge^{4+} may also enter into this position.

The occupancies of the sites (Al,Si) and Al* of the 3:2(Al,Si)-mullite approaches 86.3 and 13.7% respectively.[44] Due to the similar atomic scattering factors of Al^{3+} and Si^{4+} a more exact distinction is not possible.

The (Al,Ge)3:2-mullite also has a defect structure. Therefore sharp, very broad and weak diffuse reflections are observed. No distinct satellites are visible as in the case of the (Al,Si)-mullites. Electronmicroscopy work is in progress to clarify a possible domain structure.

Incorporation of Large Ions in (Al,Ge)-Mullite

Recently the synthesis of an unusual (Al,Ge)-mullite with incorporation of large cations in the crystal lattice was reported by Saalfeld and Klaska.[45] The synthesis was performed in a flux containing Pb- and Nd-ions. The 1–2mm long needles had an Al/Ge ratio of nearly 5:1 (exactly $2.6Al_2O_3{\cdot}GeO_2$). Microprobe analyses revealed the following composition: Al_2O_3 64.06, GeO_2 25.34, PbO 8.20, Nd_2O_3 2.40 wt%. The resulting chemical formula is:

$$[Al_{5.03}Ge_{0.97}]\ Pb_{0.15}Nd_{0.06}O_{9.71}$$

X-Ray Work

Single crystal photographs confirmed the mullite structure. However, they also showed weak reflections which are not allowed in the usual space group of Pbam. Thus the structure determination was performed on the basis of the space group $P2_12_12$. Some crystallographic data are listed in Table X. Beginning with the atomic coordinates of $3Al_2O_3{\cdot}2GeO_2$-mullite,[42] some cycles of Fourier-difference-syntheses revealed the (Pb,Nd)-positions in the vacancies of the mullite structure. Further cycles of refinement which included Pb and Nd were calculated and converged to R_w

Table IX. Interatomic Distances of the Coordination Polyhedra of 3:2(Al,Ge)-Mullite (Neumann[43])

				Distance Å
Octahedron	Al(1)	-O(2)	4 ×	1.902(1)
$\|AlO_6\|$		-O(1)	2 ×	1.930(1)
		mean	:	1.911(14)
Tetrahedron I	(Al,Ge)	-O(1)	2 ×	1.769(1)
$\|Al,GeO_4\|$		-O(2)	1 ×	1.750(1)
		-O(3)	1 ×	1.693(1)
O(3) with c.n. 2		mean	:	1.745(36)
Tetrahedron II	(Al,Ge)	-O(1)	2 ×	1.769(1)
$\|Al,GeO_4\|$		-O(2)	1 ×	1.750(1)
		-O*	1 ×	1.733(17)
O* with c.n. 3		mean	:	1.755(19)
Tetrahedron III	(Al,Ge)	-O(1)	2 ×	1.769(1)
$\|Al,GeO_4\|$		-O(2)	1 ×	1.750(1)
		-O*	1 ×	1.809(16)
O* with c.n. 3		mean	:	1.774(25)
Tetrahedron IV	Al	-O(1)	2 ×	1.816(2)
$\|Al^*O_4\|$		-O(2)	1 ×	1.820(4)
		-O*	1 ×	1.858(10)
		mean	:	1.827(20)

Standard deviations are given in parentheses

Table X. Crystallographic Data of the (Pb,Nd)-Containig (Al,Ge)-Mullite (Saalfeld and Klaska[45])

Lattice constants	a=7.6648 b=7.7914 c=2.9213 Å
Space group	$P2_12_12$
Z	$1\times[Al_{5.03}Ge_{0.97}]Pb_{0.15}Nd_{0.06}O_{9.71}$
D_x	3.79 $g\cdot cm^{-3}$

Table XI. Positional Parameters of the Average Structure of the (Pb,Nd)-Containing (Al,Ge)-Mullite (Saalfeld and Klaska[45])

Atom	Mult	x(Δ x)	y(Δ y)	z(Δ z)
		× 10^4		
Al(1)	0.5	0	0	52(15)
Al(2)	0.55	3500(1)	1618(1)	5012(10)
Al*	0.2	2315(6)	2987(5)	4866(52)
Ge	0.243	3500(1)	1618(1)	5012(10)
Pb(1)	0.03063	1182(6)	3666(6)	4810(59)
Pb(2)	0.01937	4543(12)	621(14)	4934(96)
O(1)	1.0	3717(2)	2878(2)	-90(34)
O(2)	1.0	3621(2)	5788(2)	4913(25)
O(3)	0.23066	5000	0	4339(55)
O*	0.19684	472(16)	4429(15)	5373(86)

Standard deviations are given in parentheses

= 0.021. The positional parameters are given in Table XI. Figure 14 shows a part of the mullite lattice with the (Pb,Nd)-positions. However, a distinction between the two cations was not possible.

There are two kinds of vacancies located on the a- and b-axis, both with irregular (Pb,Nd)-coordination. The eight $Pb^{(1)}$-O distances vary from 2.25 to 3.19 Å and the nine $Pb^{(2)}$-O distances from 2.37 to 3.37 Å.

This is the first time that occupation of the vacancies by cations was demonstrated. The (Pb,Nd)-ions take part in the charge balance and thus the Al_2O_3/GeO_2-ratio is shifted to the high value of 2.6:1.

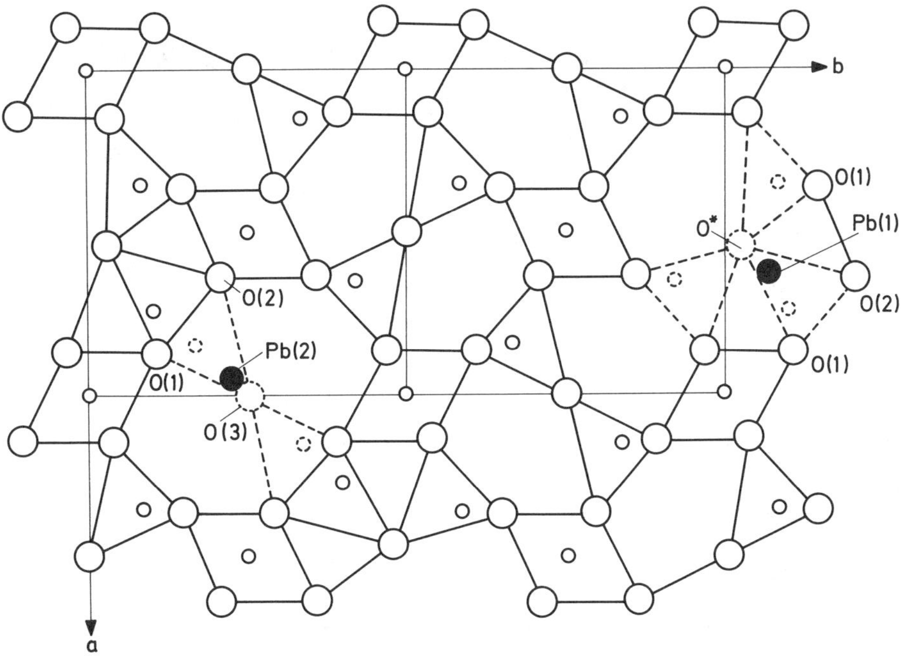

Fig. 14. Part of the mullite structure with the locations of (Pb,Nd)-ions.

Optical Data

Crystal fragments of the 3:2(Al,Ge) and 2.6:1(Al,Ge,Pb,Nd)-mullites were measured on a spindle stage of a microrefractometer to determine the influence of the large ions on the optical data. Table XII shows the results. It is clearly seen that the large ions cause an increase of the refractive indices. Axial angle and birefringence remain nearly unchanged.

Electron Diffraction

Electron diffraction studies revealed the sharp reflections of the average cell and in some planes a complicated arrangement of satellites due to a possible domain structure. This investigation is in progress and will be published soon.

Table XII. Optical Data of the Two (Al,Ge)-Mullites (λ_D = 589.3nm)

	3:2(Al,Ge)	2.6:1(Al,Ge;Pb,Nd)
Orientation	X = a Y = b	Z = c
Refractive indices	n_x = 1.6941 n_y = 1.6978 n_z = 1.7096	n_x = 1.7169 n_y = 1.7191 n_z = 1.7261
Birefringence	0.0155	0.0092
Axial angles	(+)2V = 48.53	(+)2V = 52.48
Optical axial plane	(010)	(010)

Conclusions

Czochralski-grown mullites ($2Al_2O_3 \cdot 1SiO_2$) have been the subject of our investigations. Many important properties were measured and—if possible—compared with those of synthetic mullites produced by others than the Czochralski-technique. Obviously the properties of mullite strongly depend on the composition and the growing conditions.

Another important point is the fact, that mullite has a defect structure depending on the number and distribution of vacancies. The influence of the mullite properties by chemical variations was understood so far to have its origin in replacing cations either in tetrahedral or in octahedral sites alone. The incorporation of large ions like Nd^{3+} and Pb^{2+} into (Al,Ge)-mullite showed, that another mechanism has to be taken into account: Under special growing conditions the vacancies in mullite can be filled with large cations.

Here the question arises whether the silicon-mullites are also capable to react in a similar manner. Perhaps a new way is found to control the properties of mullite, which would considerably enlarge its range of application.

Acknowledgments

The authors would like to thank H. Gerlach and Dr. J. Granzin for collecting the optical data using the spindle stage.

References

[1]W. H. Bauer, J. Gordon and C. H. Moore, "Flame Fusion Synthesis of Mullite Single Crystals," *J. Am. Ceram. Soc.*, **33**, 140–143 (1950).

[2]A. Neuhaus and W. Richartz, "Über die Einkristallzüchtung und Zustandsverhältnisse von Mullit," *Ber.Dt.Keram.Ges.*, **35**, 108–116 (1958).

[3]W. Guse and D. Mateika, "Growth of mullite single crystals ($2Al_2O_3 \cdot SiO_2$) by the Czochralski method," *J. Crystal Growth*, **22**, 237–240 (1974).

[4]W. Guse, "Compositional analysis of Czochralski-grown mullite single crystals," *J. Crystal Growth*, **26**, 151–152 (1974).

[5]S. Aramaki and R. Roy, "Revised phase diagram for the system Al_2O_3-SiO_2," *J. Am. Ceram. Soc.*, **45**, 229–242 (1962).

[6]W. E. Cameron, "Mullite: A substituted alumina," *Amer. Min.*, **62**, 747–755 (1977).

[7]R. Sadanaga, M. Tokonami and Y. Takeuchi, "The Structure of Mullite, $2Al_2O_3 \cdot SiO_2$, and Relationship with the Structures of Sillimanite and Andalusite," *Acta Crystallogr.*, **15**, 65–68 (1962).

[8]S. Durović, "Refinement of the Crystal Structure of Mullite," *Chem. zvesti*, **23**, 113–128 (1969).

[9]C. W. Burnham, "Crystal Structure of Mullite," Carnegie Inst. Washington Year Book, 62 158–165 (1963).

[10]C. W. Burnham, "Crystal Structure of Mullite," Carnegie Inst. Washington Year Book, 63 223–227 (1964).

[11]D. G. W. Smith and J. D. C. McConnell, "A Comparative Electron Diffraction Study of Sillimanite and Some Natural and Artificial Mullites," *Min. Mag.*, **35**, 810–814 (1966).

[12]S. O. Agrell and J. V. Smith, "Cell Dimensions, Solid Solution, Polymorphism and Identification of Mullite and Sillimanite," *J. Am. Ceram. Soc.*, **43**, 69–76 (1960).

[13]W. Guse and H. Saalfeld, "Das diffuse Beugungsbild von Mullit, $2Al_2O_3 \cdot SiO_2$," *Z. Kristallogr.*, **143**, 177–187 (1976).

[14]M. Tokonami, Y. Nakajima and N. Morimoto, "The Diffraction Aspect and a Structural Model of Mullite, $Al(Al_{1+2x}Si_{1-2x})O_{5-x}$," *Acta Crystallogr.*, *A*, **36**, 270–276 (1980).
[15]W. M. Kriven and J. A. Pask, "Solid Solution Range and Microstructures of Melt-Grown Mullite," *J. Am. Ceram. Soc.*, **66**, 649–654 (1983).
[16]H. R. Wenk, "Mullite-Sillimanite Intergrowth from Pelitic Inclusions in Bergell Tonalite," *N. Jahrb. Min. Abh.*, **146**, 1–14 (1983).
[17]J. Ylä-Jääski and H. U. Nissen, "Investigation of Superstructures in Mullite by High Resolution Electron Microscopy and Electron Diffraction," *Physics and Chemistry of Minerals*, **10**, 47–54 (1983).
[18]H. Saalfeld, "The Domain Structure of 2:1-Mullite," *N. Jahrb. Min. Abh.*, **134**, 305–316 (1979).
[19]J. D. C. McConnell and V. Heine, "The Incommensurate Structure and Stability of Mullite," *Phys. Rev.*, *B*, **31**, 6140–6142 (1985).
[20]Y. Nakajima, N. Morimoto and E. Watanabe, "Direct Observation of Oxygen Vacancy in Mullite, $1.86Al_2O_3{\bullet}SiO_2$ by High Resolution Electron Microscopy," *Proc. Japan Acad.*, **51**, 173–178 (1975).
[21]W. T. Stacy and W. Guse, "X-ray topographic study of Czochralski-grown mullite," *J. Crystal Growth*, **35**, 153–158 (1976).
[22]A. J. Majumdar and J. H. Welch, "New Data on Synthetic Mullite," *Trans. Brit. Ceram. Soc.*, **62**, 603–613 (1963).
[23]R. Bárta and M. Bartuska, "Mullite Studies," *Tech. Sil.*, **4**, 146–185 (1957).
[24]G. K. Kirillova, "Elektrische Eigenschaften von Mullit," *Keram. Z.*, **11**, 638–640 (1959).
[25]U. Büchner and W. Guse, "The Dielectric Tensor Components of Czochralski-Grown 2:1 Mullite Single Crystals," phys. stat. sol., (b) 80 615–618 (1977).
[26]J. Grofcsik and F. Tamás, in "Mullite, its structure formation and significance (p.12)," Publishing House Hungarian Acad. of Sciences, Budapest 1961.
[27]W. Guse, "Über die Mikrohärte des 2:1-Mullits," *Ber. Dt. Keram. Ges.*, **52**, 339–343 (1975).
[28]K. S. Mazdiyasni and L. M. Brown, "Synthesis and Mechanical Properties of Stoichiometric Aluminium Silicate (Mullite)," *J. Am. Ceram. Soc.*, **55**, 548–552 (1972).
[29]J. Gahm, "Neuere Erkenntnisse zur Mikro-Härte," *VDI-Berichte*, **160**, 25–41 (1972).
[30]W. v. Engelhard and S. Haussühl, "Festigkeit und Härte von Kristallen," *Fortschr. Min.*, **42**, 5–49 (1965).
[31]G. Gelsdorf, H. Müller-Hesse and H. E. Schwiete, "Einlagerungsversuche an synthetischem Mullit und Substitutionsversuche mit Galliumoxyd und Germaniumoxyd," *Arch. Eisenhüttenwesen*, **29**, 513–519 (1958).
[32]J. Brownell, "Subsolidus Relations Between Mullite and Iron Oxide," *J. Am. Ceram. Soc.*, **41**, 226–230 (1958).
[33]A. Muan, "Phase Equilibria at Liquidus Temperatures in the System Iron Oxide Al_2O_3-SiO_2 in Air Atmosphere," *J. Am. Ceram. Soc.*, **40**, 121–133 (1957).
[34]M. K. Murthy and F. A. Hummel, "X-Ray Study of the Solid Solution of TiO_2, Fe_2O_3 and Cr_2O_3 in Mullite ($3Al_2O_3{\bullet}2SiO_2$)," *J. Amer. Ceram. Soc.*, **43**, 267–273 (1960).
[35]S. N. Razumovski, T. A. Tunik, O. N. Fisher and S. P. Shmitt-Fogelvich, "X-ray analysis of solid solution of ferric oxide in mullite," *Ogneupory*, **18**, 46–48 (1977).
[36]H. Schneider and H. Rager, "Occurrence of Ti^{3+} and Fe^{2+} in Mullite," *J. Am. Ceram. Soc.*, C–248–C–250 (1984).
[37]H. Schneider, "Temperature-Dependent Iron Solubility in Mullite," *J. Am. Ceram. Soc.*, **70**, C–43–C–45 (1987).

[38]H. Schneider and H. Rager, "Iron incorporation in mullite," *Ceramics Intern.*, **12**, 117–125 (1986).

[39]C. R. Green and J. White, "Solid solubility of TiO_2 in mullite in the system Al_2O_3-TiO_2-SiO_2," *Trans. Brit. Ceram. Soc.*, **73**, 73–75 (1974).

[40]J. L. Miller, G. R. McCormick and S. G. Ampian, "Phase Equilibriain the System GeO_2-Al_2O_3," *J. Am. Ceram. Soc.*, **50**, 268–269 (1967).

[41]N. A. Toropov, I. F. Andreev, V. A. Orlov and S. P. Shmitt-Fogelvich, "Synthesis of Germanium Mullite and Study of its Properties," *Zhur. Prikl. Khimii*, **10**, 2143 (1970).

[42]S. Durovic and P. Fejdi, "Synthesis and crystal structure of germanium mullite and crystallochemical parameters of D-mullites," Silikáty, c 2 97–112 (1976).

[43]Chr. Neumann, "Schmelzsynthesen von Mulliten unterschiedlicher Zusammensetzung und ihre röntgenographische Charakterisierung," Dissertation Hamburg (1984).

[44]H. Saalfeld and W. Guse, "Structure refinement of 3:2-mullite ($3Al_2O_3{\cdot}2SiO_2$)," *N. Jahrb. Min. Mh.*, 145-150 (1981).

[45]H. Saalfeld and K. H. Klaska, "A Pb/Nd-stabilized mullite of the composition $Al_{5.03}Ge_{0.97}Pb_{0.15}Nd_{0.06}O_{9.71}$," *Z. Kristallogr.*, **171**, 129–133 (1985).

The Melting Nature and Growth of Mullite Single Crystals By the Modified Floating Zone Method

ISAMU SHINDO

Tsukuba Asgal Co., Ltd.
Tohkohdai 5-5, Tsukuba City, Ibaraki, 300-26 Japan

Abstract

The phase relations in the binary system Al_2O_3-SiO_2 were investigated by the slow cooling float zone (SCFZ) method. The single crystal growth of mullite of various compositions was also investigated according to the resulted phase diagram.

It is clearly demonstrated that mullite melts incongruently and their solid solution range is measured between 2:1 and 3:2 of mole ratio of Al_2O_3-SiO_2. Good quality single crystals of mullite have been successfully grown by the travelling solvent float zone (TSFZ) method using a lamp image furnace. In the solidification from the melt of this system, the constitutional supercooling phenomenon occurred easily and the characteristic cellular structure resulted. The influence of this phenomenon to phase research or to single crystal growth are discussed.

Introduction

To grow single crystals of oxide compounds, the growth method has to be selected according to melting nature of the individual materials. In the case of incongruently melting compounds or solid solutions of homogeneous composition, the travelling solvent method must be used instead of normal melt growth techniques. Thus, one must determine the correct melting characteristics of the materials to be grown.

In this Al_2O_3-SiO_2 binary system, various phase diagrams have been reported by many workers.[1–6] These results have been summarized by Pask et al.[7] However, no one has determined the correct melting nature of mullite. Various experimental reasons have been indicated by many investigators for these opposing results. The problems related to the crucible have been considered to be the important factor which has restricted the accuracy of the experimental results. For example, it is difficult to avoid the chemical reactions between the crucible materials and the melts at elevated temperatures in many cases. The following two problems have considerable influence on the growth of mullite single crystals.

1. the melting characteristics of mullite.
2. the solid solution range of the mullite.

A few investigations have been reported regarding the growth of mullite single crystals. Bauer et al.[8] have grown "opaque" mullite single crystals using the Vernuil method. Their explanation for their success, even though mullite presumably melted

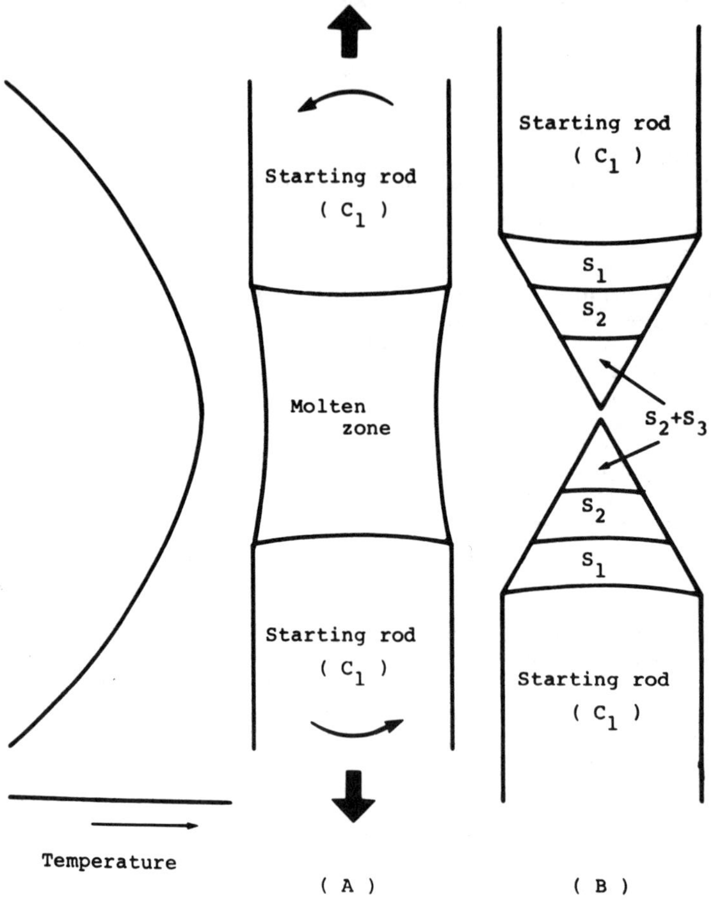

Fig. 1. Schematic diagram of the SCFZ method.

incongruently based on the Bowen's report, was derived from the deep temperature gradient across the melt zone. Mateika et al.[9] reported the results of mullite growth experiments using the Czochralski (pulling) method. Their crystals were transparent and of the composition $2Al_2O_3$-SiO_2, however, many "cleavage cracks," containing an amorphous impurity phase were also observed. The texture of these cleavage cracks appeared similar to which occur as a result of exsolution or cellular growth. It is well known that cellular growth texture is derived from constitutional supercooling formed in the boundary layer at the growing solid-liquid interface. In both Bauer's and Mateika's experiments, the formation mechanism is unknown. However, because it is an important factor in mullite growth, careful and well-controlled experiments are needed in order to determine how to avoid the formation of this structure.

If the floating zone method can be applied for studying the phase equilibria or single crystal growth of mullite, the experimental limitations and difficulties would be drastically diminished, as no container is required.

The author and his colleagues have demonstrated the usefulness of modified floating zone techniques using the lamp image furnace for determining phase relations by the slow cooling float zone (SCFZ) method[10,11] and also for growing single crystals of various incongruently melting materials by the travelling solvent float zone (TSFZ) method.[12–15] In this paper, the experimental results on the phase relations in the Al_2O_3-SiO_2 system by the SCFZ method and on the growth of mullite single crystals by the TSFZ method are reported. Moreover a new diagram has been determined.

The Slow Cooling Float Zone Method

The slow cooling float zone (SCFZ) method is a modified floating zone method and has been successfully applied to the system MgO-TiO_2.[10] The theoretical background of the SCFZ method is fundamentally based on that of the maximum fractional crystallization principle, as suggested by Bowen[16] and is similar to that of the normal freezing method.[17] The detailed procedures of this method have been described elsewhere,[10] only a brief description is given here.

In the ideal maximum fractional crystallization process, the following conditions are maintained; (1) the melt is completely homogeneous, (2) diffusion in the solid is negligible, and (3) the solid and the liquid are in equilibrium at their interface during solidification.

To realize these conditions, it is convenient to use a floating-zone technique. The continuous counter-rotation of the upper and lower shafts allows the melt to be completely mixed and kept homogeneous. Equilibrium between the solid and liquid phases can be achieved by controlling the rate of advance of the solidification front. Figure 1 is a schematic illustration of a floating melt system before the cooling process in a SCFZ experiment (A) and the resulting zonal structure in the solidified ones (B). The notations C_1, S_1, S_2, S_3 all correspond to those in Fig. 2, which is a phase diagram of a hypothetical binary system containing a peritectic compound. The solidification sequence that is determined by the EPMA technique may be plotted on a composition-temperature diagram where the temperature scale is arbitrary. In order to complete a phase diagram, the temperatures necessary for delineating the liquidus and the solidus curves must be determined for the specimen fragments with selected compositions, which are separated from the solidified cones.

Experimental Procedure

Powders of Al_2O_3 (99.9% pure) and SiO_2 (99.9% pure) were used as the starting materials. A small amount (0.1 wt percent) of Fe_2O_3 powder (99.9% pure) was also added for the absorption of the infrared light. The powders were mechanically

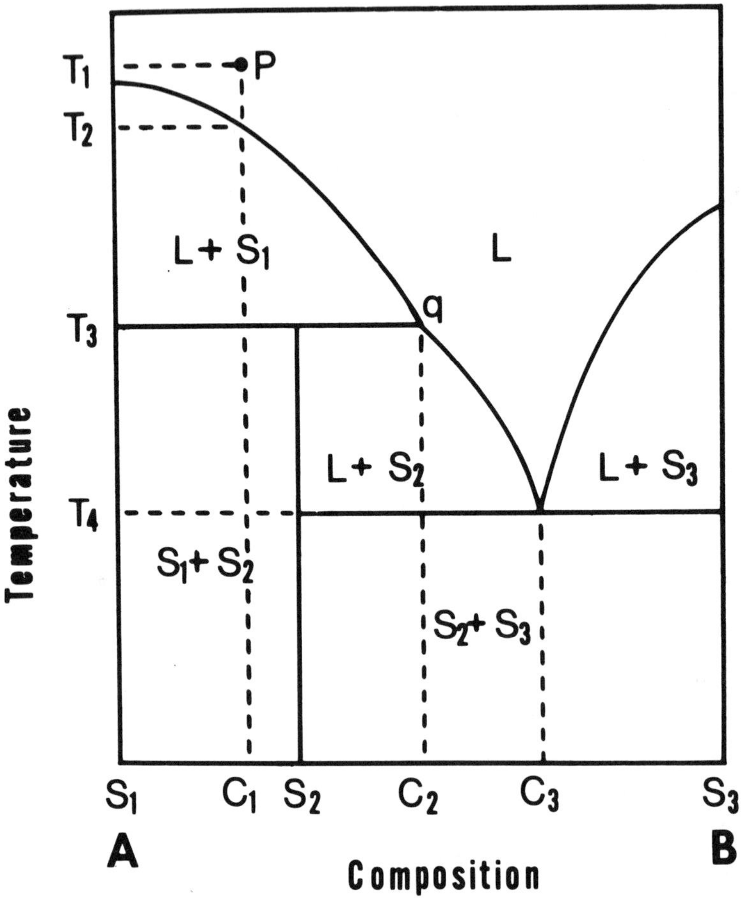

Fig. 2. Hypothetical diagram of a binary system.

mixed in various ratios, placed in a sealed rubber tube and hydrostallically pressed at a pressure of approximately 700 Kg/cm^2 to form a rod of the desired diameter (5–7 mm) and length (40–50 mm). The rods were then sintered at 1500–1600°C in air.

The floating zone apparatus used was of an infrared radiation convergence type with two 3.5KW quartz halogen lamps or a 6.5KW xenon lamp as the radiation source.

Phase Research By the SCFZ Method

Two sintered rods were used. One rod was suspended from the lower end of the upper shaft and the other was fixed at the top of the lower shaft. The facing ends of the upper and the lower sintered rods were melted and connected. The shafts were counter-rotated at 30 rpm. Approximately 30 minutes later, the electric power supply was gradually reduced at a constant rate, while the two shafts were gradually separated simultaneously. When the melt was completely condensed, two boules, each with a cone shape, were obtained.

The boules were bisected parallel to the longitudinal direction and polished mechanically. The polished sections were examined by means of optical microscopy and the EPMA technique.

Single Crystal Growth By the TSFZ Method

The apparatus and the procedures were almost the same as those in the SCFZ method. However, in the case of the TSFZ method, the melt zone was moved gradually along the rod and the electric power supply maintained at constant level.

Results and Discussion

Phase Relations By the SCFZ Method

Each SCFZ run gave a pair of boules, each having a cone-shaped recrystallized portion at one end. These recrystallized bodies have been analyzed by means of optical microscopy and the EPMA technique.

Starting Composition; Al_2O_3:SiO_2 = 3:1

Figures 3 and 4 show magnified views of the initial and the final solidified regions of the cone-shaped boule, respectively. As can be seen in Figs. 3 and 4, four

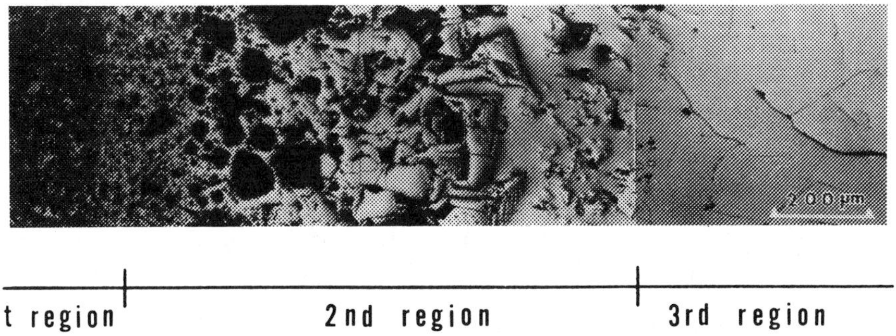

Fig. 3. Magnified view of an initial solidified body.

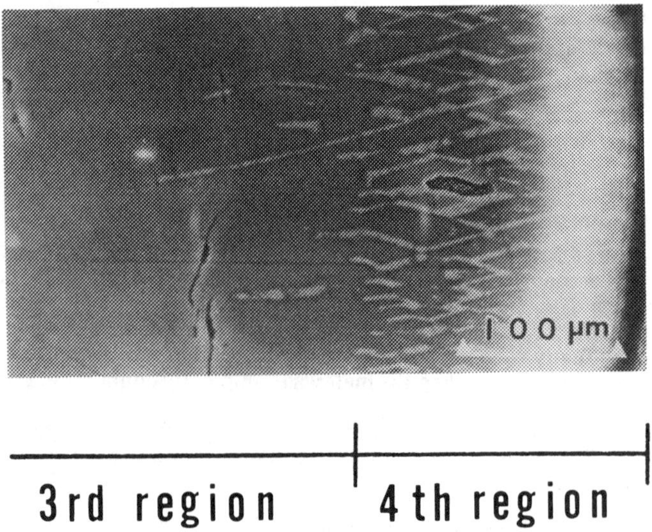

Fig. 4. Magnified view of a final solidified portion.

regions forming a zonal structure are distinguishable. The first region was the originally sintered rod, the second region corresponds to the region once melted and solidified with co-precipitation of a foreign phase. The third region is a single phase. In the fourth region, the typical lamellar structure is observed. By means of EPMA, the composition of the precipitated phase in the 2nd region was identified to be pure Al_2O_3 and that of the matrix to be $2Al_2O_3$-SiO_2. In the third region, the composition varies depending on the position. The content of Al_2O_3 in this region decreased gradually as are passed into the 4th region. At the boundary with the 4th region, the content of Al_2O_3 exhibited the smallest value, approximately $3Al_2O_3$-$2SiO_2$. The X-ray powder diffraction pattern revealed that the solid in the third region was mullite. The composition of each lamellar in the fourth region was analyzed by means of EPMA, and the results revealed that the lamellar texture consists of two phases, mullite and silica-rich glass. The composition of the latter phase was identified as approximately Al_2O_3-$9SiO_2$.

Starting Composition: 2:1

In this case almost similar results as in the case of 3:1 were obtained, although the amount of the initially solidified Al_2O_3 was less than that of the 3:1 case.

Starting Composition: 3:2

In this case the solid Al_2O_3 was not detected throughout the solidified bodies and the composition of the initially solidified body was found to be approximately $2Al_2O_3$-SiO_2. A typical lamellar structure was also observed in the final solidified region.

In both cases where the starting molar ratios of Al_2O_3 and SiO_2 were 3:1 and 2:1, the initial solidified bodies were identified as Al_2O_3. These results revealed the following facts:

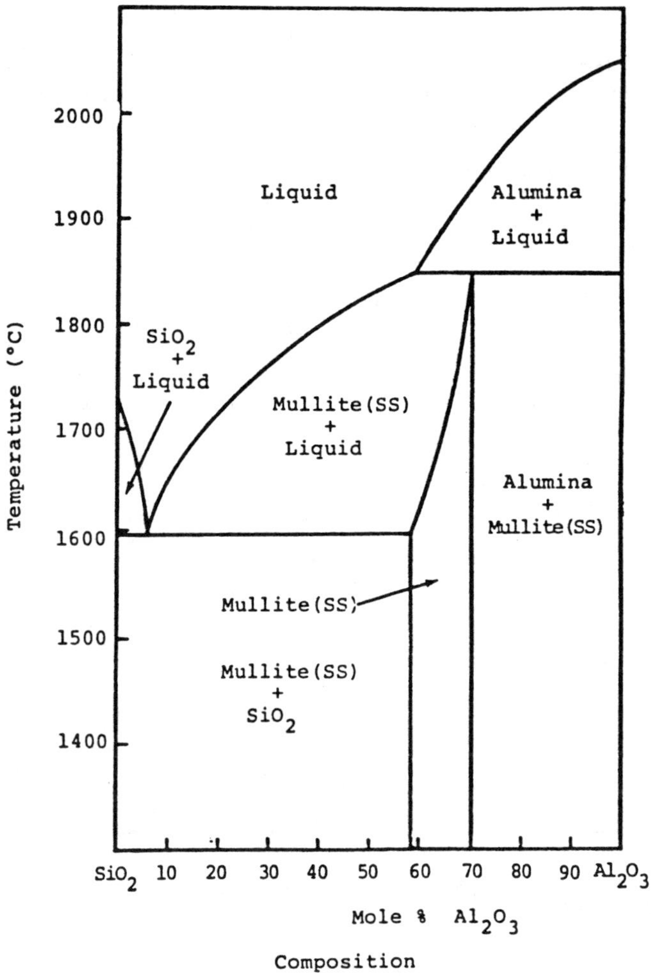

Fig. 5. Phase diagram of the system, Al_2O_3-SiO_2, from the results of the SCFZ method.

1. the solid phases encountered were Al_2O_3, mullite, and SiO_2-rich glass.
2. the solidification occurred in the following order; Al_2O_3, mullite, and the mixture of mullite and SiO_2-rich glass.
3. mullite was the peritectic compound.
4. the solid solution of Al_2O_3 in mullite extended from $3Al_2O_3 \cdot 2SiO_2$ to $2Al_2O_3 \cdot SiO_2$.

From these data a revised phase diagram of the system Al_2O_3-SiO_2 was obtained, as shown in Fig. 5, using the results of the solidus temperature measurements on the separated samples.

In the ideal SCFZ experiments, the final solidification stage should represent the eutectic lamellar structure of mullite and silica. Though the lamellar structure shown in Fig. 4 is considered to be the cellular growth lamellae because silica-rich glass is observed instead of pure silica, the compositional change between the silica-rich glass and pure silica is not so large.

Since these cellular growth structures are derived from the constitutional supercooling phenomenon formed in the liquid bordering the moving solid-liquid interface, the solidification speed plays an important role in the avoidance of this phenomenon. These problems are again discussed in below.

Single Crystal Growth of Mullite by the TSFZ Method

According to the melting nature of mullite shown in Fig. 5, a proper travelling solvent method must be used to grow a single crystal of homogeneous composition using a single crystal seed. However, the first growth experiment was made on a polycrystalline sintered rod instead of a single crystal seed, and no solvent zone was used.

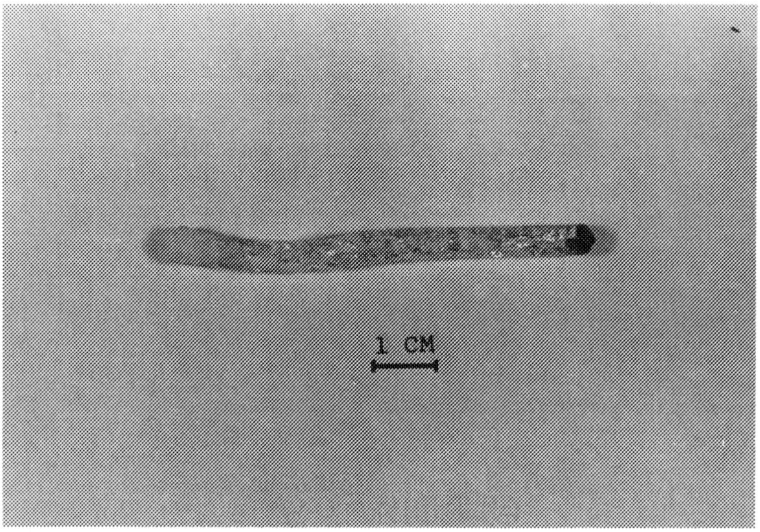

Fig. 6. A mullite crystal grown by the floating zone method.

The Starting Composition: $2Al_2O_3{:}SiO_2$

The crystal growth experiments were conducted using various growth rates. The resulting crystalline boules were studied by X-ray diffraction, optical microscopy and electron probe micro analysis.

A transparent crystal boule was obtained using a growth rate lower than 1 mm/hr. When the growth rate exceeded 1 mm/hr, the crystal tended to be "opaque." A typical crystal is shown in Fig. 6. During the first stage of growth, the alumina phase was solidified. Approximately 5 mm later, the mullite phase solidified as the

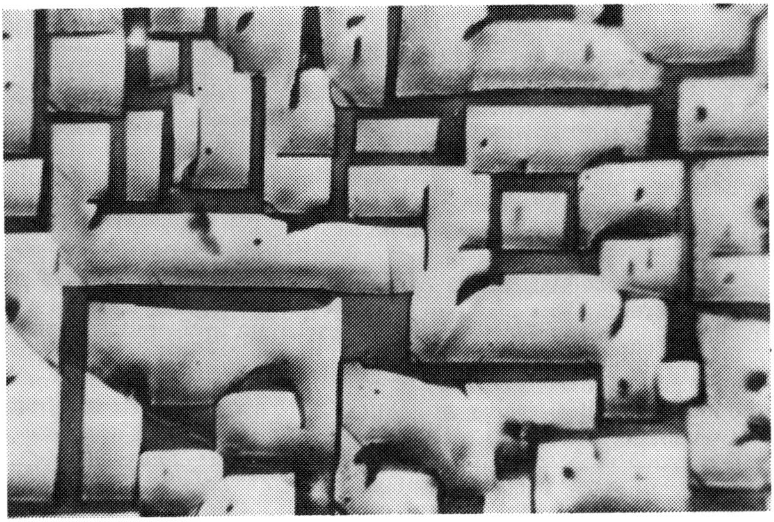

Fig. 7. Typical feature of the two phase structure shown in mullite crystals.

single phase. Even though many grain boundaries were observed at the initial stage of the mullite phase, they gradually decreased and large single grain crystals were obtained. In this experiment, a sintered rod was used as the seed, therefore, the growth process should lead to single crystals having preferred orientations. The longitudinal directions of large grains were examined as the c-axis of mullite.

Starting Composition: $3Al_2O_3$:$2SiO_2$

In this case, the transparency of the grown crystal varied with the stage of growth. In the first stage, a relatively transparent crystal was obtained, although it became gradually "opaque." No transparent crystals could be grown even under the growth rate of 0.2 mm/hr except in the first stage of growth.

The texture and the composition in the sliced and polished thin sections of the grown crystal were examined using an electron microscope and/or an electron probe microanalyzer. The Al_2O_3 concentration in the transparent crystal region grown in the first stage was measured to be 65–66 mole percent. The characteristic two phase texture was observed in the following opaque region as shown in Fig. 7. This specimen shows a microstructure consisting of mullite crystals and another phase. Mullite crystals grow parallel to the c-axis and are prismatic with well-developed {110} surfaces. Another phase was examined under an electron microscope and was confirmed to be a glass substance by the observation of its electron diffraction pattern. Al_2O_3 concentrations in the mullite and the glass phase were determined as 60.9 and 6.4 mole percent, respectively. Elemental iron was not detected in either the mullite or the glass phase.

These two-phase structures were typically seen in many samples solidified from the melt of the Al_2O_3-SiO_2 system. Thus, investigations concerned with the formation and origin of these structures were deemed one of the most important factors in the study of the phase diagram or single crystal growth related to mullite.

Various parameters control the formation of these two-phase microstructures. In the case of the cellular growth structure, the formation depends strongly on the growth rate or the solidification rate from their melt. By contrast, the formation of the eutectic, the pseudo eutectic and the exsolution structures is independent of the growth rate, however, it is very difficult to distinguish these two-phase structures using their own structure form or shape.

The cellular growth structure was caused by the formation of constitutional supercooling in the boundary layer at the solid liquid interface. Tiller et al.[17] showed that a large growth rate, a higher solute concentration, poor stirring and a large difference from unity of the effective segregation coefficient tend to produce constitutional supercooling. In the Al_2O_3-SiO_2 system, the formation of a silica-rich boundary layer at the solid-liquid interface at the solidification front is thought to be the most important factor in producing variations in the experimental results.

Thus, controlling the solidification speed was found to be very important to avoid the formation of the cellular structure. In the case of the growth of $2Al_2O_3$-SiO_2 mullite crystals, this cellular structure did not appear when the solidification speed was maintained lower than 1 mm/hr. On the other hand, in the case of the composition $3Al_2O_3 \cdot 2SiO_2$, the cellular structure was observed even at the speed of 0.3 mm/hr. In any experiments on the phase research in the system Al_2O_3-SiO_2 or the single crystal growth of mullite, the control of the solidification speed is thought to be the most important factor.

Acknowledgments

The author wishes to thank Drs. S. Kimura and F. Okamura of the National Institute for Research in Inorganic Materials for their critical reading of the manuscript and also Dr. K. Yamana of Industrial Research Institute of Ishikawa Prefecture for providing photographs.

References

[1]N. L. Bowen and J. W. Greig, "The System Al_2O_3-SiO_2," *J. Am. Ceram. Soc.*, 7 [4] 238–54 (1924).

[2]S. Aramaki and R. Roy, "Revised Phase Diagram for the System Al_2O_3-SiO_2," *J. Am. Ceram. Soc.*, **45** [5] 229–42 (1962).

[3]N. A. Toropov and F. Ya. Galakhov, "New Data for the System Al_2O_3-SiO_2," *Dokl. Akad. Nauk S.S.S.R.*, **78** [2] 299–302 (1951).

[4]J. H. Welch, "A New Interpretation of the Mullite Problem," *Nature*, **186**, 545–46 (1960).

[5]T. Horibe and S. Kuwabara, "Thermo-Analytical Investigation of Phase Equilibria in The Al_2O_3-SiO_2 System," *Bull. Chem. Soc. Japan*, **40** [4] 972–82 (1967).

[6]I. A. Aksay and J. A Pask, "Stable and Metastable Equilibria in The System Al_2O_3-SiO_2," *J. Am. Ceram. Soc.*, **58** [11-12] 507–12 (1975).

[7]R. F. Davis and J. A. Pask, "High Temperature Oxides," Vol. 5-IV, pp. 37–76 (Edited by Allen M. Alper, Academic Press, New York and London 1971).

[8]W. H. Bauer, I. Gordon, and C. H. Moore, "Flame Fusion Synthesis of Mullite Single Crystals," *J. Am. Ceram. Soc.*, **33** [4] 140–43 (1962).

[9]W. Guse and D. Mateika, "Growth of Mullite Single Crystals ($2Al_2O_3 \cdot SiO_2$) by the Czochralski Method," *J. Crystal Growth*, **22**, 237–240 (1974).

[10]I. Shindo, "Determination of The Phase Diagram by The Slow Cooling Float Zone Method: The System MgO-TiO_2," *J. Cryst. Growth*, **50** [4] 839 (1980).

[11]I. Shindo, S. Takekawa, K. Kosuda, T. Suzuki, and Y. Kawata, "Phase Relations in The Ternary System ZrO_2-Al_2O_3-SiO_2 by The Slow Cooling Float Zone Method.

"Science and Technology of Zirconia II," pp. 181–89 (Edited by Nils Claussen et al., The American Ceramic Society, Inc., Columbus, Ohio (1983).
[12]I. Shindo, N. Kimizuka, and S. Kimura, "Growth of YFe_2O_4 Single Crystals by Floating Zone Method," *Mat. Res. Bull.*, **11** [6] 637–644 (1976).
[13]I. Shindo and S. Kimura, "Single Crystal Growth of YIG by The Floating Zone Method, *J. Crystal Growth*, **41** [2] 192–98 (1977).
[14]I. Shindo, N. Ii, K. Kitamura, and S. Kimura, "Single Crystal Growth of Substituted Yttrium Iron Garnets $Y_3Fe_{5-x}(Ga,Al)_xO_{12}$ by The Floating Zone Method, *J. Crystal Growth*, **46** [3] 307–13 (1979).
[15]I. Shindo, S. Kimura, and K. Kitamura, "Growth of Mg_2TiO_4 Single Crystals by The Floating Zone Method," *J. Mat. Sci.*, **14**, 1901–06 (1979).
[16]N. L. Bowen, "Certain Singular Points on Crystallization Curves of Solid Solutions," *Proc. Nat. Acd. Sci.*, **27**, 301–09 (1941).
[17]W. A. Tiller, "Use of Controlled Solidification in Equilibrium-Diagram Studies," *Trans. AIME*, **215**, 555–9 (1959).

Modulated Structure and Vacancy Ordering in Mullite

NOBUO MORIMOTO†, YOSHIHARU NAKAJIMA‡, AND MASAO KITAMURO*

Abstract

The ordered arrangement of oxygen vacancies in mullite results in a complex modulated, or, incommensurate structure of antiphase domains. The modulated structure of mullite is constructed on the clusters of double Si-tetrahedra and triple Al-tetrahedra, which crosslink the edge-sharing AlO_6 octahedral chains running parallel to the *c*-axis. The possible composition range and stability of the modulated structure of mullite have been critically reviewed. The modulated structure of mullite has also been compared with that of plagioclase feldspar.

Historical

The chemical composition of mullite is usually denoted as $3Al_2O_3{\cdot}2SiO_2(Al_3SiO_4)$. In fact, it is a nonstoichiometric material. The structural formula is represented by $Al_2(Al_{2+2x}Si_{2-2x})O_{10-x}$ where x varies from 0.17 to 0.50.

Since the exploratory studies of mullite by Wyckoff et al.[1] and Taylor,[2] the close structural relationship between mullite and sillimanite (Al_2SiO_5) has been emphasized on the basis of the remarkable similarity in their X-ray powder diffraction patterns. Taylor[2] and Warren[3] proposed a structure model of mullite which was deduced from the structure of sillimanite by introducing oxygen vacancies. These vacancies produced charge balance due to substitution of Si^{4+} by Al^{3+} and explained smaller density of mullite (~3.16g/cm^3) compared with that of sillimanite (~3.22g/cm^3).

An average or statistical structure was determined independently by Durovic[4] and Sadanaga et al.[5] using X-ray single crystal diffraction data. Sadanaga et al. explained clearly the structural relationship between mullite and sillimanite. Although several studies were subsequently conducted by Durovic,[6] Burnham,[7,8] Nakajima[9] and Angel and Prewitt,[10] all of the results are in good agreement with the earlier studies.

As noted above, the structural formula of mullite is represented by $Al_2(Al_{2+2x}Si_{2-2x})O_{10-x}$. Therefore, the $2Al_2O_3{\cdot}SiO_2$ mullite (2:1 mullite) studied by Sadanaga et al.[5] and by Burnham[7,8] has the composition of $Al_{4.8}Si_{1.2}O_{9.6}$ (x=0.40). The average structure of this mullite is projected on (001) in Fig. 1. The structure is orthorhombic (space

†Now with the Osaka Sangyo University, Nakagaito, Daito, Osaka, 574 Japan.

‡Engineering Center, Sharp Co., 2613-1, Ichinomoto, Tenri, Nara 632, Japan.

*Department of Geology and Mineralogy, Kyoto University, Sakyo, Kyoto 606, Japan.

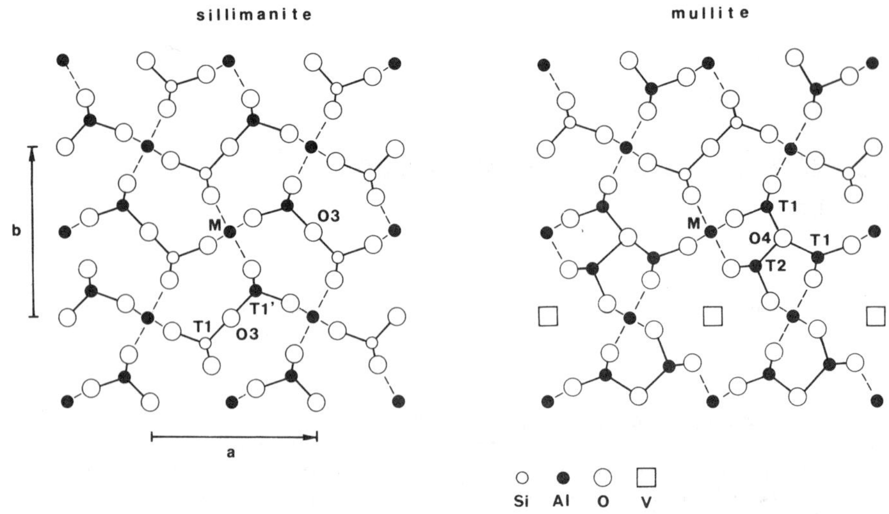

Fig. 1. Schematic drawings of the crystal structures of sillimanite and mullite. The atomic arrangement around oxygen vacancies, which are separated from each other by the unit translation of *a*, is specially shown in the mullite structure.

group Pbam) with the cell dimensions of *a*~7.6Å, *b*~7.7Å and *c*~2.9Å (Table 1). The edge-sharing AlO_6-octahedral chains run parallel to the *c*-axis and are crosslinked by doublets of Si and Al tetrahedra.

In sillimanite (Fig. 1), Al and Si atoms are completely ordered in the tetrahedral chains forming the S (sillimanite)-chain[11] with the unit length of 5.8Å (Fig. 2). However, to obtain the mullite structure from the sillimanite one, the substitution of Si by Al atoms is necessary and described by the formula $2Si^{4+}+O^{2-}\rightarrow 2Al^{3+}+$. Because the increase of the Al content gives an apparent long range disorder of Al and Si in the tetrahedral sites, the *c*-dimension for the average structure becomes about 2.9Å, half of the *c*-axis of sillimanite.

In the structure of mullite, some of the O3-atoms (see sillimanite structure in Fig. 1) at the shared corners of the doublets of the tetrahedra are missing and the related cations of T1 sites are shifted to new positions at the T2 sites (Fig. 1). The shift of these cations forms an amphibole-like or MS (modified sillimanite)-chain (Fig. 2), which was called M (mullite) chain by Saalfeld[11] based on his structure model. By the appearance of the new tetrahedra, the adjacent 03 atoms are coordinated by three tetrahedra. These are also displaced from the symmetry center toward to a less symmetrical position, 04 (Fig. 1). The long range ordering of the oxygen vacancies together with the shift of T1 cations results in the modulated structure of mullite.

It must be noticed that by the omission of the 03 atoms in the sillimanite structure, many of the neighboring 03 atoms to the vacancy are shared by the T tetrahedra (T1 and T2) forming a triplet cluster (Fig. 1). In order to avoid a concentration of electric charge into the bridging oxygen, these triple tetrahedra must be occupied by Al atoms. On the other hand, the wide solid solution range of mullite (x=0.16–0.50) requires some remaining 03 atoms to be shared by only two Si tetrahedra which form a doublet. A mixed chain consisting of the triplet and doublet

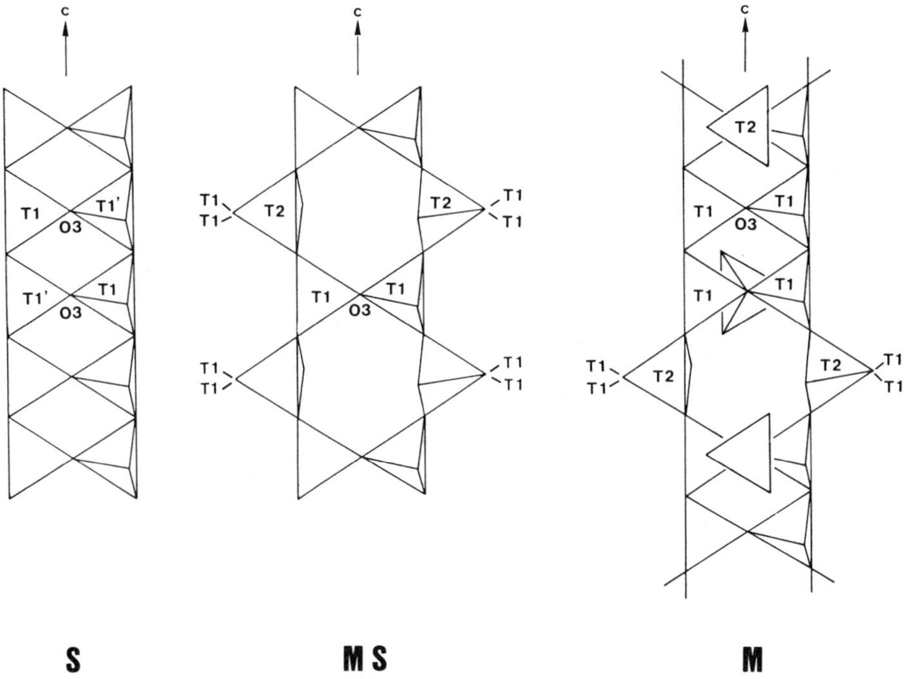

Fig. 2. S (sillimanite)-chain, MS (modified sillimanite)-chains with every other vacancy of O3 atoms elongated along the *c*-axis, and M (mullite)-chain with doublets and trefoiled clusters of Si and Al tetrahedra.[22]

clusters of tetrahedra based on a new structural model[22] is called M (mullite)-chain (Fig. 2)

Modulated Structure

Extra Reflections of Mullite

In contrast to sillimanite, mullite shows many different types of extra reflections in the X-ray diffraction patterns which cannot be explained by the average structure described above. These extra reflections were described initially by Taylor[2] and later by Kurylenko,[12] Scholze,[13] Agrell and Smith,[14] Tokonami et al.,[15] and Nakajima.[9] The same kinds of extra reflections were observed in the electron diffraction patterns by Nakajima et al.,[16] Cameron,[17] and Nakajima and Ribbe.[18]

The extra reflections in mullite are classified into two groups (Fig. 3): (1) satellite reflections and (2) diffuse streaks. The satellite reflections are further subdivided into two kinds, depending upon their diffuseness. Agrell and Smith[14] used S-, and D-mullite to represent mullites which show sharp and diffuse satellite reflections, respectively. The intensity maximum of the satellites appears as a pair around h = odd, l = odd on the hkl reciprocal planes where k is an integer. The satellite reflections changes their sharpness and position according to the Al-content (x).[18] When the Al content becomes higher, the extra reflections becomes stronger and

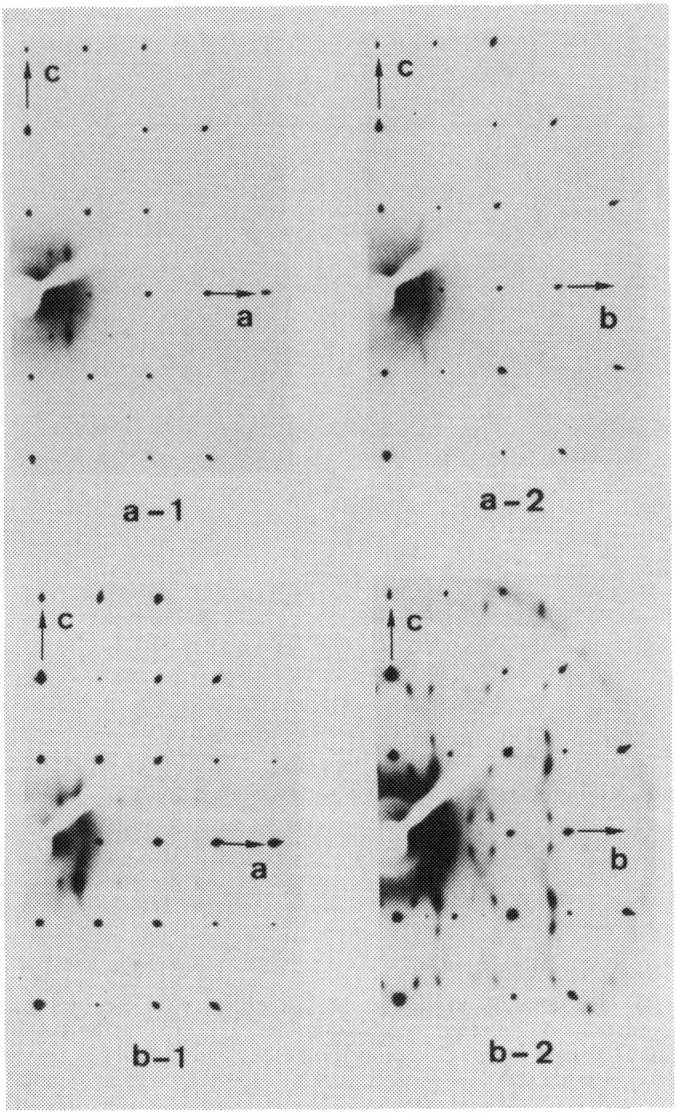

Fig. 3. X-ray diffraction patterns (precession photographs) of mullite.[9] (a) S-mullite, (b) D-mullite.

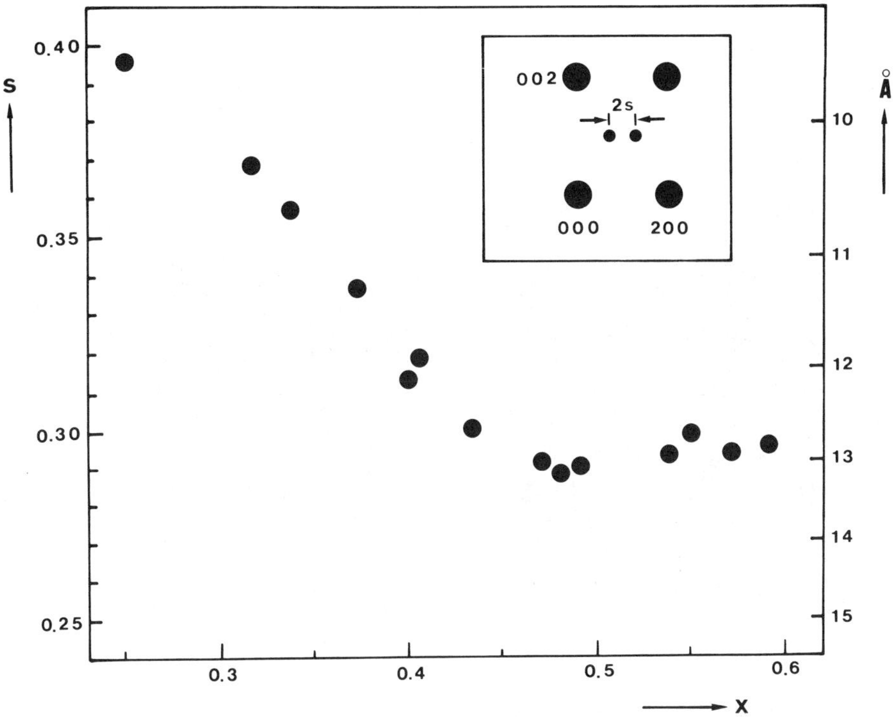

Fig. 4. Variation of the s-value with the Al content in mullite.[19] The maximum value of x affecting the s-value is apparently 0.5 or less.

sharper. The vector(s) joining a pair of the satellite reflections change(s) its (their) length as in Fig. 4. The orientation also changes from the direction parallel to (100) for X<0.5 to one slightly inclined against {601} for X>0.5.[18]

The diffuse streaks were described by Nakajima[9] and Tokonami et al.[15] Their maxima are at l = L±/2/9 and extend along the direction of either h+k=K±/4/7 or h-k=K±/4/7, where L and K are integers including 0. The intensity varies along the streaks and becomes stronger when K approaches an integer.

Modulated Structure of S-Mullite

Antiphase Domain Structure

A transmission electron microscope was first used by Nakajima et al.[16] and later by others, to elucidate the superstructure of mullite. The high resolution lattice images were taken along the *b*-axis for S-mullite of X~0.4 (Fig. 5a). In the images, elongated domains or slabs with white spots arranged in a centered lattice intergrow with each other, forming a modulated structure. The intergrowth of the slabs is slightly different in orientation according to the Al-content of mullite. The intergrowth is running parallel to (001) in Fig. 5a. However, in the case of mullite of X~0.5 or more (Fig. 5b), the intergrowth runs to a nonrational orientation slightly inclined against (601) or (601). The intergrowth in these two different orientations often coexist in a small portion of a specimen of x=5.[19]

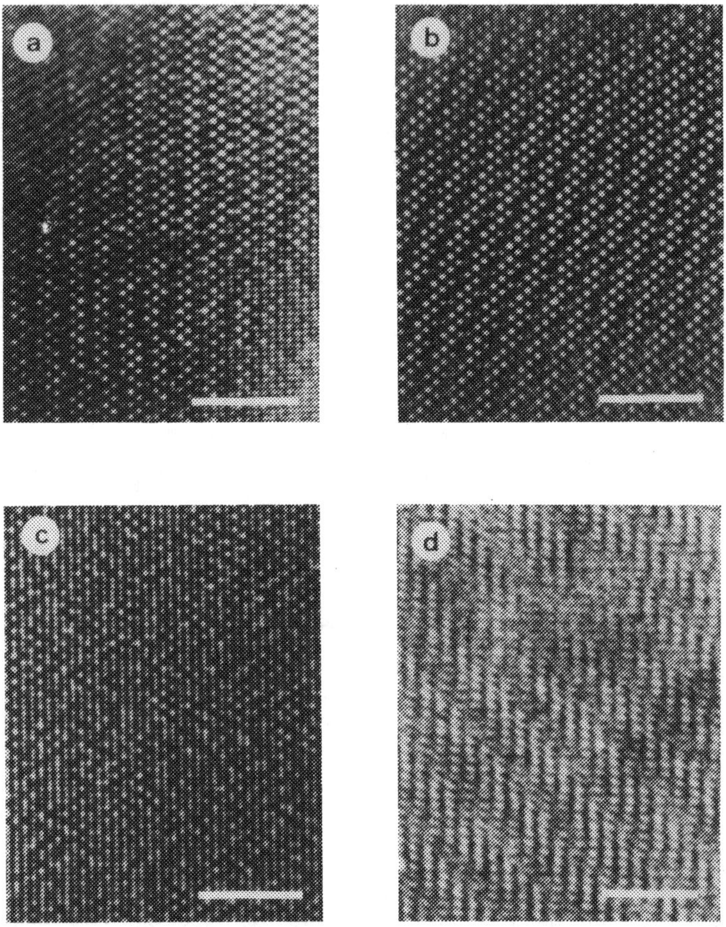

Fig. 5. Lattice images of hOl of M- and S-mullite (the *a* and *c*-axes are horizontal and vertical, respectively) and a lattice image of plagioclase feldspar. (a) S-mullite with x~0.4, (b) S-mullite with x~0.6, (c) D-mullite, (d) plagioclase feldspar (An_{50}).[23]

Two adjacent slabs are in an antiphase relation with respect to the white spots in the centered lattice with the shift of 1/2•[001] to each other. Areas with small and faint white spots are observed between these slabs, forming an antiphase domain boundaries (APB's).

In the APB's, where the large white spots are hardly recognized, the vacant oxygens are considered either to be absent or, if any, very few in number. This boundary is considered to consist of a structure very similar to that of sillimanite with no ordering of the Al and Si tetrahedra.

Thus, the modulated mullite is the intergrowth of two kinds of slabs elongated exactly or approximately along (100), slabs with oxygen vacancies on the centered lattice in (010) projections and those which do not show any oxygen vacancies with the sillimanite composition.

Structure Models of S-Mullite

Saalfeld[11] proposed a model of the mullite superstructure based on the X-ray diffraction intensity of the satellite reflections. However, this model consisting of the S- and MS-chains gives a quite different arrangement of the oxygen vacancies in the (010) projection from that in the lattice image of the defect domains. Furthermore, this model gives the intensity distribution of the satellite reflections which is different from the observed one.[21]

McConnell and Heine[20] applied a new scheme of symmetry analysis for an incommensurate modulated structure to the mullite modulated structure. They gave two pure component structures for the mullite structure; one have an oxygen vacancy and the other the Si-Al ordering. Based on this theoretical analysis, Angel and Prewitt[21] analyzed the two pure component structures using the observed intensities of the satellites. The result shows that the component structures are basically the same as those by McConnell and Heine.

The modulated structure can be also expressed by the modulation wave of the probability function of the vacancy. This concept gives a better representation of the actual arrangement of the vacancy,[22] which can be directly compared with the lattice images. Schematic drawings of the modulation wave and the vacancy arrangement in this manner are given in Fig. 6.

Superstructure of D-mullite

A typical high resolution electron microscope image for the (h0l) plane of D-mullite has been obtained at 200kV (Fig. 5c). Although the regularity is not as clear as in the S-mullite (Figs. 5a and b), the white spots corresponding to the oxygen vacant sites are locally observed to form clusters. The ordered arrangement of the white spots in the clusters is essentially the same as in the S-mullite. Moreover, the white spots on the centered lattice in the clusters are in an antiphase relation to those of the adjacent clusters. The clusters are dispersed in the matrix where the white spots are less remarkable.

Comparison of the Modulated Structures of Mullite and Plagioclase

The modulated structure of mullite described above is not uncommon for silicate crystals. Plagioclase feldspar is one of the most important rock-forming minerals and an important ceramic material. It forms an almost complete solid solution between An ($CaAl_2Si_2O_8$) and Ab ($NaAlSi_3O_8$) at high temperatures. However, most of the natural plagioclase of composition between An_{75} and An_{25} show a very complex modulated structure, giving many satellites, which are very similar to those of mullite.

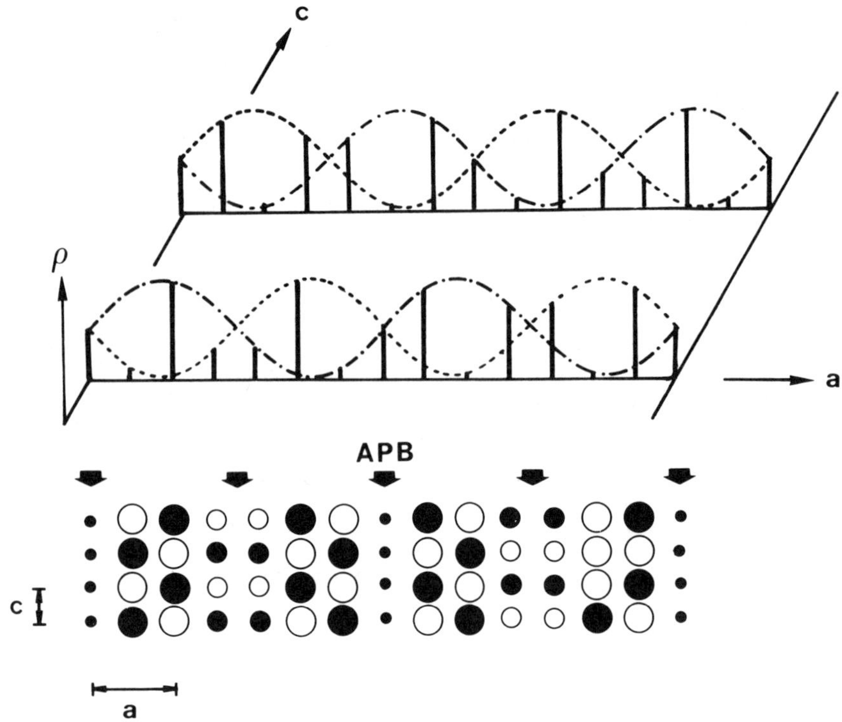

Fig. 6. (Upper): Schematic drawing of the density of vacancies (ρ) in the modulated structure of mullite. Sinisoidal curves shown by dashed and dash-dot lines represent the modulations of ρ separated by 1/2·[110]. (Lower): Schematic drawing of the density of the vacancies projected on the (010) plane. APB represents the antiphase domain boundary in mullite.

A lattice image of An_{50} plagioclase[23] is shown in Fig. 5d. The modulated structures obtained for plagioclase with different compositions are intergrowths of two kinds of slabs.[24] One kind of slabs is in an antiphase relation to the neighbors, and the other forms antiphase domain boundaries as in the modulated mullite structure. The orientation of both kinds changes continuously with composition.

Conclusion

The modulated structure in an S-mullite is comprised of intergrowths of two different sets of slabs with and without an antiphase relationship. The broader set having a width of several unit cells is in an antiphase relation to each other, corresponding to the region with an ordered distribution of oxygen vacancies, and the other to the region with no oxygen vacancies. Therefore, the broader lamellae represent the Al and vacancy-rich area and the other the Si and oxygen rich area.

Satellite and diffuse reflections are observed from the crystals which have been synthesized by zone melting and quenched to the room temperature. Their structure is preserved from their melting or decomposition temperatures. When the mullite crystals are synthesized by solid reaction from oxides or kaolinite, the products usually show a rather disordered arrangement of oxygen vacancies of the D-mullite type. Therefore, the modulated structure with intergrowths in antiphase relation to one another is considered to be stable at high temperature instead of the imaginary phase having a structure containing vacancy disorder.

References

[1]R. W. G. Wyckoff, J. W. Greig and N. L. Bowen, The X-ray diffraction patterns of mullite and sillimanite. *Am. J. Sci.*, **11**, 459–472 (1926).

[2]W. H. Taylor, The structure of sillimanite and mullite. *Z.Krist*, **68**, 503–521 (1928).

[3]B. E. Warren, The role of silicon and aluminum in complex silicates. *J. Amer. Ceramic Soc.*, **16**, 412 (1933).

[4]S. Durovic, A statistical model for the crystal structure of mullite. *Kristallografiya*, **7**, 339 (1962).

[5]R. Sadanaga, M. Tokonami, and Y. Takeuchi, The structure of mullite, $2Al_2O_3{\cdot}SiO_2$, and relationship with the structure of sillimanite and andalusite. *Acta Crystallogr.*, **15**, 65–68 (1962).

[6]S. Durovic, Refinement of the crystal structure of mullite. *Slovak Acad. Sci., Cheicke Zvesti*, **23**, 113–128 (1969).

[7]C. W. Burnham, Crystal structure of mullite. *Carnegie Institution of Washington Year Book*, **62**, 223–227 (1963).

[8]C. W. Burnham, Composition limits of mullite, and the sillimanite-mullite solid solution problem. *Carnegie Institution of Washington Year Book*, **62**, 227–22 (1963).

[9]Y. Nakajima, Dr. Thesis, Osaka University, Japan (1975).

[10]R. J. Angel and C. T. Prewitt, Crystal structure of mullite: A re-examination of the average structure. *Amer. Mineral.*, **71**, 1476–1482 (1986).

[11]H. Saalfeld, The domain structure of 2:1-mullite ($2Al_2O_3{\cdot}1SiO_2$). *Neu. Jahr. Mineral., Abh.*, **134**, 305–316 (1979).

[12]C. Kurylenko, Essay on the mullite structure. *Ceram. Abstr., 1954 June*, 1175 (1954).

[13]H. Scholze, Zum Sillimanit-Mullit problem. *Ber. Dtsch. Keram. Ges.*, **32**, 381–385 (1955).

[14]S. O. Agrell and J. V. Smith, Cell dimensions, solid solution, polymorphism, and identification of mullite and sillimanite. *J. Amer. Ceram. Soc.*, **43**, 69–76 (1960).

[15]M. Tokonami, Y. Nakajima, and N. Morimoto, Diffraction aspect and a structural model of mullite, $Al(Al_{1+2x}Si1\text{-}2x)O_{5\text{-}x}$. *Acta. Crystallogr.*, **A36** 270–276 (1980).

[16]Y. Nakajima, N. Morimoto, and E. Watanabe, Direct observation of oxygen vacancy in mullite $1.86Al_2O_3{\cdot}SiO_2$, by high resolution electron microscopy. *Proc. Japan Academy*, **51**, 173–178 (1975).

[17]W. E. Cameron, Mullite: a substituted alumina. *Amer. Mineral.*, **62**, 747–755 (1977).

[18]Y. Nakajima and P. H. Ribbe, Twinning and superstructure of Al-rich mullite. *Amer. Mineral.*, **66**, 142–147 (1981).

[19]J. Yla-Jaaski and H.-U Nissen, Investigation of superstructures in mullite by high resolution electron microscopy and electron diffraction. *Phys. Chem. Minerals*, **10**, 47–54 (1983).

[20]J. D. McConnell and V. Heine, Incommensurate structure and stability of mullite. Physical Review, B, **31**, 6140–6142 (1985).

[21]R. J. Angel and C. T. Prewitt, The incommensurate structure of mullite by Patterson synthesis. (Submitted to Acta Cryst.).

[22]M. Kitamura, N. Morimoto, and Y. Nakajima, Short range order of vacancy in mullite. *Collected Abstracts at 1988 Meeting of Japan Crystallogr. Assoc.* 9–B2 (1988).

[23]Y. Nakajima, N. Morimoto and M. Kitamura, The superstructure of plagioclase feldspar. Electron microscopic study of anorthite and laboradorite. *Phys. Chem. Minerals*, 1, 213–225 (1977).

[24]M. Kitamura and N. Morimoto, The modulated structure of the intermediate plagioclases and its change with composition. in "Feldspars and Feldspathoids" Edited by W. L. Brown, D. Reidel Publishing Company, pp. 99–119 (1984).

Structural Analysis of Glasses and Pyroxene Melts

H. MORIKAWA, F. MARUMO, T. MABUCHI, Y. SHIMIZUGAWA, AND Y. TABIRA

Research Laboratory of Engineering Materials
Tokyo Institute of Technology
Nagatsuta, Midori-ku, Yokohama 227, Japan

M. OKUNO

Department of Earth Sciences
Kanazawa University
Marunouchi, Kanazawa 920, Japan

M. MIYAKE

Department of Applied Chemistry
Yamanashi University
Takeda, Kofu 400, Japan

Abstract

Cr ions have a significant effect on a spinel primary field of some phase diagrams of silicates near diopside composition. On the other hand, Ti ions hardly change the phase diagrams. In order to clarify the roles of minor additives in crystallization of diopside, an X-ray diffraction study of silicate glasses and melts in the join diopside ($CaMgSi_2O_6$)-Ti pyroxene ($CaTiAl_2O_6$) was carried out. First, a small furnace attachable to an X-ray diffractometer was designed and constructed to investigate melt structures of silicates. As concerns a melt or glass near diopside composition, Al ions as well as a part of Ti ions seem to occupy 4-fold sites in the structure and they slightly modify the overall linkage mode of SiO_4 tetrahedra. On the other hand, Cr ions seem to occupy 6-fold sites and modify the linkage mode of SiO_4 tetrahedra.

Introduction

The Cr bearing Mg_2SiO_4 (forsterite)-$CaAl_2Si_2O_8$ (anorthite)-$CaMgSi_2O_6$ (diopside) system[1] has a wide spinel-primary-field compared with that without Cr, indicating a significant effect of Cr on phase relations in some silicate systems. Chrome spinel appears as a primary phase on the addition of 1.5 wt% $MgCr_2O_4$ (pichrochromite, Pi) to $CaMgSi_2O_6$ (diopside, Di).[1] In the join $CaMgSi_2O_6$-$CaCrCrSiO_6$ spinel appears as a primary phase on the addition of 2 wt% $CaCrCrSiO_6$.[2] A large spinel primary field is present in the join $CaMgSi_2O_6$-$CaAl_2SiO_6$-$CaCrAlSiO_6$.[3] On the other hand, the addition of Ti has only a small effect on the crystallization of some silicate systems. In the join $CaMgSi_2O_6$ (Di)-$CaTiAl_2O_6$ (Ti-pyroxene molecule, Tp), spinel appears as a primary phase on the addition of Ti-pyroxene more than 36 wt%.[4]

In this study, a small furnace attachable to an X-ray diffractometer was designed and constructed to investigate melt structure of silicates. Short-range structures of melts in the system Di-Tp were studied by the X-ray radial distribution method and the results were compared with short-range structures of the corresponding glasses. The role of Cr and Ti ions in crystallization of silicate melts is discussed, referring to the local structures around Cr and Ti of the glasses[5] in the systems Di-Pi and Di-Tp investigated by the EXAFS method.

Experimental

A small Keramax furnace which heated a silicate sample under atmospheric condition was designed and constructed since silicates under reducing condition[6] are ready to volatilize at high temperatures. The furnace is attachable to a θ-θ type goniometer of an X-ray diffractometer. Four Keramax heating elements ($LaCrO_3$ doped with CaO) of 10 mm in diameter and 140 mm in length were mounted in alumina tubes of 80 mm in length and 1 mm in thickness. They were horizontal and parallel to the X-ray beam. The horizontal and vertical distances between the centers of the heating elements were 40 and 20 mm, respectively. A sample container made of 30Rh-Pt was 10 mm in width, 20 mm in length and 3 mm in depth, and was mounted in the middle of the furnace. Above the sample container a hemicircular (100 mm in diameter) slit of 10 mm in width was made for the passage of X-rays. Kapton film of 0.075 mm in thickness was used for window materials of the furnace. The frame of the furnace was water-cooled (1 liter/min) to prevent the Kapton film from decomposing thermally. Temperatures were monitored by a Pt20-40Rh thermocouple which was located at the middle of the upper and the lower heating elements. The maximum temperature measured by the thermocouple was 1750°C. The maximum temperature of the sample in the sample container was about 1500°C. The schematic diagram of the furnace is shown in Fig. 1.

Homogeneous glasses with compositions listed in Table 1 were prepared by melting pure chemicals ($CaCO_3$, MgO, SiO_2, Al_2O_3 and TiO_2) in a Pt crucible at 1500–1550°C and quenching the crucible in water.

The glass sample was placed in the sample container and was melted by a furnace in advance. X-ray scattered intensities of melt in the small Keramax furnace were measured on a diffractometer with an asymmetrical diffraction geometry[7] using a synchrotron radiation source of the Photon Factory, the National Laboratory for High Energy Physics at Tsukuba, Japan. The power supplied to the furnace was about 1.6 kW and the liquidus temperatures of the Di70•Tp30 and Di60•Tp40 samples are 1250 and 1350°C, respectively. Monochromatic beams with λ=0.0726 nm were taken out of synchrotron radiations by a Si (111) monochromator and used for incident beams. The scattered intensities were measured by a step scanning mode at intervals of 0.5° in 2θ from 14° to 128° (from 21 to 155 in $S=4\pi\sin\theta/\lambda$). The storage ring was operated at 2.5 GeV with a ring current between 250 and 150 mA.

X-ray scattered intensities of glasses in the system Di-Tp were measured on a conventional powder diffractometer using Mo K α radiation. The scattered intensities were measured by a step scanning mode at intervals of 0.5° in 2θ from 5° to 130° (from 8 to 160 in S).

Results

Figure 2 shows X-ray scattered intensity curves of melts. After correction for absorption, the intensity data were normalized by Krogh-Moe and Norman's method[8]

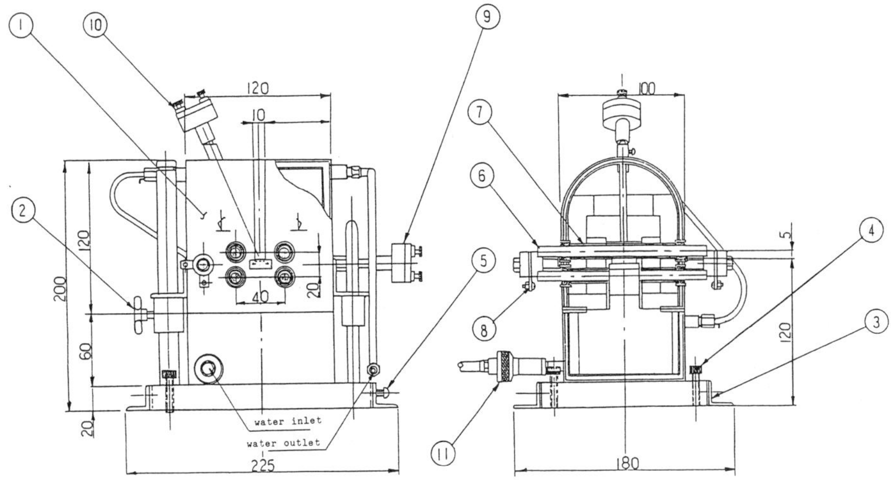

Fig. 1. Small Keramax furnace attachable to X-ray goniometer (size in millimeter). 1:Furnace frame, 2:Lock handle, 3:Guide for furnace, 4:Furnace fixing bolt, 5:Furnace positioning bolt, 6:Keramax heating element, 7:Protective alumina tube, 8:Connector for power supply, 9:Thermocouple for control, 10:Thermocouple for sample, 11:Connector for cooling water.

Table 1. Chemical composition of samples prepared in the system diopside-Ti pyroxene.

	Wt % (Mol%)				
Sample	SiO_2	CaO	MgO	Al_2O_3	TiO_2
100wt%Di	55.5(50.0)	25.9(25.0)	18.6(25.0)	- (-)	- (-)
90wt%Di.10wt%Tp	49.9(46.5)	25.7(25.6)	16.8(23.2)	4.3 (2.4)	3.4 (2.4)
80wt%Di.20wt%Tp	44.4(42.7)	25.4(26.2)	14.9(21.4)	8.6 (4.9)	6.7 (4.9)
70wt%Di.30wt%Tp	38.8(38.7)	25.2(26.9)	13.0(19.3)	12.9 (7.6)	10.1 (7.5)
60wt%Di.40wt%Tp	33.3(34.4)	25.0(27.6)	11.2(17.2)	17.1(10.4)	13.4(10.4)
50wt%Di.50wt%Tp	27.8(29.7)	24.7(28.4)	9.3(14.9)	21.4(13.5)	16.8(13.5)

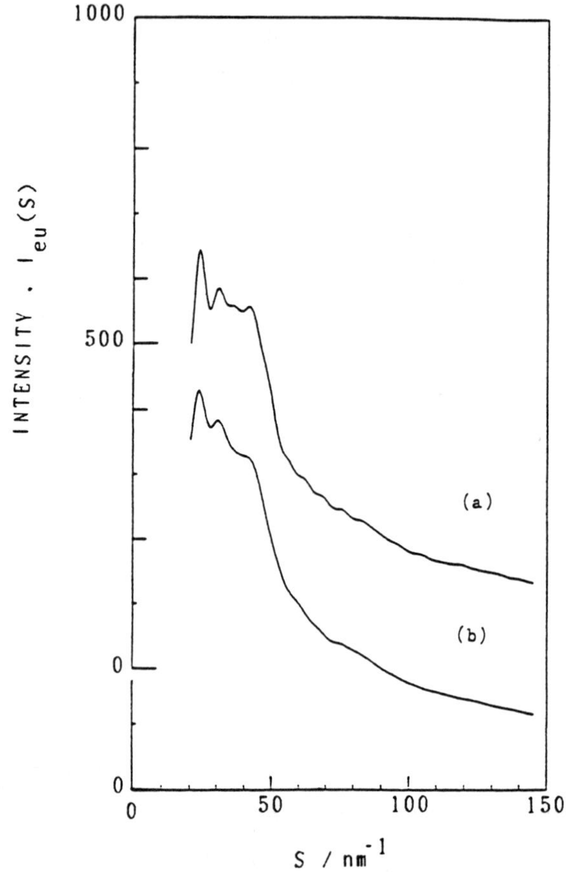

Fig. 2. X-ray scattered intensity curves of (a) Di60·Tp40 and (b) Di70·Tp30 melts.

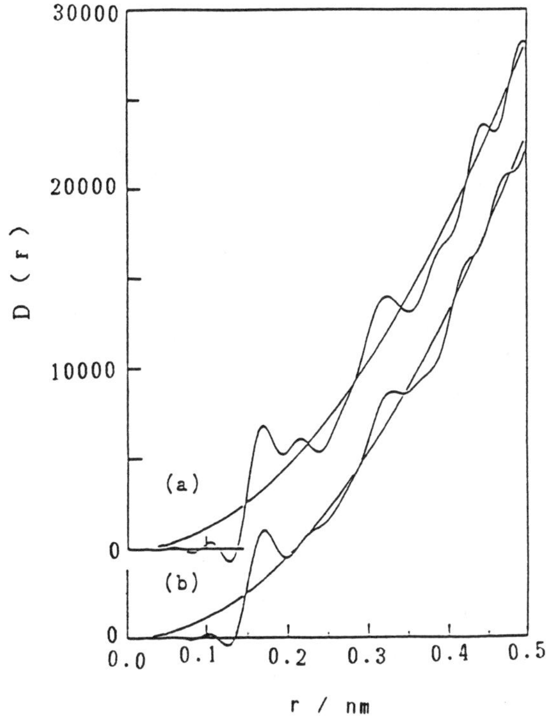

Fig. 3. Radial distribution curves of (a) Di60•Tp40 and (b) Di70•Tp30 melts.

and radial distribution curves[9,10] were calculated and are shown in Fig. 3. The radial distribution function D(r) is given by expressions (1) and (2).

$$D(r) = 4\pi r^2 \rho_e \sum_{uc} K_j + \frac{2r}{\pi} \int_0^{\infty} S \cdot i(S)\sin(rS)ds \qquad (1)$$

$$i(S) = (I_{eu}/N - \sum_{uc} f_j^2)/f_e^2 \qquad (2)$$

where the sum over uc means a sum over the atoms in the unit of composition and the subscript j represents the kind of atom. ρ_e is the average electron density, f the atomic scattering factor, f_e average scattering factor per electron, K the effective electron number, N the number of units of composition in the sample and I_{eu} the total coherent intensity.

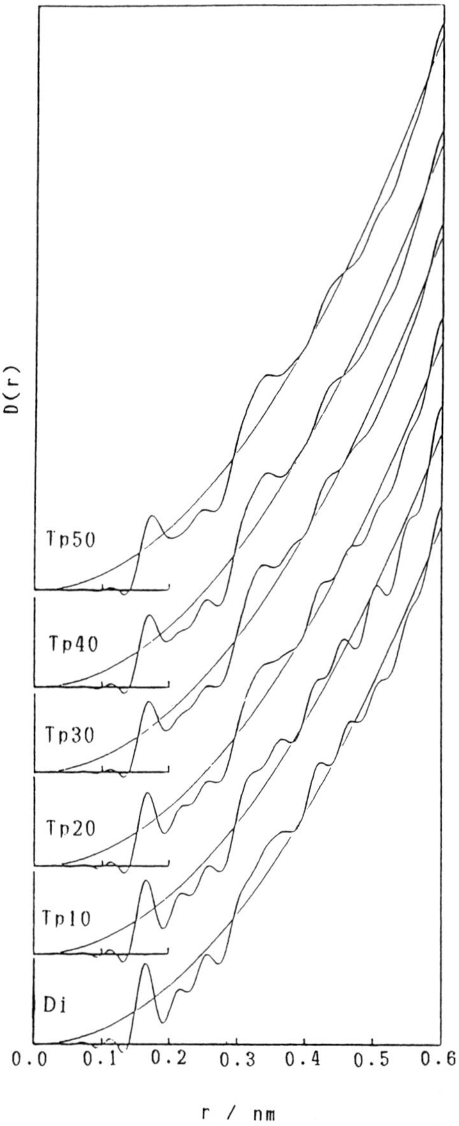

Fig. 4. Radial distribution curves of Di-Tp glasses.

The first peaks of the Di70•Tp30 and Di60•Tp40 melts were observed at 0.167 nm. Radial distribution curves of Di-Tp glass samples are shown in Fig. 4. The first peaks of the radial distribution curves were located in the range 0.164–0.172 nm, where T-O (T:tetrahedral cation) distance increases with an increase of Tp concentration. The first peaks of the Di70•Tp30 and Di60•Tp40 glasses were observed at 0.168 and 0.169 nm, respectively.

Discussion

The first peak on the radial distribution curves is assignable to the tetrahedral T-O distance. Assuming that tetrahedral Si-O and Al-O distances are 0.164 and 0.174 mn,[11] T-O distances for the Di70•Tp30 and the Di60•Tp40 glasses or melts are estimated to be 0.167 and 0.168 nm, respectively. These values are close to observed ones for the glasses or the melts in the present study and Al ions in the Di-Tp glasses seem to occupy 4-fold sites.

The first peak of the diopside glass is observed at 0.164 nm on the radial distribution curve and the distance is close to 0.163 nm of the $MgSiO_3$ glass[12] and 0.164 nm of the $CaSiO_3$ glass.[13] The second peak at 0.21 nm on the radial distribution curve is assignable to the Mg-O pair.[12] The third peak at 0.25 nm is assignable to the Ca-O and the O-O pairs,[13] and the longer Mg-O pair also contributes to the third peak.

Al ions at tetrahedral sites in the Di-Tp glasses seem to act as network former in a framework structure. However, part of Al ions in the SiO_2-Al_2O_3 glasses occupies octahedral sites[14] and this is necessary to compensate charge imbalance due to the substitution of Si^{4+} by Al^{3+}. In the Di-Tp glasses, there are Mg, Ca and Ti ions which act as network modifier to compensate the charge of AlO_4 tetrahedra.

Part of Ti ions in the Di-Tp glasses seems to occupy tetrahedral site though the presence of tetrahedraly coordinated Ti ions was uncertain on the basis of the first peak distances of the radial distribution curves. An average Ti-O distance in the Di50•Tp50 glass is reported to be 0.190 nm by the Ti K EXAFS analysis[5] and it is very close to an average Ti-O distance of 0.189 nm in an SiO_2-TiO_2 glass containing 10 mol% TiO_2.[15] Since average Ti-O distances for tetrahedral and octahedral coordinations are calculated from the sum of the effective ionic radii[11] to be 0.177 and 0.196 nm, respectively, about 30% of Ti ions in the glass seem to occupy tetrahedral sites. In the Di-Tp system, part of Ti ions seems to form tetrahedral structure units in the glass or melt, and they little modify the overall linkage mode of TO_4 tetrahedra. The presence of tetrahedraly coordinated Ti ions in pyroxene has been confirmed by the structure refinement of a Di72mol%•Tp28mol% crystal.[16]

On the other hand, an average Cr-O distance in the Di98wt%•Pi2wt% glass is reported to be 0.203 nm by the Cr K EXAFS analysis.[5] Since the Cr-O distance is close to the average Cr-O distance of 0.200 nm in Cr_2O_3 crystals (6-fold coordination), Cr ions in the Di-Pi glass seem to occupy 6-fold sites and seem to act as network modifier.

In a melt near diopside composition, part of Ti ions as well as Al ions seem to act as network former. Since they are cooperative with SiO_4 tetrahedra, diopside is ready to crystallize from the melt. On the other hand, Cr ions modify the linkage mode of SiO_4 tetrahedra. As a consequence, Cr ions prevent the crystallization of diopside. As concerns the crystallization of spinel, octahedraly coordinated Cr ions are ready to form spinel structure if there exist Mg ions nearby. On the other hand, tetrahedraly coordinated Al ions are not ready to form spinel structure because they occupy octahedral sites in the $MgAl_2O_4$ spinel structure. Ti ions have no partners among the co-existing ions to form spinel structure. Therefore, neither Al not Ti ions favor the crystallization of spinel phase.

Although the small Keramax furnace is not sufficient to study the crystallization process of mullite, some crystal-chemical features during the crystallization of pyroxene are obtained. As concerns mullite, it seems that impurities soluble in the mullite structure do not have a significant effect on the primary field of mullite.

Acknowledgments

This study was supported by the Joint Research Program (No. 85-085) of the National Laboratory for High Energy Physics, Tsukuba, Japan. A part of this study was made on Grant-in-Aid for Scientific Research (Grant No. 62103009), Ministry of Education, Science and Culture of Japan.

References

[1]K. Onuma and T. Tohara, "Effect of Chromium on Phase Relations in the Join Forsterite-Anorthite-Diopside in Air at 1 Atm," *Contrib. Mineral. Petrol.*, **84** [2-3] 174-181 (1983).

[2]K. Ikeda and K. Yagi, "Experimental Study on the Phase Equilibria in the Join $CaMgSi_2O_6$-$CaCrCrSiO_6$ with Special Reference to the Blue Diopside," *Contrib. Mineral. Petrol.*, **61** [1] 91-106 (1977).

[3]K. Onuma and T. Tohara, "The Join $CaMgSi_2O_6$-$CaAl_2SiO_6$-$CaCrAlSiO_6$ with Special Reference to Chrome Clinopyroxene and Chrome Spinel," *Lithos*, **17**, 289-298 (1984).

[4]K. Yagi and K. Onuma, "An Experimental Study on the Role of Titanium in Alkalic Basalts in Light of the System Diopside-Akermanite-Nepheline-$CaTiAl_2O_6$," *Am. J. SCi.*, Schairer Vol. **267A**, 509-549 (1969).

[5]T. Mabuchi, Y. Tabira, K. Tanaka, H. Morikawa, F. Marumo, M. Okuno, and K. Onuma, "Local Structures around Cr and Ti in Glasses of the Systems $CaMgSi_2O_6$-$MgCr_2O_4$ and $CaMgSi_2O_6$-$CaTiAl_2O_6$," Photon Factory Activity Report 1986, p. 190, National Laboratory for High Energy Physics, Tsukuba, Japan (1986).

[6]S. Iwai, T. Watanabe, I. Minato, K. Okada, and H. Morikawa, "Decomposition of Mullite by Silica Volatilization," *J. Am. Ceram. Soc.*, **63** [1-2] 44-46 (1980).

[7]H. Morikawa, F. Marumo, M. Miyake, T. Suzuki, T. Fukamachi, M. Yoshizawa, and S. Sasaki, "Diffractometer for Structural Studies of High Temperature Melts," Photon Factory Activity Report 1983/84, p. VI-11, National Laboratory for High Energy Physics, Tsukuba, Japan, (1984).

[8]N. Norman, "The Fourier Transform Method for Normalizing Intensities," *Acta Crystallogr.*, **10** [5] 370-373 (1957).

[9]H. A. Levy, M. D. Danford, and A. H. Narten, "Data Collection and Evaluation with X-Ray Diffractometer Designed for the Study of Liquid Structure," ORNL-3960, Oak Ridge National Laboratory Contract No. W-7405-eng-26, 1966.

[10]B. E. Warren, "X-Ray Diffraction," Addison-Wesley Publishing Co., Reading, Massachusetts, pp. 116-150, 1969.

[11]R. D. Shannon, "Revised Effective Ionic Radii and Systematic Studies of Interatomic Distances in Halides and Chalcogenides," *Acta Crystallogr. Sect. A*, **32** [5] 751-767 (1976).

[12]C. D. Yin, M. Okuno, H. Morikawa, and F. Marumo, "Structural Analysis of $MgSiO_3$ Glass," *J. Non-Cryst. Solids*, **55** [1] 131-141 (1983).

[13]C. D. Yin, M. Okuno, H. Morikawa, F. Marumo, and T. Yamanaka, "Structural Analysis of $CaSiO_3$ Glass by X-Ray Diffraction and Raman Spectroscopy," *J. Non-Cryst. Solids*, **80** [1-3] 167-174 (1986).

[14]H. Morikawa, S. Miwa, M. Miyake, F. Marumo, and T. Sata, "Structural Analysis of SiO_2-Al_2O_3 Glass," *J. Am. Ceram. Soc.*, **65** [2] 78-81 (1982).

[15]H. Morikawa, T. Osuka, F. Marumo, A. Yasumori, M. Yamane, and M. Nomura, "Changes in Ti Coordination Number during Pyrolysis of a SiO_2-TiO_2 Gel," *J. Non-Cryst. Solids*, **82** [1–3] 97–102 (1986).
[16]P. Fejdi, "The Role of Ti^{4+} in the Crystal Structure of Pyroxene Isomorphous Series: Diopside ($CaMgSi_2O_6$)-$CaTiAl_2O_6$," *Silikaty*, **26** [1] 7–13 (1982).

Transition Metal Distribution in Mullite

Hartmut Schneider

Forschungsinstitut der Feuerfest-Industrie
An der Elisabethkirche 27, D-5300 Bonn 1, Germany

Abstract

Dependent on synthesis temperatures and atmospheres, mullite incorporates Ti^{3+}, Ti^{4+}, V^{3+}, V^{4+}, Cr^{3+}, Mn^{2+}, Mn^{3+}, Fe^{2+}, Fe^{3+} and Co^{2+} though in strongly differing amounts. The upper solubility limit is controlled by radii and oxidation states of transition metal ions: Highest degrees of incorporation are observed for V^{3+}, Cr^{3+} and Fe^{3+} followed by Ti^{4+}. Only a very low number of Mn^{2+}, Fe^{2+} and Co^{2+} ions can enter the mullite structure.

The transition metal ions are preferably incorporated into the oxygen octahedral in mullite. Fe^{3+} also enters the oxygen tetrahedra in small amounts, while Cr^{3+} is distributed over the oxygen octahedra and the structural channels running along to the crystallographic *c* axis. Mn^{3+} substituted mullites are characterized by strong changes of lattice spacings and an associated increased lattice strain. Both observations are explained with Jahn-Teller distortion of Mn^{3+} in an octahedral environment.

Introduction

Though mullite* occurs very rarely in natural rocks, perhaps no mineral has been regarded as so basic in ceramics. This accounts for classical as well as for advanced ceramics: Mullite is one of the promising ceramic materials for high temperature applications, due to its low thermal expansion and its high creep resistance. The immense importance of mullite is documented by a large number of scientific and technical studies, which were carried out within recent years. Of particular value is the paper by Davis and Pask[1] reviewing important details of mullite synthesis and its thermodynamical, chemical, mechanical, and electrical properties. Another review paper, which deals with crystal structure, composition, formation and properties of mullite was recently presented by Schneider.[2]

The present review deals with the incorporation of transition metal cations in mullite. Because of the similarity between the crystal structures of mullite and sillimanite and andalusite, respectively, a comparison of corresponding data turned out to be useful in some cases. Kyanite, the polymorph of sillimanite and andalusite, has a different crystal structure and will not be considered.

***Mullite is named after the Isle of Mull (Scotland), where it was first found in nature.**

Crystal Structure and Mixed Crystals of the Composition $Al_{4+2x}Si_{2-2x}O_{10-x}$

Mullite has orthorhombic symmetry (space group Pbam). Chains of aluminum oxygen octahedra (=aluminum in M(1) position) with common oxygen-oxygen edges run parallel to the crystallographic *c* axis. Octahedra are linked by double chains containing aluminum oxygen and silicon oxygen tetrahedra (=aluminum and silicon in M(2) position) in a random sequence.[3] The tetrahedral chains of mullite are disturbed in a manner that some pairs of tetrahedra are misorientated with removal of some O(C)-oxygen atoms from the structure. Thereby two adjacent tetrahedral T-cation sites displace to T^+ positions. Combined with these displacements, substitutional replacement of tetrahedral silicon by aluminum takes place (Fig. 1). Resultant site occupancies are: O on O(C): 1-3/2 x; O on $O(C)^+$: 1/2x; Si: 1/2-1/2 x; and Al: 1/2+1/2 x, where x is the chemical variable of the general composition formula of mullite $Al_{4+2x}Si_{2-2x}O_{10-x}$,[4] ($0 \leq x \leq 1$).[5] The connection between octahedral and tetrahedral chains generates relatively wide structural channels running parallel to the *c* axis.[3-4,6-8]

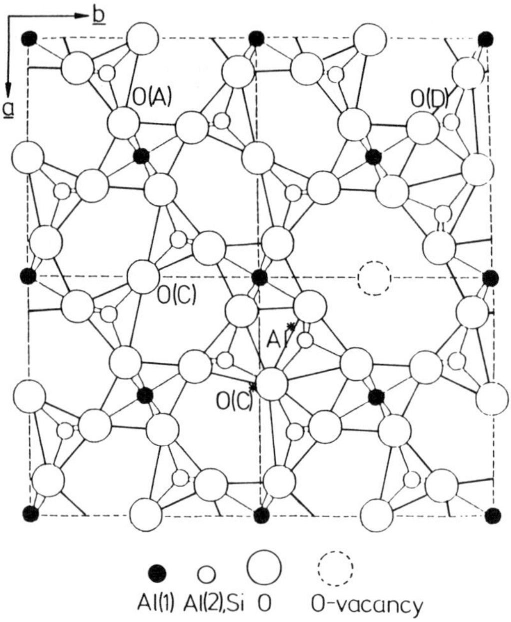

Fig. 1. Idealized projection of the crystal structure of mullite down the *c*-axis. One O(C) oxygen vacancy is indicated by a broken circle. The disturbed tetrahedra double chains adjacent to the O(C) vacancies are characterized by shifted aluminum (Al*) and oxygen (O(C)*) ions.

Two main types of mullite have been distinguished in the literature dependent on the formation conditions:

1. *Sinter-mullites* have Al_2O_3 contents ranging between about 66 and 60 mole% Al_2O_3 (78 and 72 wt% Al_2O_3) with a frequency maximum at about 60 mole%

Al_2O_3 (72 wt% Al_2O_3). These mullites have been designated as *3/2-mullites* (=$3Al_2O_3$•$2SiO_2$, x=0.25) or stoichiometric mullite. The term sinter-mullite describes a product which has been produced from the starting materials essentially by solid state reactions. The Al_2O_3 content of sinter-mullites is influenced by sintering temperatures, the duration of heat treatment, the initial bulk composition, the nature, grain size and efficiency of mixing of the starting materials, and whether corundum is nucleated.

2. *Fused-mullites* have Al_2O_3 contents ranging between about 74 and 57 mole% Al_2O_3 (82 and 70 wt% Al_2O_3) with a frequency maximum at about 66 mole% Al_2O_3 (78 wt% Al_2O_3). These mullites have been designated as *2/1-mullites* (=$2Al_2O_3$•SiO_2, x=0.40). The term fused-mullite describes a product which has been produced by melting the raw materials in an electric arc furnace above 2000°C with subsequent crystallization of mullite during the cooling of the bath. Higher alumina contents can be achieved by rapid quenching and lower alumina bulk composition, or alternatively, by a very slow cooling process. The composition of mullites crystallized from liquids at lower temperatures is primarily a function of temperature and to a lesser degree of the initial composition.

Considering only crystallographic aspects, the mullite structure theoretically fits any composition between sillimanite (x = 0) and alumina (x = 1) and there is no obvious reason why the composition should be limited to 3/2-mullite (x=0.25) and 2/1-mullite (x=0.40).[9] Actually, complete solubility between mullites does occur between about 58 and 74 mole% Al_2O_3 only at zero pressure conditions, while miscibility gaps extend between sillimanite and siliceous mullite, and between alumina rich mullites and alumina phases with mullite structure.[5,10–15]

It has been shown that the *a* lattice constant of mullite depends strongly on the Al_2O_3 content, while the simultaneous change of *b* is less intense[11,16–20] (Fig. 2). Continuing the *a* and *b* curves of the mullite solid solution series towards higher Al_2O_3 contents yields a cross of both lines at about 79 mole% Al_2O_3.[5] The observation implies that Al_2O_3-rich mullites are of tetragonal symmetry. The assumption is actually supported by the occurrence of tetragonal alumina phases with mullite structure.[5,14–15] Schneider and Rymon-Lipinski[21] recently prepared a mullite phase with tetragonal average symmetry by means of a modified sol-gel technique. However, it has been suggested that the tetragonal symmetry of this mullite phase was the result of twinning and/or domain structure of orthorhombic structural units in a unit cell scale. This compound, therefore, does not fit in Cameron's solid solution scheme.

Incorporation of Transition Metal Cations in Mullite

The data on the incorporation of transition metal cations in mullite are based on recent studies on mullites, synthesized either by annealing sample specimens of mixed-up Al_2O_3, SiO_2, and transition metal oxide powders, or by the sol-gel process.[2,22–25] Furthermore, commercial mullite bearing ceramics were studied (Table 1). Experimental results on the incorporation of transition metal cations in mullite have been compared with data published in the literature.[27–31] However, the temperature dependence of the transition metal solubility in mullite[25,32] will not be discussed in this paper.

One main aim of the present study was to describe type, oxidation state, and amount of transition metals in mullite. Another important intention was to contribute to the understanding of the structural positions of the transition metals in mullite. Microchemical, X-ray diffractometric and spectroscopic studies have been

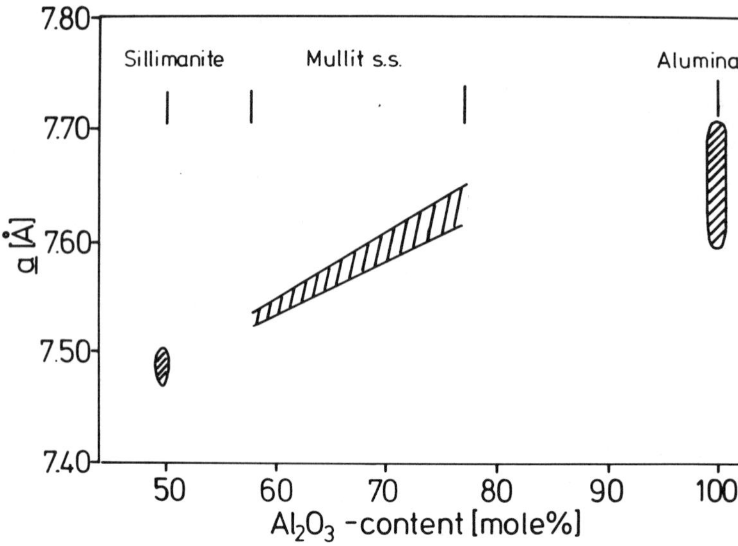

Fig. 2. Schematic drawing of *a* cell constants of mullite plotted as a function of the Al_2O_3 content, including values for sillimanite and alumina with mullite type structure. The dotted areas represent the occurrence of mullite solid solution (s.s.) or of phases with mullite type structure at atmospheric pressure and in the absence of impurities.[2]

performed in order to contribute to the understanding of the incorporation of transition metals in mullite. Neutron and electron diffractometric studies and further spectroscopic analyses are planned.

Oxidation State and Amount of Transition Metal Cations Incorporated Into Mullite

Recent knowledge on the oxidation state and the amount of transition metals cations entering the mullite structure is summarized in Table 1 and Fig. 3. Mullites synthesized from oxide powder mixtures under air furnace atmosphere incorporate Ti^{4+}, V^{5+}, Cr^{3+}, Mn^{2+}, Fe^{3+}, Co^{2+} and Cu^{2+}, though in extremely different concentrations. Ti^{3+}, V^{4+} and V^{3+} are incorporated in mullite under reducing formation atmospheres, whereas Mn^{3+} entry in mullite requires special oxidizing synthesis conditions. It turns out that the amount of transition metals in mullite is strongly dependent on the radius and oxidation state of the cation: The highest incorporation is observed if the radius and oxidation state of a given the cation correspond to those of Al^{3+}. Figure 3 shows that V^{3+}, Cr^{3+} and Fe^{3+} have the strongest tendency to be incorporated into mullite, which is similar to that of the non transition metal element gallium. The maximum amount of Ti^{4+} and V^{4+} ions entering the mullite structure is about half of that determined for V^{3+}, Cr^{3+}, and Fe^{3+}. Only low or very low amounts of Fe^{2+}, Co^{2+}, Mn^{2+} and Zr^{4+} ions can enter the mullite structure.

Knowledge on equilibrium data for the oxygen—transition metal systems with several states of oxidation may be helpful for the understanding of the incorporation processes. A graph comparing the stabilities of Mn_2O_3/Mn_3O_4 and Mn_3O_4/MnO reaction pairs with those of Fe_2O_3/Fe_3O_4, CrO_2/Cr_2O_3 and V_3O_5/V_2O_3 as a function of temperature and oxygen pressure is drawn in Fig. 4. It is shown that Cr^{3+} is the only

Table 1. Recent knowledge on maximum solid solubility and oxidation state of transition metal cations entering the mullite structure, and on synthesis conditions of mullite.

Sample key	Transition metal cation in mullite	Maximum amount of transition metal cation incorporated into mullite (given as cation oxide content)	Upper limit of solid solution known	Main oxidation states of cations observed in mullites	Cation radii in comparison to that of Al^{3+}:0.53 Å§	Synthesis conditions of mullite: Synthesis temperature	Synthesis conditions of mullite: Duration of synthesis experiments	Synthesis conditions of mullite: Remarks	References
	Sc	n.i.a.		Sc^{3+}	0.75 Å				
M-Ti(R)	Ti	$\gtrsim$ 2.1 wt% Ti_2O_3#	no	Ti^{3+}	0.64 Å	*	*	Commercially fused-mullite material produced under strongly reducing conditions	22
M-Ti(A)		~ 5.2 wt% TiO_2	yes	Ti^{4+}	0.59 Å	1550 °C	3 d	Standard (air atmosphere) laboratory synthesis	2
M-V(H)	V	$\gtrsim$ 9.4 wt% V_2O_3##	no	V^{3+}	0.64 Å	1500 °C	3 d	Strong reducing (95 % N_2 + 5 % H_2 gasflow) laboratory synthesis	23
M-V(N)		$\gtrsim$ 3.5 wt% V_2O_4###	no	V^{4+}	0.59 Å	1450 °C	3 d	Slightly reducing (N_2 gasflow) laboratory synthesis	
M-V(A)		~ 1.5 wt% V_2O_5####	yes	V^{5+}	0.54 Å	1550 °C	10 d	Sample synthesis in a sealed Pt-tube in order to avoid vanadium evaporation	
M-Cr(A)	Cr	~11.5 wt% Cr_2O_3	yes	Cr^{3+}	0.62 Å	1650 °C	10 d	Standard (air atmosphere) laboratory synthesis	23
M-Mn(A)	Mn	~ 0.5 wt% MnO	yes	Mn^{2+}	0.83 Å§§	1550 °C	10 d	Standard (air atmosphere) laboratory synthesis	23
M-Mn(O)		~ 5.7 wt% Mn_3O_4**	no	Mn^{3+}(Mn^{2+})	0.64 Å§§	890 °C	5 h	Modified sol-gel formation process with subsequent annealing in O_2-gas flow	
M-Fe(R)	Fe	< 0.1 wt% FeO	yes	Fe^{2+}	0.78 Å§§	*	*	Commercially fused-mullite material produced under strongly reducing conditions	22
M-Fe(A)		~10.3 wt% Fe_2O_3	yes	Fe^{3+}	0.65 Å§§	1300 °C	28 d	Standard (air atmosphere) laboratory synthesis	24,25

Table 1. Continued.

M-Co(A)	Co	~ 1.0 wt% CoO n.i.a.	yes	Co^{2+} Co^{3+}	0.75 Å§§ n.i.a.§§§	1550 °C	10 d	Standard (air atmosphere) laboratory synthesis	23
	Ni	n.i.a.		Ni^{2+} Ni^{3+}	0.69 Å 0.60 Å§§				
M-Cu(A)	Cu	< 0.1 wt% CuO	yes	Cu^{2+}	0.73 Å	1650 °C	10 d	Standard (air atmosphere) laboratory synthesis	23
	Zn	n.i.a.		Zn^{2+}	0.75 Å				

§ Cation radii are from Shannon and Prewitt[26] basing on a coordination number of VI with respect to oxygen ($r_{O^{2-}}$:1.40 Å).

§§ High spin cation radii (source see §).

§§§ Cation radii basing on a coordination number of V(!), (source see §).

* There exists no information on the cooling down conditions of the fused-mullite material.

** Calculated as Mn_3O_4, since according to microchemical and optic absorption spectra studies most of the manganese occurs as Mn^{3+} though, according to EPR studies, there also exists a considerable amount of Mn^{2+}.

\# Though according to microchemical and EPR studies most of the titanium occurs as Ti^{3+}, some of the transition metal occurs as Ti^{4+}.

\## Though according to microchemical studies most of the vanadium occurs as V^{3+} some of the transition metal occurs as V^{4+}.

\### Though according to microchemical and EPR studies most of the vanadium occurs as V^{4+} some of the transition metal occurs as V^{3+}.

\#### Though there was no experimental means to detect V^{5+}, the synthesis conditions suggest that this oxidation is present.

n.i.a. no information available.

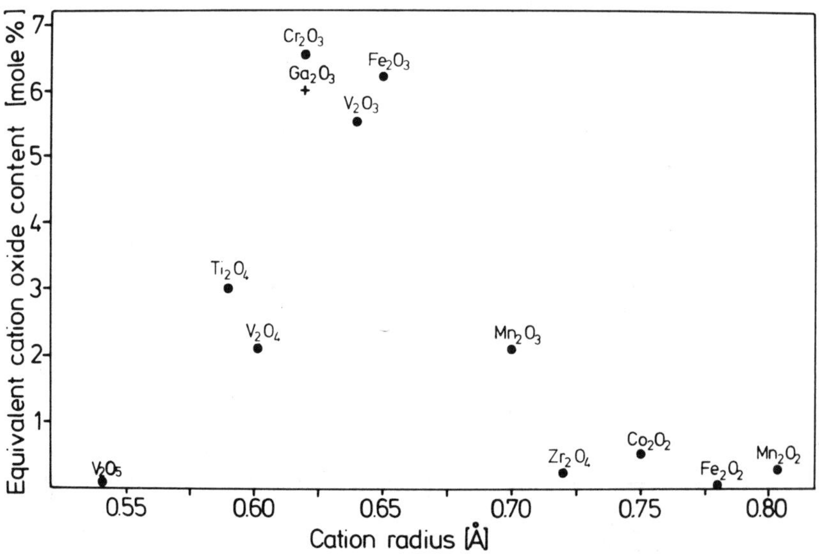

Fig. 3. Recent knowledge on maximum transition metal ions contents in mullite (given as equivalent cation oxides) plotted versus the radii of the cations.[26] The upper limit of Ti^{3+}, V^{4+} and Mn^{3+} incorporation is not known yet. Data of gallium are given for comparison. All data except those for Ga^{3+},[2] and Zr^{4+},[33] are from Table 1.

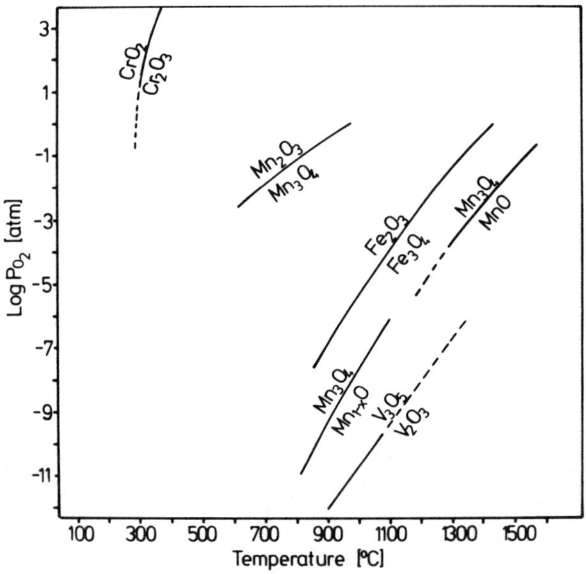

Fig. 4. Stability relations among various oxides of chromium[34] manganese,[35,36] iron[37,38] and vanadium[39] as a function of temperature and oxygen pressure.

stable oxidation state of chromium, at least about 1650°C, regardless of the oxygen pressure, while the stability field of Fe^{3+} extends up to about 1400°C at P_{O_2} 0.2 atm. V^{3+} occurs above about 1600°C at P_{O_2} 0.2 atm. but is stable under a reducing atmosphere at lower temperatures. The experimental results on vanadium, chromium, and iron incorporation into mullite obtained under varying reaction temperatures and atmospheres are in good accordance with the thermodynamic data (Table 1 and Fig. 4). Our experimental data, moreover, suggest that the TiO_2/Ti_2O_3 reaction pair displays a similar temperature and oxygen pressure dependent redox behaviour as V_3O_5/V_2O_3. The equilibrium data of the manganese oxides show that Mn^{3+} is stable only below about 1000°C. This may explain the impossibility of Mn^{3+} incorporation into mullite by sintering oxide mixtures, since this would require temperatures above 1200°C, which is far beyond the stability field of Mn^{3+}. One possible way to incorporate Mn^{3+} in mullite, therefore, is the synthesis of mullite from highly reactive metal organic compounds with the sol-gel method at low temperatures (<950°C). The same synthesis technique could enable incorporation of Co^{3+} and Ni^{3+} into the mullite structure. Another possible method to produce Mn^{3+}, Co^{3+}, and Ni^{3+} rich mullites may be synthesis with the hydrothermal technique, using oxidizing buffer systems.

Shape and size of the mullite crystals are controlled by the synthesis procedure, the synthesis temperature and atmosphere and by the type and amount of transition metal incorporated Mullites prepared by sol-gel techniques exhibit extremely small crystallites (<0.10 μm,[23] Fig. 5). The size of the mullite crystals prepared from oxide powder mixtures increases within the incorporation sequence vanadium, chromium, titanium and iron. Within the same series the shape of the mullite crystals change from reniform contours without end faces to tabular—like contours (Figs. 6a to c). Mullite crystal growth conditions are probably controlled by changing viscosities of

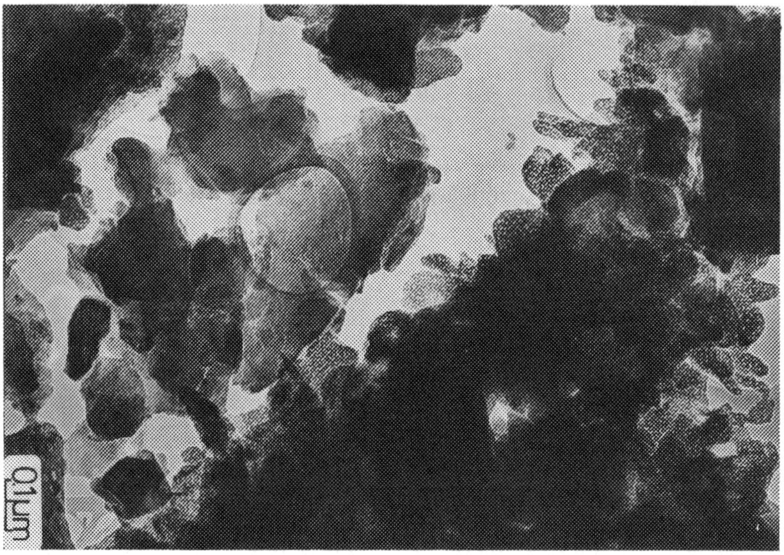

Fig. 5. Electron microscopic photograph of manganese rich mullite M-Mn(O) (Table 1) synthesized by a modified sol-gel formation and subsequent annealing at 890°C. Note the occurrence of small crystallites with coarse surfaces (≤0.1 µm[23]).

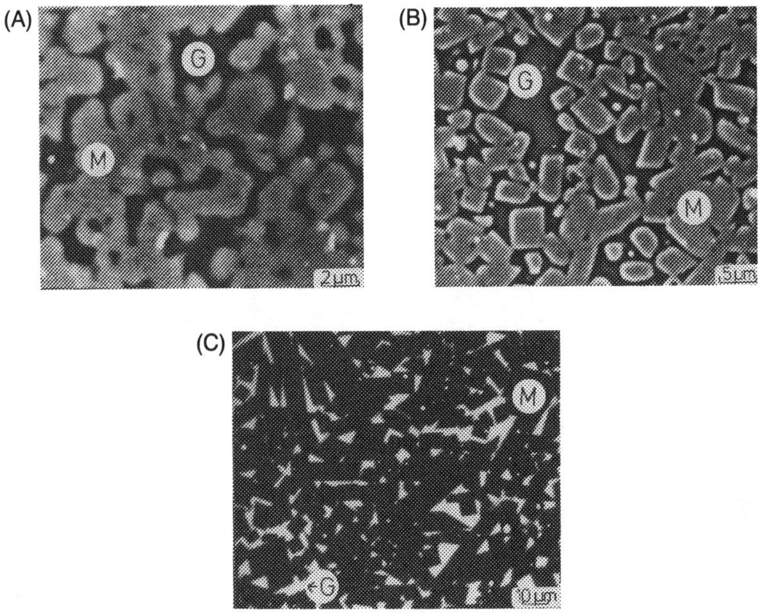

Fig. 6. Electron microprobe photographs of synthetic mullite samples consisting of Mullite (M) and coexisting glass phase (G). (a) Vanadium rich mullite sample M-V(H). Note the occurrence of very small mullite crystallites with reniform contours. (b) Chromium rich mullite sample M-Cr(A). Note the occurrence of small mullite crystals with rounded-off contours. (c) Iron rich mullite sample M-Fe(A). Note the occurrence of relatively large tabular mullite crystals with pseudo-monoclinic cross sections. Sample description is given in Table 1.

the partial melts coexisting with mullite. Going from vanadium to iron rich oxide mixtures these viscosities, apparently decrease, inducing higher velocities of the diffusion species.

Lattice Site Occupancies of Transition Metals in Mullite

The entry of transition metal cations into the mullite structure can take place by substitution of aluminum in the oxygen octahedra or tetrahedra, by substitution of silicon in the oxygen tetrahedra, by incorporation in the structural channels running parallel to the crystallographical *c* axis, or by incorporation into the structural voids produced by removal of oxygen atoms from O(C) sites (Fig. 1).

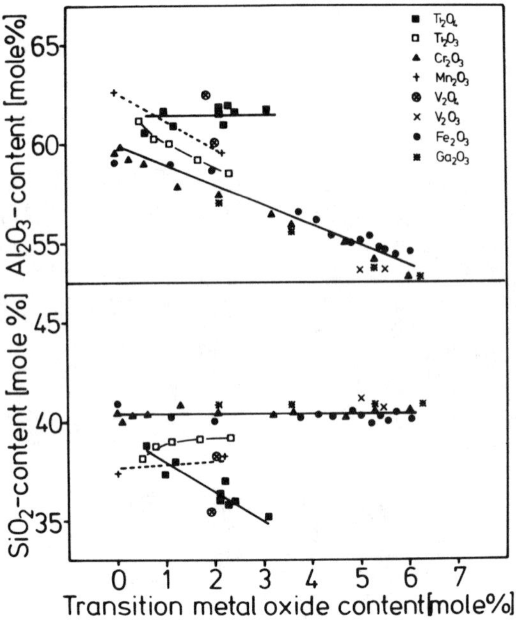

Fig. 7. Titanium, vanadium, chromium, manganese and iron incorporation given as Ti_2O_4 (M-Ti(A)), Ti_2O_3 (M-Ti(R)), V_2O_4 (M-V(N)), V_2O_3 (M-V(H)), Cr_2O_3 (M-Cr(A)), Mn_2O_3 (M-Mn(O)) and Fe_2O_3 (M-Fe(A)) in comparison to gallium incorporation,[2] given as Ga_2O_3, in mullite, plotted versus Al_2O_3 and SiO_2 contents. Sample key see Table 1.

Microchemical Studies

Plots of concentrations of SiO_2 and Al_2O_3 in mullite versus increasing concentrations of transition metal oxide yielded reciprocal and equimolar dependencies between Al_2O_3 and Cr_2O_3 and Fe_2O_3 but not between SiO_2 and Cr_2O_3 and Fe_2O_3 (Fig. 7, samples M-Cr(A) and M-Fe(A), Table 1). This indicates that the entry of Cr^{3+} and Fe^{3+} into mullite is associated with a removal of the same number of Al^{3+} from the structure. Similar microchemical relationships as for chromium and iron were obtained for titanium, vanadium and manganese containing samples M-Ti(R), M-V(H), and M-Mn (O)(Table 1). This should be interpreted in terms of a preferred occurrence of Ti^{3+}, V^{3+}, and Mn^{3+} ions in these mullites. The dependencies between titanium oxide and Al_2O_3 and SiO_2, respectively, in mullite sample M-Ti(A)(Table 1) are contrary to those described above (Fig. 7). A suitable interpretation of the data is a substitution of Al^{3+} by Ti^{4+} and simultaneous replacement of Si^{4+} by Al^{3+} in order to compensate the excess positive charge. The same explanation may work in the case of vanadium mullite M-V(N)(Table 1) providing evidence for a preferable occurrence of V^{4+} at the Al^{3+} positions.

Spectroscopic Studies

Electron paramagnetic resonance (EPR) spectroscopy has been used to determine oxidation states and structural environments of the paramagnetic transition metal cations Ti^{3+}, V^{3+}, V^{4+}, Cr^{3+}, Mn^{2+}, Fe^{3+} and Co^{2+}. Though being paramagnetic, Mn^{3+}

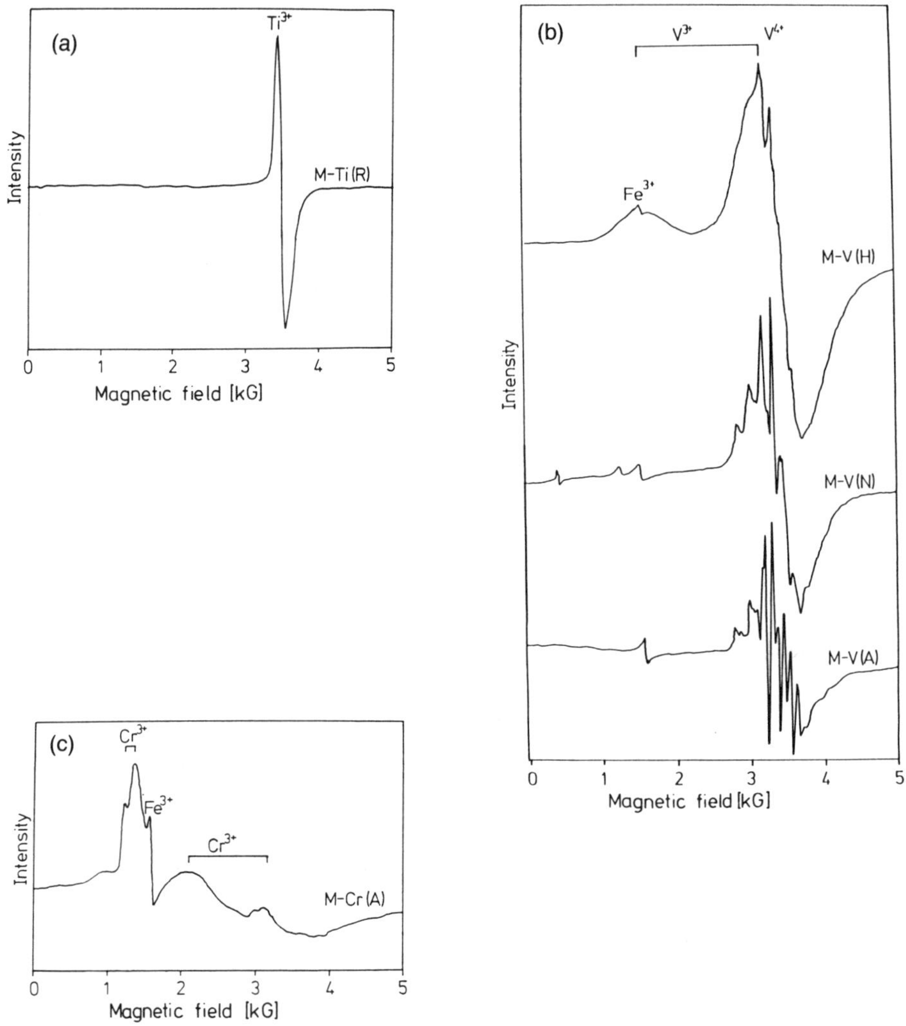

Fig. 8. Electron paramagnetic resonance (EPR) spectra of mullites exhibiting EPR signals of (a) Ti^{3+}, (b) V^{3+}, V^{4+}, (c) Cr^{3+}, (d) Mn^{2+} (see above), (e) Fe^{3+} and (f) Co^{2+}. Sample description and details of the formation conditions of the mullites are given in Table 1 for mullite samples M-Ti (R), M-V(A), M-V(N), M-V(H), M-Mn(O), M-Fe(A) and M-Co(A). In order to get a better resolution an EPR pattern of a chromium poor mullite (0.5 wt% Cr_2O_3) instead of the chromium rich mullite M-Cr(A) quoted in Table 1 (11.5 wt% Cr_2O_3) was plotted.

The EPR spectra were taken at 9.175 GHz at 290 K (Ti^{3+}(a), V^{4+}(b), Mn^{2+}(d), Fe^{3+}(e)), at 150 K (Cr^{3+}(c)), and at 25 K (V^{3+}(b), Co^{2+}(f)) using a 1mW

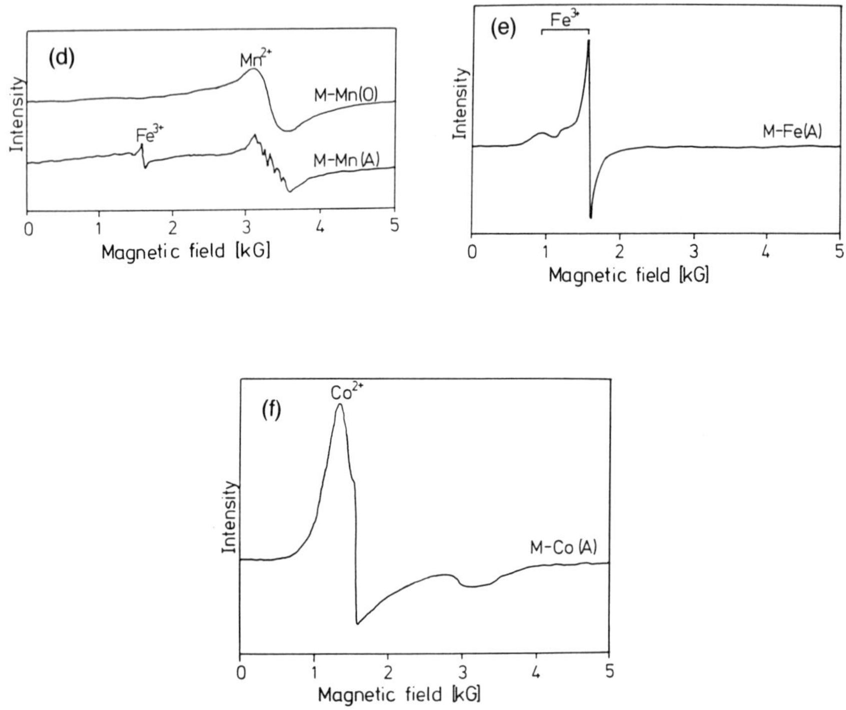

microwave power and 5 G modulation amplitude at 100 KHz. The observed EPR signals were labelled by their effective g-values, designated as g_{eff}. These values are defined by the equation $h\nu = g_{eff} \cdot \beta \cdot B$, where B is the external magnetic field at which the EPR signal appears, ν is the microwave frequency used, β is the Bohr magneton, and h is the Planck constant.

gives no EPR signals in the mullite matrix, probably due to the extreme Jahn-Teller distortion of the structure (see below). Mössbauer and optical absorption spectra were recorded from iron and manganese rich mullites, respectively, in order to get additional information on the oxidation state and the structural environment of these ions in mullite. All investigations were carried out with powdered samples, since transition metal rich mullite single crystals are not available.

Titanium Substituted Mullites

EPR spectra of mullites containing Ti^{3+} (sample M-Ti(R), Table 1) display a strong, sharp line at g_{eff}=1.9 (Fig. 8a). The shape of the EPR-signal suggests an axial crystal field symmetry at the Ti^{3+} center, probably due to the occurrence of Ti^{3+} in octahedral lattice positions.[22]

Vanadium Substituted Mullites

Vanadium rich mullites synthesized at air and in a moderately reducing atmosphere (samples M-V(A) and M-V(N), Table 1) exhibit EPR signals near g_{eff}=2.0 with characteristic hyperfine splitting (Fig. 8b). The EPR bands can be assigned to isolated V^{4+} ions probably localized in octahedral coordination.

Vanadium mullite M-V(H) synthesized at strong reducing atmospheres (Table 1) yield spectra with a weak and broad signal near g_{eff}=4.1 and a strong signal near g_{eff}=1.9. In accordance with microchemical and X-ray diffractometric data, the EPR signal group can be attributed to V^{3+} in an octahedral environment. The weak fine structure of the 1.9 band may be due to the occurrence of a small amount of V^{4+}.[40]

Chromium Substituted Mullites

Mullites containing chromium (sample M-Cr(A), Table 1) yield EPR spectra with two relatively sharp signals near g_{eff}=5, and a broad band near g_{eff}=2.5 (Fig. 8c). The sharp signals near g_{eff}=5 are similar to those effected by Cr^{3+} in silicate glass matrix and are probably due to Cr^{3+} ions in a relatively strong crystal field. These EPR signals can be attributed to isolated Cr^{3+} probably in octahedral environment. The broad EPR bands with effective g-values near 2.5 may be caused by Cr^{3+} ions occurring in locally changing crystal fields, possibly due to Cr^{3+} ions in less defined interstitial lattice positions. The intensities of the EPR signal may indicate preferred chromium incorporation at the place of octahedral aluminum at low bulk chromium contents of mullite, but preferred chromium incorporation at interstitial lattice sites at high bulk chromium contents[40] (see page 152).

Manganese Substituted Mullites

The EPR spectrum of mullite sample M-Mn(A)(Table 1) shows a signal near g_{eff}=2.0 with hyperfine splitting (Fig. 8d). The shape of the signal is very similar to that of Mn^{2+} in glass matrix, and can be assigned to Mn^{2+} in octahedral coordination.

The EPR spectrum of the manganese-rich sample M-Mn(O)(Table 1), synthesized with the sol-gel technique exhibits a signal at a similar g_{eff} value as in the M-Mn(A) spectrum (Table 1)(Fig. 8d). However, the lack of hyperfine structure even at helium temperature indicates exchange interactions between Mn^{2+} ions, which may be interpreted in terms of manganese clusters in mullite or with the occurrence of another manganese-rich phase.[40]

In the optical absorption spectrum of sample M-Mn(O)(Table 1) weak and broad peaks are identified near 14,500 and 21,500 cm^{-1}.[40] The bands correspond to those observed in manganese-rich andalusites† centered near 15,000 and 21,800 cm^{-1}, respectively, and which have been attributed to spin-allowed Mn^{3+} transitions in an octahedral crystal field.[41]

Iron Substituted Mullites

Electron paramagnetic resonance spectra of iron-containing mullites (sample M-Fe(A), Table 1) show a group of signals with effective g-values near 6.8 and 5.1 belonging to a common EPR center and a signal near an effective g-value of 4.2 (Fig. 8e). Temperature shifts and distortion indexes of signals near 6.8 and 5.1 and comparison with EPR spectra of sillimanite‡ indicate that isolated Fe^{3+} ions in octahedral coordination are present in mullite. The EPR signal near g_{eff}=4.2 is assigned to a low quantity of isolated Fe^{3+} ions in tetrahedral position.[42] According to the intensities of the signal the amount of Fe^{3+} ions entering the tetrahedra slightly increases at high annealing temperature.[24]

Mössbauer spectra recorded from iron-rich mullites (sample M-Fe(A), Table 1) consist of two relatively broad symmetric lines with a quadrupole splitting of about 1.09 mm/sec, an isomer shift of about 0.41 mm/sec., and a line width of about 0.59 mm/sec (Fig. 9). The isomer shift value is appropriate for Fe^{3+} in an octahedral environment§. The moderately large quadrupole splitting of the Fe^{3+} doublet suggests a slightly distorted octahedral iron site in mullite, in comparison to the low distorted octahedral Fe^{3+} sites in pyroxenes, garnets, amphiboles and micas (ca. 0.4 to 0.6 mm/sec), and the more distorted Fe^{3+} sites in epidote (2.0 mm/sec or higher).[43]

Cobalt Substituted Mullites

Cobalt incorporation into mullite (sample M-Co(A), Table 1) yields EPR spectra with signals near g_{eff}=4.9 and 2.2 (Fig. 8f). The EPR signals can be attributed to the occurrence of Co^{2+} in octahedral coordination.[40]

Recent spectroscopic data can be summarized in terms of

1. Incorporation of Ti^{3+}, V^{3+}, V^{4+}, Cr^{3+}, Mn^{2+}, Mn^{3+}, Fe^{3+} and Co^{2+} in octahedral structure sites of mullite, though in strongly differing amounts.
2. Incorporation of a small amount of Fe^{3+} in tetrahedral structure sites of mullite.
3. Incorporation of a considerable amount of Cr^{3+} in interstitial structure sites of mullite.

†Andalusite has a similar crystal structure as mullite, with M(1)-O_6 (M(1)=Al^{3+}) octahedra chains parallel to *c*. In a comparison of both crystal structures, it should be kept in mind that the *a* and *b* axes are interchanged in andalusite and mullite.

‡Sillimanite has a crystal structure which is very similar to that of mullite, containing chains of M(1)-O_6 (M(1)=Al^{3+}) octahedra and M(2)O_4(M(2)=Si^{4+}, Al^{3+}) tetrahedra, with both running parallel to the crystallographic *c*-axis.

§Mössbauer, microchemical, and X-ray diffractometric studies reveal evidence for the Substitution of Fe^{3+} ions in octahedral lattice positions only, while EPR provide data for tetrahedral Fe^{3+} incorporation, as well. However, this is not in contradiction, because the sensitivity of the EPR method is much higher than that of the other three methods.

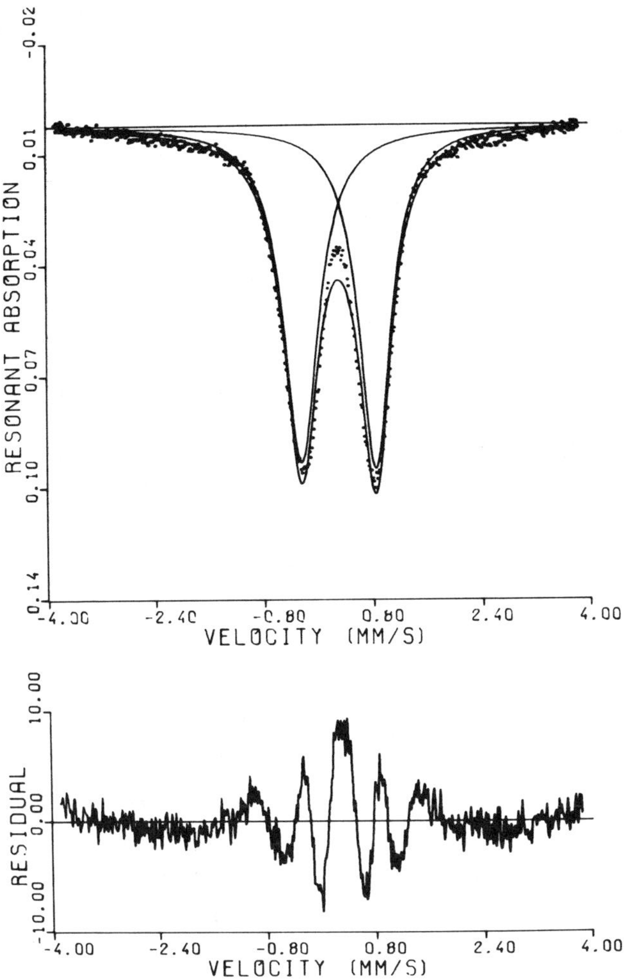

Fig. 9. Mössbauer spectrum of iron-rich synthetic sinter-mullite (sample M-Fe(A), Table 1). One doublet computer fit, which can be correlated with Fe^{3+} in octahedral coordination. Quadrupole splitting: 1.09 mm/sec, isomer shifting: 0.41 mm/sec, line width: 0.59 mm/sec.

Mössbauer spectra were obtained from about 200 mg HF/HCL-washed mullite powders at about 80 K. (Spectrum courtesy Prof. Amthauer, Salzburg).

X-ray Diffractometric Studies

The comparison of lattice constants of mullites containing transition metals with those of mullites containing other foreign cations and with impurity-free mullite yields three types of incorporation-induced expansions (Tables 2 and 3):

Table 2. Chemical composition and lattice parameters of mullites containing high amounts of transition metal cations.

Sample key (see Table 1)	Chemical composition wt%				Lattice parameters a [Å]	b [Å]	c [Å]	v [Å³]	References
Mullites containing transition metal cations									
M-Ti(A)	Al_2O_3:72.0(4),	SiO_2:24.5(3),	TiO_2 : 4.2(1),	EMA	7.5637(5)	7.7009(6)	2.8931(3)	168.51(2)	23
M-V(H)	Al_2O_3:63.0,	SiO_2:28.2,	V_2O_3 : 8.7,	XFA	7.555(2)	7.711(3)	2.8995(9)	168.92(2)	23
M-V(N)	Al_2O_3:72.5,	SiO_2:24.0,	V_2O_4 : 3.5,	XFA	7.551(1)	7.698(1)	2.8936(5)	168.19(7)	23
M-Cr(A)	Al_2O_3:60.0(4),	SiO_2:28.4(4),	Cr_2O_3:11.5(2),	EMA	7.5697(5)	7.7117(8)	2.9025(2)	169.43(2)	23
M-Mn(O)	Al_2O_3:68.4,	SiO_2:25.9,	Mn_3O_4: 5.7,	XFA	7.563(2)	7.721(2)	2.8828(7)	168.33(6)	23
M-Fe(A)	Al_2O_3:62.1(2),	SiO_2:27.4(1),	Fe_2O_3:10.3(1),	EMA	7.574(1)	7.726(1)	2.9004(5)	169.73(4)	25
Reference mullite without transition metal cations									
M-Ref(1)	Al_2O_3:71.2(2),	SiO_2:28.6(2),		EMA	7.5461(8)	7.6918(9)	2.8829(4)	167.33(3)	23
M-Ref(2)	Al_2O_3: 74	SiO_2: 26		XFA	7.587(2)	7.681(2)	2.8776(7)	167.70(6)	21

Spectroscopic data and microchemical analyses suggets that samples M-V(H), M-V(N) and M-Mn(O) contain essentially V^{3+}, V^{4+} and Mn^{3+} plus minor amounts of Mn^{2+}. That way chemical compositions are given as V_2O_3, V_2O_4 and Mn_3O_4, respectively.

M-Ref(1) was used as reference mullite for sample M-Ti(A), M-V(H), M-V(N), M-Cr(A) and M-Fe(A), (see Table 1). It was synthesized from powder sample specimens (Al_2O_3:62 wt%, SiO_2:38 wt%) by annealing 10 d at 1650 °C. After annealing the material was treated with a HF/HCl solution in order to leach the coexisting silicate glass.

M-Ref(2) was used as reference mullite for sample M-Mn(O). It was synthesized by a modified sol-gel formation process in a similar way as M-Mn(O). After synthesis the material was treated with a HF/HCl solution in order to leach coexisting amorphous phases.

EMA = Electron microprobe analyses, XFA = X-ray fluorescence analyses

Table 3. **Relative structural expansions** *a*, *b*, **and** *c*, **and** *b*/ *c*-**ratios of transition metal cation substituted mullites calculated for equivalent cation oxide contents of 5 mole% (see Table 1) in comparison to values of gallium and germanium substituted mullite, and to relative thermal expansion values (between 25 and 1000°C) of mullite, sillimanite and andalusite.**

	Incorporation - induced expansion								Thermal expansion			
Sample key transition metal Cation in mullite	M-Ti(A) Ti	M-V(H) V	M-V(N) V	M-Cr(A) Cr	M-Mn(O) Mn	M-Fe(A) Fe	M-Ga(A) Ga	M-Ge(A) Ge	3/2 mul-lite	2/1 mul-lite	Silli-manite	Anda-lusite
Δ a %	0.38	0.12	0.16	0.24	0.22 -0.73(1)	0.32	0.16	1.12	0.38	0.46	0.33	1.32
Δ b %	0.22	0.25	0.20	0.20	0.38 1.18(1)	0.38	0.29	1.14	0.64	0.61	0.83	0.98
Δ c %	0.58	0.58	0.92	0.53	0.00 0.43(1)	0.50	0.22	1.86	0.49	0.81	0.58	0.98
Δ b/Δ c	0.38	0.43	0.22	0.38	n.d.	0.76	1.30	0.61	1.31	0.75	1.43	1.35§
References	23	23	23	23	23	23	2	49	44,45	45	45-47	45-48

Incorporation induced expansions were calculated relative to reference mullite M-Ref(1), those indicated by (1) were calculated relative to reference mullite M-Ref(2), (Table 2).

§ Δa/Δc value was calculated, because a and b are interchanged in andalusite with respect to the crystallographical setting of sillimanite and mullite.

M-Ga(A): Gallium substituted mullite (5 mole% Ga_2O_3)
M-Ge(A): Germanium substituted mullite ($3Al_2O_3 \cdot 2GeO_2$)

1. Mullites containing V^{3+} and Fe^{3+} with relatively high *c*, lower *b*, and very low *a* expansion.†
2. Mullites containing Ti^{4+} and Cr^{3+} with high *c*, low *a* and even lower *b* expansion.†
3. Mullites containing Mn^{3+} with extremely high *b* and low *c* expansion, and with *a* contraction.

The intense *b* cell edge expansion of V^{3+} and Fe^{3+}-rich mullites (Tables 2 and 3) is similar to that of Ga^{3+} substituted mullites,[2] and corresponds to thermal expansion data of mullite, sillimanite and andalusite[44–48] which has been explained with the strong lengthening of the long and 'elastic' octahedral (M(1)-O(D) bond lying to either side of *b* in mullite (Fig. 1) and sillimanite and to either side of *a* in andalusite, respectively[47,48] (see footnotes on page 147). According to Hazen's principle of the structural analogous variables of composition and temperature,[50] the strong V^{3+} and Fe^{3+} incorporation-induced *b* expansion can be explained in the same manner.

The Ti^{4+} and Cr^{3+} incorporation-induced structural expansion of mullite which is stronger along *a* and *b* (Tables 2 and 3) does not fit the above mentioned model. Assuming octahedral Ti^{4+} incorporation in mullite, substitution of tetrahedral silicon by aluminum which is required for excess charge compensation, may determine the anomalous expansion behaviour. Other expansion mechanisms must be taken into account in the case of chromium incorporation into mullite: Recent X-ray Rietveld refinements on the structural positions of chromium in mullite M-Cr(A)(Table 1) in agreement with EPR data (see above) yield evidence for incorporation of about 40% of the Cr^{3+} ions in the centers of the oxygen octahedra, while about 60% occur in the structural channels along *c*, with x 0.1, y 0.5, and a distribution maximum of z near 0.[51] Possibly the anomalous incorporation-induced expansion behaviour of chromium rich mullites is influenced by the interstitial entry of Cr^{3+} into the structure.

The strong *b* expansion (but *a* contraction) of the mullite structure caused by Mn^{3+} incorporation (Tables 2 and 3) corresponds to the strong *a* length increase in Mn^{3+} substituted andalusites[41,52,53] (see footnote † above). The strong anisotropic incorporation induced structural deformation of mullite can be explained with the electron configuration of Mn^{3+} in an octahedral crystal field which causes a crystal field splitting of the $3d^4$ electrons from an energetically degenerate ground state to the high spin state $(t_{2g})^3(e_g)^1$. Considering occupation of dz^2 with one electron and dx^2-y^2 with no electron, this has the effect of strong lengthening of the octahedral M(1)-O(D) bond along *z*, caused by repulsion of the dz^2 electron and the electrons of the oxygen ligands (Jahn-Teller distortion). In addition to the M(1)-O(A) bonds of the octahedron parallel to *x* and *y*, respectively, much lower expansion or even contraction can be expected (Figs. 1 and 10). Taking into account that the *b* axis of mullite lies about 30° to either side of the M(1)-O(D) bond (= *z*) of the octahedron within the (001) plane, the Jahn-Teller distortion should effect intense *b* but much lower *a* and *c* lengthening or even shortening. The experimental results (Tables 2 and 3) actually fit these considerations.

The strong change in lattice spacings of mullite due to octahedral Mn^{3+} substitution produces considerable lattice strain. Voigt X-ray diffraction profiled analyses of sample M-Mn(O) provide high strain values for hhO (220: $=3.10\times10^{-3}$), lower values for hkl (331: $=2.49\times10^{-3}$), and the lowest values for 001 reflections (002:

†No sufficient data on the maximum incorporation of Ti^{3+} and V^{4+} are available. There is no indication for Mn^{4+} entry in the structure. The amount of Mn^{2+}, Fe^{2+}, and Co^{2+} incorporation into mullite was too small to produce a significant change in any of the lattice spacings.

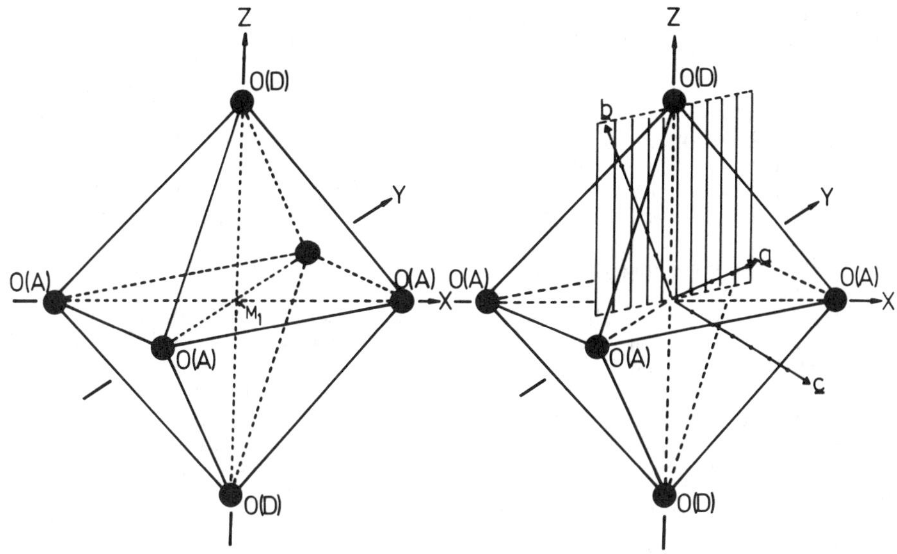

Fig. 10. Idealized version of the M(1)-O_6 octahedral in mullite. The approximate orientations of the crystallographic *a, b,* and *c* axes with respect to the *x, y* and *z* of the octahedra is shown at the right side. The dashed area represents (001).

$=1.50\times10^{-3}$).[23] Obviously, strongest incorporation-induced lattice deformations occur in the (001) plane parallel to the octahedral M(1)-O(D) bond, and parallel to the *x*, *y* angle bisector of the octahedron (Fig. 10). Strain values parallel to *c* are much lower.

Considering cation radii alone, the octahedral incorporation of impurities in mullite should produce larger increases in *b* than *a* or *c*. This 'normal' expansion behaviour has actually been observed for gallium substituted mullites,[2] and in heat-treated 3/2-mullite, sillimanite and andalusite[44–48] (Table 3). However, stronger *c* than *a* or *b* expansion occurs in titanium, vanadium, chromium and iron substituted mullites. The same effect has been described, if germanium replaces tetrahedral silicon[49] in mullite, while *c* shortening without significant change of *a* and *b* is caused by tetrahedral incorporation of the small boron and beryllium ions.[28,54] Thus the value of *c* of mullite might be sensitive to changes of the dimension of the tetrahedra double chains connecting octahedral chains along *c*. It should be mentioned, however, that such an expansion mode is difficult to understand, since tetrahedral incorporation of larger amounts of transition metals can be neglected and oxygen tetrahedra behave rather rigidly in silicates.

Conclusions

The relatively large cation sizes and high octahedral but lower tetrahedral crystal field splitting parameters of most transition metal ions may explain their preference for octahedral coordination. Fe^{3+} and Mn^{2+} ions in the middle of the 3d transition metal series make an exception: These cations have a stable d^5 electron configuration with a spherical symmetric charge distribution, similar to those of noble gases. Consequently Mn^{2+} and Fe^{3+} exhibit no site preference and their incorporation behaviours is solely controlled by the sizes of the cations. Actually a small amount of Fe^{3+} does enter the oxygen tetrahedra in mullite at high temperature.[24] On the other hand, Mn^{2+} is obviously too large to be tetrahedrally incorporated in mullite.

The entry of the relatively large transition metal ions into the oxygen octahedra at the place of aluminum should produce stronger *b* than *a* expansion, because the most elastic bond of the octahedron (M(1)-O(D)) lies to about 30°to either side of *b* (Figs. 1 and 10). Stronger *b* than *a* expansion has actually been observed for V^{3+}, Mn^{3+} and Fe^{3+} substituted mullites (Table 3, see also footnote [1] on page 152). We believe that the extreme *b* lengthening (and *a* shortening) of Mn^{3+} substituted mullites is due to repulsion of singly occupied dz^2 electron orbitals of Mn^{3+} by the electrons of the O(D) ligands along the octahedral M(1)-O(D) bonds (Jahn-Teller distortion). The intense *a* expansion of Ti^{4+} substituted mullites may be explained by substitution of tetrahedral silicon by aluminum, which does occur along with the octahedral incorporation of Ti^{4+} for charge compensation. The anomalously high *a* expansion of chromium-substituted mullite can possibly be explained with the entry of Cr^{3+} into the structural channels parallel to c. Interstitial Cr^{3+} incorporation into mullite must be a complicated process, which requires removal of Al^{3+} to form tetrahedral and/or octahedral vacancies for excess charge compensation. A large cation size not really fitting into the oxygen octahedra, which has been described to be the reason for the interstitial sodium[55] (Na^+:1.16 Å[26]) and zirconium[33] (Zr^4:0.84 Å[26]) incorporation in mullite, cannot be a main driving force in the case of chromium, since other 3d transition metal ions with larger cation sizes have not yet been localized interstitially. Possibly, the unique tendency of Cr^{3+} to occupy interstitial lattice positions in mullite can be explained with its capability to produce metastable octahedral oxygen environments in glass and silicate matrices.

While the structural position of 3d transition metals in mullite seems to be controlled essentially by electron configuration and size of cations, the amount of cation incorporation is also dependent on the oxidation state of the ions. Obviously maximum incorporation can be observed, if the transition metal occurs as an M^{3+} ion. This can be understood, taking into account that octahedral transition metal incorporation is associated with a removal of aluminum from the structure (Fig. 7 and pp. 8 to 9). The entry of cations even with radii closer to that of Al^{3+}but with deviating oxidation states (e.g. Ti^{4+} and V^{4+}) is less favourable; since it requires simultaneous tetrahedral substitution of silicon by aluminum to compensate for the excess positive charge.

Acknowledgments

Special acknowledgment is made to Dr. H. Rager (Marburg) for carrying out electron paramagnetic resonance investigations and for many valuable discussions. Thanks are also due to Profs. G. Amthauer (Salzburg) who performed Mössbauer spectra; E. Eberhard (Hannover) for transmission electron microscopy, and H. Saalfeld (Hamburg) for providing unpublished thermal expansion data on sillimanite and mullite. The excellent research assistance of Mrs. H. Krause (Bonn) was very helpful for the progress of the work.

I am greatly indebted to my wife Dipl.-Min. S. Schneider, for a patient critical review and to Mrs. H. Vogel for carefully typing the manuscript.

The 'Deutsche Forschungsgemeinschaft' (DFG) supported the research work with a grant (Schn 297/2). This is gratefully acknowledged.

References

[1]R. F. Davis and J. A. Pask, "Mullite," A. M. Alper (ed): Refractory Materials. High Temperature Oxides, Part IV, Refractory Glasses, Glass-Ceramics, and Ceramics, pp. 37–76, Academic Press, New York and London (1971).

[2]H. Schneider, "Formation, Properties and High-Temperature Behaviour of Mullite," Habiliationsschrift, Faculty of Chemistry, University of Münster (F.R.G.), pp. 1–148, (1986).
[3]C. W. Burnham, "Crystal Structure of Mullite," *Carnegie Inst. Wash. Year Book*, **63**, 223–227, (1964).
[4]J. R. Angel and C. T. Prewitt, "Crystal Structure of Mullite: A Re-examination of the Average Structure," *Amer. Min.*, **71**, 1476–1482, (1986).
[5]W. E. Cameron, "Mullite: A Substituted Alumina," *Amer. Min.*, **62**, 747–755, (1977).
[6]R. Sadanaga, M. Tokonami, and Y. Takeuchi, "The Structure of Mullite $2Al_2O_3{\cdot}SiO_2$ and Relationship with the Structures of Sillimanite and Andalusite," *Acta Cryst.*, **15**, 65–68, (1962).
[7]S. Durovic, "Die Kristallstruktur der Mullitmischkristallreihe und ihre Beziehung zur Struktur des Sillimanits," *Ber. Deutsche Keram. Ges.*, **40**, 304–315, (1963).
[8]H. Saalfeld and W. Guse, "Structure Refinement of 3:2-Mullite ($3Al_2O_3{\cdot}2SiO_2$)," *N. Jb. Mineral. Mh.*, 145–150, (1981).
[9]C. W. Burnham, "Composition Limits of Mullite and the Sillimanite-Mullite Solid Solution Problem," *Carnegie Inst. Wash. Year Book*, **63**, 227–228, (1964).
[10]H. Saalfeld, "A Modification of Al_2O_3 with Sillimanite Structure," Transaction VIIIth Int. Ceram. Congress Copenhagen, pp. 71–74, (1962).
[11]Y. Hariya, W. A. Dollase, and G. C. Kennedy, "An Experimental Investigation of the Relationship of Mullite to Sillimanite," *Amer. Min.*, **54**, 1419–1441, (1969).
[12]P. H. Duvigneaud, "Existence of Mullite without Silica," *J. Amer. Ceram. Soc.*, **57**, 224, (1974).
[13]A. P. Grigor'ev, "Crystallization of the Intermediate Phases in the Series Sillimanite-Mullite," Refractories, 448–450, (1976).
[14]P. A. Foster, Jr., "Nature of Alumina in Quenched Cryolite Alumina Melts," *J. Electrochem. Soc.*, **106**, 971–975, (1959).
[15]A. J. Perotta and J. E. Young, Jr., "Silica-free Phases with Mullite-type Structures," *J. Amer. Ceram. Soc.*, **57**, 405–407, (1974).
[16]G. Trömel, K.-H. Obst, K. Konopicky, H. Bauer, and I. Patzak, "Untersuchungen im System $Al_2O_3{\cdot}SiO_2$," *Ber. Deutsche Keram. Ges.*, **34**, 397–402, (1957).
[17]G. Gelsdorf, H. Müller-Hesse, and H. E. Schwiete, "Untersuchungen zur Frage der Mischkristallbildung in Mulliten und mullitähnlichen Verbindungen," *Sprechsaal*, **94**, 502–508, (1961).
[18]S. Durovic, "Isomorphism between Sillimanite and Mullite," *J. Amer. Ceram. Soc.*, **45**, 157–161, (1962).
[19]W. E. Cameron, "Composition and Cell Dimensions of Mullite," *Amer. Ceram. Soc. Bull.*, **56**, 1003–1011, (1977).
[20]H. Schneider and K. Wohlleben, "Microchemical Composition and Cell Dimensions of Mullites from Refractory-grade South-American Bauxites," *Ceramics Intern.*, **7**, 130–136, (1981).
[21]H. Schneider and T. Rymon-Lipinski, "Occurrence of Pseudo-Tetragonal Mullite," *J. Amer. Ceram. Soc.*, in press.
[22]H. Schneider and H. Rager, "Occurrence of Ti^{3+} and Fe^{2+} in Mullite," *J. Amer. Ceram. Soc.*, **67**, C248–250, (1984).
[23]H. Schneider, "Unpublished Results."
[24]H. Schneider and H. Rager, "Iron Incorporation in Mullite," *Ceramic Intern.*, **12**, 117–125, (1986).
[25]H. Schneider, "Temperature-dependent Iron Solubility in Mullite," *J. Amer. Ceram. Soc.*, **70**, C43–C45, (1987).

[26]R. D. Shannon and C. T. Prewitt, "Effective Ionic Radii in Oxides and Fluorides," *Acta Cryst.*, **B25**, 925-946, (1969).
[27]J. Brownell, "Subsolidus Relations between Mullite and Iron Oxides," *J. Amer. Ceram. Soc.*, **41**, 226-230, (1958).
[28]G. Gelsdorf, H. Müller-Hesse, and H. E. Schwiete, "Einlagerungsversuche an synthetischem Mullit und Substitutionsversuche mit Galliumoxid und Germaniumoxid. Teil II," *Archiv Eisenhüttenwesen*, **29**, 513-519, (1958).
[29]M. K. Murthy and F. A. Hummel, "X-ray Study of the Solid Solution of TiO_2, Fe_2O_3, and Cr_2O_3 in Mullite ($3Al_2O_3{\cdot}2SiO_2$)," *J. Amer. Ceram. Soc.*, **43**, 267-273, (1960).
[30]C. R. Green and J. White, "Solid Solubility of TiO_2 in Mullite in the System Al_2O_3-TiO_2-SiO_2," *Trans. J. Brit. Ceram. Soc.*, **73**, 73-75, (1974).
[31]S. N. Razumovskii, T. A. Tunik, O. N. Fisher, and S. P. Shmitt-Fogelevich, "X-ray Analysis of Solid Solutions of Ferric Oxide in Mullite," *Ogneupory*, **18**, 46-48, (1977).
[32]H. Schneider, "Solubility of TiO_2, Fe_2O_3 and MgO in Mullite," *Ceramics Intern.*, **13**, 77-82, (1987).
[33]H. Schneider, "Zirconium Incorporation in Mullite," *N. Jb. Miner. Mh*, **4**, 172-180, (1986).
[34]K. A. Wilhelmi, Quoted in: E. M. Levin, H. F. Mc Murdie, "Phase Diagrams for Ceramists," 1975, Supplement, p. 6, Fig. 4152, The American Ceramic Society, Columbus, Ohio, (1975).
[35]K. Schwerdtfeger and A. Muan, Quoted in: E. M. Levin, H. F. Mc Murdie, "Phase Diagrams for Ceramists," 1975 Supplement, p. 8, Fig. 4156, The American Ceramic Society, Columbus, Ohio, (1975).
[36]J. S. Hübner and M. Sato, "The Oxygen Fugacity-Temperature Relationships of Manganese Oxide and Nickel Oxide Buffers," *Amer. Min.*, **55**, 934-952, (1970).
[37]L. S. Darken and R. W. Gurry, "The System Iron-Oxygen. I. The Wustite Field and Related Equilibria," *J. Amer. Chem. Soc.*, **67**, 1398-1412, (1945).
[38]L. S. Darken and R. W. Gurry, "The System Iron-Oxygen, II. Equilibrium and Thermodynamics of Liquid Oxide and other Phases," *J. Amer. Chem. Soc.*, **68**, 798-816, (1946).
[39]J. S. Anderson and A. S. Kahn, Quoted in: E. M. Levin, H. F. Mc Murdie, "Phase Diagrams for Ceramists," 1975 Supplement, p. 16, Fig. 4170, The American Ceramic Society, Columbus, Ohio, (1975).
[40]H. Schneider and H. Rager, "Unpublished results."
[41]I. Abs-Wurmbach, K. Langer, F. Seifert, and E. Tillmanns, "The Crystal Chemistry of (Mn^{3+}, Fe^{3+})-Substituted Andalusites (Viridines and Kanonaite), ($Al_{1-x-y}Mn^{3+}$, $Fe^{3+}_y)_2(O/SiO_4)$: Crystal Structure Refinements, Mössbauer, and Polarized Optical Absorption Spectra," *Z. Krist.*, **155**, 81-113, (1981).
[42]J. Le Marshall, D. R. Hutton, G. J. Troup, and J. R. Thyer, "A Paramagnetic Resonance Study of Cr^{3+} and Fe^{3+} in Sillimanite," *Phys. Stat. Sol.* (a), **5**, 769-773, (1971).
[43]G. R. Rossmann, E. S. Grew, and W. A. Dollase, "The Colors of Sillimanite," *Amer. Min.*, **67**, 749-761, (1982).
[44]G. Oehlschlegel, A. Kockel, and A. Biedl, "Anisotrope Mischkristallbildung einiger Verbindungen des ternären Systems BaO-Al_2O_3-SiO_2. Teil II. Messungen an Strukturen mit dreidimensionaler Verknüpfung von (Si, Al) O_4-Tetraedern und Modellvorstellungen von deren Wärmedehnungsanisotropie," *Glastechn. Ber.*, **41**, 31-41, (1974).
[45]H. Saalfeld, "Personal communication."

[46]B. J. Skinner, S. P. Clark, and D. E. Appleman, "Molar Volumina and Thermal Expansions of Andalusite, Kyanite and Sillimanite," *Amer. J. Sci.*, **259**, 651–668, (1961).

[47]J. K. Winter and S. Ghose, "Thermal Expansion and High-temperature Crystal Chemistry of the Al_2SiO_5 Polymorphs," *Amer. Min.*, **64**, 573–586, (1979).

[48]H. Schneider, "Thermal Expansion of Andalusite," *J. Amer. Ceram. Soc.*, **62**, 307, (1979).

[49]H. Schneider and H.-D. Werner, "Synthesis and Crystal Chemistry of Andalusite and Mullite-type Structure in the System Al_2GeO_5-Ga_2GeO_5," *N. Jb. Miner. Abh.*, **143**, 223–230, (1982).

[50]R. M. Hazen, "Temperature, Pressure and Composition: Structurally Analogous Variables," *Phys. Chem. Minerals*, **1**, 83–94, (1977).

[51]H. Schneider, S. Jantsch, and J. Ihringer, "Rietveld Refinement of the Structural Positions of Chromium in Mullite," To be published.

[52]I. Abs-Wurmbach and K. Langer, "Synthetic Mn^{3+} Kyanite and Viridine $(Al_{2-x}Mn^{3+})SiO_5$ the System Al_2O_3-MnO-MnO_2-SiO_2," *Contrib. Mineral. Petrol.*, **49**, 21–38, (1975).

[53]M. Gunter and F. D. Bloss, "Andalusite-Kanonaite Series: Lattice and Optical Parameters," *Amer. Min.*, **67**, 1218–1228, (1982).

[54]A. Dietzel and H. Scholze, "Untersuchungen im System B_2O_3-Al_2O_3-SiO_2," *Glastechn. Ber.*, **49**, 47–51, (1955).

[55]H. Schneider, "Solid Solubility of Na_2O in Mullite," *J. Amer. Ceram. Soc.*, **67**, C130–C131, (1984).

+Aluminum, Gallium and Iron (III) Germanates With Aluminum Silicate Structures

D. MICHEL, S. ABOLHASSANI, A. KAHN, V. AGAFONOV, AND M. PEREZ Y JORBA

Laboratoire de Chimıe Appliquée de l'Etat Solide
C.N.R.S. UA302
C.E.C.M. 15 Rue Georges Urbain 94407 VITRY (France)

Abstract

The stability ranges of phases with aluminum silicate structures (mullite, andalusite or kyanite) have been studied in the systems GeO_2-M_2O_3 (M=Al, Ga, Fe^{III}). Structural features and site occupancy are discussed in comparison with homologous silicates.

Introduction

Mullite is the only compound stable at atmospheric pressure in the silica-alumina phase diagram. In addition, three aluminum silicates occur as natural minerals with the Al_2SiO_5 composition, but they are stable at high pressures only: sillimanite, andalusite and kyanite.

The skeleton of the structure of mullite and of the three Al_2SiO_5 polymorphs consists of similar infinite chains of AlO_6 edge-sharing octahedra running parallel to the crystallographic c axis. Owing to these structural relations, andalusite, for instance, is transformed at high temperature into 3/2 mullite by a topotactic reaction during which the aluminum-oxygen octahedral chains are preserved.[1,2] However, octahedral chains are linked by Al and Si oxygen coordination polyhedra which are different in each Al_2SiO_5 polymorph and in mullite.

In sillimanite, these polyhedra are AlO_4 and SiO_4 tetrahedra which define "einer" double chains (see reference 3 for nomenclature) with an ordered sequence of aluminum and silicon atoms along the c axis. In andalusite, AlO_5 and SiO_4 polyhedra are grouped in modified "zweier" double chains. Octahedral chains are linked by AlO_6 octahedra and SiO_4 tetrahedra in kyanite which has the densest structure because oxygen atoms are close packed on a distorted fcc sublattice.

Atomic arrangements in mullite are more complex due to oxygen vacancies and occupancy of two kinds of tetrahedral sites by aluminum atoms. This results in a mixture of "einer" and "zweier" double chains of tetrahedra along which Al and Si atoms are not ordered at long range.

In the binary systems between silica and sesquioxides comparable to alumina like Fe_2O_3 or Ga_2O_3, no phase isostructural with aluminum silicates is known at atmospheric pressure. Depsite a fraction of aluminum atoms can be replace by iron or gallium in mullite, there is no compound stable in the Fe_2O_3-SiO_2 or in the Ga_2O_3-SiO_2 phase diagrams. On the contrary, previous work by one of us[4–8] revealed the existence of several phases in the systems between germania and oxides of trivalent elements (aluminum, gallium and iron III).

Some of these germanates were identified as isostructural with mullite, andalusite and kyanite but other phases have no corresponding silicates. The stability range, composition and structure of various compounds have been recently determined.[9–14] Investigation on single crystals grown by chemical vapour transport led to complete structure determinations, and in particular, new structural types were found for iron,[12] aluminum[13] and gallium[14] germanates.

The present paper will be restricted to aluminum, gallium and iron III germanates having the same structure as aluminum silicates. Mullite phases are stable at high temperature in the Al_2O_3-GeO_2 and Ga_2O_3-GeO_2 systems but are not found with Fe_2O_3. In the Fe_2O_3-GeO_2 system, kyanite is the only silicate-related phase. However, as for silicon mullites, a significant rate of iron atoms can substitute aluminum in germanium mullites. The respective existence ranges of mullite, andalusite and kyanite germanates are compared with those of homologous silicates. Unit-cell variations with composition and possible site occupancy are discussed from X-ray diffraction results.

Experimental

In order to obtain homogeneous and reactive powders, coprecipitation of hydroxides was used. Starting materials were high-purity GeO_2 (quartz form) soluble in hot water, hydrated aluminum or gallium nitrate and ferric chloride. The dissolved ions were precipitated simultaneously by adding ammonium hydroxide.

Control of the solution pH is essential to maintain an equal yield from the starting composition. Previous studies determined the experimental conditions for obtaining the desired composition.[15] Usually a final pH value of 9–10 is satisfactory. After filtration and drying out, pressed powders were calcined below the crystallization temperatures which were between 800 and 900°C depending on composition. Samples were annealed at various temperatures between 950 and 1550°C. Experiments at high temperatures were performed in sealed platinum tubes to limit germanium dioxide loss.

Phases were identified by X-ray powder diagrams and their unit-cell parameters are determined from least squares refinement. DTA runs (heating rate 20°/mn) were performed from room temperature to 1400°C to determine crystallization temperatures and to detect phase transitions.

Results

Results of diffraction studies on sintered samples and on crystals grown from chemical vapour transport are presented in Tables 1, 2 and 3. Room temperature lattice constants of Al_2SiO_5 polymorphs from Winter and Ghose[16] are given for comparison.

For kyanites, a multiple overlap of the diffraction lines results from low symmetry. The diagram complexity does not allow an univocal assignment of the powder diagram at high Bragg angles and limits the accuracy of unit-cell parameter refinements.

Germanates With A Mullite-Type Structure

Mullite (space group Pbam, Z=2) and sillimanite (space group Pbnm, Z=4) have very closely related orthorhombic structures.[17] The a and b unit-cell parameters of sillimanite and mullite are nearly the same and the c parameter of sillimanite is about twice that of mullite.

Table 1. Unit-cell parameters of sillimanite and of mullite-type germanates (annealed at 1350°C)

	a(Å)	b(Å)	c(Å)	V(Å^3)
Al_2SiO_5 25°C sillimanite	7.4883(7)	7.4932(9)	5.7774(5) 2x 2.8887	332.29(5) 2x 166.1
Al_2GeO_5	7.630(1)	7.779(1)	2.9286(2)	173.8(1)
$Fe_{0.5}Al_{1.5}GeO_5$	7.702(1)	7.861(1)	2.9546(2)	178.9(1)
$Fe_{0.7}Al_{1.3}GeO_5$	7.723(1)	7.902(1)	2.9560(2)	180.3(1)
Ga_2GeO_5 (8)	7.881	8.157	3.014	193.6

Table 2. Unit-cell parameters of andalusite[15] and of isostructural germanates (annealed at 1200°C)

	a(Å)	b(Å)	c(Å)	V(Å^3)
Al_2SiO_5 25°C	7.7980(1)	7.68779(1)	5.5724(4)	342.45(6)
$Fe_{0.5}Al_{1.5}GeO_5$	8.012(1)	8.150(1)	5.757(1)	375.3(1)
$Fe_{0.7}Al_{1.3}GeO_5$	8.033(1)	8.159(1)	5.761(1)	377.6(1)
$Fe_{0.9}Al_{1.1}GeO_5$	8.046(1)	8.182(1)	5.776(1)	380.3(1)
Ga_2GeO_5	8.105(1)	8.275(1)	5.819(1)	390.3(1)

Table 3. Unit-cell parameters of kyanite[15] and of isostructural germanates (annealed at 980°C)

	a(Å) α(°)	b(Å) β(°)	c(Å) γ(°)	V(Å^3)
Al_2SiO_5 25°C	7.1262(12) 89.99(2)	7.8520(10) 101.11(2)	5.5724(10) 106.03(1)	293.60(9)
Al_2GeO_5	7.213 89.9	8.010 101.2	5.673 105.8	308.9
$Fe_{0.5}Al_{1.5}GeO_5$	7.27 90.0	8.08 101.2	5.72 105.9	316.7
$Fe_{0.7}Al_{1.3}GeO_5$	7.32 90.0	8.14 101.2	5.76 105.9	323.3
$Fe_{0.9}Al_{1.1}GeO_5$	7.35 90.0	8.17 101.2	5.79 105.9	329.5
Fe_2GeO_5	7.518 90.4	8.354 101.2	5.91 106.0	349.7

The high-temperature forms of Al_2GeO_5 and Ga_2GeO_5 have a mullite-type structure. In addition, it is possible to replace about one third of aluminum atoms in Al_2GeO_5 by iron atoms with the resulting increase in unit-cell volume shown in Table 1.

In M_2GeO_5 compounds, no evidence of cation ordering on tetrahedral sites can be drawn either from X-ray diffraction or infrared spectroscopy.

These phases can be thus regarded as "disordered sillimanites." We have verified by a TEM study that samples having a 50 mol% Al_2O_3 global composition only consist of a crystalline phase and are not mixtures of 3:2 mullite and silica-rich glass. Analysis with X-ray dispersive energy spectroscopy on single crystals confirmed the 1:1 composition of the mullite phase.

The existence range of aluminum-germanium mullite is larger than that of the corresponding silicate.[6,8] It extends from 50 to 67 mol.% (2:1 mullite) at 1550°C. This has been checked by EDX analysis.[18] As for Si-mullite, the a parameter noticeably increases whereas b slightly decreases with growing Al content. Contrary to Si-mullite, the c parameter decreases with Al content. The resulting unit-cell volume of germanium mullite is nearly independent of the composition within its large existence domain.[18] This weak variation can be explained by the similar lengths of Al-O and Ge-O bonds for a given environment. For instance, the average Al-O and Ge-O octahedral distances in α-Al_2O_3 and GeO_2 rutile are 1.92 and 1.90 Å, respectively. In the same way, the average Al-O tetrahedral distance found in alumino-silicates 1.75 Å[19] is comparable to the Ge-O distance in the quartz form of germania (1.74 Å).

To our knowledge, the only structural study of germanium mullite is that of Durovic and Fedji[20] on a single crystal with a composition close to the 3:2 stoichiometry. Germanium mullite was found to be isostructural with the aluminum silicate and germanium atoms are located in tetrahedral sites only. Although the average site occupancy is known in silicon and germanium mullites, the local order of cations on tetrahedral sites and of oxygen vacancies is still an outstanding problem. Cation ordering phenomena would be easier to detect by X-ray studies in germania homologs because of the larger difference of Al and Ge scattering factors. Growth experiments are now under way to prepare germanium mullite crystals at various compositions.

Germanates With An Andalusite-Type Structure

Ga_2GeO_5 (low-temperature form) and $(Al_xFe_{1-x})_2GeO_5$ mixed compounds have an andalusite-type structure. The temperature and composition range of the andalusite domain is shown on Fig. 1 and values of lattice constants are given in Table 2.

In a recent study of the compound Ga_4GeO_8,[14] it was found that the tetrahedral sites of the structure were occupied by gallium atoms and that germanium atoms were only in octahedra. Therefore, it was not a priori evident that germanium atoms occupy the tetrahedral sites of andalusite-type compounds as silicon atoms do in Al_2SiO_5.

In order to determine atomic positions in Ga_2GeO_5 andalusite, the structure refinement was carried out on a single crystal. As germanium and gallium ions are isoelectronic, it was impossible to distinguish them in the first refinement steps. Nevertheless, tetrahedral positions were finally assigned to germanium atoms in order to obtain satisfactory values of the thermal parameters. In addition, the resulting M-O tetrahedral bond lengths (average value 1.75 Å) corresponds rather to Ge-O distances than to Ga-O distances which are significantly longer (1.82 Å in β-Ga_2O_3). A full description of the structure of Ga_2GeO_5 is to be published elsewhere.[21]

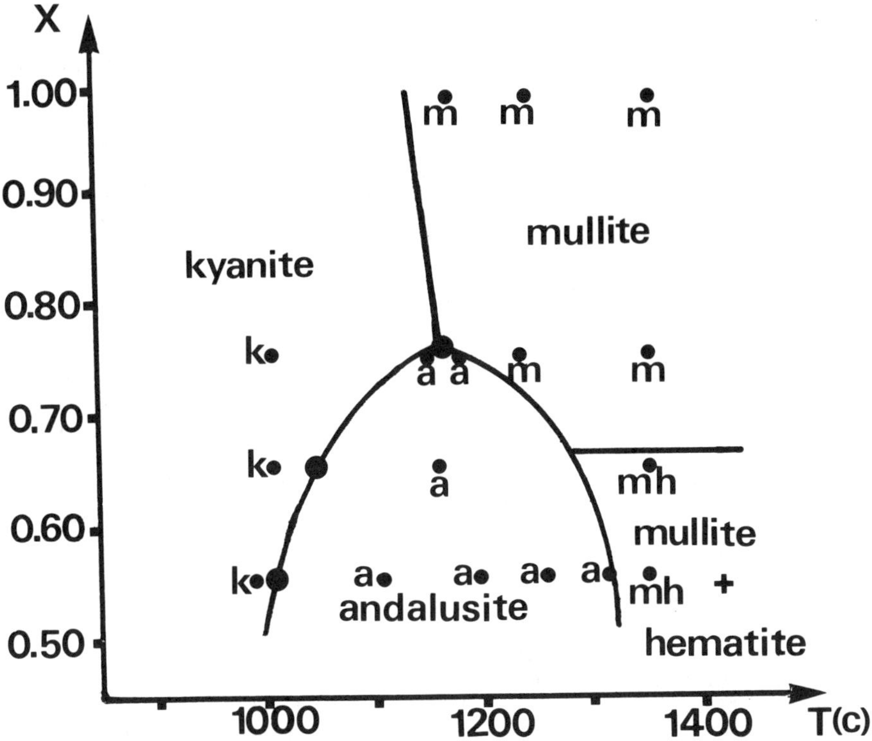

Fig. 1. Stability range of $(Al_xFe_{1-x})_2GeO_5$ phases. Small circles represents the phases observed by X-ray and large circle indicate the phase transition temperature observed by DTA.

Gallium atoms have a stronger tendency to occupy tetrahedral sites than aluminum and iron atoms. For instance, in the β stable form of gallia, half of the gallium atoms are in tetrahedral sites contrary to the corundum and hematite structures where all aluminum and iron atoms are located in octahedral sites. Therefore, in mixed $(Al_xFe_{1-x})_2GeO_5$ andalusites, Al and Fe atoms probably occupy MO_5 and MO_6 polyhedra only. From comparison of calculated and observed X-ray intensities, Schneider and Werner found that the best agreement for x=0.50 is achieved for a random distribution of the trivalent cations on these sites.[22] Similar calculations were performed on our samples for x=0.55, 0.65 and 0.75. Structure factors were obtained using the atomic positions determined by Burnham and Buerger on Al_2SiO_5[23] and by ourselves on Ga_2GeO_5.[21] Due to the limitations of powder intensity measurements, it is difficult to settle whether iron atoms are completely disordered or give preference to octahedral sites. Best fits are obtained for a slightly preferential occupancy of octahedral sites by iron atoms (~60% in octahedra and 40% in trigonal bipyramids), but studies on single crystals are needed to determine more precisely the occupancy factor of Al and Fe on each site.

Germanates With A Kyanite-Type Structure

At low temperature, $(Al_xFe_{1-x})_2GeO_5$ kyanite-type solid solutions are obtained between Al_2GeO_5 and Fe_2GeO_5 (Fig. 1).

For Al_2GeO_5, crystallization of amorphous coprecipitates results in a metastable mullite phase which transforms into stable kyanite after successive grinding and annealing treatments. On the contrary, the crystallization of aluminum-iron compounds with x>0.2 directly leads to kyanites.

A Mössbauer study on single crystals of Fe_2GeO_5 clearly showed that Fe atoms only occupy octahedral sites and consequently that Ge atoms are located on the tetrahedral sites as Si atoms in kyanite.[15] Owing to several octahedral sites in kyanite, the Mössbauer lines are broadened. The chemical shifts range between 0.34 and 0.37 mm/s which characterizes Fe^{III} in octahedra.

Discussion and Conclusions

The present results on the structure and the stability domains of mullite, andalusite and kyanite-type germanates generally are in good agreement with previous work.[6,8,22,24] The phase transformations between the Al_2SiO_5 polymorphs are particularly sluggish because they are reconstructive and require cation diffusion. It is also difficult to reach equilibrium conditions in germanates. Kyanite-andalusite transformations were observed only upon heating and appeared irreversible under DTA experimental conditions. Some minor differences between our results (obtained from coprecipitation route) and those of Schneider and Werner[24] (using direct solid state reaction) arise likely from kinetic effects.

At atmospheric pressure, germanates are often isostructural with high pressure forms of silicates. For instance, GeO_2 (rutile-type) and $ZrGeO_4$ correspond to high-pressure varieties of SiO_2 (stishovite) and $ZrSiO_4$ (scheelite-type). The similarity between the effect of pressure on silicates and cation size on germanates is illustrated by comparison of Figs. 1 and 2, which give the existence domain of $(Al_xFe_{1-x})_2GeO_5$ phases and the pressure-temperature Al_2SiO_5 diagram[25] respectively. Substituting iron to aluminum atoms in the germanium compound has the same effect as increasing pressure on Al_2SiO_5. The correspondence between the domains of sillimanite and germanate mullites clearly appears.

Another common feature of sillimanite and M_2GeO_5 mullites is the density of these structures with respect to andalusites and kyanites having the same composition. As shown in Table 4, kyanites are much more compact than mullites or sillimanite (mullite or sillimanite density~0.88 kyanite density).

Table 4. Ratio between the unit-cell volume of andalusites or kyanites and that of mullites (or sillimanite for Al_2SiO_5) with the same composition

	Andalusite	Kyanite
Al_2SiO_5 25°C(15)	1.03	0.88
Al_2SiO_5 800°C(15)	1.04	0.89
Al_2GeO_5		0.88
$Fe_{0.5}Al_{1.5}GeO_5$	1.05	0.89
$Fe_{0.7}Al_{1.3}GeO_5$	1.04	0.89
Ga_2GeO_5	1.01	

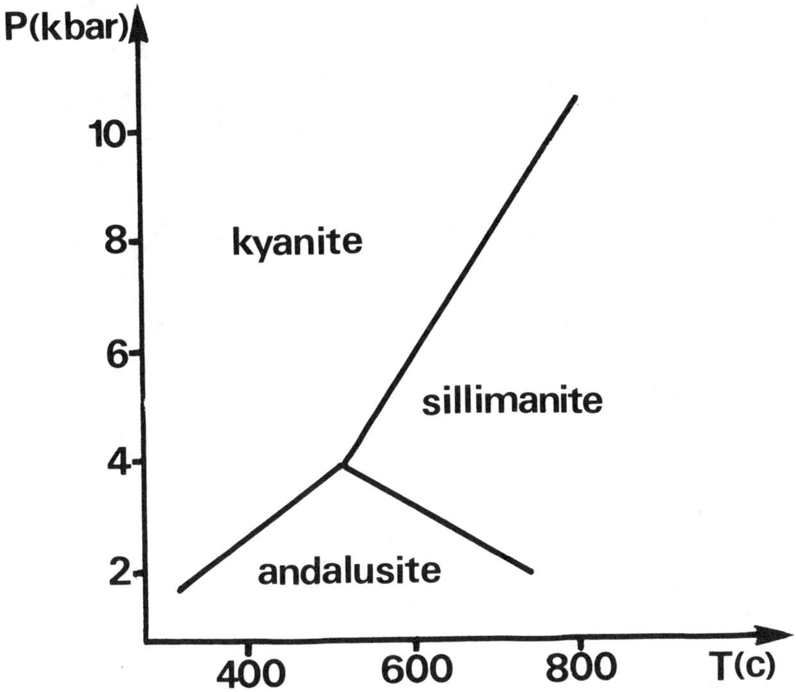

Fig. 2. Al_2GeO_5 pressure-temperature diagram from Holdaway.[25]

On the contrary, andalusites are slightly less dense than mullites or sillimanite (mullite or sillimanite density~1.03 andalusite density).

Acknowledgments

The authors are indebted to Drs. M. Harmelin for DTA experiments, E. Tronc for Mössbauer spectroscopy and L. Mazerolles for EDX analyses.

References

[1]W. Pannhorst and H. Schneider, "High-temperature Transformation of Andalusite into 3/2 Mullite and Vitreous Silica," *Mineralogical Mag.*, **42**, 195–198 (1978).

[2]H. Schneider and A. Majdic, "Kinetics and Mechanism of the Solid-State High-temperature Transformation of Andalusite into 3/2 Mullite and Silica," *Ceramurgia Int.*, **5**, 31–36 (1979).

[3]F. Liebau, "Structural Chemistry of Silicates," pp. 77–135, Springer-Verlag (Berlin), 1985.

[4]M. Perez y Jorba, P. Tarte, and R. Collongues, "Structure and Properties of Alumina-Germanium Dioxide Compounds," *C.R. Acad. Sci.*, **257**, 3417–3420 (1963).

[5]M. Perez y Jorba and R. Collongues, "Structure and Properties of Iron Oxide-Germanium Dioxide Compounds," *C.R. Acad. Sci.*, **262**, 275–277 (1966).

[6]M. Perez y Jorba, "GeO_2-Al_2O_3 and GeO_2-Fe_2O_3 Systems. Comparison with the Silica Corresponding Systems," *Silicates Industriels*, **1**, 11–17 (1968).

[7]M. Perez y Jorba and R. Collongues, "The GeO_2-Ga_2O_3 System. Comparison with the GeO_2-Al_2O_3 and GeO_2-Fe_2O_3 Systems," *C.R. Acad. Sci.*, **262**, 275–277 (1966).
[8]M. Perez y Jorba, "Study of Phases between Germanium Dioxide and some Oxides of Trivalent Elements," *Rev. Int. Hautes Temp. Refract*, **6**, 283–298 (1969).
[9]V. Agafonov, D. Michel, M. Perez y Jorba, and M. Fedoroff, "Chemical and Crystallographic Characterization of Crystals Grown by Chemical Vapour Transport in the Fe_2O_3-GeO_2 System," *Mat. Res. Bull.*, **19**, 233–239 (1984).
[10]V. Agafonov, D. Michel, A. Kahn, and M. Perez y Jorba, "Crystal Growth by Chemical Vapour Transport in the GeO_2-Ga_2O_3 System," *J. Cryst. Growth*, **71**, 12–16 (1985).
[11]V. Agafonov, A. Kahn, D. Michel, M. Perez y Jorba, and M. Fedoroff, "Growth and Structural Features of $Al_2Ge_2O_7$ Crystals," *J. Cryst. Growth*, **71**, 256–258 (1985).
[12]V. Agafonov, A. Kahn, D. Michel, and M. Perez y Jorba, "Structural Investigation of a New Iron Germanate $Fe_8Ge_3O_{18}$," *J. Solid State Chem.*, **62**, 397–401 (1986).
[13]V. Agafonov, A. Kahn, D. Michel, and M. Perez y Jorba, "Crystal Structure of a New Digermanate $Al_2Ge_2O_7$," *J. Solid State Chem.*, **62**, 402–404 (1986).
[14]A. Kahn, V. Agafonov, D. Michel, and M. Perez y Jorba, "New Germanates with Tunnel Structures: α-Ga_4GeO_8 and $Ge_4Ga_3O_{12}$," *J. Solid State Chem.*, **65**, 377–382 (1986).
[15]V. Agafonov, "Comparison between Germanates of Trivalent Elements (Al, Ga, Fe) and the Corresponding Silicates," Thesis, Paris 1985.
[16]J. K. Winter and S. Ghose, "Thermal Expansion and High-temperature Chemistry of the Al_2SiO_5 Polymorphs," *Am. Mineral.*, **64**, 573–586 (1979).
[17]S. Durovic, "Isomorphism Between Sillimanite and Mullite," *J. Am. Ceram. Soc.*, **45**, 157–161 (1962).
[18]A. Kahn, S. Abolhassani, D. Michel, L. Mazerolles, and M. Perez y Jorba, to be published.
[19]J. B. Jones, "Al-O and Si-O Tetrahedral Distances in Aluminosilicate Framework Structures," *Acta Cryst.*, **B34**, 355–358 (1968).
[20]S. Durovic and P. Fejdi, "Synthesis and Crystal Structure of Germanium Mullite and Crystallochemical Parameters of D-Mullites," *Silicaty*, **2**, 97–112 (1976).
[21]A. Kahn, S. Abolhassani, D. Michel, and M. Perez y Jorba, "Crystal structure of Ga_2GeO_5 andalusite," to be published.
[22]H. Schneider and H. D. Werner, "Synthesis and Crystal Chemistry of Andalusite-, Kyanite- and Mullite-type Structures in the System Al_2GeO_5-Fe_2GeO_5," *N. Jb. Miner. Abh.*, **140**, 153–164 (1981).
[23]C. W. Burnham and M. J. Buerger, "Refinement of the Structure of Andalusite," *Z. Kristallogr.*, **118**, 269–290 (1961).
[24]H. Schneider, "Infrared Spectroscopic Investigations on Andalusite-, and Mullite-type Structures in the System Al_2GeO_5-Fe_2GeO_5 and Al_2GeO_5-Ga_2GeO_5," *N. Jb. Miner. Abh.*, **142**, 111–123 (1981).
[25]M. J. Holdaway, "Stability of Andalusite and the Alumina Silicate Phase Diagram," *Amer. J. Sci.*, **271**, 97–131 (1971).

A Review of Powder Preparation Methods and Densification Procedures For Fabricating High Density Mullite

MICHAEL D. SACKS AND HAE-WEON LEE

Ceramics Division
Department of Materials Science and Engineering
University of Florida
Gainesville, Florida 32611

JOSEPH A. PASK

Department of Materials Science and Mineral Engineering
University of California at Berkeley
Berkeley, California 94720

Introduction

Mullite has attracted a great deal of interest in recent years due to its excellent high temperature strength and creep resistance, good chemical and thermal stability, low thermal expansion coefficient, and good dielectric properties. Consequently, mullite and mullite-containing composites have been considered for a variety of high performance applications, such as high temperature structural components,[1–5] infrared transparent windows,[6–8] substrates for microelectronic packages,[9] high temperature protective coatings,[10] radomes,[11] etc. In order to realize the full potential of mullite for technical applications and to achieve optimum physical properties, a thorough understanding of the processing behavior is essential.

In this paper, we discuss the various powder preparation techniques, consolidation (i.e. shape-forming) methods, and densification procedures for the fabrication of mullite bodies with high relative density (i.e. low porosity). Although a comprehensive (but by no means exhaustive) review of the literature is presented, our primary objective is to provide a critical analysis of the factors affecting the processing of mullite. The discussion focuses on the preparation of high purity materials of interest for high performance applications.

A variety of powder preparation methods will be reviewed, including conventional mixing of oxides, colloidal mixing techniques, and solution-precipitation techniques. We will discuss various shape-forming methods, such as dry pressing, slip casting, and sol-gel processing. The influence of powder and green compact characteristics on densification behavior and microstructure evolution will be discussed in detail. The effects of processing variables, such as chemical composition (i.e. $Al_2O_3SiO_2$ ratio) and gas atmosphere, on sintering behavior will also be discussed.

Powder Preparation

Conventional Mixing of Powders

Mullite is a common phase in many traditional ceramics, e.g. refractories, whitewares, porcelains, etc. For these applications, mullite is generally formed by heat treatment of naturally occurring aluminosilicates, such as clays and sillimanite minerals. Since these materials have overall $Al_2O_3SiO_2$ ratios that are much lower than in mullite, high temperature heat treatment results in the formation of a siliceous phase along with the mullite. In order to form single phase mullite, it is necessary to mix the aluminosilicate mineral with an aluminum-rich material. For example, mullite has been formed from clay/bauxite[12,13] clay/aluminum hydroxide,[14] clay/aluminum fluoride,[15] and clay/α-Al_2O_3[12,16,17] mixtures. The materials used in these processes are relatively coarse, so that the scale of mixing for the reaction components is, at best, on the order of a micrometer. Thus, high temperatures are required to achieve sufficiently high chemical interdiffusion coefficients for reactions to be completed in a reasonable time. Even at temperatures above 1700°C, mullitization is often incomplete, as indicated by the presence of α-Al_2O_3 and silicate glass in reacted mixtures.[12,17]

Another problem with using naturally occurring minerals to form mullite is that a considerable amount of impurities are present (e.g. alkali oxides, iron oxide, titanium oxide, etc.). These impurities are generally undesirable in technical ceramic applications. Therefore, another approach to forming mullite is to react higher purity oxides (or hydroxides) of aluminum and silicon. Aluminum-bearing compounds used to form mullite include gibbsite, $Al(OH)_3$,[18] diaspore, $AlO(OH)$,[18] and α-Al_2O_3,[17–26] while silicon-bearing compounds include quartz,[16–23] cristobalite,[18,23] fused silica,[23–26] and silicic acid.[18,19] Although higher purity is often achieved with these methods, reaction temperatures remain high because starting powders have particle sizes in the micrometer (or larger) range. Complete reaction generally requires temperatures in the range ~1600–1750°C.

It should be noted that the intimacy of mixing of the aluminum- and silicon-bearing compounds is not necessarily determined by the particle sizes of the components. Fine particles tend to form agglomerates (i.e. clusters of "primary" particles), so the scale of mixing tends to approach the agglomerate size. This point is illustrated by results reported by McAtee and Milligan.[27] They prepared aluminum- and silicon-bearing compounds by solution-precipitation techniques, but subsequently mixed the components by a conventional processing method. In one case (series I), an "alumina" (i.e. actually aluminum hydroxide) was precipitated from an aluminum nitrate solution by adding concentrated ammonium hydroxide. A "silica" gel was separately precipitated from a sodium metasilicate solution by adding concentrated hydrochloric acid. Both gels were washed by centrifugation (i.e. to remove unwanted ions, such as nitrate, chloride, sodium, etc.) and subsequently dried at room temperature. In another case (series II), an alumina gel was prepared from aluminum isopropoxide, $Al(OC_3H_7)_3$, by hydrolysis/condensation reactions. A silica gel was separately prepared in a similar manner using silicon ethoxide, $Si(OC_2H_5)_4$, dissolved in ethanol. Both gels were washed to remove alcohol and subsequently dried at room temperature. Al_2O_3-SiO_2 mixtures were prepared for both series by grinding the dried gel powders together in a motor-driven mortar for 2 h. Although gel preparation methods tend to produce powders with fine primary particle size, the scale of mixing attained in these samples was undoubtedly dominated by the size of the gel powder agglomerates. In grinding or milling operations, it is generally difficult to break down agglomerates to sizes below ~1 μm. Although heat treatment of the mixtures resulted in some mullite formation as low as 1300°C, much higher temperatures were required for extensive reaction. In samples with the stoichiometric composition of mullite (i.e. 60 mol% Al_2O_3/40 mole% $SiO_2 \approx 71.8$ wt% Al_2O_3/28.2 wt% SiO_2), the reaction was still incomplete after heat treatment at 1750°C for 2 h, as indicated by observations of small amounts of α-Al_2O_3 and α-cristobalite in X-ray diffraction (XRD) patterns. These results reflect the difficulty in achieving intimate mixing of the reaction components by conventional powder processing methods.

Colloidal Techniques

In this section, we will use the term "colloidal techniques" to refer to mullite powder preparation methods in which suspensions of fine particles—often called sols or dispersions—are utilized. These techniques are, in fact, quite similar to the conventional wet mixing of aluminum- and silicon-bearing oxide/hydroxide powders except that the size of the primary particles are much smaller (i.e. typically in the range ~5–50 nm). The advantage of using fine particles is that the distances for chemical interdiffusion are generally much shorter and, hence, lower reaction temperatures for mullite formation would be anticipated. It should be emphasized, however, that the reaction components must be well-mixed to ensure that the diffusion distances are comparable to the size of the primary particles (i.e. as opposed to the size of agglomerates).

Ghate and co-workers[28] used colloidal suspensions of γ-Al_2O_3 (~100 m^2/g specific surface area) and Ludox® silica (~230 m^2/g specific surface area) to produce mullite powders with stoichiometric composition (i.e. ~71.8/28.2 weight ratio of Al_2O_3/SiO_2). They prepared a dilute aqueous suspension of γ-Al_2O_3 which was partially stabilized against flocculation by hydrochloric acid additions. (At low pH, alumina particles develop a positive surface charge.) The silica sol was slowly added to the alumina suspension with sufficient acid to maintain a pH of ~5. The silica particles have a slight negative charge at this pH and, thus, intimate mixing of alumina and silica may have been promoted by a heteroflocculation process (i.e. due to electrostatic attraction of the oppositely charged particles). To prevent segregation of components, the mixture was subsequently "gelled" by adjusting the pH to the range of ~6–7 with ammonium hydroxide additions. The term "gelation" refers to the phenomenon in which fine particulate (or polymeric) species link up in a liquid medium to form a semi-rigid, three-dimensional, network structure. Hence, this type of powder preparation technique is often referred to as a "sol-gel" method. In Ghate's study, gelation resulted from the decrease in stability (against flocculation) of both the alumina and silica suspensions in the pH range of ~6–7. The "wet" gel was subsequently dried, ground up, and calcined to form mullite. Due to the fine size of the reactants, mullite formation was observed (in X-ray diffraction patterns) at temperatures as low as 1200°C, although complete reaction required an extended time (20 h) at temperatures $\gtrsim$ 1400°C.

Several other researchers have employed similar colloidal techniques to prepare mullite powders. Sacks and Pask[21,29] mixed aqueous suspensions of γ-Al_2O_3 (~100 m^2/g specific surface area) and Ludox® silica (~140 m^2/g specific surface area), but produced gelation by solvent evaporation instead of pH adjustment. The results were similar to Ghate's study in that complete mullitization required 24 h at 1450°C for a powder with ~73/27 weight ratio of Al_2O_3/SiO_2. Metcalf and Sant[30] also used a colloidal technique in which the mixed sols were dried (instead of chemically gelled). Powders with stoichiometric composition (~71.8 wt% Al_2O_3) were calcined to form mullite at 1550°C for 3h.

Ismail and co-workers[31,32] modified the colloidal technique by mixing aqueous silica (~130 m^2/g specific surface area) and boehmite (aluminum monohydroxide) sols under acidic conditions (pH $\approx$ 1.8), followed by gelling through solvent evaporation. The boehmite sol was prepared by hydrolysis of γ-Al_2O_3 (~68 m^2/g specific surface area) using a vigorously agitated, nitric acid solution (HNO_3/γ-Al_2O_3 molar ratio of 0.06) heated at 95°C. It should be noted that most commercially available γ-Al_2O_3 powders have a substantial amount of *hard* agglomerates. The hydrolysis conditions employed by Ismail and co-workers probably resulted in extensive breakdown of hard agglomerates through dissolution of the γ-Al_2O_3. Thus, it is likely that these workers produced a finer particle size for the alumina source (i.e. the boehmite sol) and thereby achieved a more intimate mixture of alumina and silica. The finer particle size of the boehmite was indicated by the high specific surface area (277 m^2/g) of the dried gel mixture. In addition, more intimate mixing of alumina and silica was indicated by X-ray diffraction (XRD) results which showed that extensive mullitization had occurred after 1 h at 1300°C and that the reaction was complete after 1 h at 1400°C, i.e. considerably shorter reaction times than required for Ghate's process. Similar results were reported by Roy and co-workers[33,34] using commercially available boehmite and silica sols. They observed substantial mullitization at 1330°C, but the time/temperature schedule for complete reaction was not specified.

Hamano et al.[35] also utilized boehmite and silica sols to prepare mullite powders with stoichiometric composition. The boehmite sol was prepared from aluminum ethoxide by hydrolysis/condensation reactions, using HCl as a catalyst. The silica sol was prepared by hydrolysis/condensation of methyl silicate. The two sols were then mixed and subsequently dried by slowly adding the mixture to a hot (160°C) petroleum bath. XRD results indicated the presence of a weakly crystalline Al-Si spinel phase in powders calcined in the range ~700–1250°C, with the initial formation of mullite at ~1250°C. However, this process apparently did not result in homogeneous mixing of the alumina and silica; XRD showed that calcined powders contained cristobalite at ~1250–1600°C and α-Al_2O_3 at ~1300–1700°

Sōmiya et al.[36] prepared mullite precursors with stoichiometric composition by hydrothermal treatments of a colloidal mixture of alumina and silica sols. The mixed sol was treated for 2 h at temperatures in the range 300–600°C with a pressure of 50 MPa. Powders formed at lower temperatures (300–400°C) contained a weakly crystalline aluminosilicate phase which was described as "pseudoboehmite-like" due to its similar XRD pattern to the aluminum monohydroxide. This phase decomposed during conventional calcination and another weakly crystalline aluminosilicate phase (identified as an Al-Si spinel) formed at temperatures in the range 800–1200°C (1 h). Mullite formation was not observed until 1300°C. When the hydrothermal treatment temperature was increased to 600°C, the "pseudoboehmite-like" phase was not detected. Instead, several other aluminosilicate phases with sharper XRD peaks were identified, including small amounts of mullite. Electron microscopy and chemical analysis by energy dispersive spectroscopy (EDS) indicated that the product contained agglomerates with two types of particles: (1) acicular particles (several micrometers in length) with a higher Al_2O_3/SiO_2 ratio and (2) more equidimensional particles with lower Al_2O_3/SiO_2 ratio. These morphologies and compositional inhomogeneities persisted after subsequent heat treatment at 1300°C (1 h). The calcination temperature needed to form mullite depended on the temperature of the initial hydrothermal treatment. The amount of mullite formed after calcination at 1200°C (1 h) increased as the hydrothermal treatment temperature increased from 400–600°C.

Solution Techniques

From the results discussed in the previous section concerning colloidal techniques, it is evident that substantial reductions in mullitization temperature can be achieved if the alumina and silica reactants are mixed on a fine scale. The ultimate processing technique, in terms of minimizing the chemical interdiffusion path lengths, would provide homogeneous mixing at the molecular level. This is usually the goal of researchers utilizing solution-precipitation techniques. Molecular-scale mixing may be achieved readily *in solution*; for example, the appropriate aluminum- and silicon-bearing salts can be dissolved together in a suitable solvent. However, the initial molecular-scale mixing is not necessarily maintained during subsequent processing, especially during the precipitation step. In the ideal case, it would be possible to co-precipitate a single phase, multi-cation (e.g. aluminum and silicon) precursor. (The term "precursor" refers to a compound which under-goes decomposition reactions and/or phase transformations during calcination before yielding the desired single phase compound, such as mullite). In many cases, however, the precipitated powder is multiphase and, thus, calcination conditions must be chosen not only to complete decomposition reactions and phase transformations, but also to insure that phases react to form the desired single phase compound. In some cases, sequential precipitation of single-cation precursor particles may occur, i.e. the final precipitated powder would be a mixture of single-cation precursor particles. In terms of the scale of mixing, powders produced under such conditions would probably have a similar character to powders produced by the colloidal techniques discussed in the previous section. Precautions would be required to avoid long-range segregation of the different precipitate particles during the solvent removal step (e.g. evaporation, filtration, centrifugation, etc.).

As discussed below, a large number of solution-precipitation techniques have been used to prepare mullite powders. It is of particular interest to identify processing conditions which result in the finest scale of mixing of aluminum and silicon (i.e. molecular-level mixing). There are several criteria that can be used to assess the intimacy of mixing. First, the temperature for mullitization should decrease as the mixing becomes more intimate. We noted earlier that temperatures on the order of ~1600–1750°C are required to achieve complete mullite formation (in a practical time frame) when the alumina and silica particles are in the micrometer size range. For reactant particle sizes in the range of several hundred Angstroms (i.e. as in colloidal techniques), mullitization temperatures are generally reduced to ~1300–1450°C. Therefore, reaction temperatures on the order of ~1000–1100°C might be anticipated if molecular-level mixing was achieved. In some instances, X-ray diffraction (XRD) analysis on the precipitated powder may also be used to determine if intimate mixing is achieved through a solution technique. For example, the appearance of boehmite (an aluminum hydroxide) in a precipitated precursor to mullite would indicate that segregation had occurred during the process, i.e. the powder was *not* formed by a true co-precipitation process. It should be noted, however, that segregation effects cannot always be gauged from XRD on the precipitated powders since the powders may be amorphous. However, the appearance of crystalline alumina or silica phases during heat treatment, i.e. prior to mullitization, would also indicate that segregation or sequential precipitation had occurred. For example, amorphous alumina precursors crystallize to "transition aluminas" (such as γ-, η-, or δ-Al_2O_3) at temperatures ≳500°C. Amorphous silica generally transforms to cristobalite at temperatures ≳1000°C. (The latter transformation was observed, for example, in powders prepared by some of the colloidal techniques discussed in the previous section.[28,34])

A variety of aqueous-soluble and nonaqueous-soluble compounds have been used to prepare mullite precursors by solution-precipitation techniques. Aluminum-containing compounds have included inorganic salts (e.g. chloride,[30,37–40] nitrate,[33–35,41–44] sulfate,[35,37,44–46] and chlorate,[47]) as well as organometallic compounds (e.g. alkoxides[6–8,35,36,48–54]). Similarly, the source for silicon has included the chloride,[38–40] acetate,[44] and several alkoxides.[6–8,11,30,33–36,41–43,45,47–54]

In most of the solution techniques used to prepare mullite precursors, the aluminum- and silicon-containing compounds undergo hydrolysis and condensation reactions. Initially, let us consider these reactions for a silicon compound by itself. For example, a silicon alkoxide, such as tetraethylorthosilicate (TEOS), can undergo partial hydrolysis (eq. 1) or complete hydrolysis (to form monosilicic acid, eq. 2):

$$Si(OC_2H_5)_4 + H_2O \rightarrow Si(OC_2H_5)_3OH + C_2H_5OH \quad (1)$$

$$Si(OC_2H_5)_4 + 4H_2O \rightarrow Si(OH)_4 + 4C_2H_5OH \quad (2)$$

Similarly, a metal salt, such as silicon tetrachloride, can undergo a hydrolysis reaction:

$$SiCl_4 + 4H_2O \rightarrow Si(OH)_4 + 4HCl \quad (3)$$

In addition to hydrolysis, there is a tendency to form siloxane (-Si-O-Si-) bonds by condensation reactions between two hydroxyl (i.e. silanol) groups:

$$Si(OH)_4 + Si(OH)_4 \rightarrow (OH)_3Si\text{-}O\text{-}Si(OH)_3 + H_2O \quad (4)$$

It should be noted that complete hydrolysis (eq. 2 or 3) is not required for condensation to occur; partially hydrolyzed silicate species (eq. 1) also undergo condensation reactions. The hydrolysis and condensation rates depend on a variety of factors, including the reactant concentrations, solution pH, temperature, etc. If hydrolysis and condensation were carried to completion, anhydrous silica would be formed:

$$2Si(OH)_4 \rightarrow 2SiO_2 + 4H_2O \quad (5)$$

In reality, silica particles prepared by solution techniques generally contain a considerable amount of silanol groups on the surface and in the bulk of the precipitate.

The aluminum compounds used to prepare mullite precursors can also undergo the hydrolysis and condensation reactions described above. In the ideal case, molecular mixing of aluminum and silicon can occur through a condensation reaction involving Si-OH and Al-OH groups:

$$Si(OH)_4 + Al(OH)_3 \rightarrow (OH)_3Si\text{-}O\text{-}Al\text{-}(OH)_2 + H_2O \quad (6)$$

A true co-precipitation process would involve continued condensation growth of species with aluminosiloxane (-Si-O-Al-) bonds. It is clear, however, that reactions leading to -Si-O-Si- and -Al-O-Al- bonds can also occur in solution-precipitation processes used to prepare mullite precursors. The rates of these competing reactions would have to be minimized to avoid segregation effects. Furthermore, hydrolysis rates must be carefully controlled since the rates of condensation are dependent upon the extent of hydrolysis.[55,56] The aluminum and silicon compounds used in solution processes can have widely varying hydrolysis rates. An aluminum alkoxide, for example, tends to hydrolyze much faster than the corresponding silicon alkoxide.[52] In order to avoid segregation effects, it may be desirable to carry out a prehydrolysis step for the compound with the slower hydrolysis rate. In addition, it may be helpful to control hydrolysis rates by avoiding large excesses of water (beyond the stoichiometric amount), i.e. at least during the early stages of hydrolysis. Aside from controlling the reactant concentrations, it is also common to vary the hydrolysis and condensation rates by acid or base catalyst additions. For example, the condensation reaction between silanol groups (illustrated in eqs. 4 and 5) is greatly accelerated as the pH increases.[55,56] Rapid growth and precipitation of aluminum hydroxide compounds (e.g. boehmite, bayerite, etc.) also occurs under moderately basic conditions.[50,57] Therefore, in preparing mullite precursors with molecular-level mixing of aluminum and silicon, it may be desirable to avoid base additions during the early stages of the process.

Solution techniques have been used for over 50 years to prepare aluminosilicate precursors to mullite. The initial techniques were developed to provide higher purity materials for studying the sequence of reactions during the heat treatment of clays. Insley and Ewell[37] dissolved aluminum sulfate, $Al_2(SO_4)_3 \cdot 18H_2O$, in water and then added dropwise a sodium silicate solution under conditions of vigorous stirring. The mixed solution was subsequently precipitated with NaOH additions. The gelatinous precipitate was filtered, dried, crushed up, re-washed with hot water, and re-dried. Differential thermal analysis (DTA) of a mixture with a 2:1 mole ratio of Al_2O_3:SiO_2 showed a relatively sharp exothermic peak in the range of ~980°C. Mullite formation was observed in this temperature range according to XRD analyses and the diffraction peaks became more intense at higher temperatures. Despite the washing procedure used in the above solution-precipitation technique, a considerable amount of Na_2O remained in the final powders. Therefore, Insley and Ewell also carried out a solution-precipitation process using aluminum chloride, $AlCl_3$, instead of aluminum sulfate. The gel product was subsequently electrodialyzed for 10 h to reduce the sodium ion concentration. The DTA and XRD results were similar to those obtained with gels of higher sodium content, although experiments were only carried out on a sample with an Al_2O_3:SiO_2 mole ratio of 1:2.

Ossaka[46] used a similar process in which sodium silicate and potassium aluminum sulfate, $KAl(SO_4)_2 \cdot 12H_2O$, were dissolved in a dilute aqueous solution of sulfuric acid. The mixture was co-precipitated with an excess of hexamethylenetetramine. The precipitate was subsequently filtered, washed (to remove potassium, sodium, and the amine), and dried. A mixture with a 3:2 mole ratio of Al_2O_3:SiO_2 showed a "mullite-like" phase in XRD patterns of samples heated between 910–1100°C for 5 h. DTA showed a sharp exothermic peak in the same temperature range. Well-crystallized mullite was observed in XRD patterns of powders calcined at 1250°C for 5 h.

To avoid alkali impurities, Horte and Wiegmann[38] prepared aluminosilicate precursors (overall compositions in the range ~20–85 wt% Al_2O_3) from mixed aluminum and silicon chloride solutions. Gelatinous precipitates were obtained by mixing the chloride salts with ammoniated water. The gels were subsequently washed in order to remove ammonium chloride. XRD indicated that the gels were initially amorphous. DTA showed a sharp exothermic peak in the range ~980–1010°C, but the temperature for mullite formation was not specified. Crofts and Marshall[39] also used aluminum and silicon chloride solutions to prepare aluminosilicate precursors with compositions in the range 56.5–60.9 mole% Al_2O_3. In an effort to minimize selective hydrolysis, they initially dissolved the $AlCl_3$ and $SiCl_4$ in diethyl ether (i.e. instead of water). Hydrolysis/precipitation was carried out by injecting the mixed salt solution into a vigorously stirred aqueous ammonia solution which had been heated to 50°C in order to "flash" off the ether. The precipitated powder was collected by centrifugation and subsequently filtered, dried, and calcined. Despite the turbulent conditions during the "flash-hydrolysis" process, segregation occurred to some extent during precipitation. XRD results showed that the dried powder was amorphous. However, calcined powders showed γ-Al_2O_3 at 800°C and α-cristobalite at 1200°C. Mullite formation was observed at temperatures between 1000 and 1200°C, although the reaction was not complete after 2h at 1200°C. McGee and Wirkus[40] also used a chloride process to prepare mullite. $AlCl_3$ and $SiCl_4$ were initially dissolved in absolute methanol. The mixed salt solution was then co-precipitated with ammonium hydroxide. Dried and calcined powders with stoichiometric composition (~72 wt% Al_2O_3) were amorphous to X-rays up to temperatures 1050°C (40 h). Crystallization to mullite was observed at 1100°C (40 h), although the temperature for complete reaction was not specified.

Organometallic compounds have been used extensively in preparing mullite by solution-precipitation techniques. West and Gray[48] reported the preparation of aluminosilicate precursors containing less than 5 ppm metallic impurities, but experimental conditions were not specified. Mazdiyasni and Brown[49] prepared a mullite precursor with stoichiometric composition (~71.8 wt% Al_2O_3) by initially synthesizing aluminum isopropoxide, $Al(OC_3H_7)_3$, and silicon isoproproxide, $Si(OC_3H_7)_4$. The alkoxides were then mixed and refluxed in excess isopropyl alcohol prior to hydrolysis/condensation reactions. The latter reactions were carried out by slowly adding the alkoxide solution to ammoniated water. The precipitate was subsequently washed repeatedly, dried, and calcined. XRD showed that the powder was amorphous up to ~1080°C. Extensive mullite formation was observed above ~1185°C, although higher temperatures were required for complete reaction.

Prochazka and Klug[6] also used metal alkoxides to prepare mullite powders with overall compositions of 72.3 and 76 wt% Al_2O_3. Aluminum isopropoxide and TEOS were dissolved separately in cyclohexane and subsequently mixed at the desired ratio. Hydrolysis/condensation reactions were carried out by slowly adding a water/butyl alcohol solution to the mixed alkoxide solution. The resulting gels were filtered, dried, and calcined at temperatures in the range 850–1200°C. Powders were amorphous to X-rays below ~850°C, while a spinel phase was detected at higher temperatures. Well-crystallized mullite was observed at 1150°C.

Hirata and co-workers[50,51] used two different processes—designated I and II—to prepare mullite powders from metal alkoxides. Method I was similar to the technique used by Mazdiyasni and Brown and was used to prepare powders with compositions of 75.0–76.6 wt% Al_2O_3. Excess ammoniated water (pH>7) was added slowly to a stirred mixture of aluminum isopropoxide and TEOS. After 4 h reaction time, precipitated powders were collected by centrifugation and subsequently dried at 80°C for 24 h. When the pH of the ammoniated water was ≲9.1, the dried powders were amorphous. At higher pH values, powders contained crystalline aluminum hydroxides, indicating that extensive segregation occurred during the precipitation process. (An aluminum monohydroxide, boehmite, was observed by XRD when the pH was ≳10, while aluminum trihydroxides, i.e. bayerite and nordstrandite, were also observed at pH = 11.9.) Each powder showed a relatively weak exothermic peak in DTA, but the temperature for the maximum peak height decreased from ~1010°C to ~970°C as the pH used in the precipitation process decreased from 11.9 to 7. The authors attributed the exotherm to the formation of an Al-Si spinel and suggested that the decrease in peak temperatures reflected improved chemical homogeneity. For powders precipitated at pH = 8.7, XRD showed the development of several broad peaks (attributed to the Al-Si spinel phase) after calcination at temperatures in the range of 1000–1200°C (1 h). Poorly crystallized mullite was observed by XRD at 1200°C (1 h), while extensive crystallization occurred by 1300°C (1 h).

In the second powder preparation method (II) used by Hirata and co-workers, the metal alkoxides (aluminum isopropoxide and TEOS) were first dissolved in benzene and refluxed (at ~82°C) for 4 h under flowing dry N_2. A distilled water (pH≈5.5)/isopropanol mixture (~20 wt% water) was slowly added to the benzene solution and the mixture was refluxed for 4 h. In method II, the amount of water was sufficient for complete hydrolysis of the alkoxides, but was *not* added in excess. The DTA curves for all powder II compositions (46.5–76.6 wt% Al_2O_3) showed a sharp exothermic peak at ~980°C which was attributed to the formation of both Al-Si spinel and mullite. XRD showed that the powders were amorphous after calcination treatments up to 900°C (3 h). In powders close to the stoichiometric composition, mullite and Al-Si spinel were detected after calcination at 1000 and 1100°C (3 h), while at higher temperatures only mullite was observed.

The low temperature (~1000°C) formation of mullite and the sharp DTA exotherm at ~980°C indicate that method II provides a very homogeneous (molecular-scale) mixing of aluminum and silicon in the precipitated powder. It is not clear which processing step(s) are the most critical for achieving this homogeneity. However, an important factor is probably the use of "controlled" hydrolysis/condensation conditions: (1) Excess water was *not* added, i.e. the amount of water was just sufficient for complete hydrolysis of the alkoxides. (2) Conditions were chosen to avoid high local concentrations of water during the hydrolysis. (The water was diluted in isopropanol and added slowly.) (3) Hydrolysis/condensation reactions were not accelerated with catalysts, i.e. there were no acid or base additions. Another factor contributing to the homogeneous mixing may be the initial refluxing operation; refluxing may result in some development of a molecularly mixed (Al-Si) alkoxide compound. However, a recent study by Paulick et al.[52] suggests that the formation of a double alkoxide complex is unlikely for the refluxing conditions that were used by Hirata et al.

Hamano et al.[35] prepared mullite powders with stoichiometric composition using processes similar to Mazdiyasni and Brown[49] and Hirata et al.[50,51] They initially prepared an isopropanol solution of aluminum isopropoxide and methyl silicate. In one process (co-precipitate A), the mixed alkoxide was added dropwise to a dilute NH_4OH solution at 60°C. In a second process (co-precipitate B), an isopropanol/water solution was added to the mixed alkoxide solution. In each process, precipitates were collected by centrifugation and subsequently vacuum dried. XRD showed that both powders were initially amorphous. Precipitate A showed a strong exothermic peak at ~960°C and XRD revealed the development of an Al-Si spinel phase. Mullite formation was observed at temperatures above ~1250°C. Precipitate B also showed a strong exothermic peak at ~970°C. However, XRD showed that both mullite and the spinel phase formed in this temperature range, indicating that more intimate mixing was achieved in method B. This process was similar to method II of Hirata et al.[50,51] in that the (1) water for the hydrolysis reactions was added as a diluted solution and (2) hydrolysis/condensation reactions were not accelerated by acid or base catalysts.

Paulick et al.[52] studied the effects of pH, temperature, and reaction time in preparing mullite precursors from alkoxide solutions. Aluminum isopropoxide was dissolved in excess isopropyl alcohol and added to a sufficient amount of TEOS to give an overall Al_2O_3/SiO_2 ratio equivalent to stoichiometric mullite. Hydrolysis/condensation reactions were initiated by adding water/methanol (3:1 ratio) solutions to the mixed alkoxide solutions. The solution pH was varied with nitric acid or ammonium hydroxide additions. Precipitates with the best chemical homogeneity were obtained under acidic conditions (pH≈2), while extensive segregation occurred when using solutions with no catalyst (pH≈6.5) and with base additions (pH≈10). As noted earlier, one of the difficulties in achieving co-precipitation of species with aluminosiloxane bonds (-Si-O-Al-) is that the aluminum and silicon compounds can have widely varying hydrolysis rates. Several researchers have reported that aluminum alkoxides tend to hydrolyze much faster than silicon alkoxides.[52,58] However, hydrolysis rates can be varied substantially by catalyst additions. In particular, the hydrolysis rate for TEOS increases as the pH decreases below ~7.[55,56] Thus, the acidic conditions used by Paulick et al. maximized the TEOS hydrolysis rate and this is probably one of the reasons for the good chemical homogeneity at pH≈2. Furthermore, the condensation rate for TEOS is minimized at pH≈2,[55,56] so rapid growth of polysilicate species was avoided. In addition, the low pH condition also prevented growth of large aluminum hydroxide particles.

Sōmiya et al.[36] prepared mullite and mullite precursors by hydrothermal treatments on mixed alkoxide solutions. Aluminum isopropoxide and TEOS were mixed, dissolved in benzene, and refluxed for 5 h at ~80°C under a nitrogen atmosphere. Hydrolysis/condensation of the mixed alkoxides was carried out hydrothermally at temperatures in the range 300–600°C and pressures in the range 10–100 MPa. Powders prepared at low temperatures (300–400°C) contained the same weakly crystalline "pseudoboehmite-like" phase that was formed in the hydrothermally treated colloidal sols discussed in the previous section. The product consisted of ~1–5 μm agglomerates which were composed of acicular (~100 nm long) primary particles. Analysis by EDS showed that the material was more homogeneous chemically than the powders prepared by hydrothermal treatment of colloidal sols. Conventional calcination at 1000–1200°C (1 h) resulted in the formation of a weakly crystalline Al-Si spinel phase, while mullite formation occurred at 1300°C (1 h). When the hydrothermal treatment was carried out at higher temperatures (500-600°C) and pressures (50–100 MPa), several other aluminosilicate phases were formed, including small amounts of Al-Si spinel and mullite. In contrast to the results obtained with the colloidal sols, increases in hydrothermal treatment temperature did not enhance mullitization during subsequent calcination at 1200°C (1 h).

Sōmiya et al.[36] modified the above process by first washing the hydrothermally treated product to remove organic materials (i.e. alcohols and benzene) and subsequently carrying out a second hydrothermal treatment. They observed extensive mullite formation in products treated at 600°C under 20 MPa pressure. The amount of mullite increased with time, but the particle size also increased greatly. Powders treated 72 and 168 h contained agglomerates with relatively large (~1–2 μm) acicular particles.

Sōmiya et al.[36] also carried out hydrothermal treatments on alkoxide-derived powders, i.e. instead of alkoxide solutions. A mixture of aluminum isopropoxide and TEOS was dissolved and refluxed in isopropanol. Hydrolysis/condensation reactions were initiated by adding distilled water at pH=10 to the alkoxide solution. The precipitate was dried at 120°C (24 h) and calcined at 600°C (1 h). The powder was amorphous, but EDS analysis indicated that inhomogeneities in chemical composition (Al_2O_3/SiO_2 ratio) existed (i.e. at least on the scale of tens of nanometers). Hydrothermal treatment at 600°C under 20 MPa resulted in extensive crystallization to mullite. The amount of mullite increased with treatment time (2–72 h) and EDS indicated that the chemical homogeneity was greatly improved. The crystallite size also increased rapidly with treatment time, as the specific surface area decreased from an initial value of 370 m^2/g to 31 m^2/g after 24 h. The powder was a mixture of rounded and acicular particles after short treatment times, while micrometer-size acicular particles became predominant for long treatment times.

The solution techniques discussed to this point have involved mixing either two inorganic compounds or two organometallic compounds. However, many methods involve mixing a metal salt and an organometallic compound. In a study of the phase equilibrium relationships in the Al_2O_3-SiO_2-H_2O system, Roy and Osborn[41] prepared a variety of aluminum silicate compositions using aluminum nitrate and TEOS as starting materials. The components were dissolved in ethyl alcohol, mixed, and the solvent was evaporated to form gels. Aluminum silicate compounds were synthesized by hydrothermal treatments of the gel materials. Kanzaki et al.[42,43] prepared a mullite powder with stoichiometric composition using a mixed solution of aluminum nitrate, $Al(NO_3)_3{\cdot}9H_2O$, and ethyl silicate in a water-alcohol (1:1 in volume) solvent. This solution was sprayed into a preheated quartz reaction tube at temperatures in the range of 350–650°C. The powder collected after the "spray pyrolysis" operation was amorphous. Differential thermal analysis showed a single sharp exothermic peak at ~970°C and XRD revealed extensive formation of well-crystallized mullite at 980°C. These results indicate that excellent homogeneity was achieved during the precipitation process. As in the study of Hirata et al.[50,51] it is not clear which processing step(s) were most important in preventing segregation. However, atomization of the mixed solution and rapid solvent removal during the spray pyrolysis process were probably important factors. Another factor may have been the absence of acid or base catalysts in solution. This would have resulted in minimal condensation growth in solution prior to the spray pyrolysis operation.

Hamano et al.[35] formed mullite powders with stoichiometric composition using a "spray drying" method which was very similar to the spray pyrolysis process of Kanzaki et al.[42,43] Aluminum nitrate and ethyl silicate were dissolved in ethanol and the mixed solution was sprayed into a furnace at 250°C. The powder collected after spray drying was amorphous. DTA showed a strong exothermic peak at ~980°C and XRD showed that mullite formation occurred at the same temperature. These results again suggest the importance of avoiding rapid hydrolysis/condensation conditions in order to maintain chemical homogeneity in the precipitated powders. Acid or base catalysts were not used and water (for the hydrolysis reactions) was apparently available only from the hydration water of the aluminum nitrate. Hamano et al.[35] also spray dried an ethanol-water solution of aluminum sulfate and ethyl silicate at 350°C. Molecular-level mixing was not maintained in this method, however, as XRD analysis of the powder collected after drying revealed that aluminum sulfate had re-precipitated. A weak exotherm was observed at ~980°C in DTA and the Al-Si spinel phase was present in powders calcined between ~980–1200°C. Mullite formation was observed at temperatures above 1200°C.

Roy and co-workers[33,34] also mixed an aluminum nitrate and TEOS and obtained similar results to Kanzaki et al.[42,43] and Hamano et al.[35] Components were dissolved in absolute ethanol and subsequently gelled by heating the mixed solution at 60°C for several days. Additional drying at 60°C produced powders which were amorphous. DTA showed a single sharp exotherm at ~960°C and XRD revealed poorly crystallized mullite formation at 1015°C. These results again suggest that slow hydrolysis/condensation conditions are helpful in achieving precipitated powders with molecular-scale mixing. Water was available only from the hydration water of the aluminum nitrate and the reactions were not catalyzed by acids or bases. (Since base additions were not made, it is likely that aluminum nitrate was reprecipitated during the solvent evaporation step. In the processes by Kanzaki et al.[42,43] and Hamano et al.,[35] aluminum nitrate was not observed in the precipitated powders, but drying was carried out at temperatures above the nitrate decomposition temperature.)

The importance of using controlled hydrolysis conditions to avoid segregation was also indicated in experiments reported by Meng and Huggins.[44] They initially prepared an aluminum salt solution (using either aluminum sulfate or aluminum nitrate), then added silicon acetate, $Si(C_2H_3O_3)_4$, while stirring continuously, and subsequently dried the mixture on a hot plate. When water was used as the solvent, a coarse precipitate quickly formed upon addition of the silicon acetate. Extensive segregation was indicated by XRD results which showed that powders calcined at 1300°C contained α-Al_2O_3 and α-quartz, in addition to mullite. The best results were obtained by dissolving the aluminum nitrate in methanol. Powders were amorphous at ~760°C (10 h) and mullite was the only phase detected at temperatures of ~1300–1325°C (2–18 h).

It is obvious from the above discussion that most solution techniques to prepare mullite precursors are *not* true co-precipitation processes. Sequential (or consecutive) precipitation apparently occurs in some cases and the segregation of aluminum- and silicon-containing species is extensive. There are also techniques for preparing mullite precursors in which a solution-precipitation process for one component is carried out in the presence of a previously prepared colloidal sol of the other component. (Thus, the process may be described as a "solution/colloidal" technique.) For example, Perry[11] reported hydrolyzing TEOS (with ammoniated water) in the presence of aluminum hydroxide. Calcination at 1400°C was required for complete mullitization. Metcalf and Sant[30] used a similar process in which the aluminum hydroxide was first formed by mixing an NH_4OH solution with an $AlCl_3$ solution. TEOS was subsequently added and the hydrolysis/condensation reactions were carried out by heating to 95°C for 4 h. XRD showed that mullite formation occurred after calcination at 1200°C, but complete reaction required heat treatment at 1400°C for 4 h. Yoldas[53] formed an aluminum monohydroxide sol by hydrolysis/condensation of aluminum sec-butoxide under acidic conditions. TEOS was added to form mixed sols with various Al_2O_3/SiO_2 ratios and samples were subsequently gelled. Extensive mullitization occurred at 1300°C. Klug et al.[59] carried out a similar process by adding TEOS to an aqueous suspension of colloidal boehmite. After many hours of stirring, the mixture was gelled by adding concentrated NH_4OH solution. Mullitization required temperatures above 1300°C. Roy et al.,[33,34] Shinohara et al.,[7] and Sonuparlak[8] also used colloidal boehmite and TEOS to form mullite. Similar results were obtained in that extensive mullite formation required temperatures above 1200°C in each case. Roy and co-workers[33,34] also prepared mullite by a "solution/colloidal" technique in which an aqueous Ludox® sol was mixed with an aluminum nitrate solution.

Other Techniques

The powder preparation techniques described to this point could be classified as conventional powder mixing methods, colloidal techniques, or solution-precipitation techniques. There are several procedures for preparing mullite precursors which do not fall into any of the preceding categories. For example, Moya et al.[16] used a dissolution procedure to form a "pre-mullite" powder from the clay mineral halloysite. The raw material was initially crushed and milled to <60 μm. The powder was then heated to 1000°C for 10 min to remove structural water (i.e. chemically bound hydroxyl groups) and to form a two phase mixture of Al-Si spinel and silica. The silica was dissolved by mixing the powder with a 10% NaOH solution at 90°C for 60 min. The suspension was subsequently filtered and washed with distilled water and slightly acidic water to remove impurity ions. The final composition was close to stoichiometric mullite, as chemical analysis showed 70.4% Al_2O_3, 29.0% SiO_2, 0.26% CaO, and 0.17% Fe_2O_3. The powder specific surface area was 24 m^2/g.

Gani and McPherson[60] used a flame-spraying process to produce aluminosilicate glass powders with overall Al_2O_3/SiO_2 mole ratios of 3:2 (i.e. ~71.8 wt% Al_2O_3) and 1:1 (i.e. ~63 wt% Al_2O_3). They initially prepared mullite by a conventional technique; alumina and silica were slurry mixed in acetone, dried, compacted, and heated to ~1750°C for 30 min to form mullite. The product was ball milled to produce a powder which was then sprayed into an oxyacetylene flame. Particles less than ~19

μm were completely amorphous, while larger particles were partially crystalline. The fraction of the powder which was crystalline increased with particle size. Amorphous powders crystallized to mullite at ~990°C.

Densification

Effect of Powder Characteristics and Green Microstructure on the Sintering of Mullite Powders

In this section, we review the sintering behavior of mullite powders, while reaction sintering is considered in the next section. As in any sintering process, the densification kinetics and microstructure evolution will be largely determined by the powder and green microstructure characteristics. These characteristics will, in turn, be dependent upon a number of factors, including the properties of the mullite precursor, calcination (mullitization) conditions, post-calcination treatments of the powder (e.g. grinding), and the green body forming method.

In the previous section, it was noted that the calcination temperatures required for mullitization range from ~1000°C for molecularly-mixed precursors to >1700°C for alumina and silica powders with micron-scale mixing. High temperatures are required, of course, to attain chemical interdiffusion coefficients (and mass transport rates) large enough to allow mullite formation to occur in a reasonable period of time. However, elevated temperatures also activate other diffusion-dependent phenomena, such as particle growth (i.e. coarsening) and aggregate (hard agglomerate) formation via neck growth (i.e. sintering). Particle growth is indicated by large decreases in powder specific surface area during calcination. For example, Mazdiyasni and Brown[49] prepared a mullite precursor which decreased in specific surface area from 550 m^2/g after vacuum drying (60°C for 16 h) to 280 m^2/g after calcination (600°C for 1 h). Ismail et al.[32] made a similar observation, measuring a decrease in specific surface area from 277 m^2/g for a dried mullite precursor to 1.7 m^2/g after calcination at 1400°C for 1 h. Furthermore, it was evident that aggregates (hard agglomerates) formed during calcination since the average particle size determined by light scattering was 7.5 μm, i.e. much larger than the "equivalent diameter" (~1.1 μm) calculated from the measured surface area.

As a result of particle growth and hard agglomerate formation, the calcined product generally requires extensive grinding to produce mullite powders which sinter to high relative density. Perry[11] carried out ball milling for 1-4 days on calcined powders with overall chemical composition (i.e. Al_2O_3/SiO_2 ratio) in the range ~72-78 wt% Al_2O_3. Dry pressed (138 MPa) samples required sintering temperatures of 1700-1750°C for 3 h in order to achieve relative densities in the range of 95-98%. In most cases, higher sintered densities were obtained with powders milled for 4 days. Metcalf and Sant[30] dry milled calcined powders with stoichiometric composition (~71.8 wt% Al_2O_3) for times up to 144 h and classified the ground powders into different size fractions. Powders were dry pressed at 155 MPa and compacts were subsequently sintered at 1700°C. The sintered density increased as the fraction of powder <4 μm increased. Compacts could be sintered to ~98% relative density in 1 h if all the powder was <4 μm. Sacks and Pask[21] vibratory milled calcined powders (~73 wt% Al_2O_3) in isopropanol using Al_2O_3 grinding media and a teflon-lined container. Powders were isostatically pressed (~170 MPa) to form green compacts with relative densities in the range ~54-59%. Powders milled for 5 h typically had specific surface areas in the range 2-3 m^2/g and temperatures in the range of 1700-1730°C for 12-24 h were required to achieve sintered relative densities of ~95-98%. Longer milling time (36 h) resulted in a higher specific surface area (~6.6 m^2/g) and sintered densities in the range ~95-98% could be obtained in less than 1 h at 1700°C. However, the overall chemical composition was shifted to a significantly higher Al_2O_3/SiO_2 ratio due to extensive contamination from the Al_2O_3 grinding media. Prochazka and Klug[6] wet milled an agglomerated, high specific surface area (~122 m^2/g) powder (overall composition ~76 wt% Al_2O_3) for 16 h using an alumina container and mullite grinding media. The milled powder was isostatically pressed at 800 MPa to ~50% relative green density. Samples had a relative density of ~98% and a grain size <2.0 μm after sintering at 1650°C for 3 h. Ismail et al.[31,32] used attrition milling* for 10 h to prepare a powder (~71.8 wt% Al_2O_3) having a specific surface area of 12 m^2/g and an average particle size of ~1 μm. The powder was isostatically pressed (200 MPa) to a relative green density of 51%. Samples sintered to ≥98% relative density at 1650°C in 1-3 h. Kanzaki et al.[42,43] vibratory milled* (for 50 h) powders (with stoichiometric composition) which were produced by spray pyrolysis and subsequent calcination (at 1000°C for 1 h). The powder specific surface areas were initially in the range ~2-4.5 m^2/g (depending on the spray pyrolysis temperature) and increased upon milling to the range ~12.5-17.5 m^2/g. Electron micrographs indicated that the ground powders still contained agglomerates as large as several microns. Powders were dry pressed uniaxially at 30 MPa and subsequently isostatically pressed at 200 MPa, yielding relative green densities in the range ~46-50%. Samples sintered at 1650°C for 4 h had relative densities in the range ~92-96%. Microstructural observations revealed a duplex grain structure, i.e. a mixture of submicrometer and several micrometer grains.

Sacks and Pask[29] carried out a detailed study on the effect of green microstructural inhomogeneities on the sintering behavior of mullite. They used a gel-derived powder which was extensively aggregated after calcination at 1450°C for 24 h. The initial powder had a specific surface area (determined by nitrogen gas adsorption) of 3.8 m^2/g and an average particle size (determined by an air permeability method) of 5.4 μm (see Table I). The as-calcined powder was vibratory milled for times ranging from 0.5-12 h. Table I shows that the particle size decreased rapidly with grinding, but the specific surface area increased relatively slowly. (This indicates that milling primarily results in the breakdown of porous particles, i.e. hard agglomerates). The as-calcined and milled powders were dry pressed to approximately the same relative density (~54% ± 1%) by varying the compaction pressure (140-310 MPa). The upper left micrograph in Fig. 1 shows that extremely large (>30 μm) aggregates, as well as large interaggregate pores, were present in green compacts formed with the unground (as-calcined) powder. With decreasing aggregate size (i.e. increasing grinding time), the size of the largest interaggregate pores in the green compact also decreases (Fig. 1, upper micrographs). In contrast, the bottom micrographs in Fig. 1 show that the size of the primary particles (i.e. the particles which comprise the hard agglomerates) and the intraaggregate pores are not affected by the milling process. The primary particles have sizes in the range ~0.05-0.5 μm. In most milling processes, communition becomes extremely difficult and inefficient below ~0.5 μm. The lower right micrograph in Fig. 1 shows that the 12 h ground powder still contains hard agglomerates in the several micron size range. While milling clearly results in samples

---

***It is interesting to note that the highest flexural strengths for mullite were reported by Ismail et al.[31,32] and Kanzaki et al.[42,43] Powders were ground extensively in both cases, with the latter workers using Si_3N_4 media and the former group using ZrO_2 media. The question arises as to whether the flexural strengths were affected by the introduction of second phase particulates, i.e. Si_3N_4 or ZrO_2 during grinding. (For example, several researchers[61,62] have shown that ZrO_2 additions increase the strength and toughness of mullite.) Surprisingly, Ismail et al. reported that the ZrO_2 concentration was only 0.02%, while Kanzaki et al. indicated that the highest impurity concentration was 0.22% Ni.**

Table 1. **Effect of Grinding on Powder Characteristics of Mullite**[*]

Grinding Time (h)	Average Particle Diameter (μm)	Specific Surface Area (m^2/g)
0	5.4	3.8
0.5	1.7	4.8
1	1.4	5.1
5	0.8	6.8
12	0.7	7.4

[*]From Ref. 29.

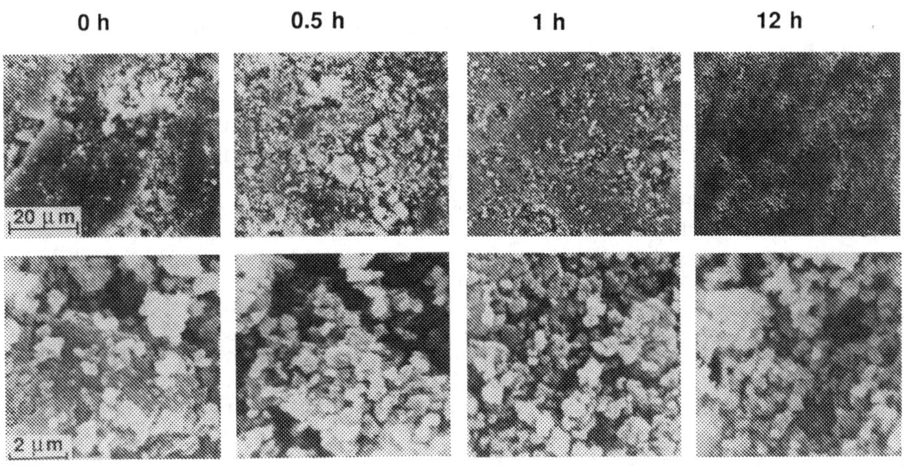

Fig. 1. Top micrographs show pressed surfaces of green compacts formed from powders ground for 0–12 h. Large aggregates break down rapidly on grinding and interaggregate pores decrease in size with increased grinding. Bottom (higher magnification) micrographs show that the "primary" particle size is the same for all grinding times. (From ref. 29.)

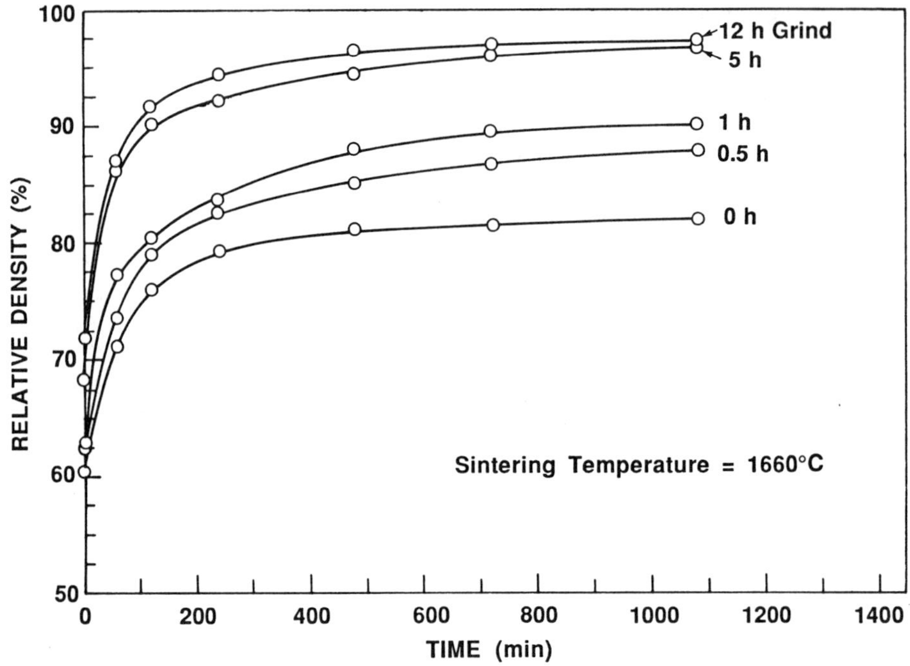

Fig. 2. Plots of relative density vs. sintering time at 1660°C using powders ground 0–12 h. (From ref. 29.)

with more uniform packing of particles, it is evident that long-range (i.e. tens of microns) packing heterogeneities still exist. Some of these heterogeneities, such as those observed in the 12 h ground powder compact at lower magnification (upper right Fig. 1), can also be attributed to the inherent difficulties associated with the dry pressing method. For example, powders tend to fill the die in a highly irregular manner, so packing variations exist even before the powder is compacted. Furthermore, substantial stress gradients arise during the compaction process, resulting in new gradients in packing density.

Although the green compacts in Fig. 1 have the same green density and primary particle size, the differences in packing behavior have a profound effect on the densification kinetics and microstructural evolution. Densification is significantly enhanced with milled powders, as shown by the plots of relative density vs. sintering time at 1660°C in Fig. 2. Micrographs for samples sintered for 18 h at 1660°C (Fig. 3) show that aggregated/agglomerated areas of the compacts sinter to full density (i.e. a pore-free state), whereas interaggregate/interagglomerate regions remain porous. Hence, the "final" density achieved during sintering increases with grinding time, i.e. as the interaggregate/interagglomerate pores decrease in size. With 12 h grinding time, powder compacts could be sintered to ~98% relative density. Nevertheless, residual non-uniformly distributed porosity is still observed in the micrographs (Fig. 3, far right). This porosity can be attributed to the initial heterogeneities in particle packing shown in Fig. 1 (far right micrographs). Detailed kinetic studies have shown that these packing heterogeneities also cause densification substages to occur.[29] The

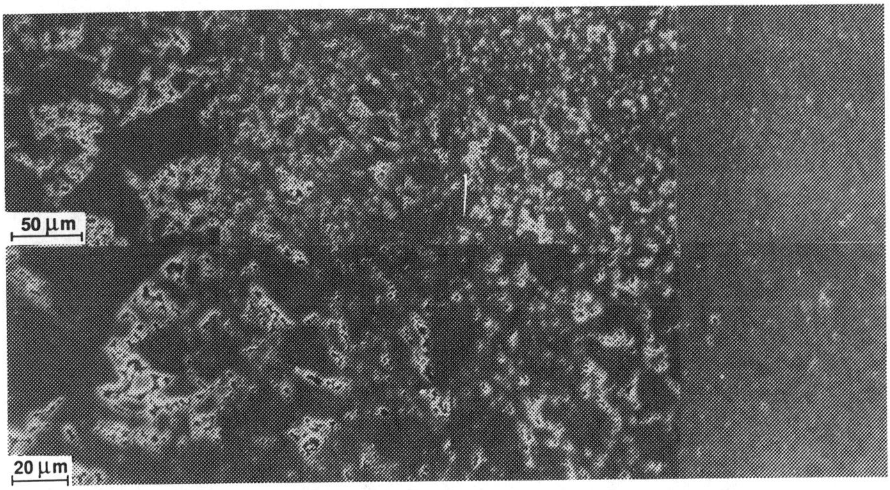

Fig. 3. Micrographs of samples from Fig. 1 after sintering at 1660°C for 18 h (From ref. 29.)

initial stage involves the elimination of very small intraaggregate pores in the partially sintered aggregates. Subsequently, larger intraaggregate pores and small interaggregate pores are eliminated. Finally, densification primarily involves the elimination of the largest interaggregate/interagglomerate pores. The micrograph in Fig. 4 illustrates the microinhomogeneities that exist in a 12 h ground sample sintered (1580°C, 6 h) to ~76% relative density, i.e. prior to the elimination of the inter-aggregate/ interagglomerate pores. The microstructure shows aggregated regions which are already undergoing final-stage sintering (i.e. grain growth has occurred and small pores are entrapped within grains) together with large porous regions (that were interagglomerate regions in the green compact) in which only initial-stage sintering (i.e. neck growth) has occurred.

Mullite is similar to other crystalline ceramics with a high degree of covalent bonding in that relatively high temperatures are required for densification. This can be attributed to the relatively low bulk and grain boundary diffusion coefficients for mullite. In principle, lower sintering temperatures for mullite could be realized by reducing diffusion distances (e.g. by reducing the average particle size). However, despite the fact that a wide range of processing techniques have been used to prepare mullite powders, there is relatively little variation in the time/temperature schedule used to obtain high sintered density (i.e. ≥95%). In the studies discussed above, sintering temperatures and times were all in the range ~1650–1750°C for 1–24 h. This indicates that the diffusion distances necessary for densification (i.e. pore elimination) did not vary substantially in these studies. This, in turn, suggests that the calcination and grinding steps tended to produce powders with a similar "effective" particle size (i.e. on the order of one micron).

It is clear from our previous discussion that the calcination temperatures used to prepare mullite powders can be reduced substantially by using techniques which provide a more intimate mixing of alumina and silica. Furthermore, mullite powders

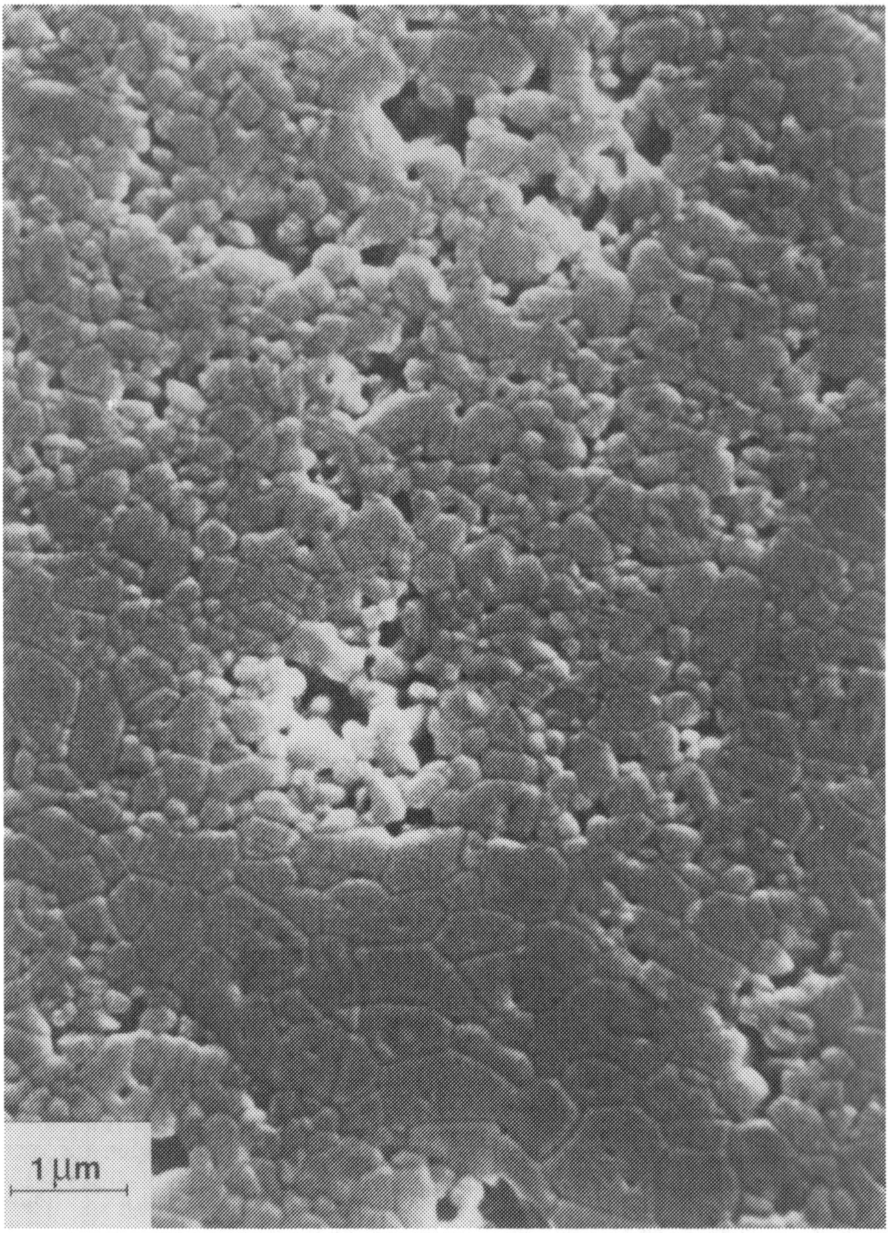

Fig. 4. Micrograph of sample sintered for 6 h at 1580°C which illustrates microinhomogeneity of gel-derived materials despite 12 h grinding of powders. (From ref. 29.)

with small primary particle size can be readily prepared due to the small size of the precipitated precursors and the low reaction temperatures. However, the lower calcination temperatures for powder preparation do not necessarily result in lower temperatures for sintering of bulk compacts. Just as the diffusion distances for mullitization are reduced by more intimate mixing of components, the diffusion lengths for neck growth processes (i.e. for aggregate formation) are shortened when the primary particles are small. Thus, aggregate formation is often as serious a concern in mullite powders formed at low temperatures using precursors prepared by solution or colloidal techniques as it is in powders prepared at high temperatures using mixtures prepared by conventional techniques. (As noted in Fig. 1, powders prepared from a colloidal technique can result in aggregates >30μm. Clearly, the "effective" particle size for bulk compact sintering is much larger than the primary particle size.) Therefore, it is important to use the minimum temperature and time necessary to complete mullitization,† since higher temperatures and longer times only contribute to aggregate formation and particle coarsening. Although this strategy appears to be followed by many investigators, some amount of aggregation appears to be unavoidable. Thus, procedures designed to eliminate and/or reduce the size of aggregates are required.

As discussed earlier, milling is used extensively to break down hard agglomerates. Aside from the fact that grinding processes are energy intensive, most milling methods are limited in that average particle sizes below ~0.5–1.0 μm are difficult to achieve. (Furthermore, prolonged milling may introduce high concentrations of undesirable impurities.) However, milling can be combined effectively with classification techniques to prepare powders with smaller average particle size. A fine-sized fraction of a milled powder can be separated by centrifugal or gravity sedimentation in a fluid. For example, Fig. 5 shows particle size distributions (determined by X-ray sedimentation) for the 12 h ground mullite powder shown in Fig. 1 and a fine fraction of the same powder in which larger particles were removed by the gravity sedimentation. (It should be noted that the average primary particle size is not changed much in this case, since sedimentation mostly results in the removal of coarse aggregates.) As discussed below, the fine powder shows significantly enhanced densification kinetics.

In addition to reducing the average particle size, sintering can also be enhanced by improving the packing characteristics of the green compact, i.e. increasing the green density, decreasing the average pore size, and eliminating large defects. We noted earlier that dry pressed samples tend to have relatively large inhomogeneities as a result of the initial packing irregularities created during die filling and the nonuniform stress gradients generated during compaction. In contrast, a number of recent studies[63–65] have shown that suspension processing techniques can be used to prepare homogeneous, high relative density green bodies. This was demonstrated with mullite by using the classified (sedimented) powder (12 h ground, fine fraction) from Fig. 5. The powder was dispersed in water at pH≈3.5 and subsequently consolidated by slip casting. Figure 6 shows the pore size distribution obtained by mercury porosimetry on the green compact. The green density is ~66% and the median pore radius is ~20 nm. This is contrasted to a dry pressed compact which has a green density of ~53% and a median pore radius ~58 nm. The larger pore sizes for

†The discussion in this section is based on the assumption that mullite powders are desired. In the next section, the advantages of sintering mullite precursors (e.g. reaction sintering) are discussed.

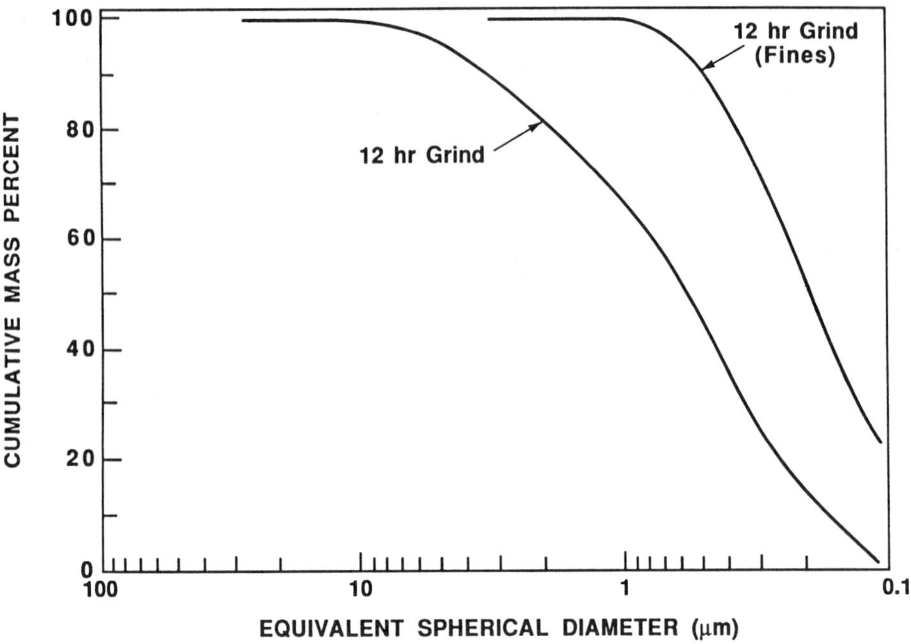

Fig. 5. Particle size distributions for the 12 h ground mullite powder shown in Fig. 1 and a fine fraction of the same powder in which larger particles were eliminated by gravity sedimentation.

the dry pressed sample result not only from the larger particle sizes (Fig. 5), but also from the packing inhomogeneities introduced during the pressing operation.

The differences in green microstructure for the dry pressed and slip cast samples result in substantial differences in the sintering kinetics. Figure 7 shows plots of relative density vs. sintering time at 1540°C. The slip cast sample sinters to ~93% relative density in 15 min and a relative density ≥99% is obtained with long sintering times. In contrast, the dry pressed sample remains below 75% relative density after long sintering times. The enhanced densification of the slip cast sample results from the smaller "effective" particle size of the sedimented powder and the more homogeneous, higher relative density green microstructure. Figure 8 shows an electron micrograph of the slip cast sample sintered to ~99% relative density (3.17 g/cm^3). The microstructure consists mainly of small, equiaxed grains with ~1.1 μm average size. However, a small percentage of extremely large elongated grains and rectangular-shaped grains (not shown in Fig. 8) were also observed. These large grains may be the reason for the slow densification rate during final stage sintering. The reason for their growth is not clear, but it may indicate that small amounts of glass phase are present at some grain boundaries.

Sintering of Mullite Precursors and Reaction Sintering

There have been many studies in which mullite bodies are formed by "reaction sintering." A wide variety of materials systems have been used, ranging from

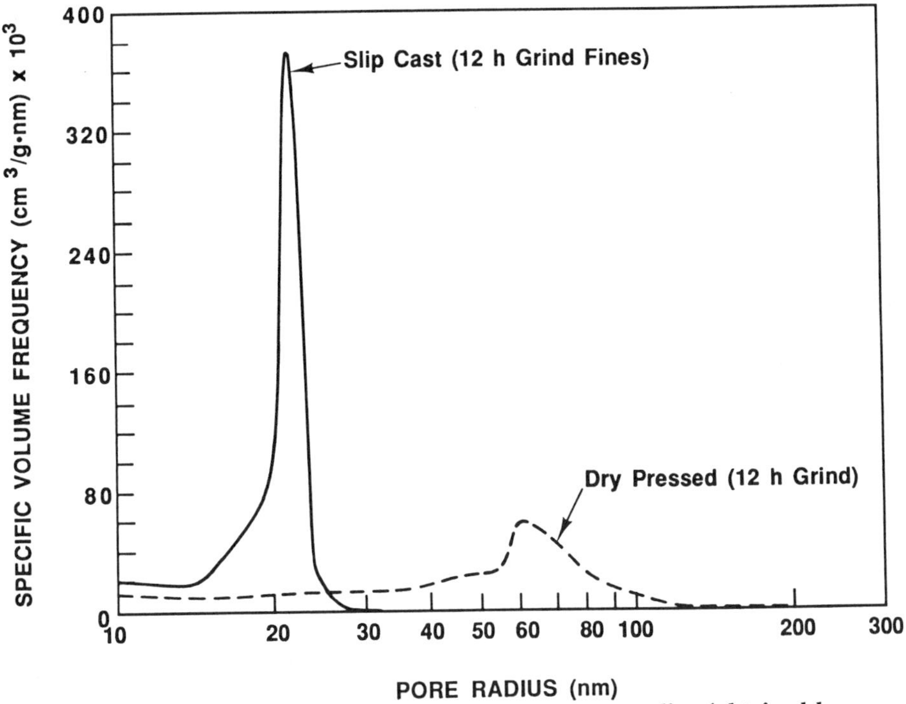

Fig. 6. Plots of specific pore volume frequency vs. pore radius (obtained by mercury porosimetry) for a dry pressed sample prepared with the 12 h ground powder and for a slip cast sample prepared with the fine fraction of the 12 h ground powder.

clay/alumina powder compacts to gel monoliths formed from organometallic compounds. In some cases, the process involves sintering a green compact formed from a mixture of two distinct powders—one alumina-rich and the other silica-rich. In other cases, green bodies are formed from a single powder (or gel) which is a mullite precursor, e.g. a material prepared by a colloidal or solution technique which is either uncalcined or calcined at a premullitization temperature. In the first process, powders must clearly undergo chemical interdiffusion and reaction to form mullite. The second method may also involve interdiffusion and reaction between aluminum- and silicon-containing phases. However, if the precursor is molecularly mixed, only decomposition reactions (e.g. dehydroxylation) and phase changes (e.g. crystallization) occur.

Mullite bodies have been prepared by reaction sintering of mixtures of clay with an aluminum-rich powder, e.g. bauxite,[12,13] α-Al_2O_3,[12,16,17] aluminum fluoride,[15] etc. The powders used in these processes are relatively coarse, so mixing is no better than micron-scale. Hence, high temperatures are required for reaction as well as densification. Hawkes[12] reported that clay/α-Al_2O_3 and clay/bauxites bodies sintered in the range ~1720–1790°C still contained small amounts of α-Al_2O_3 and silicate glass. Sacks[17] sintered clay/α-Al_2O_3 compacts (with overall composition of stoichiometric mullite) to ~94% relative density at ~1725°C. Reaction was not

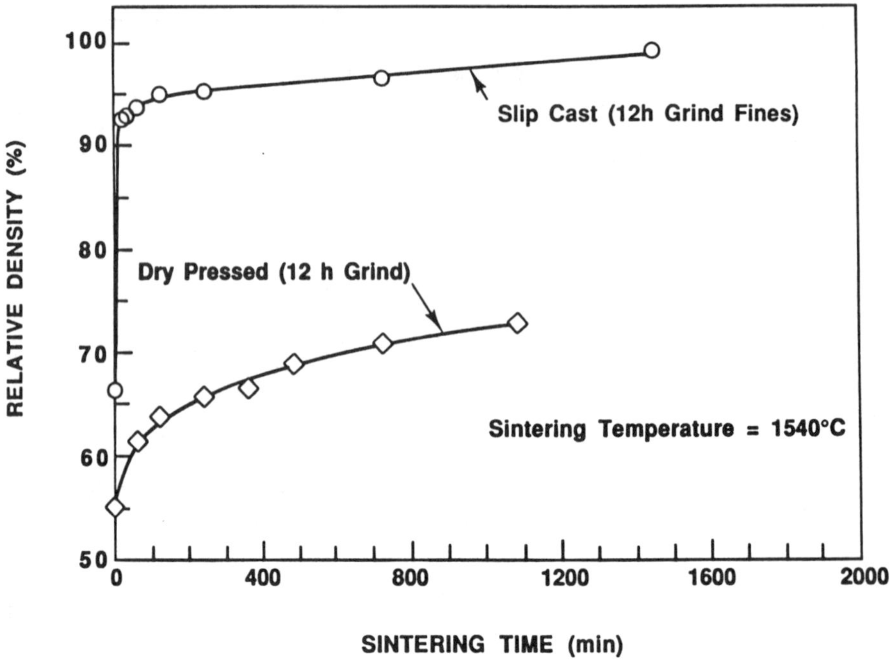

Fig. 7. Plots of relative density vs. sintering time at 1540°C for the compacts characterized in Fig. 6.

complete, however, as α-Al_2O_3 was detected by XRD and silicate glass was observed in the sintered microstructure. It was also noted that complete mullitization may have been precluded by the presence of several percent of glass-promoting impurities in the clay (i.e. Na_2O, MgO, CaO, TiO_2, Fe_2O_3). Moya et al.[16] sintered mixtures of α-Al_2O_3 and calcined halloysite with the overall composition of stoichiometric mullite. (The halloysite was initially heat treated at 1000°C to remove structural water.) The reaction to form mullite was not complete at 1600°C and the relative density was only ~77% after sintering 2 h at 1700°C in a helium atmosphere. Moya et al.[16] also sintered "premullite" powders which were prepared by high pH dissolution of the excess silica in calcined halloysite. Powders had the approximate composition of stoichiometric mullite, but also contained ~0.5% cation impurities (e.g. Fe_2O_3, CaO, Na_2O, K_2O). Dry pressed compacts (~45% relative density) were sintered in a helium atmosphere for 2 h at temperatures in the range 1350–1700°C. Sintered densities were ~93% and ~95% at 1600 and 1700°C, respectively, but microstructure observations showed residual pockets of glassy phase.

To avoid the relatively large amount of impurities introduced by using clay minerals, reaction sintering has also been carried out with mixtures of aluminum oxides/hydroxides and silicon oxides/hydroxides. Fenstermacher and Hummel[20] wet ball milled for 48 h a mixture of α-Al_2O_3 and quartz powders. Semi-dry pressed samples with the composition of stoichiometric mullite were sintered at 1710°C in a gas-fired kiln. The sintered density was ~87% and a small amount of α-Al_2O_3 was detected in the microstructure. Sacks[66] ball milled a mixture of α-Al_2O_3 and quartz powders (overall composition ~73 wt% Al_2O_3). Samples were dry pressed and sintered

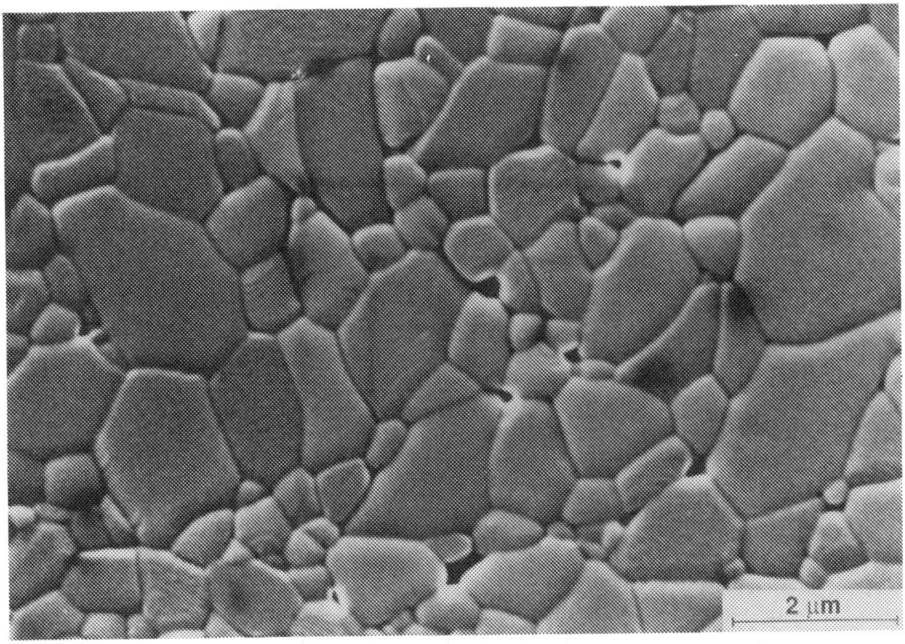

Fig. 8. Micrograph of slip cast mullite after sintering to ~99% relative density.

at 1700°C for 12 h. Complete reaction was indicated by the absence of α-Al_2O_3 in XRD patterns, but the sintered density was only ~70%. This was compared to the sintering behavior of powders which were prepared by calcining the original Al_2O_3/SiO_2 powder mixture and subsequently grinding the calcined product to approximately the same average particle size as the starting mixture. Samples were dry pressed and sintered under the same conditions as before. Calcination time/temperature schedules resulting in complete mullitization (i.e. no α-Al_2O_3 detected by XRD) produced sintered densities of ~90%. Sintered densities were lower when the calcination schedule resulted in incomplete reaction.

Rodrigo and Boch[25,26] attrition milled mixtures of a fine α-Al_2O_3 (~0.6 μm average size) and an amorphous SiO_2 (~3.6 μm average size). Powders milled for 2 h and 5 h had average sizes of ~0.6 and ~0.45 μm, respectively. Compacts with ~58% relative density were prepared by cold pressing (at 150 MPa) with ultrasonic vibration. Samples with stoichiometric composition (~71.8 wt% Al_2O_3) showed no α-Al_2O_3 in XRD after sintering for 2 h at 1600°C. As expected, enhanced shrinkage rates were observed with longer milling times. Compacts formed with 3 h ground powders could be sintered at 1600°C to relative densities of ~93% in 2 h and to ~97% in 10 h. Rodrigo and Boch also made some interesting observations regarding the effects of crystallization on reaction sintering of Al_2O_3/SiO_2 mixtures. They carried out dilatometric measurements on samples heated at 300°C/h. A sharp decrease in shrinkage rate was observed in the range ~1250–1300°C and XRD measurements showed that this coincided with extensive formation of cristobalite (i.e. crystallization of the amorphous silica). At slightly higher temperatures (~1350–1400°C), an increase in shrinkage rate was observed. The authors attributed this behavior to the formation of a metastable liquid phase, in accordance with the metastable phase diagram proposed by Davis and Pask.[67] Pask and co-workers[23,24] had previously made similar observations in their studies of the initial-stage sintering of α-Al_2O_3/amorphous silica and α-Al_2O_3/cristobalite compacts. With continued heating above 1400°C, Rodrigo and Boch observed the initial formation of mullite and this resulted in a complete cessation of shrinkage. (In fact, a slight expansion occurred between ~1400–1550°C due to the decrease in true density as mullite formed.) No further shrinkage was observed until temperatures were above 1550°C. Pask and co-workers[23,24] had also observed that mullite formation severely retards shrinkage in Al_2O_3/SiO_2 compacts.

A number of studies have shown that densification can be enhanced by using amorphous aluminosilicate powders. The most direct evidence was provided by Gani and McPherson.[60] They hot pressed (at 21 MPa) large (~13–19 μm) aluminosilicate glass powders (~71.8 wt% Al_2O_3) and monitored the shrinkage during heating (at 50°C/min). Shrinkage commenced at ~920°C and ended abruptly (after ~9% linear shrinkage) at ~1010°C. The temperature for the cessation of shrinkage coincided precisely with the crystallization (i.e. to mullite) temperature of the glass particles. It is interesting to note that this powder showed no further shrinkage at temperatures up to 1450°C, while a similar powder with the overall composition of ~63 wt% Al_2O_3/37 wt% SiO_2 showed some densification above 1400°C. According to the phase equilibrium diagram,[68] the latter composition results in a two phase mixture after crystallization (i.e. mullite and siliceous phase) and, therefore, liquid-phase (i.e. viscous) densification could occur. The mullite bodies produced by hot pressing the powder with stoichiometric composition were still highly porous (i.e. the open porosity was 33%). Gani and McPherson suggested that high density mullite could be obtained by using higher pressures during heating and/or isothermal holds below the crystallization temperature, but no experiments were carried out.

In studying the effects of powder calcination temperature on subsequent sintering behavior, Hirata et al.[50] also presented evidence that amorphous powders show enhanced densification. Experiments were carried out using powders which were precipitated (at pH=8.2) by the solution technique (method I) described earlier. The powders were initially amorphous and remained so until ~900°C. Powders calcined in the range ~900–1200°C contained a weakly crystalline Al-Si spinel phase. The initial mullite formation occurred at 1200°C, but sharp XRD peaks were not observed until 1300°C. Since the overall composition was ~76 wt% Al_2O_3, free alumina (theta or alpha) was also observed at higher temperatures. Calcined powders were isopressed (3000 kg/cm^2) and subsequently sintered at 1550°C for 2 h. Figure 9 shows the effect of calcination temperature on the green and sintered densities of powder compacts, as well as the weight loss of the samples after sintering. Powders calcined between 700°C–1100°C (i.e. amorphous or weakly crystalline powders) showed the highest sintered densities (~95–96%). A sharp decrease in relative density (to ~84%) was observed with powders calcined at 1200°C, i.e. powders which contained mullite. The sintered density continued to decrease slightly (to ~81%) with powders calcined at higher temperature, despite an increase in the compact green density. This probably reflected the detrimental effects of particle growth (coarsening) and aggregate formation, as well as additional mullitization.

Enhanced densification with amorphous (or weakly crystalline) powders is also illustrated by comparing the sintering behavior that Hirata et al.[50,51] obtained with powders prepared by methods I and II (described earlier in Solution Techniques). In one set of experiments, both powders were calcined at 1000°C for 3 h and similar average particle sizes were obtained. However, powder II showed substantial mullitization after calcination, while powder I showed no mullite formation. Table

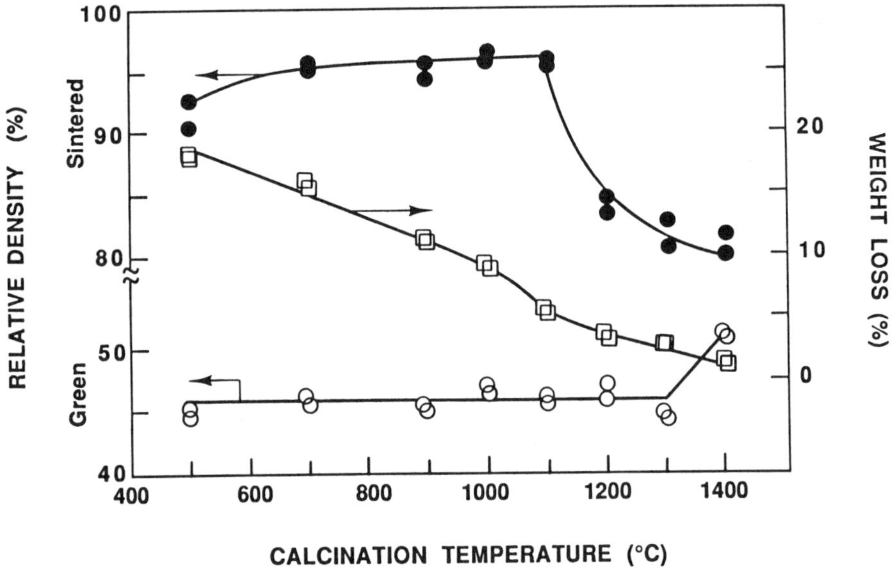

Fig. 9. Plots showing the effect of powder calcination temperature on the relative density of green compacts, the relative density after sintering (1550°C, 2 h), and the weight loss after sintering. (Redrawn from ref. 50.)

Table 2. Effect of Mullite Formation on Sintering Behavior*

Powder	Sintering Conditions	Relative Density (%)
I	1500°C, 2 h	90.1
II	1500°C, 3h	70.8
I	1550°C, 2	95.9
II	1550°C, 3 h	74.6
I	1600°C, 2h	99.0
II	1600°C, 3 h	97.5

*Both powders were calcined to 1000°C for 3 h. XRD showed that mullite had formed in powder II, but not in powder I. The results are from the study by Hirata et al.[50,51]

2 shows the sintered densities obtained at various temperatures (1500–1600°C) and times (2–3 h). It is evident that compacts prepared with powder II, which forms mullite during calcination, have a substantially lower densification rate.

One of the concerns that arises when sintering compacts prepared with mullite precursors is that relatively large weight losses may occur. This is illustrated by the results of Hirata et al. in Fig. 9. Although powders calcined in the range ~700–1100°C did sinter to high relative density, significant weight losses (i.e. ~5–16%) were observed due to dehydroxylation and the removal of residual organic groups. When such large weight losses occur, the sintering time/temperature schedule is often restricted to lower heating rates in order to avoid the development of high internal gas pressures during volatile removal. (If volatiles are removed too quickly, the resulting gas pressures are sufficient to cause catastrophic fracture of the samples.) Large weight losses during heating also tend to increase the pore volume of the compact and this may adversely affect densification. This is suggested in Fig. 9 by the decrease in sintered relative density (to ~91%) for the sample prepared with the 500°C calcined powder. Samples undergoing large weight losses and large shrinkages may also present difficulties in achieving the desired dimensional tolerances of the sintered bodies. Therefore, although pre-mullitized powders provide enhanced densification rates, it is still important to utilize sufficiently high calcination temperatures to avoid deleterious effects associated with large weight losses.

It is also noted that dilatometric effects arise during the sintering of mullite precursors and in reaction sintering due to changes in the true densities of the materials, i.e. due to chemical reactions, crystallization, polymorphic transformations, etc. This is another potential problem in that these volume changes may result in the formation of microcracks or other flaws in the compact. These defects may be difficult to eliminate during subsequent densification stages.

In all of the studies discussed in this section and in the preceding section, samples were shape-formed either by dry compaction (i.e. dry pressing, isopressing) or by slip casting. Another shape-forming method is the sol-gel process. A precursor sol is cast into a mold of the desired shape and subsequently gelled, e.g. by liquid evaporation, chemical additions, etc. The process is essentially the same as some of the colloidal or solution techniques discussed earlier for powder preparation. The main difference is in the drying behavior. In preparing powders, the gel is allowed to break up into many fragments during drying, while it is obvious that the intention in shape-forming processes is to retain a monolithic (bulk) object for subsequent sintering. The ability of a gel to remain monolithic during drying depends on a variety of factors, such as the initial reactant concentrations, aging conditions, etc. Detailed studies have been carried out for silica sol-gel materials.[69–73]

A number of researchers have formed and sintered aluminosilicate gel monoliths. Yoldas[53] prepared gels using organometallic precursors. An aluminum monohydroxide sol was initially formed by hydrolysis/condensation of aluminum sec-butoxide. TEOS was added to the sol to form gels with a wide range of Al_2O_3/SiO_2 ratios, but the reaction and gelation conditions were not specified. Wet gels were dried at room temperature and subsequently heated to 400–500°C to remove some of the residual hydroxyl and organic groups. Although Yoldas indicated that an oxide network with extensive -Si-O-Al- bonding forms in the gelled material, it seems unlikely since the processing method involved consecutive precipitation. This is supported by Yoldas' XRD results on 63 wt% Al_2O_3/37 wt% SiO_2 samples which show that mullite does not form until temperatures above 1200°C. (Furthermore, samples heated to 1200°C contained α-Al_2O_3.) Samples with various Al_2O_3/SiO_2 compositions were subsequently sintered at temperatures up to 1400°C. Microstructure observations indicated that much of the porosity was eliminated for 65 and 80 wt% Al_2O_3 samples, although bulk densities were not reported.

Roy et al.[74,75] sintered gels (~71.8 and ~80 wt % Al_2O_3) which were prepared by the colloidal and solution techniques discussed earlier.[33,34] Relative densities in the range ~92–99% were reported for "diphasic" gel pellets sintered at 1200 and 1300°C (100 min). However, for several reasons, these values must be viewed with caution. First, relative densities were calculated from apparent densities,[74] instead of bulk densities. Second, the relative density calculations were based on the true density of mullite,[75] yet the extent of mullitization in sintered samples was not reported. (Earlier work by the authors[34] indicates that complete mullitization requires higher temperatures than used in sintering.) Nevertheless, substantial densification apparently occurred at quite low temperatures. The authors also observed that "diphasic" gels (i.e. gels prepared using colloidal particles) showed enhanced densification compared to "single-phase" gels (i.e. gels prepared by solution/co-precipitation techniques). This was attributed to the "higher metastable energy" of the former material, but the effect was probably due to the fact that mullite formed at much lower temperatures (~1000°C) in the "single-phase" gel.[34]

Shinohara et al.[7] formed monolithic gels using a similar procedure to Yoldas[53] and Roy et al.[74,75] A colloidal boehmite (i.e. aluminum monohydroxide) suspension was mixed with a partially hydrolyzed TEOS solution at pH=2.5 for ~4 days. The mixture was then heated at 60°C and cast into a plastic mold just prior to gelation. Samples were dried at room temperature for several days and subsequently sintered at temperatures in the range 1100–1600°C. Samples with stoichiometric composition (~71.8 wt% Al_2O_3) sintered to ~94% relative density after 1 h at 1250°C, but no additional densification was observed with longer sintering times (up to 8 h). The cessation of densification coincided with the formation of mullite. XRD showed that a sample sintered at 1200°C for 2 h was primarily Θ-Al_2O_3, with small amounts of mullite and cristobalite. In contrast, only mullite was observed in a sample sintered at 1250°C for 2 h. These results clearly illustrate the detrimental effect of mullitization on the densification process. Even after sintering at 1600°C, the relative density increased only to ~97%.

Sonuparlak[8] also used colloidal boehmite suspensions and TEOS to prepare mullite by the sol-gel process. Boehmite powders were initially dispersed in water under acidic conditions (pH≤2). TEOS was then added to the sol in the amount necessary to obtain stoichiometric mullite. The sol was mixed at 25°C in air until gellation occurred. In order to prevent cracking, gel monoliths were air dried for more than two months at room temperature in covered molds with pinhole vents. Dried samples sintered to ~96% relative density at 1250°C (1 h), but no additional densification occurred at temperatures up to 1700°C (1 h). XRD showed that extensive mullite formation occurred in sintered (1250°C, 1 h) samples which were heat treated at 1300°C (1 h).

Sonuparlak[8] modified the above process in order to avoid the long drying times required to form crack-free monoliths. After mixing the boehmite/TEOS sol for ~1 day, the solution pH was adjusted to ~6–7 with ammonium hydroxide additions. This resulted in de-stabilization of the sol, as large flocs were observed to form immediately. After air drying at 25°C, the gel was uniaxially compacted at pressures up to 441 MPa. The sample relative density after sintering (at 1250°C, 1 h) increased with increasing compaction pressure, reaching a maximum value of ~97%. XRD showed that extensive mullitization occurred after further heat treatment at 1300°C (1 h).

Rahaman et al.[54] prepared gels by hydrolysis/condensation of aluminum sec-butoxide and TEOS. Wet gel discs were supercritically dried in an autoclave at 40°C and 10 MPa pressure, followed by conventional drying at 150°C (24 h). Sintering experiments were carried out on dried gel discs and on gel pellets formed by uniaxial compaction (20 MPa) of the dried gels. The former samples were extremely porous (~0.13 g/cm^3) and the relative density after sintering at 1350°C for 1 h was only 51%. In contrast, the gel pellets (with ~45% relative green density) could be sintered to ~99% relative density at 1200°C for 1 h. The low temperature for densification is again associated with the absence of mullite formation during sintering. XRD on gels heated to 1000°C (1 h) showed a weakly crystalline phase, while samples heated for 1 h at 1350°C (i.e. above the sintering temperature) showed extensive mullite formation.

It is apparent from the studies of Yoldas,[53] Roy et al.,[74,75] Shinohara et al.,[7] Sonuparlak,[8] and Rahaman et al.[54] that the formation of bulk mullite bodies at low sintering temperatures requires an intermediate scale of mixing of the reactants. If the aluminum- and silicon-bearing species are mixed on the molecular-level, then mullitization will occur at low temperatures (~1000°C) and subsequent densification becomes difficult. In contrast, if the level of mixing is too coarse (i.e. micrometer scale), then both mullitization and densification require high temperatures. Mixing on the scale of ~5–50 nm increases the temperature required for extensive mullitization (i.e. to ~1200–1400°C), while amorphous (or weakly crystalline) particles (i.e. alumina, silica, and/or aluminosilicate) in this size range can sinter at slightly lower temperatures.

Although the gel approach does allow lower sintering temperatures, there are obviously several difficulties associated with the method. "Wet" gels undergo large shrinkages during drying and substantial stress gradients can arise. Gel monoliths tend to warp and/or crack unless drying is carried out at slow rates or by using special conditions such as supercritical drying. To some extent, these difficulties can be overcome by compaction of gels dried at low temperatures, as demonstrated by Roy et al.,[74,75] Sonuparlak,[8] and Rahaman et al.[54] Although dry pressing operations generally produce compacts with relatively large interagglomerate pores, gels dried at lower temperatures are sufficiently deformable to produce compacts with fine pore size and relatively homogeneous structure.[8] Nevertheless, these samples still tend to have low green density, so shrinkage during sintering is large. Furthermore, large weight losses may occur during sintering due to dehydroxylation and the elimination of residual organic groups. For example, Rahaman et al.[46] observed a 13% weight loss during sintering, even though gels were calcined at 800°C for 4 h prior to sintering. The large shrinkages of gel monoliths make it difficult to achieve precise dimensional tolerances on fabricated parts. At present, the gel approach may be best suited to applications for films and coatings, i.e. two-dimensional geometries in which the difficulties associated with drying and removal of volatiles at high temperatures are less severe. However, with advances in the practical[76] and theoretical aspects[77] of drying control, the sol-gel route to fabrication of bulk mullite bodies may become a viable technology.

Effect of Chemical Composition (Al_2O_3/SiO_2 Ratio) on Sintering Behavior

Densification behavior and microstructure evolution are highly dependent on the chemical composition, i.e. the Al_2O_3/SiO_2 ratio. Sacks and Pask[21,22] studied the sintering behavior of compositions in the range 60–90 wt% Al_2O_3. Powder mixtures of Al_2O_3 and SiO_2 were calcined at 1700°C (8 h) to form mullite, mullite and aluminosilicate glass, or mullite and Al_2O_3 (i.e. depending on the overall Al_2O_3/SiO_2 ratio). The calcined materials were crushed and milled to produce powders with approximately the same particle size characteristics for each composition (~2 μm size as determined by an air permeability method and ~2 m^2/g surface area as determined by a gas adsorption method). Powders were pressed to approximately the same green relative density (~57 ± 1%) and compacts were sintered at 1700°C for 1, 4, and 12 h.

Figure 10 shows plots of the sintered relative density as a function of the overall Al_2O_3/SiO_2 ratio. Samples with 60 and 65 wt% Al_2O_3 sinter relatively quickly to their maximum density. The overall compositions for these samples fall in the mullite/liquid (two phase) region of the Al_2O_3-SiO_2 equilibrium phase diagram. Thus, the rapid densification results from liquid phase sintering. Figure 11 shows typical glass/mullite microstructures for these samples.

Slower densification rates were observed for samples with compositions in the range 71.8–74 wt% Al_2O_3. This can be attributed to a much smaller amount of liquid phase in these samples. It should be noted that several phase equilibria studies[68,78] indicate that the 71.8–74 wt% Al_2O_3 compositions lie in the mullite solid solution range and, hence, the slower densification rates might be attributed to the *absence* of liquid phase. However, microstructure studies show that the samples still contain small amounts of glassy phase. This is illustrated in Fig. 12A which shows a scanning electron micrograph for a 71.8 wt% Al_2O_3 sample sintered at 1700°C for 8 h. With longer sintering times (100 h), extensive grain growth occurs and the glass phase coalesces into large "pockets" (Fig. 12B). The amount of glass phase decreases as the overall Al_2O_3 composition decreases. For samples with 73 wt% Al_2O_3, second phase regions at grain corners could only be detected by transmission electron microscopy (Fig. 12C). A variety of direct and indirect evidence confirm that the second phase observed in samples with 71.8–74 wt% Al_2O_3 is a silicate glass: (1) Noncrystallinity was indicated from electron diffraction patterns on areas such as shown in Fig. 12C. (2) Microstructures showed elongated grains and rectangular-shaped grains that form unusual dihedral angles. With long sintering times, a small percentage of grains grew to extremely large sizes (>50 μm in length). The presence of elongated grains in aluminosilicate microstructures has generally been associated with the presence of liquid phase.[79] (3) The densification rate sharply decreases as the composition shifts from 74 to 75 wt% Al_2O_3 (Fig. 10). Glassy "pockets" were not detected in samples with 75 wt% Al_2O_3. The only second phase detected was a small amount of α-Al_2O_3 particles. The grain structure for this sample was equiaxed. (4) High temperature mechanical properties were dependent on the composition.[22,66] Compression strengths

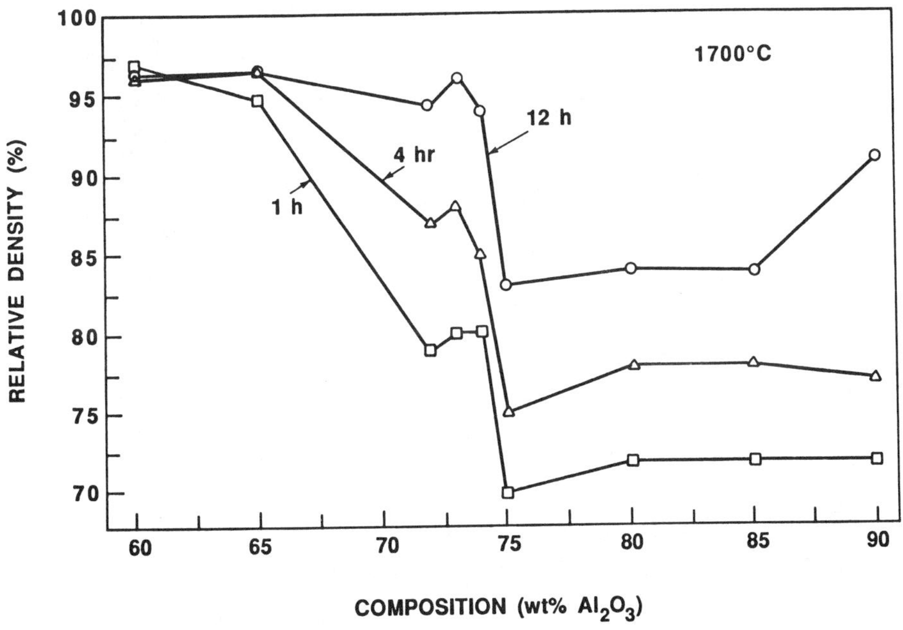

Fig. 10. Plot of relative density vs. overall chemical composition (wt% Al_2O_3) for samples sintered at 1700°C for 1, 4, and 12 h. (From ref. 22.)

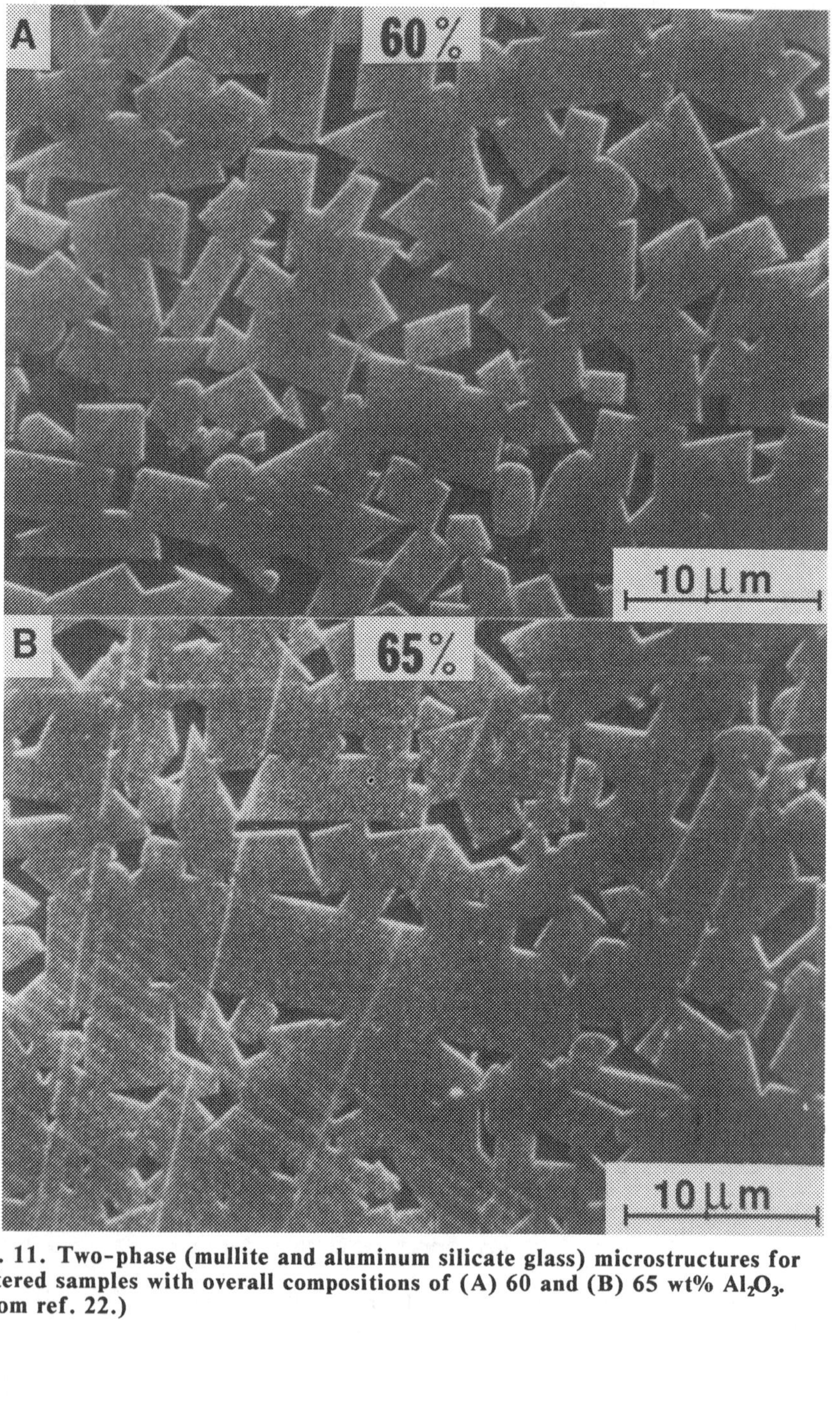

Fig. 11. Two-phase (mullite and aluminum silicate glass) microstructures for sintered samples with overall compositions of (A) 60 and (B) 65 wt% Al_2O_3. (From ref. 22.)

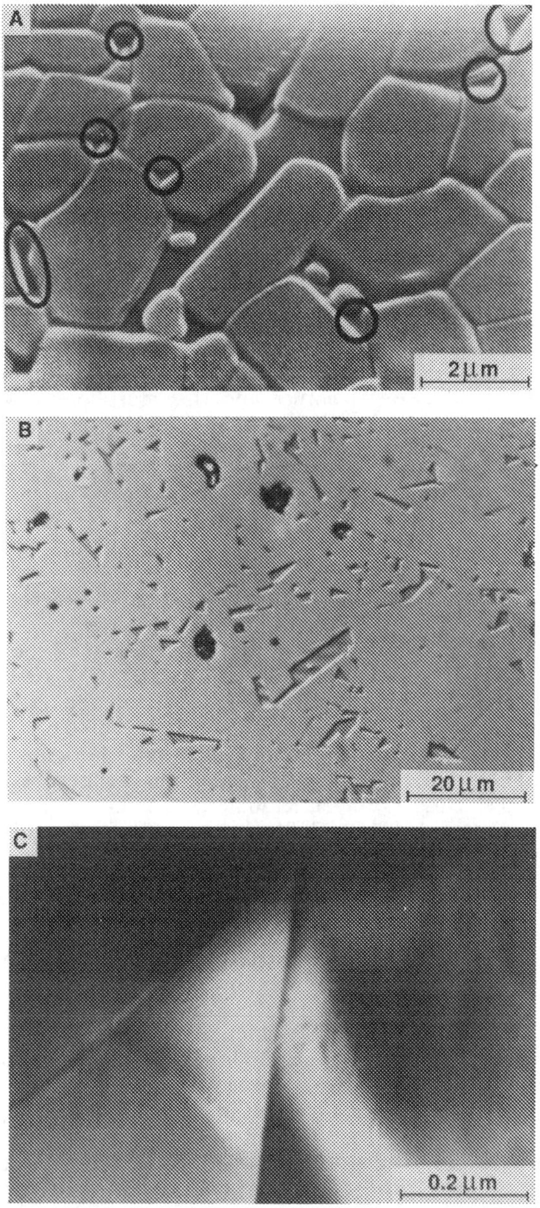

Fig. 12(A). Scanning electron micrograph of 71.8 wt% Al_2O_3 sample, sintered at 1700°C for 8 h, showing small glassy areas (circled) and occasional larger glassy regions (center). (From ref. 22.) (B): Optical micrograph of 71.8 wt% Al_2O_3 sample, sintered for 100 h at 1700°C, showing coalescence of glass phase into large "pockets." (From ref. 22.) (C): Transmission electron micrograph of a 73 wt% Al_2O_3 sample, sintered at 1700°C for 12 h, showing a small amount of glass phase. (From ref. 22.)

at 1200 and 1400°C increased as the overall composition shifted from 71.8 to 74 wt% Al_2O_3, i.e. as the amount of glass phase decreased. (5) The second phase "pockets" in Fig. 12B were readily etched with a hydrofluoric acid solution. It is well known that silicate glass is much more susceptible to HF etching than mullite.[17,80]

A number of other researchers have also shown that sintered samples with the stoichiometric composition (~71.8 wt% Al_2O_3) contain small amounts of glass phase. Prochazka and Klug[6] hot pressed samples with ~72 wt% Al_2O_3 at 1630°C. Microstructure studies showed occasional elongated grains (several micrometers in length) and small amounts of second phase at some grain corners. At higher hot pressing temperatures (~1710°C), elongated grains grew extensively in size and number and large glass pockets (~1–2 μm) could be observed. Quantitative microscopy measurements indicated that the volume fraction glass was ~2%. Samples hot pressed at 1630°C were subsequently annealed at 1780°C and the glass pockets grew to ~2–6 μm. The volume fraction also increased to ~6%. In contrast to these results, glass phase was not detected in samples with ~76 wt% Al_2O_3. Samples hot pressed at 1680°C had an equiaxed grain structure and this morphology was retained after extensive grain growth during annealing at 1800°C. Also consistent with the absence of glassy phase was the observation that higher hot pressing temperatures were required to achieve near theoretical density for ~76 wt% Al_2O_3 samples. Microstructure observations on hot pressed samples also showed a small amount of Al_2O_3 particles. These particles disappeared upon annealing at 1800–1850°C and samples became translucent. Samples with 75 wt% Al_2O_3 could be sintered to translucency at 1845–1860°C (1 h).[59]

Prochazka et al.[6,59] were able to account for the above observations by proposing that the mullite solid solution range shifts to compositions with higher Al_2O_3 content as the temperature increases. Their study indicated that the composition for stoichiometric mullite (~71.8 wt% Al_2O_3) is very close to the phase boundary between the mullite solid solution region and the mullite/liquid region at ~1600°C. With increasing temperature, the stoichiometric composition would clearly fall in the two-phase region and, hence, stoichiometric mullite would decompose to a mixture of liquid and a mullite with higher Al_2O_3 content. In most cases, residual glassy phase is expected in samples close to the stoichiometric composition because sintering is usually carried out above 1600°C. This is consistent with results in a variety of studies. For example, Perry[11] sintered compositions in the range ~72–78 wt% Al_2O_3 at ~1650–1750°C and found that relative densities were generally lower in samples with higher Al_2O_3 content (i.e. samples with less likelihood of residual glass phase). In addition, microstructure observations revealed elongated grains in lower Al_2O_3 content samples and equiaxed grains in higher Al_2O_3 content samples. Similar results were obtained by Metcalf and Sant for compositions in the range ~72–78 wt% Al_2O_3. Ismail et al.[31] reported that the bulk density for an ~72.5 wt% Al_2O_3 sample decreased slightly as the sintering time at 1650°C increased from 1.5 to 3 h. This can be explained by an increase in the amount of (lower-density) glass phase with longer sintering time. These researchers also observed the development of elongated grains in the sample sintered for 3 h. Kanzaki et al.[43] also reported decreases in density upon prolonged sintering at 1650°C and with higher sintering temperatures. Elongated grains were observed in samples with compositions in the range ~68–73 wt% Al_2O_3, while samples with ≥74 wt% Al_2O_3 had equiaxed grain structures. In contrast to the results of Sacks and Pask,[22] Kanzaki et al. did not observe significant differences in sintered densities for samples with varying composition (68–78 wt% Al_2O_3). However, this was probably due to the fact that the sintering schedule (1630°C, 4 h) was sufficient for all compositions to approach their "final" density. This interpretation is supported by the study of Hirata et al.[50,51] They found that the

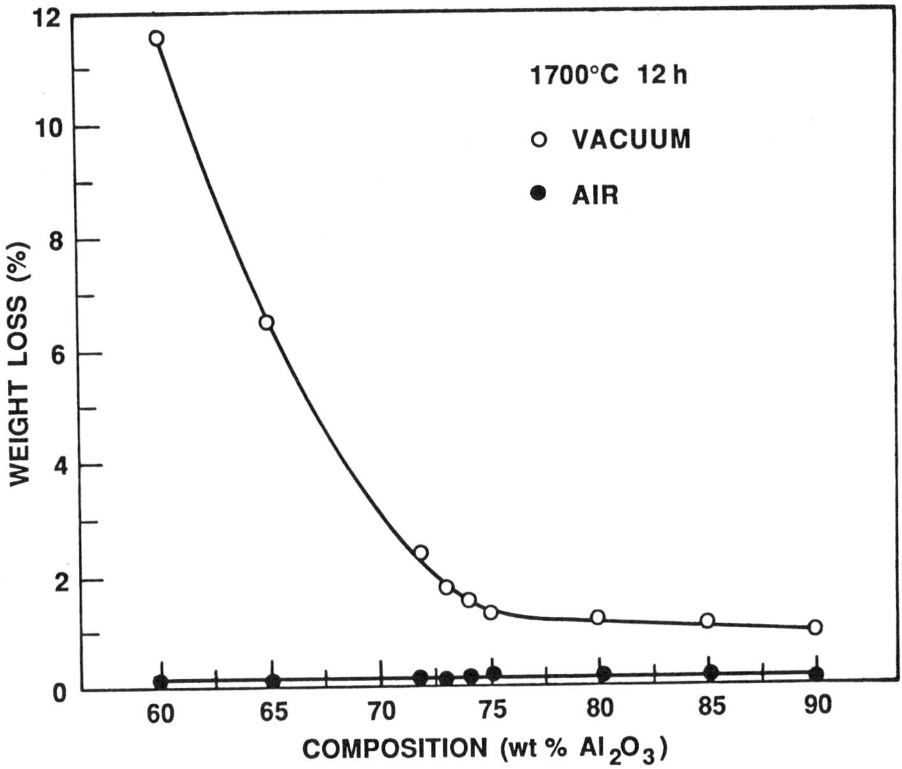

Fig. 13. Plots of percentage weight loss vs. overall chemical composition (wt% Al_2O_3) for samples sintered in air and vacuum at 1700°C for 12 h.

relative density was nearly independent of composition (~46.5–76.6 wt% Al_2O_3) for samples sintered at 1600°C for 3 h. (All of the relative densities were in range ~96–98%.) However, relative densities were much higher with the lower Al_2O_3 content samples when the sintering temperature was reduced to 1500 or 1550°C. (At these temperatures, only the samples with 46.5 wt% Al_2O_3 had high relative densities, i.e. ≳95%).

Effect of Sintering Atmosphere

The effect of atmosphere on the sintering of mullite has not been studied much. However, it is known that processing of mullite-containing materials is sensitive to the partial pressure of oxygen. Figure 13 shows the weight loss for samples with varying compositions (60–90 wt% Al_2O_3) which were sintered in air and vacuum (~10^{-6} torr) at 1700°C for 12 h.[66] In air, the weight loss was ~0.2 wt% or less for each composition. These small losses can be attributed to volatilization of physisorbed molecular water and chemisorbed hydroxyl groups. In contrast, weight losses in vacuum sintering were much larger and were highly dependent on the composition.

The large weight losses for the 60 and 65 wt% Al_2O_3 samples can be attributed to the large amount of silicate glass phase. At low partial pressure of oxygen, pure silica decomposes readily at elevated temperatures:

$$2SiO_{2(s)} \rightarrow 2SiO_{(g)} + O_{2(g)} \quad (7)$$

The amount of silicate glass phase decreases with increasing overall Al_2O_3 concentration, so the weight loss also decreases. It is interesting to note that the weight loss decreases more quickly for samples in the composition range 71.8–74 wt% Al_2O_3 compared to 75–90 wt% Al_2O_3. This is consistent with the evidence presented in the previous section that samples in the 71.8–74 wt% Al_2O_3 composition range still contain small, but decreasing, amounts of glass phase.

Although samples with 75–90 wt% Al_2O_3 do not contain glass phase, it is evident that the weight loss in vacuum is still significantly greater than in air. Part of the weight loss for all compositions (60–90 wt% Al_2O_3) can be attributed to decomposition of mullite[81]:

$$3Al_2O_3{\cdot}2SiO_{2(s)} \rightarrow 3Al_2O_{3(s)} + 2SiO_{(g)} + O_{2(g)} \quad (8)$$

For the samples with 75–90 wt% Al_2O_3, the weight loss decreases slightly with increasing overall Al_2O_3 concentration because the amount of mullite decreases (and the amount of α–Al_2O_3 increases) in this range. (α–Al_2O_3 is stable in vacuum at the temperatures used in this study.)

As a result of the mullite decomposition reaction, vacuum sintered samples develop a porous Al_2O_3 layer on the surface. This layer was removed by grinding prior to bulk density measurements.‡ Figure 14 shows a comparison of the relative densities for 60–90 wt% Al_2O_3 samples sintered in air and in vacuum. Vacuum sintered samples had higher densities for all compositions. Kinetic data was also collected for 71.8 and 80 wt% Al_2O_3 samples and these results indicated that densification rates were higher during vacuum sintering.[66] The enhanced densification may be due to increased grain boundary diffusion coefficients. During the early stages of vacuum sintering, the surface structure of the mullite particles is likely to develop an oxygen deficiency due to partial decomposition at the low oxygen partial pressure. Thus, it is possible that grain boundaries develop with a high concentration of oxygen vacancies and, in turn, this could enhance grain boundary diffusion. Etching experiments provided some evidence that the grain boundaries developed an oxygen deficiency. Vacuum sintered samples were cross-sectioned, polished, and etched in *air* at ~1550°C for ~30 min. As shown in Fig. 15A, "ridges" developed at the grain boundaries instead of "grooves" that normally form during thermal etching. The formation of ridges, a positive volume change at the grain boundaries, suggests

‡Although the porous surface layer is removed, the calculated relative densities in Fig. 14 are somewhat inaccurate because some of the weight loss also occurs from the interior of the sample, i.e. at least until the pore channels close during sintering. Therefore, the calculated relative densities for vacuum sintering may be somewhat high since theoretical density values were based on the initial sintering compositions that were used. This inaccuracy is not very significant for samples with 71.8–90 wt% Al_2O_3 because the total weight loss is relatively small. In contrast, the relative density values for the 60 and 65 wt% Al_2O_3 samples are less certain due to the larger weight losses. Nevertheless, the values in Fig. 14 are reasonable estimates. For example, calculations indicate that the 60 wt% Al_2O_3 sample is nearly fully dense. This was confirmed by microstructure observations which show very few pores in the sample.[66]

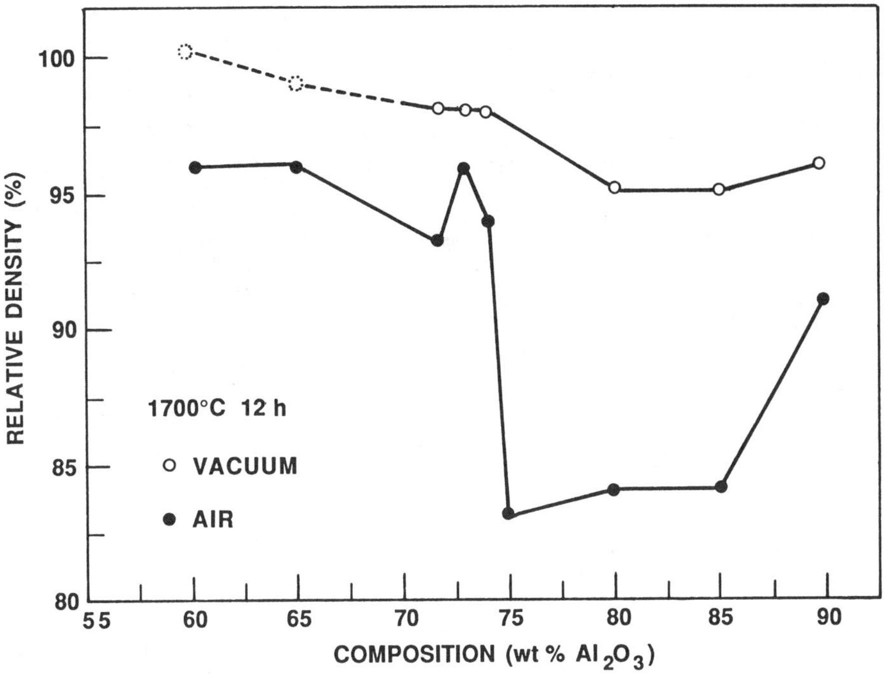

Fig. 14. Plots of relative density vs. overall chemical composition (wt% Al_2O_3) for samples sintered in air and vacuum at 1700°C for 12 h. Uncertainty in the values for the vacuum-sintered, lower Al_2O_3 content samples (indicated by the dashed line and dotted data points) arises from the large loss of silica during vacuum sintering.

that oxidation occurred during etching in air. With extended annealing in air (~100 h at ~1600°C), the ridges disappear (presumably by surface diffusion) and grain boundary grooves form (Fig. 15B).

For samples with a large amount of glass phase (i.e. 60 and 65 wt% Al_2O_3), the maximum density was reached too quickly to determine if the atmosphere affected the densification rate. The lower "endpoint" density in air (Fig. 14) is probably due to entrapped gas in the voids. This is also suggested by observations that samples sintered in an argon atmosphere have a slightly lower endpoint density.[66] (Presumably, the oxygen in air sintering diffuses out of the sample faster than argon.) Microstructure observations show that the residual pores in argon-sintered samples are larger compared to air-sintered samples, also suggesting that more gas is entrapped. Also characteristic of samples with entrapped gas is the observation that both air- and argon-sintered samples show decreases in the relative density with prolonged sintering times.[66]

Entrapped gases may also affect the final density in samples with higher Al_2O_3 content. For samples with 71.8 and 80 wt% Al_2O_3, vacuum sintering resulted in higher endpoint densities compared to air sintering. It was also observed that 80 wt% Al_2O_3 samples sintered in argon had slightly lower endpoint densities compared

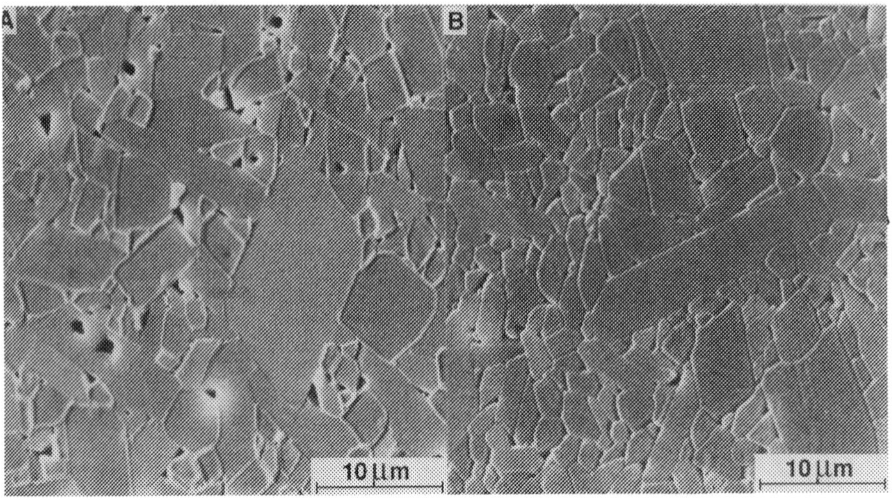

Fig. 15(A). Micrograph of vacuum-sintered 71.8 wt% Al_2O_3 sample which, after cross-sectioning and polishing, was thermally etched for ~30 min at 1550°C in *air*. (B): Micrograph of the sample in Fig. 15(A) after annealing for 100 h at 1600°C in *air*.

to air. Once again, this suggests that the oxygen in air sintering can diffuse out of the sample more readily. The study by Klug et al.[59] also indicated that oxygen can diffuse out of mullite readily during the final (i.e. closed pore) stage of sintering. They were able to sinter samples to translucency using a flowing oxygen atmosphere.

Summary

Powder preparation methods and densification procedures for the fabrication of mullite bodies with high relative density (i.e. low porosity) have been reviewed.

The calcination temperature required to prepare mullite powders depends on the scale of mixing of the aluminum- and silicon-bearing reactants. In conventional powder preparation methods, the scale of mixing of the components is, at best, on the order of a micrometer and, therefore, very high reaction temperatures (>1600°C) are required for mullitization. Colloidal techniques provide mixing on the scale of 5–50 nm, so calcination temperatures can be reduced to ~1200–1400°C. Solution-precipitation techniques have the potential to produce mullite precursors with molecular-level mixing of aluminum and silicon. In most solution-precipitation processes for preparing mullite, the aluminum and silicon-bearing components undergo hydrolysis and condensation reactions. The rates of these reactions must be carefully controlled to avoid segregation of the components and to obtain a true co-precipitated product. Mullite formation occurs at ~1000°C when the precipitated material has molecular-level mixing of aluminum and silicon. These low temperatures result in powders with fine particle size and high specific surface area.

Sintering of mullite powders to high densities requires high temperatures due to low bulk and grain boundary diffusion coefficients. To reduce the diffusion distances and to increase the driving force for sintering, it is important to use powders with fine particle sizes (<1 μm). However, such powders usually contain hard and soft agglomerates. These agglomerates must be broken down (e.g. by milling) and/or eliminated (i.e. by classification techniques) to avoid residual porosity and large grain size after sintering. Suspension processing can be used to prepare green compacts with high relative density and homogeneous packing of particles. These characteristics results in enhanced densification, less overall shrinkage during sintering, and smaller final grain size. With optimum processing, mullite bodies can be sintered to high relative density (≳99%) and fine grain size (~1 μm) at ~1540°C.

High density mullite bodies can be prepared at even lower temperatures (~1200–1300°C) by reaction sintering of amorphous or weakly crystalline materials having high specific surface area. In these techniques, it is important to complete densification prior to the mullitization reaction. This is accomplished by mixing the aluminum- and silicon-bearing components on the scale of ~5–50 nm. This delays mullite formation until ≳1250°C, but allows the high specific surface area materials to densify at slightly lower temperatures.

Densification kinetics and microstructure evolution in mullite bodies are also dependent on the overall chemical composition (i.e. the Al_2O_3/SiO_2 ratio). Higher densification rates are observed in samples which show residual glassy phase in the sintered microstructure, presumably due to liquid phase sintering. Siliceous glass is also associated with the development of microstructures having elongated (or acicular) grains and exaggerated grains. An equiaxed grain structure is observed when the glassy phase is eliminated (i.e. by using compositions with higher Al_2O_3/SiO_2 ratio).

The sintering atmosphere also affects densification behavior and microstructure development of mullite bodies. Mullite decomposes at high temperatures when the partial pressure of oxygen is low. Available evidence also indicates that densification rates are higher at low oxygen partial pressures. This may be due to enhanced diffusion at oxygen-deficient grain boundaries.

Acknowledgment

This work was supported in part by the National Science Foundation, Division of Materials Research, Ceramics and Electronic Materials Program (DMR-8451916). The authors thank P. Tewari and B. Sonuparlak for providing preprints (refs. 54 and 8, respectively).

References

[1]P. C. Dokko, J. A. Pask, and K. S. Mazdiyasni, "High-Temperature Mechanical Properties of Mullite Under Compression," *J. Am. Ceram. Soc.*, **60** (3) 150–155 (1977).

[2]T.-I. Mah and K. S. Mazdiyasni, "Mechanical Properties of Mullite," *J. Am. Ceram. Soc.*, **66** (10) 699–703 (1983).

[3]S. C. Samanta and S. Musikant, "SiC Whiskers-Reinforced Ceramic Matrix Composites," *Ceram. Eng. Sci. Proc.*, **6** (7–8) 663–672 (1985).

[4]N. Claussen and G. Petzow, "Whisker-Reinforced Oxide Ceramics," *J. Physique*, **47**, C1-693 - C1-702 (1986).

[5]P. F. Becher and T. N. Tiegs, "Toughening Behavior Involving Multiple Mechanisms: Whisker Reinforcement and Zirconia Toughening," *J. Am. Ceram. Soc.*, **70** (9) 651–654 (1987).

[6]S. Prochazka and F. J. Klug, "Infrared-Transparent Mullite Ceramic," *J. Am. Ceram. Soc.*, **66** (12) 874–880 (1983).

[7]N. Shinohara, D. M. Dabbs, and I. A. Aksay, "Infrared Transparent Mullite Through Densification of Monolithic Gels at 1250°C"; pp. 19–24 in Infrared and Optical Transmitting Materials. SPIE Vol. 683 (1986).

[8]B. Sonuparlak, "Sol-Gel Processing of Infrared Transparent Mullite," accepted by the Am. Ceram. Soc.

[9]B. H. Mussler and M. W. Shafer, "Preparation and Properties of Mullite-Cordierite Composites," *Bull. Am. Ceram. Soc.*, **63** (5) 705–710 (1984).

[10]A. G. Williams and L. V. Interrante, "Design of Metal-Organic Precursors to Aluminosilicates"; pp. 151–156 in Better Ceramics Through Chemistry, Mat. Res. Soc. Symp. Proc., Vol. 32. Edited by C. J. Brinker, D. E. Clark, and D. R. Ulrich. North-Holland, New York, 1984.

[11]G. S. Perry, "Microwave Dielectric Properties of Mullite," *Trans. Brit. Ceram. Soc.*, **72**, 279–283 (1973).

[12]W. H. Hawkes, "The Production of Synthetic Mullite," *Trans. Brit. Ceram. Soc.*, **61**, 689–703 (1962).

[13]N. B. Chatterjee and B. N. Panti, "Mullite Refractories from Clay-Bauxite or Quartz-Bauxite Mixes," *Trans. Indian Ceram. Soc.*, **24**, 116–121 (1965).

[14]A. R. Rossini, S. C. Arazi, and T. G. Krenkel, "Mullitization of Mixtures of Kaolinitic Clay and Aluminum Hydroxides," *Bol. Soc. Espan. Ceram.*, **9**, 579–591 (1970).

[15]B. P. Locsei, "Possible Technical Developments in the Manufacture of Fireclay Refractory Materials: III, Mechanisms of the Synthesis of Mullite in the System Kaolinite-Aluminum Fluoride," *Keram. Z.*, **20**, 362–367 (1968).

[16]J. S. Moya, J. Valle, and S. de Aza, "The Sintering Behavior of an Active Premullite Powder Obtained from Kandites"; pp. 409–415 in Sintering-Theory and Practice, Proc. of the 5th Int. Round Table Conf. on Sintering, Materials Science Monographs, Vol. 14. Edited by S. Pejovnik and M. M. Ristic. Elsevier, Amsterdam, The Netherlands, 1982.

[17]M. D. Sacks, "Correlation of Microstructure of Sintered Mullite Bodies with Their Stress-Strain Behavior in Compression Loading at 1200°C," M. S. Thesis, University of California, Berkeley, CA 1977.

[18]F. M. Wahl, R. E. Grim, and R. B. Graf, "Phase Transformations in Silica-Alumina Mixtures as Examined by Continuous X-ray Diffraction," *Am. Mineralogist*, **46**, 1064–1076 (1961).

[19]P. P. Budnikov, T. N. Keshishyan, and A. V. Volkova, "Kinetics of Mullite Formation from Technical Aluminum Oxide and Silica," *Zhurnal Prikladnoi Khimii*, **35** (6) 1171–1175 (1962).

[20]J. E. Fenstermacher and F. A. Hummel, "High-Temperature Mechanical Properties of Ceramic Materials: IV, Sintered Mullite Bodies," *J. Am. Ceram. Soc.*, **44**, 284–289 (1961).

[21]M. D. Sacks and J. A. Pask, "Sintering of Mullite"; pp. 193–203 in Processing of Crystalline Ceramics, Materials Science Research, Vol. II. Edited by H. Palmour III, R. F. Davis, and T. M. Hare. Plenum, New York, 1978.

[22]M. D. Sacks and J. A. Pask, "Sintering of Mullite-Containing Materials: I, Effect of Composition," *J. Am. Ceram. Soc.*, **65** (2) 65–70 (1982).

[23]A. P. S. Rana, O. Aiko, and J. A. Pask, "Sintering of α-Al_2O_3/Quartz and α-Al_2O_3/Cristobalite Related to Mullite Formation," *Ceram. Int.*, **8**, 151–153 (1982).

[24]Y. Nurishi and J. A. Pask, "Sintering of α-Al_2O_3-Amorphous Silica Compacts," *Ceram. Int.*, **8**, 57–59 (1982).

[25]P. D. D. Rodrigo and P. Boch, "High Purity Mullite Ceramics by Reaction Sintering," *Int. J. High Technol. Ceram.*, **1**, 3–30 (1985).

[26]P. D. D. Rodrigo and P. Boch, "Preparation of High Purity Mullite Ceramics," *J. Physique*, **47**, C1-411 - C1-416 (1986).

[27]J. L. McAtee and W. O. Milligan, "X-ray Diffraction Examination of Synthetic Mullite," *Tex. J. Sci.*, **2**, 200–205 (1950).

[28]B. B. Ghate, D. P. H. Hasselman, and R. M. Spriggs, "Synthesis and Characterization of High Purity, Fine Grained Mullite," *Bull. Am. Ceram. Soc.*, **52** (9) 670–672 (1973).

[29]M. D. Sacks and J. A. Pask, "Sintering of Mullite-Containing Materials: II, Effect of Agglomeration," *J. Am. Ceram. Soc.*, **65** (2) 70–77 (1982).

[30]B. L. Metcalf and J. H. Sant, "The Synthesis, Microstructure and Physical Properties of High Purity Mullite," *Trans. Brit. Ceram. Soc.*, **74** (6) 193–201 (1975).

[31]M. G. M. U. Ismail, Z. Nakai, and S. Sōmiya, "Microstructure and Mechanical Properties of Mullite Prepared by the Sol-Gel Method," *J. Am. Ceram. Soc.*, **70** (1) C-7 - C-8 (1987).

[32]M. G. M. U. Ismail, Z. Nakai, K. Minegishi, and S. Sōmiya, "Synthesis of Mullite Powder and Its Characteristics," *Int. J. High Technol. Ceram.*, **3**, 123–134 (1986).

[33]R. Roy, S. Komarneni, and D. M. Roy, "Multi-Phasic Ceramic Composites Made by Sol-Gel Technique"; pp. 347-359 in Better Ceramics Through Chemistry, Mat. Res. Soc. Symp. Proc., Vol. 32. Edited by C. J. Brinker, D. E. Clark, and D. R. Ulrich. North-Holland, New York, 1984.

[34]D. W. Hoffman, R. Roy, and S. Komarneni, "Diphasic Xerogels, A New Class of Materials: Phases in the System Al_2O_3-SiO_2," *J. Am. Ceram. Soc.*, **67** (7) 468–471 (1984).

[35]K. Hamano, Z. Nakagawa, G. Cun-ji, and T. Sato, "Formation Process of Mullite from Kaolin Minerals and Various Mixtures"; pp. 37–49 in Mullite. Edited by S. Sōmiya. Uchida Rokakuho Publishing Co., Tokyo, Japan, 1985.

[36]S. Sōmiya, M. Yoshimura, M. Suzuki, and T. Yamaguchi, "Hydrothermal Processing of Mullite Powders from Alkoxides"; pp. 63–87, ibid.

[37]H. Insley and R. H. Ewell, "Thermal Behavior of the Kaolin Minerals," *J. Res. Nat. Bur. Standards*, **14**, 615–627 (1935).

[38]V. C.-H. Horte and J. Wiegmann, "Study on the Reaction Between Amorphous SiO_2 and Al_2O_3," *Naturwiss*, **43**, 9–10 (1956).

[39]J. D. Crofts and W. W. Marshall, "A Novel Synthesis of Alumino-Silicates and Similar Materials," *Trans. Brit. Ceram. Soc.*, **66** (3) 121–126 (1967).

[40]T. D. McGee and C. D. Wirkus, "Mullitization of Alumino-Silicate Gels," *Bull. Am. Ceram. Soc.*, **51** (7) 577–581 (1972).

[41]R. Roy and E. F. Osborn, "The System Al_2O_3-SiO_2-H_2O," *Am. Mineral*, **39** (11–12) 853–885 (1954).

[42]S. Kanzaki, H. Tabata, T. Kumazawa, and S. Ohta, "Sintering and Mechanical Properties of Stoichiometric Mullite," *J. Am. Ceram. Soc.*, **68** (1) C-6 - C-7 (1985).

[43]S. Kanzaki and H. Tabata, "Sintering and Mechanical Property of Spray Pyrolyzed Mullite Powder"; pp. 51–61 in Mullite. Edited by S. Sōmiya. Uchida Rokakuho Publishing Co., Tokyo, Japan, 1985.

[44]G.-Y. Meng and R. A. Huggins, "A New Chemical Method for Preparation of Both Pure and Doped Mullite," *Mat. Res. Bull*, **18**, 581–588 (1983).

[45]J. Grofcsik and E. Vago, "Modern Methods of Clay Investigation," *Epitoanyag*, **43** (1952).

[46]J. Ossaka, "Tetragonal Mullite-like Phase from Co-precipitated Gels," *Nature*, **191**, 1000–1001 (1961).

[47]Y. M. M. Al-Jarsha, K. D. Biddle, A. K. Das, T. J. Davies, H. G. Emblem, K. Jones, J. M. McCullough, M. A. Mohd. Abd. Rahman, A. N. A. El-M. Sharf El Deen, and R. Wakefield, "Mullite Formation from Ethyl Silicate and Aluminum Chlorides," *J. Mat. Sci.*, **20**, 1773–1781 (1985).

[48]R. R. West and T. J. Gray, "Reactions in Silica-Alumina Mixtures," *J. Am. Ceram. Soc.*, **41** (4) 132–136 (1958).

[49]K. S. Mazdiyasni and L. M. Brown, "Synthesis and Mechanical Properties of Stoichiometric Aluminum Silicate (Mullite)," *J. Am. Ceram. Soc.*, **55** (11) 548–552 (1972).

[50]Y. Hirata and K. Shimada, "Preparation and Sinterability of Fine Mullite Powder from Mixed Alkoxides"; pp. 89–122 in Mullite. Edited by S. Sōmiya. Uchida Rokakuho Publishing Co., Tokyo, Japan, 1985.

[51]Y. Hirata, K. Sakeda, Y. Matsushita, and K. Shimada, "Preparation of Fine SiO_2-Al_2O_3 Powders by Hydrolysis of Mixed Alkoxides," *Yogyo-Kyokai-Shi*, **93** (9) 101–105 (1985).

[52]L. A. Paulick, Y.-F. Yu, and T.-I. Mah, "Ceramic Powders from Metal Alkoxide Precursors"; pp. 121–129 in Ceramic Powder Science, Advances in Ceramics, Vol. 21. Edited by G. L. Messing, K. S. Mazdiyasni, J. W. McCauley, and R. A. Haber. American Ceramic Society, Westerville, OH, 1987.

[53]B. E. Yoldas, "Microstructure of Monolithic Materials Formed by Heat Treatment of Chemically Polymerized Precursors in the Al_2O_3-SiO_2 Binary," *Bull. Am. Ceram. Soc.*, **59** (4) 479–483 (1980).

[54]M. N. Rahaman, L. C. DeJonghe, S. Shinde, and P. H. Tewari, "Sintering and Microstructure of Supercritically Dried, Gel-Derived Mullite," submitted to the Am. Ceram. Soc.

[55]R. K. Iler, The Chemistry of Silica, John Wiley, New York, 1979.

[56]K. D. Keefer, "The Effect of Hydrolysis Conditions on the Structure and Growth of Silicate Polymers"; pp. 15–24 in Better Ceramics Through Chemistry. Mats. Res. Soc. Symp. Proc., Vol. 32. Edited by C. J. Brinker, D. E. Clark, and D. R. Ulrich. North-Holland, New York, 1984.

[57]W. H. Gitzen, Alumina as a Ceramic Material, American Ceramic Society, Columbus, OH, 1970.

[58]B. E. Yoldas, "Preparation of Glasses and Ceramics from Metal-Organic Compounds," *J. Materials Sci.*, **12**, 1203–1208 (1977).

[59]F. J. Klug, S. Prochazka, and R. H. Doremus, "Alumina-Silica Phase Diagram in the Mullite Region," General Electric Technical Information Series Report No. 86CRD077, August, 1986.

[60]M. S. J. Gani and R. McPherson, "Crystallization of Mullite from Al_2O_3-SiO_2 Glasses," *J. Aust. Ceram. Soc.*, **13** (2) 21–23 (1977).

[61]N. Claussen and J. Jahn, "Mechanical Properties of Sintered, In Situ-Reacted Mullite-Zirconia Composites," *J. Am. Ceram. Soc.*, **63** (3–4) 228–229 (1980).

[62]J. S. Moya and M. I. Osendi, "Microstructure and Mechanical Properties of Mullite/ZrO_2 Composites," *J. Mater. Sci.*, **19**, 2909–2914 (1984).

[63]M. D. Sacks and T.-Y. Tseng, "Preparation of SiO_2 Glass from Model Powder Compacts: I, Formation and Characterization of Powders, Suspensions, and Green Compacts," *J. Am. Ceram. Soc.*, **67** (8) 516–532 (1984).

[64]M. D. Sacks and G. W. Scheiffele, "Properties of Silicon Suspensions and Slip Cast Bodies," *Ceram. Eng. Sci. Proc.*, **6** (7–8) 1109–1123 (1985).

[65]E. A. Barringer and H. K. Bowen, "Formation, Packing, and Sintering of Monodisperse TiO_2 Powders," *J. Am. Ceram. Soc.*, **65** (12) C-199 - C-201 (1982).

[66]M. D. Sacks, "Sintering Behavior of Mullite-Containing Materials," Ph.D. Thesis, University of California, Berkeley, CA, 1979.

[67]R. F. Davis and J. A. Pask, "Diffusion and Reaction Studies in the System Al_2O_3-SiO_2," *J. Am. Ceram. Soc.*, **55** (10) 525–531 (1972).

[68]I. A. Aksay and J. A. Pask, "Stable and Metastable Equilibria in the System SiO_2-Al_2O_3," *J. Am. Ceram. Soc.*, **58** (11–12) 507–512 (1975).

[69]M. Yamane, S. Aso, and T. Sakaino, "Preparation of a Gel from Metal Alkoxide and Its Properties as a Precursor of Oxide Glass," *J. Mat. Sci.*, **13**, 865–870 (1978).

[70]M. Nogami and Y. Moriya, "Glass Formation Through Hydrolysis of $Si(OC_2H_5)_4$ with NH_4OH and HCl Solution," *J. Non-Crystalline Solids*, **37**, 191–201 (1980).

[71]P. Yu, H. Liu, and Y. Wang, "Hydrolysis and Gelling Process of Silicon Ethoxide," *J. Non-Crystalline Solids*, **52**, 511–520 (1982).

[72]S. Sakka, "Gel Method for Making Glass"; pp. 129–167 in Glass, III, Treatise on Materials Science and Technology, Vol. 22. Edited by M. Tomozawa and R. Doremus. Academic Press, New York, 1982.

[73]L. C. Klein and G. J. Garvey, "Monolithic Dried Gels," *J. Non-Crystalline Solids*, **48**, 97–104 (1982).

[74]S. Komarneni, Y. Suwa, and R. Roy, "Application of Compositionally Diphasic Xerogels for Enhanced Densification: The System Al_2O_3-SiO_2," *J. Am. Ceram. Soc.*, **69** (7) C-155 - C-156 (1986).

[75]R. Roy, Y. Suwa, and S. Komarneni, "Nucleation and Epitaxial Growth in Diphasic (Crystalline + Amorphous) Gels"; pp. 247–258 in Science of Ceramic Chemical Processing. Edited by L. L. Hench and D. R. Ulrich. Wiley, New York, 1986.

[76]L. L. Hench, "Use of Drying Control Chemical Additives (DCCAs) in Controlling Sol-Gel Processing," pp. 52–64, ibid.

[77]G. W. Scherer, "Theory of Drying Gels," to be published in the proceedings of the 3rd International Conference on Ultrastructure Processing, San Diego, CA, Feb. 1987.

[78]S. Aramaki and R. Roy, "Revised Phase Diagram for the System Al_2O_3-SiO_2," *J. Am. Ceram. Soc.*, **45** (5) 229–242 (1962).

[79]W. Von Lohre and H. Urban, "Contribution to the Morphology of Mullite," *Ber. Deut. Keram. Ges.*, **37**, 249–251 (1960).

[80]M. D. Sacks, R. B. Langston, S. T. Tso, and J. A. Pask, "Corrosion and Mechanical Behavior Correlations with Composition and Microstructure of Aluminum Silicate Refractories," *Bull. Am. Ceram. Soc.*, **58** (7) 691–697 (1979).

[81]R. F. Davis, I. A. Aksay, and J. A. Pask, "Decomposition of Mullite," *J. Am. Ceram. Soc.*, **55** (2) 98–101 (1972).

Mullite Derived From Diphasic Nanocomposite Gels*

SRIDHAR KOMARNENI, PROFESSOR OF CLAY MINERALOGY† AND RUSTUM ROY, EVAN PUGH PROFESSOR OF THE SOLID STATE MATERIALS RESEARCH LABORATORY

The Pennsylvania State University
University Park, PA 16802

Abstract

Mullite composition gels were prepared as single phases and as diphasic nanocomposites. Bulk and apparent densities of 96 and 97% respectively were measured for the nanoheterogeneous diphasic mullite gels made by the mixing of two sols after sintering at 1300°C. Another type of diphasic gel made from silica sol and aluminum nitrate solution and the single phase gels made from tetraethoxysilane and aluminum nitrate showed bulk densities of 72 and 79% respectively after identical sintering treatments at 1300°C. Results of densification studies of the different gels sintered at 1400°C closely parallel those that were sintered at 1300°C. The truly diphasic gels which were prepared by mixing boehmite (AlOOH) sol and silica sol resulted in much greater densification than the other type of nominally diphasic gel because the compositional heterogeneity on a nanometer scale was preserved in the former but not in the latter. This enhanced densification of compositionally diphasic gels compared to the single-phase gels is attributed to the heat of reaction available in the former and not in the latter. Neither the use of isostructural mullite seeds nor the use of non-isostructural seeds such as anatase and kaolinite significantly affected the densification behavior of the different mullite gels. The densification behavior of the diphasic gels made from two sols was not significantly affected by the nature of the catalyst (HCl or HNO_3) used.

Introduction

The concept of diphasic ceramic + ceramic gels as a new class of materials with interesting potential was first introduced by one of us.[1] This is a new direction for the sol-gel processing science and is now well established in our laboratory.[2–8] The goal of ceramic materials processing via the diphasic ceramic gels to exploit the thermodynamics of metastable materials and in particular to utilize the heat of reaction of the two discrete phases. Here, this concept is applied to the processing of mullite which has numerous technological applications such as infrared transmit-

---

***Invited paper.**

†Also with the Department of Agronomy.

ting materials,[9] refractory materials,[10] substrate materials[11] and high-temperature structural materials.[12,13] Although the processing of mullite from different starting materials has been studied extensively by numerous workers,[14–20] the diphasic approach to mullite processing has just begun.[7,8,13] The difference between diphasic composites and single-phase gels was earlier investigated (see Hoffman, Roy and Komarneni, 1984)[7] by differential thermal analysis and x-ray powder diffraction, the latter to determine phase change with heating. Seeding of the gels with the final equilibrium phase has not only led to the lowering in crystallization temperatures[21–23] but also resulted in enhanced densification of the sintered bodies.[24,25] The seeded gels have been classified as structurally diphasic.[21] The objective of this research was to show that the diphasic nanocomposites are superior to single-phase gels in the sintering of mullite.

Experimental Procedure

Stoichiometric mullite (71.83 wt% Al_2O_3-28.17 wt% SiO_2) gels were prepared by the diphasic and single phase routes. Two varieties of diphasic gels were prepared as follows: (1) using two sols and (2) using one solution and one sol. Diphasic gels of the first variety, the truly diphasic, were made by initially peptizing boehmite (AlOOH)[‡] with HNO_3 or HCl to make an alumina sol and mixing with a silica sol, Ludox[§] and gelling at room temperature. Diphasic gels of the second variety were made by mixing aluminum nitrate nonahydrate with the silica sol and gelling. The single phase gels were made by first dissolving aluminum nitrate nonahydrate in absolute ethanol, adding tetraethoxysilane [$Si(OC_2H_5)_4$], and mixing and then gelling at 60°C in a water bath.

Both the diphasic and single phase gels were also seeded with mullite which is the final equilibrium phase as well as non-isostructural seeds such as anatase (TiO_2) and kaolinite [$Si_4Al_4O_{10}(OH)_8$]. The different seed crystals were introduced and mixed with the different components as fine sols prior to gelation.

The diphasic gels of the first variety (with and without seeds) made from boehmite and silica sols were heated at 60°C to remove water, ground in an agate mortar with a pestle and pelletized prior to sintering. The second variety of diphasic gels (with and without seeds) made from aluminum nitrate and silica sol and the single-phase gels were initially dried at 60°C, ground and calcined between 400°-500°C in a pot furnace to decompose nitrates and organics, ground in an agate mortar with a pestle and pelletized prior to sintering. All the xerogel powders were mixed with approximately 1% polyvinyl alcohol (PVA) and pressed into pellets at 15,000 psi and sintered in a programmed furnace at 1300° and 1400°C in air. The heating rates were ~2°C/min up to 225°C (heated for 2h at this temperature), ~2°C/min between 225° and 500°C (heated for 2h at this temperature), ~4.5°C/min between 500° and 1300°C, and 1400°C. The samples were heated for 3h at the final sintering temperatures. Apparent densities of the sintered pellets after evacuation were measured by the Archimedes technique. Bulk densities of the sintered pellets were measured after coating them with wax. Microstructural characterization of the fractured surfaces of the sintered pellets was carried out by scanning electron microscopy (SEM).[¶] Powder x-ray diffraction (XRD)[**] using graphite mono-

---

‡Dispural, Remet Chemical Corp., Chadwicks, NY.

§Ludox, "AS," E. I. du Pont de Nemours and Co., Inc., Wilmington, DE.

¶Model DS-130, International Scientific Instruments, Santa Clara, CA.

chromated CuKα radiation of the crushed pellets was carried out to determine phase purity of the samples.

The diphasic gel made from boehmite and silica sol and the single-phase gel were characterized by ^{29}Si magic angle spinning nuclear magnetic resonance (MASNMR) spectroscopy. Solid-state ^{29}Si MASNMR spectra were obtained by using a high-resolution, narrow-bore commercial spectrometer†† with a home-built MAS broad-banded probe fitted with a standard spinning assembly.[26] The ^{29}Si MASNMR spectra were obtained at 79.5 MHz using the above spectrometer and the ^{29}Si chemical shifts were recorded with respect to tetramethylsilane. No resolution enhancement was undertaken and the spectra were presented with the appropriate line broadening.

The diphasic gels of the first variety were also characterized by transmission electron microscopy using a Philips 420 microscope.

Results and Discussion

^{29}Si MASNMR Spectra

The ^{29}Si MASNMR spectra of single and diphasic mullite gels are shown in Fig. 1. The diphasic mullite gel made from two sols shows two resonances, one at -100 ppm and another at -111.8 ppm from TMS which is indicative of silica gel and may be assigned to HO $Si(OSi)_3$ and $Si(OSi)_4$, respectively.[27] This is an expected result because the two discrete phases i.e., silica and alumina sols do not react at room temperature and thus the above ^{29}Si MASNMR spectrum is due to the presence of the discrete silica sol. The single-phase gel, on the other hand, shows a broad ^{29}Si resonance at about -103.3 ppm (Fig. 1). The low-field shift of the ^{29}Si resonance of the single-phase gel compared to the diphasic gel is an indication of the substitution of the aluminum in the nearest neighbor environment of the silicon in the former since the low-field chemical shift, in general, is proportional to the extent of Al substitution in the nearest neighbor environment of Si in crystalline aluminosilicates.[28,29] The presence of Al in the nearest neighbor environment of the Si clearly demonstrates that the Al and Si components are mixed almost on an atomic scale unlike the case of diphasic gels.

Transmission Electron Microscopy

Transmission electron micrograph (Fig. 2) of the diphasic mullite gel using two sols clearly shows that the silica and alumina sols are intimately mixed even though they remained as separate entities. The spherical silica sol and the fibrillar nature of the boehmite sol are clearly seen (Fig. 2). Thus the diphasic mullite gel is nano-heterogeneous i.e., heterogeneous on a nanometer scale. The single-phase gel showed no distinct morphology.

Densification Behavior

The geometrical green densities of single phase, diphasic (one sol + one solution) and truly diphasic (two sols) gels are 51, 51 and 54% of theoretical density, respectively. The bulk and apparent densities of the unseeded and seeded diphasic and single-phase mullite gels sintered at 1300°C are presented in Table I. The diphasic gels made from two sols show enhanced densification compared to either the single-phase gel or the other variety of the diphasic gel made from one sol and one

---

**Philips Electronic Instruments, Inc., Mahwah, NJ.

††WH-400, Bruker Instruments, Incs., Billerica, MA.

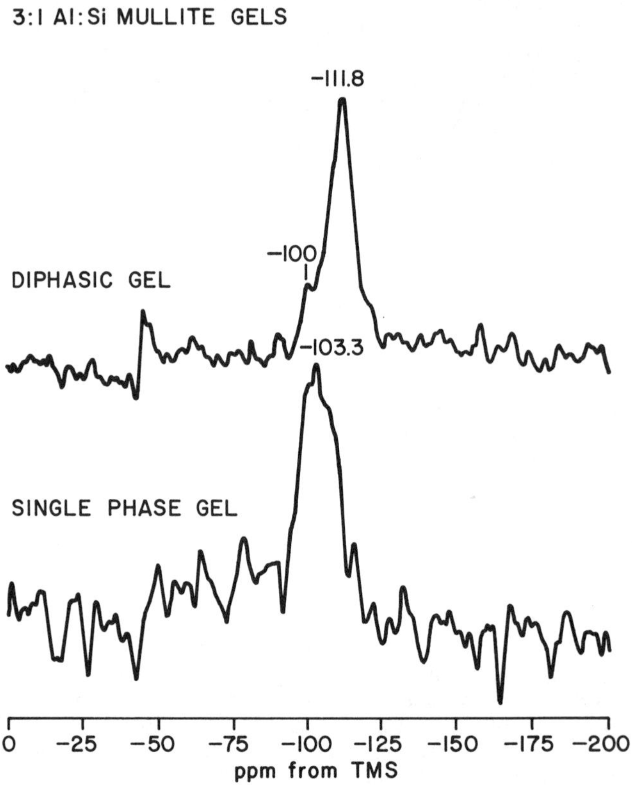

Fig. 1. **^{29}Si MASNMR spectra of single and diphasic mullite composition gels.**

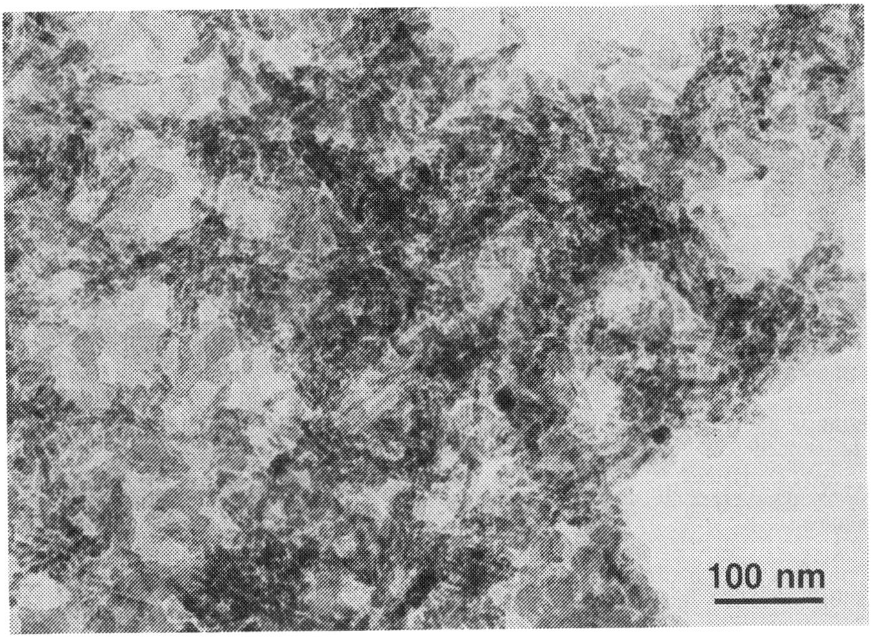

Fig. 2. Transmission electron micrograph of diphasic mullite gel made from two sols showing spherical silica sol particles and fibrillar boehmite particles mixed intimately on a nanometer scale.

solution as evidenced by higher bulk densities in the former compared to the latter (Table 1) even though their green densities are about the same. The bulk density of the diphasic gel (3.02 g/cm^3) made from two sols is almost the same as its apparent density (3.03 g/cm^3) which shows that there is little or no interconnected porosity. On the other hand, the bulk densities of the other diphasic gel and the single-phase gel are much lower than their apparent densities (Table 1) which show that there is a lot of interconnected porosity in these sintered gels. Microstructural characterization of the fractured surfaces of the different gels corroborate the density results. The diphasic gels made from two sols show good densification upon sintering at 1300° and 1400°C with a few micropores (Fig. 3A, 3B) which are not interconnected while the single-phase gels and the other diphasic gel made from one sol and one slution show little or no densification with large cracks (Fig. 3C, 3D). The compositionally diphasic gels which were prepared by mixing boehmite sol and silica sol resulted in much greater densification (Table 1, Figure 3) than the other type of diphasic gels which were made by $Al(NO_3)_3$ and silica sol because the heterogeneity on a nanometer scale was preserved in the former but not in the latter. The enhanced densification of the diphasic gels made from two sols may be attributed to the heat of reaction between the two sols. The diphasic gels store much higher metastable energy than the single-phase gels (Fig. 4) and thus the enhanced densification of the diphasic gels may be attributed to the additional energy provided during the exothermic heat of reaction. Results of the bulk and apparent densities (Table 2) and the microstructures of the unseeded and seeded diphasic and single-phase gels

Table 1. Bulk and apparent densities[1] of unseeded and seeded diphasic and single phase mullite gels sintered at 1300°C

Sample	Densities, g/cm^3	
	Bulk	Apparent
Diphasic gel from silica sol and boehmite	3.02 (95.2)*	3.03 (95.6)
Diphasic gel from silica sol and boehmite + mullite seeds	2.97 (93.8)	3.00 (94.8)
Diphasic gel from silica sol and boehmite + anatase seeds	3.02 (95.1)	3.05 (96.2)
Diphasic gel from silica sol and boehmite + kaolinite seeds	3.00 (94.8)	3.04 (95.8)
Diphasic gel from silica sol and $Al(NO_3)_3$	2.28 (71.9)	3.07 (96.7)
Diphasic gel from silica sol and $Al(NO_3)_3$ + mullite seeds	2.11 (66.5)	3.03 (95.6)
Diphasic gel from silica sol and $Al(NO_3)_3$ + anatase seeds	2.17 (68.3)	3.04 (96.0)
Diphasic gel from silica sol and $Al(NO_3)_3$ + kaolinite seeds	2.27 (71.7)	2.94 (92.8)
Single phase gel	2.41 (76.0)	2.97 (93.5)
Single phase gel + mullite seeds	2.34 (73.9)	2.97 (93.7)
Single phase gel + anatase seeds	2.39 (75.4)	2.97 (93.7)
Single phase gel + kaolinite seeds	2.31 (72.9)	2.99 (94.4)

[1] = Average for duplicate samples
* = Numbers in parentheses are percentages of theoretical density

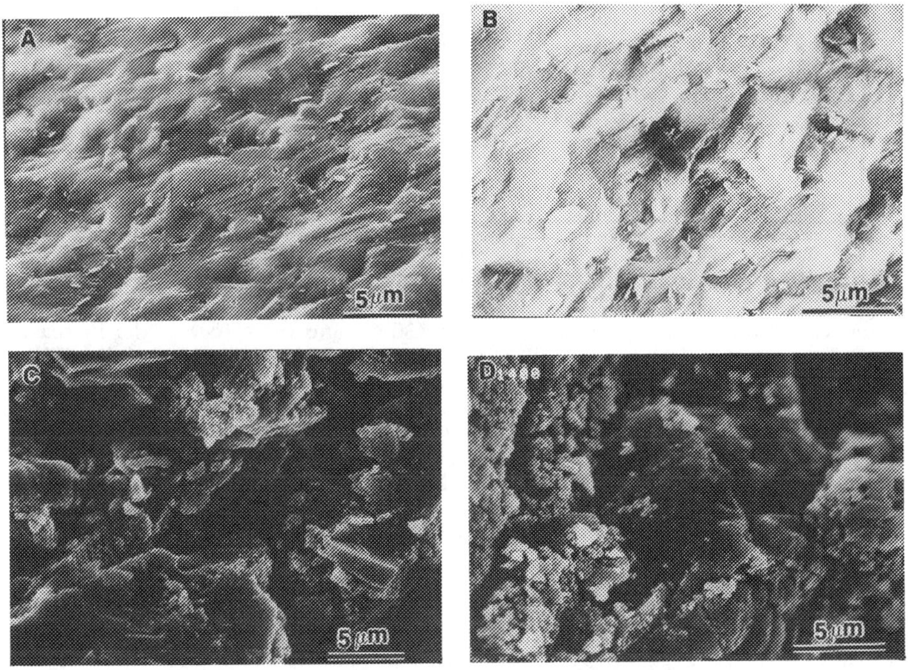

Fig. 3. Scanning electron micrographs of fractured surfaces of sintered mullite gels: A. Diphasic mullite gel made from two sols sintered at 1300°C showing good densification with a few micropores; B. Same as A but sintered at 1400°C; C. Single phase gel sintered at 1400°C showing large cracks with little or no densification; D. Diphasic gel made from silica sol and $Al(NO_3)_3$ solution sintered at 1400°C showing little or no densification.

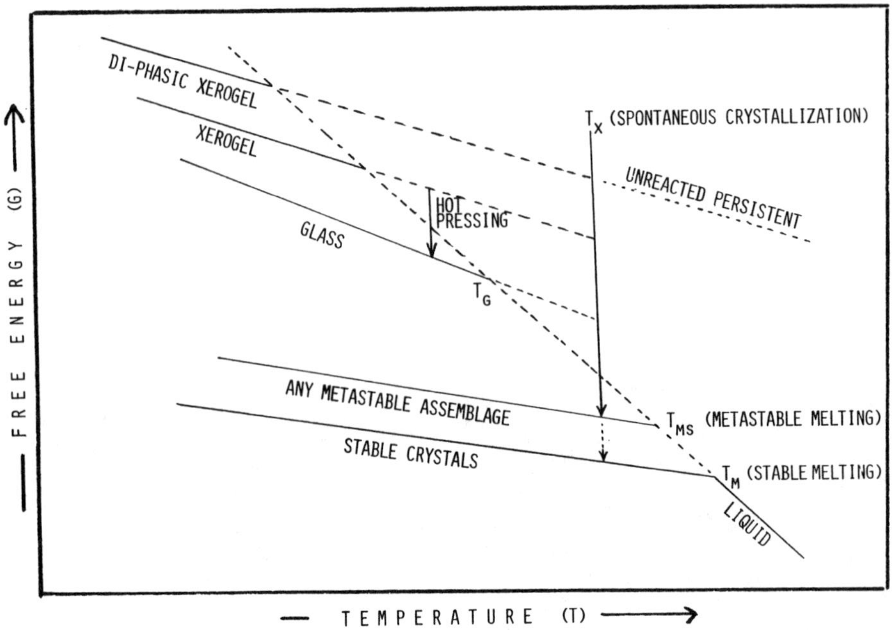

Fig. 4. Representation of G-T relations among isoplethal phases.

sintered at 1400°C closely parallel those at 1300°C and the effect of the increased temperature appears to be insignificant on densification (Tables 1 and 2). These results further confirm out earlier predictions[1] and findings[8] that the diphasic gels are superior to single-phase gels for enhanced densification.

The densities of the different gels were not affected upon isostructural seeding with mullite (Tables 1 and 2) which indicates little or no effect of seeding. Seeding of the different mullite composition gels with non-isostructural seeds such as anatase and kaolinite also showed no effect on their densification as measured by the densities (Tables 1 and 2). Characterization of the microstructures did not show any difference between the unseeded and mullite seeded gels sintered at 1300° or 1400°C.

Effect of Catalyst on Densification Behavior

The bulk and apparent densities of the diphasic gels made from silica sol and boehmite sol with HNO_3 and HCl catalysts and sintered at 1300° and 1400°C are presented in Table 3. The densification behavior of the diphasic gels is the same irrespective of the acid catalyst used (Table 3). Both HNO_3 and HCl appear to be equally effective when they were used to peptize boehmite during the mullite gel-making process.

Conclusion

The truly diphasic nanocomposite approach or the nanoheterogeneous sol-gel processing using two sols (or more for a multicomponent gel) is advantageous for enhancing densification for two reasons: (1) there are no organics or major amounts of anions to decompose or volatilize during the sintering process unlike the one sol

Table 2. Bulk and apparent densities[1] of unseeded and seeded diphasic and single phase mullite gels sintered at 1400°C

Sample	Densities, g/cm^3	
	Bulk	Apparent
Diphasic gel from silica sol and boehmite	3.04 (95.9)*	3.07 (96.8)
Diphasic gel from silica sol and boehmite + mullite seeds	2.99 (94.3)	3.02 (95.3)
Diphasic gel from silica sol and boehmite + anatase seeds	3.04 (95.8)	3.06 (96.6)
Diphasic gel from silica sol and boehmite + kaolinite seeds	3.01 (94.9)	3.05 (96.3)
Diphasic gel from silica sol and $Al(NO_3)_3$	2.29 (72.3)	3.05 (96.1)
Diphasic gel from silica sol and $Al(NO_3)_3$ + mullite seeds	2.16 (68.2)	3.08 (97.3)
Diphasic gel from silica sol and $Al(NO_3)_3$ + anatase seeds	2.18 (68.7)	3.05 (96.3)
Diphasic gel from silica sol and $Al(NO_3)_3$ + kaolinite seeds	2.29 (72.3)	3.04 (95.9)
Single phase gel	2.49 (78.5)	3.04 (95.9)
Single phase gel + mullite seeds	2.37 (74.8)	3.03 (95.6)
Single phase gel + anatase seeds	2.43 (76.6)	3.05 (96.3)
Single phase gel + kaolinite seeds	2.39 (75.4)	3.03 (95.5)

[1] = Average for duplicate samples
* = Numbers in parentheses are percentages of theoretical density

Table 3. Bulk and apparent densities* of diphasic gels made from silica sol and boehmite sol with different catalysts and sintered at two different temperatures

Sintering temperature (°C)	HNO_3 catalyzed gel		HCl catalyzed gel	
	Bulk density (g/cm^3)	Apparent density (g/cm^3)	Bulk density (g/cm^3)	Apparent density (g/cm^3)
1300	3.02 (95.2)†	3.03 (95.6)	3.03 (95.5)	3.07 (96.8)
1400	3.04 (95.9)	3.07 (96.8)	3.04 (95.9)	3.07 (96.8)

*Average for duplicate samples
†Numbers in parentheses are percentages of theoretical density

and one solution diphasic gels or the single-phase gels which require a calcining step and may lead to non-uniform segregation of components; and (2) the discrete phases of the diphasic gel, at least in part, provide additional energy by the exothermic heat of reaction during sintering to enhance densification.

Acknowledgment

This research was supported by the Air Force Office of Scientific Research under Contract No. F49620-85-C-0069.

References

[1]R. Roy, "Ceramics from Solutions: Retrospect and Prospect," Abstracts, Annual Meeting of the Materials Research Society, p. 370 (1982).

[2]D. Hoffman, R. Roy and S. Komarneni, "Diphasic Ceramic Composites Via a Sol-Gel Method," *Mat. Letters*, **2** [3] 245–247 (1984).

[3]D. Hoffman, S. Komarneni and R. Roy, "Preparation of a Diphasic Photosensitive Xerogel," *J. Mat. Sci. Letters*, **3** [5] 439–442 (1984).

[4]R. Roy, S. Komarneni and D. M. Roy, "Multiphasic Ceramic Composites Made by Sol-Gel Technique," in the Proceedings of the Symposium on "Better Ceramics Through Chemistry" (Eds. C. J. Brinker, D. E. Clark and D. R. Ulrich), Elsevier, North-Holland, pp. 347–359 (1984).

[5]D. W. Hoffman, R. Roy and S. Komarneni, "New sol-Gel Strategies for Making Ceramic-Ceramic Composites," Abstracts, 1984 American Ceramic Society Meeting, *Am. Ceram. Soc. Bull.*, **63** [4] 459 (1984).

[6]R. A. Roy and R. Roy, "Diphasic Xerogels: I. Ceramic-Metal Composites," *Mat. Res. Bull.*, **19** [1] 169–177 (1984).

[7]D. W. Hoffman, R. Roy and S. Komarneni, "Diphasic Xerogels, A New Class of Materials: Phases in the Al_2O_3-SiO_2 System," *J. Am. Ceram. Soc.*, **67** [7] 468–471 (1984).

[8]S. Komarneni and R. Roy, "Application of Compositionally Diphasic Xerogels for Enhanced Densification: The System Al_2O_3-SiO_2, *J. Am. Ceram. Soc.*, **69** [7] C-155-C-156 (1986).

[9]S. Prochazka and F. J. Klug, "Infrared-Transparent Mullite Ceramic," *J. Am. Ceram. Soc.*, **66** [12] 874–880 (1983).

[10]P. D. D. Rodrigo and P. Boch, "High Purity Mullite Ceramics by Reaction Sintering," *Int. J. High Tech. Ceramics*, **1** [1] 3–30 (1985).

[11]P. E. Debely, E. A. Barringer and H. K. Bowen, "Preparation and Sintering Behavior of Fine-Grained Al_2O_3-SiO_2 Composite," *J. Am. Ceram. Soc.*, **68** [3] C-76-C-78 (1985).

[12]S. Kanzaki, H. Tabata, T. Kumazawa and S. Ohta, "Sintering and Mechanical Properties of Stoichiometric Mullite," *J. Am. Ceram. Soc.*, **68** [1] C-6-C-7 (1985).

[13]M. G. M. U. Ismail, Z. Nakai and S. Somiya, "Microstructure and Mechanical Properties of Mullite Prepared by the Sol Gel Method," *J. Am. Ceram. Soc.*, **70** [1] C-7-C-8 (1987).

[14]S. Aramaki and R. Roy, "Revised Equilibrium Diagram for the System Al_2O_3-SiO_2, *Nature*, **184**: 631–632 (1959).

[15]R. F. Davis and J. A. Pask, "Diffusion and Reaction Studies in the System Al_2O_3-SiO_2," *J. Am. Ceram. Soc.*, **55** [10] 525–531 (1972).

[16]I. A. Aksay and J. A. Pask, "Stable and Metastable Equilibria in the System SiO_2-Al_2O_3," *J. Am. Ceram. Soc.*, **59** [11–12] 507–512 (1975).

[17]A. P. S. Rana, O. Aiko and J. A. Pask, "Sintering of α-Al_2O_3/quartz and α-Al_2O_3/Cristobalite Related to Mullite Formation," *Ceram. Int.*, **8** [2] 151–161 (1982).

[18]Y. Nurishi and J. A. Pask, "Sintering of α-Al_2O_3-Amorphous Silica Compacts," *Ceram. Int.*, **8** [1] 57–63 (1982).

[19]R. R. West and T. J. Gray, "Reactions in Silica-Alumina Mixtures," *J. Am. Ceram. Soc.*, **41** [4] 132–136 (1958).

[20]B. B. Ghate, D. P. H. Hasselman and R. M. Spriggs, "Synthesis and Characterization of High Purity, Fine Grained Mullite," *Bull. Am. Ceram. Soc.*, **52** [9] 670–672 (1973).

[21]Y. Suwa, R. Roy and S. Komarneni, "Lowering Crystallization Temperatures by Seeding in Structurally Diphasic Al_2O_3–MgO Xerogels," *J. Am. Ceram. Soc.*, **68** [9] C-238-C-240 (1985).

[22]Y. Suwa, S. Komarneni and R. Roy, "Solid-State Epitaxy Demonstrated by Thermal Reactions of Structurally Diphasic Xerogels: The System Al_2O_3," *J. Mat. Sci. Lett.*, **5** [1] 21–24 (1986).

[23]G. Vilmin, S. Komarneni and R. Roy, "Lowering Crystallization Temperature of Zircon by Nanoheterogeneous Sol-Gel Processing," *J. Mat. Sci.*, **22** [9] 3556–3560 (1987).

[24]M. Kumagai and G. L. Messing, "Enhanced Densification of Boehmite Sol-Gels by α-Al_2O_3 Seeding," *J. Am. Ceram. Soc.*, **67** [11]: C-230-C-231 (1984).

[25]S. Komarneni, Y. Suwa and R. Roy, "Enhancing Densification of 93% Al_2O_3-7% MgO Triphasic Xerogels with Crystalline α-Al_2O_3 and $MgAl_2O_4$ Seeds," *J. Mat. Sci. Lett.*, **6** [5] 525–527 (1987).

[26]C. A. Fyfe, G. C. Gobbi, J. S. Hartman, R. E. Linkinski, J. H. O'Brien, E. R. Beange, and M. A. R. Smith, "High Resolution Solid-State MAS Spectra of ^{29}Si, ^{27}Al, ^{11}B, and Other Nuclei and Inorganic Systems Using a Narrow-bore 400 MHz High Resolution NMR Spectrometer," *J. Magn. Reson.*, **47** [1] 168–173 (1982).

[27]G. Maciel and D. Sindorf, "Silicon-29 Nuclear Magnetic Resonance Study of the Surface of Silica Gel by Cross-Polarization and Magic Angle Spinning," *J. Am. Chem. Soc.*, **102** [25] 7606–7607 (1980).

[28]E. Lippmaa, M. Magi, A. Samoson, G. Engelhardt, and A. R. Grimmer, "Structural Studies of Silicates by Solid-State High Resolution ^{29}Si NMR," *J. Am. Chem. Soc.*, **102** [15] 4889–4893 (1980).

[29]E. Lippmaa, M. Magi, A. Samoson, M. Tarmak, and G. Engelhardt, "Investigation of the Structure of Zeolites by Solid-State High-Resolution ^{29}Si NMR Spectroscopy," *J. Am. Chem. Soc.*, **103** [17] 4992–4996 (1981).

Surface Chemistry Effects on Homogeneity and Crystallization of Colloidal Mullite Sol-Gels

J. C. HULING AND G. L. MESSING

Department of Materials Science and Engineering
The Pennsylvania State University
University Park, PA 16802

Abstract

Crystallization and microstructural development of mullite from colloidal sol-gels is directly related to colloid surface chemistry. Heterocoagulation between boehmite and silica particles yields homogeneous gels that transform directly to mullite between 1270–1300°C. Clustering of silica particles at intermediate pHs results in heterogeneously mixed gels that require longer-range diffusion and preferential crystallization of cristobalite and α-alumina at ~1250°C. The presence of the less reactive cristobalite and α-alumina inhibits complete mullite formation to 1450°C. Furthermore, an amorphous grain boundary phase, not observed in homogeneous gels, forms in the heterogeneous gels and is present until 1550°C.

Introduction

Sol-gel processes for the preparation of high-purity mullite may be broadly classified as either polymeric or colloidal (or, in some cases, a combination of these), depending on the nature of the alumina and silica precursors. Polymeric sol-gel systems are produced[1–3] by mixing metal alkoxides such as aluminum isopropoxide or sec-butoxide (or salts such as aluminum nitrate, sulfide or chloride) with tetraethoxysilane (TEOS) or tetramethoxysilane (TMOS). Acid-catalyzed hydrolysis of the mixture yields a homogeneous polymerized oxide network in which molecularly-mixed components crosslink.[4] But, due to the relatively slow hydrolysis rate of silicon alkoxides compared to those of aluminum, experimental conditions such as the ratio of water to alkoxide, catalyst concentration and reaction temperature must be carefully controlled to promote simultaneous reaction and homogeneous incorporation of the components.

In colloidal sol-gel systems the scale of component mixing is necessarily limited by the maximum size of the discrete particles involved (typically 10–30 nm). Colloidal or fumed silica is widely used, and the source of alumina is usually boehmite (γ-Al00H) or γ-Al_2O_3.[5,6] The gel structure in colloidal systems is primarily influenced by the surface chemistries of the precursor particles and the resulting electrostatic particle interactions that occur both in suspension and as the aqueous solvent is removed. Because H^+ and OH^- are the potential-determining ions for oxide and hydroxide colloids, the solvent pH and ionic strength control particle interactions.

Polymeric and colloidal mullite-precursor gels exhibit distinctly different mullitization behavior because of the direct relationship between diffusion distance and reaction kinetics. Polymeric gels have homogeneous, molecular-scale mixing and mullite formation requires only short-range diffusion through the amorphous mixed precursor. In such gels, crystallization of a cubic spinel-type phase and/or mullite occurs rapidly at 980°C, as evidenced by a sharp exothermic peak during differential thermal analysis.[3] However, the covalent bonding of crystalline mullite severely reduces diffusion rates and thus low bulk densities are obtained unless sintering temperatures >1600°C or external pressure are applied.

In contrast to polymeric gels, colloidal gels containing discrete yet reactive precursor particles require long-range diffusion. Hence, mullitization does not occur from mixtures of amorphous silica and transitional aluminas until 1250–1300°C. The increased crystallization temperature of the colloidal gels allows the amorphous colloidal phase to effect densification by viscous flow. The ensuing crystallization proceeds slowly, as evidenced by very broad exotherms during thermal analysis.[5,6]

The shortest interparticle diffusion distance and lowest crystallization temperature in colloidal gels corresponds to an ideal particle arrangement in which the two precursors' particles have the maximum amount of adjacent surface. In a mixed colloidal sol in which the two precursors' particles significantly differ in size, this arrangement corresponds to completely discrete smaller particles homogeneously surrounding and separating completely discrete larger particles. For boehmite and silica particles, ~100–150 nm and 20 nm diameters, respectively,[7,8] this configuration can be most nearly obtained experimentally by utilizing the component particles' opposite surface charges. As shown by the zeta potential-pH curve[9] in Fig. 1, there is a Coulombic attraction between negatively-charged silica and positively-charged alumina in the pH range 2–9 due to acid-base surface reactions. Although the data in Fig. 1 is for α-alumina, boehmite, γ-AlOOH, is nearly identical in surface character due to the overwhelming effect of surface hydration. Electrostatic attraction, combined with clustering of fine particles by van der Waals' interactions, is expected to lead to heterocoagulation and maximum pairing of unlike particles in the gel[10–12] as schematically shown in Fig. 2A. Outside the pH range 2–9, solubilities of the components increase sharply and adsorption of hydrated Al^{3+} and Si^{4+} ions from solution can significantly alter the particle surface chemistries.[13,14]

Heterocoagulation between discrete unlike particles does not necessarily occur whenever the pH is in the range where the components' surfaces are oppositely charged. If sol processing conditions are such that the particles of a given component are initially coagulated (generally an irreversible condition), homogeneous colloid mixing in the combined sol will be prevented. The effective diffusion distance will, in these cases, be increased from the average individual particle size to the average size of the initially-formed clusters. Segregation of components and gel heterogeneity expected under such conditions, as illustrated in Fig. 2B, necessitates longer times and higher temperatures for complete mullitization. In addition, it becomes more likely that independent crystallization of the separate components will occur, decreasing reactivities of the individual components and further hindering complete mullitization.

This paper presents an investigation of the role of colloidal sol homogeneity in mullite crystallization, crystallization of the precursors and microstructural development.

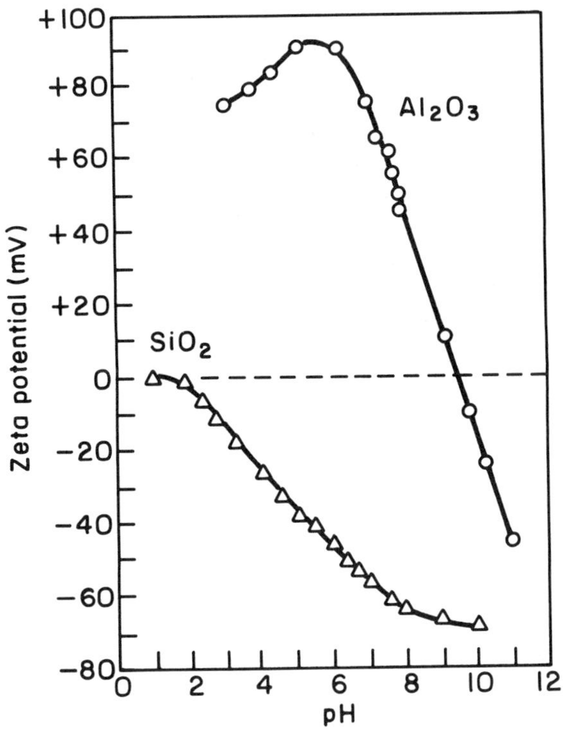

Fig. 1. Effect of pH on the zeta potential of silica and alumina particles (from ref. 9).

Experimental Procedure

Two gels differing only in precursor homogeneity were obtained by using extremes in colloid surface chemistry. The gels contained identical boehmite* and silica† and were 73.0 weight percent Al_2O_3 and thus were within the single-phase mullite phase field.

Gel A was designed to promote heterocoagulation of the boehmite and silica particles to obtain maximum homogeneity of precursor mixing (Fig. 2A) and thus the minimum diffusion distance for reaction. The silica sol was first diluted with deionized water such that upon addition of the boehmite powder the combined sol would be 20 weight percent solids. During dilution, the sol pH decreased from about 9 to slightly above 7, where strong electrostatic repulsive forces continue to dominate silica particle interactions. The boehmite powder was slowly added to the vigorously stirred silica sol at approximately one gram per minute to avoid forming clumps of powder in the mixture. During 90 minutes of stirring the pH returned to near 9. Concentrated HNO_3 was then added dropwise to the mixture to lower the pH to near

---

*Catapal SB boehmite, Vista Chemical, Ponca City, OK.

†Ludox HS collodial silica, Dupont, Wilmington, DE.

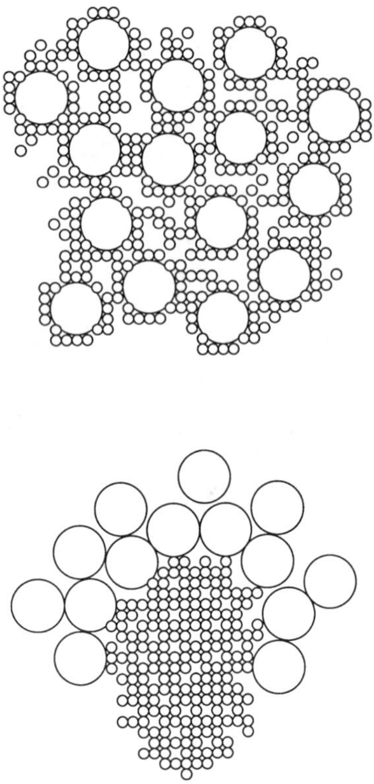

Fig. 2. Schematic diagrams of the expected precursor particle arrangements in gels A and B (small circles = silica, large circles = boehmite).

3 and allow gelation. While lowering the pH, a maximum in viscosity was observed in the pH range 5–6. According to Fig. 1, the alumina and silica surface charges would give a maximum product in this range, resulting in maximum electrostatic attraction between the unlike particles. The pH of the combined sol was maintained near 3 by further periodic additions of HNO_3 and the combined sol was stirred overnight prior to sample preparation (see below).

Gel B was designed to result in heterogeneous precursor mixing (Fig. 2B) and an increase in the diffusion distance for mullite crystallization. In this case the silica sol was diluted with deionized water from its original 40 weight percent solids content to the 20 weight percent solids content used in the final mixed sol. Concentrated HNO_3 was then added, prior to boehmite addition, to rapidly (<15 seconds) lower the silica sol's pH to near 2. At this pH, according to Fig. 1, the surface charge of the silica particles would be near zero, eliminating repulsive electrostatic forces and allowing coagulation of most colloidal systems. However, massive coagulation and settling of colloidal silica is generally not observed below pH 3 due to the formation of a substantial hydration layer that surrounds the particles and provides short-range steric protection.[8,15] This phenomenon, which is unique to silica, is the basis for the temporary "acidic stabilization" of colloidal silica. After stirring for 90 minutes, the acidic silica sol was added to a 20 weight percent boehmite sol which was pre-stabilized at pH 3 with concentrated HNO_3. The sol mixture was again stirred overnight prior to sample preparation.

In gel B, heterocoagulation between discrete particles does not occur. However, contrary to most reports the silica particles are probably not discrete but are instead small clusters. Silica cluster formation results from irreversible coagulative particle collisions during lowering of the pH of the 20 weight percent silica sol. Between the pH regimes of high electrostatic stability (pH>7) and temporary "hydration layer induced" stability (pH<3), colloidal silica rapidly coagulates and settles.[8] This coagulation process is generally irreversible[16] and thus clusters formed during passage through the intermediate pH regime will not be redispersed at pH 2. This is especially true at low pH, since silica surface charge here is negligible. Hydration layer formation within clusters may tend to further "bind" the individual particles together, while hydration layers on the surface of the clusters would sterically prevent further cluster growth.

Based on the above argument, the proposed structure of gel B contains more heterogeneous boehmite-silica mixing (expected to be similar to Fig. 2B) and fewer boehmite-silica contacts due to silica clustering.

Gel samples in the form of 1–2 μm thick unsupported films were produced by spin-coating the sol mixtures onto polished camphor discs, as described in detail elsewhere.[17] Following sublimation of the camphor discs' surfaces, the gel films were removed and sintered for two hours at temperatures from 1250°C to 1550°C. Microstructural analysis was performed on the sintered film surfaces using SEM imaging. Crystalline phase identification was performed using X-ray powder diffraction of sol mixtures that were dried, crushed and fired simultaneously with the gel films.

Results and Discussion

The crystalline phases detected in the fired A and B gels using X-ray diffraction are listed in Table 1 in order of their predominance at each temperature. The mullitization of gel A begins near 1270°C and is complete within two hours at 1300°C. At 1250°C, prior to mullitization, gel A contains a mixture of δ- and θ-alumina stabilized at this temperature by silica.[18] Mullite becomes the major phase at 1270°C, but the continued presence of δ and θ-alumina shows that the reaction is not complete within two hours at this temperature. At and above 1300°C, however, only mullite peaks are detected. This smooth transition of phases over a small temperature range is consistent with the expected crystallization behavior of a homogeneous mixture of the colloidal precursors requiring long-range (i.e., nanometer-scale rather than atomic-scale) diffusion. Differential thermal analysis of gel A during rapid (20°C/min) heating reveals a small exothermic peak at 1350°C. This peak temperature is displaced from the isothermal transformation temperature (~1270°C) due to the kinetics-limited mullitization.

The intimate boehmite-silica particle-particle configuration allows homogeneous high-temperature pre-crystallization dissolution of alumina throughout the viscous silica phase. Such incorporation of alumina into the amorphous silica structurally prevents independent crystallization of silica to cristobalite or of the transitional alumina to α-alumina.

The difference in phase formation between the two gels is first observed at 1250°C. In gel B silica forms cristobalite as a major phase. This supports the claim that the silica particles in gel B are not as homogeneously distributed as those in gel A. The clustering of silica particles in the gel B configuration necessarily allows less high-temperature dissolution of alumina into the viscous, amorphous silica phase.

Table 1. Crystalline Phases Present in the Fired Gels

Temperature	Gel A	Gel B
1250°C	δ & θ-alumina	δ & θ-alumina cristobalite
1270°C	mullite δ & θ-alumina	mullite δ & θ-alumina cristobalite α-alumina
1300°C	mullite	mullite α-alumina cristobalite δ & θ-alumina
1350–1450°C	mullite	mullite and α-alumina
1550°C	mullite	mullite

During sintering, dissolved alumina cannot fully penetrate the larger silica regions, hence leaving silica pockets in which formation of cristobalite can occur.

The amount of cristobalite in gel B is maximized at 1270°C, and mullite begins to form by reaction of the δ- and θ-alumina with the amorphous silica-alumina phase surrounding the cristobalite in the silica pockets. The mullite formation is identical to that occurring to a greater extent in gel A at this temperature. For the same reasons that silica is able to crystallize to cristobalite below 1250°C, the transitional δ- and θ-aluminas partially transform to α-alumina at 1270°C in the alumina-rich regions of gel B that result from the silica segregation.

At 1300°C only traces of δ- and θ-alumina and cristobalite are detected due to increased mullite and α-alumina formation. The disappearance of δ- and θ-alumina at 1350°C corresponds to the maximum formation of α-alumina, which then decreases slowly with increasing temperature as mullitization proceeds sluggisly from this stable form of alumina. After two hours at 1550°C only crystalline mullite is detected.

The complete mullitization of gel B requires higher temperatures due to longer-range diffusion of components and the crystallization of relatively inert phases that do not form in gel A. The independent formation of these more stable crystalline phases increases the difficulty of mullitization beyond that due to the component segregation by decreasing the components' reactivities. Although the cristobalite can react relatively easily to form mullite, the introduction of the very stable α-alumina subsequently requires higher temperatures and longer times for full reaction. The effects of such decreased component reactivity are seen in the retention of α-alumina after two hours at 1450°C in gel B. The kinetic effects resulting from the increased gel heterogeneity are shown by gel B's differential thermal analysis, in which the exothermic peak is 20°C above that for gel A. The longer-range component diffusion required for mullite formation in the heterogeneous gel B increases the time lag of the exothermic peak during rapid heating.

It is appreciated and should be noted that a contributing factor to the formation of cristobalite (and its consequences) in gel B is the presence of sodium as a stabilizing ion in the colloidal silica (Ludox HS) used in preparing both gels. This ion lowers the viscosity of the amorphous silica during sintering and facilitates the ordering of silica tetrahedra during cristobalite formation. The presence of dissolved alumina in the silica during sintering structurally hinders cristobalite formation. Therefore, the effect of the sodium on silica's independent crystallization is enhanced in gel B.

The differences in the mullitization of A and B gels, evidenced by the presence of intermediate phases in Table 1 and the differences in peak temperatures during differential thermal analysis, are also apparent from the microstructures of the sintered gel films in Figs. 3–5. At 1250°C, prior to mullitization, the films have indistinguishable microstructures (Fig. 3) because the scale of precursor segregation in gel B is below the resolution limit of the SEM. At 1550°C, however, where only mullite is detected by X-ray diffraction of both fired gels, it is apparent that a significant amount (i.e., <5%) of an unreacted amorphous phase remains at the gel B grain boundaries (Fig. 4B). No such phase is apparent in the microstructure of gel A (Fig. 4A) at any stage of transformation.

It is proposed that the microstructure in Fig. 4B results from alumina-rich mullite initially forming from the heterogeneous component distribution of gel B. This is possible due to the composition range of the single-phase mullite field, which extends from 71.8 to ~74 weight percent Al_2O_3. It would also be probable considering

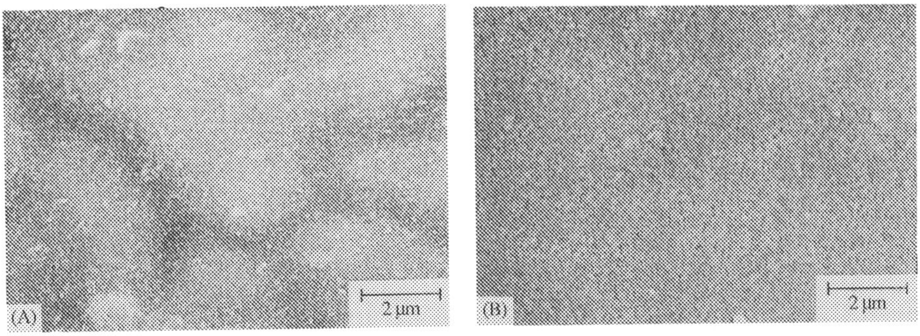

Fig. 3. Surfaces of films fired at 1250°C for 2 hours, (A) gel A, (B) gel B.

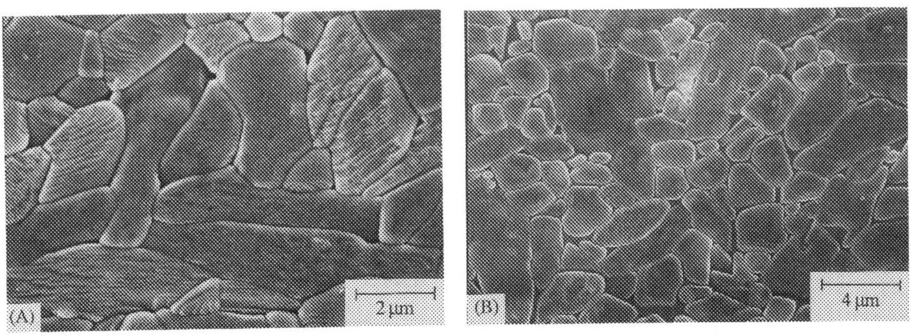

Fig. 4. Surfaces of films fired at 1550°C for 2 hours, (A) gel A, (B) gel B.

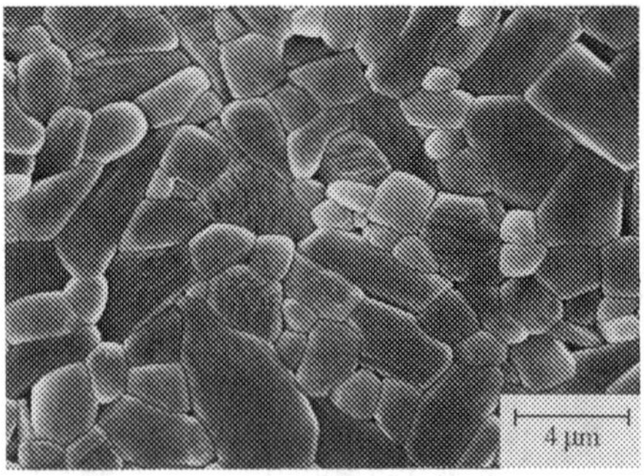

Fig. 5. Surface of fired film of gel B, 1550°C for 10 hours.

the effective "removal" of silica as a reacting phase by cristobalite formation at 1270–1300°C. By increasing the sintering time to ten hours at 1550°C the amorphous phase reacts with the Al_2O_3-rich mullite and the grain boundary phase disappears (Fig. 5).

Summary and Conclusions

Colloidal boehmite-silica gels begin mullite crystallization near 1270°C, a temperature high enough to allow prior viscous densification of the gel due to the presence of amorphous silica. The times and temperatures required to complete mullitization are dependent upon the gel's homogeneity. The homogeneity is directly related to adjustments in precursor sol pH and precursor particle surface chemistries. Heterocoagulation of discrete precursor particles yields a homogeneously mixed gel that transforms fully to mullite between 1270°C and 1300°C without individual component crystallization. Initial clustering of one precursor yields a heterogeneous, component-segregated gel in which longer-range diffusion is required for mullite formation. The segregation results in individual precursor crystallization and thus decreases component reactivities. These transformation sequence differences are clearly reflected in microstructural development, with no grain boundary phase formation in homogeneous gel-derived mullite and amorphous grain boundary phase formation in heterogeneous gel-derived mullite.

Acknowledgment

This work was performed at The Pennsylvania State University's Center for Advanced Materials with funding from the Gas Research Institute; additional support for J. C. Huling was provided by The Pennsylvania State University under the Graduate School Fellowship program.

References

[1]K. S. Mazdiyasni and L. M. Brown, "Synthesis and Mechanical Properties of Stoichiometric Aluminum Silicate (Mullite)," *J. Am. Ceram. Soc.*, **55** [11] 548–552 (1972).

[2]B. E. Yoldas, "Microstructure of Monolithic Materials Formed by Heat Treatment of Chemically Polymerized Precursors," *Am. Ceram. Soc. Bull.*, **59** [4] 479–483 (1980).

[3]K. Okada and N. Otsuka, "Characterization of the Spinel Phase from SiO_2-Al_2O_3 Xerogels and the Formation Process of Mullite," *J. Am. Ceram. Soc.*, **69** [9] 652–656 (1986).

[4]L. A. Paulick, Y. Yu and T. Mah, "Ceramic Powders from Metal Alkoxide Precursors," pp. 121–129 in Advances in Ceramics, Vol. 21, Ceramic Powder Science. Edited by G. L. Messing, K. S. Mazdiyasni, J. W. McCauley and R. A. Haber. American Ceramic Society, Westerville, Ohio, 1987.

[5]D. W. Hoffman, R. Roy and S. Komarneni, "Diphasic Xerogels, A New Class of Materials: Phases in the System Al_2O_3-SiO_2," *J. Am. Ceram. Soc.*, **67**, [7] 468–471 (1984).

[6]M. G. M. U. Ismail, Z. Nakai, K. Minegishi and S. Somiya, "Synthesis of Mullite Powder and its Characteristics," *Int. J. High Tech. Ceram.*, **2**, 123–134 (1986).

[7]Vista Chemical Company Technical Information, "Preparation and Rheology of CATAPAL Alumina Dispersions," 1987.

[8]R. K. Iler, The Chemistry of Silica. Wiley-Interscience, New York, 1979.

[9]R. Snel, "Control of the Porous Structure of Amorphous Silica-Alumina: 2. The Effects of pH and Reactant Concentration," *App. Catalysis*, **12**, 189–200 (1984).

[10]R. K. Iler, "Adsorption of Colloidal Silica on Alumina and of Colloidal Alumina on Silica," *J. Am. Ceram. Soc.*, **47** [4] 194–198 (1964).

[11]R. D. Harding, "Heterocoagulation in Mixed Dispersions," *J. Colloid Interface Sci.*, **40** [2] 164–173 (1972).

[12]P. E. Debely, E. A. Barringer, and H. K. Bowen, "Preparation and Sintering Behavior of Fine Grained Al_2O_3-SiO_2 Composite," *J. Am. Ceram. Soc.*, **68** [3] C-76-C-78 (1985).

[13]E. Matijevic, F. J. Mangravite, Jr., and E. A. Cassell, "Stability of Colloidal Silica, IV. The Silica-Alumina System," *J. Colloid Interface Sci.*, **35** [4] 560–568 (1971).

[14]R. K. Iler, "The Effect of Surface Aluminosilicate Ions on the Properties of Colloidal Silica," *J. Colloid Interface Sci.*, **55** [1] 25–34.

[15]J. A. Kitchener, *Disc. Faraday Soc.*, **52**, 379–400 (1971).

[16]J. T. G. Overbeek, "How Colloid Stability Affects the Behavior of Suspensions," pp. 25–44 in Materials Science Research, Vol. 17, Emergent Process Methods for High-Technology Ceramics. Edited by R. F. Davis, H. Palmour III, and R. L. Porter. Plenum Press, New York, 1984.

[17]J. C. Huling and G. L. Messing, "A Preparation Method for Unsupported Sol-Gel Thin Films," *J. Am. Ceram. Soc.*, **71** [4] C-222-C-224 (1988).

[18]R. K. Iler, "Effect of Silica on Transformations of Fibrillar Colloidal Boehmite and Gamma Alumina," *J. Am. Ceram. Soc.*, **47** [7] 339–341 (1964).

Sintering of Mullite Prepared By Sol-Gel Method

M. G. M. U. ISMAIL AND ZENJIRO NAKAI

Research and Development Division
Chichibu Cement Co. Ltd.
5310 Mikajiri
Kumagaya, Saitama, Japan

SHIGEYUKI SŌMIYA*

Research Laboratory of Engineering Materials
Tokyo Institute of Technology
4259 Nagatsuda
Midori, Yokohama, Japan

Abstract

The influence of Al_2O_3 content in mullite on physical, thermal and electrical properties was studied. Four compositions containing 70.0, 71.8, 73.0 and 75.0% by wt. of Al_2O_3 were prepared by sol-gel method. The main crystalline phase present on calcination of these gel powders was mullite. Their sinterability decreases with increasing Al_2O_3 content. The 70.0 wt% composition decomposes at 1700°C for 3 hours to high alumina mullite and silica which changed the grain morphology from equiaxed to elongated. These alumino-silicate compositions show a flexural strength of 350–400 MPa at room temperature. High alumina mullite shows low strength at temperature between 1000°C and 1300°C.

Introduction

Mullite $3Al_2O_3 \cdot 2SiO_2$) is considered as a suitable material for use in high temperature applications. The low dielectric constant and thermal expansion also permit mullite to be used as an electronic package material. Different methods of preparation[1–4] of mullite were tried recently. The main difficulty observed was homogeneous mixing of Al_2O_3 and SiO_2. Presence of a liquid phase along the grain boundaries in sintered specimens influences the high temperature properties. In the present investigation a sol-gel method was used to prepare mullite. Four compositions having different

***Present Address:**
The Nishi Tokyo University
3-7-19 Seijo
Setagaya, Tokyo, 157 Japan

proportions of Al_2O_3 and SiO_2 were selected. The main crystalline phase present in these four compositions was mullite. The compositions selected contained 70.0, 71.8, 73.0 and 75.0% by wt. of Al_2O_3. These will be hereafter referred to as A70, A72, A73 and A75, respectively. A70 and A75 mullites are outside the stable solid-solution limit proposed by Aksay et al.[5]

Experimental Procedure

Mullite contain 70.0, 71.8, 73.0 and 75.0% by wt. of Al_2O_3 were prepared by the sol-gel method. The details of the method of preparation are described elsewhere.[6] The gels formed were dried, powdered and calcined at 1400°C for 1 hour for mullitization. The phase identification was carried out by XRD method. The calcined powders were attrition-milled for 8 hours, and the slurry was dried at 120°C for 18 hours. The powders were uniaxially pressed into 50x30x8 mm blocks and followed by isostatic pressing under 200 MPa. The green compacts were sintered under ambient conditions between 1500°C and 1700°C for different durations. The bulk density of the sintered specimens was determined by liquid displacement method.

The sintered blocks were cut into 40x4x3 mm test pieces and the surface finishing was carried out in accordance with the JIS R-1601-1981 for Testing Method for Flexural Strength (Modulas of Rupture) of High Performance Ceramics. The flexural strength was determined by 3-point bend method. The fracture toughness was determined by measuring the crack length obtained by indentation method.

For SEM (model JSM-25S JEOL Ltd. Japan) observations the specimens were polished and thermally etched at 1500°C. For TEM (model JEM-2000FX, JEOL Ltd. Japan) observations the specimens were ion-thinned. The energy-dispersive microanalysis (EDX) was carried out using a probe size of =10nm. Dielectric constant was determined by using specimens of 50mm in diameter and 1.5mm in thickness at a frequency of 1 MHz at 25°C. Thermal diffusivity was measured by laser flash method and thermal conductivity was calculated. Thermal expansion was measured using a Dialatometer with Al_2O_3 as a standard.

Results and Discussion

Powder Characteristics

The XRD patterns of A70, A72, A73 and A75 powders calcined at 1400°C for 1 hour show only mullite as the crystalline phase. The powder characteristics of these compositions are summarized in Table 1. The larger surface area of A75 compared to others is due to incomplete solid-solution of Al_2O_3 even after calcination at 1400°C for 1 hour. The crystallite size also increases with increasing Al_2O_3 content. The average particle sizes of A70, A72, A73 and A75 after attrition milling were 1.08μm, 1.04μm, 1.05μm and 1.06μm respectively. The TEM observations of theses powders shows similar particle morphology.

Sintering

The powder compacts of A70, A72, A73 and A75 were cold isostatically pressed under 200MPa and sintered at temperatures varying from 1600°C to 1700°C for durations ranging from 1 to 3 hours. Typical sintering curves are given in Fig. 1. A70 composition could be sintered at a temperature of 1600°C to a density >98% of theoretical density. A slight decrease in sintered density was observed at 1700°C after a slight increase at 1650°C, which was caused by the decomposition of mullite to form high alumina mullite and silica. For A70 the condition selected for sintering

Table 1. Powder characteristics of Mullite compositions.

	A70	A72	A73	A75
Surface area (BET) m^2/g	12.7	12.6	13.9	15.9
Crystallite size (nm)	26.0	27.1	26.8	28.9
Lattice parameters (nm)				
a	0.7587	0.7567	0.7571	0.7574
b	0.7679	0.7690	0.7683	0.7683
c	0.2886	0.2886	0.2889	0.2886

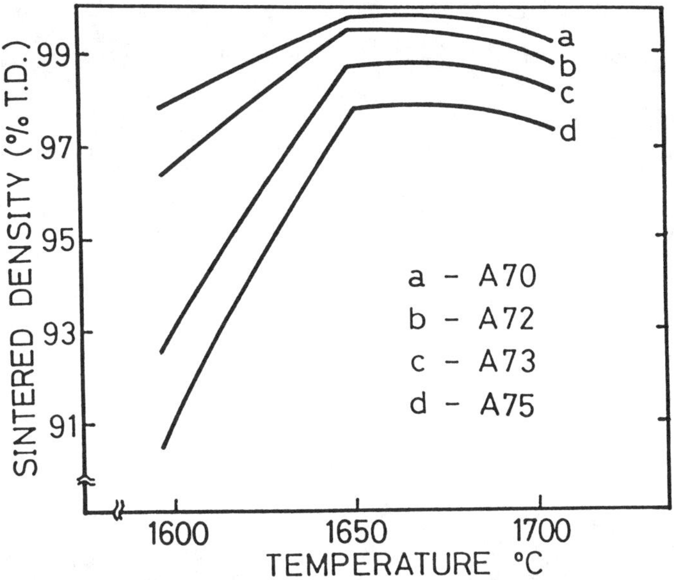

Fig. 1. Sintering curves of mullite compositions.

is 1640°C for 1.5 hours and the sintered density was 99.1% of theoretical. The SEM microphotographs of A70 sintered at 1640°C for 1.5 h consists of equiaxed grains; no elongated grains are observed. The average grain size was 2.0 μm (Fig. 2). With increase of sintering temperature to 1700°C the grain morphology changed from equiaxed to elongated which is typical for a matrix containing a glassy phase. Mullite A70 composition, where the silica content is higher than the stoichiometric mullite decomposes to alumina-rich mullite at temperatures above 1650°C. This resulted in excess liquid SiO_2 in the matrix and the grain morphology was changed. The triple point of A70 sintered at low temperature shows only a thin layer of grain boundary phase. The SEM microphotographs of A72 sintered at 1650°C for 1.5 h and 1700°C for 3 h are shown in Fig. 3. The grain morphology in both case was equiaxed. This

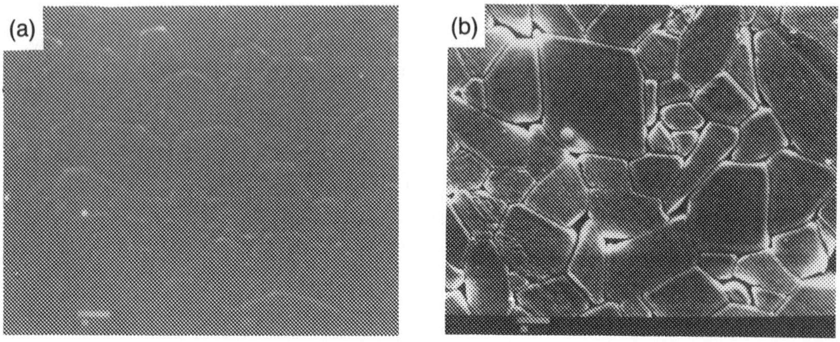

Fig. 2. SEM micrographs of A70 mullite sintered at (a) 1640°C for 1.5 h (b) 1700°C for 3 h (bar=1μm).

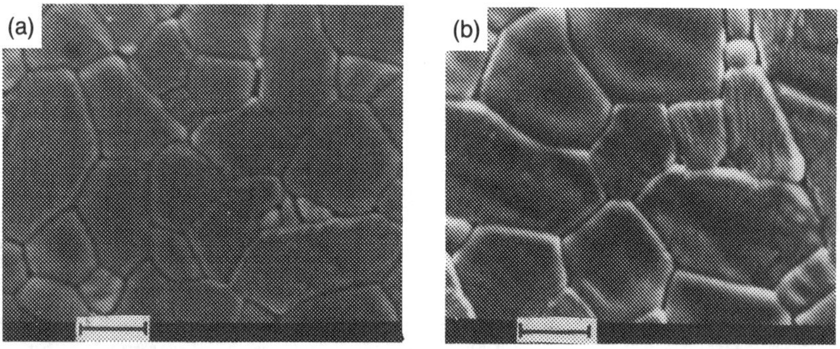

Fig. 3. SEM micrographs of A72 mullite compositions sintered at (a) 1650°C for 1.5 h (b) 1700°C for 3 h (bar=1μm).

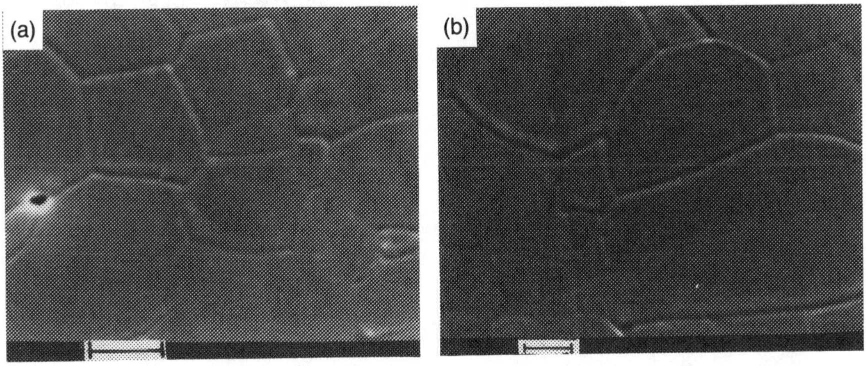

Fig. 4. SEM micrographs of A73 mullite compositions sintered at (a) 1650°C for 2 h (b) 1700°C for 3 h (bar=1μm).

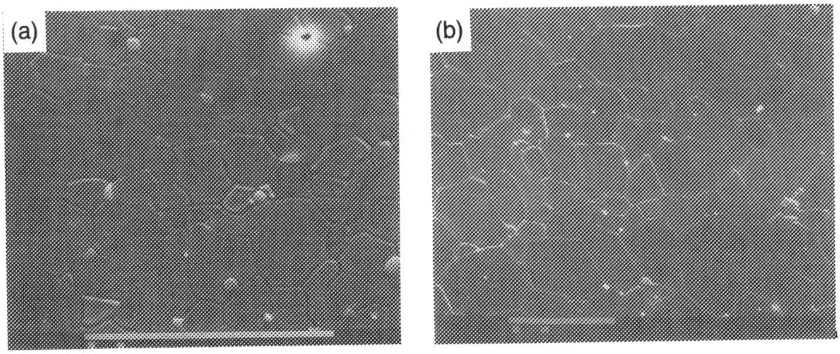

Fig. 5. SEM micrographs of A75 mullite compositions sintered at (a) 1650°C for 3 h (b) 1700°C for 3 h (bar=10μm).

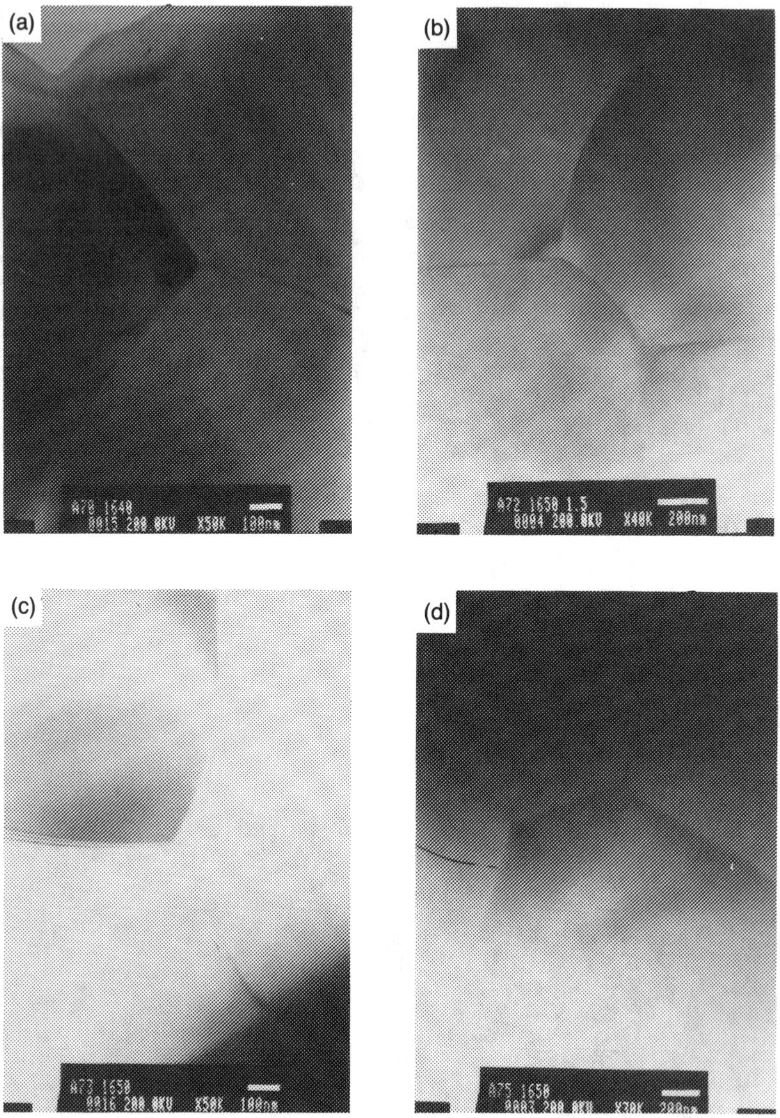

Fig. 6. TEM micrographs of triple grain junction of mullite compositions sintered at (a) A70–1640°C/1.5 h (b) A72–1650°C/1.5 h (c) A73–1650°C/2 h (d) A75–1650°C/3 h.

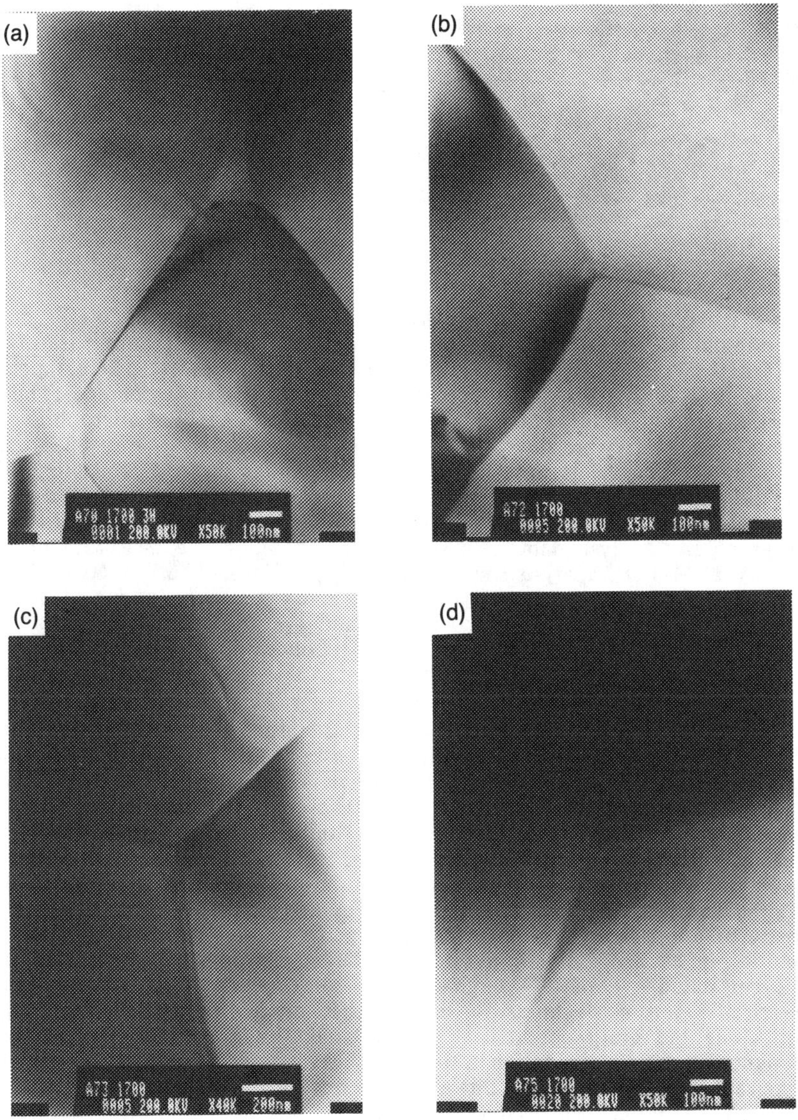

Fig. 7. TEM micrographs of triple grain junction of mullite compositions sintered at 1700°C for 3 h.

	Al (atom%)	Si (atom%)
a -	68.9	30.9
b -	62.5	37.3
c -	64.9	34.7
d -	70.3	29.6
e -	63.8	35.7

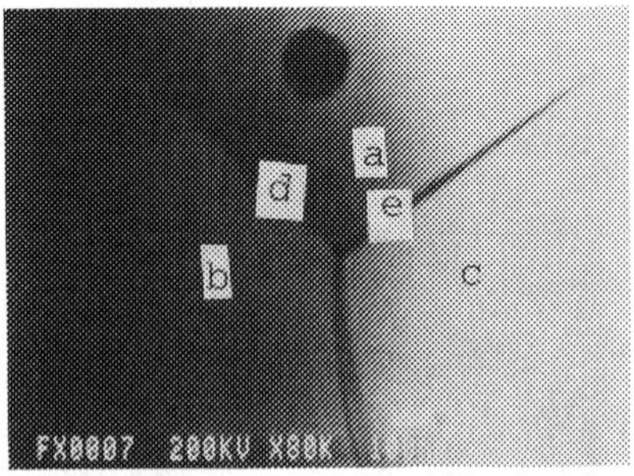

Fig. 8. EDX microanalysis of A72 mullite sintered at 1650°C for 1.5 h.

shows that the stoichiometric mullite A72 does not decompose at 1700°C. The SEM microphotographs of A73 and A75 sintered at different temperatures are given in Figs. 4 and 5, respectively. The microstructure of A73 and A75 is composed of equiaxed grains. The grain size increases with increasing Al_2O_3 content. The grain size of A75 mullite sintered at 1650°C for 3 h is 10 μm. The TEM microphotographs of the grain boundary are shown in Figs. 6 and 7, respectively. No secondary phase was observed along the grain boundary. Mullites containing Al_2O_3 in stoichiometric proportion or more do not decompose at around 1700°C. The energy dispersive microanalysis (EDX) was carried out on stoichiometric mullite (A72), and the TEM micrograph and analysis data are given in Fig. 8. The points analyzed within the grains by TEM-EDX analysis show similar compositions. The thin layer along the grain boundary shows the presence of zirconia and yttria with alumina silica. There was no free silica observed along the grain boundary. These impurities were a result of using TZP balls as the grinding media in attrition milling.

Physical Properties

The variation of flexural strength with temperature of A70, A72, A73 and A75 is shown in Fig. 9. The flexural strength decreases at 500°C for all compositions and increases again. The reason for this behaviour is not clearly understood. But the sudden increase in thermal expansion at 500°C creates an expansion gradient along the specimen which can lower the strength. Mullite of stoichiometric composition maintains a strength of 350 MPa from 1000°C to 1300°C and decreases to 300 MPa at 1400°C. With increase of Al_2O_3 content in mullite the high temperature flexural strength decreases. The fracture toughness of A72 is 2.5 MPa•$m^{1/2}$ and decreases with increase of temperature up to 1200°C and increases slightly again at 1300°C. As shown in Fig. 8 there was no liquid phase at the triple grain junction and only a very thin precipitated layer of ZrO_2 and Y_2O_3 was observed along the grain boundary. The increase of fracture toughness above 1200°C is not due to the presence of glassy phase along the grain boundary. As explained by Mah and Mazdiyasni,[7] the dominance of energy dissipation at the crack tip over grain boundary sliding results in apparent strengthening of mullite at high temperature. In the case of high alumina mullite (A75) similar behaviour was observed.

Electrical Properties

The electrical resistivity of A72 was determined using specimens of 50 mm in diameter and 1.5 mm in thickness. The average value obtained for electrical

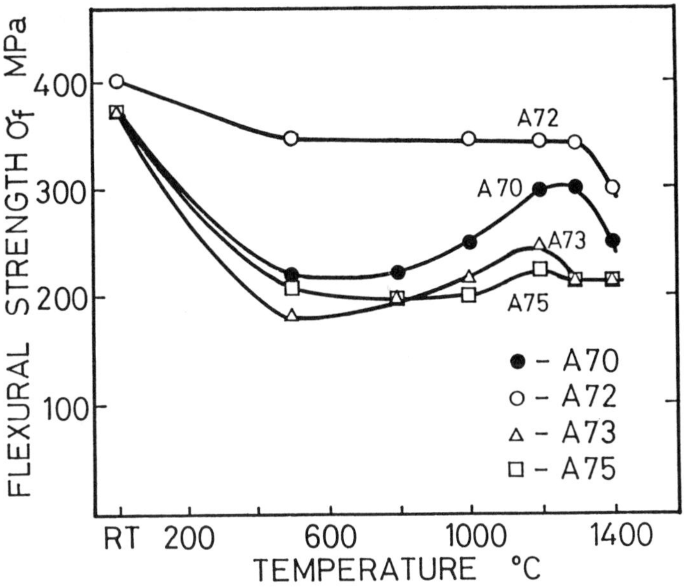

Fig. 9. Variation of flexural strength with temperature for different mullite compositions (sintering conditions A70–1640°C/1.5 h, A72–1650°C/1.5 h, A73–1650°C/2 h and A75–1650°C/3 h).

resistivity was $>10^{15}$ Ω•cm. Which is greater than the values ($>10^{13}$ Ω•cm) reported for commercial mullite. The presence of glassy phase along the grain boundary reduces the electrical resistivity. Whereas the increase is an evidence of absence of glassy phase and impurities such as Na_2O and K_2O. The values determined for dielectric constant of A70, A72 and A75 are 7.60, 7.63 and 7.65, respectively. Within the mullite composition range studied the dielectric constant did not vary remarkably. Increase of SiO_2 beyond 30% by wt. influenced the dielectric properties. This is due to the presence of glassy phase in the matrix.

Thermal Properties

Thermal expansion data of compositions studied are given in Table 2. With increase of Al_2O_3 content the coefficient of linear expansion increases. Mullite also shows a sharp increase of thermal expansion between 400°C and 800°C. This is the cause of low flexural strength between 500°C and 800°C.

Variation of thermal diffusivity of A70, A72, A73 and A75 with temperature are given in Fig. 10. Thermal diffusivity of A70 and A72 decreases with increase of temperature, whereas A73 and A75 show a completely different behaviour. Thermal diffusivity of A73 and A75 decreases up to 600°C and increases again. The reason for this behaviour is not known.

Summary

The four compositions selected in this study contain Al_2O_3 from 70 to 75% by wt. The main phase present in four compositions was mullite. The sinterability decreases with increasing Al_2O_3 content. The high Al_2O_3 mullite (A75) could be sintered only

Table 2. Thermal expansion data of Mullite compositions.

	$\alpha_{20-400°C}/°C \times 10^6$
A70	4.30
A72	4.30
A73	4.37
A75	4.48

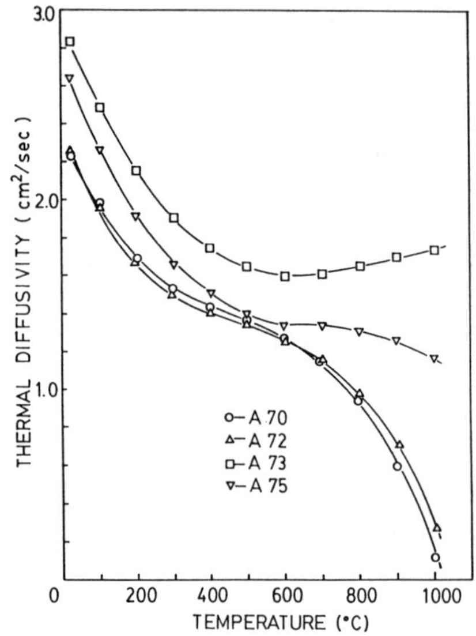

Fig. 10. Thermal diffusivity versus Temperature for A70, A72, A73 and A75 mullite compositions.

up to 97% of theoretical density whereas stoichiometric mullite A72 could be sintered to >99% of T.D. at 1650°C for 1.5h. The microstructure of all the compositions consisted of equiaxed grains, but the grain size increased with the Al_2O_3 content. After sintering these compositions at 1700°C for 3 hours, A70 mullite underwent decomposition leaving a glassy phase in the matrix. Stoichiometric mullite A72 shows the highest flexural strength of 400 MPa at room temperature. Other compositions show flexural strength of 350 MPa at room temperature. With increase of Al_2O_3 content high temperature flexural strength decreases. The electrical resistivity and dielectric constant are nearly constant within the solid solution limit.

Acknowledgment

Thanks are due to Takao Kumaki and Hideo Tsunatori for technical assistance.

References

[1]K. S. Mazdiyasni and L. M. Brown, "Synthesis and Mechanical Properties of stoichiometric alumino silicate (Mullite)," *J. Am. Ceram. Soc.*, **55** (11), 548–52, 1972.

[2]B. B. Ghate, D. P. H. Hasselman, and R. M. Spriggs, "Synthesis and characterization of high purity fine grained mullite," *Am. Ceram. Bull.*, **52** (9), 670–72, (1973).

[3]S. Kanzaki, H. Tabata, T. Kumazawa, and S. Ohta, "Sintering and Mechanical Properties of stoichiometric Mullite," *J. Am. Ceram. Soc.*, **68** (1), C_6-C_7, (1985).

[4]T. Yamaguchi, M. Suzuki, S. Hiraishi, M. Yoshimura, and S. Sōmiya, "Hydrothermal crystallisation of alumino silicate fine powder by hydrolysis of mixed alkoxides," *J. Mater. Soc. Jpn.* **21**, 241– , (1984), (in Japanese).

[5]I. A. Aksay and J. A. Pask, "Stable and metastable equilibria in the system SiO_2-Al_2O_3," *J. Am. Ceram. Soc.*, **58** (11–12), 507–12, (1975).

[6]M. G. M. U. Ismail, Zenjiro Nakai, Keiichi Minegishi and Shigeyuki Sōmiya, "Synthesis of Mullite powder and its characteristics," *Int. J. High Technology Ceramics*, **2**, 123–134, (1986).

[7]T. Mah and K. S. Mazdiyasni, "Mechanical Properties of Mullite," *J. Am. Ceram. Soc.*, **66** (10), 699–703, (1983).

Preparation and Characterization of Mullite Powders From Alkoxides and Other Chemical Routes

K. S. MAZDIYASNI

General Atomics
San Diego, California 92138

Abstract

Hydrolytic and thermal decomposition of metal alkoxides is a convenient method for preparing highly active, high purity, homogeneous and ultra-fine particle size bulk ceramic powder for characterization and consolidation studies. This method has been successfully applied to the preparation of many ceramic materials including aluminum silicate, mullite, $3Al_2O_3{\cdot}2SiO_2$ systems. Hot pressing of the mullite powders prepared by this technique resulted in near theoretically dense bodies with uniform microstructure containing acicular and polygonal grains 1 to 3 μm in size. Transmission electron microscopy, high temperature X-ray, infrared spectrascopy are used to observe nucleation, growth, crystallite morphology of the synthesized powders, and microstructural features. Mechanical properties at room and elevated temperatures are discussed.

Introduction

Phase Relationships

High purity mullite has excellent potential for high temperature applications, since it exhibits excellent stability, low thermal expansion, and good high temperature strength and creep resistance. Phase relationships in the vicinity of mullite are very sluggish and considerable research has been devoted toward their clarification, often with contradicting results. A recent review of the literature has been accomplished by Klug[1]; key points relative to the current study will be noted here. Originally, Bowen and Greig[2] described mullite as the compound $3Al_2O_3{\cdot}2SiO_2$ (72% Al_2O_3) with no solid solution and incongruent melting at 1820°C. In contrast, Toropov and Galakhov[3] reported mullite to exist over the 72% to 78% Al_2O_3 composition range and melting congruently at 1910°C. Aramaki and Roy[4] also reported congruent melting, but at 1850°C and stated that mullite existed over the 71.8% to 74.3% Al_2O_3 region. Further work by Aksay and Pask,[5] and others, confirmed the incongruent melting of mullite and defined a 70.5% to 74.0% composition range. More recently, Prochazka and Klug[6] defined the mullite range at 72% to 75% Al_2O_3 at 1600°C, changing to 74% to 76% at 1800°C and melting of a 76% composition at 1905°C. The latest study of Klug[1] confirmed this mullite range and reported the incongruent melting of the 77% composition at 1890°C.

Synthetic Mullite

Considerable effort has been directed in recent years toward the synthesis of polycrystalline stoichiometric mullite with the composition $3Al_2O_3 \cdot 2SiO_2$. In the usual method of preparing mullite, kaolinite, and related materials are thermally decomposed to mixed oxides.[7] Mixtures of various kinds of Al_2O_3 and SiO_2 have also been commonly used.[8] The coprecipitation of Al and Si salts is a convenient method for preparing an amorphous, highly active material.[9] A variation of this method is the slow injection of a solution of the metal halides in organic solvents through a rubber diaphragm into a vigorously stirred NH_4OH solution at 50°C.[10] Several materials other than clay and gels, e.g., mixtures of α-quartz, silicic acid, or β-cristobalite with diaspore ($HAlO_2$), gibbsite [$Al(OH)_3$], or α-alumina, have been reported.[11] Thermal decomposition of methylsiloxyaluminum compounds has also been used to prepare mullite.[12]

Current Study

Emphasis in this paper is placed on the use of metal alkoxides to prepare high-purity, highly active, homogeneous high temperature polycrystalline mullite powders. Alkoxides are $M(OR)_n$ compounds, where n is the valence of the metal, M, and R is an organic group. Thermal and hydrolytic decomposition of metal alkoxide $M(OR)_n$ can be employed to obtain hydroxide, sol, gels, and submicron size <50 nm single- and mixed-phase amorphous and or crystalline oxide powders simply by employing an alkoxide or mixed alkoxides as the starting materials.

Synthesis of Alkoxy Derived Mullite

Typically, mixed metal alkoxides in a mutual solvent, such as isopropyl alcohol, are decomposed hydrolytically to yield the corresponding mixed metal oxides, hydroxides, and/or sols.

Example of synthesis of polycrystalline ceramic powders, such as mullite, is as follows:

$$6Al(OC_3H_7)_3 + 2Si(OC_2H_5)_4$$

$$+ \text{ Excess } H_2O \xrightarrow[NH_3]{\text{Trace HNO3}} 2[(3AlSi)(OH)_{13}] + \text{alcohols} \qquad (1)$$

$$2[(AlSi)(OH)_{13}] \xrightarrow[\Delta]{\text{Vacuum}} 3Al_2O_3 \cdot 2SiO_2 + 13H_2O \qquad (2)$$

According to Paulick, Yu, and Mah,[13] the above hydrolytic reaction is very sensitive to experimental conditions, such as the presence of acidic or basic catalysts, reaction temperature, and molar ratio of alkoxides to H_2O. Three critical experimental parameters, pH, temperature, and reaction time, which affected the rate of hydrolysis, were systematically studied. For all of the experiments, the volume ratio of 100 ml of mixed alkoxide solution to 500 ml of a H_2O/CH_3OH (3:1 ratio) solution was used. To ensure reproducibility of each experiment, a large batch of mixed alkoxide solution was prepared in excess isopropyl alcohol and aliquots were withdrawn from this.

The same authors claim that in an acidic environment (pH = 2), the hydrolytic decomposition was completed within 7 hr at room temperature. Stoichiometric mullite was the only product obtained as evidence by both elemental analysis and an

X-ray diffraction (XRD). Under the alkaline condition (pH = 10), the degree of hydrolytic decomposition was both time and temperature dependent. The TEOS was only partially hydrolyzed even after heating at 70°C for 20 hr. The oxide powder obtained was composed of 65% Al_2O_3 and 35% mullite. In the neutral condition, pH = 6.5, the oxide powder obtained was mainly alumina. The degree of hydrolysis is only moderately affected by temperature and it is not affected by the reaction time. The result is consistent with the reported IR spectra.[13] Therefore, the hydrolytic decomposition of mixed TOES and $Al(OC_3H_7)_3$ alkoxides is controlled mainly by the rate of TEOS decomposition, since $Al(OC_3H_7)_3$ can be readily hydrolyzed in water without any catalysts.

The difference between acid versus base catalyzed hydrolysis of mixed TEOS and $Al(OC_3H_7)_3$ can be accounted for by the different reaction mechanisms of TEOS hydrolysis. Nucleophilic hydrolysis and condensation in alkaline solutions tend to produce highly cross linked species, which may not be completely hydrolyzed. In contrast, in acidic solutions, the electrophilic reaction mechanism favors the production of weakly cross linked species which tend to be completely hydrolyzed.[15]

Acid Catalyzed

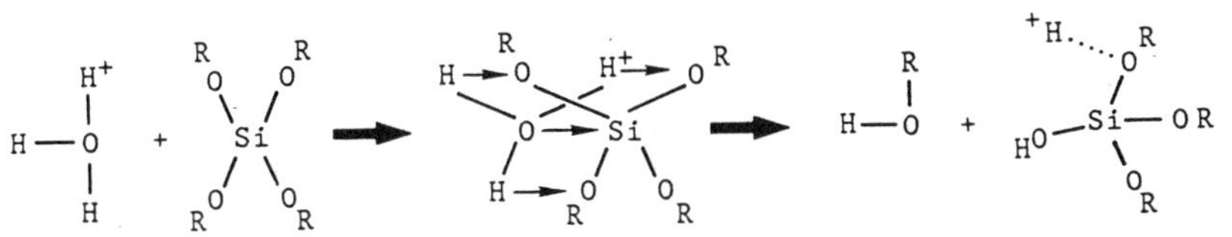

(3)

Base Catalyzed

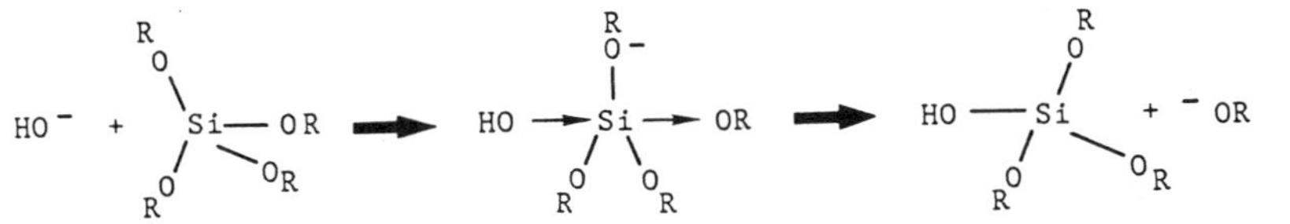

(4)

The specific surface area of the as-calcined mullite powder obtained from acidic hydrolysis condition as reported by Paulick, et al.[13] was 197 m^2/g. It was noted that clusters of particles developed during concentration of the hydroxide slurry. These particles polymerized upon further drying, causing bound -OH groups in the clusters, thereby resulting in hard agglomerates instead of fine powder. Ultra-fine mullite powder, with a surface area of 416 m^2/g, was obtained by hydrolytic decomposition of mixed TEOS and $Al(OC_3H_7)_3$ under an alkaline condition.

Purity and Particle Size

Spectrographic analysis for impurities in the mixed oxide calcined at 600°C for 1 hr showed <230 ppm impurities present in two representative batches of metal alkoxides decomposed to the mixed oxides. Separate analyses for the oxides demonstrated the maintenance of high-purity levels on a continuing basis. The most interesting aspect of oxide preparation in this manner, however, is not the very high purity but the unusually small particle size.

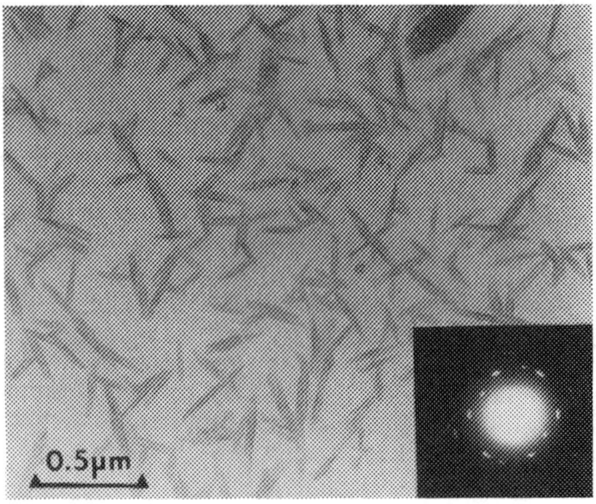

Fig. 1. Electron micrograph of as prepared mullite.

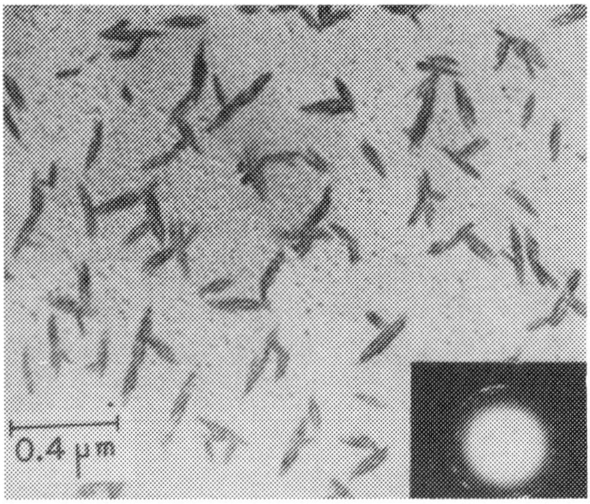

Fig. 2. Electron micrograph of mullite powders calcined at 600°C for 1 hr.

Figures 1 and 2 show electron micrographs of as prepared and calcined at 600°C for 1 hr in air particles of dispersed mullite powders. Samples were prepared by dispersal in ethyl alcohol using ultrasonic vibration, followed by spraying the dispersion on thin carbon substrates. Other techniques, such as dispersal in distilled water or xylene, resulted in agglomerated samples and unsatisfactory determinations. The as-prepared powders are extremely active with a surface area of 550 ±10% m^2/g; however, the surface area is reduced to half when the powders are calcined at 600°C for 1 hr. This reduction in surface area is attributed to nucleation and growth of the finer particles to larger crystallites.

The needlelike morphology of the crystallites is retained in the absence of a liquid phase, in the alkoxy-derived mullite powder prepared near-room-temperature conditions and calcined to higher temperatures, contrary to the findings of previous investigators. The intimate mixing of highly active fine particulates in the alkoxide decomposition process described is apparently responsible for formation of the acicular morphology which is characteristic of crystalline mullite.

High Temperature X-Ray

The X-ray diffraction pattern of alkoxy-derived mullite was observed from room temperature to 1705°C on a Pt40Rh substrate using CuKα radiation. Temperatures were measured with a Pt-Pt10Rh thermocouple. Initially the powder was amorphous, with a very broad peak extending from 24° to 31° 2θ. No significant changes were observed up to 435°C. As the temperature was increased at ~200°C intervals between 435° and 1100°C the aforementioned peak disappeared completely. However, at 1185° to 1200°C peaks corresponding to orthohombic mullite, with the exception of a peak at 2θ = 16.4°, began to appear. From 1200° to 1700°C all peaks corresponding to mullite were observed and increased in intensity with temperature. This phase was retained during cooling to room temperature and recycling of the same sample, indicating complete solid solution of the $3Al_2O_3 \cdot 2SiO_2$ under these experimental conditions.

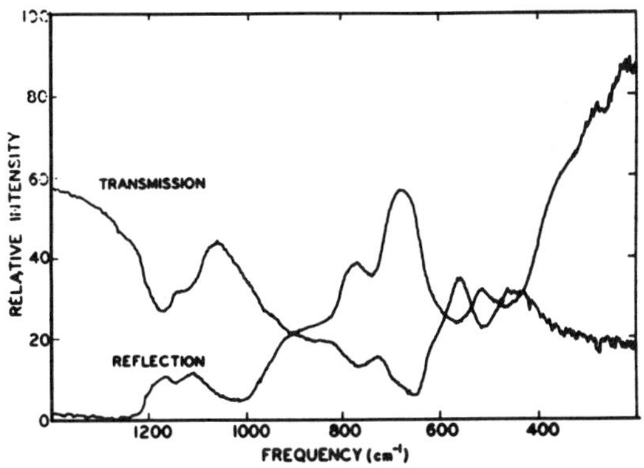

Fig. 3. Infrared spectra of alkoxy-derived mullite.

Infrared Absorption Frequencies

The mullite powder transmission spectrum and corresponding reflection spectrum of a polycrystalline solid sample are shown in Fig. 3. A 3-mg sample of the oxide was dispersed in 400 mg of anhydrous spectrographic grade CsI powder and pressed into disks for IR studies. The IR transmission spectrum was obtained using a grating spectrophotometer over the region 1400 to 200 cm^{-1}, where the spectral slit width seldom exceeded 2 cm^{-1}. This instrument was also used to measure specular reflection of the solid sample at 15° off normal incidence. Polarization effects were uncorrected and uncompensated for in this spectra. The IR spectrum of alkoxy-derived mullite differs slightly from that reported for standard mullite,[16] although the overall contour of the bands is in agreement and a one-to-one correlation exists between transmission minima. There are minor frequency differences observed in the 900, 500, and 450 cm^{-1} regions assigned to ν(SiO), ν(AlO), and ∂_d(OSiO), respectively, by MacKenzie.[17] These frequency variations may in part result from the thermal history of the powder as well as the inherent broadness of the observed peaks.

The IR spectrum of a material depends on its crystal structure; if it is inhomogenous, a different spectrum is associated with each local structure. There is evident that additional IR-active modes may be activated or obliterated in inorganic spectra by thermal history and mechanical damage to the surfaces of the particulates.

Agglomeration

Hoch and Nair[18] and Haberko[19] have demonstrated that calcination of gels washed with alcohol resulted consistently in a powder composed of agglomerates which were much more porous with low powder density and mechanically weaker than agglomerates obtained from gels washed with water. Apparently these weak agglomerates could be easily crushed under compaction pressure, in contrast to the behavior of the strong agglomerates obtained from gels washed with water. Haberko[19] suggests that, in compacts of the strong agglomerates two populations of pores very different in size are present; this characteristics of the strong agglomerates is perhaps related to their resistance of crushing. The presence of the two different inter- and intra-agglomerates pore size lead to severe fluctuations of the sintering driving force across the body and the resultant densification of the body is poor. However, the collapse of weak agglomerates observed on compaction results in a very fine uniform microstructure of the compact.

Consolidation

The as-prepared $3Al_2O_3{\bullet}2SiO_2$ powder was calcined at 600°C for 1 hr and ground in a B_4C mortar to effect comminution of the larger agglomerates before processing. The very fine, dry powder is quite active and readily becomes statically charged because of the very high surface area of the particulates and their clean surfaces. Specimens 3/4-in. to 2-1/2-in. in diameter were uniaxially vacuum-hot-pressed at ~0.013 Pa in graphite dies with Ta, Mo, or graphfoil liners. Some specimens were also subjected to reactive hot-pressing (more commonly known as pressure calcintering) in graphite dies.

Completely white specimens with densities 99% to 99.5% of theoretical (3180 to 3190 kg/m^3) were readily vacuum-hot-pressed at 35 MPa at 1400° to 1600°C for 15 to 60 min. Results were best when powders were consolidated at 1500°C for 30 min. The concentration of SiO_2 did not change significantly under these experimental conditions. Electron microprobe analysis of the vacuum-hot-pressed samples indicated a homogeneous dispersion of SiO_2 in the Al_2O_3 matrix.

Microstructure

The unique microstructural features of thermally etched bodies prepared by this technique, shown in Fig. 4 is interlocking needlelike, acicular, and polygonal grains, depending on both the composition of the powder used and firing schedule.

Additionally, minor compositional changes and different thermal histories of mullite samples result in a significant change in microstructure as well as variation in the observed cell dimensions of the orthorhombic crystal structure. These variations result mainly from the many possibilities for Al-Si order-disorder.[16]

The TEM of the alkoxy-derived mullite are shown in Fig. 5. In most part, the grain boundaries are clean and intergranual porosity is absent. Most of the mullite grains have a size less than 1 μm. However, the elongated polygonal grains are 2 or 3 times larger in grain size. Figure 6 shows that some pores are trapped in grains and exhibit faceted features. Also a few glassy-like boundary layers are observed. The sample appears to be free from other type defects, except dislocations as shown in Fig. 7.

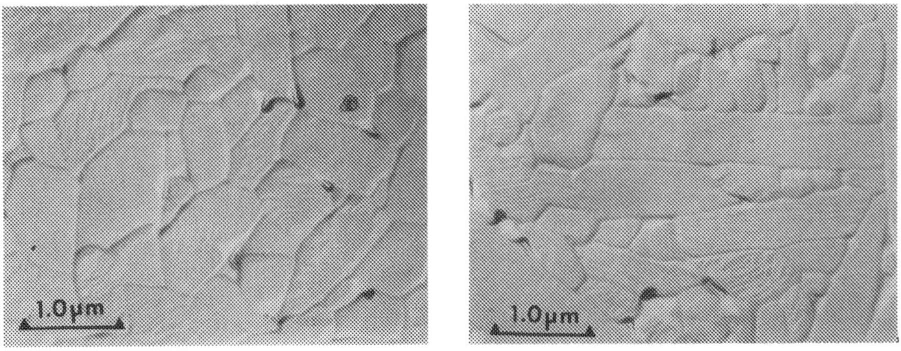

Fig. 4. Electron micrographs of mullite vacuum hot-pressed at 3.5 MPa and 1500°C for 1/2 hr.

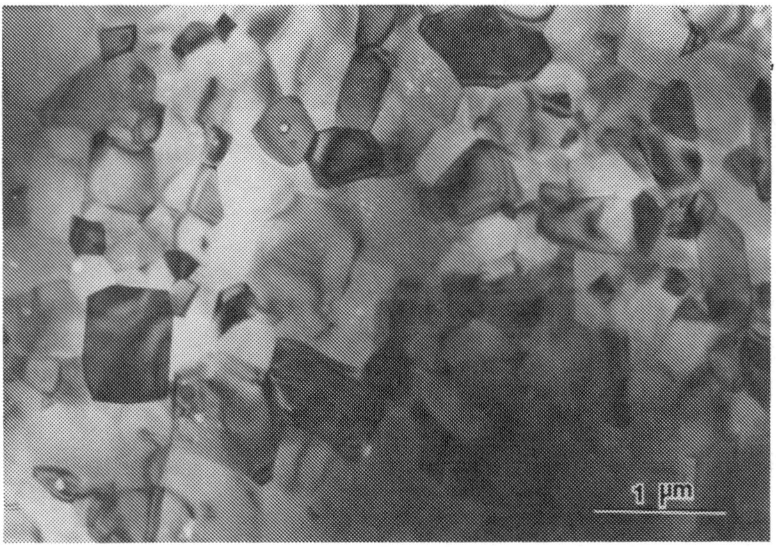

Fig. 5. Transmission electromicrograph of alkoxy-derived mullite.

Fig. 6. Transmission electromicrograph of elongated grains with some pores entrapped within grains.

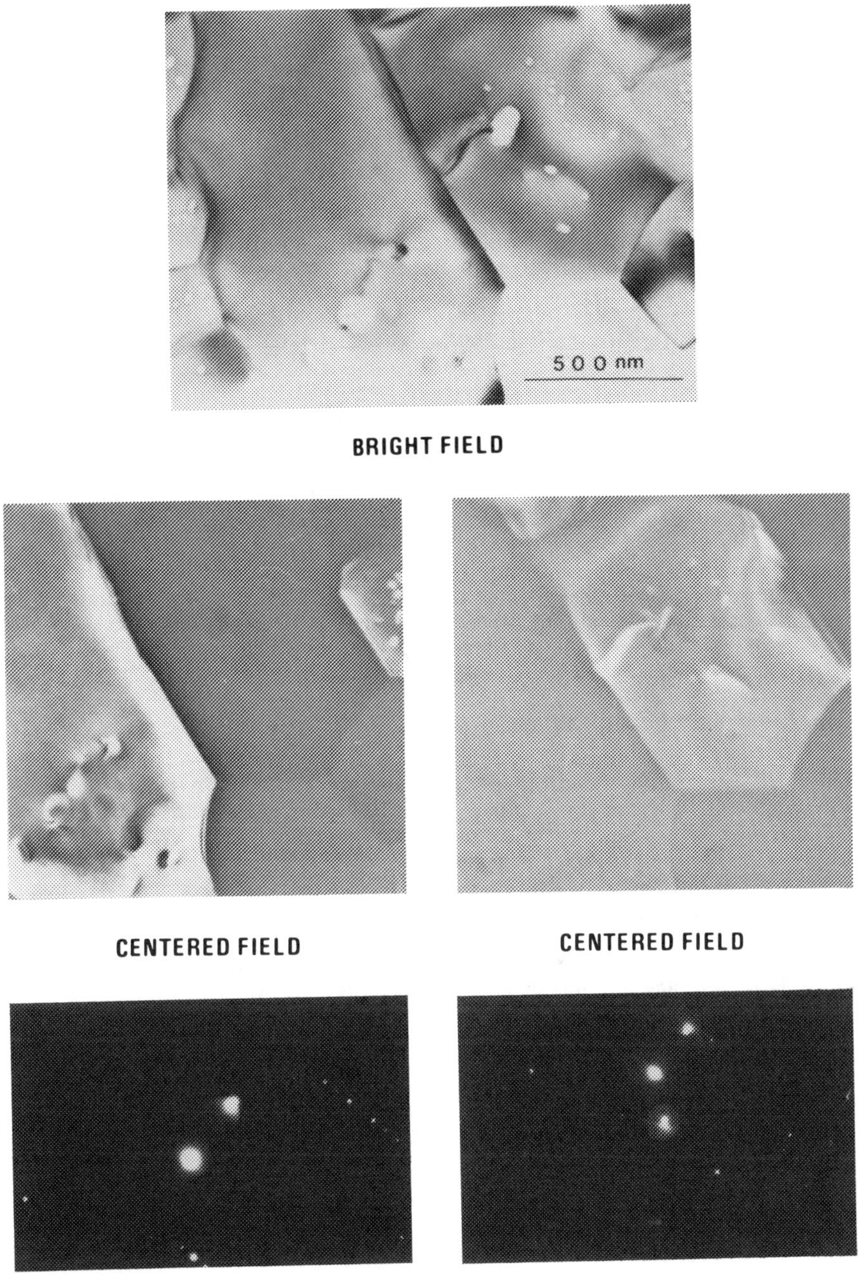

Fig. 7. Mullite grains with dislocation.

Mechanical Properties

Mechanical property data of mullite have been reported by a number of investigators. Mazdiyasni and Brown[20] found a four-point flexural strength of 273 MPa for hot-pressed material. Penty[21] reported a value of ~170 to 240 MPa in three-point bending for hot-pressed material, while Metcalfe and Sant[22] reported 152 MPa in three-point bending for sintered material. Mah and Mazdiyasni[23] reported a four-point value of 130 MPa at room temperature with a slight increase to ~145 MPa at 1400°C and a fracture toughness of 1.7 $MPam^{1/2}$ at room temperature. Kanzaki, et al.,[24] reported a three-point bend strength of 360 MPa with a slight decrease at 1400°C, and a room temperature fracture toughness of 2.8 $MPam^{1/2}$. In a later article[25] the strength and fracture toughness of mullite are reported at room temperature and 1300°C as a function of composition. Room temperature strength did not vary significantly (360 to 400 MPa) but at 1300°C the 68% Al_2O_3 composition had a strength of 600 MPa, and 71.8% composition had a strength of ~400 MPa and the 74% composition had a strength of ~250 MPa. Fracture toughness at room temperature also did not vary with composition (2.4 to 2.8 $MPam^{1/2}$), but at 1300°C the 68% Al_2O_3 composition was ~4.7 $MPam^{1/2}$ whereas the other compositions did not change.

Dokko, et al.[26] observed compressive fracture stress of ≈8000 kg/cm^2 (≈110,000 psi) at 1400°C which appeared to exceed all reported values for polycrystalline oxide bodies. Also the creep resistance of the mullite measured by Dokko, et al.[26] at 1400°C was similar to that measured by Lessing, et al.[27] in a four-point bending. Both bending and compressive creep data exhibited a linear stress dependence. The apparent activation energy for creep was determined to be ≈170 Kcal/mol. This value agrees not only with those reported previously,[26,27] but with activation energy for lattice diffusions (≈168 Kcal/mol) obtained by Aksay and Pask.[5]

Conclusion and Summary

The alkoxide decomposition process for the preparation of single phase high purity and fine grain size mullite has been demonstrated. The extremely high purity in excess of 99.95% can be obtained. Particle size ranges below 50 nm are easily achieved while maintaining exceptionally high level of purity. The oxide powders have extremely high activity during subsequent solid state reactions making feasible low-temperature firing processes to prepare high density oxide ceramics. The alkoxy technique is applicable to all mixed oxides where the metal alkoxides are readily prepared or commercially available. Advantage of the process, lie in the intimate mixing of the highly active powders as prepared facilitating further processing steps while maintaining stoichiometry and purity throughout the forming process.

Hydrolytic decomposition of metal alkoxides provide an inexpensive, economical route toward the production of high purity fine particulate bulk powders and fine grained ceramics on a large scale.

References

[1]F. J. Klug, "Alumina-Silica Phase Equilibria in the Mullite Region," Ph.D. Dissertation, Rensselaer Polytechnic Institute, Troy, New York, December 1984.

[2]N. L. Bowen and J. W. Greig, "The System Al_2O_3-SiO_2," *J. Am. Ceram. Soc.*, 7 (4) 238–254 (1924).

[3]N. A. Torpov and F. Y. Galakhov, "The SiO_2-Al_2O_3 System," *Dokl, Akad. Nouk. SSSR*, **78** (2) 299–302 (1951).

[4]S. Aramaki and R. Roy, "Revised Phase Diagram for the System Al_2O_3-SiO_2," *J. Am. Ceram. Soc.*, **45** (5) 299–242 (1962).

[5]I. A. Aksay and J. A. Pask, "Stable and Metastable Equilibria in the System SiO_2-Al_2O_3," *J. Am. Ceram. Soc.*, **58** (11-12) 507-512 (1975).
[6]S. Prochazka and F. J. Klug, "Infrared-Transparent Mullite Ceramic," *J. Am. Ceram. Soc.*, **66** (12) 874-880 (1983).
[7]G. W. Brindley and M. Nakahira, "The Kaolinite-Mullite Reaction Series: I, II, and III," ibid., **42** (7) 311-324 (1959).
[8]L. B. Pankratz, W. W. Weller, and K. K. Kelley, "Low-Temperature Heat Content of Mullite," U.S. Bur. Mines, Rep. Invest., No. 6287, p. 7 (1963).
[9]R. Roy, "Aid in Hydrothermal Investigation: II," *J. Am. Ceram. Soc.*, **39** (4) 145-146 (1956).
[10]J. D. Crofts and W. W. Marshall, "Novel Synthesis of Alumino-Silicates and Similar Materials," *Trans. Brit. Ceram. Soc.*, **66** (3) 121-126 (1967).
[11]F. M. Wahl, R. E. Grim, and R. B. Graf, "Phase Transformations in Silica-Alumina Mixtures as Examined by Continuous X-Ray Diffraction," *Am. Mineral.*, **46** (9-10) 1064-1076 (1961).
[12]S. Otani and A. Kojima, "Thermal Transformation of SiO_2-Al_2O_3 Gels Prepared from Methylsiloxyaluminum Compounds," *Kogyo Kagaku Zasshi*, **67** (10) 1509-1512 (1964).
[13]L. A. Paulik, Y.-F. Yu, and T.-I. Mah, "Ceramic Powders from Metal Alkoxide Precursors," Advances in Ceramics, Vol. 21, Ceramic Powder Science, Edited by Messing, Mazdiyasni, McCauley, and Haber, pp. 121-129, (1987).
[14]R. K. Iler, The Chemistry of Silica, John Wiley, New York (1979).
[15]K. D. Keefer, "The Effect of Hydrolysis Conditions on the Structure and Growth of Silicate Polymers," Better Ceramics Through Chemistry, P. J. Brinker, D. E. Clark, and D. R. Ulrich, etds. (North-Holland), p. 15 (1984).
[16]S. Aramaki and R. Roy, "Revised Phase Diagram for the System Al_2O_3-SiO_2," *J. Am. Ceram. Soc.*, **45** (5) 229-242 (1962).
[17]K. J. D. MacKenzie, "Infrared Frequency Calculations for Ideal Mullite ($3Al_2O_3 \cdot 2SiO_2$)," ibid., **55** (2) 68-71 (1972).
[18]M. Hoch and K. M. Nair, Ceramurgia, International 2, p. 88 (1976).
[19]K. Haberko, "Characteristics and Sintering Behavior of Zirconia Ultrafine Powders," Ceramurgia, International 5, p. 148 (1979).
[20]K. S. Mazdiyasni and L. M. Brown, "Synthesis and Mechanical Properties of Stoichiometric Aluminum Silicate," *J. Am. Ceram. Soc.*, **55** (11), 548-592 (1972).
[21]R. A. Penty, Ph.D. Dissertation, Lehigh University, Bethlehem, Pennsylvania, 1972.
[22]B. L. Metcalfe and J. H. Sant, "The Synthesis, Microstructure, and Physical Properties of High Purity Mullite," *Transactions and Journal of the British Ceramic Society*, **74** (6) 193-201 (1975).
[23]T.-I. Mah and K. S. Mazdiyasni, "Mechanical Properties of Mullite," *J. Am. Ceram. Soc.*, **66** (10) 699-703 (1983).
[24]S. Kanzaki, H. Tabata, T. Kumazawa, and S. Ohta, "Sintering and Mechanical Properties of Stoichiometric Mullite," *J. Am. Ceram. Soc.*, **68** (1) C-6-C-7 (1985).
[25]S. Kanzaki, O. Abe, M. Ohashi, and H. Tabata, pp. 625-632 in Ceramic Materials and Components of Engines, Edited by W. Bunk and H. Hausner, Verlag Deutsche Keramlsche Gesellschaft, Bad Honnef, 1986.
[26]P. C. Dokko, J. A. Pask, and K. S. Mazdiyasni, "High Temperature Mechanical Properties of Mullite Under Compression," *J. Am. Ceram. Soc.*, **60**, No. 3-4 (1977).
[27]P. A. Lessing, R. S. Gordon, and K. S. Mazdiyasni, "Creep of Polycrystalline Mullite," *J. Am. Ceram. Soc. Discussion and Note*, **58**, March-April 1975.

Mullite Formation From Aluminum and Silicon Alkoxides

Bulent E. Yoldas

Advanced Research
PPG Industries, Inc.
P. O. Box 11472
Pittsburgh, PA 15238-0472

Abstract

Mullite can be prepared from precursor materials produced by hydrolytic polycondensation of silicon and aluminum alkoxides as well as from mixing sols separately derived from these compounds. Homogeneity and the molecular networks of these precursor materials are fundamentally affected by the condensation reactions. These differences in the precursors in turn affect the crystallization, sintering and microstructure. The crystallization of mullite from the stoichiometric polymer $Al_6Si_2O_{13}$ at ~980°C appears to be related to the intimacy and homogeneity of aluminum-silicon bonding. The homogeneous condensate, $Al_6Si_2O_{13}$, also shows an enhanced densification at this crystallization regime. In this paper, the effect of preparation methods on the crystallization of mullite is presented.

Introduction

Use of metal-organic compounds, particularly metal alkoxides, as precursors to ceramic and glass formation recently became widespread. The initial attraction of metal-organic derived materials was largely based on their high purity, reactivity, and submicron particle size. Further advances in ceramic processing was made by forming porous monolithic glass and ceramic gels without the aid of thermal consolidation,[1,2] which can later be sintered. These processes are often referred to as "sol-gel" techniques.

Investigation of materials in the Al_2O_3-SiO_2 binary by various "sol-gel" methods have been reported.[3–9] In this paper hydrolytic polycondensation of silicon and aluminum alkoxides is reviewed within the context of forming precursor condensates to mullite and modification of these condensation reactions to produce an intimate and homogeneous aluminum-silicon bonding.

Mullite Preparation Methods

The hydrolytic polycondensation rates of aluminum and silicon alkoxides are substantially different:

$$\equiv \text{Si-OR} + \text{HO-M} \rightarrow \equiv \text{Si-O-M} + \text{ROH} \qquad (1)$$

$$= Al\text{-}OR + HO\text{-}M \rightarrow = Al\text{-}O\text{-}M + ROH \quad (2)$$
(where M is H, Al = or Si ≡)
Reaction rate (s) >> (1)

Thus a homogeneous aluminosilicate cannot readily be formed by a standard hydrolysis of a mixture of these true alkoxides. Hydrolytic polycondensation of such a mixture gives gross inhomogeneities in the inorganic network since condensation alumina immediately occurs. The first of the following three preparation methods is an aqueous system which has been described in detail elsewhere.[8] The second and third methods differ significantly even though they are both aqueous systems. In the third method a precipitate is formed initially with a well defined amount of water and redissolved. These methods are briefly described below.

Method I: Use of Alkoxide Derived Sols

The first technique is to utilize an AlO(OH) sol[10] and react the alkoxy bonds of $Si(OR)_4$ with the hydroxyl bonds of alumina colloids:

$$= Al\text{-}OH + RO\text{-}Si \equiv \rightarrow = Al\text{-}O\text{-}Si \equiv + ROH \quad (3)$$

This technique provides a scale of mixing on the order of ~10–100 nanometers.

The second technique involves the partial hydrolysis of $Si(OR)_4$ followed by the addition of unhyrolyzed aluminum alkoxide:

$$(RO)_3Si\text{-}OH + RO\text{-}Al(OR)_2 \rightarrow (RO)_3Si\text{-}O\text{-}Al(OR)_2 + ROH \quad (4)$$

In this technique, the degree of condensation of the siloxane, obviously plays an important role in the resultant homogeneity. Preparation of materials using $Si(OR)_4$ and aqueous colloidal AlO(OH) solutions presents several difficulties. The immiscibility may be dealt with by the introduction of alcohol as a mutual solvent. To reduce the amount of alcohol required, the alumina sol can be initially concentrated. The AlO(OH) sols most amenable to concentration are those having 0.07 moles acid per mole AlO(OH) which can be concentrated to 15 to 20% equivalent Al_2O_3 by weight.[11] This significantly reduces the amount of alcohol required to facilitate the miscibility. Secondly, mixing of these two components leads to gelling within a short period of time, the exact time depending upon solution concentration pH, temperature, etc. The mixture must therefore be rapidly homogenized before the system solidifies. In the process, the hydroxy groups of the AlO(OH) colloids react with $Si(OR)_4$, creating chemically-bonded silica around and between the alumina colloids.

This method creates silica-rich molecular grain boundaries at the several hundred Angstrom level. And mullite precursors obtained by his method do not show ~ 980°C DTA peak.

Method 2: Polymerization by Modifying Chemical Reaction Rates

It is found that the alkoxide mixtures form a more homogeneous oxide network when the chemical encounter rates leading to condensation are significantly reduced. The chemical encounter rates in the condensation reaction can be reduced first by diluting the alkoxides in a nonaqueous solvent, SLV and then by introducing water in very small quantities.

$$x\ Al(OR')_3,\ y\ Si(OR)_4,\ z\ SLV + w\ H_2O \rightarrow = Al\text{-}O\text{-}Si \equiv + ROH,\ SLV \quad (5)$$
where w << x,y << z and SLV is a solvent, e.g. alcohol

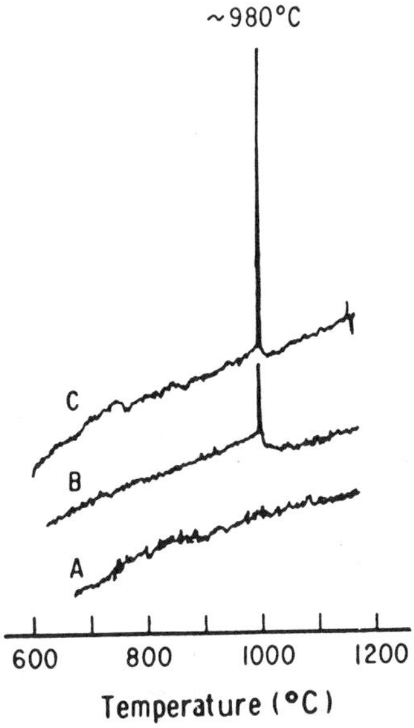

Fig. 1. The occurrence and intensity of the ~980°C DTA peak of amorphous $Al_6Si_2O_{13}$ of the rate of hydrolysis of the base solution prepared from a stoichiometric mixture of $Al(OC_4H_9)_3$ and $Si(OC_2H_5)_4$ in alcohol. (A) The solution is hydrolyzed vigorously with excess water, (B) The solution is thinly poured over a tray and hydrolyzed in humid air within hours. (C) The solution is hydrolyzed in humid air over 90 days while mixing.

The importance of the rate of hydrolysis on the molecular-structural makeup and on the mullite crystallization is demonstrated by the experimental results shown in Fig. 1. Here a stoichiometric mixture of $Al(OC_4H_9)_3$ and $Si(OC_2H_5)_4$ corresponding to the mullite composition was prepared in dry ethyl alcohol at a concentration of five weight percent equivalent oxide. This solution was divided into three portions and hydrolyzed at three different rates. The first portion (A) was mixed directly with excess water, the second portion (B) was spread thinly over a tray and allowed to hydrolyze and dry in humid air, and the last portion (C) was put in an open flask and stirred for three months, providing a very slow and methodical hydrolysis with humid air. The DTA peaks of these three materials during the heat treatment through 980°C are shown in Figure 1. This experiment shows that the vastly different rates of hydrolysis between aluminum and silicon alkoxides can be largely compensated by using a slow restricted hydrolysis.

Method 3: Copolymer Method

This method differs from the previous one in the sense that 0.5–1.0 moles of H_2O can be added into $Si(OR)_4$, $Al(OR)_3$ mixture at once. Unlike the previous method, a precipitation first occurs and later dissolves.[12]

This method is based on the fact that polyorganoaluminum condensates formed by 0.5 to 1.0 mole of water per mole aluminum alkoxide dissolve in alcohols to form clear polymer solutions.[12] In this method, stoichiometric quantities of the aluminum and silicon alkoxides are first mixed and then hydrolyzed with 0.5 to 1.0 mole of

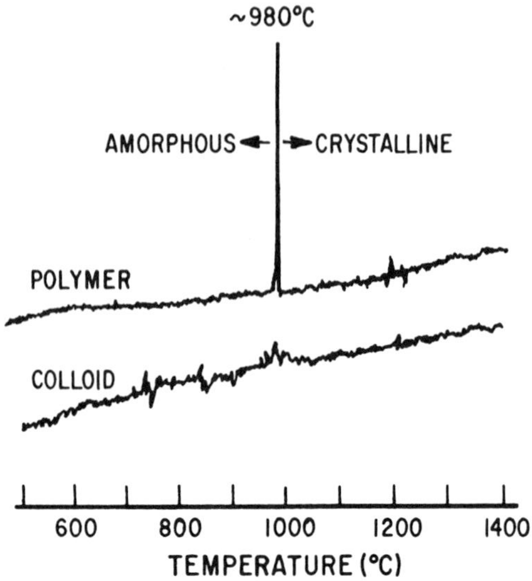

Fig. 2. A DTA peak at approximately 980°C and crystallized to mullite occurs when $Al_6Si_2O_{13}$ composition is formed by the polymeric method. When the same composition is formed by colloidal precursors, this DTA peak does not occur.

water per mole of total alkoxides in an alcohol medium to cause an initial polymerization of silicon with aluminum in a small molecular scale first, and followed by a secondary hydrolysis to complete the reactions. In a preferred variation of this theme, a stoichiometric amount of silicon alkoxide is first introduced into the water/alcohol liquor and then given up to 10 minutes to react before the aluminum alkoxide is introduced. The hydrolysis and self-condensation rate of $Si(OR)_4$ is much slower than that of $Al(OR)_3$. When the aluminum alkoxide is introduced into the liquor, a white precipitate forms, but this condensate dissolves, forming a water-clear solution within hours at 60°C. The total equivalent oxide concentration should be kept below 10% by weight in the liquor. After the solution clears, it may be hydrolyzed further by excess water, which causes either immediate gelation or fine powder formation, depending on variations in the hydrolysis conditions such as pH etc.

Pyrolysis of dried condensate to an oxide state shows that the oxide content varies considerably, depending on the hydrolysis conditions. For example, the oxide content of the condensate produced by method 3 increases from ~69% to ~79% by weight when nitric acid is introduced during the final hydrolysis. This figure goes up to 88–90% when hydrogen peroxide instead of water is used. The oxide content gives a revealing insight into the nature of molecular-structural variations in these condensates, since an extended polymerization leads to a higher oxide content by the elimination of terminal groups.[13]

These precursor solutions may also be used for coating and deposition of thin films, where the final hydrolysis is done in a non-liquid state.

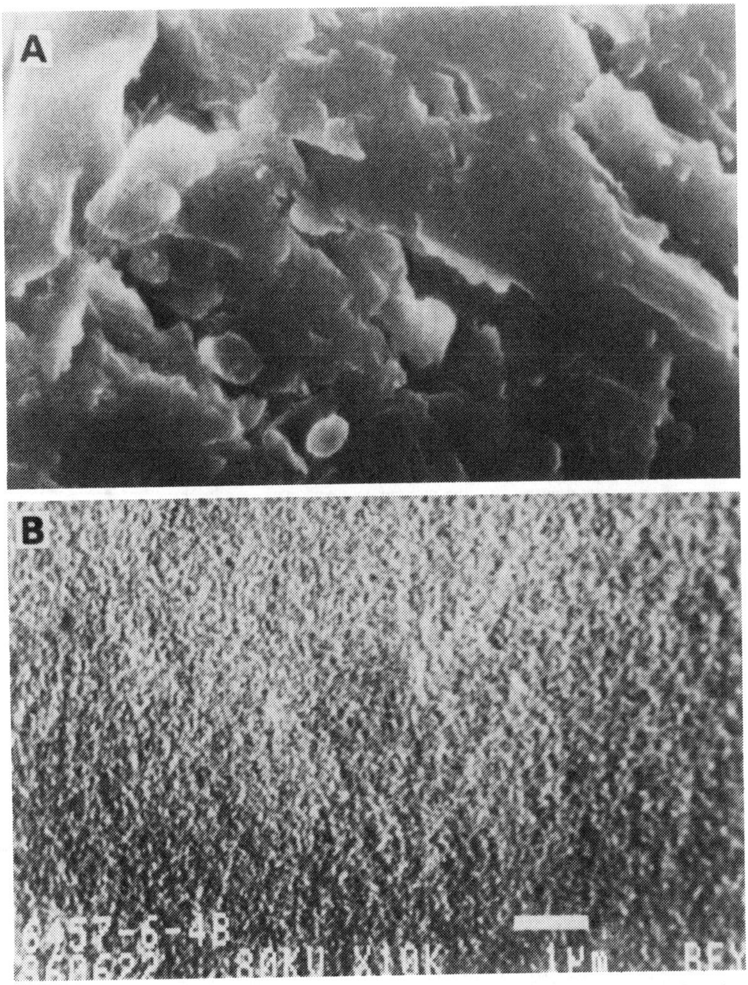

Fig. 3. Change in the ultrastructure of amorphous polymeric $Al_6Si_2O_{13}$ during the 980°C crystallization. (A) Below the 980°C. (B) Above 980°C.

High Temperature Effects of Condensation Reactions On Crystallization

In these experiments, the occurrence and intensity of the 980°C DTA peak is found to be closely related to the hydrolysis conditions which effects the intimacy and homogeneity of the aluminum-silicon bonding in the precursor networks. For example, mullite prepared from colloidal aqueous sols shows a very small or non existent exothermic peak at 980°C when similarly treated (Figure 2). Thus, the 980°C exothermic peak might be used for monitoring the nature of the precursor ultrastructure. The crystallization of mullite from the amorphous $Al_6Si_2O_{13}$ polymer at 980°C is also associated with a spontaneous densification and a dramatic change

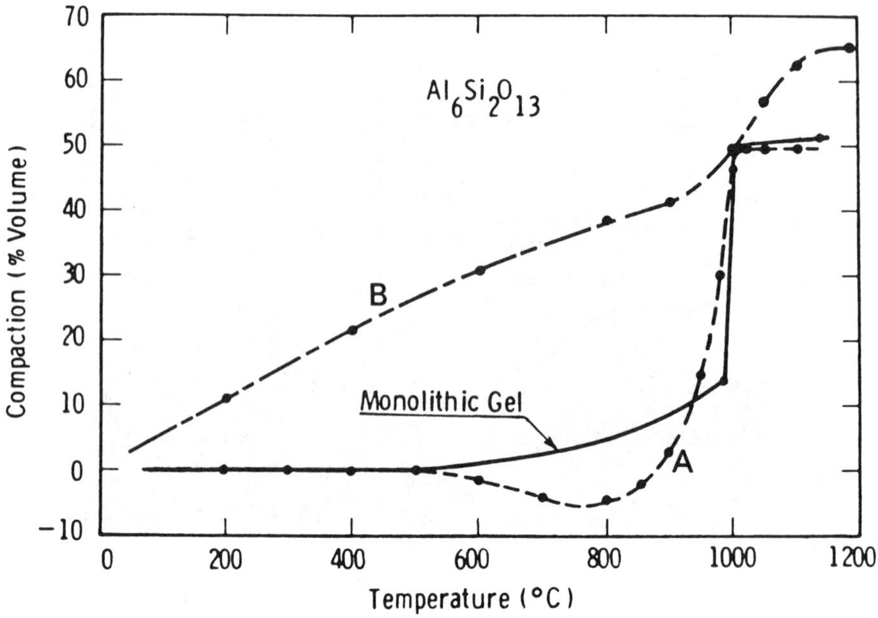

Fig. 4. The sintering behavior of $Al_6Si_2O_{13}$ powder condensate around 980°C also reflects the intimacy of the aluminum and silicon polymerization in the network. (A) Powder produced by slow hydrolysis of a stoichiometric alkoxide mixture in alcohol using humid air. (B) Powder produced from a precursor solution where $Si(OC_2H_5)_4$ was initially hydrolyzed.

in the microstructure (Figure 3). In monolithic gels, (Figure 4) approximately 35% linear shrinkage occurs. The amorphous material crystallizes into ~50 nm crystalline domains. The material attains the appearance of fusion. The powders produced from the same solution also exhibit similarly accelerated sintering in this temperature regime (curve A in Figure 4). In comparison, a sample produced from the reaction of hydrolyzed $Si(OC_2H_5)_4$ with $Al(OC_4H_9)_3$ powders shows a completely different sintering behavior (B in Figure 4). In the latter case, the initial hydrolysis of $Si(OC_2H_5)_4$ clearly leads to formation of free SiO_2, which is reflected in the temperature dependence of the sintering curve. The surface activity, nature of the terminal bonds, stoichiometry, and morphology of the network structure are affected by the condensation conditions, and all appear to play a significant role on the sintering and crystallization in this system.

The microstructure of mullites produced by polymeric and colloidal condensation techniques discussed here has been investigated.[14] Those investigations showed that there were significant differences in grain geometry, orientation, and porosity between mullite samples prepared by the two different methods. Mullite produced from the reaction of $Si(OC_2H_5)_4$ with colloidal AlO(OH) sol tends to show elongated crystals with considerable intergranular void space, e.g. 15%, and the presence of a liquid phase. Mullite produced from the alcohol-based polymer solutions, on the other hand, has a dense microstructure with equiaxed grains with no indication of a liquid phase presence.[14]

Summary

The crystallization, sintering and microstructure of mullite obtained via chemical polymerization of alkoxides are fundamentally affected by the condensation reactions. Crystallization of mullite from the amorphous $Al_6Si_2O_{13}$ phase at ~980°C appears to be closely related to the intimacy and homogeneity of alumina-silica bonding in the ultrastructure. Mullite prepared from colloidal aqueous alumina shows a very small or no exothermic peak at ~980°C. Occurrence of 980°C is also accompanied by a large spontaneous densification in this system. These studies also indicate that considerable ultra and microstructural design can be performed in chemically condensed ceramic materials.[14]

References

[1]B. E. Yoldas, "A Transparent Porous Alumina," *Am. Ceram. Soc. Bull.*, **54** (3), 286–88 (1975).

[2]B. E. Yoldas, "Preparation of Glasses and Ceramics from Metal-Organic Compounds," *J. Mater. Sci.*, **12**, 1203–8 (1977).

[3]A. K. Chakraborty and D. K. Ghosh, "Comment on Diphasic Xerogels, A New Class of Materials: Phases in the System Al_2O3-SiO_3," *J. Am. Ceram. Soc.*, **69** (8), C-202-C-203 (1986).

[4]Y. M. M. Al-Jarsha, K. D. Biddle, A. K. Das, T. J. Davies, H. G. Eimblem, K. Jones, J. M. McCullough, M. A. Mohd, A. B. D. Rahman, A. N. A. El-M Sharf el deen, and R. K. Wakefield, "Mullite Formation from Ethyl Silicate and Aluminum Chlorides," *J. Mater. Sci.*, **20**, 1773–81 (1985).

[5]B. E. Yoldas, "Microstructure of Monolithic Materials Formed by Heat Treatment of Chemically Polymerized Precursors in Al_2O_3-SiO_2, Binary," *Am. Ceram. Soc. Bull.*, **59** (4), 479–83 (1980).

[6]B. E. Yoldas, D. Partlow, "Formation of Mullite and Other Alumina-Based Ceramics via Hydrolytic Polycondensation of Alkoxides," to be published in J. Mater. Sci.

[7]D. W. Hoffman, R. Roy, and S. Komarneni, "Diphasic Xerogels, A New Class of Materials: Phases in the System Al_2O_3-SiO_2," *J. Am. Ceram. Soc.*, **67** (7), 468–71 (1984).

[8]S. Prochazka and F. J. Klug, "Infrared Transparent Mullite Ceramic," *J. Am. Ceram. Soc.*, **66** (12), 874–80 (1983).

[9]K. S. Mazdiyasni and L. M. Brown, "Synthesis and Mechanical Properties of Stoichiometric Aluminum Silicate (Mullite)," *J. Am. Ceram. Soc.*, **55** (11), 548–52 (1972).

[10]B. E. Yoldas, "Alumina Sol Preparation from Alkoxides," *Am. Ceram. Soc. Bull.*, **54** (3), 289–90 (1975).

[11]B. E. Yoldas, "Alumina Gels that Form Porous Transparent Al_2O_3," *J. Mater. Sci.*, **10**, 1856–60 (1975).

[12]B. E. Yoldas, U.S. Patent No. 4,535,026, 8/13/85.

[13]B. E. Yoldas, "Effect of Variations in Polymerized Oxides on Sintering and Crystalline Transformations," *J. Am. Ceram. Soc.*, **65** (8), 387–93 (1982).

[14]J. A. Pask, X. W. Zhang, A. P. Tomsia, B. E. Yoldas, "Effect of Sol-Gel Mixing on Mullite Microstructure and Phase Equilibria in the α Al_2O_3-SiO_2 System," *J. Am. Ceram. Soc.*, **70** (10), 704–7 (1987).

Processing of Ultrafine Mullite Powder Through Alkoxide Route

H. SUZUKI AND H. SAITO

Department of Mechanical System Engineering
Toyota Technological Institute
2-12-1 Hisakata, Tenpaku, Nagoya 468 Japan

Y. TOMOKIYO

HVEM Lab. Kyushu University
6-10-1 Hakozaki, Higashi, Fukuoka 812 Japan

Y. SUYAMA

Japan Fine Ceramics Center
2-4-1 Mutsuno, Atsuta, Nagoya 456 Japan

Abstract

The mullite precursor with a stoichiometric composition was prepared by a controlled hydrolysis and subsequent controlled copolymerization of Si and Al alkoxides. The precursor was amorphous and had a pseudoböehmite-like structure below 800°C and crystallized above 900°C to form ultrafine mullite powders via an Al, Si-spinel phase. The composition of the Al, Si-spinel phase was similar to the stoichiometric mullite composition. The powders have very large surface area and crystallization of mullite on calcination was completed above 1200°C.

Introduction

Mullite is promising as a structural material as well as a substrate material for integrated circuits because of its low dielectric constant and similar thermal expansion to silicon. Mullite powder for sintering use is prepared by various methods such as the coprecipitation method using various salts,[1,2] the sol-gel process[3,4] and the spray-pyrolysis technique.[5] Until now, it has been difficult to prepare fine and homogeneous stoichiometric mullite powder because of incomplete mixing of the raw materials and their low diffusivities. From Si and Al mixed alkoxides, Mazdiyasni and Brown[6] succeeded in the preparation of a high-purity submicron aluminum silicate with an acicular, needle-like morphology and a nominal composition of $3Al_2O_3{\cdot}2SiO_2$. These powders could be vacuum-hot-pressed into highly dense, translucent mullite polycrystalline bodies. Prochazka and Klug[7] also prepared mullite ceramics, transparent in the infrared, by hot-pressing and hot-isostatically pressing starting materials derived from the mixed alkoxides. However, they did not investigate the homogeneity in the composition of the powders and details of the formation process of mullite. Hirata et al.[8,9] also examined some properties of the SiO_2-Al_2O_3 powders with a nominal composition of $3Al_2O_3 \cdot 2SiO_2$ and concluded that the powders prepared by the hydrolysis of Si and Al mixed alkoxides crystallized via the Al, Si spinel phase into mullite with a composition of $2Al_2O_3 \cdot SiO_2$, θ-and α-alumina above 1200°C. Mullite has also been synthesized by Yamaguchi et al.[10] and Suzuki et al.[11] under hydrothermal conditions from mixed Si and Al alkoxides. In these processes, the mixed alkoxides were treated under 50 MPa at 300°-400°C for 2 hours and the products were calcined at 1300°C for 1 hour to yield mullite. All the powders prepared in these processes were thought to be inhomogeneous due to the difference in the hydrolysis rates between Si and Al alkoxides. In the present work, processing of the ultrafine mullite powder through an alkoxide route using the partial hydrolysis method was investigated and the formation process of mullite is discussed in detail.

Experimental Procedure

Silicon ethoxide, $Si(OC_2H_5)_4$, and aluminum isopropoxide $Al(OC_3H_7^i)_3$, were used as starting materials. $Si(OC_2H_5)_4$ was partially hydrolyzed with acidic water before mixing and reaction with aluminum alkoxide, while $Al(OC_3H_7^i)_3$ was refluxed in isobutylalcohol to partially substitute $-OC_4H_7^i$ groups for $-OC_3H_7^i$ groups of $Al(OC_3H_7^i)_3$. These two solutions were mixed and reacted at room temperature for 24 hours. The mullite precursor solution prepared in this way was hydrolyzed by drop-wise addition of the excess amount of distilled water. A schematic diagram for the powder preparation is shown in Fig. 1. The composition of the powders prepared

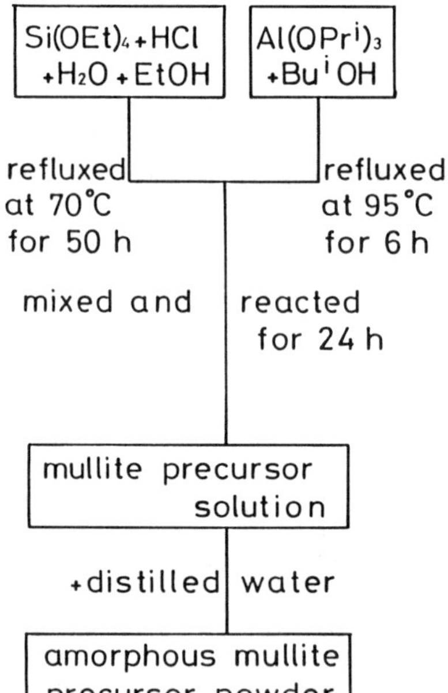

Fig. 1. A schematic diagram for the partial hydrolysis method.

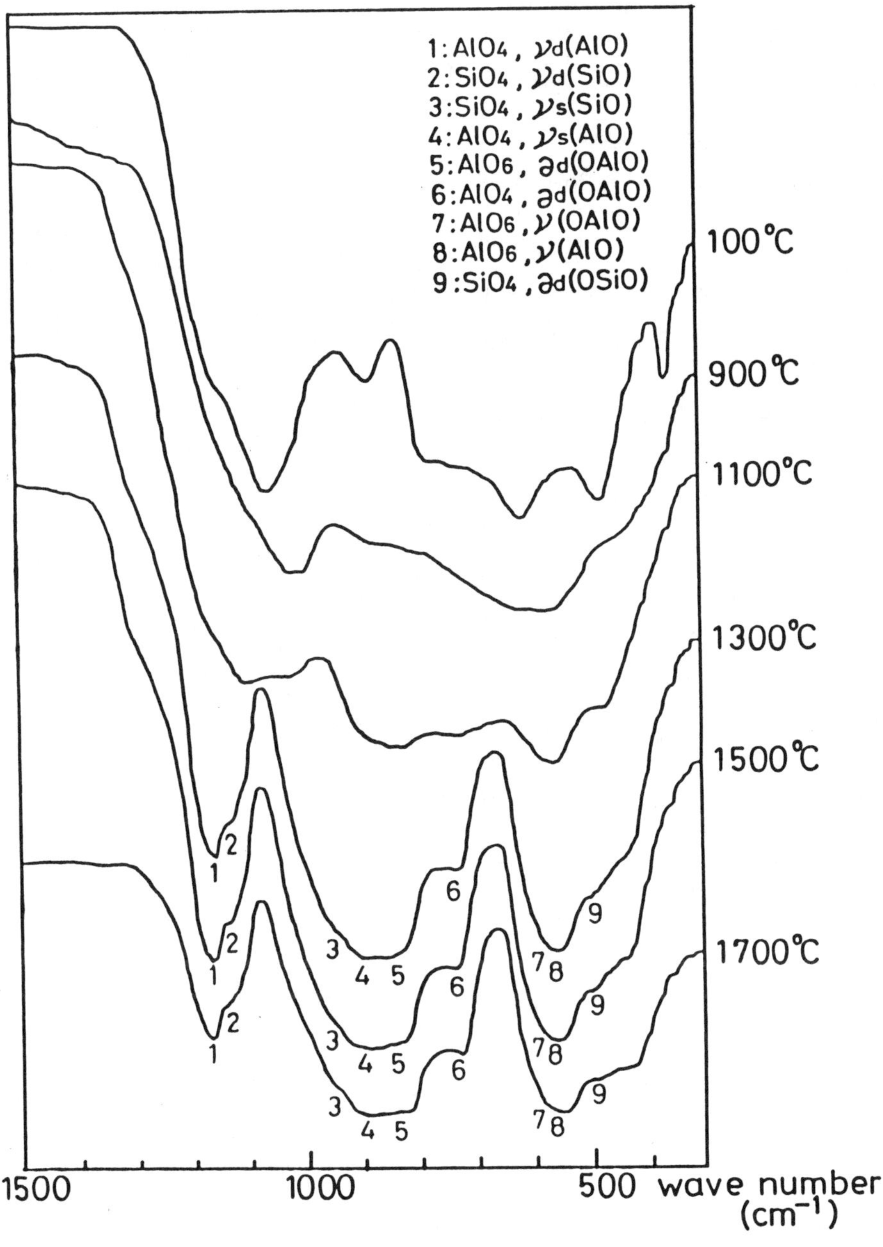

Fig. 2. IR spectra of calcined powders.

in this procedure was quite similar to the stoichiometric composition of mullite.[12] The powders were calcined for 2 hours at various temperatures up to 1700°C and examined by infrared spectroscopy, X-ray powder diffraction technique, transmission electron microscopy and energy dispersive X-ray spectroscopy.

Results and Discussion

Characteristic Changes of Powders on Heating

Changes in the infrared (IR) spectra of the powders on heating is shown in Fig. 2. The IR spectrum of as prepared mullite precursor powder desiccated at 100°C is similar to that of a pseudoböehmite-like crystal[13] prepared by the hydrothermal reaction of Al and Si alkoxides. However, mullite precursor powder is amorphous in this stage and its IR spectrum suggests the presence of Al^{3+} with tetrahedral coordination in the structure. Formation reaction of the mullite precursor is thought to be as follows:

$$SiO_2(OR)_{6-x}(OH)_x + 6Al(OR)_{3-y}(OH)_y \longrightarrow Si_2Al_6O_7(OR,OH)_{12} + H_2O + ROH \text{ (R; Alkyl group)}$$

The mullite phase was identified at 1100°C by X-ray diffraction (Table-1) but the IR spectrum of the powder calcined at 1100°C for 2 hours does not coincide with that of ideal mullite calculated by Mackenzie et al.[14] because an Al, Si-spinel phase coexists in the calcined powders. The IR spectra of powders calcined above 1300°C coincide with that of ideal mullite. The two peaks between 1100 and 1200 cm^{-1} (1 and 2 peaks in Fig. 2) vary with the composition of mullite sensitively. That is, as the alumina content of mullite increases, the peak at higher wave numbers moves to lower wave numbers and decreases in intensity compared with the second peak. The peak at 1165 cm^{-1} of the powders calcined above 1300°C have higher intensity compared with the second one, suggesting the powders are of stoichiometric composition.

Table 1. Phase determined by X-ray diffraction

Calcination Temp. (°C)	Phase
800	Amorphous
900	Amorphous
1000	Al,Si-spinel > Amorphous
1100	Mullite > Al,Si-spnel > Amorphous
1200	Mullite
1700	Mullite

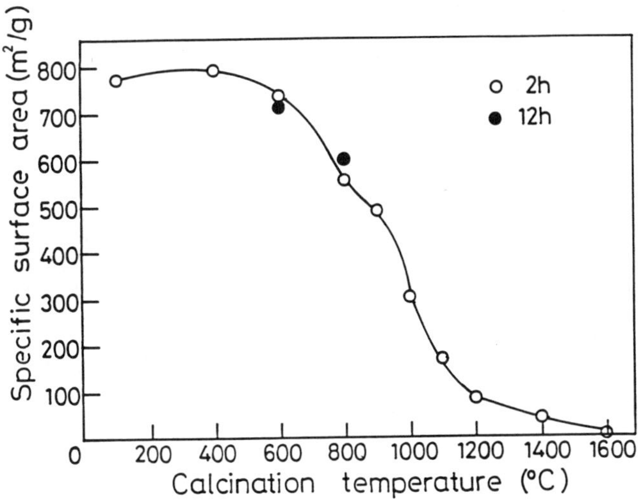

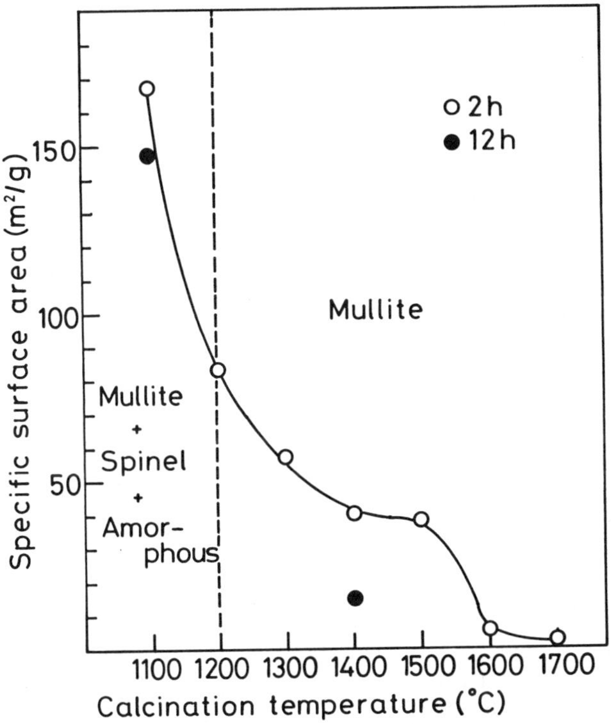

Fig. 3. Change in specific surface areas of calcined powders (a) and change in specific surface areas of powders calcined above 1100°C (b).

As shown in Fig. 3-(a), the powders calcined below 1100°C have large surface areas of more than 167 m^2/g. Especially, the amorphous mullite precursor powders calcined below 800°C have very large surface areas of more than 550 m^2/g and the surface area abruptly decreases above 900°C because amorphous powders crystallize via an Al, Si-spinel phase into mullite at temperatures ranging from 900 to 1200°C. The decrease in the surface areas continues up to 1400°C and levels off temporarily and then decreases again (Fig. 3-(b)). This change in the surface areas may correspond to the grain growth of the particles. Average sizes of the particles calculated from the surface areas are shown in Table 2. Below 1500°C, the average particle sizes are smaller than 50 nm and ultrafine mullite powders are obtainable. Above 1600°C, particles of about 50 nm are incorporated into the larger particles with about 400 nm in diameter because of the enhanced diffusivity of atoms.

Table 2. Average grain size calculated by specific surface area

Calcnation Temp.(°C)	Surface area (m^2/g)	Averga grain size (nm)
1100	167	11
1200	83	23
1300	57	33
1400	40	47
1500	38	50
1600	5	380
1700	2	470

Lattice parameters of the "a" axis are also sensitive to the composition. As shown in Fig. 4, lattice parameters of "b" and "c" axes are almost constant and coincide with the value reported in JCPDS 15-776 below 1500°C. The lattice parameters were determined from (401), (041) and (002) reflections of mullite using high purity silicon powders as an internal standard. However, in the "a" axis, the lattice parameter is slightly larger than that of stoichiometric mullite. In the case of prolonged heating at 1400°C and 1500°C (Fig. 4), the lattice parameter decreased to give a stoichiometric value. This result and those of the phase and compositional analyses indicate the possibility of an order-disorder transformation around 1500°C, such as α to β transformation of cordierite.[15] The Al/Si ordering had been suggested by Aramaki and Roy[16] but further investigation is indispensable to explain this phenomenon.

Transmission Electron Microscopy

From the results of X-ray powder diffraction, the mullite precursor powders were considered to be amorphous up to 900°C. However, the electron diffraction patterns of the powders calcined at 900°C indicate the presence of an Al, Si-spinel (Fig. 5). The observed particle sizes of mullite were from 6 to 11 nm at 1100°C.[12] These ultrafine mullite powders were incorporated into the large particles consuming the Al, Si-spinel and amorphous phase. Above 1200°C, no crystalline phase of the Al, Si-spinel but mullite was identified by the electron diffraction and the chunky morphology of mullite suggests the absence of a liquid phase below 1600°C (Fig. 6).

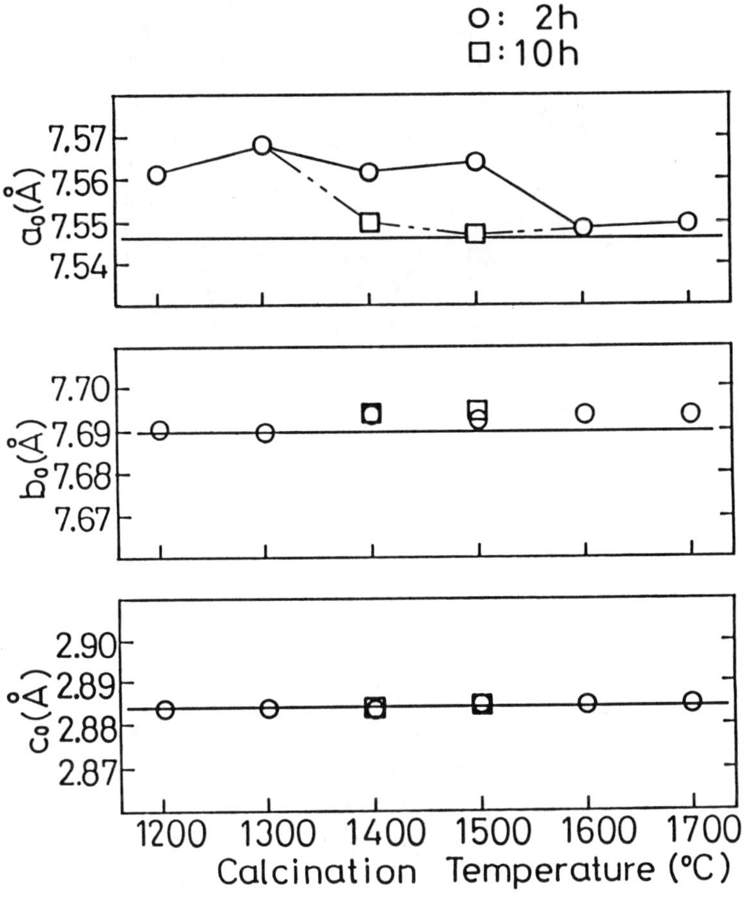

Fig. 4. Lattice parameters of mullite calcined at different temperatures.

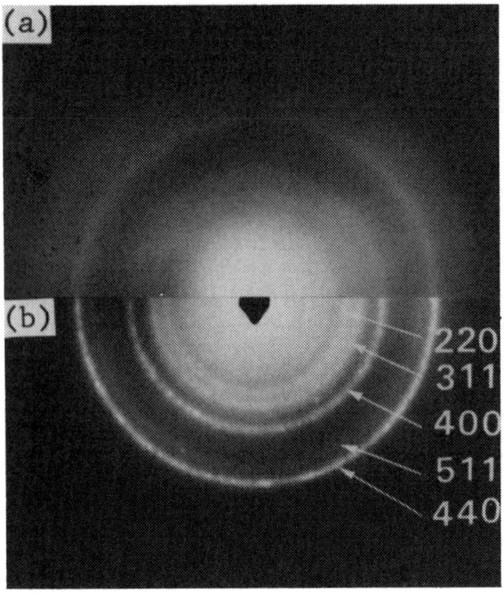

Fig. 5. Electron diffraction pattern of the powder calcined at 900°C for 2 hours.

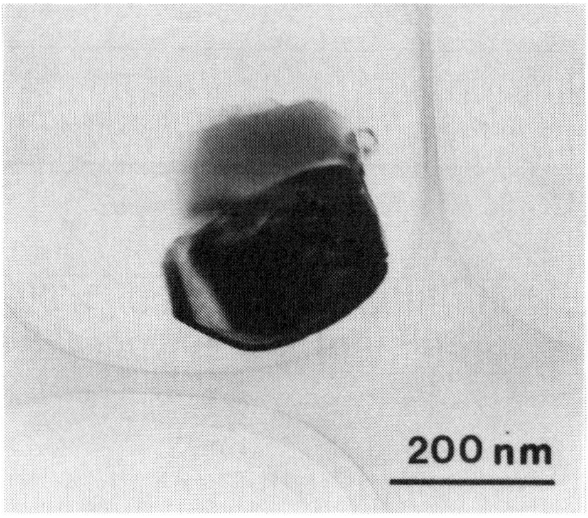

Fig. 6. Transmission electron micrograph of the powder calcined at 1600°C for 2 hours.

Table 3. Alumina content determined by EDS in TEM. (Average of 10 points or more.)

Calcination temperature (°C)	900	1000	1100	1200	1300	1500	1600	1700
wt% Al_2O_3	71.3	71.0	71.7	72.0	72.0	71.9	71.8	73.1

The compositions of these powders determined by an energy dispersive X-ray spectroscopy (Table 3) indicate that the particles have a stoichiometric composition up to 1600°C and the fluctuation of the compositions in each particles decreased with the increase in the temperature. At 900°C, the average composition of powders was about 71.3 wt% Al_2O_3 (Table 3). If the inhomogeneous polycondensation of Si and Al alkoxides had occurred in the precursor solution, Al_2O_3 would have to be detected at an elevated temperature. On the other hand, the Al_2O_3 content of the powders calcined at 1700°C is higher than that of stoichiometric mullite (Table 3), indicating that the mullite-Al_2O_3 solvus bends over toward the alumina side. This result corresponds well with the measurement of the lattice parameters of the powders (Fig. 4) and also with the phase diagram proposed by Prochazka and Klug.[7]

Formation Process of Mullite

As already described above, the mullite precursor powders with the stoichiometric composition were amorphous below 800°C and have a pseudoböehmite-like structure. However, prolonged heating of the powders at 800°C for 50 hours gave diffuse but quite similar diffraction patterns to Al, Si-spinel's (Fig. 5). The powders heated at 900°C for 12 hours gave the same diffraction patterns. The observed diffuse patterns suggest the presence of the fine Al, Si-spinel smaller than 10 nm. Okada and Ōtsuka[17] claimed that the chemical composition of the spinel phase was around $SiO_2 \cdot 6Al_2O_3$ but our present results indicate the presence of an Al, Si-spinel phase with a stoichiometric mullite composition (Fig. 7). The reason for this compositional difference is ascribed to the formation process of mullite. The presence of a spinel phase with a composition of $2Al_2O_3 \cdot 3SiO_2$ had been predicted by Brindley and Nakahira.[18] Considering the composition of the spinel phase and the fact that the amorphous precursor contains 4-coordinated Al, the unit cell of the spinel phase observed in this work is probably represented by $(Al_x,\square_{5.33-x})(Si,Al)_8 Al_{10.67}O_{32}$. The spinel phase seems to be metastable such as μ-phase of cordierite.[15]

The formation process of alkoxy-derived ultrafine mullite with a stoichiometric composition is considered to be as follows:

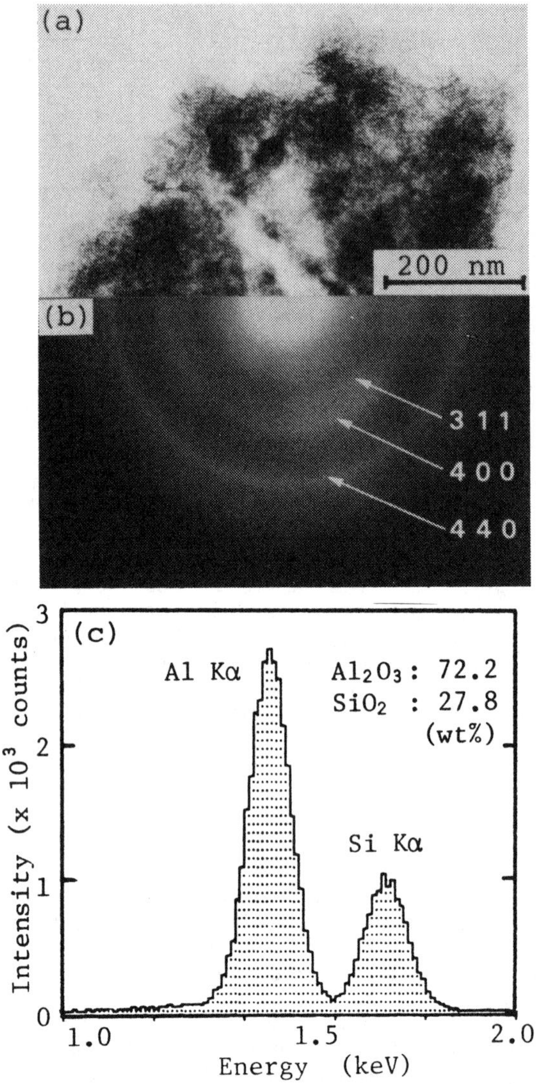

Fig. 7. An example of EDS analysis by TEM. (a) Bright field image of powders calcined at 900°C for 2 hours. (b) Diffraction pattern from (a). (c) EDS spectrum from the area of (a).

1. Copolymerization of partially hydrolyzed $Si_2O(OR)_6$ and $Al(OR)_3$ occurs to yield an amorphous mullite precursor of a pseudoböehmite-like structure with a stoichiometric mullite composition. (at ambient temperature)
2. Crystallization of the amorphous mullite precursor into an Al, S-spinel phase with a stoichiometric mullite composition (around 800°C).
3. Grain growth of the Al, Si-spinel phase (800°-1100°C).
4. Conversion of the spinel phase into mullite (around 1100°C).
5. Grain growth of mullite (above 1200°C).

Conclusions

Homogeneous and ultrafine mullite precursor powder with a stoichiometric composition was prepared through the alkoxide route. The precursor prepared by the partial hydrolysis method of Si and Al alkoxides had a pseudoböehmite-like structure and was amorphous below 800°C. The precursor crystallized into an Al, Si-spinel phase with a stoichiometric mullite composition. The spinel to mullite conversion was completed at 1200°C to yield ultrafine mullite powder with a stoichiometric composition.

References

[1]G. Y. Meng and R. A. Huggins, "A New Chemical Method for Preparation of both pure and doped Mullite," *Mat. Res. Bull.*, **18** [5] 581–88 (1983).

[2]K. Hamano, T. Sato, and Z. Nakagawa, "Properties of Mullite Powder Prepared by Coprecipitation and Microstructure of Fired Bodies," *Yogyo-Kyokai-Shi*, **94** [8] 818–22 (1986).

[3]M. G. M. U. Ismail, Z. Nakai, K. Minegishi, and S. Sōmiya, "Synthesis of Mullite Powder and its Characteristics," *Int. J. High Technology Ceramics*, **2**, 123–34 (1986).

[4]M. G. M. U. Ismail, Z. Nakai, and S. Sōmiya, "Microstructure and Mechanical Properties of Mullite Prepared by the Sol-Gel Method," *J. Am. Ceram. Soc.*, **70** [1] C-7–C-8 (1987).

[5]S. Kanzaki, T. Kumazawa, J. Asaumi, O. Abe, and H. Tabata, "Dependency of Mechanical Property of Sintered Mullite on Chemical Composition," *Yogyo-Kyokai-Shi*, **93** [7] 407–08 (1985).

[6]K. S. Mazdiyasni and L. M. Brown, "Synthesis and Mechanical Properties of Stoichiometric Aluminum Silicate (Mullite)," *J. Am. Ceram. Soc.*, **55** [11] 548–52 (1972).

[7]S. Prochazka and F. J. Klug, "Infrared-Transparent Mullite Ceramic," *J. Am. Ceram. Soc.*, **66** [12] 874–80 (1983).

[8]Y. Hirata, H. Minamizono and K. Shimada, "Property of SiO_2-Al_2O_3 Powders Prepared from Metal Alkoxide," *Yogyo-Kyokai-Shi*, **93** [1] 36–44 (1985).

[9]Y. Hirata, K. Sakeda, Y. Matsushita, and K. Shimada, "Properties of Fine SiO_2-Al_2O_3 Powders by Hydrolysis of Mixed Alkoxides," *Yogyo-Kyokai-Shi*, **93** [9] 577–80 (1985).

[10]T. Yamaguchi, M. Suzuki, S. Hiraishi, M. Yoshimura, and S. Sōmiya, "Hydrothermal crystallization of alumino silicate fine powder prepared by hydrolysis of mixed alkoxides," *J. Mat. Soc. Japan*, **21**, 241–44 (1984).

[11]M. Suzuki, S. Hiraishi, M. Yoshimura, and S. Sōmiya, "Preparation of mullite powder by calcination of the products hydrothermally treated from mixed alkoxides or mixed sols," *Yogyo-Kyokai-Shi*, **92** [6] 320–27 (1984).

[12]H. Suzuki, Y. Tomokiyo, Y. Suyama, and H. Saito, "Preparation of ultrafine mullite powder from metal alkoxides," *Yogyo-Kyokai-Shi*, **96** [1] 67–73 (1988).

[13]S. Sōmiya, M. Yoshimura, M. Suzuki, and S. Hiraishi, "Preparation of Powders in the Al_2O_3-SiO_2 System by Hydrothermal Treatment of Alkoxide," *J. Japan Soc. Powder and Powder Metallurgy*, **31** [6] 215–20.
[14]K. J. D. Mackenzie, "Infrared Frequency Calculations for Ideal Mullite ($3Al_2O_3 \cdot 2SiO_2$)," *J. Am. Ceram. Soc.*, **55** [2] 68–71 (1972).
[15]H. Suzuki, K. Ota, and H. Saito, "Preparation of Cordierite Ceramics from Metal Alkoxides (Part 2) Sintering," *Yogyo-Kyokai-Shi*, **95** [2] 170–75 (1987).
[16]S. Aramaki and R. Roy, "Revised Phase Diagram for the System Al_2O_3-SiO_2," *J. Am. Ceram. Soc.*, **45** [5] 229–42 (1962).
[17]K. Okada and N. Ōtsuka, "Characterization of the Spinel Phase from SiO_2-Al_2O_3 Xerogels and the Formation Process of Mullite," *J. Am. Ceram. Soc.*, **69** [9] 652–56 (1986).
[18]G. W. Brindley and M. Nakahira, "The Kaolinite-Mullite Reaction Series: III, The High-Temperature Phases," *J. Am. Ceram. Soc.*, **42** [7] 319–24 (1959).

Characterization of SiO_2-Al_2O_3 Powders Prepared From Metal Alkoxides

SENSHU MITACHI, MASAMI MATSUZAWA, AND KIMIYOSHI KANEKO

Hokko Chemical Industry Co., Ltd.
2165 Toda, Atsugi-shi, Kanagawa 243, Japan

SHUZO KANZAKI AND HIDEYO TABATA

Government Industrial Research Institute, Nagoya
Hirate-cho, Kita-ku, Nagoya 462, Japan

Abstract

Aluminum silicate powders ranging from 69 to 79 wt% Al_2O_3 were synthesized by hydrolysis of mixed benzene solutions of $Al(Oi\text{-}C_3H_7)_3$ and $Si(OCH_3)_4$ with water (pH=7) [(A)-series], or with aqueous ammonia (pH=10) [(B)-series].

In the case of the stoichiometric composition, the starting amorphous powder of the (A)-series transformed directly into crystalline mullite at >980°C. The (B)-series powder converted to mullite above 1250°C by way of a spinel phase. In both cases, the specific surface area of the powders decreased rapidly during mullitization. There was a large difference in the homogeneity and morphology of the particles, and also in the sinterability between the as-prepared powders of the (A) and the (B)-series. Both powders densified to >98% of theoretical density by sintering at 1650°C for 4 h. The flexural strength for the samples from the (B)-series powder was 340 MPa (r.t.) and 430 MPa (1300°C). The sintered density depended markedly on the chemical composition; it decreased with increasing alumina content. The silica-rich specimens showed higher strength values at 1300°C than at room temperature.

Introduction

Mullite ($3Al_2O_3{\cdot}2SiO_2$) has been recently recognized as a promising engineering ceramic for high temperature applications because of its high strength at high temperatures.[1,2] In the present study, hydrolysis of metal alkoxides[3] has been adopted for the preparation of mullite powders. This paper discusses crystallization, sintering and the mechanical properties of ceramics made from these powders.

Experimental Procedure

The starting materials in this work were benzene solutions of $Al(Oi\text{-}C_3H_7)_3$ (1mol/l) and $Si(OCH_3)_4$ (2mol/l). The alumina-silica powders contain 69 to 79 wt%

Al_2O_3 were prepared by the hydrolysis of the corresponding mixed benzene solutions of metal alkoxides. A mixed benzene solution of 1000ml was rapidly added by vigorous stirring to a mixture of 750ml of methanol and 1500ml of distilled water [(A)-series], or to a mixture of 750ml of methanol and 1500ml of aqueous ammonia (pH=10) [(B)-series]. Heated under reflux for 3 h with stirring, the resulting precipitate was then filtered and dried at 80°C. The obtained powders were calcined at various temperatures. The calcined powders were ground in methanol using a vibratory mill with silicon nitride balls as milling media for 50 h or 80 h. The powder compacts were formed by die pressing at 30 MPa, followed by cold isostatic pressing at 200 MPa, and were then sintered at 1650°C for 4 h (heating rate; 10°C/min). The sintered bodies were sliced and ground with a #200 and a #400 diamond wheel to prepare test pieces of 3mm by 4mm by 40mm. The flexural strength was measured by three-point bending with a span length of 30mm and a cross head speed of 0.5mm/min. High temperature strength was also measured at 1300°C in air. For each test, two or three bars were used.

Results and Discussion

Stoichiometric Mullite Powders

Formation Process of Mullite from Precursors

The exact chemical composition and purity of powders calcined at 1250°C for 1 h (without grinding), at which temperature the powders fully mullitized, were determined by means of X-ray fluorescence analysis and atomic absorption spectroscopy, respectively. The mullite powder had a chemical composition of 71.6 wt% Al_2O_3/28.4 wt% SiO_2 for the (A)-series and 71.5 wt% Al_2O_3/28.5 wt% SiO_2 for the (B)-series. The major impurities in the powders were 30 ppm Na_2O, 40 ppm K_2O, and <10 ppm CaO irrespective of the hydrolysis method.

The DTA curves illustrated in Fig. 1 were obtained using a heating rate of 10°C/min for the hydrolysis products dried at 80°C for 24 h. Only one sharp exothermic peak at ~980°C was observed in the (A)-series powder. In the (B)-series powder, two weak exothermic peaks were identified at 980°C and 1250°C. From the XRD patterns (Fig. 2) of the powders heated at various temperatures (heating rate; 10°C/min), the exothermic peak at 980°C was attributed to the transformation from an amorphous phase to crystalline mullite in the (A)-series powder. In the case of (B)-series powder, the exotherm at 980°C was attributed to the formation of the so-called spinel phase.[4,5] At 1250°C the exotherm was attributed to the transformation of the spinel to mullite.

Powder Characteristics

Changes of specific surface area with calcination temperature for powders (A) and (B) without grinding are shown in Fig. 3. In both cases, the specific surface area decreased rapidly and was accompanied by mullitization. Rapid grain growth associated with mullitization was also observed by TEM (Fig. 4). Aggregated and porous particles formed during mullitization at 1000°C for 1 h in the (A) powder (Fig. 4). In the case of the (B)-series, the granular particles and the needle-like particles co-existed before mullitization; they changed into well-dispersed and polyhedrally-shaped particles after mullitization at 1200°C for 1 h (Fig. 4).

The following possible mechanism is suggested to account for differences in homogeneity of the mullite precursor and the morphology of the obtained mullite particles between the (A) and (B)-series. When the starting solution was hydrolyzed with water (pH=7) [(A)-series], fairly large chains and networks of siloxane bonds containing Al (-O-Si-O-Al-O-Si-) were formed by partial hydrolysis. On the

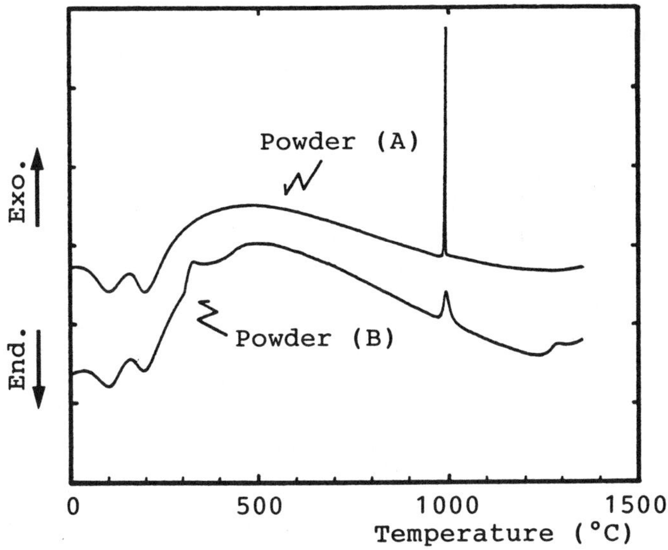

Fig. 1. DTA curves at heating rate of 10°C/min for uncalcined powders (A) and (B).

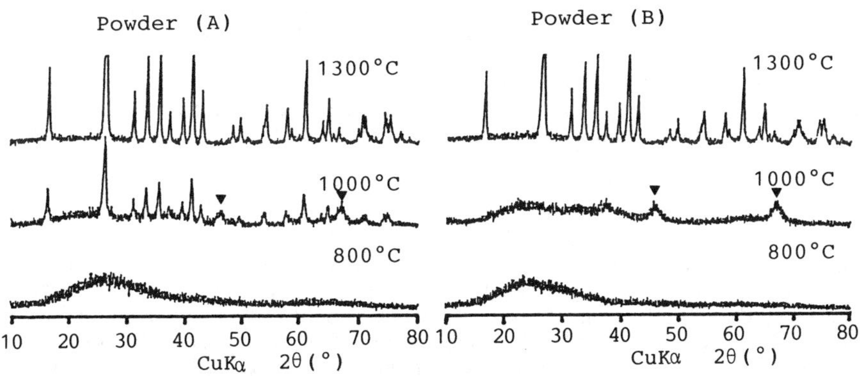

Fig. 2. X-ray diffraction patterns of hydrolyzed powders (A) and (B) calcined at various temperatures. (heating rate; 10°C/min) (▼; spinel).

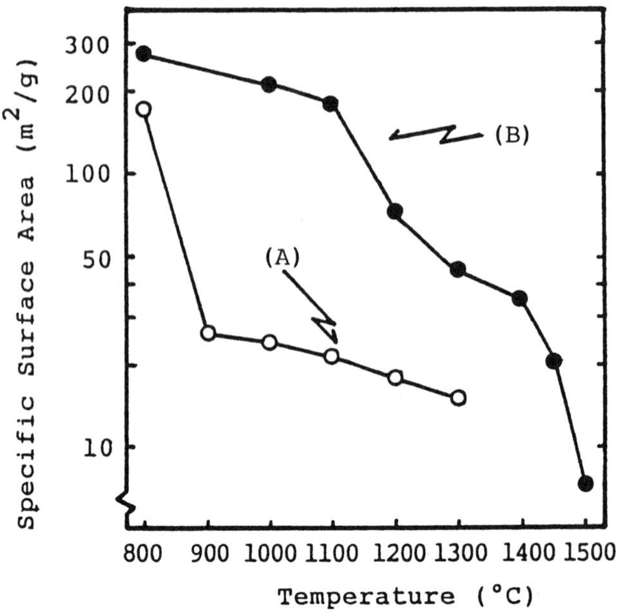

Fig. 3. Changes of specific surface area with calcination for the hydrolyzed powders (A) and (B). (heating rate; 5°C/min, soaking time; 1 h).

contrary, when the solution was hydrolyzed with aqueous ammonia (pH=10) [(B)-series], hydrolysis was accelerated by hydroxyl groups[6] and resulted in the formation of short chains of siloxane bonds. Hydroxides were also partially formed. The conclusive evidence which accounts for the above differences has not been obtained. However, if this is the case, then the Si and Al elements are distributed homogeneously on an atomic level in the case of the (A)-series, and thus, the precursor transformed directly into crystalline mullite on heating. In the case of powder (B), since the distribution of Si and Al is less homogeneous than that of (A), the precursor transformed to the spinel phase followed by the formation of mullite on further heating. It suggests that the resultant morphology of the obtained mullite particles shown in Fig. 4 is also influenced by the dimension of the chains and the networks.

Sinterability

To obtain approximately the same surface areas after calcination and grinding for 50 h (about 35m^2/g), powders (A) and (B) were calcined for 1 h at 950°C and 1450°C, respectively (Fig. 5). The compositions of the powders after grinding were the same as before grinding, within the limits of resolution of X-ray fluorescence. The green density of compacts (A) and (B) after cold isostatic pressing at 200 MPa were 1.45g/cm^3 and 1.55g/cm^3, respectively. Linear thermal shrinkage curves obtained at a heating rate of 10°C/min for these powder compacts are illustrated in Fig. 6. The (B) powder was more sinterable than the (A) powder. Moreover, the bulk densities of the sintered bodies produced from powders (A) and (B) at 1650°C for 4 h were 3.08g/cm^3 and 3.14g/cm^3, respectively.

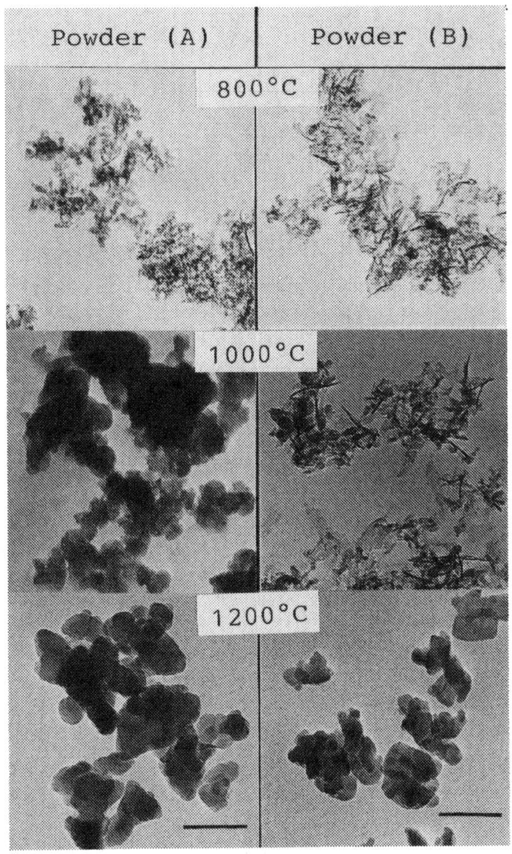

Fig. 4. Transmission electron micrographs of the powders (A) and (B) calcined at 800°C, 1000°C and 1200°C for 1 h. (heating rate; 5°C/min) (Bar=0.1μm).

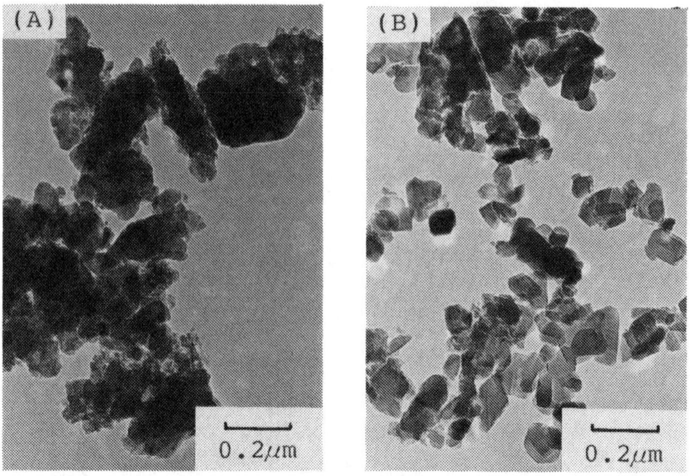

Fig. 5. Transmission electron micrographs of the calcined powders (A) and (B) with the same specific surface area. [Powder (A); calcined at 950°C for 1 h, S. A. = 35.8m^2/g, Powder (B); calcined at 1450°C for 1 h, S. A. = 33.9m^2/g] (heating rate; 5°C/min) (Bar=0.1μm).

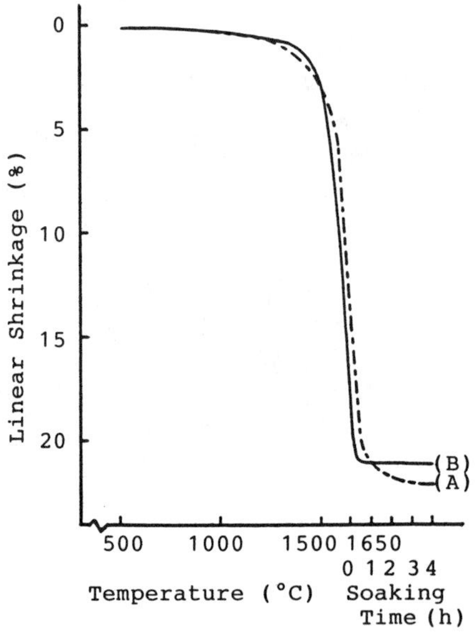

Fig. 6. Sintering behavior of the compacts of (A) and (B) powders. (heating rate; 10°C/min).

SiO_2-Al_2O_3 Powders [(B)-series] with Various Chemical Compositions

The SiO_2-Al_2O_3 powders containing 69 to 79 wt% Al_2O_3 were synthesized in the same way as the (B)-series to study the effect of composition on the powder characteristics and mechanical properties of the sintered bodies.

Powder Characteristics

The SiO_2-Al_2O_3 powders with six different chemical compositions were synthesized. Chemical analysis of these powders after calcination at 1200°C for 1 h without grinding are shown in Table 1. These calcined powders were fully mullitized. Before mullitization, the silica-rich powders were only composed of granular particles. By contrast, needle-like particles co-existed with granular ones in the alumina-rich powders. Both a reduction in particle size and better dispersibility among the particles were observed with increasing Al_2O_3 content after mullitization (Fig. 7).

Table 1. Chemical composition analysis of powders synthesized by route B. (a) Calcined at 1200°C for 1 h. (without grinding) (b) Grinding for 80 h after calcination and then sintered at 1650°C for 4 h.

	(a) calcined at 1200°C, 1h		(b) sintered at 1650°C, 4h	
	Al_2O_3(wt%)	SiO_2(wt%)	Al_2O_3(wt%)	SiO_2(wt%)
NO.1	68.7	31.3	68.9	31.1
NO.2	69.4	30.6	69.7	30.3
NO.3	70.4	29.6	70.5	29.5
NO.4	71.5	28.5	71.5	28.5
NO.5	73.4	26.6	73.6	26.4
NO.6	78.6	21.4	78.9	21.1

Sinterability and Flexural Strength

The powders were calcined and ground at 1200°C and for 80 h, respectively. Neither SiO_2 nor Si_3N_4 was introduced by milling with Si_3N_4 media (Table 1). Figure 8 shows the relative density of the sintered body as a function of composition. The powders having a composition within the silica-rich region were more easily sintered owing to the participation of a liquid phase sintering mechanism. The microstructures of thermally etched specimens obtained from the bodies sintered at 1650°C for 4 h are shown in Fig. 9. In the silica-rich specimens, prismatic mullite grains are observed, while corundum grains are found to be dispersed around mullite grains in the alumina-rich specimens. At the composition of 73.6 wt% Al_2O_3, mullite grains have equiaxed morphology, and neither a glassy phase nor corundum grains were identified. As expected from the phase diagram proposed by Prochazka and Klug,[7] single phase crystalline mullite can be obtained in this compositional region.

Figure 10 shows the relationship between flexural strength of the sintered bodies and their chemical composition. The flexural strength of stoichiometric samples (71.8 wt% Al_2O_3) was 340 MPa at room temperature and 430 MPa at 1300°C. It is evident that silica-rich specimens exhibit higher flexural strength at 1300°C than at room

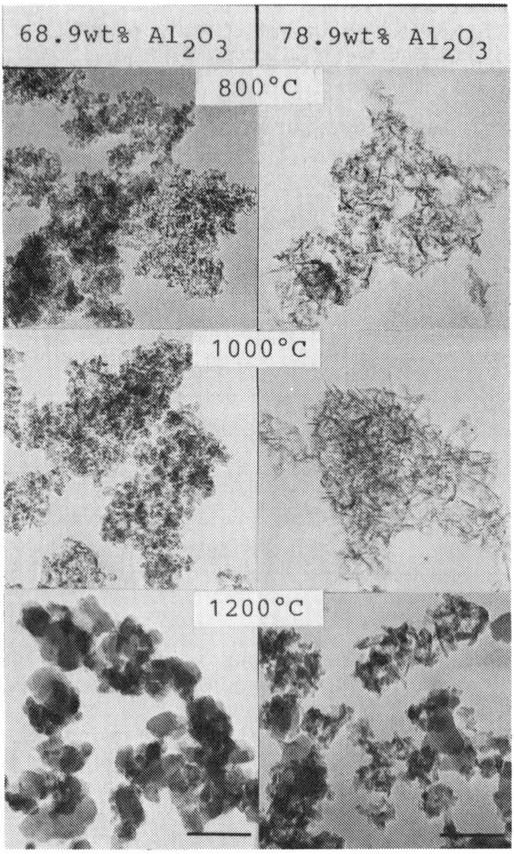

Fig. 7. Transmission electron micrographs of the powders calcined at 800°C, 1000°C and 1200°C for 1 h (heating rate; 5°C/min) with various chemical compositions. (Bar=0.1μm).

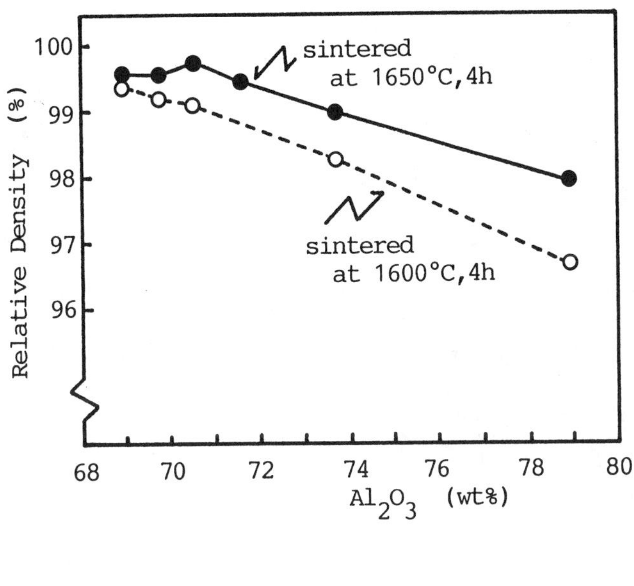

Fig. 8. Changes of bulk density with chemical composition for samples sintered at 1600°C and 1650°C for 4 h. (heating rate; 10°C/min).

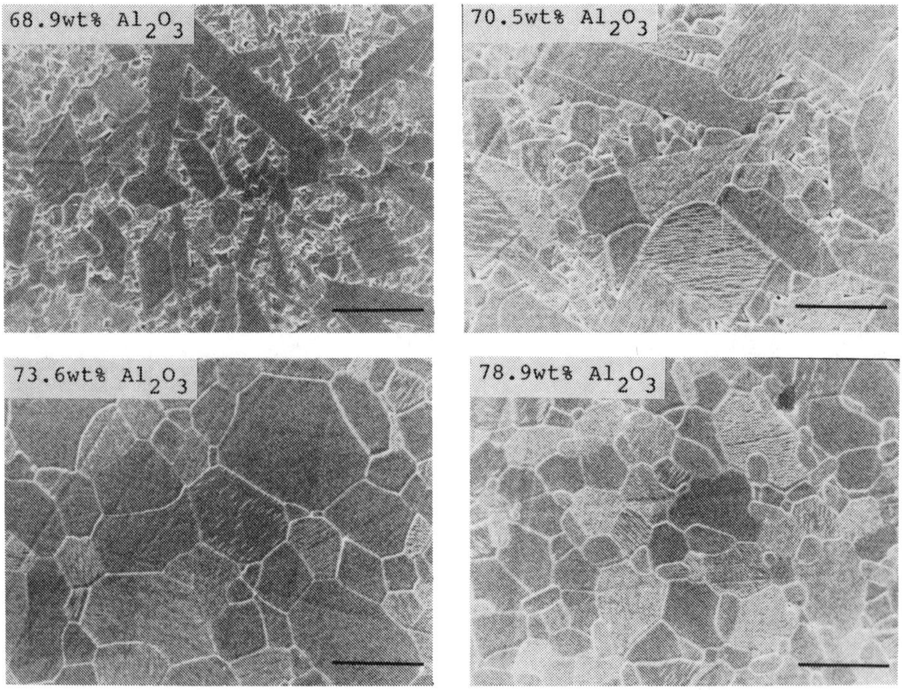

Fig. 9. Microstructure of samples sintered at 1650°C for 4 h with various compositions. (thermal etching) (Bar=2μm).

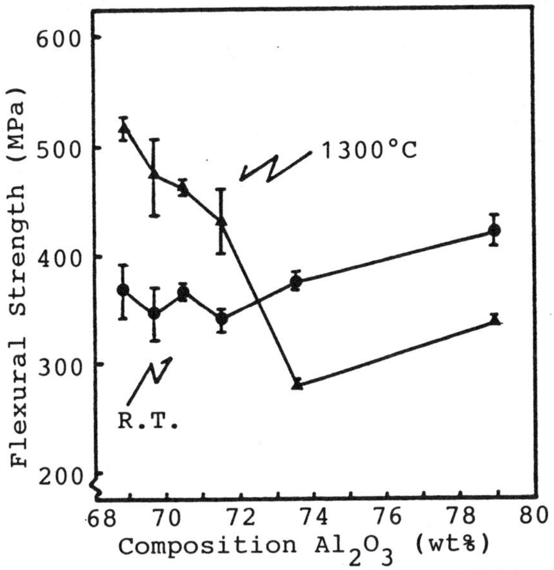

Fig. 10. Flexural strength at room temperature and at 1300°C in air vs. chemical composition for samples sintered at 1650°C for 4 h.

temperature. This behavior can be explained by the stress relaxation mechanism of glassy grain boundary phases at elevated temperatures.[3]

Conclusions

Two kinds of stoichiometric mullite precursors were synthesized from mixed alkoxide solutions by different hydrolysis conditions. Upon heating, one powder [(A)-series] was transformed directly into crystalline mullite above 980°C, whereas the other [(B)-series] was converted to mullite above 1250°C through a spinel intermediate phase.

Since the (B)-series powder was more easily sintered than that of the (A)-series, SiO_2-Al_2O_3 powders with different composition (69 to 79 wt% Al_2O_3) have been synthesized by the method of (B). Powders having an alumina-rich composition consist of finer and more easily dispersed particles than the silica-rich powders. However the former powders required higher firing temperature than that of the latter to obtain a fully densified body. The flexural strength for the sintered body of stoichiometric composition was 340 MPa at room temperature and 430 MPa at 1300°C. The silica-rich specimens showed higher strength values at 1300°C than at room temperature.

References

[1]S. Kanzaki, H. Tabata, T. Kumazawa, and S. Ohta, "Sintering and Mechanical Properties of Stoichiometric Mullite," *J. Amer. Ceram. Soc.*, **68** (1) C6-7 (1985).

[2]P. C. Dokko, J. A. Pask, and K. S. Mazdiyasni, "High-Temperature Mechanical Properties of Mullite Under Compression," ibid., **60** (3-4) 150-155 (1977).

[3]K. S. Mazdiyasni and L. M. Brown, "Synthesis and Mechanical Properties of Stoichiometric Aluminum Silicate (Mullite)," ibid, **55** (11) 548-552 (1972).

[4]A. K. Chakraborty, "Formation of Silicon-Aluminum Spinel," ibid., **62** (3–4) 120–124 (1979).

[5]G. W. Brindley and M. Nakahira, "The Kaolinite-Mullite Reaction Series: I, II, III," ibid., **42** (7) 311–324 (1959).

[6]D. C. Bradley and C. Prevedorou-Demas, "Metal Oxide Trialkylsilyloxide Polymer. Part IV. Titanium Oxide Triethylsilyloxide Polymers," *J. Chem. Soc. (A)*, 43–44 (1967).

[7]S. Prochazka and F. J. Klug, "Infrared-Transparent Mullite Ceramics," ibid., **66** (12) 874–880 (1983).

Mullite Powder From Hydrothermal Processing

SHIGEYUKI SŌMIYA*, MASAHIRO YOSHIMURA, MATSUO SUZUKI† AND TOSHIHIDE YAMAGUCHI†

Research Laboratory of Engineering Materials
Tokyo Institute of Technology
Nagatsuta, Midori, Yokohama 227 Japan

Abstract

Mullite powders can be prepared from alkoxides through the following hydrothermal processing; [I] Hydrothermal treatment and calcination, [II] Two-stage hydrothermal treatment, or [III] Hydrothermal crystallization of coprecipitates.

In processes [I] and [II], a pseudoboehmite-like phase containing Al and Si homogeneously was prepared by the hydrothermal treatment of the mixed solution of Al-isopropoxide and Si-tetraethoxide ($Al_2O_3/SiO_2 = 3/2$) at 300°C under 50 MPa for 2 h. The product consisted of plate-like particles of about 5 nm × 100 nm. When it was calcined in air, it yielded fine mullite powders of about 0.1 μm granular with homogeneous composition above 1300°C via Al-Si spinel phase above 600°C, process [I].

In process [II], organic solvents were washed away by distilled water after the first stage hydrothermal treatment at 300°C, 50 MPa for 2 h, then the produced pseudoboehmite-like precipitate was again hydrothermally treated at 600°C under 20 MPa for 2 to 168 h to produce needle-like mullite particles 5 nm wide and 100 nm long. Small amounts of AS(H)-II phase was also produced in these processes.

In process [III], amorphous coprecipitates, hydrolyzed from mixed alkoxide solutions and calcined at 600°C in air, were treated hydrothermally at 600°C under 20 MPa for 2 to 72 h. The coprecipitates which had inhomogeneous Al and Si distributions changed into homogeneous mullite $Al_2O_3/SiO_2 = 1.5 \pm 0.3$ after the treatment for 72 h. The mullite showed aggregated acicular shapes 0.25 μm wide and 0.5 μm long on average including a few larger ones up to 2 μm long.

Hot pressing of the acicular mullite yielded dense ceramics (d = 3.1g/cm^2, T.D. = 98%) at 1600°C under 40 MPa for 30 min.

---

*Present address
The Nishi Tokyo University
3-7-19 Seijo, Setagaya, Tokyo 157 Japan

†Present address
Dainihon Ink and Chemicals Corp.
1-3 Takasago, Takaishi, Osaka 592 Japan

Introduction

Ultra fine submicron powders are new expectant materials for various uses such as ceramics, catalysts, paints, magnetics, and fillers because they often exhibit more magnificent properties than usual powders. Such ultra fine powders are required not only to be fine but also to be pure, to have narrow distribution of particle size and not to agglomerate. And moreover it is important to control the particle's shape. Ultra fine powders have become useful in order to make reliable ceramics by controlling their microstructures, purities, homogeneity, etc. Especially to obtain ceramics with homogeneous composition and distribution, it is required to mix their component units uniformly to fine orders. For such demands, ultra fine powders are now eagerly requested.

In the Al_2O_3-SiO_2 system, mullite is the only stable compound under normal pressure. Pure mullite, however is difficult to obtain, because it is an incongruent melting compound and has a wide solid solution range including metastable state. Therefore, in order to obtain pure mullite, it is necessary to make an effort not only to exclude impurity such as alkali, but also to homogenize its composition. Until now, many studies to synthesize mullite have been achieved by using fine powders.[1] Recently, Mazdiyasni and Brown[2] have studied the alkoxide process, that is, the process of powder preparation by hydrolysis or thermal decomposition of metal alkoxide. By using fine powders prepared by alkoxide process, high temperature and high strength mullite ceramics and thin film translucent mullite were obtained. The alkoxide process has merits to be able to mix the components in liquid state and to produce fine powders of high purity.

On the other hand, hydrothermal synthesis of ceramics powder, which uses water under high temperature and high pressure as a reaction medium, has merit of being able to promote reactions at lower temperatures than usual solid state reactions, and being able to produce highly pure crystalline powders. Therefore, we examined new processes to produce mullite fine powder by combination of the alkoxide and hydrothermal processes as follows.

1. Hydrothermal treatment and calcination
2. Two-stage hydrothermal treatment
3. Hydrothermal crystallization of coprecipitates

Experimental

Raw Materials

Aluminum•isopropoxide ($(Ci\text{-}C_3H_2O)_3Al$, M_w 2042, M_p 131°C, S_g 1.035, Al cont. 13.2±0.2%, purity above 99%, commercial name AIPD, Kawaken Fine Chemicals Co., Japan) 82.2g and Silicon•tetraethoxide ($(C_2H_5O)_4Si$, M_w 208.3, S_g 0.93–0.94, SiO_2 cont. 28.8%, residual HCl less than 0.01%, commercial name Ethylisilicate 28, Nippon Colcoat Chem Co., Japan) 30.0g were weighed (Al_2O_3:SiO_2 = 3.2 mol ratio) in a 500 ml flask with agitator. After 240 ml benzene was added in the flask, the content was solved by agitating at refluxing temperature (78–80°C) for 5 h in dry nitrogen. The obtained benzene solution of mixed alkoxide was used as the starting material for hydrothermal treatment.

For comparison, alumina sol (Al_2O_3 cont. 10.14%, PH 4.99, Vis. 370 cps, S_g 1.1, commercial name Aluminasol 200, Nissan Chem. Ind. Co., Japan) and acid stabilized silica sol (SiO_2 cont. 20.12%, PH 2.55, Vis. 1.8 cps, S_g. 1.1, commercial name Snowted O, Nissan Chem. Ind. Co., Japan) were weighed to Al_2O_3:SiO_2 = 3.2 (mol ratio) and agitated for 10 min by glass rod to be used as the raw material for hydrothermal treatment.

Table 1. Products prepared by the hydrothermal treatment (for 2 h) of mixed alkoxide (Al_2O_3:SiO_2 = 3:2, mole ratio)—treated directed—

Temperature (°C)	Pressure(MPa) 10	50	100
300	B	B	B
400	B	B	B
500	B [a]	B [a]	B+H+A [a]
600	Am [a]	S [a]	A+H+S+M [a]

a) With hydrocarbon
B : Pseudoboehmite (γ-AlOOH)-like phase
H : Hydralsite (2 Al_2O_3·2 SiO_2·H_2O ; after Roy and Osborn (1954))
A : AS(H)-II (close to Al_2SiO_5 ; after Aramaki and Roy (1963))
M : Mullite (3 Al_2O_3·2 SiO_2-2 Al_2O_3·SiO_2)
Am : Amorphous phase, S : Al-Si Spinel

Results and Discussions

Products of Hydrothermal Treatment

The products of hydrothermal treatment of the mixed solution of Al-isopropoxide Si-tetra ethoxide (Al_2O_3/SiO_2 = 3/2), at various temperature under various pressures for 2 h, are shown in Table 1. The X-ray powder diffraction patterns of these products have broad peaks as shown in Figs. 1 to 3, which suggest that these products are not well crystallized or consist of fine crystals. Examination of the products by gas chromatography revealed that high molecular hydrocarbons, M_w = 100 to 300, were produced as by-products.

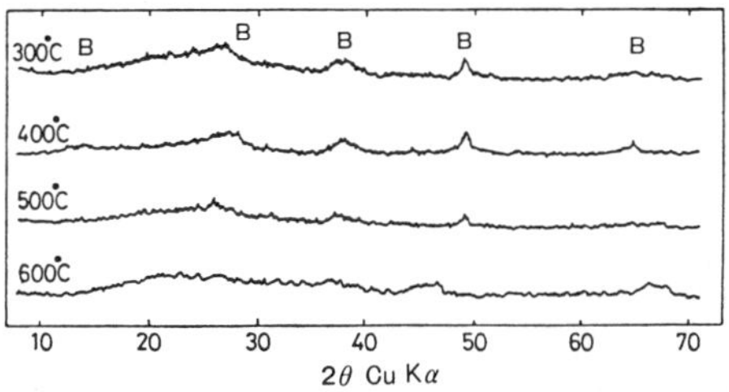

Fig. 1. X-ray powder diffraction patterns of products after hydrothermal treatment, at various temperature under 10 MPa for 2 h, of alkoxide (Al_2O_3:SiO_2 = 3:2, mole ratio) (B: Pseudoboehmite-like phase).

Influence of Hydrothermal Temperature

As shown in Fig. 1, the X-ray diffraction patterns of the products under 10 MPa at 300 to 480°C look like pseudoboehmite (Al00H•nH_2O), which were detected by Sato el al.[3] by neutralization of aluminum salt solution. We call this phase pseudoboehmite-like, because it may include Si.

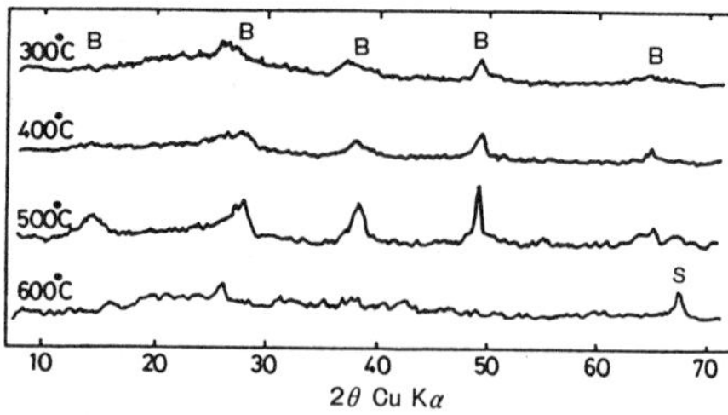

Fig. 2. X-ray powder diffraction patterns of products after hydrothermal treatment at various temperatures under 50 MPa for 2 h of alkoxide (Al_2O_3:SiO_2 = 3:2, mole ratio). B: Pseudoboehmite-like phase, S: Al-Si Spinel.

The products under 10 MPa at 300 to 400°C are not well crystallized pseudoboehmite-like phase. The X-ray patterns of the products at higher temperature became weaker, so the products at 600°C might be regarded as amorphous.

The X-ray diffraction of the products under 50 MPa at 300°C (Fig. 2) does not show a well-crystallized pseudoboehmite-like phase, and the diffractions grew sharper as the temperature rose to 400 to 500°C. At 600°C these diffractions disappeared and new broad diffractions appeared around 2θ = 25 - 26° and 2θ = 66 - 68°. The former peak is near the main diffraction (210)(120) of mullite and the latter is near the diffraction of Al-Si spinel, which may be the precursor for mullite.

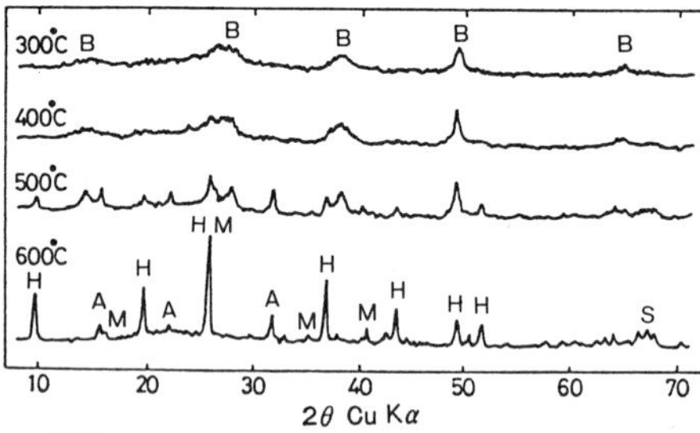

Fig. 3. X-ray powder diffraction patterns of products after hydrothermal treatment, at various temperature under 100 MPa for 2 h, of alkoxide (Al_2O_3:SiO_2 = 3:2, mole ratio) (B: Pseudoboehmite-like phase, H: Hydralsite, A: AS(H)-II, M: Mullite, S: Al-Si Spinel).

The products under 100 MPa below 400°C (Fig. 3) are almost the same as the products under 10 MPa. But, at 500°C hydralsite ($2Al_2O_3 \cdot 2SiO_2 \cdot H_2O$) which was identified by Roy and Osborn,[4] and AS(H)-II (close to Al_2SiO_5) by Aramaki and Roy,[5] were observed. At 600°C pseudoboehmite-like phase disappeared, hydralsite and AS(H)-II, however, grew.

Roy and Osborn[4] obtained hydralsite by hydrothermal treatment at 405 to 530°C under 35 to 70 MPa for 6 to 30 days from kaolinite, nacrite, dickite, halloysile or $Al_2O_3 \cdot SiO_2$ gel. Also Aramaki and Roy[5] obtained AS(H)-II by hydrothermal treatment at 500 to 700°C under 50 to 500 MPa for 7 days from $Al_2O_3 \cdot SiO_2$ gel of Al_2O_3/SiO_2 = 1.

In these past studies, hydralsite or AS(H)-II were obtained by hydrothermal treatment for more than 6 days. But, in this study, we obtained these products by hydrothermal treatment for only 2 h.

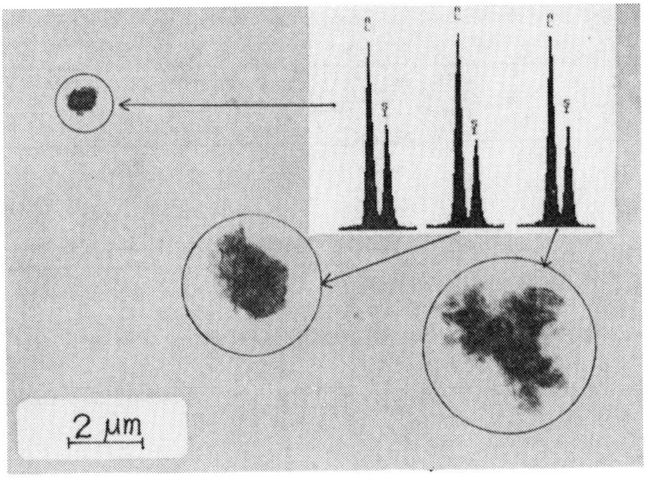

Fig. 4. TEM photograph of the product after hydrothermal treatment (300°C, 50 MPa, 2 h) of mixed alkoxide (Al_2O_3:SiO_2 = 3.2, mole ratio) and EDS analysis of each particle.

Influence of Hydrothermal Pressure

Crystallization of hydrated products were promoted as the pressure increased. That is, the pressure promoted the crystallization of pseudoboehmite-like phase at 300 to 400°C, and the products in the Al_2O_3-SiO_2-H_2O system, such as hydralsite and AS(H)-II, at 500 to 600°C.

It is noticeable that the increase of water partial pressure promoted the reaction between Al_2O_3 and SiO_2, that is, mullite, which is not a hydrated product, was produced slightly at 600°C under 100 MPa.

Pseudoboehmite-Like Phase

High crystalline boehmite (γ-AlOOH) was produced by the hydrothermal treatment of aluminum•isopropoxide at 300°C under 50 MPa for 2 h. On the other other hand, crystalline pseudoboehmite-like phase was produced from mixed alkoxide under the same condition (Fig. 2). This product (Fig. 4) consisted of particles of 1–5 μm in length, in which Al and Si were detected. Al and Si were detected with same ratio, in both 1 μm square EDS analysis and 20 nm diameter spot analysis. From these EDS analyses (Fig. 5) it can be concluded that Al_2O_3 and SiO_2 components exist homogeneously in areas larger than 20 nm. In high magnification TEM, aggregates of fine plate particles of a few nm thickness and a few hundred nm length could be observed (Fig. 6). The IR absorption spectra of pseudoboehmite-like phase (Fig. 7) was different from those of boehmite and amorphous silica which were obtained by the hydrothermal treatment under the same condition as pseudoboehmite-like phase. This suggests that this phase is not a mixture of psuedoboehmite and amorphous silica, and Al and Si are dispersed uniformly in this phase.

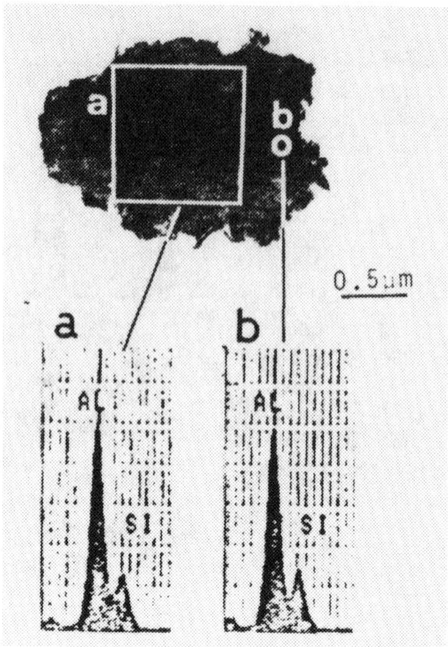

Fig. 5. TEM photograph of the product after hydrothermal treatment (300°C, 50 MPa, 2 h) of mixed alkoxide (Al_2O_3:SiO_2 = 3:2, mole ratio) and EDS analysis of the area a and b.

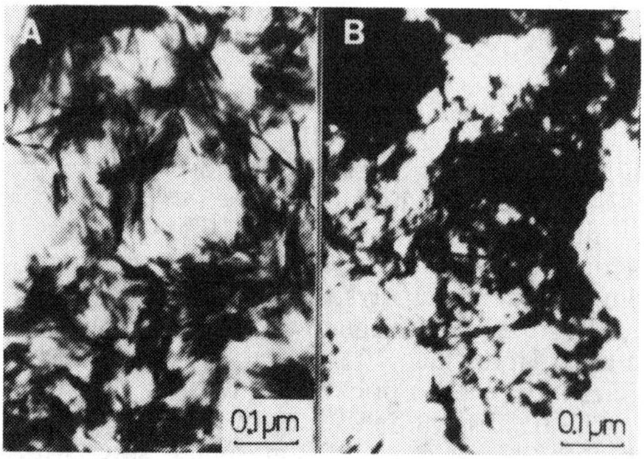

Fig. 6. TEM photographs of the products after hydrothermal treatment (300°C, 50 MPa, 2 h) of (A) mixed alkoxide and (B) mixed sol Al_2O_3:SiO_2= 3:2, mole ratio).

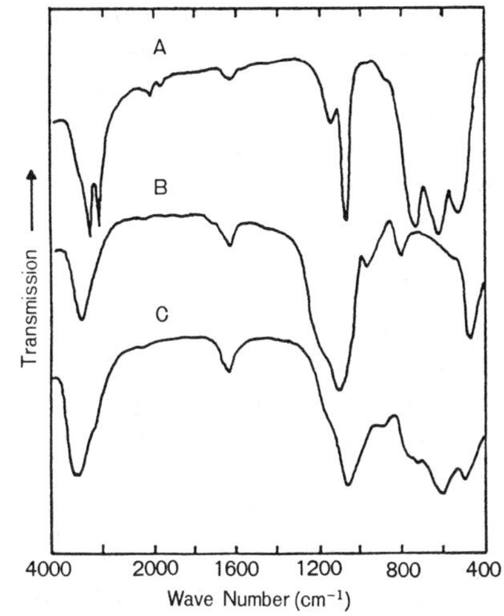

Fig. 7. IR absorption spectra of products after hydrothermal treatment (300°C, 50 MPa, 2 h) of alkoxide. A: Boehmite from Al-isopropoxide. B: Amorphous silica from Si-ethoxide. C: Pseudoboehmite-like phase from mixed alkoxide (Al_2O_3:SiO_2 = 3:2, mole ratio).

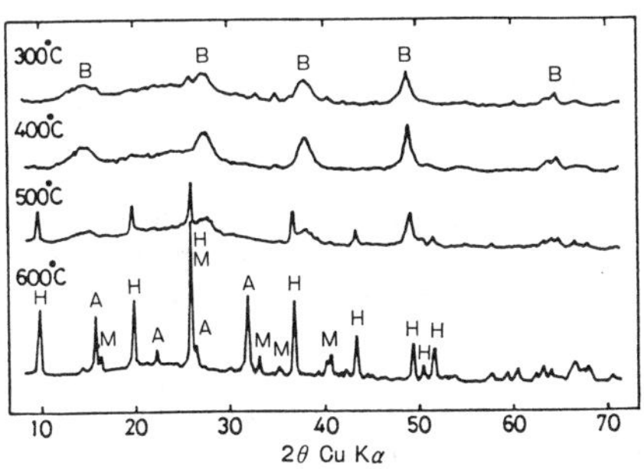

Fig. 8. X-ray powder diffraction patterns of products after hydrothermal treatment at various temperatures under 50 MPa for 2 h of sol (Al_2O_3:SiO_2 = 3:2, mole ratio). B: Pseudoboehmite-like phase, H: Hydralsite, A: AS(H)-II.

Influence of Organic Solvents

X-ray diffraction patterns of the products of hydrothermal treatment of the mixture of Alumina sol and silica sol (Al_2O_3:SiO_2 = 3:2 mol ratio), under the same condition as the mixed alkoxide, are shown in Fig. 8.

The diffraction patterns of pseudoboehmite, hydralsite and AS(H)-II were sharper than those from alkoxide (Fig. 2). This means that the crystallinity of these products was higher than those from alkoxide. In alkoxide process, organic solvent, such as benzene of alcohol which was produced by hydrolysis of alkoxide, might lower the fugacity of water and suppress production from and growth of the hydrated products such as pseudoboehmite-like phase. The solubility of silica and alumina in hot water might be lowered in order to include organic solvent. So, silica and alumina from alkoxide might be much more super-saturated to nucleate rapidly than those from sol. This might be one reason for the production of fine powders. Organic solvent might be absorbed on the surface of the particles so that the particles hardly resolved to the hot water and pseudoboehmite-like phase hardly changed to hydralsite and AS(H)-II. In such a case, adsorbed organics might be reacted by the catalytic action of fine particles. In fact, high molecular hydrocarbon was synthesized as a by-product by the hydrothermal treatment of alkoxide above 500°C, which suggests a reaction between the pseudoboehmite-like phase and organic solvent.

In Fig. 6, TEM photographs are shown of pseudoboehmite-like phase produced by the hydrothermal treatment at 300°C under 50 MPa for 2 h of the mixed alkoxide and the mixed sol. Evidently the former was finer than the latter.

Organic solvent might be effective to stabilize the fine pseudoboehmite-like phase precipitate, but might suppress the production of hydralsite, AS(H)-II and mullite. To obtain these products, a longer time or higher water: partial pressure might be needed, or an additional step, such as the hydrothermal treatment without organics, might be needed.

Roy and Osborn[5] obtained mullite in their hydrothermal equilibrium study above 485°C under 35 to 55 MPa for 3 to 30 days, using kaolinite, coprecipitated gel, etc. as starting materials. Also, DeVijnck[6] obtained mullite by the hydrothermal treatment of coprecipitated gel at 600°C under 39.9 MPa for 7 days. Now, the object of this study is to obtain mullite in relatively short times. So, we examined the other two processes. That is, 1) firing of the pseudoboehmite-like phase and 2) additional hydrothermal treatment without organic solvent.

Firing of the Hydrothermally Obtained Precipitate

The Products of Firing

In Fig. 9, TG-DTA curves of the precipitates of pseudoboehmite-like phase obtained by the hydrothermal treatment of alkoxide and sol at 300°C under 50 MPa for 2 h, which were dried at 120°C in air, are shown. Each curve had a weight loss or endotherm peak due to the dehydration of pseudoboehmite-like phase between 400 and 600°C. Above 600°C, there were no abrupt weight loss or exo-end therm peak, but the precipitates, when fired at 1400°C, were changed to mullite as shown by X-ray analysis. The DTA curves turned upwards above 1100°C and there might be broad exotherm peak around 1300°C. This might be the exotherm of mullite crystallization.

In Figs. 10 and 11, are shown the X-ray powder diffraction patterns of products after calcination, between 600 and 1300°C in air for 1 h, of pseudoboehmite-like precipitates obtained by the hydrothermal treatment of alkoxide and sol. The X-ray diffractions of pseudoboehmite-like phase disappeared after the calcination at 600°C. Between 600°C and 1200°C, new diffractions appeared around $2\theta = 66–68$ which might belong to Al-Si spinel. The diffraction patterns of the products fired at 1350°C show mullite.

Brindley and Nakahira[7] reported that kaolinite changed to Al-Si spinel after calcination at 950–1000°C. But, in this study, Al-Si spinel was already produced just after the dehydration of pseudoboehmite-like phase around 600°C. However, to obtain mullite, further firing near to 1300°C was needed.

TEM photograph and EDS patterns of mullite produced by the firing at 1300°C for 1 h are shown in Fig. 12. Mullite obtained from alkoxide had granules of about 0.1 μm dia. as shown in Fig. 12. On the other mullite from sol had no distinct

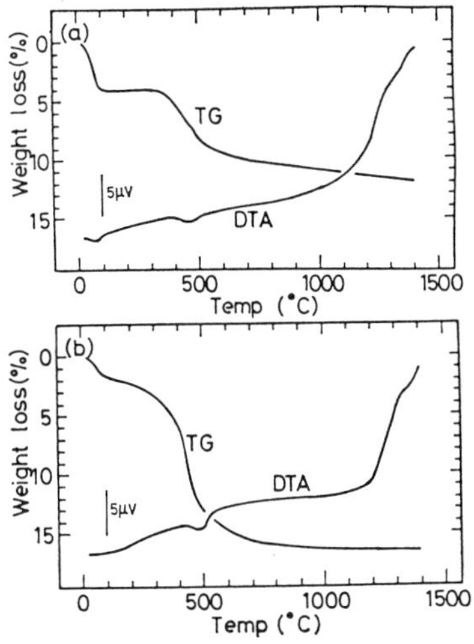

Fig. 9. TG-DTA curves of the products after hydrothermal treatment (300°C, 50 MPa, 2 h) of (a) alkoxide and (b) sol (Al_2O_3:SiO_2 = 3:2, mole ratio).

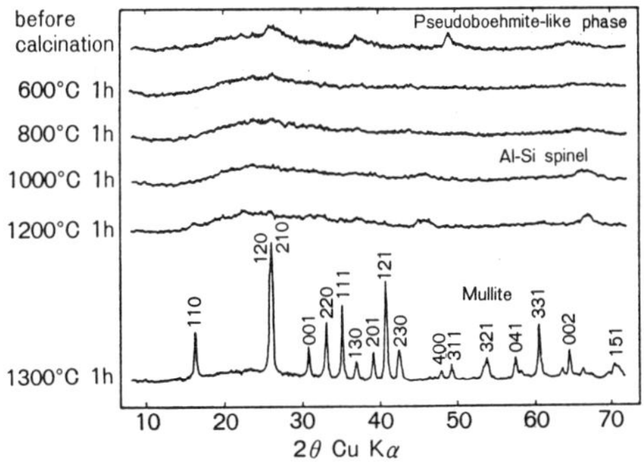

Fig. 10. X-ray powder diffraction patterns of products after calcination of the hydrothermally treated (300°C, 50 MPa, 2 h) alkoxide (Al_2O_3:SiO_2= 3:2, mole ratio).

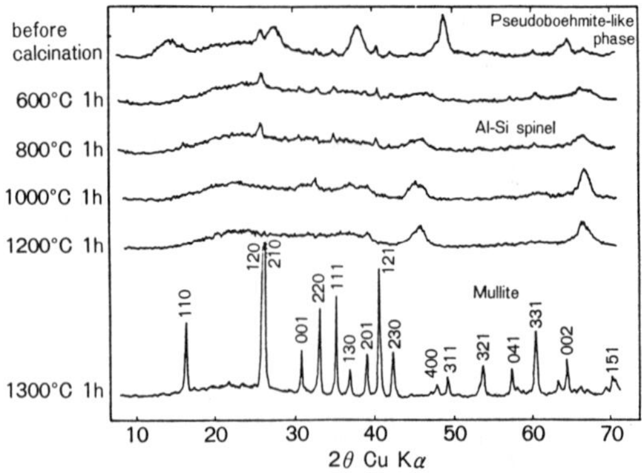

Fig. 11. X-ray powder diffraction patterns of products after calcination of the hydrothermally treated (300°C, 50 MPa, 2 h) sol (Al_2O_3:SiO_2 = 3:2, mole ratio).

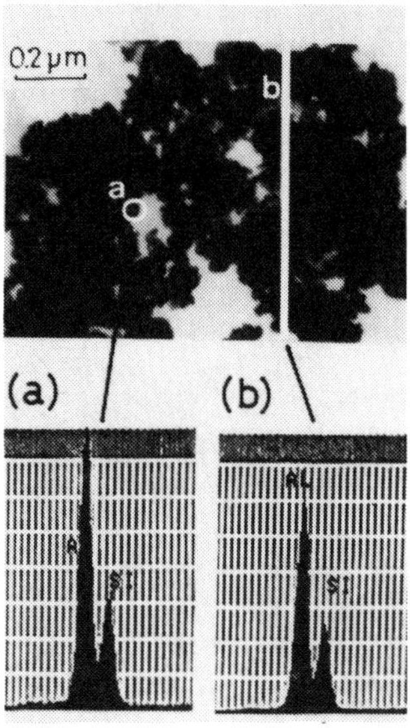

Fig. 12. TEM photograph of the product after calcination (1300°C, 1 h) of the hydrothermally treated (300°C, 50 MPa, 2 h) mixed alkoxide (Al_2O_3:SiO_2 = 3:2, mole ratio) with EDS analysis of the spot a and the line b.

shape. In Fig. 12, Al_2O_3/SiO_2 ratio from a spot analysis of about 20 nm and (b) line analysis are similar, i.e., the former 1.7 and the latter 1.5.

This means that the products from alkoxide are 3:2 mullite with homogeneous composition. On the other hand, the composition of the products from sol was not homogenous in the Al_2O_3/SiO_2 ratio.

The crystallite sizes were calculated by X-ray method using the diffraction area (201) and (220). Mullite obtained from alkoxide consisted of fine particles of 46 nm

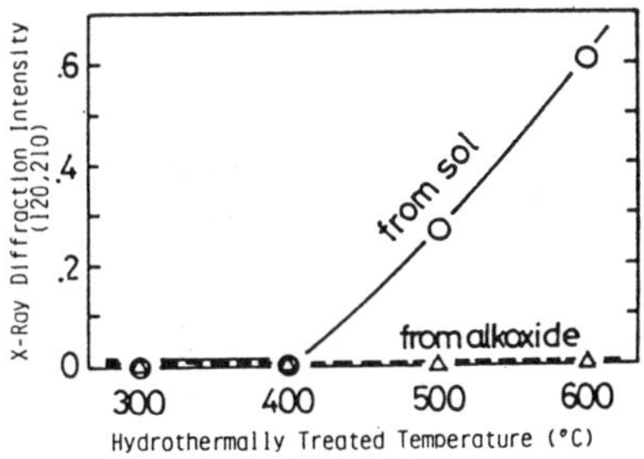

Fig. 13. Apparent amount of mullite after calcination at 1200°C for 1 h in air, against hydrothermally treated temperature.

(parallel to C axis) and 34 nm (perpendicular to C axis), which were finer than those from sol (81 nm, 60 nm).

Influence of the Temperature of Hydrothermal Treatment

Figure 13 shows the ratios between the temperature of hydrothermal treatment under 50 MPa for 2 h and the apparent yield of mullite obtained by calcination (in air at 1200°C for 1 h) of the hydrothermal products. This was indicated in the area of (120,210) diffraction peaks of mullite.

In case of using alkoxide, mullite was not formed by calcination at 1200°C. This is the same situation as that shown by Al-Si spinel which was obtained by hydrothermal treatment at 600°C and did not change to mullite by calcination below 1300°C (Figs. 9 and 10).

On the other hand, calcination (at 1200°C) of the precipitate obtained by the hydrothermal treatment of sol at 500°C and 600°C yielded mullite. This means, that hydralsite or AS(H)-II (Fig. 8), obtained by hydrothermal treatment of sol at 500°C and 600°C, acted as precursors for mullite. But, the compositions of these precursors were not just the same as mullite, so the yielded mullite did not have a homogenous composition. For example, large needle-like particles, (a) in Fig. 14, had the composition of Al_2O_3/SiO_2 ratio 2.2, and the fine granule particles (b) had Al_2O_3/SiO_2 ratio 1.5. From TEM and EDS analysis of the hydrothermal products at 600°C under 50 MPa for 2 h, the large needle-like particles (a) could be formed from AS(H)-II crystals and the fine particles (b) from hydralsite ($2Al_2O_3{\cdot}2SiO_2{\cdot}H_2O$) (Fig. 15). By the calcination at 1300°C, mullite was obtained with the same shape and composition as the precursors. The crystallite size of this mullite calculated by X-ray method, was 94 nm in length and 59 nm in width. This size was far smaller than the shape in TEM (Fig. 4). The needle-like mullite in TEM thus might be polycrystals. The composition of particles of AS(H)-II from EDS analysis had more Al_2O_3 content that Al_2SiO_5 which was proposed by Aramaki and Roy.[5]

From these results, the following was concluded. To obtain mullite with homogenous composition, firing of the pseudoboehmite-like precipitates obtained by hydrothermal treatment of the mixed alkoxide is the better process. On the other hand, to obtain mullite at lower temperature in short times, the calcination of hydrated aluminosilicate precipitate obtained by the hydrothermal treatment of mixed sol at 500–600°C under 50 MPa, is better.

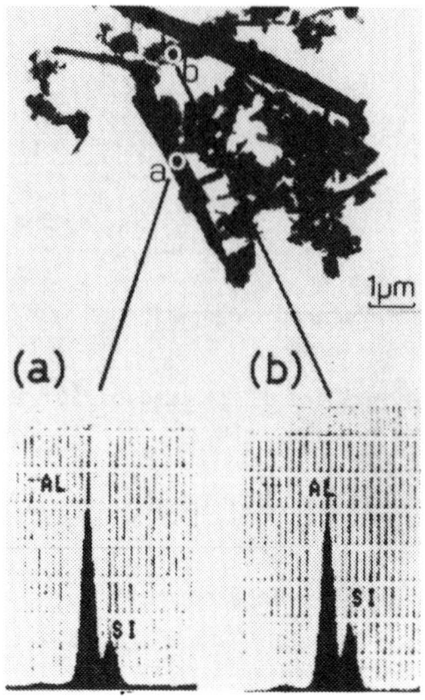

Fig. 14. TEM photograph of the product after calcination (1300°C, 1 h) of the hydrothermally treated (600°C, 50 MPa, 2 h) mixed sol (Al_2O_3:SiO_2 = 3:2, mole ratio) with EDS analysis of the particle a and the particle b.

Mullite Synthesis By Two Stage Hydrothermal Treatment

As mentioned above, fine mullite powders with homogeneous composition could be obtained by firing the pseudoboehmite-like phase precipitate produced by the hydrothermal treatment of the mixed alkoxide above 1300°C. But, it may be desirable to synthesize mullite directly by hydrothermal treatment only, because of low energy. However, in the hydrothermal treatment above 500°C of alkoxide solution, the organic solvent stabilized pseudoboehmite-like phase prevents further reaction between Al_2O_3 and SiO_2. Thus, in this study, we examined a new two-stage process. That is, at first, pseudoboehmite-like phase with homogeneous composition was obtained by the hydrothermal treatment of alkoxide. Then, in the second stage, after organic solvents are washed away by distilled water, the pseudoboehmite-like precipitates are again hydrothermally treated to produce mullite fine powders.

Washing By Distilled Water

After hydrothermal treatment of mixed alkoxide at 300°C under 50 MPa for the prescribed hours, the organic solvents, such as benzene and alcohol, were washed away by distilled water by the following procedure, i.e., Reducing–Adding pressure between 0 and 50 MPa for 5 times using a pressure control valve and pressure pump. The organic solvent was washed away by this procedure after 3 times, and 2-propanol, ethanol, benzene were not detected by gas chromatography after 4 or 5 times of this procedure.

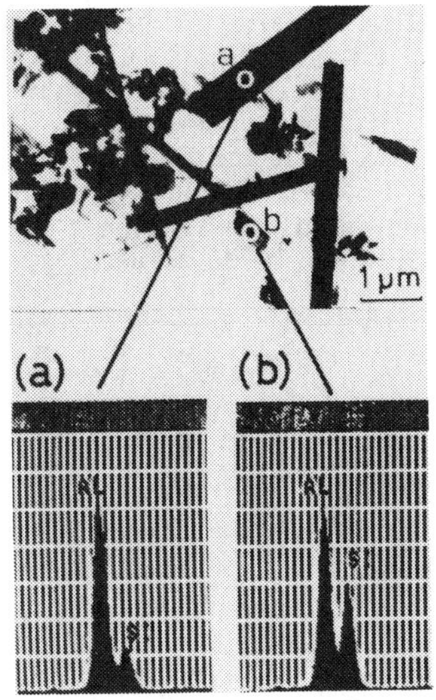

Fig. 15. TEM photograph of the product after hydrothermal treatment (600°C, 50 MPa, 2 h) of mixed sol (Al_2O_3:SiO_2 = 3:2, mole ratio) with EDS analysis of the particle a and the particle b.

Table 2. Products prepared by the hydrothermal treatment (for 2 h) of mixed alkoxide (Al_2O_3:SiO_2 = 3:2, mole ratio)—Treated after washing at 300°C—

Temperature (°C)	Pressure(MPa) 10	50	100
300	B	B	B
400	B	B	B
500	B	B+H	B+H+A
600	S	A+H+S+M	A+H+S+M

B : Pseudoboehmite (γ-AlOOH)-like phase
H : Hydralsite ($2Al_2O_3 \cdot 2SiO_2 \cdot H_2O$; after Roy and Osborn (1954))
A : AS(H)-II (close to Al_2SiO_5 ; after Aramaki and Roy (1963))
M : Mullite ($3Al_2O_3 \cdot 2SiO_2$-$2Al_2O_3 \cdot SiO_2$)
Am : Amorphous phase, S : Al-Si Spinel

Influence of the Temperature in the Second Stage Hydrothermal Treatment

Table 2 shows the results of the second stage hydrothermal treatment at 300 to 600°C under 10 to 100 MPa for 2 h after washing the precipitates produced by the hydrothermal treatment of alkoxide at 300°C under 50 MPa for 30 minutes. Comparing with the direct hydrothermal treatment (without washing) under the same conditions, there was no difference below 400°C, but above 500°C aluminosilicates such as hydralsite, AS(H)-II and mullite, were produced under the lower pressures. At 500°C under 50 MPa, mullite was produced together with hydrated products. So, to produce mullite effectively, the hydrothermal treatment at 600°C above 10 MPa below 50 MPa might be desirable.

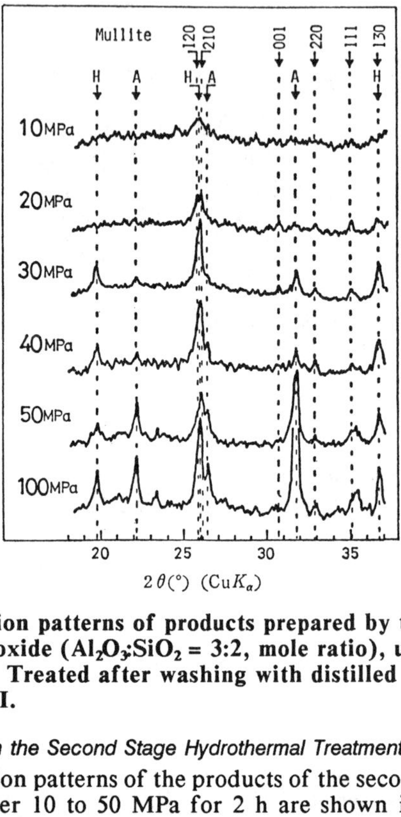

Fig. 16. X-ray diffraction patterns of products prepared by the hydrothermal treatment of mixed alkoxide (Al_2O_3:SiO_2 = 3:2, mole ratio), under various pressures at 600°C for 2 h. Treated after washing with distilled water at 300°C. H: Hydralsite, A: AS(H)-II.

Influence of the Pressure in the Second Stage Hydrothermal Treatment

The X-ray diffraction patterns of the products of the second stage hydrothermal treatment at 600°C under 10 to 50 MPa for 2 h are shown in Fig. 16. Above 30 MPa, much mullite was formed, but hydralsite and AS(H)-II were also produced together with mullite. Larger amounts of hydralsite and AS(H)-II were produced with increasing pressure. So, in order to produce mullite only, the best condition was deduced to be 20 MPa at 600°C.

Influence of Time in the Second Stage Hydrothermal Treatment

The second stage hydrothermal treatment at 600°C under 20 MPa was carried on for 2 to 168 h (Fig. 17). The yield of mullite increased with time, and hydralsite

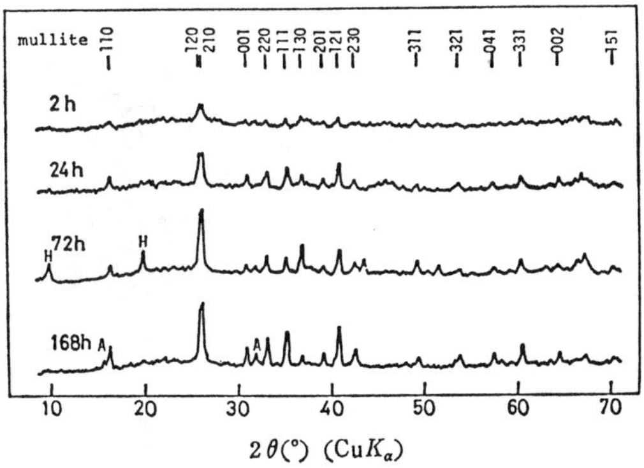

Fig. 17. X-ray powder diffraction patterns of products prepared by the hydrothermal treatment of mixed alkoxide (Al_2O_3:SiO_2 = 3:2, mole ratio), at 600°C under 20 MPa for various hours. Treated after washing with distilled water at 300°C. H: Hydralsite, A: AS(H)-II.

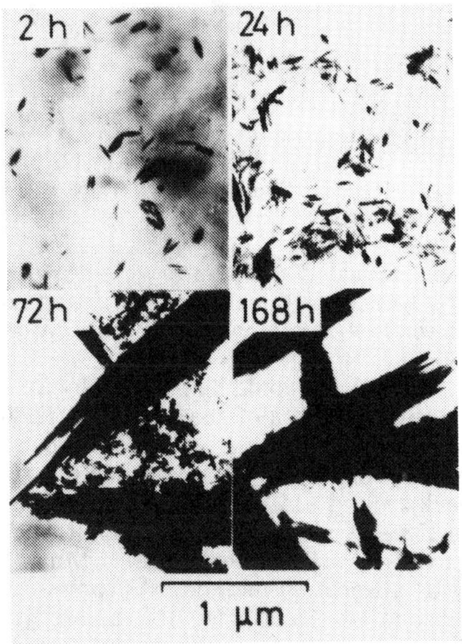

Fig. 18. TEM photographs of products prepared by the hydrothermal treatment of mixed alkoxide (Al_2O_3:SiO_2 = 3:2, mole ratio), at 600°C under 20 MPa for various hours. Treated after washing with distilled water at 300°C.

and AS(H)-II were also coprecipitated. TEM photographs of these products are shown in Fig. 18. In the products of the hydrothermal treatment for 2 h, needle-like mullite of 100 nm length and 5 nm width was observed. The needles grew to 1 μm length and 20 nm width in 168 h. But, in the hydrothermal products for 24 h and 72 h, fine planes were also observed together with the needle-like crystals. EDS

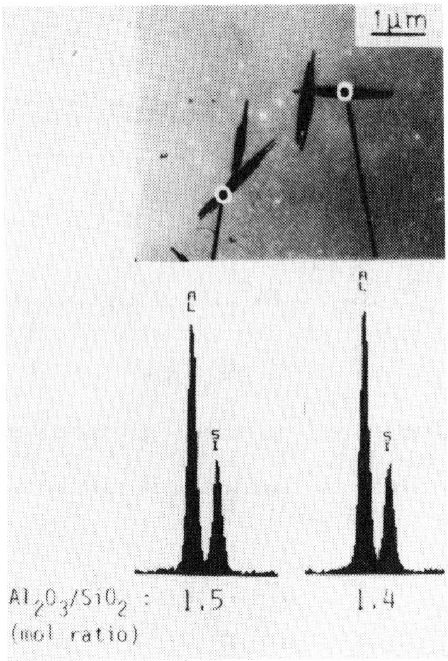

Fig. 19. TEM photograph of the product prepared by the hydrothermal treatment of mixed alkoxide (Al_2O_3:SiO_2 = 3:2, mole ratio), at 600°C under 20 MPa for 168 h. Treated after washing with distilled water at 300°C. With EDS analysis of the particle a and the particle b.

analysis of these products revealed that there were various particles of different compositions of Al_2O_3/SiO_2 ratio in the short time treatment. After the long time treatments needle-like crystals with homogeneous composition could be obtained (Fig. 19).

The process of mullite synthesis could be pursued by infrared absorption spectra (Fig. 20). In the IR absorption spectra of the product after hydrothermal treatment for 2 h, an absorption could be observed around 1050 cm^{-1} similar to pseudo-boehmite-like phase. This absorption peak decreased as the treating hours increased, and a new absorption peak around 1150 cm^{-1} developed. The spectra of the products after 168 h are just the same as mullite. It is difficult to distinguish mullite from sillimanite ($Al_2O_3 \cdot SiO_2$) by X-ray diffraction method. Muan[8] reported the difference between them by using IR absorption method. As shown in Fig. 21, IR absorption spectra of the hydrothermal products after 168 h, are just the same as mullite obtained by firing in air at 1400°C for 1 h. They could be distinguished from other aluminosilicates, such as sillimanite, kyanite and andalusite.

From the above results, it was proved that needle-like mullite of micron length with almost homogeneous composition (small amount of AS(H)-II as by product) could be obtained by the following process.

Organic solvent was washed away by distilled water after the first stage hydrothermal treatment at 300°C under 50 MPa, then the produced pseudoboehmite-like precipitate was again hydrothermally treated at 600°C under 20 MPa for many hours (168 h = 7 days) to produce needle-like mullite. DeVijnck[6] reported that only mullite phase was obtained by the hydrothermal treatment of alumina-silica

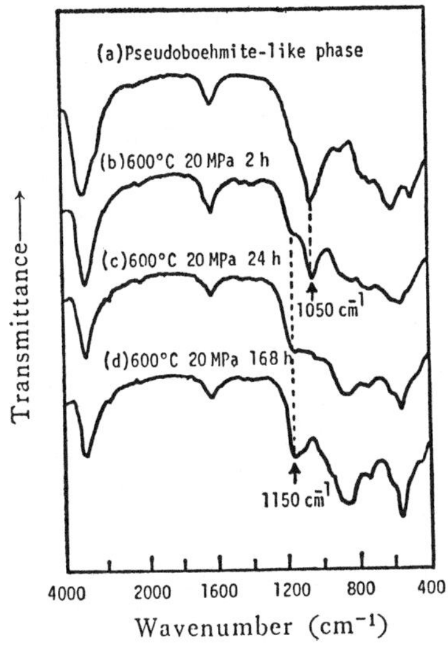

Fig. 20. IR absorption spectra of (a) Pseudoboehmite-like phase, obtained by the direct hydrothermal treatment of mixed alkoxide,* at 300°C under 50 MPa. (b)~(d) Products prepared by the hydrothermal treatment of mixed alkoxide,* at 600°C under 20 MPa for various hours. Treated after washing with distilled water at 300°C.

***Al_2O_3:SiO_2 = 3:2, mole ratio.**

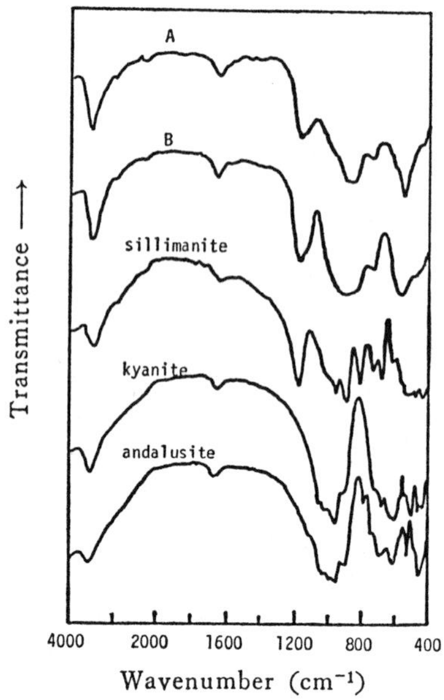

Fig. 21. IR absorption spectra of (a) mullite obtained hydrothermal treatment (600°C, 20 MPa, 168 h). (b) mullite obtained by calcination (1400°C, 1 h), from mixed alkoxide (Al_2O_3:SiO_2 = 3:2, mole ratio) and minerals in the Al_2O_3-SiO_2 system (sillimanite, kyanite, andalusite).

coprecipitate at 600°C, under 39.2 MPa, for 7 days. In this study, in the hydrothermal treatment above 20 MPa, the yield of AS(H)-II was increased. And also in the longer time (336 h = 14 days) treatment at 600°C under 20 MPa, AS(H)-II could not be excluded. Therefore, it is necessary to consider making more uniform compositions and shapes in a shorter time and also preventing the synthesis of AS(H)-II phase.

Synthesis By Hydrothermal Crystallization Processing

To get a large amount of mullite powder, it is easier to treat solids than solutions. In this section we discuss the method to get fine mullite crystals by the hydrothermal treatment of the system Al_2O_3-SiO_2 fine powder which is obtained by the hydrothermal treatment of mixed alkoxide.

Hydrolysis Under One Atmosphere

Aluminum•isopropoxide and silicon tetraethoxide with an Al_2O_3/SiO_2 mol ratio of 3/2 were weighed in a 500 ml flask. After equal weight Isopropanol was added, the content was solved by agitating. Then distilled water with Ph equal to 10 prepared by ammoniated water was slowly added, hydrized. The resulting powder was dried at 120°C for 24 h in air, then calcine at 600°C for 1 h in air.

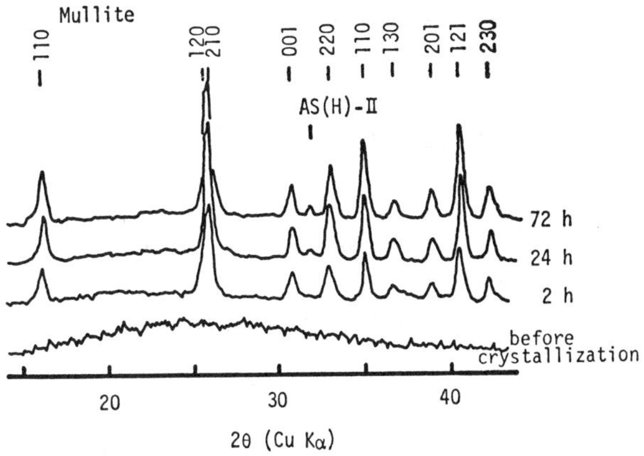

Fig. 22. X-ray powder diffraction patterns of the products before and after hydrothermal crystallization, at 600°C under 20 MPa for various hours, of amorphous powder prepared by hydrolysis of mixed alkoxides (Al_2O_3:SiO_2 = 3:2, mole ratio).

Hydrothermal Crystallization

As-calcined powder 10g and distilled water were weighed into an autoclave 30 $\phi \times 200$ h. Then, crystallized hydrothermally at 600°C under 20 MPa for 2 to 72 h. After treatment powder was dried at 120°C for 24 h in air.

Products

Figure 22 shows the comparison of X-ray powder diffraction patterns of the products before and after hydrothermal crystallization, at 600°C under 20 MPa for 2 to 72 h, of amorphous powder prepared by hydrolysis of mixed alkoxides. Powders before crystallization are obviously amorphous. They crystallized to mullite after 2 h hydrothermal treatment. With longer treatment times, the amount of mullite increased. At the same time, a slight amount of AS(H)-II crystallized and increased in amount with treatment time. These results are similar to the conclusions of the above sections, "Influence of the Pressure in the Second Stage Hydrothermal Treatment" and "Influence of Time in the Second Stage Hydrothermal Treatment."

In the absorption spectra shown in Fig. 23, an absorption around 1050 cm^{-1} could be observed in the amorphous powder before hydrothermal treatment, around 1150 cm^{-1} in hydrothermally crystallized powder over 24 h and the middle stage in powder hydrized for 2 h. So this powder is a mixture of mullite crystals and amorphous powder.

In Fig. 24 SEM photographs of products before and after hydrothermal treatment are shown. And in Table 3 are listed the particle size (D_{SEM}) estimated from the SEM photo, the specific surface area (S_{BET}) by B.E.T. method, and the crystallite size (Dx) calculated by X-ray method using the diffraction area (001).

The crystallite size does not vary with treatment time, about constant. But D_{BET} of powder treated for 2 h is smaller than Dx, 19.2 nm. D_{BET} is average size of whole powder. The powder treated hydrothermally for 2 h is mixed with fine mullite powder with particle size of 65 nm (D_{BET}/Dx) and more fine amorphous powder.

Assuming that surface area of amorphous and mullite powder is 370 m^2/g and 33 m^2/g, respectively, mullite/amorphous ratio is possible to estimate 80/20. After

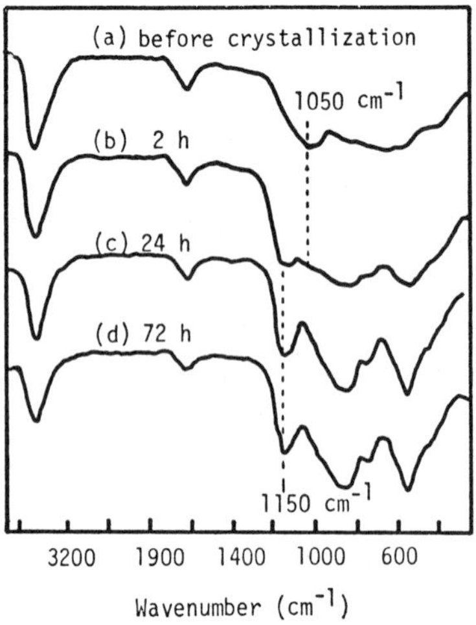

Fig. 23. Infrared absorption spectra of the products before and after hydrothermal crystallization, at 600°C under 20 MPa for various hours, of amorphous powder preprared by hydrolysis of mixed alkoxides (Al_2O_3:SiO_2 = 3:2, mole ratio).

Fig. 24. SEM photographs of products before and after hydrothermal crystallization, at 600°C under 20 MPa for various hours, of amorphous powder preprared by hydrolysis of mixed alkoxides (Al_2O_3:SiO_2 = 3:2, mole ratio).

Table 3. Characteristics of the powder prepared by hydrothermal crystallization (600°C, 20 MPa) of amorphous alumino-silicate powder from mixed alkoxides.

hydrothermal crystallization	Surface area (m^2/g)	D_{BET} (nm)	D_X (nm)	D_{SEM} (nm)
before crystallization	370	5	-	-
2 h	100	19.2	65.3	200 X 100
24 h	31	61.2	66.8	500 X 250
72 h	36	52.7	54.3	500 X 250 ~ 2000 X 400

(A) before crystallization
(B) 72 h

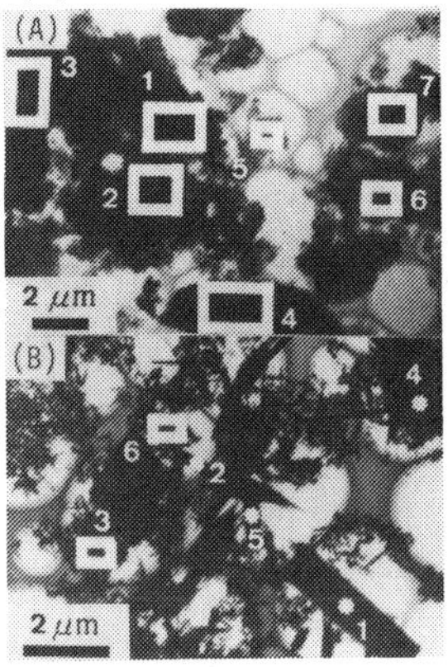

area	Al_2O_3/SiO_2 (mol ratio)
A(whole area)	1.61
A - 1	1.21
2	1.22
3	2.11
4	3.13
5	1.21
6	1.63
7	1.34
B(whole area)	1.36
B - 1	1.83
2	1.50
3	1.26
4	1.51
5	1.46
6	1.45

Fig. 25. SEM photographs and EDS analysis of the products before and after hydrothermal crystallization, at 600°C under 20 MPa for 72 h, of amorphous powder prepared by hydrolysis of mixed alkoxides (Al_2O_3:SiO_2 = 3:2, mole ratio).

hydrothermal treatment for over 24 h, D_{BET} = Dx. So hydrothermal treatment for 24 h completes crystallization to form powders as single crystals.

On the other hand, D_{SEM} is bigger than Dx and it increases with increasing time. Figure 25 shows SEM photographs and EDS analysis of the products before and after hydrothermal crystallization. In amorphous powder before hydrothermal crystallization, there are several areas far from Al_2O_3/SiO_2 = 1.5. It indicates that products homogeneous in Al_2O_3/SiO_2 mol ratio can not be prepared by hydrolysis under 1 atm. On the other hand after hydrothermal crystallization for 72 h, powders form needles like mullite crystallites. At the same time, Al_2O_3/SiO_2 mol ratio equal to 1.5 ± 0.3, so it is evident that homogenization is occurring.

In homogeneous amorphous gel which is prepared by hydrolysis of mixed alkoxides, homogeneous mullite is formed after hydrothermal crystallization. Particles prepared by this method, however, are secondary particles and about 0.1–2 μm in size.

Hot Pressing

Hot Pressing

Starting materials are amorphous powder which is hydrolyzed from mixed alkoxide solution and calcined at 600°C in air and mullite powder which is crystallized from amorphous powder by hydrothermal treatment at 600°C under 20 MPa. Specimens 100 nm in diameter were hot-pressed under 20 to 40 MPa at 1400 to 1600°C in graphite die in argon.

Evaluation Method

The surface of the cut specimens were polished by 0.5 μm diamond paste, etched at 1550°C for 10 min and examined by SEM. Bending test sample bars (3 x 4 x 30 mm^3) wore cut from the hot-pressed billet and polished through 2.5 μm diamond paste. Specimens were tested using a three-point-bending test configuration. (Span 20 mm, head-speed 0.5 mm/min).

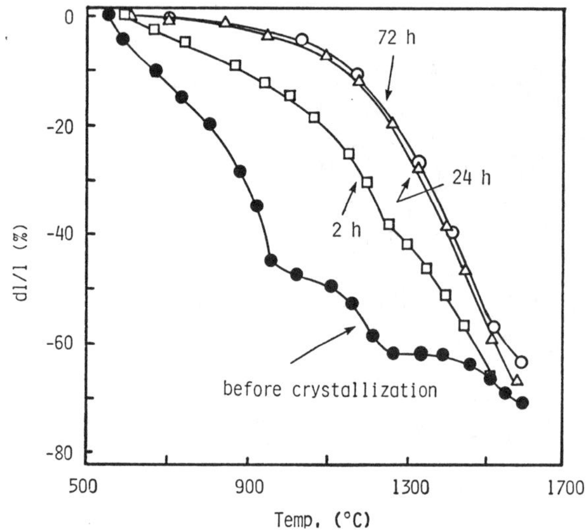

Fig. 26. Relationship between temperature and shrinkage at hot-pressing (40 MPa, 300°C/h) of compact from the powders crystallized hydrothermally at 600°C under 20 MPa.

Results

Figure 26 shows the relationship between temperature and shrinkage in hot pressing. Untreated amorphous powder changed into Al-Si spinel at 950°C and at 1250°C into mullite. Powder treated hydrothermally for 2 h. Contains some amorphous material which changed into mullite at 1250°C. On the other hand, powder treated for 24 and 72 h crystallized to mullite will sinter at over 1200°C.

Figure 27 shows density of mullite ceramics by hot-pressing. At 1600°C density is over 98% for each powder. But at lower temperatures untreated powder is the easiest to densify.

In Fig. 28 the typical microstructure of several mullite ceramics is shown. Table 4 gives their grain size. Although specific surface area and shrinkage in hot-

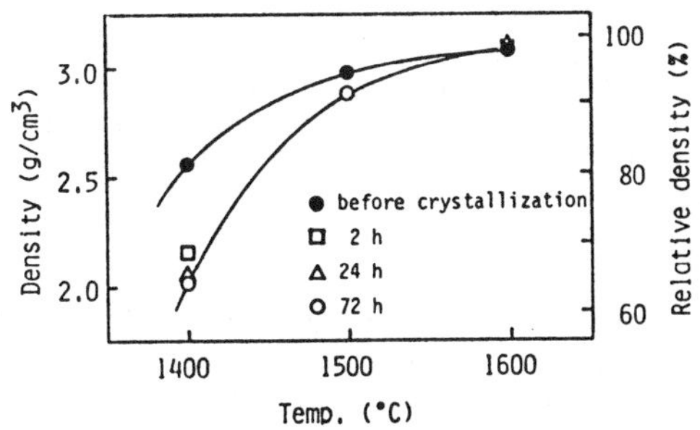

Fig. 27. Relationship between hot-pressing temperature and density of mullite ceramics under 40 MPa for 30 min from the powders crystallized hydrothermally at 600°C under 20 MPa.

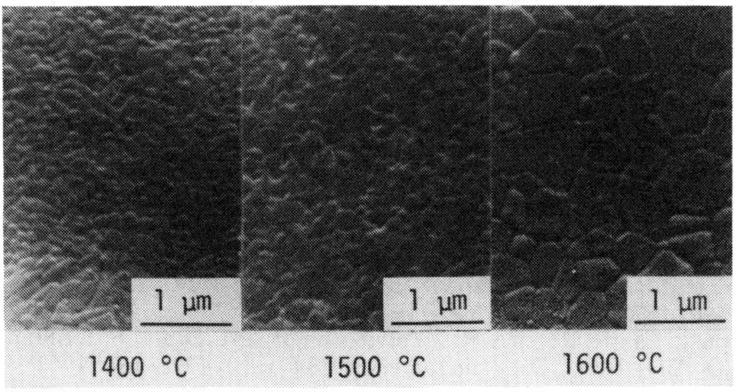

Fig. 28. Microstructures of mullite ceramics hot-pressed under 40 MPa for 30 min from the powders crystallized hydrothermally at 600°C under 20 MPa for 72 h.

Table 4. Grain size of the mullite ceramics hot-pressed at various temperatures under 40 MPa for 30 min from the powders crystallized hydrothermally at 600°C under 20 MPa.

crystallization (600°C, 20 MPa)	grain size (μm)*		
	1400°C	1500°C	1600°C
before crystallization	0.27	0.34	0.67
2 h	0.22	—	0.51
24 h	0.21	0.28	0.40
72 h	0.25	0.29	0.58

* measured by intercept method

pressing are different for each powder, the grain size of mullite ceramics made from the different powders at a given temperature is not so different.

Bending strength of mullite ceramics hot pressed at 1600°C under 22 MPa for 30 min. from the powder crystallized hydrothermally for 24 h was 195 MPa.

It is difficult to make high density and high strength mullite ceramics from needle-like powder.

Summary

We have shown the results by processing of [I] Hydrothermal treatment and calcination, [II] Two-stage hydrothermal treatment, and [III] Hydrothermal crystallization of coprecipitates. Mullite fine powders were synthesized at 600°C under hydrothermal condition.

References

[1a]Mullite, Ed. by Shigeyuki Sōmiya. Uchida Rokakuho Publishing Co., Tokyo, pp. 141, (1985) (in Japanese).

[b]Mullite 2—Raw Materials and Sintered Bodies. Ed. by Shigeyuki Sōmiya. Uchida Rokakuho Publishing Co., Tokyo, pp. 168, (1987) (in Japanese).

[c]Advances in Ceramics, Vol. 21, Ceramic Powder Science. Ed. by G. L. Messing, K. S. Mazdiyasni, J. W. McCauley, R. A. Haber, *Am. Ceram. Soc.*, pp. 825 (1987).

[2]K. S. Mazdiyasni and L. M. Brown, "Synthesis and Mechanical Properties of Stoichiometric Aluminum—Silicate (Mullite)," *J. Am. Ceram. Soc.*, **55** [11] 548–52 (1972).

[3]Taichi Sato, Fusaji Ozawa, and Hiromi Terada, "Studies on the Formation of Gelatinous Aluminum Hydroxide and Some Properties of the Derived Alumina 1, Preparation of Aluminum Hydroxide Gels," *Asahi Glass Kogyo Gijutsu Shoreikai Hokoku*, **16**, 365–378 (1970).

[4]R. Roy and E. F. Osborn, "The System Al_2O_3-SiO_2-H_2O," *Am. Mineralogists*, **39** [11/12] 853–85 (1954).

[5]S. Aramaki and R. Roy, "A New Polymorph of Al_2BiO_5 and Futur Studies in the System Al_2O_3-SiO_2-H_2O," *Am. Mineralogists*, **48** [11/12] 1322–134 (1963).

[6]Y. DeVijnck, "Transformation Hydrothermale de gels Silico-aluminewx Coprecipites," *Silicate Ind.*, **41** [11] 451–461 (1976).

[7]G. W. Brindley and M. Nakahira, "The Kaolinite-Mullite Reaction Series II Metakaolin III The High Temperature Phases," *J. Amer. Ceram. Soc.*, **42** [7] 314–318 (1959), 319–324 (1959).

[8]A. Muan, "Phase Equilibria at Liquidus Temperatures in the System Iron Oxide-Al_2O_3-SiO_2 in Air Atmosphere," *J. Am. Ceram. Soc.*, **40** [4] 121–133 (1957).

Characterization and Sintering of Al_2O_3-SiO_2 Powders Formed By Chemical Vapor Deposition

SABURO HORI AND RYUICHI KURITA

Advanced Material Systems Laboratory
Kureha Chemical Industry Co., Ltd.
3-25-1 Hyakunincho, Shinjuku-Ku, Tokyo 160 Japan

Abstract

Spherical ultrafine (Av. 40–70 nm) Al_2O_3-SiO_2 composite powders (67–80 wt% Al_2O_3) were prepared by chemical vapor deposition using a combustion flame of approximately 1900°C. The powders consisted of an amorphous phase and γ-Al_2O_3. They also exhibited two DTA mullitization peaks between 940° and 990°C, suggesting that SiO_2 was metastably contained in γ-Al_2O_3 and reacted with Al_2O_3 to form mullite before the amorphous pre-mullite crystallized. Due to the small amount of remaining amorphous phase after the mullitization, the compacts prepared from these powders did not exhibit any significant densification up to 1400°C except shrinkage accompanying the mullitization. However they could be densified to ≥ 98.8% T.D. by sintering at 1600°C for 2 h. The fair sinterability was attributed mostly to the small particle sizes. Due to the chemical homogeneity and the fair sinterability, a dense and uniform microstructure and high bending strength was achieved, especially in the compositions containing excessive Al_2O_3.

Introduction

For fabrication of high performance mullite ceramics, it is believed to be necessary to adopt chemically-processed powders because the chemical homogeneity and the purity level have significant influences on the microstructure and properties of mullite. The techniques to fabricate such mullite ceramics can be divided into two categories, depending on the fabrication route; the powder is either mullitized during sintering or before sintering by calcination. By the first fabrication route[1–6] in which amorphous pre-mullite or a uniform mixture of Al_2O_3 and SiO_2 is mullitized during sintering, the sintering temperature can be lowered. However, the microstructure may not be very uniform because of the remaining glass phase and the uneven densification and grain growth. By the second route,[7–9] in which the Al_2O_3-SiO_2 mixture is mullitized by calcination, milled, formed and then sintered, the sintering temperature required is relatively high, but the microstructure is more uniform and the bending strength is higher than in the first route. If the homogeneity is maintained in the first route, or if the sinterability is improved in the second route, the microstructure and properties of mullite can be further improved.

Cuer, Elston and Teichner[10] prepared Al_2O_3-SiO_2 composite powders by a chemical vapor deposition (CVD) method using a combustion flame. Gani and

McPherson[11] have conducted an intensive study on characterization of Al_2O_3-SiO_2 powders prepared by a CVD method using an r.f. plasma. But they did not report the sintering behavior of these powders.

The present authors showed that powders prepared by a flame CVD method can be useful to fabricate two-component oxide ceramics such as ZrO_2-toughened Al_2O_3,[12] aluminum titanate ceramics[13] and corundum-rutile composites.[14] The objectives of this study are to characterize the Al_2O_3-SiO_2 composite powders prepared by the CVD technique and to attempt to fabricate mullite of uniform microstructure and good mechanical properties by sintering the pressed compacts of the CVD powders at relatively low temperatures without calcination. Separate studies on the consolidation of these CVD powders by colloidal filtration techniques are reported elsewhere[15,16] and can be compared with this study.

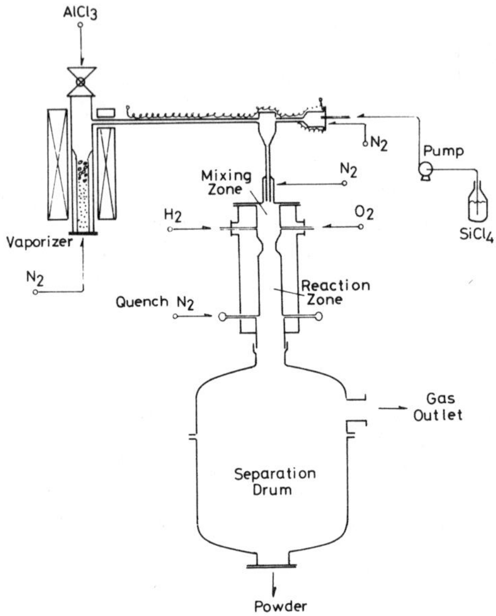

Fig. 1. CVD apparatus for producing Al_2O_3-SiO_2 powders.

Experimental Procedure

The CVD Al_2O_3-SiO_2 powders (67–80 wt% Al_2O_3) were prepared by using an apparatus whose schematic diagram is shown in Fig. 1. The mixture of $AlCl_3$ and $SiCl_4$ vapors and N_2 carrier gas was introduced into the mixing zone of the reactor in which the temperature of approximately 1900°C was maintained by a combustion flame formed by H_2 and O_2. The temperature gradually dropped in the reaction zone and became approximately 900°C at the exit before the reaction products were quenched by N_2. The total reaction time was approximately 65 ms. The produced powder was collected at the bottom of the separation drum.

The powders were characterized by transmission electron microscopy (TEM), chemical analysis, X-ray diffraction (XRD) and differential thermal analysis (DTA). Some of the powders were also calcined at various temperatures and subjected to XRD analysis to study the phase changes. The impurities contained in these powders varied; the total impurities (excluding H_2O and Cl) were initially in the range of

2000–5000 ppm, but dropped to 200–500 ppm after a reactor modification. The major impurities contained in the powders were Na_2O, CaO, MgO and Fe_2O_3. The sintering experiments and the mechanical property measurements were conducted with powders containing 200–500 ppm total impurities.

The powders were deagglomerated by ultrasonic agitation, mixed with a surfactant, dried, isostatically, pressed at 294 MPa into bar-shaped specimens (approximate sintered dimensions 3 × 3 × 20 mm^3) and sintered in air between 1550° and 1650°C for 2 h. Using shorter specimens, the firing shrinkage was directly monitored up to 1500°C using a dilatometer.

After polishing, the fired specimens were subjected to bulk density measurements, microstructural observation and bending strength measurements. For the calculation of relative densities, the compositional range of mullite solid solution was assumed to be 70.5–74.0 wt% Al_2O_3[17] and the theoretical densities of various mullite compositions were determined according to Cameron's data.[18] The microstructures were observed by scanning electron microscopy (SEM) on polished and thermally-etched surfaces. The room temperature bending strengths were measured with specimens sintered at 1600°C for 2 h.

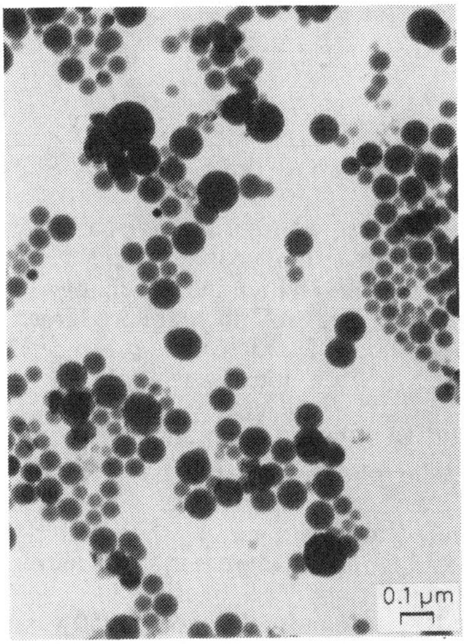

Fig. 2. Transmission electron micrograph of a typical CVD Al_2O_3-SiO_2 powder (66.8 wt% Al_2O_3).

Results and Discussion

Powder Morphology

Figure 2 shows a typical CVD Al_2O_3-SiO_2 powder (66.8 wt% Al_2O_3) produced in this study. The particles are spherical and ultrafine, and slightly agglomerated. The particle sizes of this powder are shown in Fig. 3, which exhibits a log-normal size distribution, as is typical with CVD powders.[19] The particle sizes varied noticeably from run to run with the average size changing between 40 and 70 nm. As the

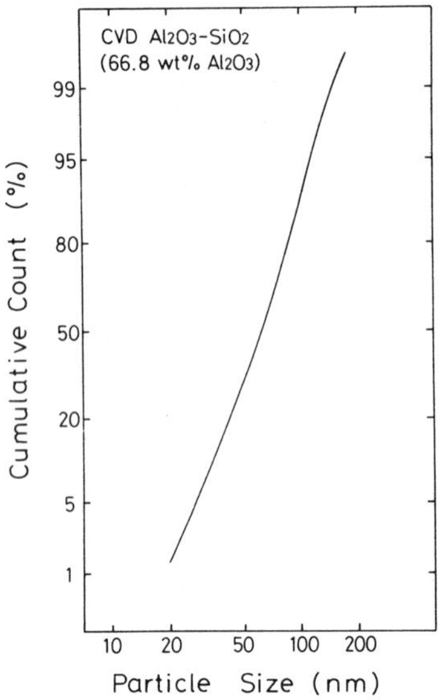

Fig. 3. Particle size distribution of the CVD Al_2O_3-SiO_2 powder shown in Fig. 2.

reactor was run at temperatures where a liquid phase was expected, a slight change of temperature and/or composition probably had a significant influence on the amount and the viscosity of liquid which in turn changed the growth rate of particles. It should be noted that the Al_2O_3-SiO_2 powders are larger than the Al_2O_3-ZrO_2[20] and the Al_2O_3-TiO_2[21] powders, both of which were produced by the same technique but at temperatures where no or little liquid was expected.

Phase Composition and Changes on Heating

Table 1 indicates the phases detected by XRD in the as-produced and the calcined powders of various compositions. In the as-produced powders, an amorphous phase appeared to be the major phase and γ-Al_2O_3 was the minor phase. The intensity of the γ-Al_2O_3 peaks increased with increasing Al_2O_3 content. Weak mullite peaks were observed only occasionally.

By calcination for 10 min, mullite formed at 950°C and became the major phase at 1000°C with all the compositions. The direct mullitization from the amorphous phase without forming an intermediate spinel phase is considered to be evidence of chemical uniformity, at least to a certain extent, as schematically shown by Okada et al.[22] At temperatures higher than 1000°C, the remaining amorphous phase gradually disappeared. The Al_2O_3 changed from the γ-phase to the more ordered δ-phase and then to the θ-phase, but did not change to α up to 1500°C. As the pure Al_2O_3 prepared by the same CVD technique transformed to α below 1400°C, the addition of SiO_2 appeared to stabilize the Al_2O_3 to its transition forms. This stabilization is probably due to the solid solution of SiO_2 to Al_2O_3, as discussed in detail by Gani et al.[11] with plasma-prepared Al_2O_3-SiO_2 powders.

Figure 4 exemplifies DTA data of two different compositions between 940° and 1000°C, which exhibited two exothermic peaks with the as-produced powders.

Table 1. Phase Changes of Various Powder Compositions by Calcination

Composition (wt% Al_2O_3)	As-Produced	900°C	950°C	1000°C	1100°C	1300°C	1500°C
65.7	AM(s)	AM(s) γ(vw)	AM(s) M(w) γ(w)	M(s) AM γ(w)	M(s) AM δ(w)	M(s) AM(w) δ(w)	M(s) AM(vw)
67.2	AM(s) γ(w)	AM(s) γ(w)	AM(s) M(w) γ(w)	M(s) AM γ(w)	M(s) AM(w) δ(w)	M(s) AM(w) δ(w)	M(s)
71.2	AM(s) M(w) γ(w)	AM(s) M(w) γ(w)	AM(s) M γ(w)	M(s) AM(w) γ(w)	M(s) AM(w) δ(w)	M(s) AM(w) δ(w)	M(s)
74.5	AM γ(w)	AM γ(w)	AM M γ(w)	M(s) AM(w) γ(w)	M(s) AM(w) δ(w)	M(s) δ(w) AM(vm)	M(s)
79.8	AM γ	AM γ	AM M γ	M(s) γ AM(w)	M(s) δ AM(w)	M(s) δ, θ AM(vw)	M(s) θ, δ

Note 1) The powders were calcined for 10 min.
2) Phases; AM: amorphous, M: mullite, γ, δ, θ: Al_2O_3 phases
3) XRD Peak Intensities; (s): strong, (w): weak, (vw): very weak

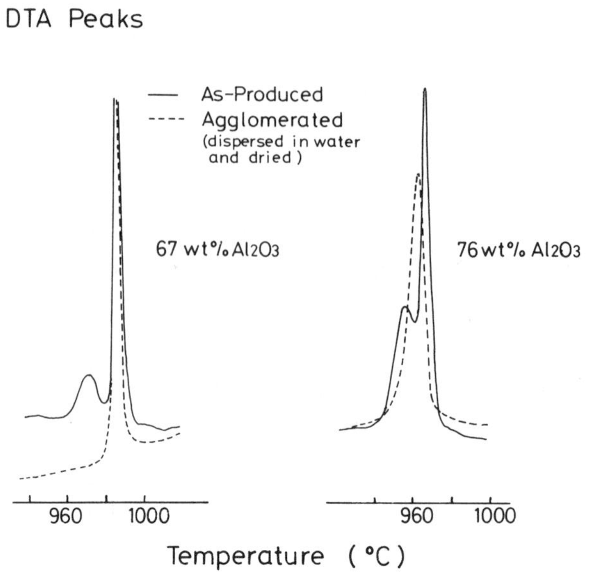

Fig. 4. DTA data of two typical compositions at 940–1000°C (heating rate 20°C/min). Both compositions exhibited double exothermic peaks with the as-produced powders, but not with the agglomerated powders. The lower temperature peak became stronger with increasing Al_2O_3 content.

With increasing Al_2O_3 content, the intensity of the first peak increased and the both peaks shifted toward lower temperatures. As Gani and McPherson[11] proved with the CVD Al_2O_3-SiO_2 powders prepared by an r.f. plasma, both of these two exothermic peaks are considered to correspond to mullitization. The lower temperature peak corresponds to the formation of mullite from a metastable solution of SiO_2 in γ-Al_2O_3, and the higher temperature peak corresponds to crystallization of an amorphous pre-mullite phase. The CVD powders, therefore, apparently consist of two phases; an amorphous pre-mullite and γ-Al_2O_3 containing a significant amount of SiO_2. The reason for this two-phase structure can be attributed either to a metastable subliquidus relation between the liquid phase and the metastable Al_2O_3 solid solution at high temperatures or to a metastable liquid immiscibility upon cooling in the Al_2O_3-rich region of the Al_2O_3-SiO_2 binary.

The double DTA peaks disappeared when the powders were dispersed into water, dried and then subjected to DTA analysis. The mullitization appeared to occur as a whole at the average temperature. The DTA peak of the agglomerated powder with 67 wt% Al_2O_3 was close to the second peak of the as-produced powder because the first peak was relatively small, while the peak of the agglomerated powder with 76 wt% Al_2O_3 emerged between the two peaks of the as-produced powder because the two peaks were comparable in size. The reason for this phenomenon is not clear, but it suggests some effect of agglomeration or particle-particle interaction on the mullitization.

The mullite crystallized from the CVD powder containing 69 wt% Al_2O_3 exhibited the following lattice parameters after calcination at 1000°C for 10 min.

a = 0.76174 ± 0.00026 nm
b = 0.76776 ± 0.00031 nm
c = 0.28886 ± 0.00002 nm

According to Cameron's data,[18] the lattice parameter **a** corresponded to 83 wt% (73 mol%) Al_2O_3. Okada et al.[23] revealed that a gel of homogeneous Al_2O_3-SiO_2 mixture crystallized directly to a mullite of Al_2O_3-rich composition containing approximately 70 mol% Al_2O_3. The occurrence of Al_2O_3-rich mullite is considered to be usual when mullite is crystallized directly from the amorphous state. This Al_2O_3-rich mullite disappeared at higher temperatures.

Densification Behavior

The sintering experiment was conducted with 4 different compositions, as shown in Table 2. The green density was low with specimens of 76.5 wt% Al_2O_3, whose particle size was smaller (av. ≈40 nm) than the others (av. ≈60 nm).

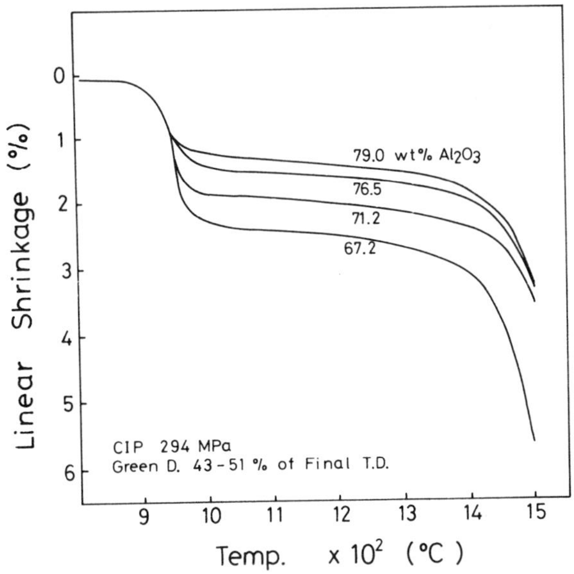

Fig. 5. Firing shrinkage of the powder compacts of various compositions up to 1500°C (heating rate 5°C/min). No significant densification was observed up to 1400°C. The shrinkage at ≈950°C was not the densification but the density change accompanying mullitization.

The direct observation of firing shrinkage, as shown in Fig. 5, revealed that considerable shrinkage accompanied mullitization. This shrinkage cannot be attributed to elimination of pores, but rather to the change of true density from the amorphous phase (≈2.8 g/cm^3) to the crystalline mullite (3.1–3.2 g/cm^3). After the mullitization, no significant densification was observed up to 1400°C, which suggests very little remaining glass phase.

The relative densities achieved by sintering these compacts between 1550° and 1650°C for 2 h did not vary from composition to composition. The densities

Table 2. Green and Sintered Densities of Various Compositions of Compacts Prepared from CVD Al_2O_3-SiO_2 Powders

	67.2 wt% Al_2O_3		71.2 wt% Al_2O_3		76.5 wt% Al_2O_3		79.0 wt% Al_2O_3	
	g/cm^3	Relative D.	g/cm^3	Relative D.	g/cm^3	Relative D.	g/cm^3	Relative D.
Green	1.537	49.1%	1.627	51.2%	1.375	42.8%	1.689	54.1%
800°C 2 h	1.561	49.9	1.658	52.2			1.736	52.8
1550°C 2 h	2.748	87.9					2.770	84.3
1600°C 2 h	3.121	99.8	3.138	98.8	3.202	99.6	3.260	99.2
1650°C 2 h	3.115	99.7	3.119	98.2	3.188	99.1	3.299	100.0
Theoretical D.	3.126		3.177		3.215		3.285	

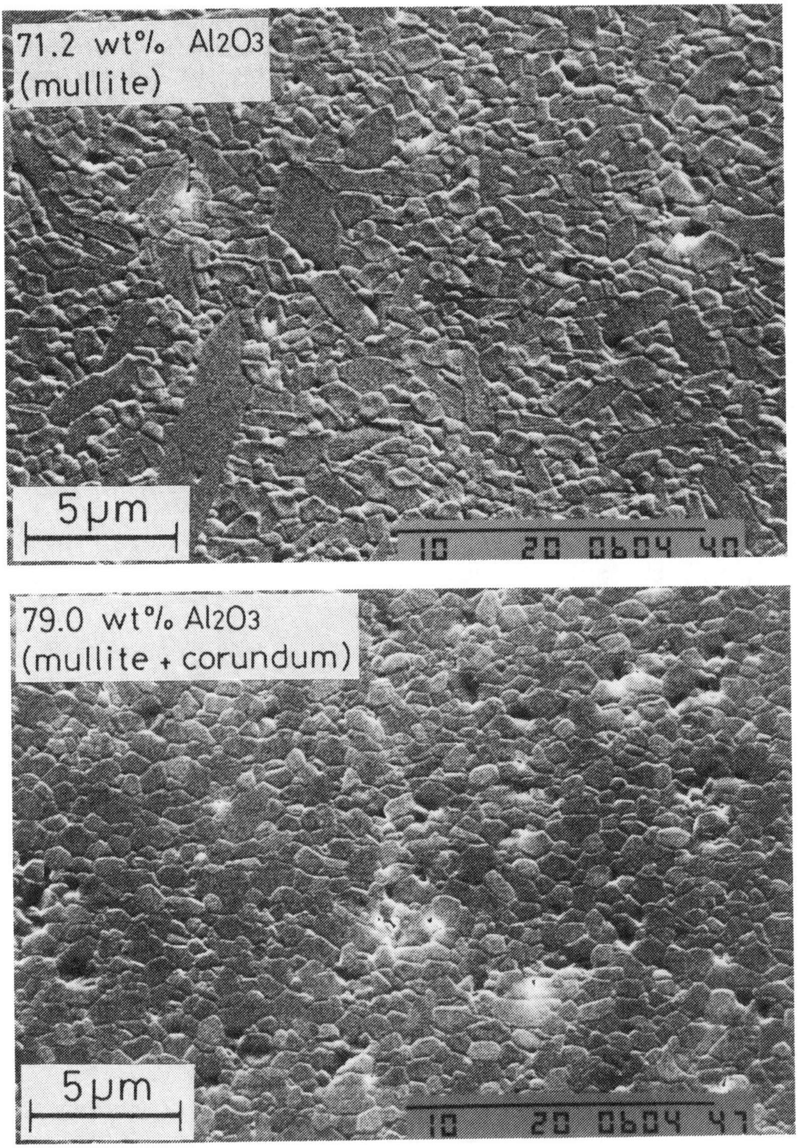

Fig. 6. Microstructures of mullite (71.2 wt% Al_2O_3) and mullite-corundum composite (79.0 wt% Al_2O_3) sintered at 1600°C for 2 h from CVD powders.

achieved at 1600°C for 2 h were ≥98.8%. The temperature required to achieve nearly full density was ≈1600°C for these CVD powders, which is ≈50°C lower than those for the mullitized/milled powders,[7–9] but much higher than those reported by Debély et al. and Komarneni et al.[6] Both Debély et al. and Komarneni et al. introduced some heterogeneity into the powders for enhancing densification. The CVD powders are, therefore, rather similar to those mullitized/milled powders in terms of homogeneity. But the CVD powders exhibited better sinterability than the mullitized powders, which can be attributed mostly to smaller particle sizes.

Microstructure and Bending Strength

Figure 6 shows the microstructures of two compositions sintered at 1600°C for 2 h; one with 71.2 wt% Al_2O_3 which exhibited only the mullite phase by XRD and the other with 79.0 wt% Al_2O_3 which exhibited mullite and corundum. With increasing Al_2O_3 content, the grains became more isotropic and uniform in size and the abnormally-grown mullite grains disappeared. The uniform microstructure was achieved in the Al_2O_3-rich compositions because there was very little liquid phase and also because the uniformly-dispersed Al_2O_3 grains controlled the grain growth.

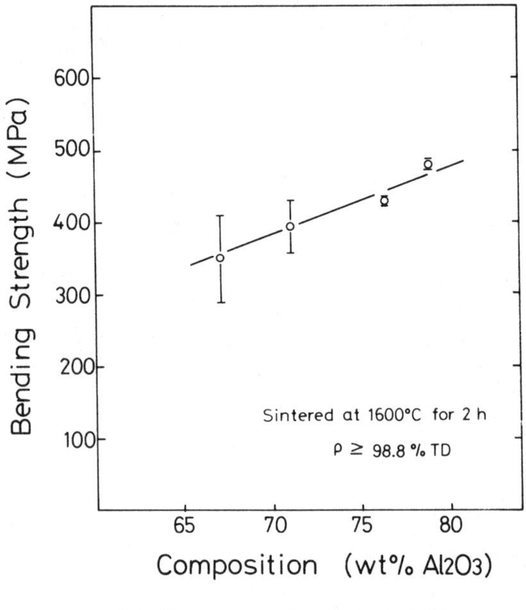

Fig. 7. Room temperature bending strength of Al_2O_3-SiO_2 ceramics sintered at 1600°C for 2 h from CVD powders as function of composition between 67 and 79 wt% Al_2O_3.

Figure 7 shows the dependence of bending strength on the composition with specimens sintered from the CVD powders at 1600°C for 2 h. The bending strength increased with increasing Al_2O_3 content. The strength was 480 MPa with 79 wt% Al_2O_3, which is much higher than those reported previously with sintered mullite or mullite-corundum composites of this compositional range. Even with 67.2 wt% and 71.2 wt% Al_2O_3, the strength data are comparable to those high values previously reported.[7–9] The reason of these high bending strengths can be attributed to the dense and fine-grained microstructure and therefore small flaw sizes, especially in the Al_2O_3-rich compositions. The chemical uniformity and the fair sinterability of the CVD powders resulted in the uniform microstructure and high strength.

Summary

The Al_2O_3-SiO_2 powders prepared by a flame CVD technique were spherical and ultrafine, and exhibited a two-phase structure consisting of amorphous pre-mullite phase and γ-Al_2O_3 containing a metastable SiO_2. The structure was essentially the same as those prepared by an r.f. plasma. The densification behavior was rather similar to the mullitized/milled powders because of the chemical homogeneity, but the temperature required to achieve nearly full density was ≈50°C lower because of the smaller particle sizes. It was demonstrated that the high bending strength could be achieved especially in the Al_2O_3-rich compositions by sintering the mostly amorphous CVD powders without calcination. Further improvement in microstructure and mechanical properties may be possible by lowering the sintering temperature through colloidal processing or by lowering the impurity level below he current 200 ppm.

References

[1]K. S. Mazdiyasni and L. M. Brown, "Synthesis and Mechanical Properties of Mullite," *J. Am. Ceram. Soc.*, **55** [11] 548–552 (1972).

[2]M. Mah and K. S. Mazdiyasni, "Mechanical Properties of Mullite," *J. Am. Ceram. Soc.*, **66** [10] 699–703 (1983).

[3]S. Prochazka and F. J. Klug, "Infrared-Transparent Mullite Ceramic," *J. Am. Ceram. Soc.*, **66** [12] 874–880 (1983).

[4]P. E. Debély, E. A. Barringer, and H. K. Bowen, "Preparation and Sintering Behavior of Fine-Grained Al_2O_3-SiO_2 Composites," *J. Am. Ceram. Soc.*, **68** [3] C-76–78 (1985).

[5]Y. Hirata, H. Minamizono, and K. Shimada, "Sinterability of Fine SiO_2-Al_2O_3 Powders Derived from Metal Alkoxide," *Yogyo-Kyokai-Shi*, **93** [8] 475–478 (1985).

[6]S. Komarneni, Y. Suwa, and R. Roy, "Application of Compositionally Diphasic Xerogels for Enhanced Densification: The System Al_2O_3-SiO_2," *J. Am. Ceram. Soc.*, **69** [7] C-155-156 (1986).

[7]S. Kanzaki, H. Tabata, T. Kumazawa, and S. Ohta, "Sintering and Mechanical Properties of Stoichiometric Mullite," *J. Am. Ceram. Soc.*, **68** [1] C-6-7 (1985).

[8]S. Kanzaki, T. Kumazawa, J. Asaumi, O. Abe, and H. Tabata, "Dependency of Mechanical Property of Sintered Mullite on Chemical Composition," *Yogyo-Kyokai-Shi*, **93** [7] 407–408 (1985).

[9]M. G. M. U. Ismail, Z. Nakai, and S. Sōmiya, "Microstructure and Mechanical Properties of Mullite Prepared by the Sol-Gel Method," *J. Am. Ceram. Soc.*, **70** [1] C-7-8 (1987).

[10]J. P. Cuer, J. Elston, and S. J. Teichner, "Methods of Production and Properties of Finely Divided Solids Prepared in a Flame Reactor III. Properties of Alumina," *Bull. Soc. Chim. Fr.*, **1961**, 89–93 (1961).

[11]M. S. J. Gani and R. McPherson, "Glass Formation and Phase Transformations in Plasma Prepared Al_2O_3-SiO_2 Powders," *J. Mater. Sci.*, **12** [5] 999–1009 (1977).

[12]S. Hori, M. Yoshimura, S. Sōmiya, R. Kurita, and H. Kaji, "Mechanical Properties of ZrO_2-Toughened Al_2O_3 Ceramics from CVD Powders," *J. Mater. Sci. Letters*, **4** [4] 413–416 (1985).

[13]S. Hori, M. Yoshimura, and S. Sōmiya, "Effects of Excess Al_2O_3 and TiO_2 on Microstructure and Properties of Al_2TiO_5 Ceramics Prepared from CVD Powders,"

pp. 87–94 in Defect Properties and Processing of High-Technology Nonmetallic Materials, Edited by Y. Chen, W. D. Kingery, and R. J. Stokes, MRS Symposia Proceeding 60, Materials Research Society, Pittsburgh, 1986.

[14]S. Hori, H. Kaji, M. Yoshimura, and S. Sōmiya, "Deflection-Toughened Corundum-Rutile Composites," pp. 283–288 in Advanced Structural Ceramics, Edited by P. F. Becher, M. V. Swain and S. Sōmiya, MRS Symposia Proceeding 78, Materials Research Society, Pittsburgh, 1987.

[15]Y. Hirata and I. A. Aksay, "Colloidal Consolidation and Sintering Behavior of CVD-Processed Mullite Powders," pp. 611–22 in Proceedings of the 22nd University Conference on Ceramics and the International Materials Symposium, Edited by J. A. Pask and A. G. Evans, Plenum Press (1987).

[16]Y. Hirata, I. A. Aksay, R. Kurita, S. Hori, and H. Kaji, "Processing of Mullite with Powders Processed by Chemical Vapor Deposition," these proceedings.

[17]I. A. Aksay and J. A. Pask, "Stable and Metastable Equilibria in the System SiO_2-Al_2O_3," *J. Am. Ceram. Soc.*, **58** [11–12] 507–512 (1975).

[18]W. E. Cameron, "Composition and Cell Dimensions of Mullite," *Am. Ceram. Soc. Bull.*, **56** [11] 1003–1011 (1977).

[19]G. D. Ulrich, "Theory of Particle Formation and Growth in Oxide Synthesis Flames," *Combus. Sci. Technol.*, **4** [2] 47–57 (1971).

[20]S. Hori, "ZrO_2 and ZrO_2-Containing Ultrafine Powders by Chemical Vapor Deposition Method," pp. 21–28 in Zirconia Ceramics 1, Edited by S. Sōmiya, Uchida Rokakuho, Tokyo, 1983.

[21]S. Hori, Y. Ishii, M. Yoshimura, and S. Sōmiya, "Preparation of Codeposited Al_2O_3-TiO_2 Powders by Vapor Phase Reaction Using Combustion Flame," *Yogyo-Kyokai-Shi*, **94** [4] 400–408 (1986).

[22]K. Okada and N. Ōtsuka, "Characterization of the Spinel Phase from SiO_2-Al_2O_3 Xerogels and the Formation Process of Mullite," *J. Am. Ceram. Soc.*, **69** [9] 652–56 (1986).

[23]K. Okada, Y. Hoshi, and N. Ōtsuka, "Formation Reaction of Mullite from SiO_2-Al_2O_3 Xerogels," *J. Mater. Sci. Letters*, **5** [12] 1315–1318 (1986).

Processing of Mullite With Powders Processed By Chemical Vapor Deposition

YOSHIHIRO HIRATA AND ILHAN A. AKSAY

Department of Materials Science and Engineering
Advanced Materials Program, Washington Technology Center
University of Washington
Seattle, WA 98195

RYUICHI KURITA, SABURO HORI AND HISATSUGU KAJI

Kureha Chemical Industry Co., Ltd.
Advanced Material System Laboratories
Iwaki-City, Fukushima 974, Japan

Abstract

The main purpose of colloidal dispersion and consolidation techniques is to control the packing density fluctuation in a green compact and the structure of the hierarchically clustered particles during consolidation. In this paper, we summarize our results on the sintering behavior of mullite compacts produced through colloidal consolidation of CVD-processed mullite powder (67.9 and 72.2 wt% Al_2O_3) with an average particle size of 38–65 nm. We show that the pore size distribution of the compacts plays the most crucial role on the densification behavior during sintering. Since at a given temperature smaller pores in a compact are eliminated faster than larger pores, it is important to decrease the pore size along with the pore volume in order to lower the sintering temperature.

The factors that affect the pore size distribution were identified as (i) the degree of dispersion in the colloidal suspension stage, and (ii) the structure of the particle clusters which are formed during casting as affected by the type of dispersant and the forming technique used. As the size of the particle agglomerates inheritated from the as-received powders is reduced through chemical leaching of the grain boundaries in electrostatically dispersed suspension (pH 4), the packing density of the green compact increases and in turn higher sintered densities are achieved at a given temperature. Reduction of the agglomerate size causes (a) increase in the packing density, (b) increase in the volume of first generation voids (contained within the first generation particle clusters), (c) decrease in the volume of second generation voids (contained among the first generation particle clusters) in a green compact, and (d) decrease in the pore size of second generation voids. The electrosteric stabilization through PMAA-NH_4 is superior to electrostatic stabilization in increasing the densification rate of green compacts. The polymer coating of particles reduces the average pore size and narrows the pore size

distribution of the second generation voids. Long range particle segregation during consolidation, due to differential mass, affects the pore size distribution in a green compact. When segregation is prevented, smaller particles fill the voids formed among larger particles, yielding narrower pore size distribution of the second generation voids. The experimental data showed that the filtration technique is superior to sedimentation in the reduction of long range particle segregation. When the pore size distribution approached a monosize state, compacts could be densified to relative densities above 97% by sintering at 1500–1600°C for 2 hours. The average grain size was smaller for the sintered body formed from electrosterically stabilized suspension than electrostatically stabilized suspension.

Introduction

Mullite ($3Al_2O_3 \cdot 2SiO_2$) is the only stable compound in the SiO_2-Al_2O_3 system.[1] Mullite also has a high melting point (1828°C), low density (3.17 g/cc), high covalent character and a low coefficient of thermal expansion (5.6×10^{-6}/°C).[2,3] Recent studies on the SiO_2-Al_2O_3 system have revealed that (i) dense mullite of high purity shows little degradation in strength above 360 MPa up to 1300°C[4–6] and infrared transparency,[3,7–9] and (ii) the strength and fracture toughness of mullite at higher temperature are very sensitive to the small amount of alumina-silicate glass[6,10,11] and the microstructure.[5,12] These properties of mullite can be compared to those of Si_3N_4 or SiC in terms of being a candidate for high temperature structural applications.

It is well known that mechanical or electrical properties of ceramic materials are influence by porosity, microstructure and chemical/phase composition, and depend upon the processing. When processing is done with submicron size powders, the colloidal suspension route is superior to conventional dry processing for the production of dense ceramics with a uniform microstructure.[13–15] The physical and chemical stability of a colloidal suspension is reflected in the character of the resulting green compact[16] which, in turn, controls the properties of the sintered body. This paper describes the colloidal processing (dispersion, consolidation and sintering) of CVD-processed ultrafine SiO_2-Al_2O_3 powders[17] (<0.1 μm) and the evaluation of the microstructure of the sintered body.

Experimental Procedure

Colloidal dispersion, consolidation and sintering studies were performed on the CVD-processed amorphous SiO_2-Al_2O_3 powders with the composition of 67.9 and 72.2 wt% Al_2O_3. The properties of these powders are listed in Table 1. They were processed by the Kureha Chemical Industry Company of Japan. The mixed vapors of $AlCl_3$ and $SiCl_4$ were introduced by N_2 flow into the combustion flame (~1900°C) of H_2/O_2 gas to form SiO_2-Al_2O_3 particles. The preparation and detailed characterization of CVD-processed SiO_2-Al_2O_3 powders will be reported by Hori et al.[18] in the proceedings of the 1st International Workshop on Mullite. Powder No. 1 was supplied in freeze-dried form and contained strongly clustered primary spherical particles. The other powder was supplied in a dilute (~3v/o) aqueous suspension form. A typical electron micrograph of the CVD-processed SiO_2-Al_2O_3 particles is shown in Fig. 1. The volume-surface mean diameter of the particles, calculated from the specific surface area and the true density, was 65 nm for Powder No. 1 and 38 nm for Powder No. 2. The total content of Na_2O, CaO and Fe_2O_3 was in the range of 2400–5200 ppm.

Stable aqueous suspensions of 2.9–31 vol% were prepared either electrostatically at pH 4 or electrosterically with an NH_4 salt of PMAA (Darvan C) at pH 6.5–7.5. Either HCl or NH_4OH were used for pH adjustment. The degree of dispersion of the

Table 1. **Properties of CVD-processed amorphous SiO_2-Al_2O_3 powders**

Powder	Composition Al_2O_3 (wt%)	Na_2O (ppm)	CaO (ppm)	Fe_2O_3 (ppm)	Density (g/cc)	Specific surface area(m^2/g)	Volume-surface mean diameter(nm)	As-received suspension pH	As-received suspension concentration wt%(vol%)
No.1	67.9	2050	83	270	2.801	33.2	65	Freeze-dried	
No.2	72.2	2110	1160	1910	2.863	54.6	38	5.6	8.0(2.9)

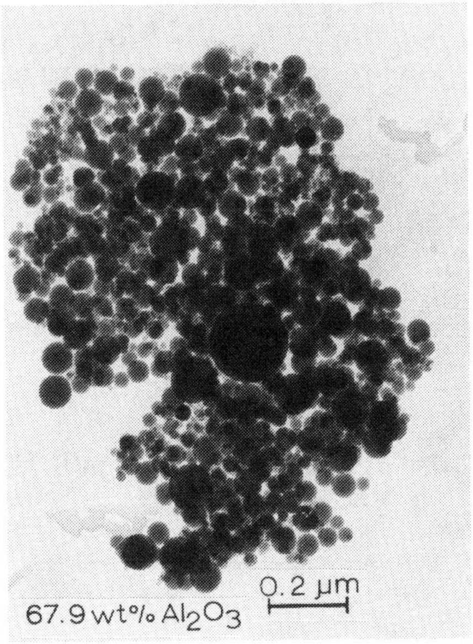

Fig. 1. Transmission electron micrograph of CVD-processed spherical SiO_2-Al_2O_3 particles with a composition of 67.9 wt% Al_2O_3.

SiO_2-Al_2O_3 particles was examined with an X-ray absorption particle analyzer. The colloidal filtration technique through gypsum molds and the sedimentation method were used to consolidate the colloidal particles. Green compacts dried at room temperature or 110°C were sintered at 1050°–1600°C for 2 hours. Densities of the green compacts dried at 500–570°C and sintered bodies were measured by the Archimedes method using kerosene and distilled water, respectively. Sintered samples were polished with 1 μm diamond paste and thermally etched at 50°C lower than the sintering temperature for one hour. The resulting microstructure was evaluated by scanning electron microscopy (SEM).

Results and Discussion

Dispersion

The zeta-potential of the dilute suspension of Powder No. 1 (67.9 wt% Al_2O_3) is shown in Fig. 2 as a function of pH. The isoelectric point of the colloidal particles was pH 7.7 and closer to that of an Al_2O_3 suspension (pH 7–10) than that of a SiO_2 suspension (pH ~2). The results of Fig. 2 suggest that the powder will be dispersed at pH 3–5 and at a pH above 9 due to the strong repulsion force among the highly charged particles. Figure 3 shows the dispersion states of the as-received SiO_2–

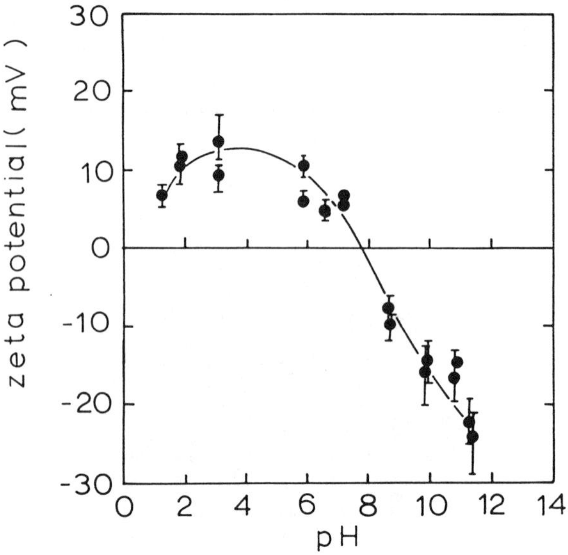

Fig. 2. Zeta potential of as-received amorphous SiO_2-Al_2O_3 powders with a composition of 67.9 wt% Al_2O_3.

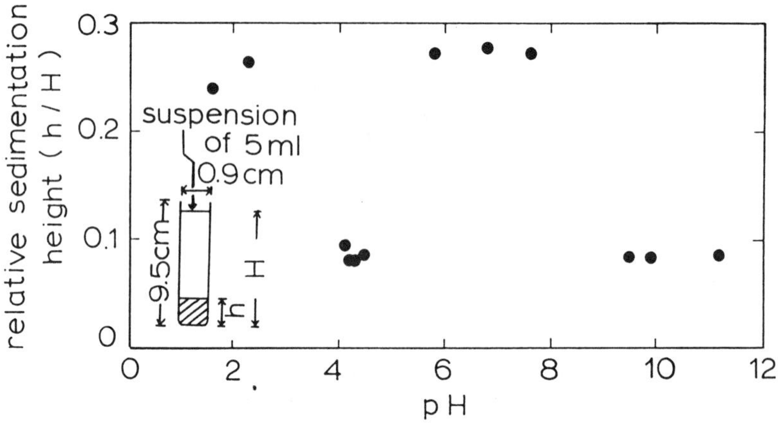

Fig. 3. Relative sedimentation height as a function of pH for the dilute suspension (~1 v/o) of amorphous SiO_2-Al_2O_3 powders with a composition 67.9 wt% Al_2O_3.

Al_2O_3 particles (Powder No. 1) in the dilute suspensions (~1v/o) of different pH. The relative sedimentation height reached a constant value after 2 hours of settling. The data in Fig. 3 correspond to the sedimentation height at the settling time of 20–24 hours. From Fig. 3, it is clear that a higher dispersion state (lower sedimentation height) is achieved in the suspension of pH ~4 and pH above 9.5. This result

Table 2. **Preparation conditions of the colloidal suspensions for the measurement of particle size distribution**

Suspension	Concentration (vol%)	pH	Stirring Time
A	2.9	11.09–10.72	1.77 h
B	2.0	9.09–9.07	2.30 h
C	2.0	4.08–4.35	2.33 h
D	2.3	3.92–3.97	17.00 h
E	2.3	4.02–3.86	3.60 day
F	2.3	3.90–4.02	4.40 day
G	2.3	4.02–3.93	5.20 day
H	2.3	4.18–4.67	7.10 day

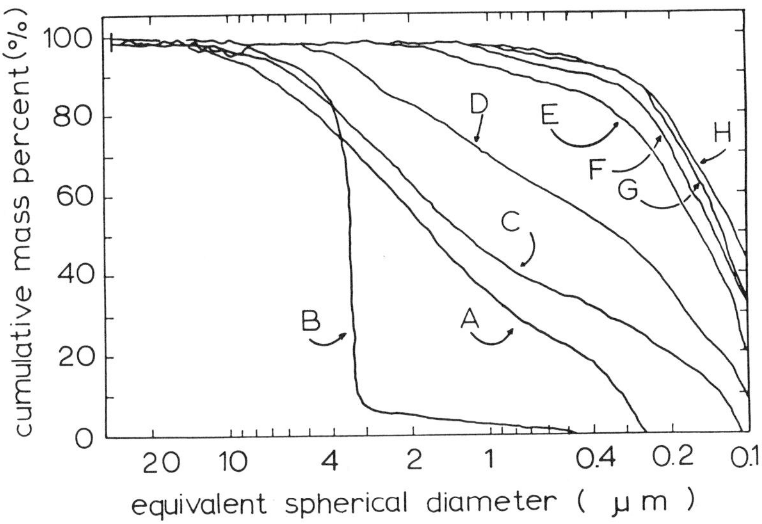

Fig. 4. Particle size distributions of colloidal SiO_2-Al_2O_3 particles with a composition of 67.9 wt% Al_2O_3. (See Table 2 for associated data and legend.)

agrees with the predictions from the data on zeta potential. More quantitative data on the degree of dispersion is presented in Table 2 and Fig. 4. A comparison of the particle size distributions of Powder No. 1 in the suspension at pH 4, 9, and 11 (stirring time 2–2.5 hours) indicates that the electrostatic stabilization at pH 4 is most effective in the dispersion of the particle agglomerates formed during powder processing. Moreover, the large agglomerates could be broken into submicron sizes (average size ~0.1 μm) through chemical leaching of the grain boundaries at pH 4 for 1 week. Since the agglomerate was weak, the ultrasonic technique also was effective for the destruction of large agglomerates.

Consolidation and Sintering

The main purpose of colloidal dispersion and consolidation techniques is to control the packing density fluctuations in a green compact by controlling not only

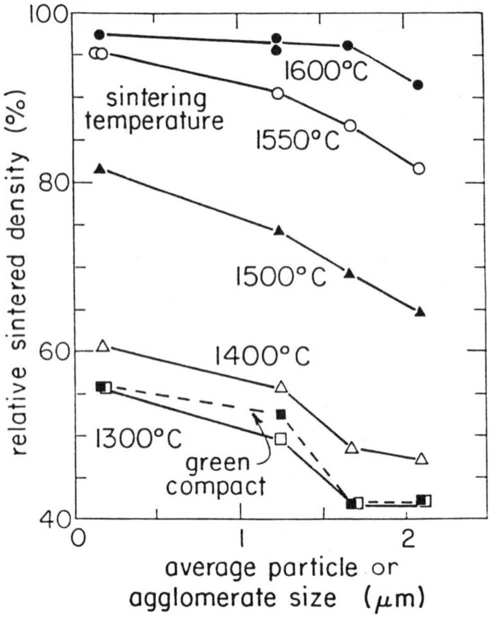

Fig. 5. Effect of average particle or agglomerate size on the densification of compacts formed by colloidal filtration technique (Powder No. 1).

the degree of dispersion in the suspension stage but also the structure of the hierarchically clustered particles during consolidation.[16,19,20] The results presented below clearly illustrate the importance of controlling packing density fluctuations on the sintering behavior. The roles of three factors:

(a) the extent of agglomeration in the as-received powders,
(b) long range particle segregation during consolidation, and
(c) the type of dispersant.

are described in the following subsections.

Effect of Agglomerate Size

We first illustrate the dependence of sintered densities on the extent of particle agglomeration inherited from the as-received powders[17] (Fig. 5). Powder No. 1 in this study displayed extensive agglomeration in the as-received state. The size of these agglomerates was reduced through chemical leaching of the grain boundaries in electrostatically dispersed suspensions of 20±3 v/o solids content. Green compacts were made by the colloidal filtration technique. The relative density measurement was based on the true density of 2.801 g/cc of the as-received amorphous SiO_2-Al_2O_3 particles for green compacts, and the density 3.141 g/cc for sintered mullite. From Fig. 5, it is evident that as the size of the agglomerates is reduced, the packing density of the green compacts increases and, in turn, higher sintered densities are achieved at a given temperature. Therefore, increasing the degree of dispersion of particles is the first key step for processing dense mullite at a low sintering temperature.

Quantitative analysis[21] of the hierarchy of pores in a green compact was also conducted using a close packing model structure (FCC). Although the detailed calculations will be published in elsewhere, the results suggest the possibility of obtained compacts with packing densities in the range of 54.8–74.1%: Reduction in the agglomerate size causes:

1. increase in the packing density,
2. increase in the volume of first generation voids (contained within first generation clusters),
3. decrease in the volume of second generation voids (contained among the first generation clusters), and
4. decrease in the pore size of second generation voids.

In the range of packing density 40.6–54.8%, the following changes are expected with reduction of the agglomerate size:

1. increase in the packing density,
2. increase in the volume of first and second generation voids,
3. decrease in the volume of third generation voids (contained among the second generation clusters), and
4. decrease in the pore size of third generation voids.

Since at a given temperature smaller pores in a compact are eliminated faster than larger pores,[22,23] it is important to decrease the pore size along with the pore volume in order to lower the sintering temperatures. Therefore, the conclusions derived from the calculation of hierarchy of pores support the relation (Fig. 5) between the agglomerate size and the green density or densification behavior of the compacts.

Long Range Particle Segregation

The influence of the powder consolidation technique on the sintering behavior of the compacts was also studied. The difference in the densification behavior for the compacts formed by filtration and sedimentation is compared in Figs. 6 and 7. The compact formed by filtration is densified faster than that formed by sedimentation. This result is associated with the degree of long range particle segregation[24,25] during consolidation due to the differences in mass. Of course, in the case of monosize particles of single phase systems this problem doesn't exist. However, the degree of long range particle segregation becomes lower in the green compact produced by filtration than that produced by sedimentation for the powder systems displaying a finite size distribution. As a result, pore size distribution becomes narrower in the compact produced by filtration than that formed by sedimentation, because smaller particles fill the voids formed among larger particles when segregation is prevented. The difference in densification behavior in Fig. 6 and Fig. 7 reflect this difference in the pore size distribution of a green compact. Therefore, it is essential to prevent segregation until consolidation is completed for the achievement of a sinterable microstructure of a green compact.

Effect of Dispersant

Figure 8 summarizes the comparison of the sintering behavior of compacts formed from electrostatically and electrosterically stabilized suspensions of Powder No. 2. In electrosteric stabilization, the amount of PMAA-NH_4 used affects the green density (Fig. 9). Low green densities results when the amount of PMAA-

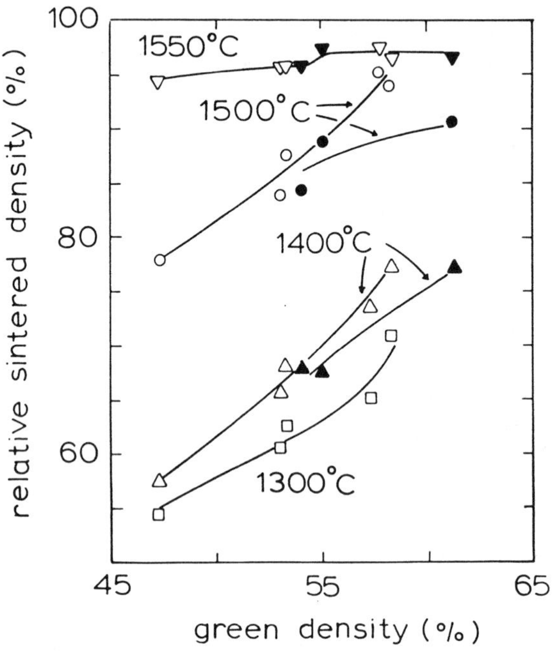

Fig. 6. Densification behavior of compacts formed from the electrostatically stabilized suspension of Powder No. 2 by filtration (□ Δ ○ ∇) and sedimentation (▲ ● ■).

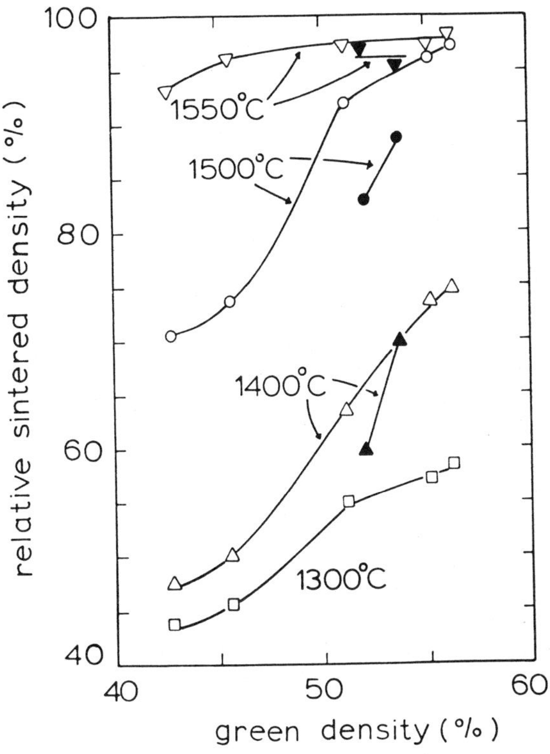

Fig. 7. Densification behavior of compacts formed from the electrosterically stabilized suspension of Powder No. 2 by filtration (□ Δ ○ ∇) and sedimentation (▲ ● ■).

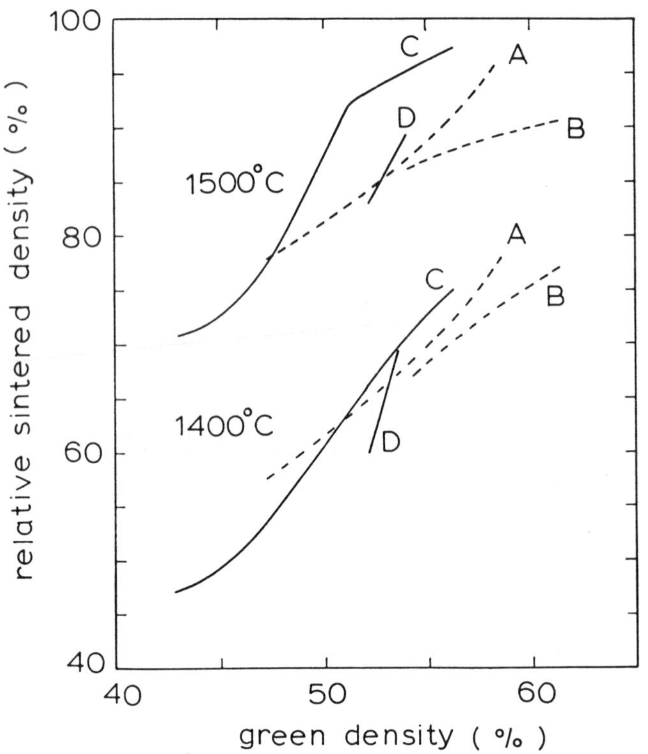

Fig. 8. Comparison of densification behavior between compacts formed from electrostatically (A,B) and electrosterically (C,D) stabilized suspension of Powder No.2 by filtration (A,C) and sedimentation (B,D).

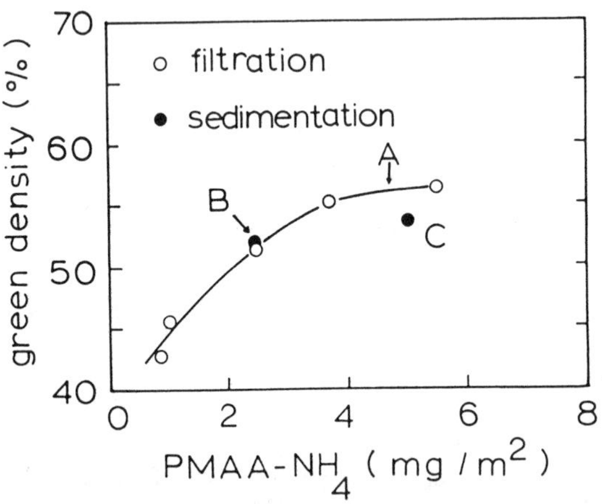

Fig. 9. Amount of PMAA-NH$_4$ and green density for Powder No. 2. (See Table 3 for associated data and legend.)

Table 3. Properties and consolidation methods of electrosterically stabilized suspensions associated with Fig. 9.

Sample	Suspension		Consolidation
	vol% vs.	pH	
A	7.78–8.95	6.51–6.93	filtration
B	7.78	6.93	sedimentation
C	9.84	7.50	sedimentation

NH_4 is not sufficient (i.e., ~1 mg/m^2) to coat all the particles. Clustering thus occurs at low suspension concentrations due to bridging flocculation.[26] The change in the green density with the amount of PMAA-NH_4 shown in Table 3 and Fig. 9 suggests that 4 mg/m^2 of polymer is required to completely coat the particles of Powder No. 2 at the concentration of 8–9 vol%.

Figure 8 shows that electrosteric stabilization is superior to electrostatic stabilization in increasing the densification rate of green compacts with densities above 50% at temperatures above 1400°C. This difference in the sintering behavior relating to the type of dispersant used in the suspension stage can be correlated to the change in pore size distribution of green compacts. In our previous studies, we have shown that transitions from a dispersed colloidal suspension to a consolidated state resemble the nucleation and growth processes observed in atomic systems.[16,18,19] Densely packed clusters of primary particles form and grow as a result of a multiple site nucleation process when a critical particle number density is exceeded at a given particle-particle interaction potential. The smallest size pores are contained within these first generation particle clusters. Second generation pores are contained between the first generation particle clusters. These second generation pores are larger than the intra cluster pores. Coating of the particles with a polymer reduces the size of the second generation voids in a green compact. This decrease in the size of second generation voids lower the sintering temperature required to attain full-densification.

Evaluation of Microstructure

Figure 10 shows the relationship between grain growth and densification during sintering of the compacts formed by filtration from the electrostatically stabilized suspension of Powder No. 1. Green density does not affect a change in the porosity of either the closed pores or the open pores with densification. Most of the open pores were transformed to closed pores at the relative density ~92%. From Fig. 10, it was found that (i) the grain size is independent of the densification until the critical density (~92%) where the porosity of the open pores suddenly approaches to 0%, and (ii) grain size rapidly increases with decreasing porosity of the closed pores in the bulk density range above the critical density. These results suggest that (i) open pores act as grain growth inhibitors, and (ii) the microstructure of small grain size is achieved together with high sintered density by shifting the critical density to near full density or suppressing the transformation of open pores to closed pores.

Figure 11 and the data of Table 4 illustrate the effect of coating the particles with a polymer in the suspension stage on the grain growth during sintering of Powder No. 2. The scanning electron micrographs of mullite sintered at 1550°C for 2 hours are shown in Fig. 12. (See also Table 5.) The average grain size and the change of grain size with sintering temperature become smaller for the compact formed from electrosterically stabilized suspension than the electrostatically stabilized suspension. That is, coating the particles produces a green compact with a microstructure which increases the densification rate and decreases the grain growth

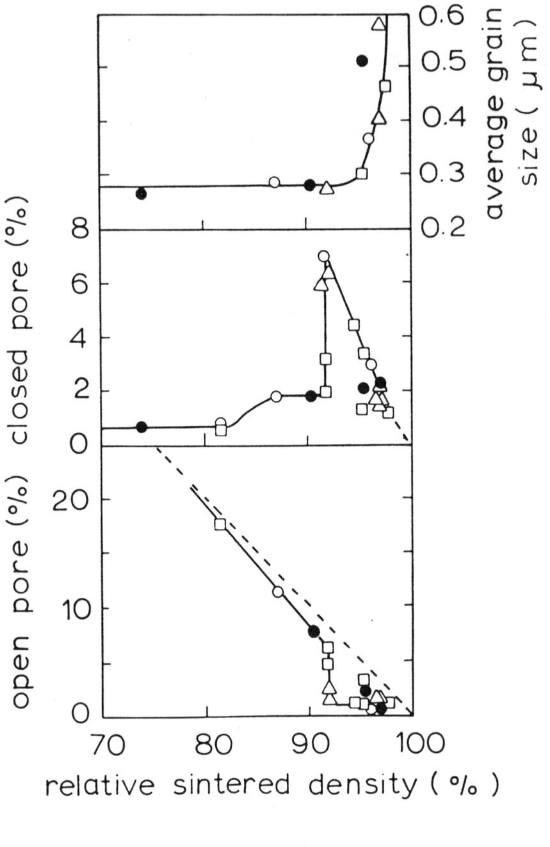

Fig. 10. Densification and growth of the compacts formed by filtration from electrostatically stabilized suspension (Powder No. 1). Green density-○: 42%, ●:52%, □: 56%, Δ: 63%.

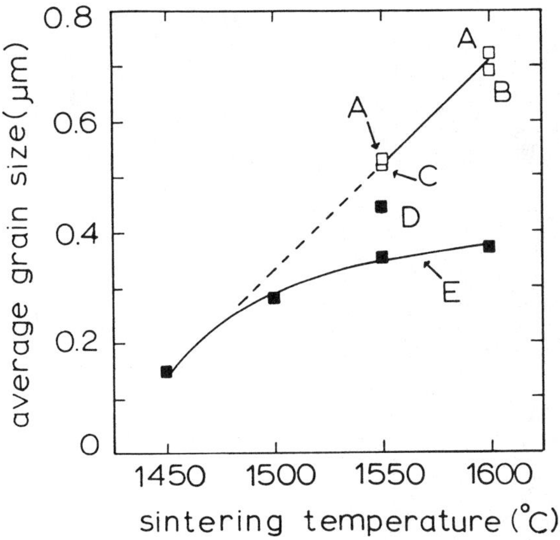

Fig. 11. Average grain size of mullite as a function of sintering temperature for the compacts formed from electrostatically (A,B,C) and electrosterically (D,E) stabilized suspension of Powder No. 2. (See Table 4 for associated data and legend.)

Table 4. Properties of the colloidal suspensions and green compacts giving the grain sizes of sintered samples shown in Fig. 11.

Sample	Consolidation	$PMAA\text{-}NH_4$ (mg/m^2)	Green Density (%)
A	filtration	0.0	53.2
B	filtration	0.0	58.3
C	sedimentation	0.0	54.0
D	filtration	2.5	51.3
E	filtration	3.7	55.2

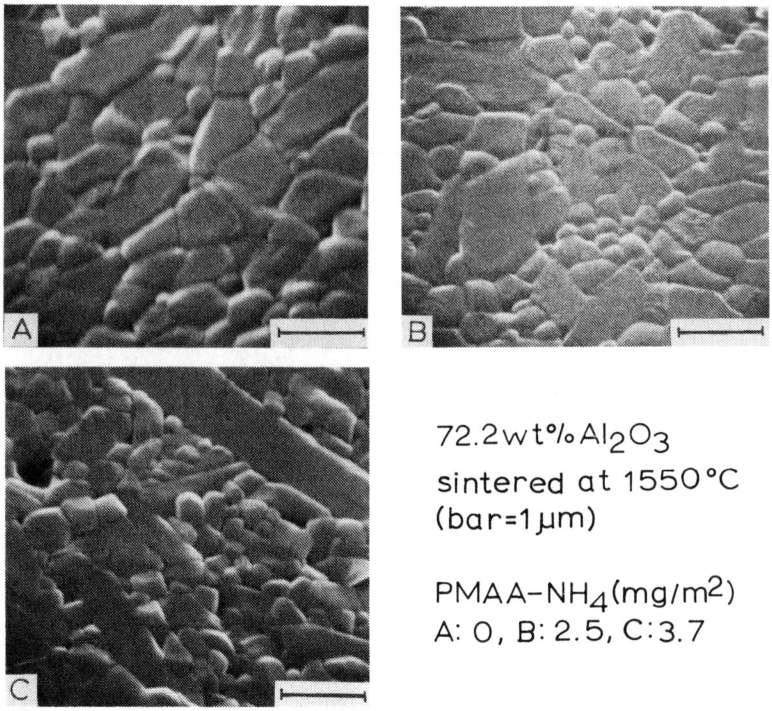

Fig. 12. Microstructure of mullite with a composition of 72.2 wt% Al_2O_3, sintered at 1550°C for 2 hours (bar = 1 μm). (See Table 5 for associated data and legend.)

Table 5. Effect of coating the particles with a polymer in the suspension stage on the average grain size of mullite sintered at 1550°C for 2 hours

Sample	PMAA-NH_4 (mg/m^2)	relative density(%) green	relative density(%) sintered	Average grain size (μm)
A	0	53.2	96.9	0.53
B	2.5	51.3	97.1	0.44
C	3.7	55.2	95.6	0.36

rate. A detailed discussion regarding this result will be presented in a subsequent paper.

Conclusions

Colloidal processing (dispersion, consolidation and sintering) and microstructures of sintered samples were studied using CVD-processed ultrafine SiO_2-Al_2O_3 powders (67.9 and 72.2 wt% Al_2O_3) with an average particle size of 38–65 nm. Electrostatic stabilization in an acid medium of pH ~4 was effective in the dispersion of the agglomerates formed during processing. Pore size distribution of the green compacts played the most crucial role on the densification behavior during sintering. Factors that affect the pore size distribution were identified as (i) the degree of dispersion in the colloidal suspension stage, and (ii) the structure of the particle clusters formed during consolidation as affected by the dispersant and the forming technique. When the pore size distribution approached a monosize system, green compacts could be densified to a relative density above 97% by sintering at 1500–1600°C for 2 hours. The mullite grain size was smaller for the sintered body formed from the electrosterically stabilized suspension than that produced from the electrostatically stabilized suspension.

References

[1]I. A. Aksay and J. A. Pask, "Stable and Metastable Equilibria in the System SiO_2-Al_2O_3," *J. Am. Ceram. Soc.*, **58** [11–12], 507–12 (1975).

[2]R. F. Davis and J. A. Pask, in "High Temperature Oxides," Part 4, A.M. Alper, ed., Academic Press, New York (1971) pp. 37–76.

[3]K. S. Mazdiyasni and L. M. Brown, "Synthesis and Mechanical Properties of Stoichiometric Aluminum Silicate (Mullite)," *J. Am. Ceram. Soc.*, **55** [11], 548–52 (1972).

[4]S. Kanzaki, H. Tabata, T. Kumazawa, and S. Ohta, "Sintering and Mechanical Properties of Stoichiometric Mullite," *J. Am. Ceram. Soc.*, **68** [1], C6-C7 (1985).

[5]S, Kanzaki, T. Kumazawa, J. Asaumi, O. Abe, and H. Tabata, "Dependency of Mechanical Property of Sintered Mullite on Chemical Composition," *Yogyo-Kyokai-Shi*, **93** [7], 407–8 (1985).

[6]M. G. M. U. Ismail, Z. Nakai, and S. Somiya, "Microstructure and Mechanical Properties of Mullite Prepared by the Sol-Gel Method," *J. Am. Ceram. Soc.*, **70** [1], C7-C8 (1987).

[7]S. Prochazka and F. J. Klug, "Infrared-Transparent Mullite Ceramic," *J. Am. Ceram. Soc.*, **66** [12] 874–80 (1983).

[8]N. Shinohara, D. M. Dabbs, and I. A. Aksay, "Infrared Transparent Mullite Through Densification of Monolithic Gels at 1250°C," *Proc. of SPIE*, **683**, 19 (1986).

[9]M. Ohashi, H. Tabata, S. Kanzaki, and S. Mitachi, "Preparation of Transparent Mullite Ceramics," Proc. of Annual Meeting of Ceram. Soc. Japan, 245–46 (1986).

[10]M. Mah and K. S. Mazdiyasni, "Mechanical Properties of Mullite," *J. Am. Ceram. Soc.*, **66** [10], 699–703 (1983).

[11]J. M. Rincon, G. Thomas, and J. S. Moya, "Microstructural Study of Sintered Mullite Obtained from Premullite," *J. Am. Ceram. Soc.*, **69** [2], C29-C31 (1986).

[12]S. Kanzaki, O. Abe, H. Tabata, T. Kumazawa, and S. Ohta, "Effect of Microstructure on Mechanical Properties of Mullite Ceramics," Proc. of Annual Meeting of Ceram. Soc. Japan, 243–44 (1986).

[13]I. A. Aksay, F. F. Lange, and B. I. Davis, "Uniformity of Al_2O_3-ZrO_2 Composites by Colloidal Filtration," *J. Am. Ceram. Soc.*, **60** [10], C190-C192 (1983).

[14]F. F. Lange, B. I. Davis, and E. Wright, "Processing-Related Fracture Origins: IV, Elimination of Voids Produced by Organic Inclusions," *J. Am. Ceram. Soc.*, **69** [1], 66–69 (1986).

[15]T. M. Shaw and B. A. Pethica, "Preparation and Sintering of Homogeneous Silicon Nitride Green Compacts," *J. Am. Ceram. Soc.*, **69** [2], 88–93 (1986).

[16]I. A. Aksay, in "Advances in Ceramics," Vol. 9, J. A. Mangles and G. L. Messing, eds., Am. Ceram. Soc., Columbus, OH (1984) pp. 94–104.

[17]Y. Hirata and I. A. Aksay, "Colloidal Consolidation and Sintering Behavior of CVD-Processed Mullite Powders," Proceedings of the 22nd University Conference on Ceramic, and the International Materials Symposium, pp. 611–22, J. A. Pask and A. G. Evans, eds., Plenum Press (1987).

[18]S. Hori and R. Kurita, "Characterization and Sintering of Al_2O_3-SiO_2 Powders Formed by Chemical Vapor Deposition Methods," to appear in the Proceedings of the 1st International Workshop on Mullite, Tokyo 1987, Am. Ceram. Soc. (1988).

[19]I. A. Aksay and C. H. Schilling, in "Advances in Ceramics," Vol. 9, J. A. Mangels and G. L. Messing, eds., Am. Ceram. Soc., Columbus, OH (1984), pp. 85–93.
[20]C. Han, I. A. Aksay, and O. J. Whittenmore, in "Advances in Materials Characterization II," R. L. Snyder, R. A. Condrate, Sr., and P. F. Johnson, eds., Plenum Publ. Co., New York (1985), pp. 339–47.
[21]Y. Hirata and I. A. Aksay, "Quantitative Analysis of Hierarchy of Pores," unpublished work.
[22]C. Herring, "Effect of Change of Scale on Sintering Phenomena," *J. Appl. Phys.*, **21** [4], 301–3 (1950).
[23]M. D. Sacks and J. A. Pask, "Sintering of Mullite Containing Materials: II, Effect of Agglomeration," *J. Am. Ceram. Soc.*, **65** [2], 70–77 (1982).
[24]C. Han, "Sintering of Bimodal Powder Compacts," M.Sc. Thesis, University of California, Los Angles, CA (1985).
[25]Y. Hirata and I. A. Aksay, "Long Range Particle Segregation in Colloidal Filtration," Extended Abstracts, Am. Ceram. Soc., 39th Pacific Coast Regional Meeting (1986), p. 6.
[26]D. H. Napper, "Polymeric Stabilization of Colloidal Dispersion," Academic Press, London (1983).

Sintering and Mechanical Properties of Mullite Derived Via Spray Pyrolysis

SHUZO KANZAKI AND HIDEYO TABATA

Government Industrial Research Institute, Nagoya
Hirate-cho, Kita-ku, Nagoya 462, Japan

TAKESHI KUMAZAWA

Technical Research Laboratory
Mino Yogyo Co., Ltd.
Kamezaki Kitaura-cho, Handa, Aichi 475, Japan

Abstract

Mullitization and densification behavior of spray pyrolyzed powders, microstructure development and mechanical properties of sintered bodies were studied with compositions ranging from 60 wt% Al_2O_3/40 wt% SiO_2 to 78 wt% Al_2O_3/22 wt% SiO_2. The as-sprayed amorphous powders were transformed directly into crystalline mullite on heating above 980°C irrespective of chemical composition. The relative density decreased with increasing Al_2O_3 content. Densification of powder compacts with composition less than 74 wt% Al_2O_3 was promoted by the presence of glassy grain-boundary phase which led to an acicular mullite grain morphology. The absence of the glassy phase resulted in poor sinterability and equiaxed grain morphology. The room temperature strength and fracture toughness of the sintered bodies were almost independent of chemical composition. The high temperature strength, however, was significantly affected by the composition, and increased with increasing SiO_2 content up to 1300°C. These trends in the mechanical properties are discussed in terms of microstructure and impurities included in the starting powders.

Introduction

Mullite ($3Al_2O_32SiO_2$) is a well known materials as one of the important constituents in refractories. Since high purity synthetic mullite has been pointed out to be a potential high-performance refractory ceramics,[1,2] considerable efforts have been directed toward the synthesis and the mechanical property evaluation of the stoichiometric mullite.[3–6] Although several studies have been conducted to evaluate mechanical properties of the stoichiometric mullite, only a few studies concerning the effect of chemical composition on mechanical properties have been reported[7] so far. As the microstructure and the sinterability of mullite ceramics strongly depend upon its composition ratio, i.e., Al_2O_3/SiO_2 ratio,[8,9] it is expected that mechanical properties of these ceramics depend on their composition as well.

Although fine-grained and high purity powder is required for obtaining the well-controlled microstructure of the fully densified bodies, the highly homogeneous distribution of the constituent elements in the starting powder is particularly important especially in the case of multi-component systems. In order to obtain high purity and fine grained mullite powders, various synthesis methods have been conducted, e.g. thermal decomposition, coprecipitation, hydrolysis etc.[3,6,10–14] Among these synthesis methods, spray pyrolysis (or EDS)[5,15] is a promising technique to prepare highly homogeneous powder, especially for multi-component systems.

In the present study, silica-alumina powders with the composition ranging from 60 wt% Al_2O_3/40 wt% SiO_2 to 78 wt% Al_2O_3/22 wt% SiO_2 were prepared by spray pyrolysis method. Mullitization process and the sinterability of the synthesized powders have been investigated. The effect of chemical composition on microstructure and mechanical properties of the sintered bodies was also examined.

Experimental Procedure

Preparation of the Starting Powders and Sintered Bodies

Nine different kinds of aluminum silicate powders with the composition ranging from 60 wt% Al_2O_3/40 wt% SiO_2 to 78 wt% Al_2O_3/22 wt% SiO_2 were prepared by spray pyrolysis method.[5] Starting materials were aluminum nitrate and ethyl silicate. Both materials were dissolved into water-methanol (1:1 in volume) mixed solvent. Concentration of the mixed solution was 0.2 mol/L in terms of the corresponding aluminum silicate compositions.

The mixed solution was sprayed with compressed air through a water cooled glass nozzle into a preheated quartz reaction tube at 650°C in air. The as-obtained powder was calcined at 950°C for 1 h to convert amorphous powder into crystalline mullite. After calcination, the powder was ground for 80 h in a nylon-line vibratory mill with silicon nitride balls using methanol as a mixing medium. After drying and screening (–150 mesh), powders were calcined at 800°C for 1 h to burn out any organic materials contained. Typical impurities in the 71.8 wt% Al_2O_3 powder after grinding were 0.0266% Na_2O, 0.0286% Fe_2O_3, <0.0167% TiO_2, 0.0241% K_2O, <0.0140% CaO and <0.0166% MgO in wt% as determined with atomic absorption spectroscopy. Powders were designated according to their alumina content, e.g. the powder with 71.8 wt% Al_2O_3 composition is referred to as 71.8A. Powder compacts were die-formed under uniaxial pressure (30 MPa) followed by isostatic pressing at 200 MPa. The compacts were sintered at 1630°C or at 1650°C for 4 h.

Property Measurement

The crystallization sequence of as-sprayed amorphous powder was determined by DTA and X-ray diffraction (XRD). Bulk densities of the sintered bodies were measured by a fluid displacement method using distilled water. Specific surface area of the powders was measured by BET method. SEM and TEM were used for the observation of microstructure of the specimens. Thermal etching (1500°C for 20 min) and chemical etching (0°C for 24 h in 1% hydrofluoric acid solution) were carried out on the polished specimens for SEM examination. Selected-area diffraction studies and electron energy loss spectroscopy (EELS) were performed to identify grain-boundary phase.

The sintered plates were sliced and ground with #400 diamond wheel to prepare test pieces of the size 3 by 4 by 38 mm. Flexural strength was measured by three-point bending test (30 mm span) from room temperature to 1400°C in air at a loading rate of 0.5 mm/min. Fracture toughness K_{IC} was determined using SENB test pieces in four-point bending (30/10 mm spans) method. The notched bars were prepared by introducing a notch of 1 mm depth and 0.1 mm with into the same bending bars as flexural strength measurement.

Results and Discussion

Powder Characteristics

The as-sprayed powder was composed of spherical hollow particles with many small pores and amorphous to X-ray diffraction irrespective of chemical composition. Figure 1 shows DTA curves for the as-sprayed powders with different composition measured at a heating rate of 10°C/min. A single sharp exothermic peak was observed at about 980°C regardless of the composition.

The XRD patterns of the 71.8A powder, quenched from various temperatures, are shown in Fig. 2. Mullite is the only crystalline phase detected above 980°C in the composition range of 60A to 75A. Corundum was detected in the 78A composition when heated above 1500°C as expected by the SiO_2-Al_2O_3 phase diagram.[16]

These results indicate that spray pyrolyzed powder transforms directly from the amorphous state to crystalline mullite. This behavior might be attributed to the powder preparation method. Since the pyrolysis temperature was high enough compared to the thermal decomposition temperature of each starting material, instantaneous decomposition both materials occurred together and a highly homogeneous mixed powder formed. Furthermore, the results suggest that the mullitization temperature is not affected by the composition, i.e., the Al_2O_3/SiO_2 ratio.

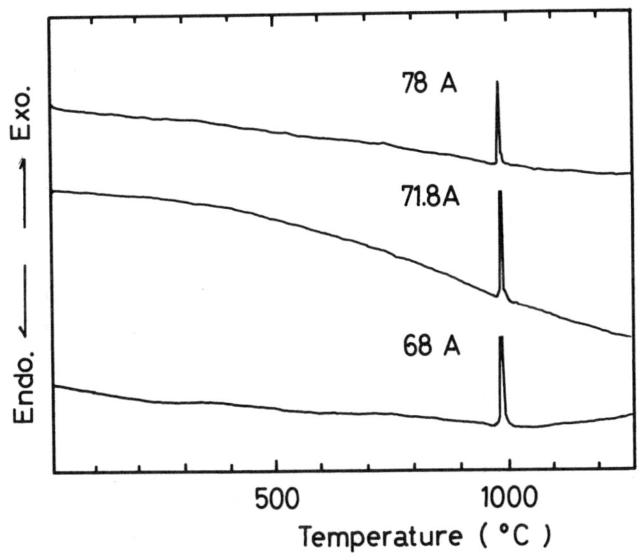

Fig. 1. DTA curves for the as-sprayed powders with different composition.

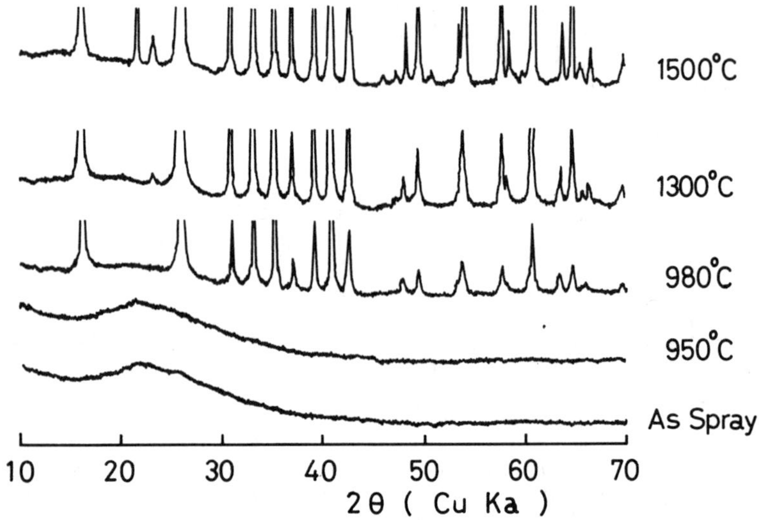

Fig. 2. X-ray diffraction patterns for 71.8A powder quenched from various temperatures indicated.

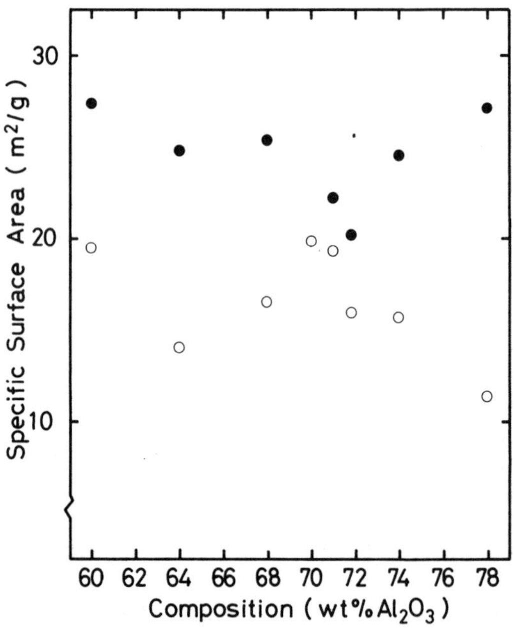

Fig. 3. Specific surface area of powders (○) as-sprayed and (●) calcined at 950°C for 1 h and ground after calcination as a function of composition.

Figure 3 shows the relationship between specific surface area and composition for powders as-sprayed, calcined at 950°C for 1 h and ground after calcination. The specific surface area of the as-sprayed powders ranged from 10 to 20m^2/g depending on composition. However, calcination without grinding results in a lowering of this value to less than 5m^2/g owing to mullitization. Spherical hollow mullite particles were easily broken down to small flakes on grinding and the value increased between 20 and 27m^2/g.

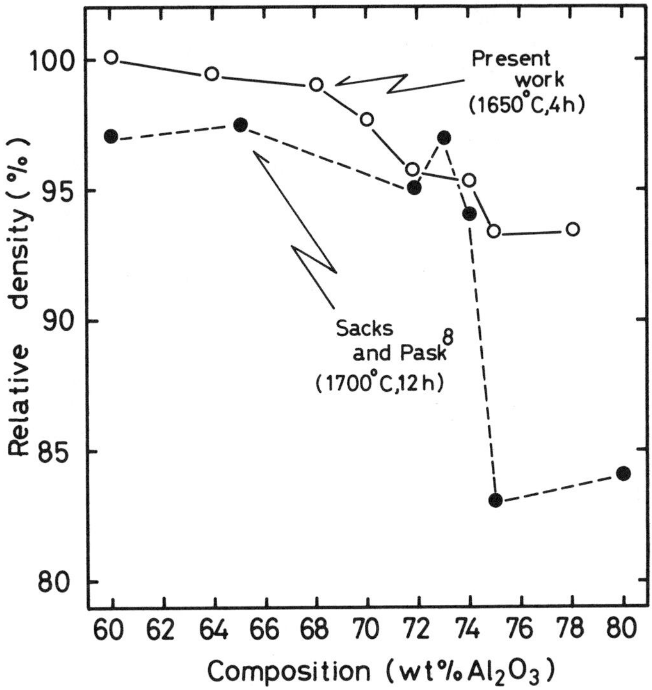

Fig. 4. Relative density of the specimens as a function of composition.

Effect of Composition on Density and Microstructure

The effect of composition on relative density of the samples sintered at 1650°C for 4 h is shown in Fig. 4. The densification behavior in relation to composition was generally in accord with the results previously reported.[6,8] The bulk density decreased with increasing Al_2O_3 content between 71.8 and 75 wt% Al_2O_3, then increased slightly at 78 wt% Al_2O_3. Compositions between 60 and 70 wt% Al_2O_3, in mullite and liquid phase region, attained maximum density due to liquid phase sintering.

Representative microstructures of compositions 68 to 78 wt% Al_2O_3 are shown in Fig. 5. The phase present in the sintered specimens determined by SEM and XRD were mullite + glassy grain-boundary phase for 60 to 71.8 wt% Al_2O_3, single phase mullite for 74 wt% Al_2O_3 and mullite + corundum for 75 and 78 wt% Al_2O_3.

In the case the compositions with 68 and 71.8 wt% Al_2O_3, acicular mullite grains and grain-boundary second phases can be seen. The morphology of mullite grains become granular when the Al_2O_3 content in the composition exceed 74 wt% and

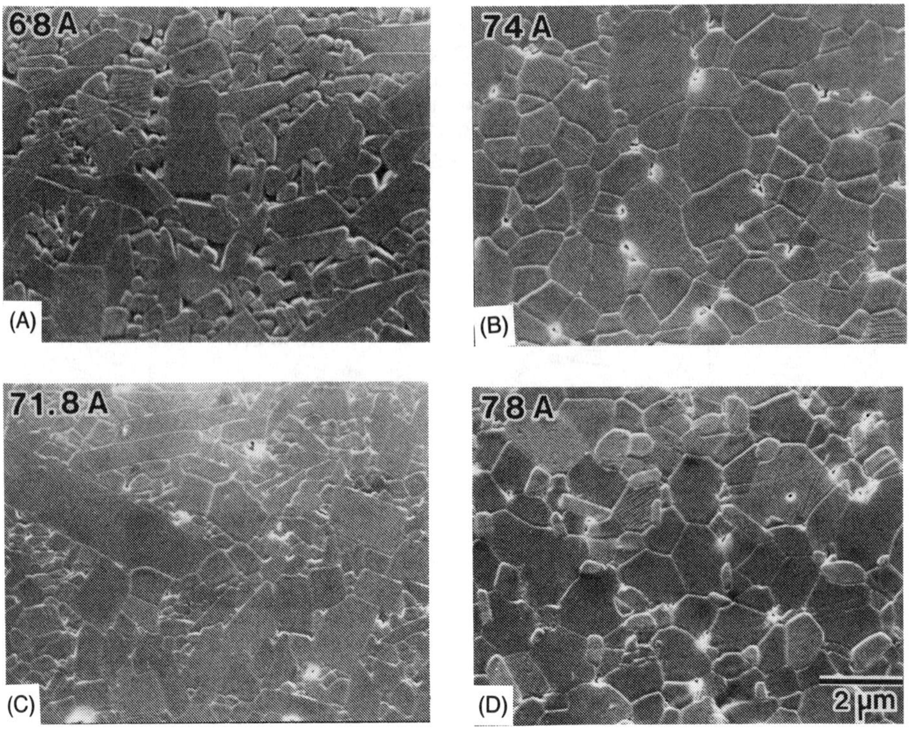

Fig. 5. Typical SEM photographs of the specimens sintered at 1650°C for 4 h.

corundum particles were observed both in the 75 and 78 wt% Al_2O_3. These results suggest that the liquid phase is necessary to attain high density and acicular grain morphology. Features of interest in the 71.8 wt% Al_2O_3 composition is the existence of glassy phase, despite the use of relatively pure and fully mullitized powder. If the 71.8 wt% Al_2O_3 composition would fall into the mullite solid solution range, the sintered sample should be composed of only crystalline mullite. Three possibilities can be considered for the existence of a glassy phase in 71.8A;

1. insufficient reaction caused by inhomogeneous distribution of Al_2O_3 and SiO_2 in the raw material
2. impurities in the raw materials cause liquid phase even though the impurity level is considerably lower
3. 3:2 mullite is not stable above a certain temperature as suggested by Prochazka and Klug[17] or Sacks and Pask.[8]

It was revealed that the amorphous spray pyrolyzed powder was directly converted into mullite at low temperature (Fig. 1) and the liquid phase was not observed in the 74 wt% Al_2O_3 composition (Fig. 5). These facts support that the formation of the glassy phase is not due to incomplete reaction nor the presence of impurities in the raw material, as already pointed out.[8] Figure 6 shows the

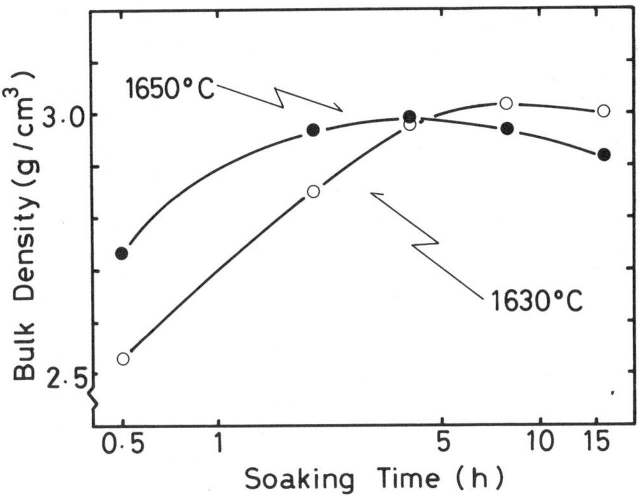

Fig. 6. Bulk density of the 71.8A sample sintered at 1630°C and at 1650°C as a function of soaking time.

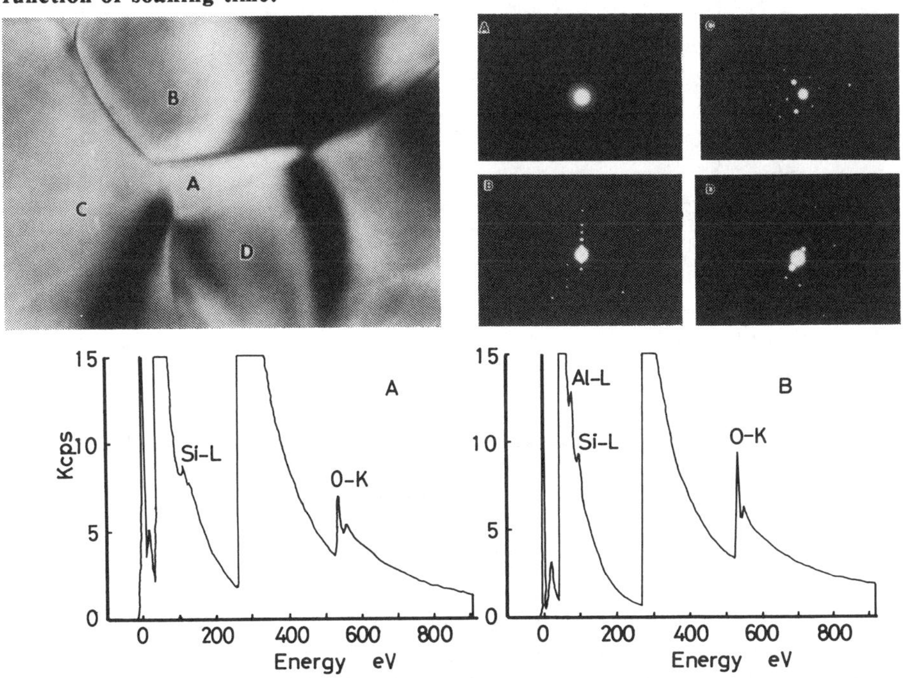

Fig. 7. TEM photograph, microbeam diffraction (70A) and EELS spectra of the 70A sample.

relationship between soaking time and bulk density for the 71.8A sample. The density increased for short soaking times and decreased with prolonged soaking time.

Expansion of lattice parameter along a-axis was also observed with decrease in density, e.g. 7.544A and 7.546A for 4 h and 16 h soaking at 1650°C, respectively.[9] It suggests that the Al_2O_3 content in mullite increases with soaking time at 1650°C. Furthermore, TEM, electron beam diffraction patterns and EELS revealed that the grain-boundary glassy phase is composed of silicon and oxygen, while the mullite grain contains aluminum as well (Fig. 7).

Therefore it is apparent that the grain-boundary glassy phase is composed of almost pure SiO_2 and does not contain any appreciable amount of Al_2O_3. Hence, it appears that the 3:2 mullite is not stable as a single phase at the temperature concerned, and phase separation occurs to form a mullite phase with a higher Al_2O_3 content than 71.8 wt% Al_2O_3 and amorphous SiO_2. Sacks and Pask[8] have pointed out that mullite with this composition exists as a metastable assemblage of high Al_2O_3 mullite and glass phase. On the other hand, Prochazka and Klug[17] have postulated that mullite solidus bends over to alumina side and 3:2 mullite does not exist as an equilibrium phase above a certain temperature. However, it is still uncertain which of the above explanations is more reasonable to explain the present experimental results.

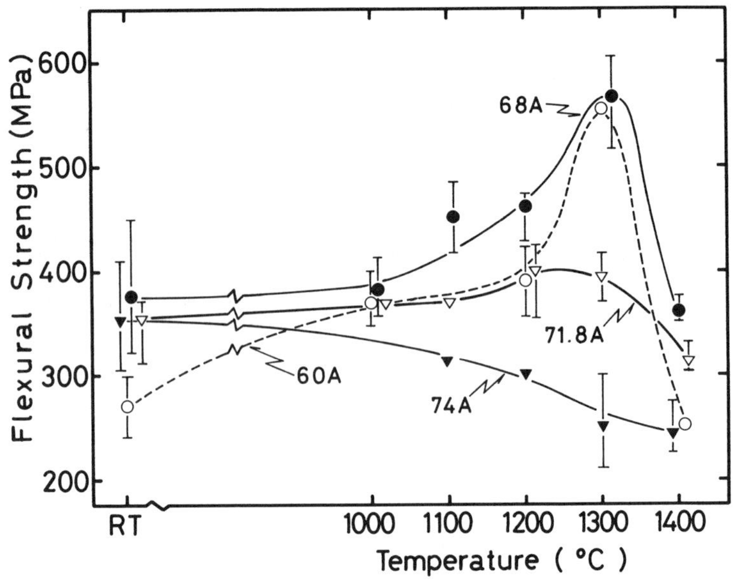

Fig. 8. Flexural strength of the specimens with different compositions as a function of temperature.

Effect of Composition on Mechanical Properties

Plots of three-point bending strength as a function of temperature for the specimens with representative compositions are shown in Fig. 8. Average strength values of the stoichiometric composition 71.8A are 352, 398 and 392 MPa at room temperature, 1200° and 1300°C, respectively. The flexural strength decreased to 310

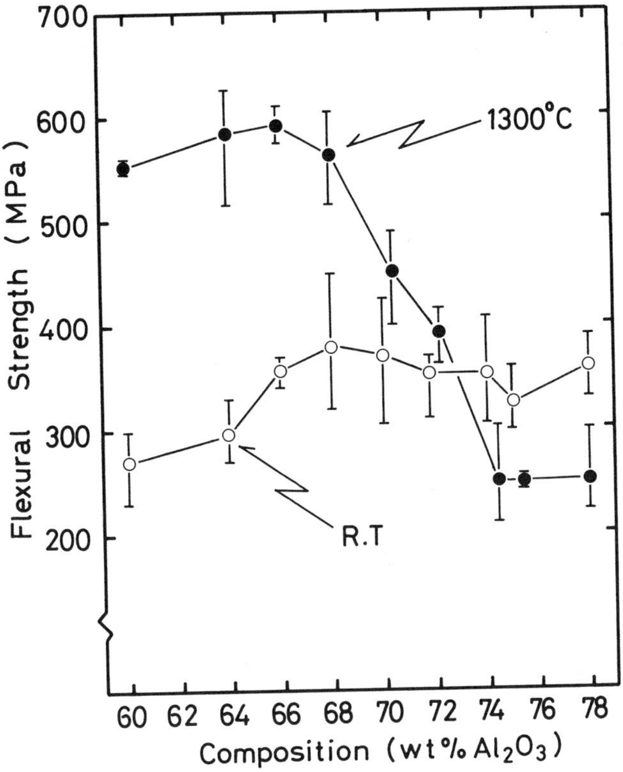

Fig. 9. Flexural strength of the specimens at room temperature and at 1300°C as a function of composition.

MPa at 1400°C. A steep increase in strength was observed between 1200° and 1300°C for SiO_2-rich compositions and a drastic drop in strength was observed at 1400°C. In contrast, the strength of the 74A composition, composed only of mullite, decreased with increasing temperature.

Figure 9 shows the strength at room temperature and at 1300°C as a function of composition. The room temperature strength is independent for compositions between 66 and 78 wt% Al_2O_3, while a slight decrease in strength was recognized for 60 and 64A. In contrast, the high temperature strength is strongly dependent on composition.[7,18]

The strength measured at 1300°C increased for the compositions with a higher SiO_2 content than the stoichiometric composition 71.8A, while the values for the higher Al_2O_3 contents decreased. An average strength of 592 MPa was obtained for 66A. These results and microstructural observations (Fig. 5, Fig. 7) suggest that the glassy grain-boundary phase plays an important role for the high temperature mechanical properties of mullite ceramics.

Figure 10 shows load-deflection curves measured at room temperature, 1200°, 1300° and 1400°C for the samples 60A and 68A. The compositions tested below 1200°C fractured in a brittle manner, while the fracture at 1300°C was accompanied

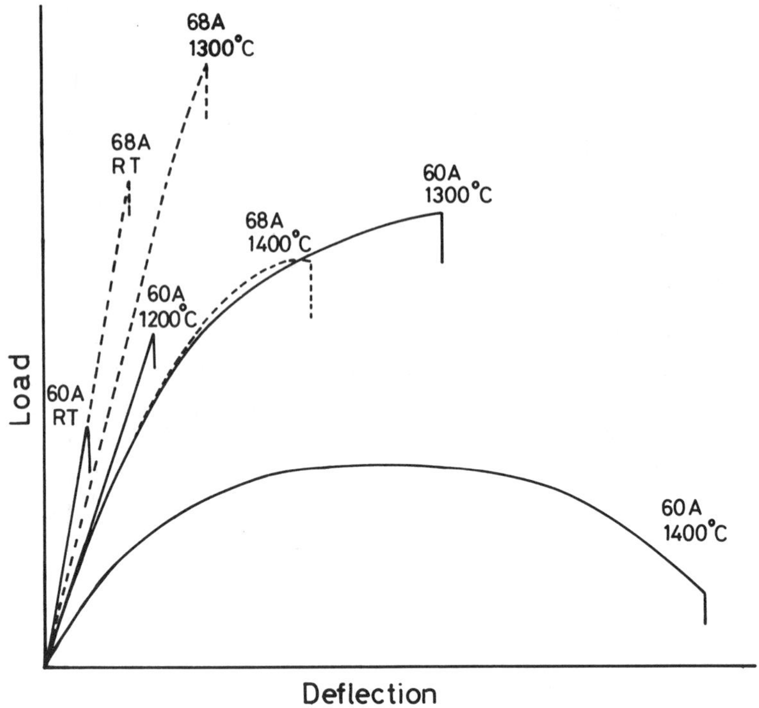

Fig. 10. Load-deflection curves of the 60A and 68A specimens at room temperature, 1200°C, 1300° and at 1400°C.

by plastic deformation. The rate of deformation increased with increasing SiO_2 content and with testing temperature. Mah and Mazdiyasni[4] studied mechanical properties of mullite with stoichiometric composition over the temperature range from room temperature to 1500°C. They suggested that a dominance of energy dissipation at the crack tip over grain-boundary sliding results in an "apparent" strengthening of mullite at high temperatures. Similar consideration can also be applied in this study. The glassy phase which is located exclusively at the triple point and composed only of SiO_2, acts as an energy absorber when its viscosity decreases at high temperature. Although the glassy grain-boundary phase is advantageous for the rapid loading, the existence of such a vitreous phase will cause disadvantages for time-dependent loading, i.e., the creep resistance at high temperatures is predicted to be low.

In addition, the effect of impurities in the raw materials on high temperature strength has been considered. Figure 11 shows the strength of samples with different impurity levels as a function of temperature. The values for the sample 68A and 71.8A are the results obtained in this study. The values for the sample 67A-C is a commercially produced mullite with 67 wt% Al_2O_3 composition, prepared from natural aluminum silicate raw material which includes relatively high amount of impurities.[19]

The sample 71.8A-A has been prepared from alkoxy-derived high purity powder with a total amount of impurity less than 0.03 wt%.[20] Hence the amount of impurities

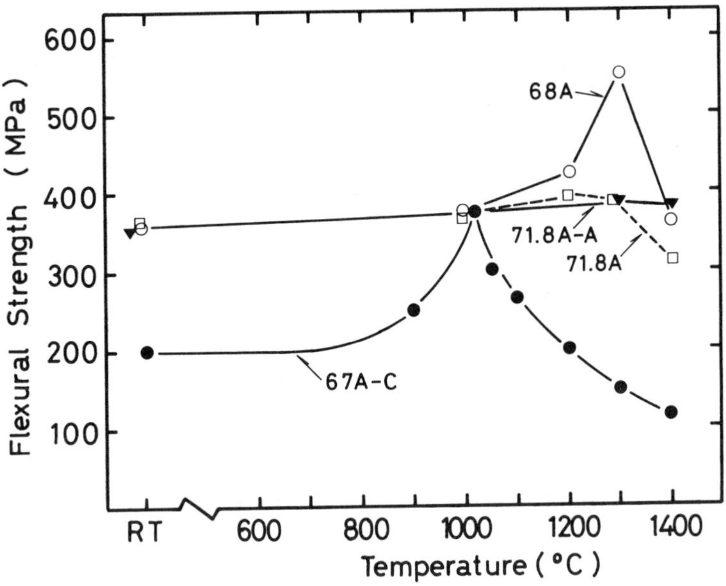

Fig. 11. Flexural strength of the specimens with different composition and amount of impurities. (68A, 71.8A; Present work, 67A-C; Ref. 19, 71.8A-A; Ref. 20)

in similar compositions are in the order of 67A-C > 68A and 71.8A > 71.8A-A. The difference in high temperature strength in each series is clearly recognizable. Although both samples 67A and 67A-C have a maximum in strength, the peak temperature of the 67A-C is about 300°C lower than that for 68A. The difference in peak temperature indicates that the temperature at which the viscosity of the grain-boundary glassy phase decreases is sufficiently different. While the 71.8 wt% Al_2O_3 composition includes a small amount of glassy phase, the high temperature strength does not decreased even at 1400°C when high purity raw powder was utilized. These results indicate that the impurities in the raw materials should be as low as possible because they seriously reduce the viscosity of the glassy phase. On the other hand, the strength of the sample without glassy phase decreased gradually with increasing temperature which may be attributed to the decrease in elastic modulus with increasing temperature.

The fracture toughness K_{IC} as a function of composition measured at room temperature and at 1300°C is shown in Fig. 12. The room temperature K_{IC} values are in the range between 2.4 and 2.8 $MPam^{1/2}$, and no significant difference was observed with compositional change. The high temperature K_{IC} increased with increasing SiO_2 content, and a maximum value of 4.7 $MPam^{1/2}$ was obtained for 68A. The 78A sample shows no significant difference with temperature. The result obtained at 1300°C is in accordance with the flexural strength for all compositions except for 78A.

Observation of the fracture surfaces showed that the surface area formed above 1200°C is larger than that at room temperature and a part of the acicular mullite grains was pulled out above 1300°C. Therefore the increase in K_{IC} with increasing

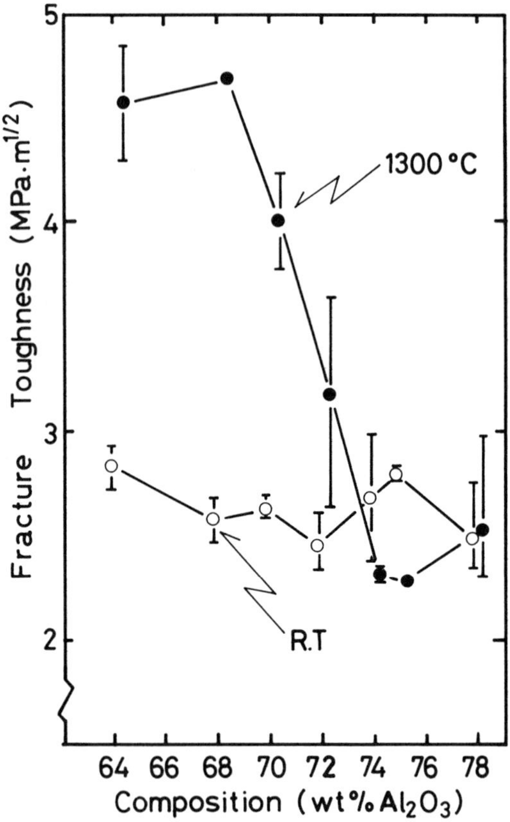

Fig. 12. Fracture toughness of the specimens at room temperature and at 1300°C as a function of composition.

temperature can be explained mainly by the pulling out of the acicular grains due to the decrease in viscosity of the glassy phase.

Conclusions

Sinterability of the powder compacts increased with increasing excess amount of SiO_2, because of liquid phase sintering. The strength values at high temperature were significantly affected by the composition, and increased with increasing SiO_2 content up to 1300°C. This behavior is mainly attributed to energy dissipation at the viscous grain-boundary phase. However, an excess amount of SiO_2 resulted in poor strength above 1300°C. It is concluded that the amount of glassy phase in the sintered bodies and the impurities contained in the starting powder are the dominant factors affecting the mechanical properties of mullite ceramics, especially at high temperatures.

References

[1]P. C. Dokko, J. A. Pask, and K. S. Mazdiyasni, "High-Temperature Mechanical Properties of Mullite under Compression," *J. Am. Ceram. Soc.*, **60** [3-4] 150-55 (1977).

[2]P. A. Lessing, R. S. Gordon, and K. S. Mazdiyasni, "Creep of Polycrystalline Mullite," *J. Am. Ceram. Soc.*, **58** [3-4] 149 (1975).

[3]B. L. Metcalfe and J. H. Sant, "The synthesis, Microstructure and Physical Properties of High Purity Mullite," *Trans. Brit. Ceram. Soc.*, **74**, 193-201 (1975).

[4]T. Mah and K. S. Mazdiyasni, "Mechanical Properties of Mullite," *J. Am. Ceram. Soc.*, **66** [10] 699-703 (1983).

[5]S. Kanzaki, H. Tabata, T. Kumazawa, and S. Ohta, "Sintering and Mechanical Properties of Stoichiometric Mullite," *J. Am. Ceram. Soc.*, **68** [1] C-6-7 (1985).

[6]P. D. D. Rodrigo and P. Boch, "High Purity Mullite Ceramics by Reaction Sintering," *Int. J. High Technology Ceramics*, **1**, 3-30 (1985).

[7]S. Kanzaki, T. Kumazawa, J. Asaumi, O. Abe, and H. Tabata, "Dependency of Mechanical Property of Sintered Mullite on Chemical Composition," *Yogyo-Kyokai-Shi*, **93** [7] 407-08 (1985).

[8]M. D. Sacks and J. A. Pask, "Sintering of Mullite-Containing Materials: I, Effect of Composition," *J. Am. Ceram. Soc.*, **65** [2] 65-70 (1982).

[9]T. Kumazawa, S. Kanzaki, J. Asaumi, O. Abe, and H. Tabata, "Sinterability of SiO_2-Al_2O_3 Powders by Spray Pyrolysis: Effect of Composition," *Yogyo-Kyokai-Shi,*, **94** [5] 485-90 (1986).

[10]K. S. Mazdiyasni and L. M. Brown, "Synthesis and Mechanical Properties of Stoichiometric Aluminum Silicate (Mullite)," *J. Am. Ceram. Soc.*, **55** [11] 548-52 (1972).

[11]B. B. Ghate, D. P. H. Hasselman, and R. M. Spriggs, "Synthesis and Characterization of High Purity, Fine Grained Mullite," *Am. Ceram. Soc. Bull.*, **52** [9] 670-72 (1973).

[12]T. A. Wheat, E. M. H. Sallam, and A. C. D. Chaklader, "Synthesis of Mullite by a Freeze-Dry Process," *Ceramurgia Int.*, **2**, 42-44 (1979).

[13]B. E. Yoldas, "Microstructure of Monolithic Materials Formed by Heat Treatment of Chemically Polymerized Precursors in the Al_2O_3-SiO_2 Binary," *Am. Ceram. Soc. Bull.*, **59** [4] 479-83 (1980).

[14]Y. M. M. Ai-Jarsha, K. D. Biddle, A. K. Das, T. J. Davis, H. G. Emblem, K. Jones, J. M. McCullough, M. A. Mohd.Abd.Rahman, A. N. A. El-m. Sharf el Deen, and R. Wakefield, "Mullite Formation from Ethyl Silicate and Aluminum Chlorides," *J. Mater. Sci.*, **20**, 1773-81 (1985).

[15]D. M. Roy, P. R. Neurgaonkar, T. P. O.'Holleran, and R. Roy, "Preparation of Fine Oxide Powders by Evaporative Decomposition of Solutions," *Am. Ceram. Soc. Bull.*, **56** [11] 1023-24 (1977).

[16]I. A. Aksay and J. A. Pask, "Stable and Metastable Equilibria in the System SiO_2-Al_2O_3," *J. Am. Ceram. Soc.*, **58** [11-12] 507-12 (1975).

[17]S. Prochazka and F. J. Klug, "Infrared-Transparent Mullite Ceramic," *J. Am. Ceram. Soc.*, **66** [12] 874-80 (1983).

[18]S. Kanzaki, O. Abe, M. Ohashi, and H. Tabata, "Mechanical Properties of Pressureless Sintered Mullite Ceramics," 625-31 in Ceramics Materials and Components for Engines, Edited by W. Bunk and H. Hausner, Verlag Deutsche Keramische Gesellschaft, 1986.

[19]T. Kawanami, "Properties of Commercial Mullite Ceramics," 123-35 in Mullite, Edited by S. Somiya, Uchida Rokakuho, 1985.

[20]T. Kumazawa, S. Kanzaki, and H. Tabata; unpublished work.

High-Purity Mullite Ceramics By Reaction-Sintering

P. BOCH AND T. CHARTIER

Ecole Nationale Supérieure de Céramique Industrielle
Limoges, France

P. D. D. RODRIGO

University of Moratuwa
Moratuwa, Sri-Lanka

Abstract

Reaction-sintering of alumina + silica mixtures is an inexpensive route to produce mullite ceramics. The processing parameters have to be tailored to favour densification prior to recrystallization of silica into cristobalite and mullitization. The use of submicron powders sintered at a fast heating rate allows sintering treatments at 1600°C to yield almost theoretical density and ceramics containing the maximum possible mullite content. The alumina-to-silica ratio is a critical parameter, and its influence has been studied in mixtures varying from 68 to 80 wt% alumina. Mullite always begins to form with the 70.4% Al_2O_3 composition. The 71.8% Al_2O_3 mixture leads to the best densification and the finest microstructure, whereas the 75% Al_2O_3 mixture exhibits poor densification and exaggerated grain growth. The silica-rich 68% Al_2O_3 composition has the higher room temperature strength, and high-temperature strength increases to 450 MPa at 1200°C.

Introduction

Alumina ceramics continue to occupy the first rank among technical ceramics. Two classes of reasons can explain this prominence. Firstly, alumina possesses interesting properties for mechanical as well as for electrical applications. Secondly, alumina powders are rather low in cost, and the fabrication processes are straight forward (sintering at temperatures below 1600°C for the "reactive" aluminas). Therefore, alumina is often chosen for applications where other ceramic compounds could be more adequate. This is particularly true for mullite ceramics whose potential has not been fully exploited because of difficulties in the techniques of preparation. The aim of the present paper is to show that reaction-sintering of alumina and silica mixtures is an efficient route to produce high-technology mullite ceramics, which could aid in the development of these materials.

Reaction-Sintering of Mullite

Many studies have been devoted to the understanding of the Al_2O_3-SiO_2 diagram, and, in particular, Pask and Aksay have shown the existence of two complementary diagrams.[1] In the stable diagram, mullite melts incongruently at about 1830°C, and its solid solution ranges from 70.5 to 74 wt% Al_2O_3. In the metastable diagram, SiO_2-ordered mullite melts congruently at about 1880°C, and its solid solution extends to 77 wt% Al_2O_3. By contrast, SiO_2-disordered mullite melts congruently at about 1900°C, and its solid solution extends to 83 wt% Al_2O_3. A cristobalite-alumina diagram also exists with a eutectic at 1260°C but without mullite.[2]

Most of the earlier studies were concerned with melt-grown mullite, whereas most of the more recent work focuses on how to sinter fine grain, high purity, high density mullite ceramics.[3–4] Mullite is characterized by having a low density (3.16 $g.cm^{-3}$), low thermal conductivity and low thermal expansion (α_{20}• c = 5. 10^{-6} K^{-1}), with a Young's modulus of about 230 GPa, a strength and toughness generally lower than those of alumina at room temperature but higher at elevated temperatures, and excellent creep resistance.[5–8]

The absence of commercially available mullite powders—at least in the range of cost and quality which corresponds to alumina powders—is a demonstration of the fact that the key point remains the preparation of starting materials. Therefore, three main routes may be chosen, namely :

1. the preparation of mullite powders from clay-based materials, for instance "premullite" obtained from halloysite,[9]
2. the preparation of mullite powders using chemical or physical methods (e.g. coprecipitation, hydrolysis, sol-gel, CVD, etc.),[10–16]
3. or the preparation or mullite ceramics by reaction-sintering of alumina plus silica mixtures.[3–8] This is the topic of the present paper.

For a mullite with the 3/2 stoichiometry, the reaction is:

$$3Al_2O_3 + 2SiO_2 \rightarrow (3Al_2O_3.2SiO_2)$$

Non-stoichiometric mullite can form when the alumina-to-silica ratio is different from 3/2. A noticeable excess of alumina or of silica yields corundum + mullite or silica + mullite materials, respectively. The case of the 3/2 stoichiometry will be discussed first, followed by examples of other alumina-to-silica ratios.

Mullite of "3/2" Composition

Reaction-Sintering of Mullite Ceramics

The preparation of mullite ceramics by reaction-sintering of mixtures of powders of alumina and silica involves chemical reactions, phase transformations, and densification. The chemical reactions are mainly due to mullitization, but even in the simplest case of a "3/2" mixture—where the end-product is stoichiometric mullite—some intermediate, non-equilibrium compounds can develop. Phase transformations are essentially related to silica transitions and to mullitization. Sintering mechanisms can be operative in a pre-reacted state, in a partially-reacted state, or in a fully-reacted state.

The previous studies on the reaction-sintering of mullite[3–4,6–8] lead to four comments, also valid for the preparation of zirconia-toughened ceramics by reaction-sintering between alumina and zircon:

1. Crystalline mullite powder is not an easily sinterable compound, which is related to the slowness of diffusion. Nearly all data on densification and grain growth during hot-pressing or on creep of mullite lead to activation enthalpies of about 700 kJ/mole.[17–19] Such a value is very similar to the activation energy for lattice diffusion of Si^{+4} (702 kJ/mole),[20] which could mean that the diffusion of silicon ions controls densification, grain growth, and creep of mullite. Other creep data such as stress exponent (about 1.5) and grain size dependence (about −2) also support the idea that the diffusion of Si^{+4} plays a major role on the creep of mullite and on its densification and grain growth.[18,19] Therefore, heat treatments at relatively high temperature are required to densify mullite ceramics. In the case of reaction-sintering, the processing treatments should be tailored to delay mullitization in relation to densification, in order that sintering develops in a pre-reacted or partially-reacted state rather than in a fully-reacted state.
2. The temperature of mullite formation and rate of mullite development depend on the crystalline forms of the Al_2O_3 and SiO_2 components. A comparison among the literature data is difficult, because (1) the differences in the crystalline forms of the silica and alumina species and (2) the effect on the reactivity of the differences of purity, particle size, form and nature of particles (e.g. single crystals or polycrystals).[3,4,6,21] However, most data show τ-Al_2O_3 or other alumina components with a defect structure to be the most reactive. For silica, there are contradictory data. Our results have shown that for the particle sizes which were studied here (from about 0.5 μm to 1.5 μm), cristobalite is more reactive than amorphous silica or quartz. However, amorphous silica and quartz generally transform to cristobalite at temperatures where mullitization is just beginning, which means that mullitization mainly develops by reaction between α-alumina and cristobalite, regardless of the starting materials. The experiments described in the present paper have been carried out using α-alumina + amorphous silica mixtures (pressed samples) or α-alumina + α-quartz mixtures (tape-cast samples).
3. The densities of the silica forms are about 2.22, 2.32, and 2.65 $g.cm^{-3}$ for amorphous silica, cristobalite, and quartz, respectively. Mullite has a low-density (3.16 $g.cm^{-3}$) in comparison with the initial "3/2" Al_2O_3-SiO_2 mixture (which has a density of 3.31 $g.cm^{-3}$ for the case of α-alumina + amorphous silica mixture). These differences in densities imply that strong dilatometric effects are expected to develop during reaction-sintering. Some of them have been demonstrated to be detrimental to densification.[21]
4. Finally, it must be pointed out that Brook and Yangyun have stressed that during reaction-sintering of powder compacts the rate of densification depends on D^{-3} or D^{-4} (D being the mean particle size), whereas the rate of reaction depends on D^{-1} or D^{-2}.[22] Hence, densification mechanisms are more strongly favoured than the reaction mechanisms by a decrease in the particle size of the reactants.

These four comments suggest that the preparation of mullite ceramics by reaction-sintering is facilitated by very fine powders, the choice of amorphous silica or quartz and α-alumina as starting materials, and a firing cycle which favours densification prior to mullitization.

Experimental

Forming Methods

Two forming methods were chosen: dry pressing and tape casting.

Table 1. Starting materials for dry-pressing

Material	Source	Purity	Mean particle size
Alumina:α-Al_2O_3	Reynolds RC172DBM*	> 99.7%	0.65 μm
Silica:amorphous SiO_2	Alfa 89709†	> 99.5%	3.60 μm

*Reynolds Chem. Co. Arkansas USA
†Alfa Research Chemicals and Materials, Karlsruhe FRG.

Table 2. Composition of the casting slurry

Material	Source	Vol%	Wt%
Alumina:α-Al_2O_3	Alcoa A16SG*	20.6	47.9
Quartz:α-SiO_2	Sifraco C800†	12.2	18.8
Deflocculant:phosphate ester	Gerland PE8312‡	1.1	0.7
Solvent:methylethylketone	Rhône-Poulenc§	35.7	16.7
and:ethanol	Rhône-Poulenc§	18.8	8.6
Plasticizer:dibutyl phtalate	Rhône-Poulenc§	3.2	1.9
and:polyethylene glycol	Rhône-Poulenc§	4.2	2.7
Binder:polyvinyl butyral	Monsanto¶	4.2	2.7

*Alcoa, Pittsburgh USA
†Sifraco, Paris France
‡Gerland Chemical, Paris France
§Rhône-Poulenc, Paris France
¶Monsanto, Saint Louis USA

1. For pressing, starting materials were α-alumina and amorphous silica (Table 1).[21] The 3/2 Al_2O_3/SiO_2 mixtures were prepared by attrition milling in ethanol, using zirconia balls, to three different particle sizes for the mixtures: 0.80, 0.60; and 0.45 μm. After drying, crushing and mixing with organic binders, powders were sieved through a 200 μm sieve and uniaxially pressed with ultrasonic assistance to improve the compaction homogeneity.[23] Samples were disc-shaped, having a 30 mm diameter and a 3 mm thickness.
2. For tape casting, starting materials were α-alumina and α-quartz (Table 2).[24] The preparation of the casting slurry was done by ball-milling in two stages. First, the ceramic powders were deagglomerated in the solvent, with the help of a dispersant. Second, the binders and the plasticizers were added to the slurry. Most of the experiments were carried out with a tape thickness of about 200 μm (in the dried state). The dried tapes were punched into square units (50 mm by 50 mm). In some cases, studies were done on these units and in other cases studies were done on 4 mm-thick multilayered laminates, prepared by stacking 20 units and thermocompressing them (54 MPa, 100°C).

Heat Treatments

Heat treatments were tailored, taking into account three conditions. First, the initial heating rate must be low, to burn out the organic components in a progressive manner. Second, the final heating rate up to the sintering temperature must be rather fast. Indeed, the apparent activation enthalpy for densification seems to be higher than the corresponding term for mullitization.[21,24] Therefore, a fast heating rate and a high temperature of firing are required to delay mullitization in comparison to densification. Third, the firing temperatures must not exceed 1550–1650°C to keep the production cost of mullite ceramics at the same level as alumina ceramics.

Details on heat treatments will be given below (see: results and discussion—influence of heating rate). In all cases, firing was done using an electric furnace with $MoSi_2$ heating elements, in air.

Characterization

The quantitative determination of crystalline phases was done by comparing the areas under the X-ray peaks of ground samples with those of mechanical mixtures of individual components (α-alumina, α-quartz, β-cristobalite, and mullite) with the same mean particle size as the ground samples.[21] The quantitative determination of amorphous silica was done by etching ground samples with hydrofluoric acid. Microstructural observations were done by SEM on samples polished and thermally etched (at 100°C below the firing temperature). The bulk densities were measured by Archimede's method, and the results were compared with data obtained from mixtures of powders of individual components, assuming a rule of mixtures. The apparent densities were obtained from sample dimensions. Shrinkage curves were recorded on a differential dilatometer. Elastic moduli were determined by ultrasonic techniques. Mechanical strength was determined by biaxial flexure of discs,[25] in the as-sintered state, without any polishing. Finally, toughness (K_{1c}) was measured on single-edge-notched-beam samples (25 mm by 4 mm by 3 mm).

Results and Discussion

Influence of Particle Size

Dilatometry and XRD experiments showed that there are three broad stages in the reaction-sintering of 3/2 Al_2O_3-SiO_2 mixtures:

1. For temperatures < 1250–1300°C, densification begins to develop in a pre-reacted state, i.e. in bodies composed of amorphous silica and alumina (pressed samples) or of quartz and alumina (tape-cast samples).
2. The shrinkage rate slows for temperatures from 1300°C to 1400°C, which is the domain where vitreous silica and quartz recrystallize into β-cristobalite. For quartz-containing mixtures, this behaviour could be explained by the fact that the expansion associated with the quartz-to-cristobalite transformation counterbalances the pore elimination. However, for amorphous silica-containing mixtures, the recrystallization into cristobalite corresponds to an overall contraction, which could have been thought to assist densification. Among the possible reasons for the decrease in the densification rate, one could suggest the formation of microflaws around the silica particles, which reduces the number of particle-to-particle contacts, or the elimination of amorphous silica, which favours densification by viscous flow.
3. For temperatures > 1400°C, mullite develops. The shrinkage rate of the sample is low, and an overall expansion is even sometimes observed. It has been verified that no porosity develops, therefore the expansion is due to the high specific volume of mullite in comparison with the 3/2 mixture (0.316 instead of 0.302 $cm^3.g^{-1}$).

Figure 1 (from Ref. 8) shows the densification curves and the XRD peaks corresponding to samples pressed from powders of different particle sizes. The reduction in the particle size leads to a slight decrease in the temperature of mullitization but to a noticeable increase in the rate of densification. This is in agreement with Yangyun and Brook's view that fine particles favour densification in comparison with mullitization, and confirms that the preparation of mullite ceramics by reaction-sintering requires fine (submicron) powders. Other experiments

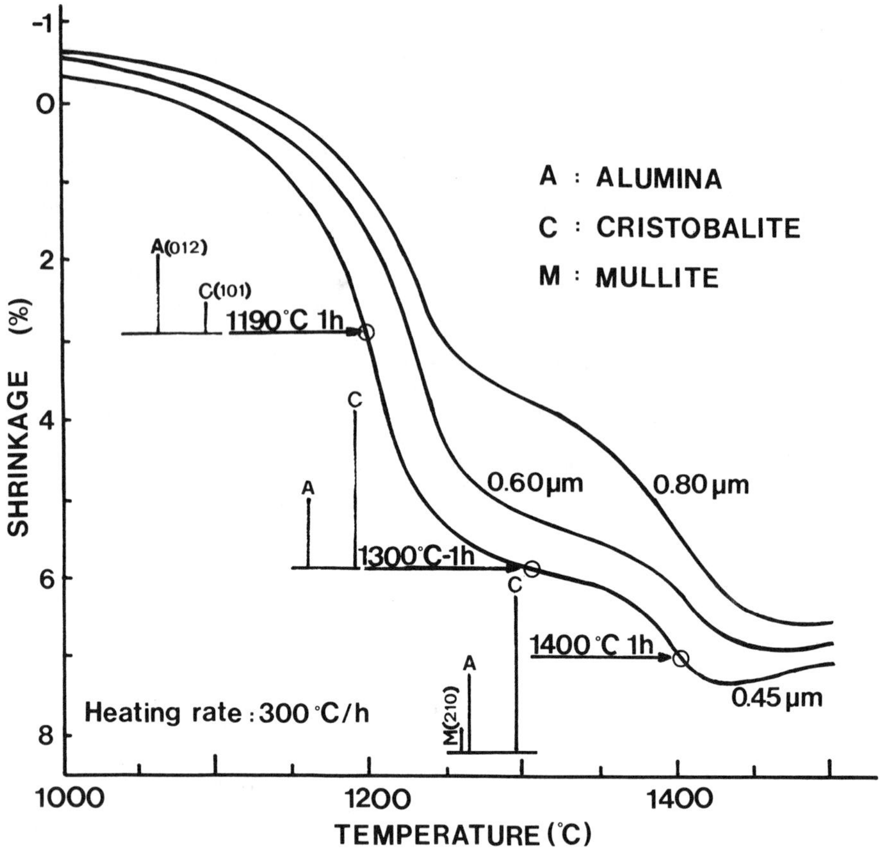

Fig. 1. Shrinkage curves and XRD peaks vs. temperature for three 3/2 Al_2O_3-SiO_2 mixtures of different fineness: 0.80, 0.60, and 0.45 μm (from Ref. 8).

on pressed or cast samples, for mullite as well as for ZrO_2-mullite, have also confirmed this point.

Influence of Heating Rate

The influence of heating rate has been discussed in Ref. 8 and 21 for pressed samples containing amorphous SiO_2 and in Ref. 24 for tape-cast samples containing crystalline quartz. The results are as follows:

1. A separation of an early densification stage from that of subsequent mullitization was never obtained, contrary to what was obtained by Claussen and Jahn in the case of ZrO_2-mullite.[26] It is believed that such a separation requires the use of ultrafine powders—below the size which was used here—and the presence of additives or impurities.

2. Low heating rates are detrimental to densification. A comparison between two firing cycles (A and B) with the same final temperature (1600°C for 2 h) but different heating rates (30°C/min from 1300 to 1600°C for A and 1°C/min for B) shows that both cycles yield fully mullitized samples. However, the relative density was lower for samples fired with cycle B (d = 90%) than for those fired with cycle A (d = 94%), although the former samples were exposed to temperatures over 1300°C for nearly 5 h more than the latter ones. Upon reaching 1600°C, the relative content of mullite was 50% and 98%, and the relative density was 83% and 77% for "A" samples and "B" samples, respectively. Thus faster heating rates, where sintering develops in the pre-reacted and partially-reacted states must be preferred to lower heating rates, where sintering develops in the partially-reacted and fully-reacted states. One reason for this is the low sinterability of mullite, as previously mentioned. Another possible reason could be the presence of amorphous silica residue in mixtures rapidly heated to high temperatures without sufficient time for complete recrystallization of silica into cristobalite. The silica residue would be in the liquid state (softening point at about 1550°C), which would help densification by viscous flow mechanisms.
3. Two-step firing cycles can improve densification in comparison with one-step cycles. Table 3 (from Ref. 24) give details on firing cycles and materials properties, respectively. Among the results:

- an increase in the final temperature from 1600 to 1620°C improves both densification and mullitization, but a further increase up to 1650°C is detrimental to densification, which could be related to the burning out of organic residues.[27,28] A bimodal microstructure (small equiaxed grains plus large elongated grains) can develop for the highest temperatures of firing,
- the choice of the intermediate temperature of the two-step cycle is rather critical. Comparison of dilatometry and XRD data suggests that the point is to promote densification before those reactions (transformation of quartz into cristobalite and/or mullitization) begin to develop.

The best densification was obtained for two-step heating (from room temperature to 1350°C at 20°C/min, then from 1350 to 1620°C at 60°C/min, with a soaking time of 10 h at 1620°C). Samples were fully mullitized, with a relative density of about 99%, which allows them to be translucent (Fig. 2).

Influence of the Al_2O_3 to SiO_2 Ratio

Experimental

Mullite with 3/2 stoichiometry contains 71.8 wt% Al_2O_3, but the mullite solid solution extends from about 70.5 to about 74.8 wt% Al_2O_3. Moreover, it is interesting to study the influence of a small excess of silica or of alumina beyond the solubility range. Therefore, five compositions were considered, namely: 68.0, 71.8, 75.0, 77.3, and 80.0 wt% Al_2O_3. Samples were prepared by pressing using the same procedure as for 3/2 mullite. Starting materials were RC172 DBM Reynolds α-alumina and Alfa Chemicals vitreous silica. Firing temperatures were 1570 and 1600°C, with soaking times of 1 min, 30 min, 1.5 h, 4 h, and 10 h. Heating rates were 30°C/min from room temperature to 1300°C, then 60°C/min up to the firing temperature.

Table 3. Firing cycles and materials properties

Heating rate and max.temp. First stage	Heating rate and max.temp. Second stage	Phases (wt%)	Apparent density	Microstruct. aspect
10°C/min 1300°C	60°C/min 1600°C 10 h	m 93.9 a 4.4 c 1.7	94.1%	Equiaxed grains
10°C/min 1300°C	60°C/min 1620°C 10 h	m 100	96.8%	Equiaxed and elongated grains
10°C/min 1350°C	60°C/min 1620°C 10 h	m 100	93.6%	Equiaxed grains
20°C/min 1350°C	60°C/min 1620°C 10 h	m 100	98.9%*	Equiaxed and elongated grains

(m = mullite ; a = alumina ; c = cristobalite)
*translucent sample

Fig. 2. Translucent mullite sample (thickness: 0.2 mm) produced by tape-casting and firing according to cycle F (Table 3).

Results and Discussion

Phase Composition

A quantitative study of the chemical composition of the crystalline phases has been carried out by XRD.[6,21] Three samples were tested simultaneously in a Huber Guinier chamber operating on the basis of Seeman-Bohlin focusing. $CuK_{\alpha 1}$ peak positions were determined to ±0.0025° relative to an α-alumina standard. Using all mullite peaks which do not superimpose with other crystalline phases (alumina or cristobalite), the unit-cell parameters were obtained and refined using a computer program. The study was carried out using data on the relative quantities of alumina, cristobalite, and mullite, as well as data on the variation of the cell parameters of mullite with its Al_2O_3 content (the "a" and "c" parameters being the most sensitive to the changes in composition).[29,30] When amorphous phases can exist, which is the case below the silica-rich end of the range of mullite solid solution, a chemical etch was conducted using a 10% hydrofluoric solution to leach amorphous silica. Using a balance of high sensitivity (0.0001 g) the weight of a ground sample immersed in HF solution was recorded at different time intervals. The weight of a 75% Al_2O_3 sample did not change significantly during 3 h in 10% HF solution, whereas the corresponding curve for a 68% Al_2O_3 sample indicated a noticeable leaching (about 3 wt%).

This study on phase composition has been detailed in Ref. 21. Its main results are:

1. The solubility range of Al_2O_3 in mullite extends from 70.4 wt% to 74.8 wt%, which is very close to the figures given by Aksay and Pask (70.5 to 74 wt%).[1] An excess of amorphous silica is found in samples with less than 70.4% Al_2O_3 and an excess of crystalline corundum is found in samples with more than 74.8% Al_2O_3.
2. For bulk samples with the 68% Al_2O_3 composition, the silica residue does not recrystallize into cristobalite during subsequent firing (1350 or 1400°C, for 2 days). However, recrystallization is obtained when heat treatments are performed on ground samples. The effect of grinding may be the facilitation of heterogeneous nucleation on surfaces.
3. The measurements of the "a" and "c" parameters of the mullite cell show that mullite begins to develop with a composition close to 70.4% Al_2O_3 (i.e. the silica-rich end of the stable mullite solid solution), regardless of the bulk composition of the starting mixture. The composition of mullite remains at this constant value as long as all silica is not consumed. When all silica is consumed, the remaining alumina (for mixtures with more than 70.4% Al_2O_3) progressively dissolves into mullite, whose composition can go up to 74.8% Al_2O_3 (i.e. the alumina-rich end of the stable mullite solid solution).

Densification

The densification kinetics are sensitive to the alumina-to-silica ratio (Fig. 3). Below about 1350°C, the shrinkage rate is higher for the silica-rich mixtures, which can be due to the densification of the silica zones (viscous flow). Above 1350°C, the shrinkage rate is maximum for the 3/2 stoichiometric mullite and minimum for the 75% Al_2O_3 composition. The final density is also sensitive to the alumina-to-silica ratio: the highest density is observed in 3/2 mullite whereas the lowest is observed in 75% Al_2O_3 samples (Fig. 4).

Microstructure

The microstructure depends on the alumina to silica ratio. For 3/2 mullite, the microstructure consists of small equiaxed grains. For 75% Al_2O_3, it consists of a mixture of small equiaxed grains (5 by 5 μm) with large, elongated grains (7 by 30 μm).

A quantitative microstructural study[31] was carried out using a procedure described by Shaw and Brook.[32] Specific surface area density (Sv^P) was evaluated by counting surface intersections (P_l), using a unit length of random line on the polished sections ($Sv^P = 2 P_l$). Sv was also evaluated from porosity measurements ($S'v = 6/\pi D$ with π being the porosity percentage and D the mean diameter of pores,

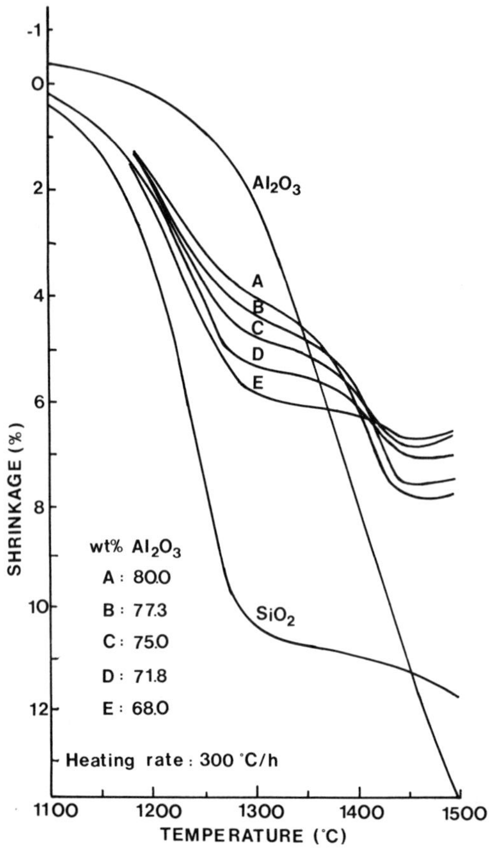

Fig. 3. Relative shrinkage vs. temperature for Al_2O_3/SiO_2 mixtures of various compositions (from Ref. 24).

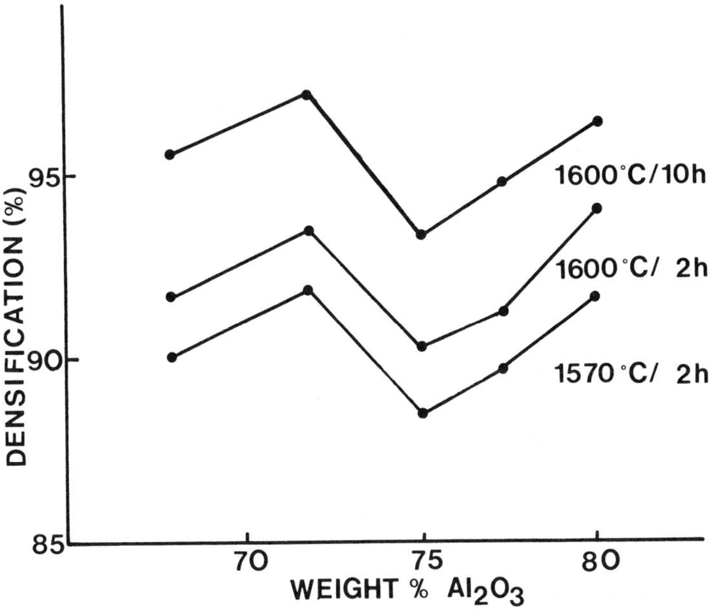

Fig. 4. Densification vs. composition for various Al_2O_3/SiO_2 mixtures (from Ref. 24).

assumed to be spherical). Specific grain-boundary area density (Sv^{gb}) was evaluated by counting grain-boundary intersections with random lines on polished sections. The coarsening-to-densification ratio can thus be evaluated.[33,34] The plot of pore surface area vs. instantaneous density gives a trajectory lying between pure coarsening (fall of Sv from its initial value toward zero with density remaining constant) and pure densification (increase from the initial density to the full density, where Sv is zero). For the grain-boundary area, the curve of Sv^{gb} vs. instantaneous density begins at a point given by the initial density and zero grain-boundary area, then rises, and finally decreases to the value for the grain-boundary area in the dense sample.

As far as the specific surface area is concerned, both 71.8 and 75% Al_2O_3 compositions exhibit similar behaviour. Figure 5 shows that Sv^P and $S'v^P$ continuously decrease when density increases. This suggests that the mechanisms which act to reduce porosity are similar in both compositions. However, the specific surface area at a given density is always lower in the 75% Al_2O_3 composition than in the 3/2 composition, which indicates that pores are more numerous but smaller in this latter composition.

As far as the grain-boundary surface area is concerned, the 71.8 and 75% Al_2O_3 compositions behave in a different manner (Fig. 6):

1. For the 3/2 composition, there are two stages. Up to a relative density of about 90%, sintering develops at the same time as Sv^{gb} increases. The grain-boundary area reaches a maximum of about $1.3 \times 10^6 m^2.m^{-3}$ for $d = 90\%$ and then decreases, which indicates that grain coarsening predominates.

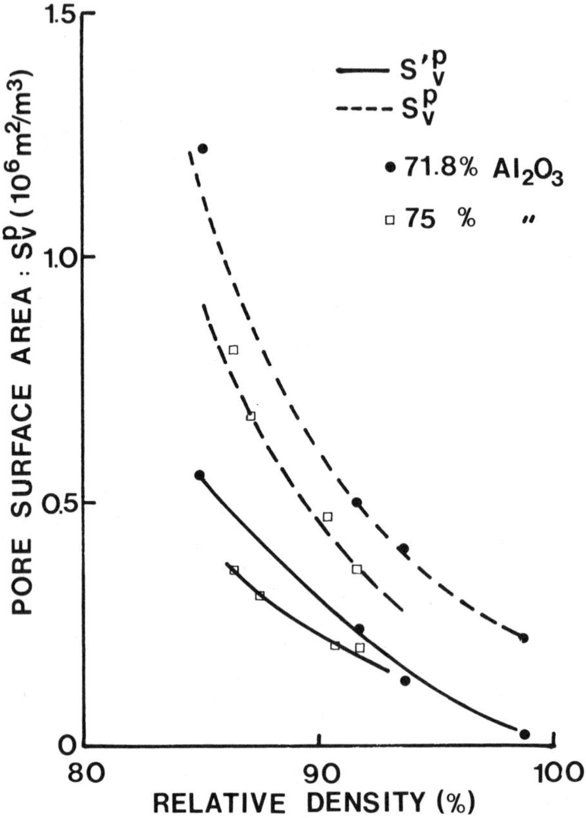

Fig. 5. Specific surface area density vs. relative density for 71.8 and 75 wt% Al_2O_3 compositions (Sv^P: stereological evaluation, $S'v^P$: porosimetry measurements).

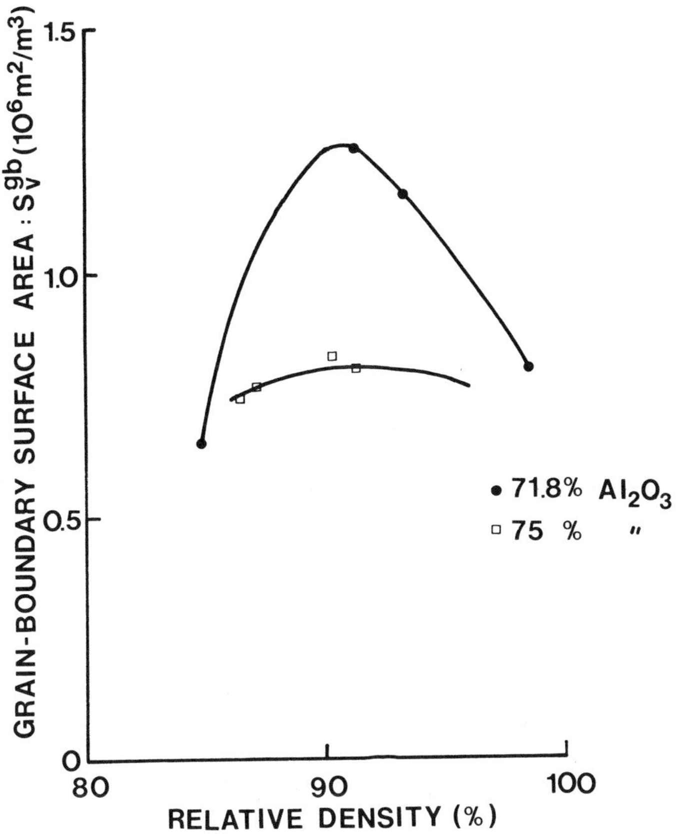

Fig. 6. Specific grain-boundary area density vs. relative density for 71.8 and 75 wt% Al_2O_3 compositions.

2. For the 75% Al_2O_3 composition, the grain-boundary surface area keeps a rather constant value of about $0.7 \times 10^6 m^2.m^{-3}$ in the whole density range. This indicates that grain coarsening develops in the early stage of sintering, which slows densification.

Figures 7, 8, and the micrographs in Fig. 9 confirm that the initial densification rate and the initial grain growth rate are faster in the 75% Al_2O_3 composition than in the 3/2 composition. Therefore, pores are trapped in intragranular positions, which prevents them from further elimination, and exaggerated grain growth develops. This behaviour can be compared with what is observed in MgO-free alumina.[35] According to Brook's theory,[36] a coarse microstructure with large grains is a result of the separation of pores from grain boundaries. Moreover, the microstructural changes during sintering depend on the relative values of mobility parameters, in particular lattice diffusion, grain-boundary diffusion, surface diffusion, pore mobility, and grain-boundary mobility.

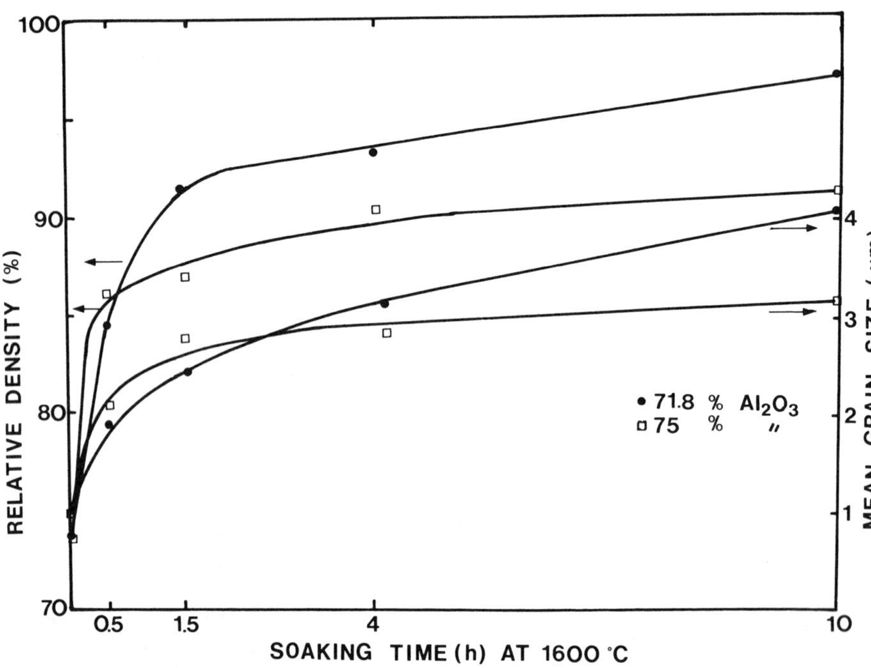

Fig. 7. Densification vs. time (sintering at 1600°C) for 71.8 and 75% Al_2O_3 compositions and grain growth vs. time (sintering at 1600°C) for 71.8 and 75% Al_2O_3 compositions.

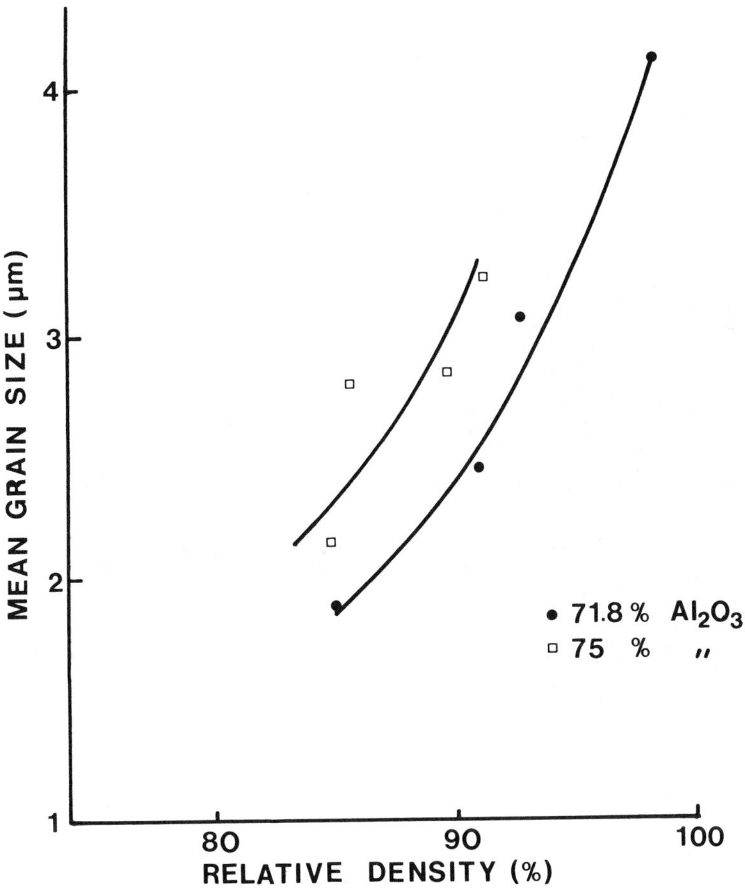

Fig. 8. Grain growth vs. relative density (sintering at 1600°C) for 71.8 and 75% Al_2O_3 compositions.

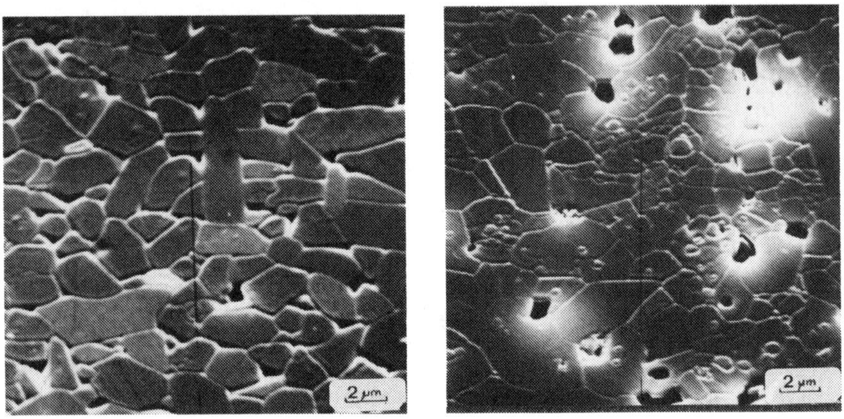

Fig. 9a. Microstructures of 71.8 and 75% Al_2O_3 compositions after sintering at 1600°C for 10 h (photos 1 and 2).

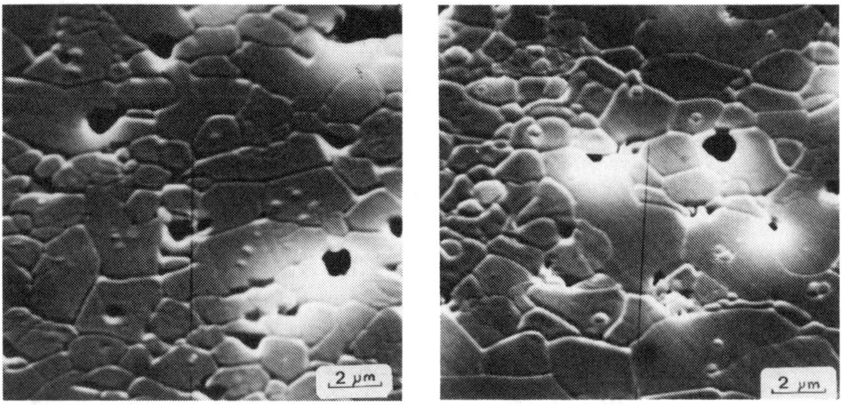

Fig. 9b. Microstructures of 71.8 and 75% Al_2O_3 compositions, for a densification of 91.5% (photos 3 and 4).

Exaggerated grain growth was observed in the product containing high alumina mullite and a small amount (1 vol%) of α-corundum precipitates, whereas such grain growth was not observed in the product which contained low-alumina mullite and a significant amount (5 vol%) of a silica-rich phase. Therefore, the abnormal grain growth of high-alumina mullite cannot be attributed to the role of a grain-boundary liquid phase. The crystal lattice structure of the mullite solid solution must undergo some modifications as its alumina-to-silica ratio varies. Indeed, the amounts of Si^{+4} and Al^{+3} cations are modified which leads to some perturbations in the crystallographic structure, and therefore in the mobility parameters. On this basis, and in the light of the information concerning the effect of MgO on the grain growth of alumina, we could consider the variation of diffusion mechanisms with composition as a possible reason for the exaggerated grain growth in the high-alumina mullite. In addition, the change in the mullite composition, which varies from 70.4% Al_2O_3

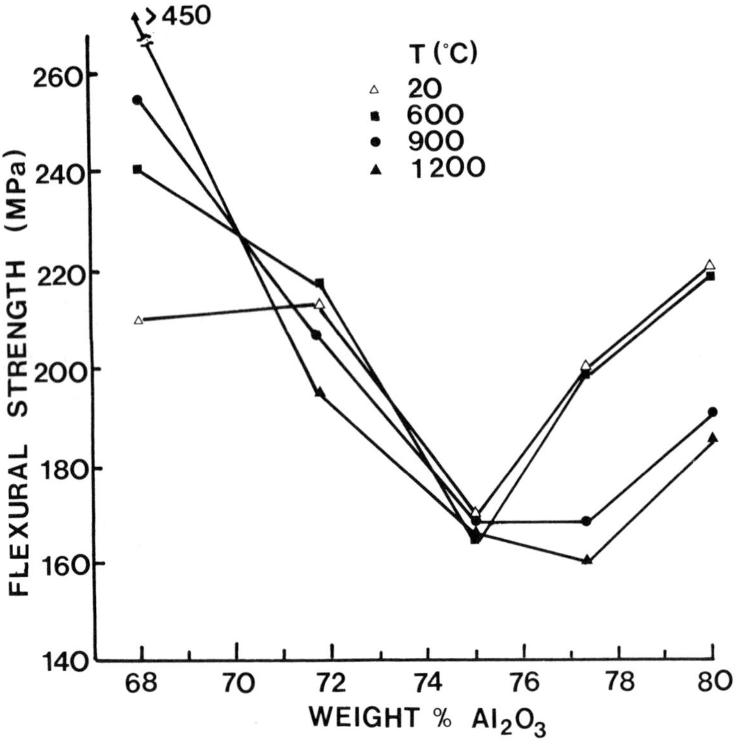

Fig. 10. Flexural strength vs. composition at 20, 600, 900, and 1200°C, for reaction-sintered (1600°C for 10 h) mullite ceramics.

to its final composition, could be involved. Further studies are needed to elucidate this point.

Mechanical Properties

Figure 10 shows the relationship between chemical composition and R.T. flexural strength of reaction-sintered mullite ceramics at 20, 600, 900, and 1200°C. The plot of strength vs. composition has a form similar to that of the plot of densification vs. composition (Fig. 4). The 75% Al_2O_3 composition, which has the lowest densification, exhibits the lowest strength, whereas the SiO_2-rich (68% Al_2O_3) and Al_2O_3-rich (80% Al_2O_3) compositions of identical densification (96%) exhibit almost identical strengths. The close relation between densification and strength exists even at 600°C. Hence, the low temperature strength of the studied materials depends more on their densification than on their composition.

The situation is different for the high-temperature behaviour. The strength of the 71.8% Al_2O_3 composition, which has neither free silica nor free alumina, does not significantly vary from R.T. to 1200°C. It is the same for the 75% Al_2O_3 composition, which has a small amount (1%) of free alumina. In contrast, the strength of the Al_2O_3-rich and SiO_2-rich compositions show significant variations with temperature. An increase in temperature from 600 to 1200°C causes a decrease of 15 to 20% in the strength of the alumina-rich materials, whereas it causes a dramatic increase in the

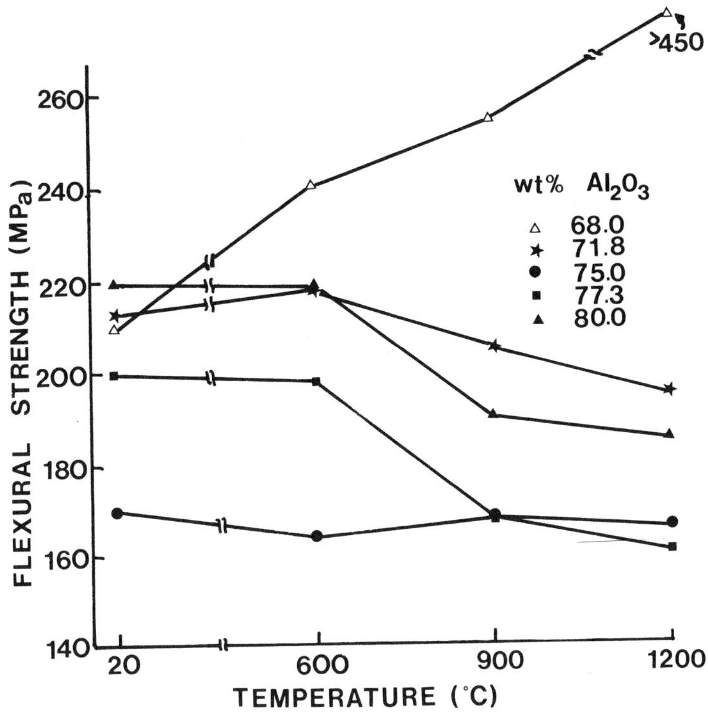

Fig. 11. Flexural strength vs. temperature for reaction-sintered (1600°C for 10 h) mullite ceramics of 68, 71.8, 75, 77.3, and 80% Al_2O_3 compositions.

strength of the 68% Al_2O_3 composition (450 MPa at 1200°C) (Fig. 11). This behaviour must be related to the presence of an excess of silica in the 68% Al_2O_3 ceramics. At room temperature, the viscosity of vitreous silica is very high and the behaviour is purely brittle. As temperature increases, viscosity decreases and micro-plasticity develops. This relaxes the stresses at crack tips and can allow some dissipation mechanisms to act.[37,38] The main parameter here is the value of the viscosity. When it is high enough, which is the case for high-purity silica, strength is improved even at rather high temperatures. When it is low, which is the case for zirconia-toughened mullite prepared from natural zircon—see related paper by Boch, Chartier and Giry in this publication—the creep rate is fast and the high-temperature strength is low.

Literature data[5,38–47] on mechanical properties of mullite give strength values varying from about 150 MPa[38] to more than 450 MPa for hot-pressed materials produced from sol-gel powders.[46] Typical values for chemically-prepared mullites are close to 400 MPa,[43,45,47] but low values (less than 150 MPa) can be found in some very pure, dense, hot-pressed mullites.[5] Toughness varies from less than 2 MPa√m to more than 3 MPa√m, whereas the results of the present study are close to 2 MPa√m. The increase in strength with temperature and the beneficial influence of a small excess of silica have been observed.[5,45] The present R.T. strength values of about 200 MPa may be considered as rather satisfactory, taking into account that the properties of ceramics prepared by reaction-sintering of commercial powders, with a firing temperature of about 1600°C, cannot be expected to compete with those of materials produced by hot-pressing or from chemically-prepared powders. Besides, the high-temperature mechanical strength of reaction-sintered mullites is at the same level as that of chemically prepared materials, which confirms the potential of reaction-sintered mullite ceramics for high-temperature applications. Finally, it must be pointed out that ZrO_2-toughened mullite ceramics could be expected to show improved mechanical properties as will be seen in the related paper noted above.

Conclusion

The main conclusions of this study can be summarized as follows:

1. Nearly dense (>97%) mullite ceramics can be prepared by reaction-sintering of mixtures of alumina and silica, for firing temperatures below or equal to 1600°C, i.e. for conditions close to those which are used to sinter "reactive" aluminas.
2. Alpha alumina and vitreous silica are the best starting materials for the reaction-sintering of high-purity mullites.
3. The alumina content of the stable mullite solid-solution varies from about 70.5 to 74.8 wt% Al_2O_3.
4. Mullite always begin to form with a composition close to the silica-rich end (i.e. 70.5 or 71.8% Al_2O_3) whatever the bulk composition is. The composition later moves toward the final one.
5. Excessive grain growth develops in the 75% Al_2O_3 materials. Quantitative microstructural observations show this behaviour can be compared with that of alumina in the absence of MgO additives.
6. The alumina-to-silica ratio affects the mechanical properties. The silica-rich material (68% Al_2O_3) exhibits the best mechanical properties. In particular, its strength continuously increases from room temperature to more than 1200°C.

References

[1]I. A. Aksay and J. A. Pask, "Stable and Metastable Equilibria in the System Al_2O_3-SiO_2," *J. Am. Ceram. Soc.*, **58** [11–12] 507–12 (1975).

[2]S. H. Risbud and J. A. Pask, "SiO_2-Al_2O_3 Metastable Phase Equilibrium Diagram Without Mullite," *J. Mat. Sc.*, **61** [1–2] 2449–54 (1978).

[3]Y. Nurishi and J. A. Pask, "Sintering of α-Al_2O_3-Amorphous Silica Compacts," *Ceram. Inter.*, **8** [2] 57–59 (1982).

[4]A. P. S. Rana, O. Aiko, and J. A. Pask, "Sintering of α-Al_2O_3/Quartz and α-Al_2O_3/Cristobalite Related to Mullite Formation," *Ceram. Inter.*, **8** [4] 151–153 (1982).

[5]T. Il Mah and K. S. Mazdiyasni, "Mechanical Properties of Mullite," *J. Am. Ceram. Soc.*, **66** [10] 699–703 (1983).

[6]P. D. D. Rodrigo and P. Boch, "High Purity Mullite Ceramics by Reaction Sintering," *Int. J. High Technology Ceramics*, [1] 3–30 (1985).

[7]P. Boch, J. P. Giry, and P. D. D. Rodrigo, "Reaction Sintering of Mullite and Zirconia-Mullite Ceramics" pp. 307–14 in Ceramic Materials and Components for Engines. Edited by W. Bunk and H. Hausner, Deutsche Keramische Geselleschaft, Bad Honnef FRG (1986).

[8]P. Boch, J. P. Giry, and P. D. D. Rodrigo, "Reaction Sintering of Mullite-Based Ceramics," Workshop on Advanced Ceramics, Tokyo Institute of Technology, Tokyo, September 4–5 (1986), to be published by Elsevier Applied Science.

[9]J. M. Rincon, G. Thomas, and J. S. Moya, "Microstructural Study of Sintered Mullite Obtained from Premullite," *J. Am. Ceram. Soc.*, **69** [2] C29–C31 (1986).

[10]S. Prochazka and F. J. Klug, "Infrared-Transparent Mullite Ceramic," *J. Am. Ceram. Soc.*, **66** [12] 874–80 (1983).

[11]I. A. Aksay, "Densification of Mullite via Colloidal Processing: a Review," First International Workshop on Mullite, Tokyo, November 9–10 (1987).

[12]K. S. Mazdiyasni, "Preparation of Mullite Powders From Alkoxides and Other Chemical Routes," Ibid.

[13]S. Somiya, M. Yoshimura, M. Suzuki, and T. Yamagushi, "Hydrothermal Synthesis of Mullite," Ibid.

[14]Y. Hirata, I. A. Aksay, R. Kurita, S. Hori, and H. Kaji, "Processing of Mullite With Powders Processed by Chemical Vapor Deposition," Ibid.

[15]S. Komarneni and R. Roy, "Mullite Derived from Diphasic Gels," Ibid.

[16]S. Kankazi, "Sintering and Mechanical Properties of Mullite Derived From Spray Pyrolysis," Ibid.

[17]B. B. Ghate, D. P. H. Hasselman, and R. M. Spriggs, "Kinetics of Pressure-Sintering and Grain Growth of Ultra Fine Mullite Powder," *Ceram. Inter.*, **1**, 105–110 (1975).

[18]P. C. Dokko, J. A. Pask, and K. S. Mazdiyasni, "High Temperature Mechanical Properties of Mullite Under Compression," *J. Am. Ceram. Soc.*, **60** [3–4] 150–55 (1977).

[19]M. D. Sacks and J. A. Pask, "Sintering of Mullite," pp. 193–203, in Processing of Crystalline Ceramics. Edited by H. Palmour, R. F. Davis, and T. M. Hare. Plenum Press, New-York USA (1978).

[20]I. A. Aksay, "Diffusion and Phase Relationship Studies in the Alumina Silica System," PhD Thesis, Univ. of California, Berkeley (April 1973).

[21]P. D. D. Rodrigo, "Reaction-Sintering of Mullite-Based Ceramics," Ph.D. Thesis, Univ. of Limoges, May 1986.

[22]S. Yangyun and R. J. Brook, "Preparation of Zirconia-Toughened Ceramics by Reaction-Sintering," *Science of Sintering*, **17**, 1, 35–47 (1985).

[23]B. Rogeaux and P. Boch, "Influence of an Ultrasonic Assistance to the Powder Compaction on the Weibull Modulus of Sintered Alumina," *J. Mat. Sc. Lett.*, **4**, 403–4 (1985).

[24]T. Chartier and P. Boch, "Preparation of Mullite Ceramics by Tape Casting and Reaction-Sintering," Science of Ceramics n•14, Canterbury, September 7-9 (1987).

[25]J. C. Glandus, "Meaning of the Biaxial Flexure Test of Discs for Strength Measurements," Science of Ceramics n•13, Les Editions de Physique, Paris France, 595–601 (1986).

[26]N. Claussen and J. Jahn, "Mechanical Properties of Sintered, in Situ-Reacted Mullite-Zirconia Composites," *J. Am. Ceram. Soc.*, **63** [3–4] 228–29 (1980).

[27]S. J. Bennison and M. P. Harmer, "Swelling of Hot-Pressed Al_2O_3," *J. Am. Ceram. Soc.*, **68** [11] 591–97 (1985).

[28]T. Chartier, J. L. Besson, and P. Boch, "Mechanical Properties of ZrO_2-Al_2O_3 Laminated Composites," to be published in the proceedings of "Zirconia 86," Am. Ceram. Soc. publ.

[29]W. E. Cameron, "Composition and Cell Dimension of Mullite," *Am. Ceram. Soc. Bull.*, **56** [11] 1003–11 (1977).

[30]P. Das and R. Choudhury, "Further X-ray Diffraction Studies on Mullite Cell Dimensions," *Interceram*, **4**, 22–23 (1987).

[31]E. E. Underwood, A. R. Colcord, and R. C. Waugh, "Quantitative Relationships for Random Microstructure," pp. 25–70 in Ceramic Microstructures. Edited by R. M. Fulrath and J. A. Pask, Wiley, New-York USA (1968).

[32]N. J. Shaw and R. J. Brook, "Structure and Grain Coarsening During the Sintering Of Alumina," *J. Am. Ceram. Soc.*, **69** [2] 107–10 (1986).

[33]M. F. Yan, R. M. Cannon, and H. K. Bowen "Grain Boundary Migration in Ceramics," pp. 276–307 in Ceramic Microstructures '76. Edited by R. M. Fulrath and J. A. Pask, Westview Press, Boulder USA (1977).

[34]J. E. Burke, K. W. Lay, and S. Prochazka, "The Effect of MgO on the Mobility of Grain Boundaries and Pores in Aluminum Oxide," *Mater. Sci. Res.*, **13**, 417–25 (1980).

[35]K. A. Berry and M. P. Harmer, "Effect of MgO Solute Development in Al_2O_3," *J. Am. Ceram. Soc.*, **69** [2] 143–49 (1986).

[36]R. J. Brook, "Controlled Grain Growth," pp. 331–64 in Treatise on Materials Science and Technology, Vol. 9. Edited by F. F. Y. Wang, Academic Press, New-York USA (1976).

[37]C. R. Cheeseman and G. W. Groves, "The Mechanism of the Peak in Strength and Toughness at Elevated Temperatures in Alumina Containing a Glass Phase," *J. Mat. Sc.*, **20**, 2614–22 (1985).

[38]K. S. Mazdiyasni and L. M. Brown, "Synthesis and Mechanical Properties of Stoichiometric Aluminium Silicate (Mullite)," *J. Am. Ceram. Soc.*, **55** [11] 548–52 (1972).

[39]B. L. Metcalfe and J. H. Hant, "The Synthesis, Microstructure and Physical Properties of High Purity Mullite," *Trans. Brit. Ceram. Soc.*, **74** [6] 193–201 (1975)

[40]S. Kankazi and H. Tabata, "Sintering and Mechanical Properties of Stoichiometric Mullite," *J. Am. Ceram. Soc.*, **68** [1] C6–C7 (1985).

[41]S. Somiya, "Mullite," Uchida Rokakuho Publ. Co., Tokyo Japan (1985).

[42]S. Kankazi, T. Kumazawa, J. Asaumi, O. Abe, and H. Tabata, "Dependency of Mechanical Properties of Sintered Mullite on Chemical Composition," *Yogyo-Kyoshai-Shi*, **93** [7] 407–408 (1985).

[43]M. G. M. U. Ismail, Z. Nakai, K. Minegishi, and S. Somiya, "Synthesis of Mullite Powder and its Characteristics," *Int. J. High Technology Ceramics*, **2** [2] 123–134 (1986).

[44]T. Sato, M. Ishizuka, and M. Shimada, "Sintering and Characterization of Mullite-Alumina Composites," *Ceram. Int.*, **12**, 61–65 (1986).

[45]S. Kanzaki, O. Abe, M. Onashi, and H. Tabata, "Mechanical Properties of Pressureless Sintered Mullite Ceramics," pp. 625–32 in Ceramic Materials and Components for Engines. Edited by W. Bunk and H. Hausner, Deutsche Keramische Gesselleschaft, Bad Honnef FRG (1986).

[46]H. Onishi, T. Kawanami, K. Miyazaki, and T. Hiraiwa, "Mechanical Properties of Mullite," Ibid. pp. 633–42.

[47]M. G. M. U. Ismail, Z. Nakai, and S. Somiya, "Microstructure and Mechanical Properties of Mullite Prepared by the Sol-Gel Method," *J. Am. Ceram. Soc.*, **70** [1] C7-C8 (1987).

Formation Process of Mullite

Kiyoshi Okada* and Nozomu Ōtsuka

Department of Inorganic Materials
Tokyo Institute of Technology
O-okayama, Meguro, Tokyo 152, Japan

Abstract

Formation processes of mullite from various starting materials were summarized, and the observed differences of their routes were interpreted with respect to their structural differences. All the starting materials crystallized from amorphous state around 1000°C firing. Difference in crystalline phase could be interpreted by the degree of mixing of SiO_2 and Al_2O_3 components in the starting materials which corresponded to the number of Si-O-Al linkages. They were divided into three groups: (1) Mullite directly crystallized from amorphous state, starting materials have infinitely mixed state of SiO_2 and Al_2O_3 components; (2) Spinel phase crystallized before crystallization of mullite, starting materials have rather poor mixing state; and (3) Both spinel and mullite phases almost simultaneously crystallized from amorphous state, starting materials have intermediate mixing state. Kaolin group minerals were one of the good examples. Mullite formed around 1000°C from the starting materials in type (1). Type (3) corresponded to primary mullite which was always richer in Al_2O_3 than $2SiO_2 \cdot 3Al_2O_3$ irregardless of the starting compositions.

Introduction

Many studies on mullitization from various starting materials have been reported. Effects of impurities, firing atmosphere, and some kinds of pre-treatment, i.e., compaction, NaOH treatment, grinding, were also examined, especially in the mullitization of kaolin. The results are summarized as follows: (1) Mullite crystallized directly from amorphous state, (2) Spinel phase crystallized from amorphous state and transformed to mullite at higher firing temperatures, (3) Both spinel and mullite phases simultaneously crystallized from the amorphous state to transformed completely to mullite with heating to higher temperatures. Mullitization temperature, therefore, widely differed due to the different formation processes.

Mullite in transitional state was first reported by Brindley and Nakahira,[1] which was considered to be $SiO_2 \cdot Al_2O_3$ composition. Schüller and Kromer[2] reported a scaly appearance for mullite formed by a reaction in the solid state, and needles for mullite crystallized in the melt. They called the former "primary" and the latter "secondary."

---

*Member, The American Ceramic Society.

Recently, some workers[3,4,5] reported that the composition of primary mullite was richer in Al_2O_3 than $2SiO_2{\cdot}3Al_2O_3$ from the data of analytical electron microscopy (ATEM) and lattice constants (LC). Okada and Ōtsuka[6] showed that mullite which formed around 1000°C from any starting materials always had a higher Al_2O_3 content than $2SiO_2{\cdot}3Al_2O_3$ independent of the bulk composition, and gradually changed to $2SiO_2{\cdot}3Al_2O_3$ as the firing temperature increased.

In this paper, reports on mullitization from various starting materials are summarized, and the differences in their formation processes are discussed, based on the structure of starting materials. Chemical composition of formed mullite is also discussed.

Mullitization

Papers concerning mullitization from various starting materials are summarized in Table 1. These starting materials were divided into several groups: (1) mixtures of sols or oxides, (2) precipitates from solutions dissolving various kinds of reagents, (3) glassy materials and xerogels prepared by melting, gelation or spray pyrolysis methods, and (4) clay minerals. Various mullitization routes can be understood by considering mixing states of SiO_2 and Al_2O_3 components in the starting materials.

In the case of mixtures of sols or oxides, SiO_2 and Al_2O_3 particles reacted to form mullite at high firing temperature. The formation temperature changed depended mainly on their particle sizes. When the sol was composed of amorphous or low crystalline state materials, γ-Al_2O_3 crystallized at fairly low temperatures and then formed mullite: amorphous $\rightarrow$ γ-Al_2O_3 + amorphous $\rightarrow$ mullite. If mixing was poor and/or particles used were relatively large, unreacted portion crystallized to α-Al_2O_3 and cristobalite which remained up to fairly high firing temperatures.

In the case of the precipitation method, mixing states of two components in the precipitates depended largely on their precipitation conditions. There are many reports using alkoxides for the starting reagents of both components. It is, however, well known that the hydrolysis speeds of Al-alkoxides are much faster than those of Si-alkoxides. If these alkoxides in the solution were hydrolyzed by H_2O and/or NH_4OH solution, hydrolysis of Al-alkoxides preceded that of Si-alkoxides. If salts listed in Table 1 were used for the starting reagents, their solutions were acidic. Precipitation occurs by changing pH of the solution from acid to neutral. In this case, the Al_2O_3 component acted as a nucleus and accelerated the precipitation of SiO_2 component with Al_2O_3. Therefore, the precipitates were considered to be composed of agglomerated particles of very small $SiO_2{\cdot}nH_2O$ and $Al_2O_3{\cdot}mH_2O$ particles, or particles with a dual structure of an inner Al_2O_3-rich core with an outer SiO_2-rich surface. Schematic models for various starting materials are shown in Fig. 1. Only a small number of Si-O-Al linkages were considered to be present in these precipitates. They showed the following scheme for mullitization: amorphous $\rightarrow$ spinel phase + amorphous $\rightarrow$ mullite. Here, spinel phase was formed at around 900–1000°C and contained several wt% of SiO_2 in it as reported elsewhere.[10,40] Almost all of the starting materials prepared by precipitation showed a similar mullitization route, though one exception was reported by Ossaka.[7] He formed a solution with a very low concentration by dissolving 0.002 M Na_4SiO_4 and NH_4-alum and synthesized precipitate by homogeneous precipitation. This starting material showed direct mullitization at around 950°C firing. In this case, very low concentration and homogeneous precipitation method using $(CH_2)_6N_4$ was certainly an important factor to obtain infinitely mixed precipitates.

There have been reported some starting materials that show direct mullitization from the amorphous state. These starting materials are glasses prepared by the conventional melting method,[16,17] xerogels prepared by slowly hydrolyzing and gelling

No	Ref.	Si-source	Al-source	Cryst. Phase	Preparation
1	7	Na_4SiO_4	(NH_4)-alum	Mu	precipitated by $(CH_2)_6N_4$
2	8	$Si(OEt)_4$	$Al(NO_3)_3 \cdot 9H_2O$	Mu	gelation at 60°C
3	9	$Si(OEt)_4$	$Al(NO_3)_3 \cdot 9H_2O$	Mu	spray pyrolysis
4	10	$Si(OEt)_4$	$Al(NO_3)_3 \cdot 9H_2O$	Mu	gelation at 60°C
5	11	$Si(OEt)_4$	$Al(O^iPr)_3$	Mu	spray pyrolysis
6	12	$(EtO)_3Si\text{-}O\text{-}Al(OBu)_2$		Mu	gelation by HCl
7	13	Kaolinite		Mu	metakaolin pressed 250kb
8	13	Kaolinite		Mu	replaced OH by F
9	14	Kaolinite		Mu	with LiF
10	15	Pyrophyllite		Mu	
11	16	SiO_2	Al_2O_3	Mu	glass by melting
12	17	SiO_2	Al_2O_3	Mu	glass by melting
13	18	$Si(OMe)_4$	$Al(O^iPr)_3$	Sp	precipitated by NH_4OH
14	19	$Si(OMe)_4$	$Al(O^iPr)_3$	Sp	precipitated by NH_4OH
15	20	$Si(OEt)_4$	$Al(OEt)_3$	Sp	precipitated by $CO(NH_2)_2$
16	21	$Si(OEt)_4$	$Al(O^iPr)_3$	Sp	precipitated by H_2O
17	22	$Si(OEt)_4$	$Al(O^iPr)_3$	Sp	precipitated by H_2O
18	23	$Si(OEt)_4$	$Al(O^iPr)_3$	Sp	precipitated by HCl, H_2O
19	24	$Si(OEt)_4$	$Al(OBu)_3$	Sp	precipitated by HNO_3
20	25	$Si(OPr)_4$	$Al(O^iPr)_3$	Sp	precipitated by NH_4OH
21	10	$Si(OEt)_4$	$Al(NO_3)_3 \cdot 9H_2O$	Sp	precipitated by NH_4OH
22	19	$Si(OEt)_4$	$Al(NO_3)_3 \cdot 9H_2O$	Sp	precipitated by NH_4OH
23	26	$Si(OEt)_4$	$AlCl_3$	Sp	precipitated by NH_4OH
24	27	$Si(OEt)_4$	$Al_2Cl(OH)_5$	Sp	precipitated by NH_4Ac
25	12	Me_3SiOAc	$Al(O^iPr)_3$	Sp	gelation by H_2O
26	28	$SiAc_4$	$Al(NO_3)_3 \cdot 9H_2O$	Sp	dry
27	29	$SiCl_4$	$AlCl_3$	Sp	
28	30	Silica sol	$Al(NO_3)_3 \cdot 9H_2O$	Sp	dry
29	31	Silica sol	$Al_2(SO_4)_3$	Sp	freeze-dry
30	32	Silica sol	$Al_2(SO_4)_3$	Sp	
31	33	Fumed silica	$Al_2(SO_4)_3$	Sp	dry
32	34	Colloidal	$Al(NO_3)_3 \cdot 9H_2O$	Sp	precipitated by NH_4OH
33	22	Silica sol	Alumina sol	Sp	hydrothermal
34	8	Silica sol	Alumina sol	Sp	
35	18	Silica sol	Alumina sol	Sp	dry
36	35	Silica sol	Alumina sol	Sp	dry
37	19	Silica sol	Alumina sol	Sp	dry
38	36	Kaolinite		Sp	
39	29	Kaolinite		Sp	
40	1	Kaolinite		Sp	
41	37	Kaolinite		Sp	
42	4	Kaolinite(syn.)		Sp	
43	38	Halloysite		Sp	
44	39	Halloysite		Sp	
45	40	Halloysite		Sp	
46	41	Allophane		Sp	grinding
47	42	$Si(OEt)_4$	$Al(O^iPr)_3$	Sp+Mu	gelation
48	19	$Si(OMe)_4$	$Al(O^iPr)_3$	Sp+Mu	precipitated at pH7
49	38	Kaolinite		Sp+Mu	
50	39	Kaolinite		Sp+Mu	
51	43	Kaolinite		Sp+Mu	
52	44	Kaolinite		Sp+Mu	
53	40	Dickite		Sp+Mu	
54	40	Allophane		Sp+Mu	
55	40	Imogolite		Sp+Mu	

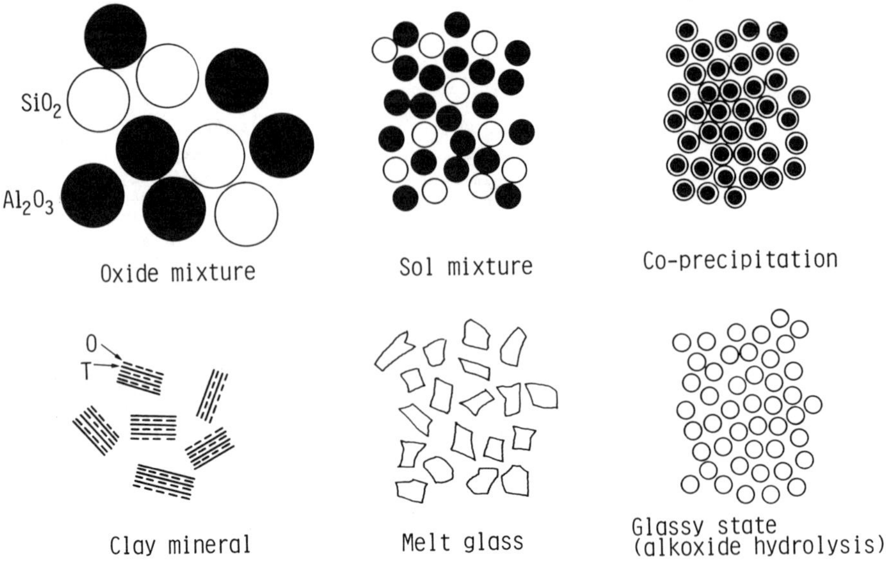

Fig. 1. Schematic models for various starting materials. T: SiO_2 tetrahedral layer, O: $Al(O,OH)_6$ octahedral layer.

the solution,[8,10,12] and xerogels prepared by a spray pyrolysis method.[9,11] Their structures were considered to be similar to the structures of their solutions. The structure of the melt glass in the SiO_2-Al_2O_3 system has been reported by Morikawa et al.[45] They proposed a mullite-like structure for the glass with a composition similar to mullite. It was, therefore, considered that the plausible structure was a random network with Si-O-Al linkages or, if it was not the case, that it was made from agglomerates of molecularly mixed $SiO_2 \cdot nH_2O$ and $Al_2O_3 \cdot mH_2O$ particles. If the mixing state in solution was not homogeneous, it did not yield similar results. When Si-alkoxide and $Al(NO_3)_3 \cdot 9H_2O$ are dissolved in alcohol, the following reactions must occur.

$$Si(OR)_4 + H_2O \rightarrow Si(OR)_3(OH) + ROH \quad (1)$$
$$2Si(OR)_3(OH) \rightarrow (RO)_3Si\text{-}O\text{-}Si(OR)_3 + H_2O \quad (2)$$
$$Al(NO_3)_3 + H_2O \rightarrow Al(NO_3)_2(OH) + HNO_3 \quad (3)$$
$$2Al(NO_3)_2(OH) \rightarrow (NO_3)_2Al\text{-}O\text{-}Al(NO_3)_2 + H_2O \quad (4)$$
$$Si(OR)_3(OH) + Al(NO_3)_3 \rightarrow (RO)_3Si\text{-}O\text{-}Al(NO_3)_2 + HNO_3 \quad (5)$$

According to Shalumov et al.,[46] the reaction rate of formula (1) is about 1000 times faster than that of formula (5) at room temperature. Therefore, the reaction of formula (2) should progress more extensively than the reaction of formula (5). This implies that polymerization of SiO_2 component progressed rapidly so that the structure of the solution corresponded to a poor mixing state as the reaction time became longer at room temperature. Okada and Ōtsuka[47] found that the constituent crystalline phase in thin films, which were prepared by the dip-coating method using a similar solution, changed from mullite to spinel phase as the time from the preparation of the solution exceeded two months at room temperature. It was considered to be caused by the polymerization of the SiO_2 and the Al_2O_3 components individually in the solution. Increase of reaction rate of formula (5) became relatively higher than that of formula (1) as the reaction temperature increased. This tendency was preferable for the formation of a precursor with Si-O-Al linkages. Xerogels reported by Refs. 8 and 10 were prepared at 60°C and gelled within two weeks. This hydrolysis condition was suitable for formation of precursors with atomistically mixed structures since these starting materials showed direct mullitization. However, precise examination by transmission electron microscopy of this specimen previously reported[48] revealed to co-existence with mullite of a small amount of spinel phase which was difficult to detect by the powder X-ray diffraction method. This may be due to the high concentration of solution used for the preparation. In the case of spray pyrolysis, the prepared solution was instantaneously dried and reacted by spraying it into a high temperature at about 600°C. The structure of these xerogels was suggested to be similar to that of the solution.

In the case of clay minerals, the situation was a little complicated and the crystallized phase varied with the mineral species. In almost all cases, i.e., kaolinite, dickite, allophane and imogolite,[40] both spinel phase and mullite were simultaneously formed from the amorphous state. The formation of mullite was not as much as that of spinel phase. The spinel phase formation slightly preceded mullite formation in halloysite but both spinel and mullite were formed by firing to a little higher temperature. Metakaolin is the state before crystallization of phases by firing of the kaolin group minerals; it is composed of a SiO_4 tetrahedral layer and largely distorted AlO_4 tetrahedral layer. This structure[49] shows large anisotropy. It has an infinitely mixed arrangement of the SiO_2 and Al_2O_3 components along the direction perpendicular to the layers but has perfectly segregated arrangements along the direction within the layers. There are considerable numbers of Si-O-Al linkage in this structure in which numbers were evaluated to be intermediate between the numbers in the structure of glassy state and those of mixture of sols. These minerals formed both spinel and mullite. The following result supports this idea that the numbers of Si-O-Al linkages were an important factor to decide the crystallizing phase. When allophane is dry ground by mortar, the Si-O-Al linkage in its structure breaks down.[41] Only spinel phase was crystallized from this specimen though not only spinel phase but also mullite were crystallized from the untreated specimen as shown in Table 1. On the other hand, the spinel phase was not detected in the thermal change of pyrophyllite.[15] This may be deduced because pyrophyllite has about twice as many Si-O-Al linkages as there are in the kaolin group minerals. The structure of anhydrous pyrophyllite which is the state just before crystallization has a tighter arrangement than metakaolin. However, the lower Al_2O_3 content in pyrophyllite may be considered to be responsible.

There are some reports examining the effects of some pretreatments of starting materials. Effects of impurities have been most widely studied.[50,51,52,53] They should act as a nucleus former, but the results obtained are not always compatible. Lemaitre et al.[52] report that Ca cation works as an accelerator of mullite formation and Mg cation works as an inhibitor. On the other hand, both of them are reported to act as inhibitors by MacKenzie.[50] We examined effects of impurities, i.e., Li, Na, K, Mg, Ca, Fe, Ti and Zr cations, on the SH and RH xerogels.[53] All of them work as inhibitors of mullite formation with Ti cation being the most effective. The impurity that was most effective on mullitization was fluorides. These accelerated the formation of mullite greatly. Examples are fluorated kaolinite,[13] LiF[14] and AlF_3.[53] Since high vapor pressure of fluorides is deduced to lower the energy barrier of the reaction from amorphous state to mullite, formation of mullite was accelerated. Effect of firing atmosphere is also reported by MacKenzie,[54] and Inoue and Okuda.[55]

Mullitization is accelerated under H_2O atmosphere and vacuum because they act as nucleus formers for mullite. Weiss et al.[13] reported that if metakaolin is pressed at 250 kb and fired, this specimen directly crystallizes to mullite from the amorphous state. On the other hand, metakaolin pressed at 125 kb showed a sequence like metakaolin → quartz + spinel phase → mullite by firing; and metakaolin pressed at 90 kb showed metakaolin → spinel phase → mullite. These drastic changes in the formation process must be connected with a structure change of metakaolin by compaction, but the details are uncertain.

From these results, different mullitization routes observed with various starting materials could be understood by considering the mixing state of SiO_2 and Al_2O_3 components, i.e., the numbers of Si-O-Al linkages in their structures.

Primary Mullite

Chemical compositions of mullite formed at various firing temperatures were examined by ATEM and LC data. Experimental conditions in this paper are the same as those reported elsewhere.[48] Starting materials examined were precipitates[10] with various bulk compositions (RH), xerogels[10] prepared by gelation with various bulk compositions (SH), glass prepared by melting and various kinds of clay minerals.[40] They were fired for 24 h at various temperatures. Relation between chemical compositions of mullite and their LC data was examined by Cameron.[56] He reported a linear relation between compositions and lengths of *a* axis, as lengthening of *a* axis corresponds to Al_2O_3-rich composition for mullite. To confirm this relation, chemical compositions of mullite in some specimens were analyzed by ATEM. Figure 2 shows the relation between measured Al_2O_3 content by ATEM and the estimated value from *a* axis data. They showed good agreement except for one

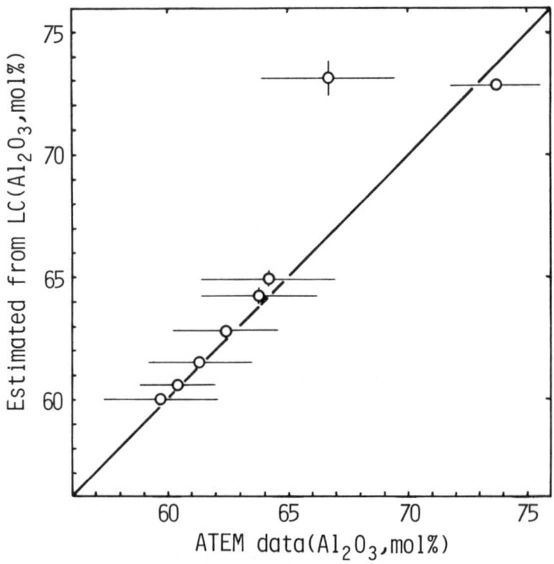

Fig. 2. Relation between measured Al_2O_3 content in mullite by ATEM and estimated one from the length of a axis. Error bars represent the standard deviations.

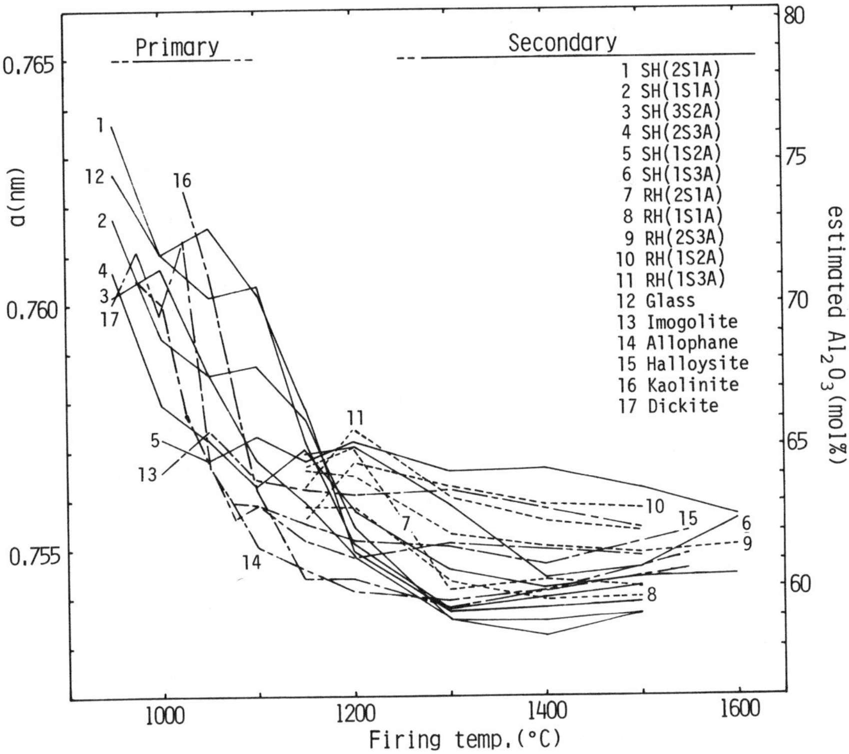

Fig. 3. Relations between the lengths of a axis and their firing temperatures for mullite formed from various starting materials. S: SiO_2, A: Al_2O_3.

datum. This corresponds to the specimen fired just above its crystallization temperature. Two reasons were considered for this discrepancy. One was due to the lattice loosening because of heating just above crystallization. Second was due to the influence of co-existence material by inappropriate pretreatment for specimen preparation. It was concluded that chemical composition of mullite was able to be estimated from the length of *a* axis except for specimen just after crystallization. Figure 3 shows relations between the lengths of *a* axis and firing temperatures for various specimens. The curves show similar trends: the *a* axis length decreases with increase of firing temperature in the lower firing temperature region and becomes almost constant with further increase in firing temperature. This indicates that mullite formed at low firing temperatures around 1000°C was Al_2O_3-rich, and the Al_2O_3 content gradually decreased with higher firing temperatures up to 1200-1300°C. Two stepped formation curves are reported which correspond with the change of the length of *a* axis.[6] From these results, mullite formed around 1000°C firing corresponded to the so-called "primary mullite."

Lattice constants of specimens fired for various times at fixed temperatures were measured to examine stability of primary mullite which was considered to be a transient state and are shown in Fig. 4. In the specimens fired at 1000°C, the length of *a* axis became constant as firing time exceeded 24 h and still remained constant

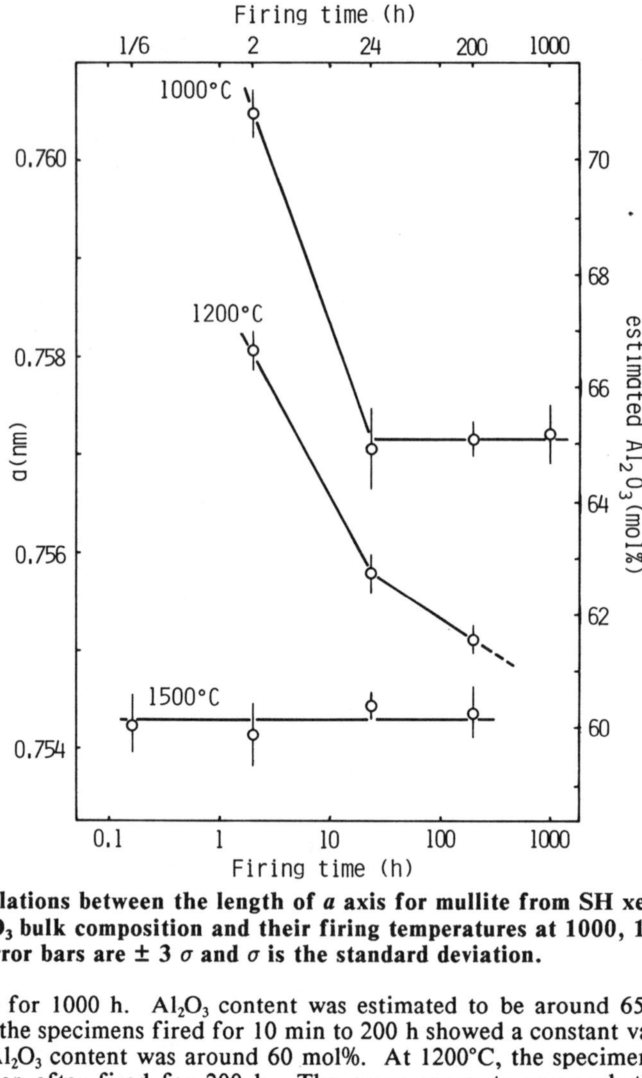

Fig. 4. Relations between the length of *a* axis for mullite from SH xerogel with $2SiO_2$-$3Al_2O_3$ bulk composition and their firing temperatures at 1000, 1200 and 1500°C. Error bars are ± 3 σ and σ is the standard deviation.

after firing for 1000 h. Al_2O_3 content was estimated to be around 65 mol%. At 1500°C, all the specimens fired for 10 min to 200 h showed a constant value and the estimated Al_2O_3 content was around 60 mol%. At 1200°C, the specimens showed a decrease even after fired for 200 h. The curve seems to approach the value of 1500°C specimens with longer firing time at 1200°C. After all, composition of mullite formed at 1200°C was anticipated to become 60 mol% with a longer firing time. On the other hand, composition of mullite formed at 1000°C kept its Al_2O_3 rich composition even after firing for 1000 h.

Conclusions

Four mullitization routes were found for thermal change from various starting materials to mullite and are shown in the SiO_2-Al_2O_3 equilibrium diagram reported by Aksay and Pask[57] (Fig. 5). These can be summarized as follows.

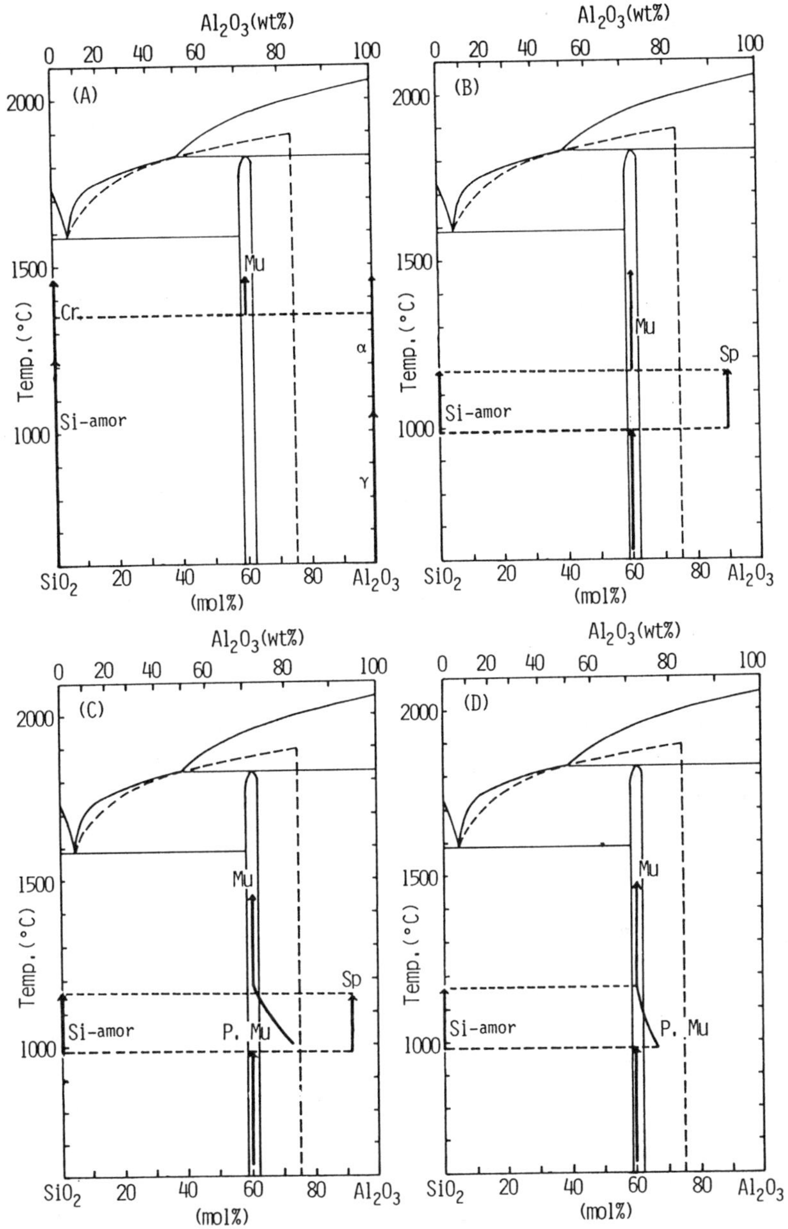

Fig. 5. Four mullitization routes shown in the SiO_2-Al_2O_3 equilibrium diagram. P. Mu: primary mullite, Mu: secondary mullite, Sp: spinel phase, γ: γ-Al_2O_3, α: α-Al_2O_3, Si-amor: amorphous SiO_2, Cr: cristobalite.

1. Mixtures of sols and/or oxides: starting materials → γ-Al_2O_3 + amorphous SiO_2 → α-Al_2O_3 + cristobalite → mullite + (α-Al_2O_3) + (cristobalite).
2. Precipitates: starting materials → spinel phase (with 7–8 wt% SiO_2) + amorphous SiO_2 → mullite.
3. Clay minerals: starting materials → spinel phase (with 7–8 wt% SiO_2) + amorphous SiO_2 + primary mullite (Al_2O_3 rich composition) → mullite.
4. Glassy states: starting materials → primary mullite (Al_2O_3 rich composition) + amorphous SiO_2 → mullite.

Acknowledgments

A part of this work is supported by The Kazuchika Ohkura Memorial Foundation and a Research Grant-in-Aid for the Encouragement of Young Scientists.

References

[1]G. W. Brindley and M. Nakahira, "The Kaolinite-Mullite Reaction Series: III, The High-Temperature Phases," *J. Am. Ceram. Soc.*, **42** [7] 319–324 (1959).

[2]K. H. Schüller and H. Kromer, "Primary Mullite as a Pseudomorph after Kaolinite," pp. 533–538 in Proceedings of the International Clay Conference, Mexico City, 1975. Edited by S. W. Bailey, Applied Publishing, Wilmette, Ill., 1976.

[3]K. Okada, N. Ōtsuka, and J. Ossaka, "Characteristics of the Thermally Decomposed Products from SiO_2-Al_2O_3-H_2O Compounds-With Respect to Allophane–," *Yogyo-Kyokai-Shi*, **93** [11] 671–678 (1985).

[4]S. Tomura, Y. Shibasaki, H. Mizuta, and T. Maeda, "Exsolution of Mullite from Spherical Kaolinite," *Yogyo-Kyokai-Shi*, **94** [8] 917–921 (1986).

[5]K. Okada, Y. Hoshi, and N. Ōtsuka, "Formation Reaction of Mullite from SiO_2-Al_2O_3 Xerogels," *J. Mater. Sci. Lett.*, **5** [12] 1315–1318 (1986).

[6]K. Okada and N. Ōtsuka, "Chemical Composition Change of Mullite During Formation Process," pp. 497–502 in Science of Ceramics 14, Canterbury, 1987. Edited by D. Taylor. The Institute of Ceramics, 1988.

[7]J. Ossaka, "Tetragonal Mullite-like Phase from Co-Precipitated Gels," *Nature*, **191** [4792] 1000–1001 (1961).

[8]D. W. Hoffman, R. Roy, and S. Komarneni, "Diphasic Xerogels, A New Class of Materials: Phases in the System Al_2O_3-SiO_2," *J. Am. Ceram. Soc.*, **67** [7] 468–471 (1984).

[9]S. Kanzaki, H. Tabata, T. Kumazawa, and S. Ohta, "Sintering and Mechanical Properties of Stoichiometric Mullite," *J. Am. Ceram. Soc.*, **68** [1] C-6–C-7 (1985).

[10]K. Okada and N. Ōtsuka, "Characterization of the Spinel Phase from SiO_2-Al_2O_3 Xerogels and the Formation Process of Mullite," *J. Am. Ceram. Soc.*, **69** [9] 652–656 (1986).

[11]O. Sakurai, N. Mizutani, and M. Kato, "Preparation of Mullite Powders from Metal Alkoxides by Ultrasonic Spray Pyrolysis," *J. Ceram. Soc. Jpn*, **96** [6] 639–645 (1988).

[12]J. C. Pouxviel, J. P. Boilot, A. Dauger, and L. Huker, "Chemical Route to Aluminosilicate Gels, Glasses and Ceramics," pp. 269–274 in Better Ceramics Through Chemistry II, Edited by C. J. Brinker, D. E. Clark, and D. R. Ulrich, Materials Research Society, 1986.

[13]A. Weiss, K.-J. Range, and J. Russow, "The Al, Si Spinel Phase from Kaolinite (Isolation, Chemical Analysis, Orientation) and Relations to Its Low-Temperature Precursors)," pp. 34–37 in Proceedings of the International Clay Conference, Tokyo, 1969, Vol. 2. Edited by L. Heller. Israel University Press, 1970.

[14]R. Oyamada, "The Mechanism of Mullite Formation in Low-Grade Kaolin at Low Temperatures," *Denki Kagaku*, **49** [5] 286–292 (1981).

[15]J. Lemaitre, A. J. Leonard, and B. Delmon, "Mechanism of the Thermal Transformations of Metakaolinite," (in Fr.), *Bull. Mineral.*, **105** [5] 501–507 (1982).
[16]T. Takamori and R. Roy, "Rapid Crystallization of SiO_2-Al_2O_3 Glasses," *J. Am. Ceram. Soc.*, **56** [12] 639–644 (1973).
[17]R. Ganz and W. Kronert, "Crystallization Behavious of High Temperature Ceramic Fibre of the Al_2O_3-SiO_2 System," *Interceram*, **31** [2] 136–144 (1982).
[18]K. Hamano, T. Sato, and Z. Nakagawa, "Properties of Mullite Powder Prepared by Coprecipitation and Microstructure of Fired Bodies," *Yogyo-Kyokai-Shi*, **94** [8] 818–822 (1986).
[19]S. Mitachi, N. Tanaka, K. Kaneko, H. Tabata, and S. Kanzaki, "Effect of Synthetic Methods on Properties of SiO_2-Al_2O_3 Powders," Annual Meeting Abstracts of Ceramic Society of Japan, pp. 239–240 (1986).
[20]H. Yamada and S. Kimura, "Studies on the Co-precipitates of Alumina and Silica Gels and its Transformations at Higher Temperatures," *Yogyo-Kyokai-Shi*, **70** [3] 87–93 (1962).
[21]S. Prochazka and F. J. Klug, "Infrared-Transparent Mullite Ceramic," *J. Am. Ceram. Soc.*, **66** [12] 874–880 (1983).
[22]M. Suzuki, S. Hiraishi, M. Yoshimura, and S. Somiya, "Preparation of Mullite Powder by Calcination of the Products Hydrothermally Treated from Mixed Alkoxide or Mixed Sol," *Yogyo-Kyokai-Shi*, **92** [6] 320–327 (1984).
[23]H. Suzuki and H. Saito, "Processing of the Fine Mullite Powder from Metal-Alkoxides and Its Sintering," *Yogyo-Kyokai-Shi*, **95** [7] 697–702 (1987).
[24]B. E. Yoldas, "Microstructure of Monolithic Materials Formed by Heat Treatment of Chemically Polymerized Precursors in the Al_2O_3-SiO_2 Binary," *Am. Ceram. Soc. Bull.*, **59** [4] 479–483 (1980).
[25]K. S. Mazdiyasni and L. M. Brown, "Synthesis and Mechanical Properties of Stoichiometric Aluminum Silicate (Mullite)," *J. Am. Ceram. Soc.*, **55** [11] 548–552 (1972).
[26]Y. Hirata, H. Minamizono, and K. Shimada, "Property of SiO_2-Al_2O_3 Powders Prepared from Metal Alkoxide," *Yogyo-Kyokai-Shi*, **93** [1] 36–44 (1985).
[27]Y. M. M. Al-Jarsha, K. D. Biddle, A. K. Das, T. J. Davies, H. G. Emblem, K. Jones, J. M. McCullough, M. A. M. A. Rahman, A. N. A. El-M. S. El Deen, and R. Wakefield, "Mullite Formation from Ethyl Silicate and Aluminum Chlorides," *J. Mater. Sci.*, **20** [5] 1773–1781 (1985).
[28]G.-Y. Meng and R. A. Huggins, "A New Chemical Method for Preparation of Both Pure and Doped Mullite," *Mat. Res. Bull.*, **18** [5] 581–588 (1983).
[29]V. J. Wiegmann and C. H. Horte, "Uber die exotherme Reaktion des Metakaolin zwischen 950 und 1000°C," *Z. anorg. allgem. Chem.*, **286** [5/6] 268–287 (1956).
[30]A. K. Chakraborty and D. K. Ghosh, "Crystallization Behavior of Al_2O_3 in the Presence of SiO_2," *J. Am. Ceram. Soc.*, **70** [3] C-46–C-48 (1987).
[31]T. A. Wheat, E. M. H. Sallam, and A. C. D. Chaklader, "Synthesis of Mullite by a Freeze-Dry Process," *Ceramurgia Intern.*, **5** [1] 42–44 (1979).
[32]A. K. Chakraborty, "Formation of Silicon-Aluminum Spinel," *J. Am. Ceram. Soc.*, **62** [3–4] 120–124 (1979).
[33]H. Saito, M. Mizuno, M. Shiraishi, and K. Yamashita, "Preparation of High-Purity Mullite Powders," Annual Meeting Abstracts of Ceramic Society of Japan, pp. 233–234 (1987).
[34]Y. Kubota and H. Takagi, "Synthesis and Properties of Mullite," Annual Meeting Abstracts of Basic Science Meeting of Ceramic Society of Japan, p. 62 (1986).
[35]T. Hiraiwa, K. Miyazaki, T. Kawanami, and H. Ohnishi, "Synthesis of High Purity, Fine Grained Mullite Powders," Annual Meeting Abstracts of Ceramic Society of Japan, pp. 103–104 (1986).

[36]J. E. Comeforo, R. B. Fischer, and W. F. Bradley, "Mullitization of Kaolinite," *J. Am. Ceram. Soc.*, **31** [9] 254–259 (1948).
[37]A. J. Leonard, "Structural Analysis of the Transition Phases in the Kaolinitie-Mullite Thermal Sequence," *J. Am. Ceram. Soc.*, **60** [1–2] 37–43 (1977).
[38]H. D. Glass, "High Temperature Phases from Kaolinite and Halloysite," *Am. Mineral.*, **39** [3–4] 193–207 (1954).
[39]Y. Tsuzuki, "Mechanism of the 980°C Exotherm of Kaolin Minerals," *J. Earth Sci.*, Nagoya Univ., **9** [2] 305–344 (1961).
[40]K. Okada, N. Ōtsuka, and J. Ossaka, "Characterization of Spinel Phase Formed in the Kaolin-Mullite Thermal Sequence," *J. Am. Ceram. Soc.*, **69** [10] C-251-C-253 (1986).
[41]K. Okada, N. Ōtsuka, and J. Ossaka, "Structural Change of Allophane by Dry Grinding and Its Influence to the Allophane-Mullite Thermal Sequence," *Clay Sci.*, **6** [6] 283–294 (1987).
[42]Y. Hirata and K. Shimada, "Preparation and Sinterability of Fine SiO_2-Al_2O_3 Powders from Mixed Alkoxides," pp. 18–19 in New Investigations of Functional Ceramics. Edited by Koizumi. Ministry of Education, Science and Culture Grant-in-Aid for Special Research. 1985.
[43]H. J. Percival, J. F. Duncan, and P. K. Foster, "Interpretation of the Kaolinite-Mullite Reaction Sequence from Infrared Absorption Spectra," *J. Am. Ceram. Soc.*, **57** [2] 57–61 (1974).
[44]A. K. Chakraborty and D. K. Ghosh, "Reexamination of the Kaolinite-to-Mullite Reaction Series," *J. Am. Ceram. Soc.*, **61** [3–4] 170–173 (1978).
[45]H. Morikawa, S. Miwa, M. Miyake, F. Marumo, and T. Sata, "Structural Analysis of SiO_2-Al_2O_3 Glasses," *J. Am. Ceram. Soc.*, **65** [2] 78–81 (1982).
[46]B. Z. Shalumov, A. M. Bessarabov, M. V. Mikheev, L. A. Gornostaeva, and B. I. Polyakov, "Kinetics of Autocatalytic Hydrolysis of Tetraethyl Orthosilicate in the Presence of Aluminum Nitrate," *React. Kinet. Catal. Lett.*, **25** [3–4] 177–179 (1984).
[47]K. Okada and N. Ōtsuka, "Preparation of Transparent Mullite Film by Dip-Coating Method," The First International Workshop on Mullite, in press.
[48]K. Okada and N. Ōtsuka, "Chemical Composition Change of Mullite Formed from $2SiO_2{\cdot}3Al_2O_3$ Xerogel During Formation Process," *J. Am. Ceram. Soc.*, **70** [10] C-245-C-247 (1987).
[49]S. Iwai, H. Tagai, and T. Shimamune, "Procedure for Dickite Structure Modification by Dehydration," *Acta Crystallogr., Sect. B*, **27** [1] 248–250 (1971).
[50]K. J. D. MacKenzie, "The Effects of Impurities on the Formation of Mullite from Kaolinite-Type Minerals I. The Effect of Exchangeable Cations," *Trans. Brit. Ceram. Soc.*, **68** [3] 97–101 (1969).
[51]K. J. D. MacKenzie, "The Effects of Impurities on the Formation of Mullite from Kaolinite-Type Minerals II. The Effect of Exchangeable Anions," *Trans. Brit. Ceram. Soc.*, **68** [3] 101–103 (1969).
[52]J. Lemaitre, M. Bulens, and B. Delmon, "Influence of Mineralizers on the 950°C Exothermic Reaction of Metakaolinite," pp. 539–544 in Proceedings of the International Clay Conference, Mexico City, 1975. Edited by S. W. Bailey. Applied Publishing, Wilmette, Ill., 1976.
[53]K. Okada and N. Ōtsuka; unpublished data.
[54]K. J. D. MacKenzie, "The Effects of Impurities on the Formation of Mullite from Kaolinite-Type Minerals III. The Effect of the Firing Atmosphere," *Trans. Brit. Ceram. Soc.*, **68** [3] 103–109 (1969).
[55]K. Inoue and S. Okuda, "Effect of Heating Atmospheres on Phase Transformation of Kaolinite," *Nendo Kagaku*, **10** [1] 11–16 (1970).

[56]W. E. Cameron, "Composition and Cell Dimensions of Mullite," *Am. Ceram. Soc. Bull.*, **56** [11] 1003–1011 (1977).
[57]I. A. Aksay and J. A. Pask, "Stable and Metastable Equilibria in the System SiO_2-Al_2O_3," *J. Am. Ceram. Soc.*, **58** [11–12] 507–512 (1975).

Mullite-Silica Ceramics For Insulating Substrate Material

SHUZO KANZAKI, MASAYOSHI OHASHI AND HIDEYO TABATA

Government Industrial Research Institute, Nagoya
Hirate-cho, Kita-ku, Nagoya 462, Japan

TAKASHI KURIHARA, SHO-ICHI IWAI AND SHIN-ICHI WAKABAYASHI

Shinko Electric Industries Co., Ltd.
711 Kurita, Nagano 380, Japan

Abstract

SiO_2-Al_2O_3 powders containing 20, 30 and 40 wt% Al_2O_3 were synthesized by hydrolysis of metal alkoxide. Sinterability of the powders and the relationship between their chemical composition and the properties of the sintered bodies, such as thermal, electrical and mechanical properties, have been studied. These powders can be densified to almost full density at 1300° to 1350°C for 4 h without the formation of cristobalite. The sintered bodies consisted of very fine prismatic mullite particles and glass matrix. Dielectric constant, thermal expansion coefficient and flexural strength decreased with increasing SiO_2 content in the starting powder and ranged from 4.3 to 5.3, 1.8 to 2.9 × 10^{-6}/°C and 124 to 201 MPa, respectively.

Introduction

Ceramic substrates have played an important role in high performance electronic products. In recent years logic devices with higher integration density and higher signal propagation rate have been required. High dielectric constant and a large thermal expansion coefficient cause a delay in signal propagation, and thermal mismatch between substrate and silicon chip in the fabrication process, respectively.[1] Alumina ceramics have been widely used as a substrate material for intergrated circuits (IC), but it has a relatively high dielectric constant (9.6) and a larger thermal expansion coefficient (7.0 × 10^{-6}/°C) than that of silicon (3.5 × 10^{-6}/°C). Therefore, materials which have a lower dielectric constant than alumina and a thermal expansion coefficient closer to that of silicon have been required. Because of its fairly lower dielectric constant and the closer thermal expansion coefficient to that of silicon, mullite ($3Al_2O_3{\cdot}2SiO_2$) has been considered to be a candidate material for application as an IC substrate. Although a number of papers have been published with regard to the application of mullite ceramics for substrate, most of the reports have been limited to the stoichiometric mullite composition.[2,3] It has been reported that mechanical properties of the mullite ceramics depend on the chemical composition due to the difference in their microstructure.[4] Other properties of mullite ceramics such as thermal and electrical properties are also expected to be influenced by their chemical composition.

In the present study, the sintering behavior of silica-alumina powders containing 20, 30 and 40 wt% Al_2O_3, and thermal, electrical and mechanical properties of the sintered bodies were examined in relation to their chemical composition.

Experimental

Starting materials* were amorphous powders containing 20, 30 and 40 wt% Al_2O_3 (hereafter referred as 20A, 30A and 40A), which were synthesized by hydrolysis of mixed solutions of $Al(OC_3H_7^i)_3$ and $Si(OCH_3)_4$ with ammonia aqueous solution (PH10). In order to compare sinterability and the properties of the sintered bodies with 20A, 30A and 40A, powders containing a larger amount of Al_2O_3 (60A-74A) were also examined. These powders were calcined at 1230°C (20A-40A), or at 1300°-1350°C (60A-74A) for 1 h. After grinding with a vibrational mill for 24 h (20A-40A), or 50 h (60A-74A), the powders were dried in a rotary evaporator and were passed through a 140 mesh screen followed by heat treatment at 800°C for 1 h to remove organic contamination. Powder compacts were formed as disks (16 mm dia. × 3 mm thick), blocks (6 × 6 × 19 mm), and plates (40 × 48 × 3 mm, 40 × 48 × 6 mm) under uniaxial pressure (30 MPa) followed by isostatic pressing (200 MPa). These compacts were sintered in air at 1300°-1450°C (20A-40A), or at 1650°C (60A-74A) for 4 h (Heating rate: 10°C/min).

Specific surface areas and crystallization process of the powders were examined by BET† method and DTA‡, respectively. The thermal shrinkage curves of the powder compacts and the thermal expansion coefficients of the sintered bodies were measured by a dilatometer§ at a heating rate of 5°C/min. The thermal expansion data were calibrated by the use of NBS-fused silica (Reference Materials 739). The amount of mullite and cristobalite formed in the sintered bodies were determined by X-ray diffraction using sodium chloride as an internal standard. The sintered plates for the measurement of the dielectric constant were ground into 0.5 mm in thickness with a #400 diamond grinding wheel. Test bars of the size 3 × 4 × 40 mm for the mechanical strength measurement were prepared from the square plates by cutting, grinding and finally polishing with 6 μm diamond paste. Flexural strength was measured by means of three-point bending¶ at room temperature with a span length of 30 mm and cross head speed of 0.5 mm/min. Microstructure of the powders and sintered bodies was examined by the use of SEM, and TEM**.

---

*Hokko Chemical Industry Co., Ltd.

†Shimazu Seisakusho, Ltd. ACCUSORB 2100-01.

‡Rigaku Denki Co., Ltd. PTC-10A.

§Shinku Riko Co., Ltd. DL-1500.

¶Shimazu Seisakusho, Ltd. DCS-2000.

**JEOL Ltd. T-200, JEM-4000FX.

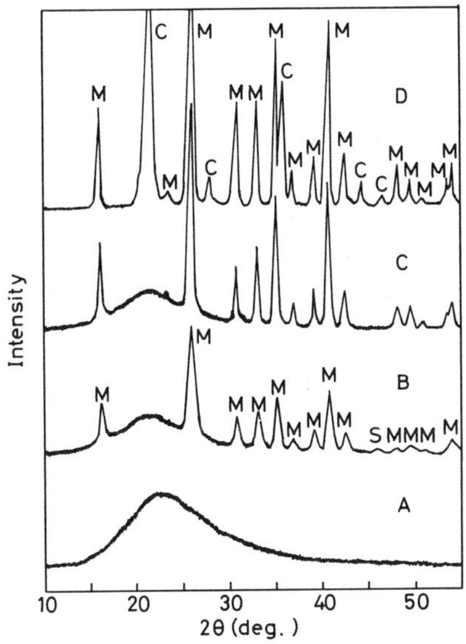

Fig. 1. X-ray diffraction patterns of 40A. A: as received B: Cristobalite S: Spinel.

Results

Characterization of Powders

The as-synthesized amorphous powders transformed into mixed crystalline phases of mullite and spinel on heating at about 980°C where a sharp exothermic peak was observed. The XRD patterns of 40A powder after heating at various temperatures are shown in Fig. 1. As the powders 20A, 30A, and 40A were calcined and ground, predominant mullite phase and a slight amount of spinel phase were observed. Upon heating to around 1300°C, the spinel phase disappeared and mullite remained as a crystalline phase. Cristobalite was found to form by further heating of the powders 20A, 30A and 40A. For calcined powders in the composition range from 60A to 74 A, only mullite was observed as a crystalline phase and cristobalite was not detected even when sintered above 1600°C. Specific surface areas of the as-synthesized 20A, 30A and 40A powders were 236, 227 and 183 m^2/g, respectively. After calcination and grinding, they were reduced down to 17, 18 and 30 m^2/g, respectively.

TEM photographs of the powders (20A, 40A) after calcination and grinding are shown in Fig. 2. The particle size of the powder 40A is smaller than that of 20A in accordance with their surface area. In both powders very fine particles (about 15 nm long) are dispersed in the matrix. The morphology of the fine particles is prismatic and the aspect ratio increased with increasing SiO_2 content.

Sinterability and Microstructure of the Sintered Bodies

Thermal shrinkage curves for the compacts of 20A, 30A and 40A are shown in Fig. 3. All the powders show more rapid densification beginning at about 1250°C. The temperature at which rapid densification occurred decreased slightly with increase in SiO_2 content of the powder. Although the formation of cristobalite was

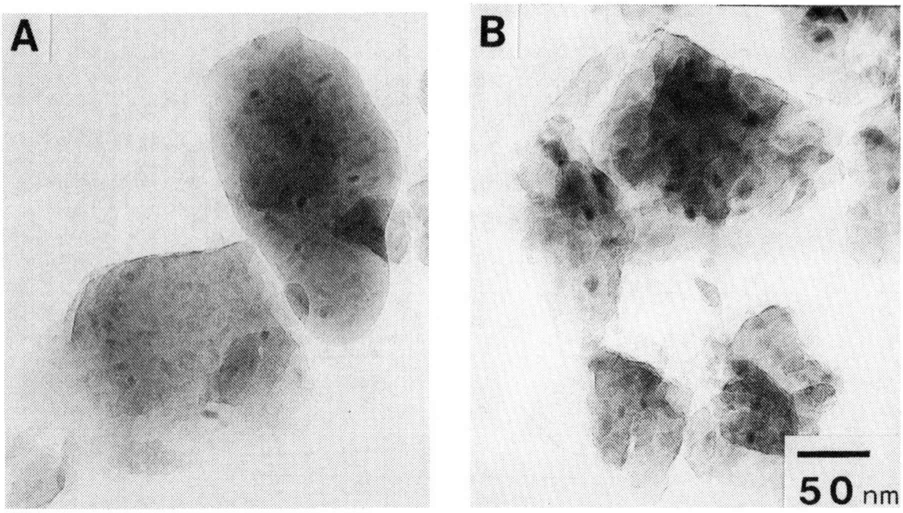

Fig. 2. TEM photographs of SiO_2-Al_2O_3 powders. A: 20A B: 40A Calcination: 1230°, 1 h Grinding: 24 h.

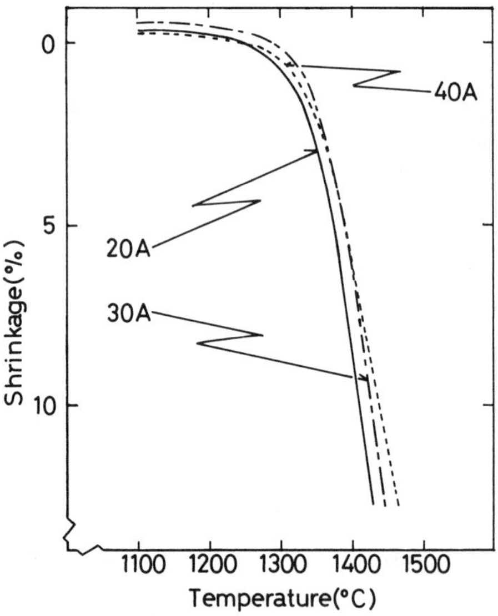

Fig. 3. Thermal shrinkage curves for compacts of SiO_2-Al_2O_3 powders. Heating rate: 5°C/min.

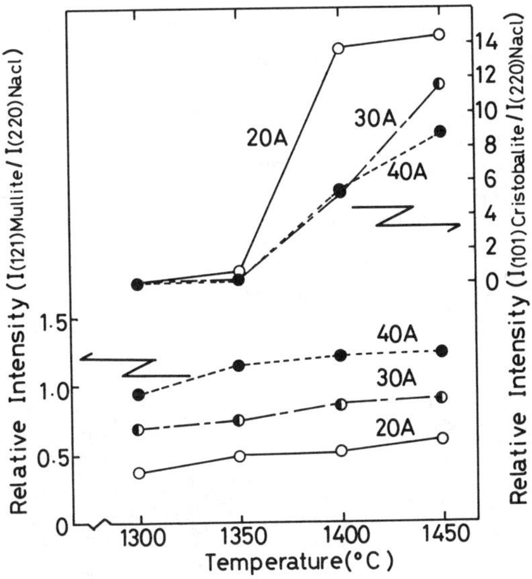

Fig. 4. The influence of temperature on relative intensity of mullite and cristobalite. Heating rate: 5°C/min.

observed at high temperature as shown in Fig. 1, the temperature at which cristobalite formed depended on the chemical composition of the powder. Figure 4 shows the amount of mullite and cristobalite formed as a function of temperature. The amount of mullite increased with the increase in Al_2O_3 content, and also slightly increased with increasing temperature. Cristobalite was not formed lower than 1300°C in any composition. However, it began to form rapidly above 1350°C, and the amount increased with increasing temperature and with increasing SiO_2 content.

Effect of temperature on bulk density is shown in Fig. 5. The bulk density increased with increasing temperature irrespective of composition. On 20A sufficient

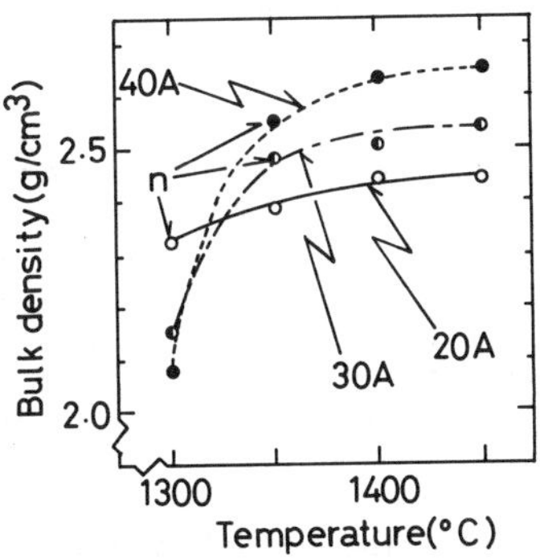

Fig. 5. Effect of temperature on bulk density. Heating rate: 10°C/min, Soaking time: 4 h.

densification was already attained at 1300°C, and the bulk density increased slightly above 1300°C. While on 30A and 40A the bulk density increased rapidly from 1300° to 1350°C, and increased gradually above that temperature.

More than 98% of theoretical density, determined by measuring the porosity on SEM photographs,[5] was attained in each sample without the formation of cristobalite (marked with "n" in Fig. 5). The increase in bulk density above the temperature marked with "n" can be accounted for by the volume change caused by the transformation from amorphous silica to cristobalite. Thus, nearly full densification is achieved at the temperature marked with "n" for each composition. The firing temperature, at which fully densified bodies were obtained, increased with increasing Al_2O_3 content and the temperature above 1600°C was necessary for compositions in the range from 60 to 74 wt% Al_2O_3.

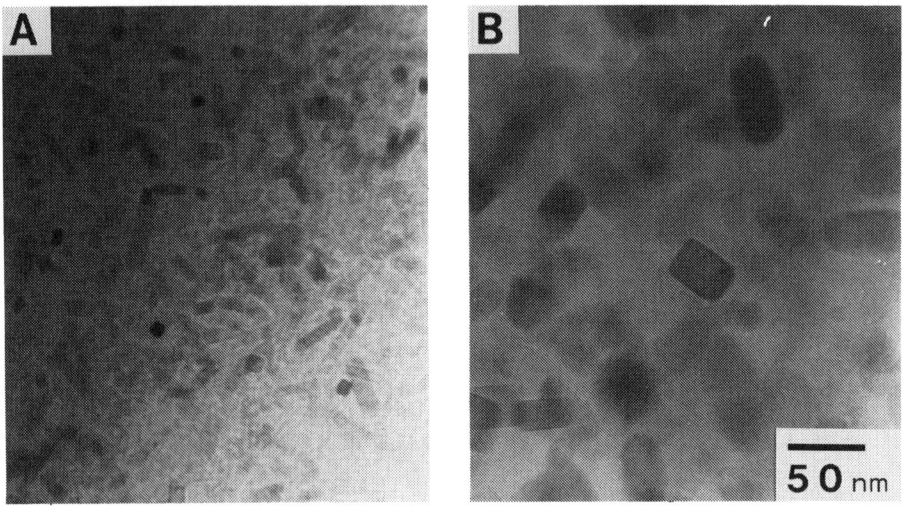

Fig. 6. TEM photographs of the sintered bodies. A: 20A, 1300°C, 4 h B: 40A, 1350°, 4 h.

Figure 6 shows representative TEM photographs of the samples sintered at the temperature marked with "n" in Fig. 5. It can be seen that fine and prismatic particles were randomly dispersed in the glass matrix for both samples. The prismatic particles might be assigned to mullite on the basis of the results with XRD. Sample 20A contained a larger amount of glassy phase than 40A as expected from the SiO_2-Al_2O_3 phase diagram.[6] The aspect ratio of the particles tend to increase with increasing SiO_2 content. Furthermore, morphology of the prismatic particles are quite analogous to that of the particles dispersed in calcined powder.

Properties of Sintered Body

The effects of composition and the amount of cristobalite in the sintered body on thermal expansion have been considered. Figure 7 shows the thermal expansion for samples with cristobalite (20A(2), 40A(2)) and without cristobalite (20A(1), 40A(1)). The thermal expansion of the samples without cristobalite increase linearly

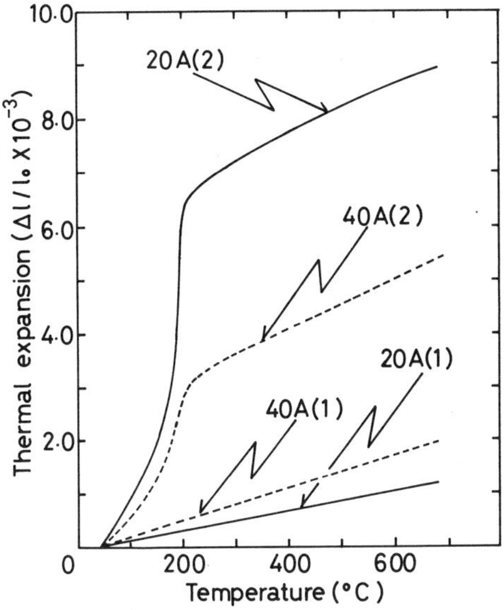

Fig. 7. The influence of composition on the thermal expansion of the sintered bodies. (1) without cristobalite (2) with cristobalite.

with temperature at a small slope and increase with increasing Al_2O_3 content. While, a large expansion caused by the α-β transformation of cristobalite was observed at about 200°C for the samples with cristobalite. Therefore, it is requisite to densify without the formation of cristobalite for the preparation of a sintered body with small and linear expansion in this system.

The following measurement of properties was performed mainly with the sintered bodies without cristobalite. The data obtained were arranged as a function of the amount of mullite phase. The amount of mullite in the composition range examined was estimated as follows. According to the equilibrium phase diagram,[6] stable mullite solid solution exists from 70.5 to 74 wt% Al_2O_3 below 1750°C. However, a glassy phase has been observed even within this composition range as reported.[7] Therefore 74 wt% Al_2O_3 is assumed to be the boundary composition between glass + mullite and mullite. This assumption was in good agreement with the experimental results of XRD and microstructural observation. On the basis of this assumption, the compositions containing larger amounts of SiO_2 than 26 wt% are composed of mullite with the Al_2O_3/SiO_2 ratio of 74 wt%/26 wt% and silica glass, for example, 40A consists of mullite with the composition 40 wt% Al_2O_3/14 wt% SiO_2 and 46 wt% SiO_2. In addition the properties of silica glass used were the values previously reported.[8,9]

Thermal Expansion Coefficient

The linear thermal expansion coefficient (30°-400°C) of the sintered bodies as a function of composition is shown in Fig. 8. The thermal expansion coefficient tends to increase with an increasing amount of mullite in the composition range from 20A (1.8×10^{-6}/°C) to 74A (4.0×10^{-6}/°C). The result obtained was approximately consistent with Turner's equation[10] (1) which is a theoretical formula for the thermal expansion coefficient of a composite material.

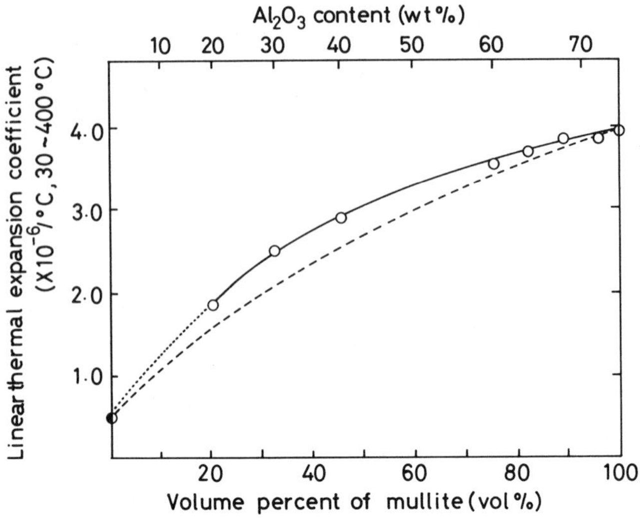

Fig. 8. Thermal expansion coefficient (30–400°C) of the sintered bodies as a function of the amount of mullite. Solid line: This study. Broken line: Turner's equation (1). ◐: cited from ref. 8.

$$\alpha_c = \frac{\alpha_M V_M E_M + \alpha_S V_S E_S}{V_M E_M + V_S E_S} \quad (1)$$

where α is the linear thermal expansion coefficient, E is Young's modulus, V is volume fraction of the phase, and subscripts C, M and S represent composite, mullite and silica glass, respectively. Young's modulus was assumed to be 180GPa for mullite and 72GPa for silica glass.

Dielectric Constant

The dielectric constant of the sintered bodies measured at the frequency of 1 MHz as a function of composition is shown in Fig. 9. The sintered bodies with cristobalite showed a slightly higher value compared with the values without cristobalite. The dielectric constant without cristobalite decreased with increasing SiO_2 content, and an approximately linear relationship between the dielectric constant and the composition was observed. The results obtained showed good agreement with one of the theoretical formulas (2) for composite material.[11]

$$\epsilon_C = \epsilon_M V_M + \epsilon_S V_S \quad (2)$$

where, ϵ is the dielectric constant, V is the volume fraction of the phase, and subscripts C, M and S represent composite, mullite and silica glass, respectively. The samples in the composition range from 20A to 40A showed fairly low dielectric constants which correspond to half the value of alumina. The low dielectric constant is advantageous for high speed propagation of signals in an electronic circuit.

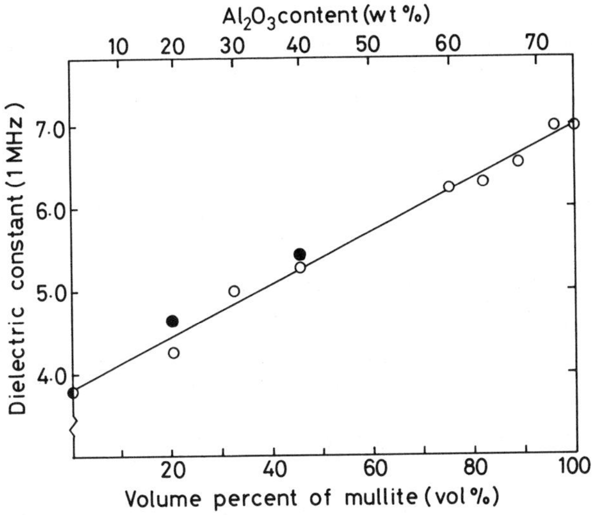

Fig. 9. Dielectric constant (1MHz) of the sintered bodies as a function of the amount of mullite. Solid line: equation (2). ◐: cited from ref. 9. ●: with cristobalite.

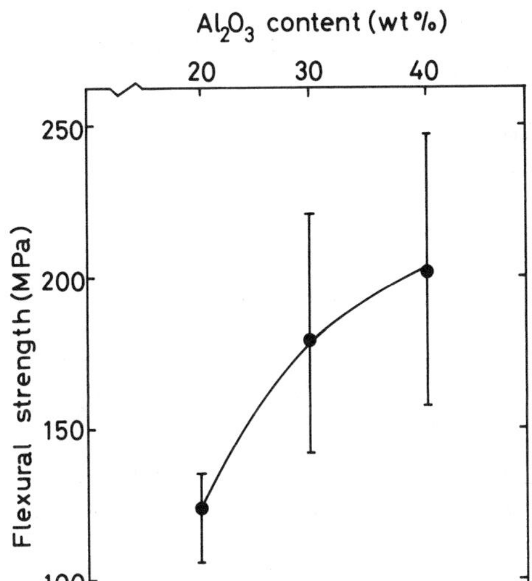

Fig. 10. Flexural strength of the sintered bodies as a function of the amount of mullite.

Flexural Strength

Dependence of composition on the flexural strength is shown in Fig. 10. The flexural strength increased with increase of the amount of mullite in the composition range from 20 to 40 wt% Al_2O_3. The average strength of 120 MPa for sample 20A is considered to be sufficient as a substrate material.

Summary and Conclusions

Sinterability of the alkoxide derived SiO_2-Al_2O_3 powders, microstructure and some properties of the sintered bodies were examined.

1. Powders with the composition range from 20 to 40 wt% Al_2O_3 were densified to more than 98% of theoretical density at a temperature range between 1300° and 1350°C without the formation of cristobalite.
2. In this composition range the sintered bodies were composed of prismatic mullite particles and glass matrix. The aspect ratio of the mullite particles decreased with increase in Al_2O_3 content.
3. The average linear thermal expansion coefficient (30°-400°C) and the dielectric constant of the sintered bodies decreased with increase in SiO_2 content.
4. The flexural strength at room temperature increased with increasing Al_2O_3 content, that is, the amount of mullite phase.

The values of the average linear thermal expansion coefficient, the dielectric constant and the flexural strength obtained for 20A specimen were 1.8×10^{-6}/°C, 4.3 and 124 MPa, respectively. These values are adequate for applications as an IC substrate material. Furthermore, it is also suggested that these values are able to be controlled by varying the Al_2O_3/SiO_2 ratio in the starting powder.

Acknowledgment

The authors wish to express their gratitude to Mr. S. Mitachi of Hokko Chemical Industry Co., Ltd, for providing the starting powder.

References

[1]S. Nishigaki, "Multilayer ceramic substrate," *Electronic ceramics*, [5] 61–71 (1985).

[2]N. Kamehara, K. Niwa and K. Murakawa, "Packaging material for high speed computers," IMC 1982 Proceedings Tokyo, 388–93 (1982).

[3]G. S. Perry, "Microwave Dielectric Properties of Mullite," *Trans. Brit. Ceram. Bull.*, **72**, 279–83 (1973).

[4]S. Kanzaki, T. Kumazawa, J. Asaumi, O. Abe and H. Tabata, "Dependency of Mechanical Property of Sintered Mullite on Chemical Composition," *Yogyo-kyokai-shi*, **93** [7] 407–08 (1985).

[5]E. E. Underwood, A. R. Colcord and R. C. Waugh, Quantitative relationships for random microstructures, 25–34, "Ceramic Microstructures," R. M. Fulrath and J. A. Pask, John Wiley & Sons, Inc., New York, 1968.

[6]E. M. Levin, C. R. Robbins and H. F. McMurdie, Metal oxide systems, 123, Phase diagrams for ceramists, American Ceramic Society, Inc., Columbus, 1964.

[7]T. Kumazawa, S. Kanzaki, J. Asaumi, O. Abe and H. Tabata, "Sinterability of SiO_2-Al_2O_3 Powders Prepared by Spray Pyrolysis: Effect of Chemical Composition," *Yogyo-kyokai-shi*, **94** [5] 485–90 (1986).

[8]W. D. Kingery, H. K. Bowen and D. R. Uhlmann, Thermal properties, 595, Introduction to Ceramics, John Wiley & Sons, Inc., New York, 1975.
[9]W. D. Kingery, H. K. Bowen and D. R. Uhlmann, Dielectric properties, 933, Introduction to Ceramics, Ibid, 1975.
[10]W. D. Kingery, H. K. Bowen and D. R. Uhlmann, Thermal properties, 604, Introduction to Ceramics, Ibid, 1975.
[11]H. C. Graham and N. M. Tallan, Microstructural-electrical properties theory, 511, Physics of electronic ceramics Part A, L. L. Hench, D. B. Dove, Marcel dekker, Inc., New York, 1971. This article is a reprint of the article which appeared in Yogyo-Kyokai-Shi (Japanese version), Vol. 95, No. 12, 1213–18, 1987.

Influence of Powder Characteristics on Microstructure and Mechanical Properties of Mullite Ceramics

TAKESHI KUMAZAWA AND SHIGETOSHI OHTA

Technical Research Laboratory
Mino Yogyo Co., Ltd.
1-46 Kamezaki, Handa 475, Japan

SHUZO KANZAKI AND HIDEYO TABATA

Government Industrial Research Institute, Nagoya
Hirate-cho, Kita-ku, Nagoya 462, Japan

Abstract

Influence of powder characteristics on microstructure and mechanical properties of stoichiometric mullite was studied with the use of raw powders synthesized by means of (S) spray pyrolysis, (Ha) hydrolysis with water, (Hb) hydrolysis with aqueous ammonia, and (C) co-precipitation methods. Mullitization temperature of the as-synthesized amorphous powders and the bulk density of the sintered bodies increased in the order of (S) < (Ha) < (Hb) = (C). The degree of homogeneity in the starting powder was closely related to the rate of grain growth and amount of glassy phase formed during sintering. Powder preparation methods do not affect the instantaneous fracture behavior of each specimen up to 1300°C. When the cross-head speed was reduced, however, the degree of plastic deformation prior to fracture increased in the order of (S) < (Ha) < (Hb) < (C) at elevated temperature, in accordance with the amount of glassy grain boundary phase.

Introduction

It is well recognized that homogeneity of chemical composition in the starting powder as well as high-purity and fine-grained particles are the most important factors to obtain a dense sintered body with precisely controlled microstructure.[1] There have been reported that the composition (SiO_2/Al_2O_3 ratio) of mullite has a great influence on the microstructure development of the sintered body and consequently affect the mechanical properties of mullite ceramics.[2,3] Besides the composition, mullitization sequence of the starting materials also has some influence on the microstructure.[4–7] However, there is still some dispute about the formation process of so-called spinel phase and mullitization sequence as well.[8–10] On the other hand, direct transformation from the amorphous phase to crystalline mullite has been observed in the powders derived from the spray pyrolysis[11] and coprecipitation method.[12]

In the present study, four different powder preparation methods have been carried out to study the influence of powder characteristics on the sintering behavior, microstructure development and mechanical properties of stoichiometric mullite.

Experimental Procedure

Four different mullite powders with stoichiometric composition (71.8 wt% Al_2O_3/28.2 wt% SiO_2) were synthesized by means of spray pyrolysis, hydrolysis and co-precipitation techniques. The spray pyrolized powder [S] was prepared by the instantaneous decomposition of mixed water-methanol solution of aluminum nitrate and ethyl silicate with the same procedure as reported previously.[11] Two different types of powders were prepared by the hydrolytic decomposition of mixed benzene solution of aluminum isopropoxide and methylsilicate with water (pH = 7)[Ha] or with aqueous ammonia (pH = 10)[Hb] followed by drying. In the case of the co-precipitation method [C], mullite precursor was coprecipitated from the same mixed solution as [S] with aqueous ammonia. The preparation sequence of each method is summarized in Table 1.

Table 1. Synthesis of various mullite precursors

Symbol	Method of synthesis		
S	$Al(NO_3)_3$ + $Si(OEt)_4$ in H_2O,MeOH	Spray Pyrolysis at 650 °C	Amorphous Powder
Ha	$Al(OPr^i)_3$ + $Si(OMe)_4$ in C_6H_6	Hydrolysis (H_2O, MeOH)	Amorphous Powder
Hb	$Al(OPr^i)_3$ + $Si(OMe)_4$ in C_6H_6	Hydrolysis (NH_4OH, H_2O, MeOH)	Amorphous Powder
C	$Al(NO_3)_3$ + $Si(OEt)_4$ in H_2O,MeOH	Co-precipitation (NH_4OH, H_2O, MeOH)	Amorphous Powder

Me:CH_3-, Et:C_2H_5-, Pr^i:iso-C_3H_7-

All the powders obtained were calcined at temperatures where full mullitization occurred: 950°C for S and Ha, and 1450°C for Hb and C. The calcined powders were ground by a vibratory mill using a nylon-lined pot and silicon nitride balls with ethanol as a grinding medium. The powder compacts were formed by use of a cold isostatic press at 200 MPa.

Thermal shrinkage for these compacts was measured by a differential type dilatometer* up to 1650°C in air. Bulk density of the specimens sintered between 1570° to 1650°C for 4 h was measured by the Archimedes method using distilled

*Rigaku Denki Co., Ltd. TMA-1700.

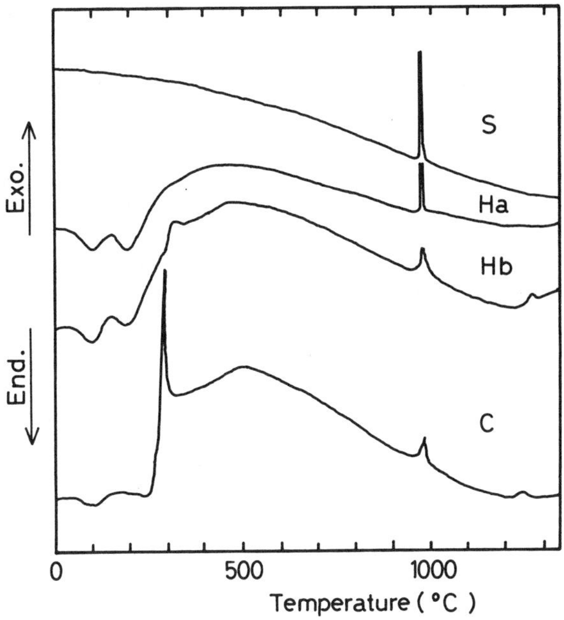

Fig. 1. DTA curves for powder prepared by different methods.

water. Microstructure of the sintered specimens was observed by means of SEM† on polished and etched surfaces. The specimens sintered at 1650°C for 4 h were machined and ground with #400 diamond wheel to prepared test pieces of the size 3 × 4 × 40 mm. The flexural strength of the test pieces was measured‡ by three-point bending from room temperature to 1400°C in air. Identification of the crystalline phases was carried out by XRD§.

Results and Discussion

Characteristics of the Powders

All the as-synthesized powders were amorphous to XRD. The DTA curves for the precursors are shown in Fig. 1. A broad endothermic and exothermic peaks were observed below 500°C, corresponding to dehydration and burn out of the volatile organic solvent residue, respectively, except for powder S. In the case of the S powder, a single and sharp exothermic peak was observed at about 980°C, which is attributed to mullite formation from amorphous phase as mentioned in previous reports.

---

†JEOL Ltd. T-300.

‡Shimadzu Seisakusho, Ltd. DSC-2000.

§Rigaku Denki Co., Ltd. Rad-ru.

The Ha powder also showed a clear single exothermic peak at about 980°C. However, a trace of Al-Si spinel was detected in addition to the predominant mullite phase by XRD. It indicates the simultaneous formation of mullite and Al-Si spinel at about 980°C. In the case of Ha powder, no other endothermic or exothermic reaction was observed at a higher temperature than 980°C.

On the other hand, two exothermic peaks were observed at above 500°C for powders Hb and C. The first sharp peak appeared at about 980°C and the second one, which is smaller and broader, was observed at about 1300°C. These peaks were attributed to the formation of Al-Si spinel from amorphous phase and the transformation into mullite, respectively.

The above results suggest that the highest degree of homogeneity in the elementary distribution of Al and Si can be achieved by spray pyrolysis method, and that the degree of homogeneity seemed to decrease in the order of Ha, Hb and C powders.

Sintering Behavior and Microstructure

Calcination and grinding condition, specific surface area after grinding and chemical composition of each powder determined by X-ray fluorescence analysis are summarized in Table 2. In order to attain full mullitization and a similar value of specific surface area, calcination and grinding were adjusted to optimal condition. The chemical composition of each obtained powder was in good agreement with the calculated value. Figure 2 shows the TEM[1] micrographs of the powders calcined and ground. The particle size distribution in S and Ha powder were between 0.2 and 0.7 um, and that of Hb and C powder were in the range of 0.1 to 0.3 um. The morphology of the particles showed that S and Ha powder are composed of aggregated and porous particles, whereas Hb and C powders were composed of fine polyhedral and well dispersed particles.

Table 2. Chemical composition, treatment condition and surface area of the starting powders.

Symbol	Al_2O_3 content	Calcination	Grinding	Surface area
S	71.2 wt%	950 °C 1h	80 h	20.1 m^2/g
Ha	71.6	950	50	35.8
Hb	71.5	1450	50	33.9
C	71.5	1450	50	28.5

Thermal shrinkage curves of rectangular bars (5 × 5 × 20 mm) made from the mullite powders are shown in Fig. 3. All the samples tended to shrink gradually with increasing temperature up to about 1500°C, and rapid shrinkage was observed for Ha, Hb and C samples between 1500° and 1650°C. The rapid shrinkage of S sample was observed to start above 1500°C. The S and Ha specimens showed further shrinkage when heated at 1650°C. Figure 4 shows bulk density of each specimen as a function of firing temperature with a soak of 4 h.

---

[1]JEOL Ltd. 4000F.

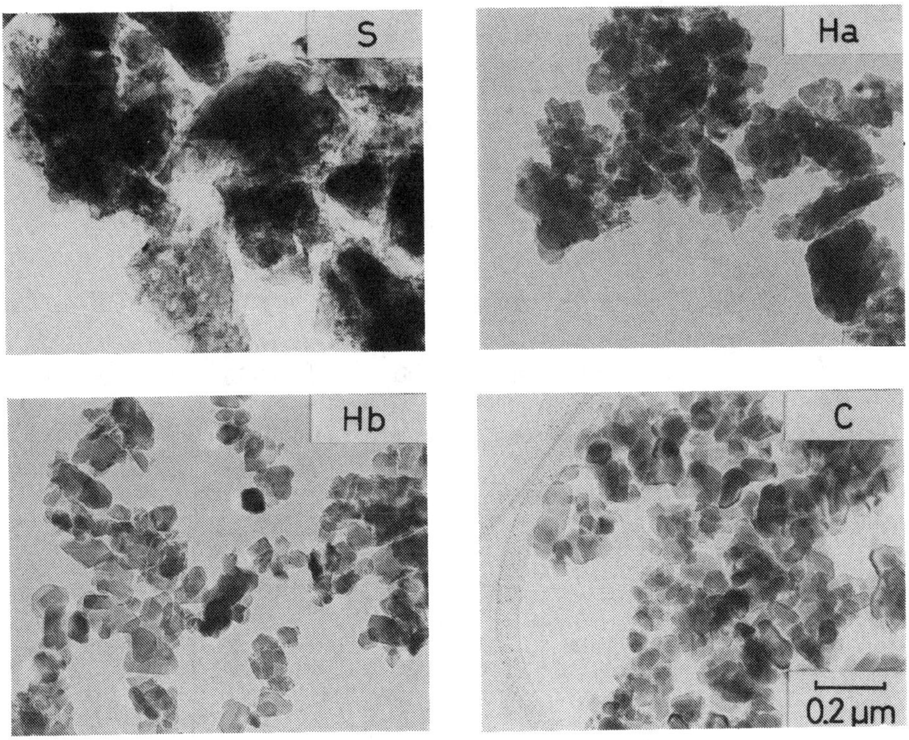

Fig. 2. Transmission electron micrographs of the powders after grinding.

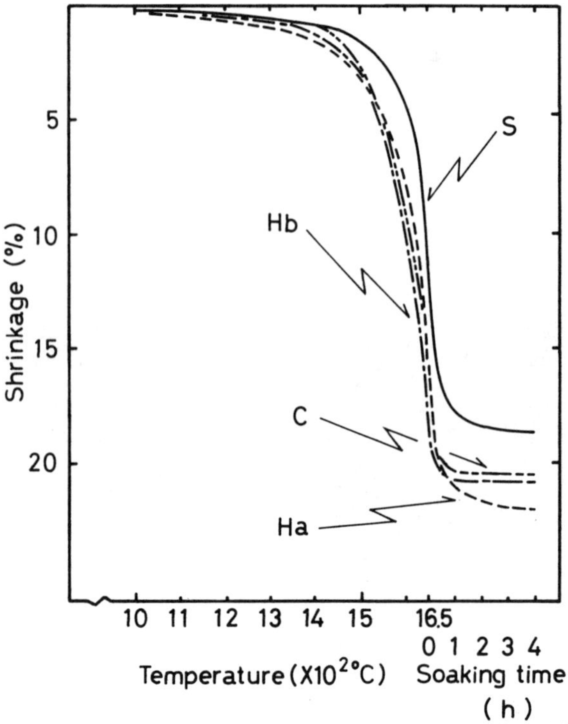

Fig. 3. Change of thermal shrinkage curves with synthesis of mullite powder.

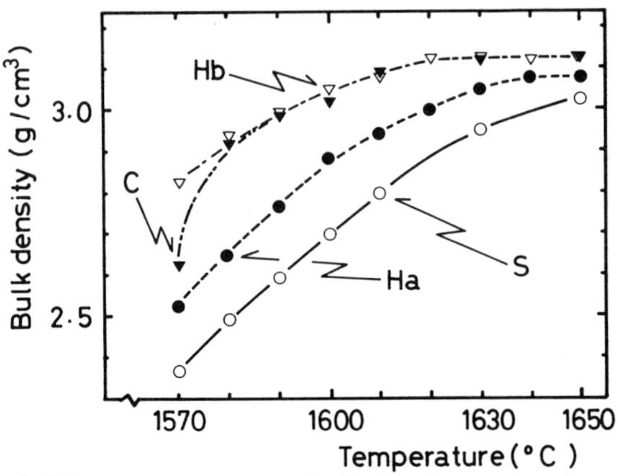

Fig. 4. Effect of firing temperature on bulk density.

It is obvious that Hb and C powders are more sinterable than the S powder, as expected from Fig. 3. The temperature at which full densification can be attained is different for each specimen. It is also suggested that firing temperatures higher

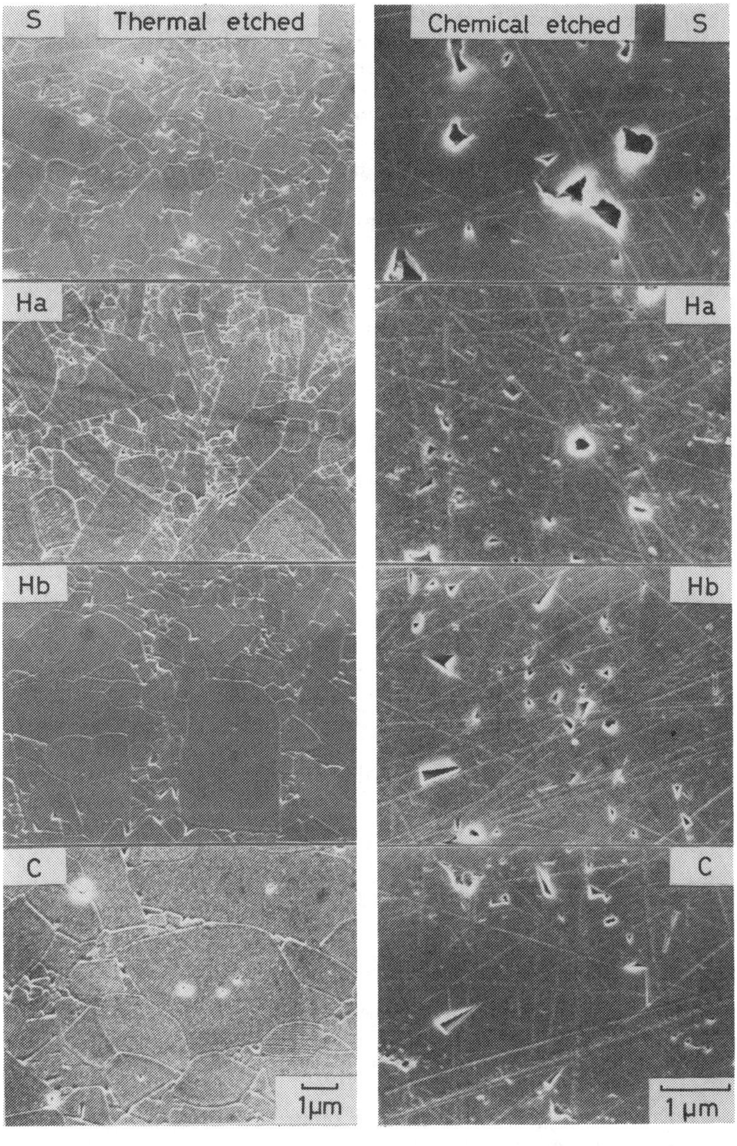

Fig. 5. Scanning electron micrographs of the specimens sintered at 1650°C for 4 h.

than 1650°, 1640°, and 1630°C are necessary, for the full densification of the S and Ha, Hb, and C powders. Figure 5 shows SEM photographs of the samples fired at 1650°C for 4 h. Each sample was thermally etched in air for 15 min at 1500°C (left) or chemically etched with a 1% hydrofluoric acid solution for 24 h at 0°C (right).

It seems that specimen S contains a large amount of pores than the other specimens. Grain growth is found to proceed along with the densification. The presence of a glassy phase in the sintered body is recognized from the pores with angular morphology formed by chemical etching, which were exclusively located at the three grain junctions. The specimens Hb and C obviously contain larger amounts of glassy phase than S and Ha. Therefore, the higher densification rate for the sample Hb and C can be attributed to the amount of liquid phase formed, which also accounts for the fraction of Al-Si spinel phase formed during calcination of the starting powders.

The amount of glassy phase in the sintered bodies is closely dependent upon the route of the mullitization process, that is, direct transformation from amorphous to mullite or mullitization by way of an Al-Si spinel phase.

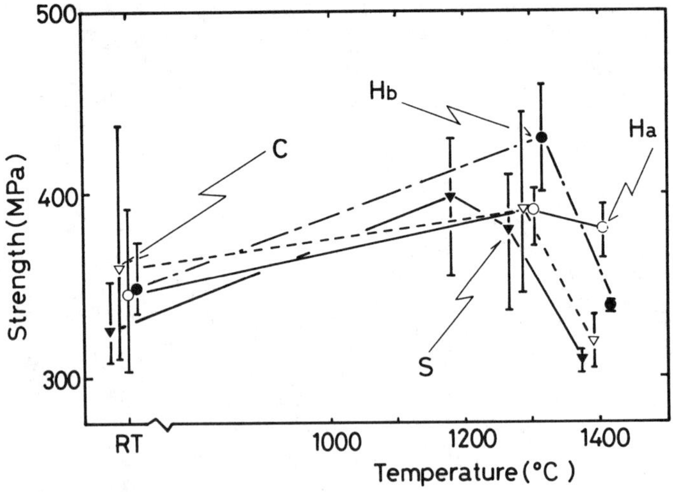

Fig. 6. Temperature dependence of the flexural strength of mullite.

Influence of the Powder Characteristics on Mechanical Properties of the Sintered Mullite

The flexural strength as a function of temperature for the specimens prepared by the different powder preparation methods is shown in Fig. 6. Little difference in the room temperature flexural strength was observed among the powders, and the strength value ranged from 320 to 360 MPa. The flexural strength increased gradually with temperature up to 1300°C, and decreased rather abruptly except for the Ha specimen. The increase in the strength with increasing temperature can be attributed to the presence of the grain boundary glassy phase.[13] The ratio of the strength values at 1300°C and at room temperature increased in the order of C, Ha, Hb and S. The amount of Al-Si spinel formed during calcination and also the amount of glassy phase which exists in the sintered body increases in the order C, Hb, Ha and S. Therefore, the increase in the flexural strength ratio at 1300°C and at room temperature is correlated to the increase of the amount of glassy phase in the sample. The decrease in the strength above 1300°C is reasonably attributable to the decrease in viscosity of the grain boundary glassy phase.[14] Although specimens Ha and S contained almost the same amount of glassy grain-boundary phase, the flexural strength of specimen S decreased more than that of Ha. This discrepancy can be explained by the different amount of metallic impurities in both specimens. Specimen S contained 240 ppm of Na_2O and specimen Ha contained only 30 ppm. Therefore, the rapid decrease in the viscosity of silica glass[15] with the increasing amount of alkaline element is considered to be responsible for the rapid decrease in the strength value at 1400°C.

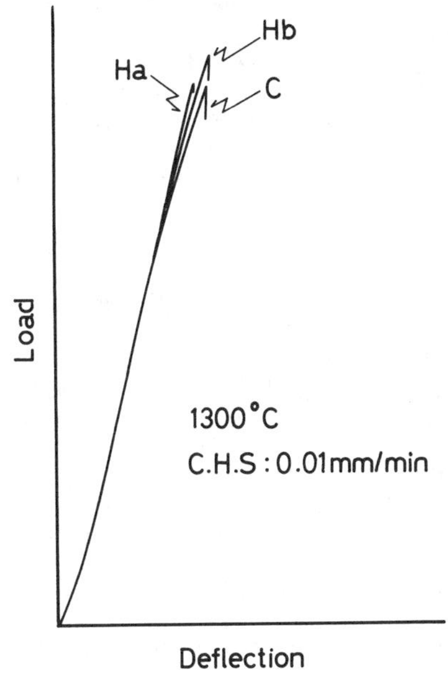

Fig. 7. Typical load-deflection curves for specimens Ha, Hb and C at 1300°C.

Even when the instantaneous fracture strength of mullite ceramics does not decrease at an elevated temperature, time-dependent fracture can decrease at high temperature because all the mullite ceramics examined contained a glassy grain boundary phase to some extent regardless of the powder preparation method. Then, three point bending test was performed at 1300°C with a very slow loading rate of 0.01 mm/min. Figure 7 shows typical load-deflection curves for the specimens Ha, Hb and C. The sample Ha ruptured in a rather brittle manner, while the fracture of specimens Hb and C was accompanied by plastic deformation to some extent. The extent of plastic deformation for sample C was highest. Consequently, it is recognized that the increase in the amount of glassy phase results in the lowering of time-dependent failure strength at elevated temperatures.

From the above results, it is concluded that the powder preparation method affects elemental homogeneity in the starting powder, which greatly influences the development of the microstructure and accordingly the mechanical properties, especially time-dependent fracture behavior of the sintered mullite ceramics.

Summary and Conclusions

Mullite powders of stoichiometric composition (71.8 wt% Al_2O_3) were prepared by different processes, e.g., spray pyrolysis [S], hydrolysis with water [Ha], hydrolysis with aqueous ammonia [Hb] and co-precipitation [C] methods. Influence of the powder characteristics on microstructure and mechanical properties was investigated and the following results were obtained:

1. Mullitization temperature of the as-synthesized amorphous powders increased in the order of S < Ha < Hb < C.
2. The bulk density of the sintered bodies increased in the order of S < Ha < Hb = C, and the rate of grain growth and the amount of grain-boundary glassy phase also increased in the same order.
3. Powder preparation methods or the amount of glassy phase in the sintered bodies do not influence very much the instantaneous fracture strength of the sintered bodies from each powder up to 1300°C. On the contrary, time-dependent failure strength is found to be considerably affected by powder characteristics.

References

[1]R. J. Brook, "Processing Technology for High Performance Ceramics," *J. Mater. Sci. and Engi.*, **71**, 305–312 (1985).

[2]M. D. Sacks and J. A. Pask, "Sintering of Mullite-Containing Materials: I, Effect of Composition," *J. Am. Ceram. Soc.*, **65** [2] 65–70 (1982).

[3]S. Kanzaki, T. Kumazawa, J. Asami, O. Abe and H. Tabata, "Dependency of Mechanical Property of Sintered Mullite on Chemical Composition," *Yogyo-Kyokai-Shi*, **93** [7] 407–408 (1985).

[4]P. D. D. Rodrigo and P. Boch, "High Purity Mullite Ceramics by Reaction Sintering," *Int. J. High Technology Ceramics*, **1**, 3–30 (1985).

[5]K. S. Mazdiyasni and L. M. Brown, "Synthesis and Mechanical Properties of Stoichiometric Aluminum Silicate (Mullite)," *J. Am. Ceram. Soc.*, **55** [11] 548–552 (1972).

[6]B. L. Metcalfe and J. H. Sant, "The Synthesis, Microstructure and Physical Properties of High Purity Mullite," *Trans. J. Brit. Ceram. Soc.*, **74**, 193–201 (1975).

[7]T. Kumazawa, S. Kanzaki, J. Asaumi, O. Abe and H. Tabata, "Sinterability of SiO_2-Al_2O_3 Powders Prepared by Spray Pyrolysis: Effect of Chemical Composition," *Yogyo-Kyokai-Shi*, **94** [5] 485–490 (1986).

[8]G. W. Brindley and M. Nakahira, "The Kaolinite-Mullite Reaction Series: III, The High-Temperature Phases," *J. Am. Ceram. Soc.*, **42** [7] 319–324 (1959).

[9]M. Bulens, A. Leonard and B. Delmon, "Spectroscopic Investigations of the Kaolinite-Mullite Reaction Sequence," *J. Am. Ceram. Soc.*, **61** [1-2] 81–84 (1959).

[10]K. Okada and N. Otsuka, "Characterization of the Spinel phase from SiO_2-Al_2O_3 Xerogels and the Formation Process of Mullite," *J. Am. Ceram. Soc.*, **69** [9] 652–656 (1986).

[11]S. Kanzaki, H. Tabata, T. Kumazawa and S. Ohta, "Sintering and Mechanical Properties of Stoichiometric Mullite," *J. Am. Ceram. Soc.*, **68** [1] C6-C7 (1985).

[12]D. W. Hoffman, R. Roy and S. Komarneni, "Diphasic Xerogels, A new Class of Materials: Phases in the System Al_2O_3-SiO_2," *J. Am. Ceram. Soc.*, **67** [7] 468–471 (1984).
[13]S. Kanzaki, T. Kumazawa and H. Tabata, "Sintering and Mechanical Properties of Mullite Derived Via Spray Pyrolysis," Proceeding of Mullite Workshop (1987).
[14]T. Mah and K. S. Mazdiyasni, "Mechanical Properties of Mullite," *J. Am. Ceram. Soc.*, **66** [10] 699–703 (1983).
[15]G. W. Morey, "The Properties of Glass," 2nd ed., Reinhold Publishing Corporation, pp. 132–165 (1954).

Microstructure and Bending Strength of Highly Pure Mullite Ceramics

MINEO MIZUNO

Research and Development Division
Kyoritu Ceramic Materials Co. Ltd.
Tsukisan-cho 2-41, Minato-ku, Nagoya, 455-91 Japan

MASANORI SHIRAISHI

STK Ceramics Laboratory Corp.
Tsukisan-cho 1-11
Minato-ku, Nagoya, 455 Japan

HAJIME SAITO

Toyota Technological Institute
Hisakata 2-12-1
Tenpaku-ku, Nagoya, 468 Japan

Abstract

Dense mullite ceramics with high bending strength were produced by hot-pressing or by pressureless sintering using highly pure fine mullite powder prepared from aluminum sulfate and fumed silica. Bending strengths of the specimens were found to have relationships to the densities and the microstructures which vary according to heating conditions. High bending strength mullite ceramics of 350 to 404 MPa require following conditions: 1. Bulk densities above 3.06 g/cm^3, 2. Maximum grain size less than 25 μm, 3. The fraction of elongated grains larger than 2 μm between 30 to 70%.

Introduction

Highly pure mullite ceramics have been studied extensively because of high bending strength and excellent creep resistance even at temperatures higher than 1200°C.[1–3] Therefore, mullite is expected to be one of candidates for high temperature structural applications.

Several kinds of aluminum salt and fumed silica were examined as starting materials of mullite powder, and the combination of aluminum sulfate and fumed silica was found to be superior to other materials for obtaining dense sintered mullite bodies with high bending strength.[4,5]

Ryshkewitch,[6] and Coble and Kingery (1956)[7] found the strength to decrease exponentially with increasing porosity because the pores decreased the load-bearing cross-section area.

Carniglia[9] reexamined the Orowan, Petch, and Knudsen equations on average grain size vs average porosity-normalized strength of a number of specimens of a brittle material. He supported the Orowan and Petch equations.

Experimentally, it is well known that strength of polycrystalline ceramics decreases with increasing average grain size. It may depend on the fact that critical flaw sizes increase with increasing grain size for many types of flaws.[10] Also, there is a strong negative correlation between fracture stress and grain size for large grains at fractures.[11]

Anisotropic grains such as pillor-shaped, fiber-like or plate-like increase the strength for some ceramics. For example a crack perpendicular to the direction of an elongated grain would require more fracture energy to develop, leading to the strength increase of the specimen.[10,12]

Dense mullite ceramics with high strength were fabricated in this study by hot-pressing or by pressureless sintering. Although many factors affect the strength of ceramics,[13,14] the bending strength is demonstrated to depend strongly on the following microstructure features:

1. Porosity of specimens.
2. Shape, size and distribution of grains.
3. Phases (mullite and glassy phases).

Experimental Procedure

Powder Preparation

The schematic diagram of mullite powder preparation by pyrolysis is shown in Fig. 1. Aluminum sulfate and fumed silica were used as starting materials. Fumed silica* was mixed with aqueous aluminum sulfate solution† to realize the stoichiometric mullite composition ($3Al_2O_3 \cdot 2SiO_2$; 71.8 wt% Al_2O_3·28.2 wt% SiO_2). The mixture was stirred for 5 h and dried at 120°C. The dried gel was heated at 1000°C to decompose the aluminum sulfate into γ-Al_2O_3. The resultant mixture of γ-Al_2O_3 and amorphous silica reacted to form mullite at 1350°C for 3 h.

The obtained mullite powder was analyzed by X-ray fluorescence and was identified by XRD. Grain morphology was observed by transmission electron microscope (TEM). And the specific surface area was determined by the B.E.T. method.

Preparation and Testing of Mullite Bodies

Mullite bodies were hot-pressed in a carbon die at temperatures from 1550 to 1700°C for 1 h under 10 MPa in vacuum (less than 10^{-3} torr). Furthermore, green compacts of mullite powder were also cold isostatically pressed at 200 MPa and sintered under atmospheric conditions at temperatures from 1630 to 1690°C for 1 to 10 h.

The bulk density was measured by the Archimedes' method. The relative density was calculated using 3.17g/cm^3 as theoretical density (ASTM XRD card 15–776).

---

*No. 200, Nippon Aerosil Co.: Specific surface area is 200 m^2/g.

†It contains 7.8 wt% of Al_2O_3.

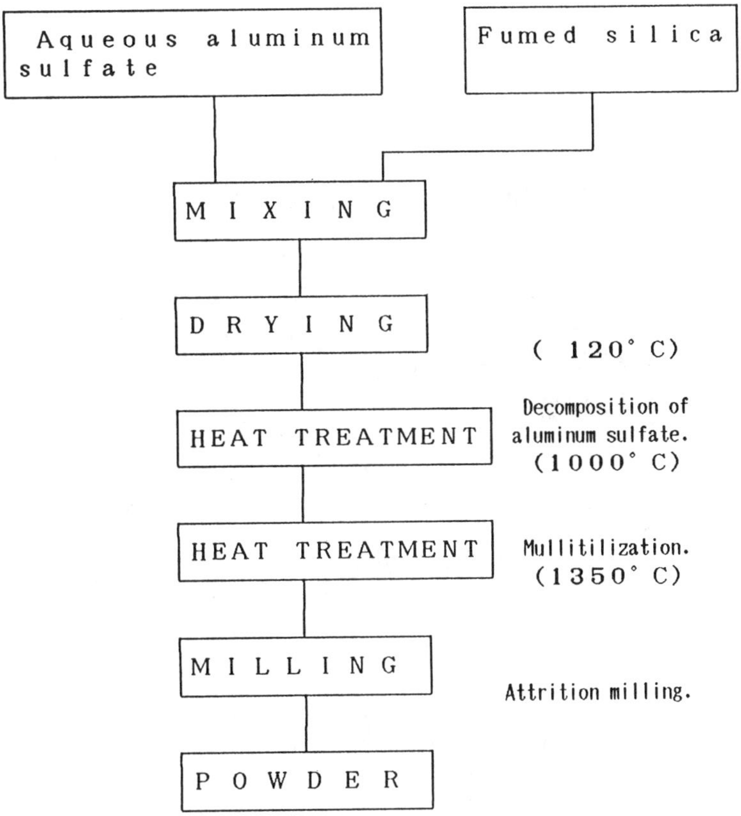

Fig. 1. Processing of mullite powder.

Specimens were cut into testbars of 3 by 4 by 40 mm. Polished specimens were tested in 3-point bending with a span of 30 mm. Microstructures of polished surfaces after thermal etching and of fracture surfaces after the bending test were observed by scanning electron microscope (SEM).

Results and Discussion

Mullite Powder

TEM observations show that the mullite powder is composed of particles from 0.1 to 0.2 μm in size (Fig. 2). The specific surface area is $24m^2/g$. Particles have a uniform shape with narrow particle size distribution. A selected area electron diffraction pattern shows that the particles are well crystallized. The crystalline phase identified by XRD is mullite.

The weight ratio of Al_2O_3 to SiO_2 (Table 1) is 2.56 which corresponds to the theoretical composition of mullite. Impurities, especially alkaline and alkaline earth metal oxides, cause the formation of glassy phases in mullite. Since the glassy phases may degrade the mechanical properties at high temperatures, the powder should

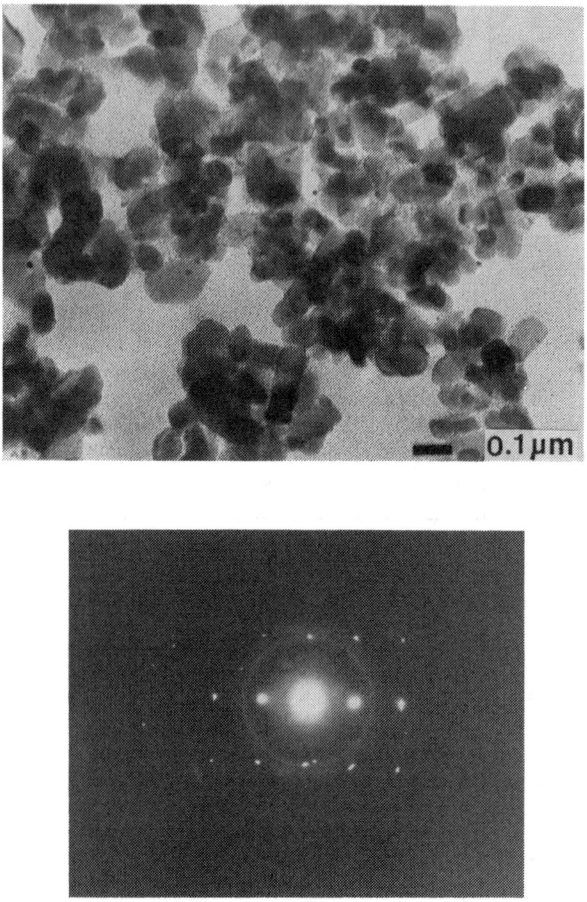

Fig. 2. Transmission electron micrographs of mullite powder.

Table 1. Chemical composition of mullite powder analyzed on XRF

SiO_2	27.90
Al_2O_3	71.41
Fe_2O_3	0.03
CaO	0.01
MgO	0.01
K_2O	0.01
Na_2O	0.03
Al_2O_3/SiO_2	2.56

contain as little an amount of impurities as possible. The total amount of Na_2O, K_2O, CaO, and MgO is 0.06 wt%, which is much lower than that of conventional mullite.[15]

Hot-Pressed Specimens

Crystalline phase of the hot-pressed samples identified by XRD is mullite, and no other phases were found.

Scanning electron micrographs of hot-pressed specimens show that the body is composed of small granular grains below 1 μm and elongated prismatic crystals (Fig. 3). The significant growth of elongated grains with increasing temperature suggests that they grew by liquid phase crystal growth. Impurities of 0.6 wt% of alkaline and alkaline earth metal oxides may have played some role for lowering the temperature of formation of a glassy phase.

A few large grains in the specimen HP-1550 (hot-pressed at 1550°C) are dispersed in the matrix of small grains under 1 μm. The maximum size of elongated grains of the specimen HP-1550 observed by SEM is 5 μm (Fig. 4), and the bulk density is only 3.00g/cm^3 (relative density of 94.6% TD).

With increasing temperature up to 1700°C, elongated prismatic grains increase in quantity and size consuming small granular grains. A small amount of pores at grain boundaries is observed in the specimen HP-1600. The maximum size of the

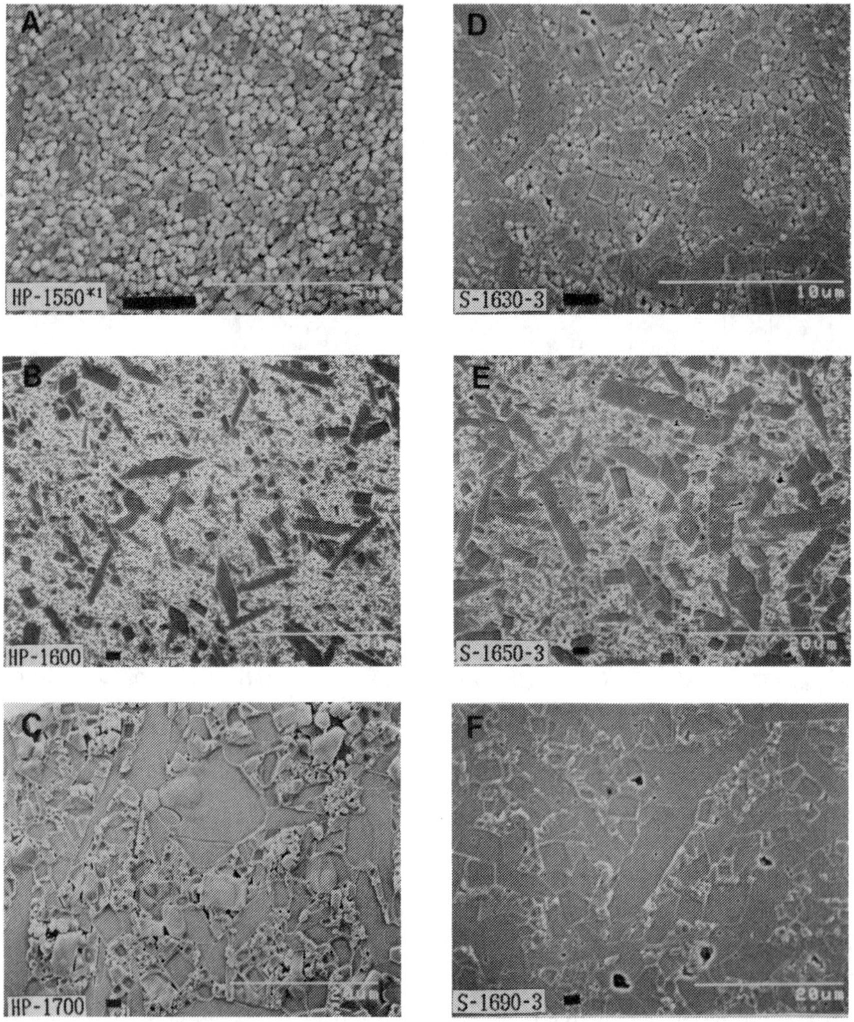

Fig. 3. Scanning electron micrographs of mullite specimens. (A) to (C); Hot-pressed for 1 h under 10 MPa in vacuum at 1550°, 1600° and 1700°C, respectively. (D) to (F); Sintered for 3 h under ordinary pressure at 1630°, 1650° and 1690°C, respectively. (Bars 2 μm) *1 Specimen designation.

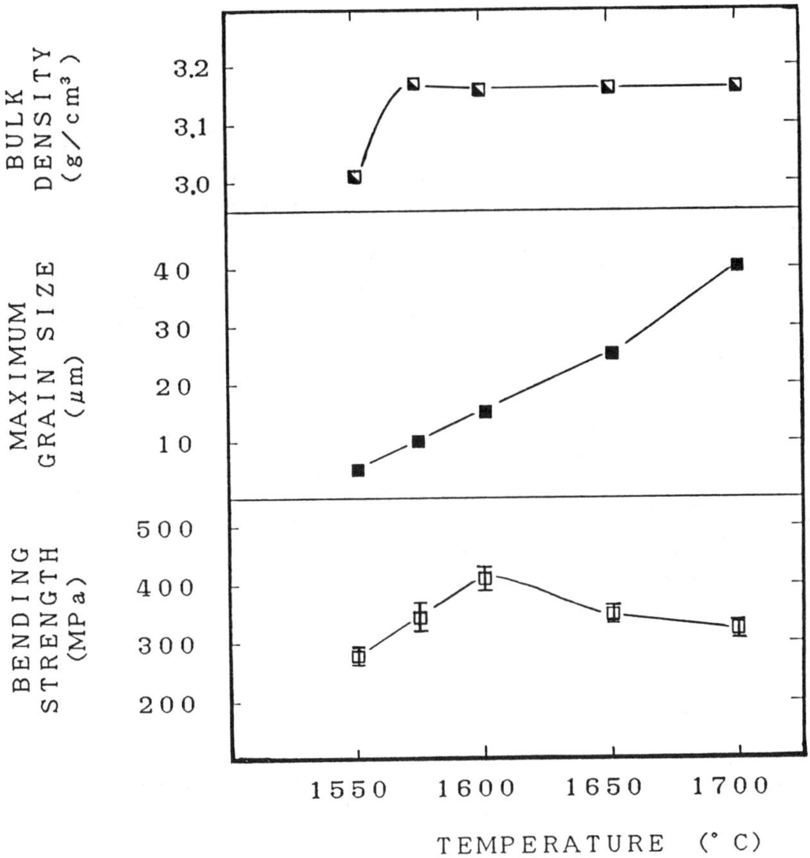

Fig. 4. Effect of hot-pressing temperature on bulk density, maximum grain size*1 and bending strength*2. *1 Maximum grain size observed on SEM. *2 3-point average (□), maximum and minimum bending strength.

elongated grains is about 15 μm. Exaggerated grain growth to 40 μm and agglomeration of pores at tripple points are observed in the specimen HP-1700.

The specimen HP-1700 shows spherical structure inside some large grains, suggesting partial melting. Mullite of stoichiometric composition is equilibrated with glassy-phase above 1600°C in the SiO_2-Al_2O_3 binary system (Prochazka and Klug[16]). Since the solidus line of mullite and mullite plus liquid phase bends over to the alumina side at higher temperatures, treatment at higher temperatures over 1600°C is suggested to cause the formation of mullite with higher alumina content and more glassy phase.

Bending strength varies with hot-pressing temperatures (Fig. 4). The low bulk density of the specimen HP-1550 leads to low bending strength. The existence of pores exponentially reduces the strength.[6,7]

Since the density of specimens hot-pressed does not change above 1600°C, the difference in bending strength is considered to be due only to the specific

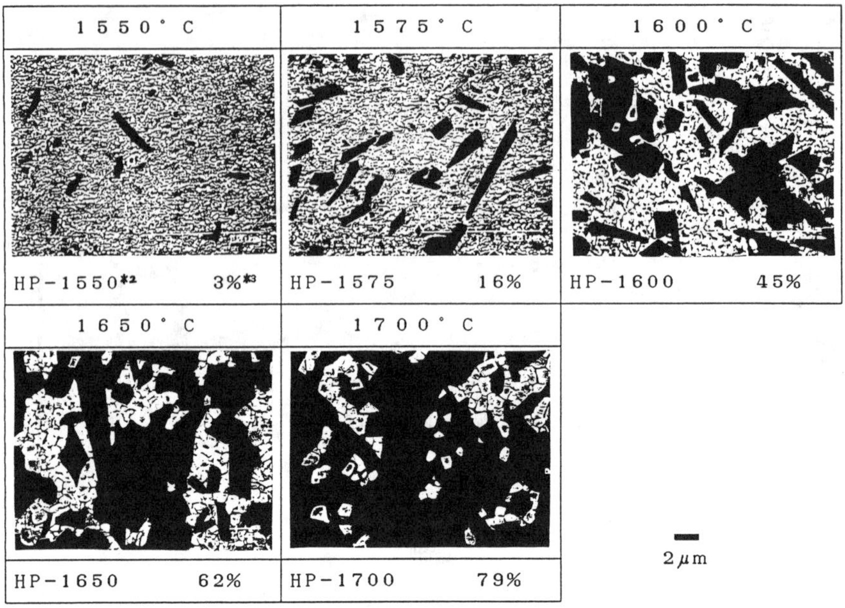

Fig. 5. Effect of hot-pressing temperature[*1] on the area (dark) of grains larger than 2 μm (SEM observation). [*1] Hot-pressed at each temperature for 1 h under 10 MPa in vacuum. [*2] Specimen designation. [*3] Percentage of the dark area.

microstructure. Aspect ratios of elongated grains are from 4 to 15 in all specimens. The maximum grain size, however, linearly increases with increasing temperature. Elongated mullite grains show intergranular fractures, and small mullite grains in the matrix show transgranular fractures and might involve also intergranular fractures (SEM observation). The increasing strength of specimens can be explained by the fracture energy of grains.[17]

The highest average strength (404 MPa) is obtained in the specimen HP-1600. The strength decreases by overheating. When the maximum grain lengths exceed about 25 μm, the strengths of specimens again degrade. Large grains themselves would become the origins of fracture[13] and they could form large flaws.[12] The decrease in the strength of specimens HP-1650 and HP-1700 is considered to depend on the exaggerated grain growth over 25 μm in size and on the extention of flaws.

Portion of elongated grains also seem to affect the strength. The areas of elongated grains larger than 2 μm, indicated as the dark areas in scanning electron micrographs (Fig. 5), increase from 3% to 79% with increasing temperature from 1550 to 1700°C. The occupancy of elongated grains in the specimen HP-1575 having relative density of 99.7% was only 16%. This may be the reason for its low strength.

Observation of microstructure and the relation between bending strength and heating temperature indicate that an optimum size and area of elongated grains are required for mullite grains to get high strength. These conditions for hot-pressed mullite ceramics are considered to be: (1) the maximum size of elongated grains about 15 μm, (2) the area of grains larger than 2 μm in size about 45% (SEM observation).

Sintered Specimens

Crystalline phase of the sintered specimens is mullite and no other phases were identified by XRD. This situation is the same as for hot-pressed specimens.

Bulk densities, maximum size of elongated grains, bending strength, and areas of elongated grains larger than 2 μm vary with increasing temperature and time (Fig. 6 and 7). These experimental results show that hot-pressed specimens are not different from sintered specimens except for density and pores. Sintered specimens involve intergranular pores with diameters up to 2 μm and intragranular pores with diameters up to 0.2 μm. Consequently, the bulk densities were a little lower than those of hot-pressed specimens. They vary from 3.00 to 3.10g/cm^3 (relative densities from 94.6 to 97.8%) according to sintering conditions.

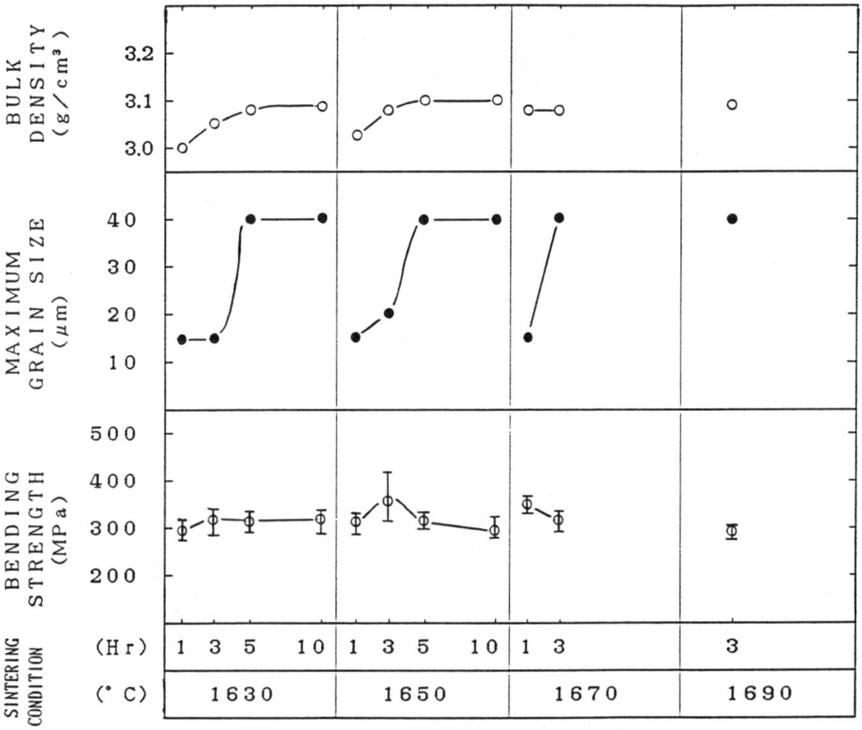

Fig. 6. Effect of sintering condition on bulk density, maximum grain size[*1] and bending strength[*2]. [*1] Maximum grain size observed on SEM. [*2] 3-point average (○), maximum and minimum bending strength.

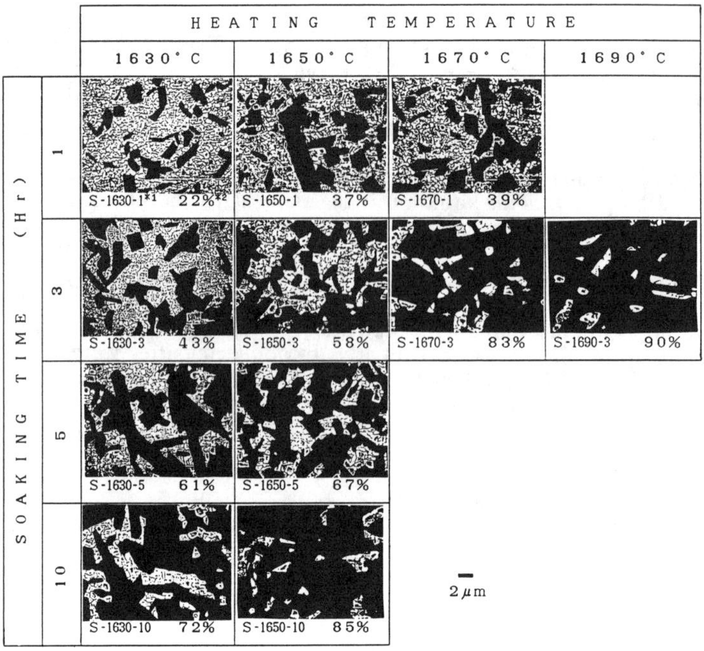

Fig. 7. Effect of sintering condition on the area (dark) of grains larger than 2 μm (SEM observation). *1 Specimen designation. *2 Percentage of the dark area.

Specimens S-1650-3 (sintered at 1650°C for 3 h) and S-1670-1 show high average bending strengths of 357 and 350 MPa, respectively. The maximum strength in all specimens in 421 MPa for S-1650-3. These two specimens exhibit the same density (3.08g/cm^3, relative density of 97.2%), and similar maximum sizes for elongated grains (from 15 to 20 μm). The portions of the area of grains larger than 2 μm estimated on SEM are 58% for S-1650-3 and 39% for S-1670-1. Their densities are lower than those of the hot-pressed specimens. The strengths, however, are almost the same.

Specimens S-1630-1, S-1630-3 and S-1650-1, having densities of 3.00, 3.05 and 3.03g/cm^3 (relative densities of 94.6, 96.2 and 95.6%), respectively, are weak in strength. Intragranular pores up to 0.2 μm in diameter in these specimens seem to be too small to act as fracture origins, but intergranular pores up to 2 μm in diameter may be large enough to act as fracture origins. Therefore, densities lower than 3.05g/cm^3 may reduce the strength.

The bending strengths of the specimens, S-1630-5, S-1630-10, S-1650-10, S-1670-3 and S-1690-3, are rather low, in spite of higher densities of 3.07 to 3.10g/cm^3. This is considered to depend on the exaggerated grain growth up to 40 μm, resulting in the strength decrease in the same fashion as for hot-pressed specimens.

Conditions for High-Strength Mullite Ceramics

High-strength mullite ceramics that are hot-pressed or pressureless-sintered can be fabricated under the following conditions:

1. Bulk density more than 3.06g/cm^3.
2. Maximum size of elongated grains less than 25 μm.
3. The areal portion of elongated grains larger than 2 μm in size from 30 to 70%.

Hot-pressing at 1600°C for 1 h and sintering at 1650°C for 3 h and at 1670°C for 1 h are included in these conditions. Their products show high bending strengths on the average.

Conclusion

1. Mullite ceramics were prepared from highly pure mullite powder by hot-pressing at temperatures from 1550 to 1700°C for 1 h under 10 MPa and by pressureless-sintering at temperatures from 1630 to 1690°C.
2. These specimens are composed of small granular mullite grains below 1 μm in size and elongated prismatic mullite crystals up to 40 μm in length.
3. Densities of specimens and maximum sizes of elongated mullite grains increase with increasing temperature and time. Bending strengths, however, decrease with overheating and prolonged sintering time. This is considered to be due to exaggerated grain growth over 25 μm in length.
4. High-strength mullite ceramics that are hot-pressed or pressureless-sintered require the following conditions:
   1. Bulk densities above 3.06 g/cm^3.
   2. Maximum grain size less than 25 μm.
   3. The fraction of elongated grains larger than 2 μm between 30 to 70%.

Acknowledgment

The writers deeply thank Dr. Yoshiko Suwa for her critical review of the manuscript and helpful discussion.

References

[1]T. Hiraiwa, "Synthesis of High-purity Mullite Powder," *Kogyo Zairyo*, **34** [13] 67–72 (1986).

[2]S. Kanzaki and H. Tabata, "Sintering and Mechanical Properties of Stoichiometric Mullite," *J. Am. Ceram. Soc.*, **68** [1] C-6–C-7 (1985).

[3]M. G. M. U. Ismail, Z. Nakai, and S. Somiya, "Microstructure and Mechanical Properties of Mullite Prepared by the Sol-Gel Method," *J. Am. Ceram. Soc.*, **70** [1] C-7–C-8 (1987).

[4]M. Mizuno and H. Saito, "High-Purity Mullite powder prepared by Pyrolysis and Its Properties," pp. 31–41 in Mullite II. Edited by S. Somiya. Uchida Rokakuho Publishing Co., Tokyo, Japan 1987.

[5]H. Saito, M. Mizuno, and M. Shiraishi, "Preparation of High-purity Mullite Powders," Abstract for Annual Meeting of Japan Ceram. Soc. 233–234 (1987).

[6]E. Ryshkewitch, "Compression Strength for Porous Sintered Alumina and Zirconia," *J. Am. Ceram. Soc.*, **36** [1] 65–8 (1953).

[7]R. L. Coble and W. E. Kingery, "The effect porosity on the mechanical properties of sintered alumina," *J. Am. Ceram. Soc.*, **39**, 377 (1956).

[8]R. H. Herron and W. J. Smothers, "Technique for Microscopic Observation of crack Propagation," *Am. Ceram. Soc. Bull.*, **46** [2] 181–185 (1967).

[9]S. C. Carniglia, "Reexamination of Experimental Strength-vs-Grain size Data for Ceramics," *J. Am. Ceram. Soc.*, **55** [5] 243–49 (1972).

[10]H. P. Kirchner, pp. 6–9 in Strengthening of Ceramics, Marcel Dekker, Inc. 1979.

[11]H. P. Kirchner, R. M. Gruver, and W. A. Sotter, "Characteristics of Flaws at Fracture Origins and Fracture Stress-Flaw Size relations in Various Ceramics," *Mater. Sci. Eng.*, **22**, 147–156 (1976).

[12]F. F. Lange, "Relation Between Strength, Fracture Energy, and Microstructure of Hot-pressed Si_3N_4," *J. Am. Ceram. Soc.*, **56** [10] 518–22 (1973).

[13]K. Hamano, pp. 25–46 in "Mechanical Properties of Ceramics," Edited by Yogyo Kyokai. Yogyo Kyokai, Japan 1979.

[14]K. Komeya, pp. 59–75 in "Mechanical Properties of Ceramics," Edited by Yogyo Kyokai. Yogyo Kyokai, Japan 1979.

[15]T. Kawanami, "Properties of Commercial Mullite Ceramics," pp. 123–35 in Mullite. Edited by S. Somiya. Uchida Rokakuho Publishing Co., Tokyo, Japan 1985.

[16]S. Prochazka and F. J. Klug, "Infrared-Transparent Mullite Ceramic," *J. Am. Ceram. Soc.*, **66** [12] 874–80 (1983).

[17]H. P. Kirchner and R. M. Gruver, "Strength-Anisotropy-Grain Size Relations in Ceramic Oxides," *J. Am. Ceram. Soc.*, **53** [5] 232–36 (1970).

Preparation of Transparent Mullite Films by Dip-Coating Method

Kiyoshi Okada* and Nozomu Ōtsuka

Department of Inorganic Materials
Tokyo Institute of Technology
O-okayama, Meguro, Tokyo, 152 Japan

Abstract

Transparent mullite thin films were prepared by the dip-coating method. The films were colorless, transparent, dense and tightly adherent to silica glass substrates. The transmittance, phase content, orientation of mullite, Vickers microhardness, bend strength, thermal stability and surface structure of the films were examined. Stability of the coating solution was also examined using solutions with different aging times.

Introduction

Formation of mullite needs a fairly high firing temperature of around 1500°C if particles of silica and alumina were used for the starting materials. However, firing to around 1000°C is good enough to yield mullite, if xerogel[1] hydrolyzed slowly at 60°C from the solution of tetraethoxysilane (TEOS) and aluminum nitrate nonahydrate in ethanol was used. It is well-known that thin films of amorphous and crystalline materials can be prepared at relatively low temperatures by the dip-coating method.[2–4] If mullite film can be prepared, it has potential use as a heat and chemical resisting film.

In this paper, preparation of mullite films was examined to put this specialty of yielding mullite at low heating temperature to practical use. Crystalline phase and microstructure of the films were investigated by powder X-ray diffraction (XRD) and scanning electron microscopy (SEM). Some properties, i.e., transmittance, hardness, bend strength and thermal stability, of the films were measured. Stability of the coating solution was also examined using solutions with different aging times.

---

*Member, The American Ceramic Society.

Experimental Procedure

Preparation of Thin Films

The coating solution was prepared by dissolving TEOS† and aluminum nitrate nonahydrate† in absolute ethanol so as to be $3Al_2O_3 \cdot 2SiO_2$ composition. Concentration of the solution was adjusted to be 0.025, 0.1, 0.15 and 0.2 M. The surface of silica glass slides 2 mm thick was ultrasonically cleaned with an acetone solution and used for the substrate after drying. Dip-coating was made by the following procedure. The substrates were dipped into the solution, withdrawn vertically with speeds of 0.5, 1, 1.5 and 3 cm/min, and dried at room temperature for 10 min. These treatments were done in dry nitrogen or air with 60% relative humidity. The dried coating substrates were heated for 5 min at 500°C and further heated for 10 min at 1050°C in air. This procedure was repeated up to twenty times to make thicker films. Dip-coating was made using the solutions with different aging times from 2 to 170 days to examine stability of this solution.

Measurement of Properties

Crystalline phase in the films was identified by XRD‡. Integrated intensity for each reflection was measured to determine formation amount and orientation relation of formed phases on the films. No internal standard was used for the measurement. Orientation index was calculated according to the Lotgering method.[5] Lattice constant of mullite was calculated using eleven reflections by the least squares program RSLC-3.[6] External standard§ method was adopted for the correction of reflection angle of the XRD instrument. Surface structure and thickness of the films were observed by SEM¶. Transmittance of the coating films was measured by spectrophotometer** with wave-length between 200 and 800 nm. Vickers microhardness was measured by Vickers microhardness tester††. Measurements were made at loads of 0.245 to 4.9 N for 30 s. Four point bend strengths‡‡ of the coated and non-coated silica glass slides were measured with a cross head speed of 0.2 mm/min. The coated films were further heated for 1 h at 1100, 1200, 1300 and 1400°C to examine thermal stability of the films.

---

†Wako Pure Chemicals Co., Osaka, Japan.

‡Geigerflex, Rigaku Denki Co., Tokyo, Japan.

§Standard reference material 640a, National Bureau of Standards, Washington, DC.

¶MSM-9, Akashi Seisakusho Co., Tokyo, Japan.

**UVIDEC-610, Japan Spectroscopic Co., Tokyo, Japan.

††MVK, Akashi Seisakusho Co., Tokyo, Japan.

‡‡DCS-R-10T-S, Shimadzu Seisakusho Co., Kyoto, Japan.

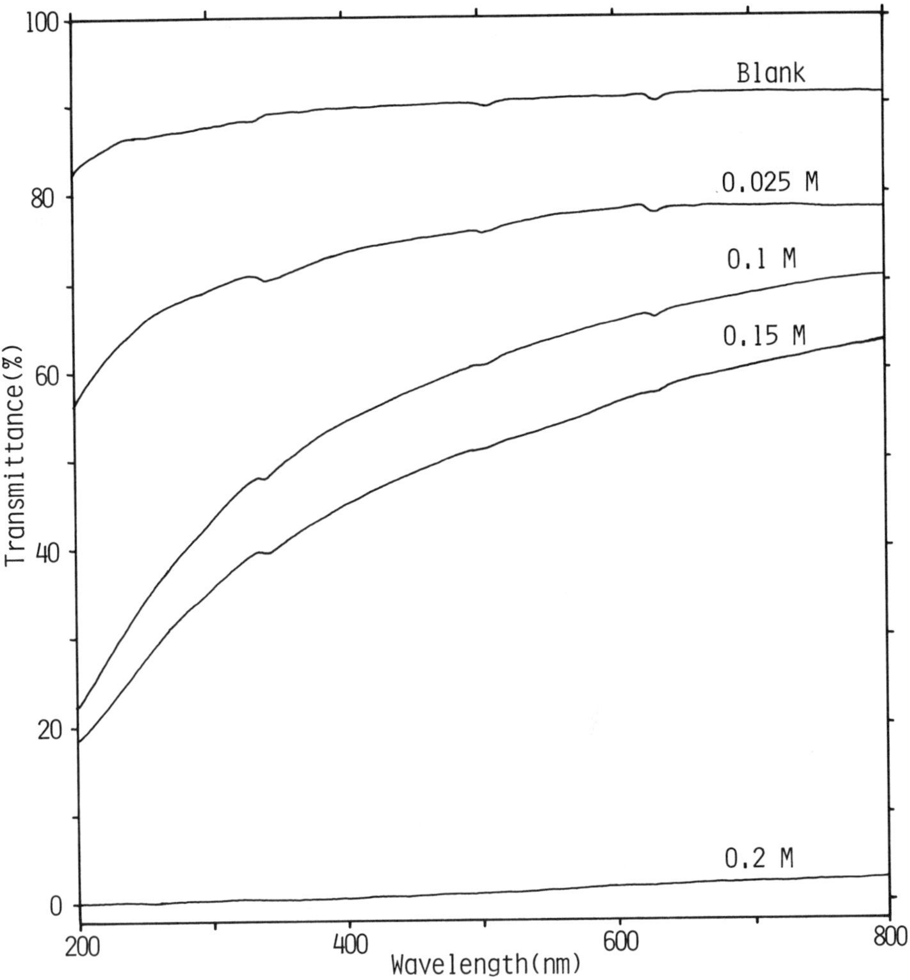

Fig. 1. Transmittance curves of coated specimens using solutions of various concentrations. "Blank" represents uncoated silica glass substrate. Experimental conditions were as follows: withdrawal speed 1.5 cm/min, aging of the solution 2 days, coated 20 times and heat treated at 1050°C for 10 min.

Results and Discussion

Coating Solution

Coating films were prepared using solutions with different concentrations. Crystalline phase observed was only mullite in 0.025, 0.1 and 0.15 M solutions, but spinel phase was also present in 0.2 M solution. Figure 1 shows transmittance curves of the specimens from the solutions of various concentrations. Transmittance decreased gradually as the concentration of the solution changed from 0.025 to 0.15

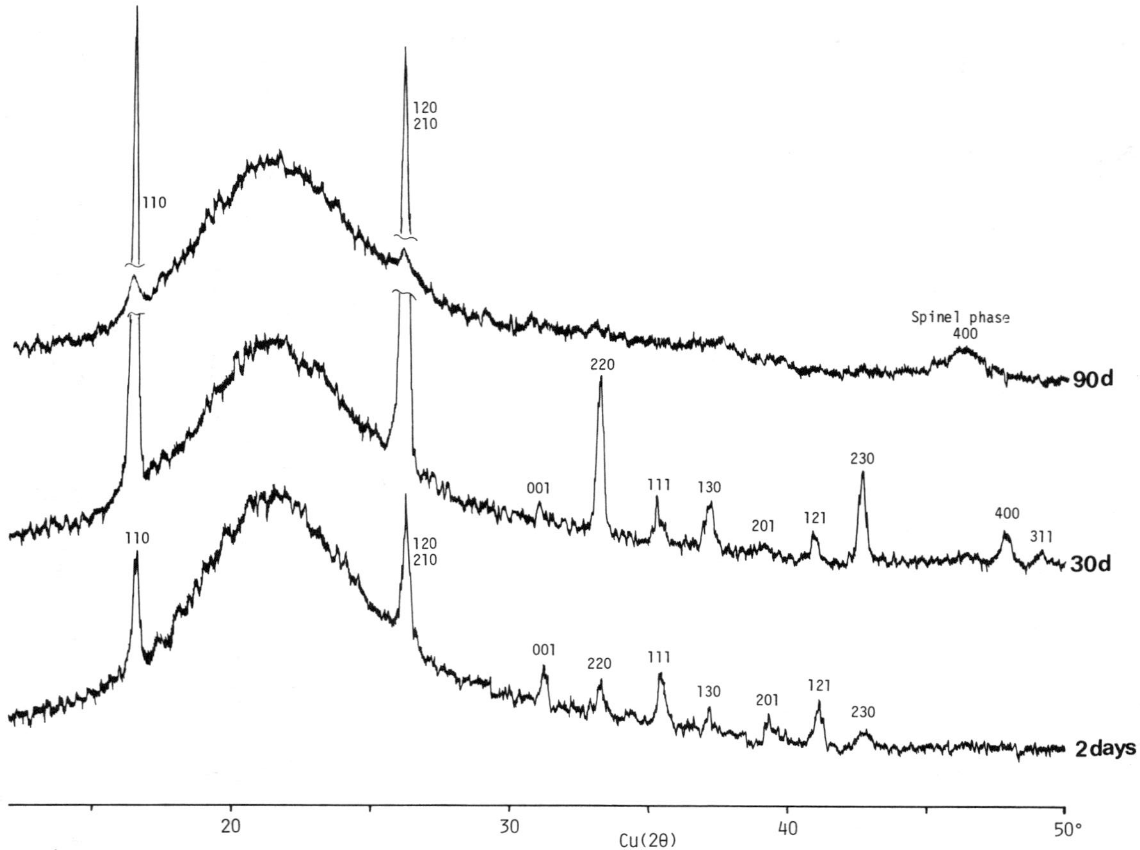

Fig. 2. XRD patterns of some coating films heat treated at 1050°C with different aging times of the solutions. Large hump around 22° in Cu 2θ is due to the substrate.

M. It should be caused by the difference of the coating film thicknesses as they were linearly increased with the concentrations from about 1 micrometer in 0.025 M, 4 micrometer in 0.1 M, 6 micrometer in 0.15 M to about 10 micrometer in 0.2 M. The drop in transmittance between 0.15 and 0.2 M specimens was very large. It should be caused by the inadequately high viscosity of the solution because of the high concentration.

Dip-coating was made under a dry nitrogen atmosphere and air with 60% relative humidity. No detectable difference was observed in obtained coating films by different atmospheres in the present experiments.

Effect of Aging of Solution

In this coating solution, hydrolysis of TEOS gradually progresses by reaction with water from aluminum nitrate nonahydrate. Therefore, it is necessary to examine this effect. Figure 2 shows XRD patterns of some specimens with different aging times of the solutions at 25°C. There were remarkable difference among them. Figure 3 shows relations between formation amounts of mullite and spinel phase, and aging time of the solution. Formation amount of mullite in the coating films increased with aging time and showed a maximum around 30 days. However, the amount of mullite decreased with longer aging time and finally the crystalline phase in the film changed from mullite to spinel phase. No crystalline phase change was observed when the solution was vigorously mixed before dip-coating. This result implied structural change of precursor compounds in the solution with aging time. Formation of spinel phase instead of mullite with longer aging time was considered to be caused by the polymerization of the silica component. Intensities of (hkO) reflections in 30 day aged specimens were clearly intensified compared with other reflections, i.e., (hkl) and (001). Since orientation for (hkO) was considered, Lotgering index was calculated (Fig. 3). Their values increased rapidly with aging time and became 1 which corresponds to perfectly orientated (hkO).

Lattice Constant of Mullite

Lattice constant of the 30 days aged specimen calculated was a = 0.7636 (5), b = 0.7682 (6) and c = 0.2891 (1) nm. (Here, numbers in the parentheses represent standard deviations.) Comparing these values with those from powders of xerogel[7] with mullite composition fired for 24 h at 1050°C (a = 0.7522 (8), b = 0.76880 (13) and c = 0.28806 (3) nm), the former showed longer *a* and *c* axes. Although the lattice constant for the specimen of the film may be lengthened due to the tension caused by the difference of thermal expansion of coating film and that of the substrate, chemical composition of mullite estimated from the length of *a* axis[8] was calculated to be 76 mol% in Al_2O_3, which is a richer Al_2O_3 composition than the ordinary one.

Microstructure

Figure 4 shows SEM photographs of surface and cross section of the coating film. Surface of the coating film looked smooth and dense, but some lines were observed running mosaicly. These must be induced by the difference of thermal expansion between coating film and substrate. Average thermal expansions of mullite[9] and silica glass[10] were reported to be around 5.6×10^{-6} and 0.5×10^{-6}/°C, respectively. Therefore, coating films should suffer tension from substrate. In fact, birefringence was observed along the lines in the film by reflection microscopy (RM). From photographs 4(B) and (C), the film was found to be composed of grains less than 1 micrometer.

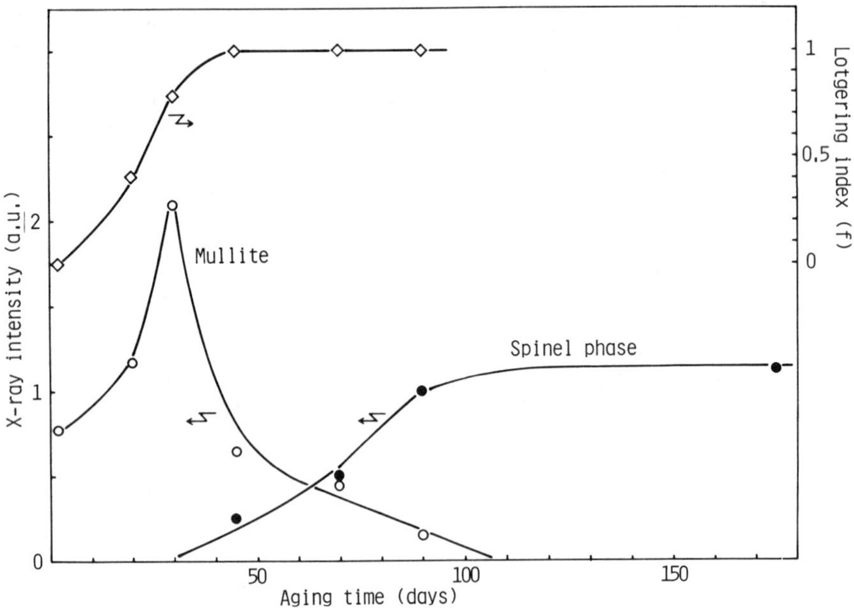

Fig. 3. Relations between formation amounts of mullite and spinel phase, Lotgering index for (hkO) reflection of mullite and aging time of the solution. Intensity data of nonorientated specimens were taken from JCPDS (Card No. 15–776).

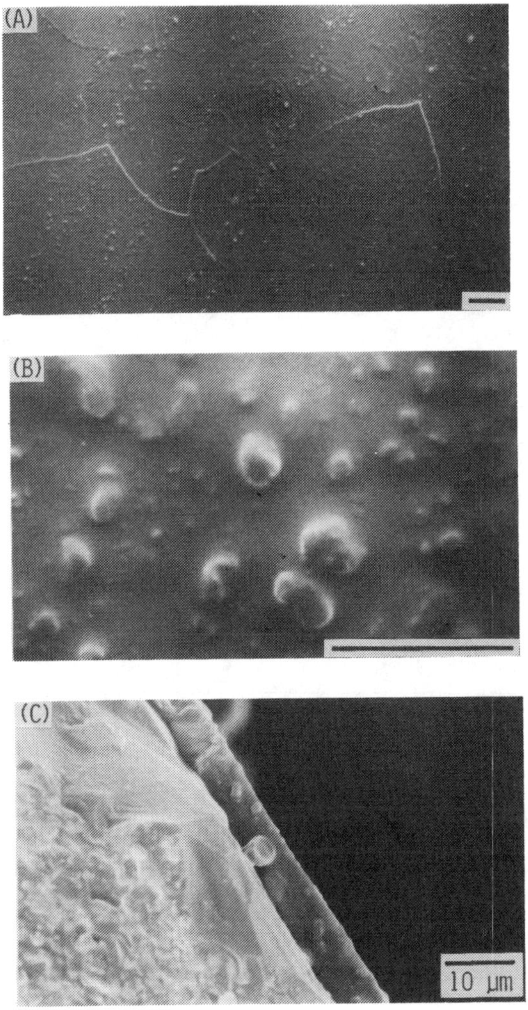

Fig. 4. SEM photographs of the coating film. (A), (B): surface, (C): cross section. Experimental conditions were same with Fig. 1 using 0.1 M solution.

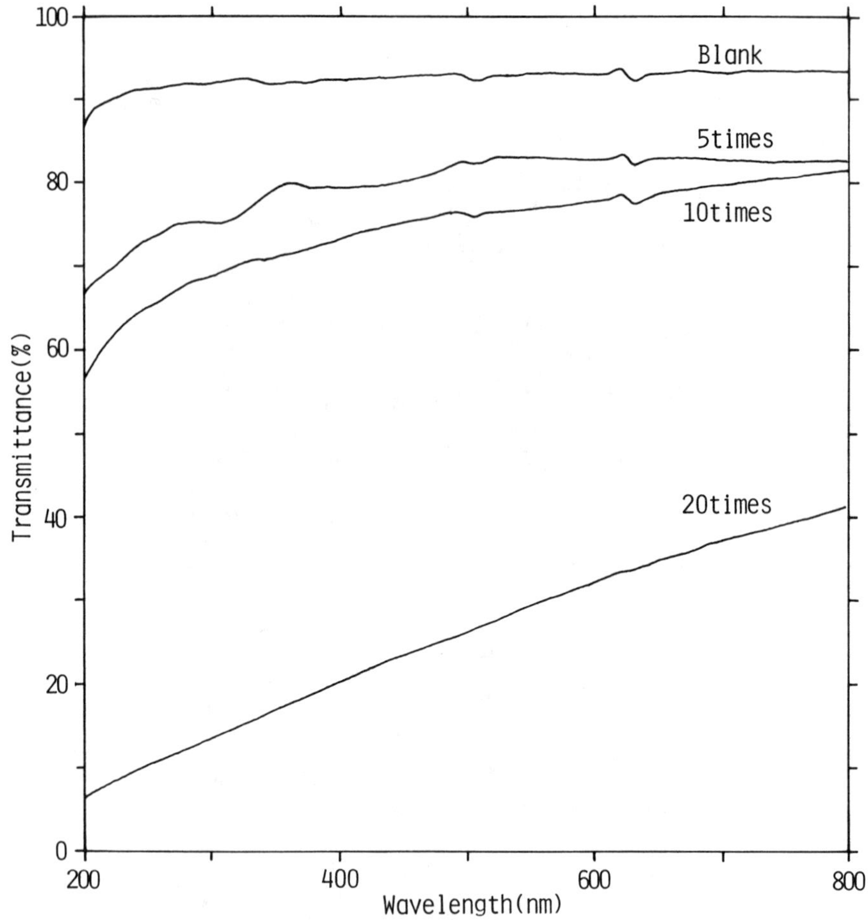

Fig. 5. Transmittance curves of the specimens coated various times using 20 days aging solution. "Blank" represents uncoated substrate. Experimental conditions were as follows: withdrawal speed 3 cm/min, concentration of the solution 0.15 M and heat treated at 1050°C for 10 min.

Transmittance

Figure 5 shows transmittance curves of uncoated and coated substrates. That for a five times coated specimen showed high transmittance of about 82% at 800 nm and showed two absorption peaks at around 300 and 450 nm by interference of thin film. Transmittance curve for a ten times coated specimen also showed relatively high transmittance of 81% at 800 nm. However, that for a twenty times coated one showed a large drop to 41% at 800 nm. From observations of these specimens by RM and SEM micrographies, numbers of white spots, consisting of some platy fragments gathered irregularly and piled up, were found to increase acceleratingly on the surface of the films with the number of dip-coatings. Increase of these white spots caused lowering of transmittance of the films.

Thermal Stability

Thermal stability of these films was examined by heating them at various temperatures. Specimens heated at 1100 and 1200°C showed no apparent change in XRD patterns. However, specimens heated above 1300°C yield cristobalite in addition to mullite. A number of cracks were observed in their films from the SEM observation. It is caused by the volume change of cristobalite at the high-low inversion.

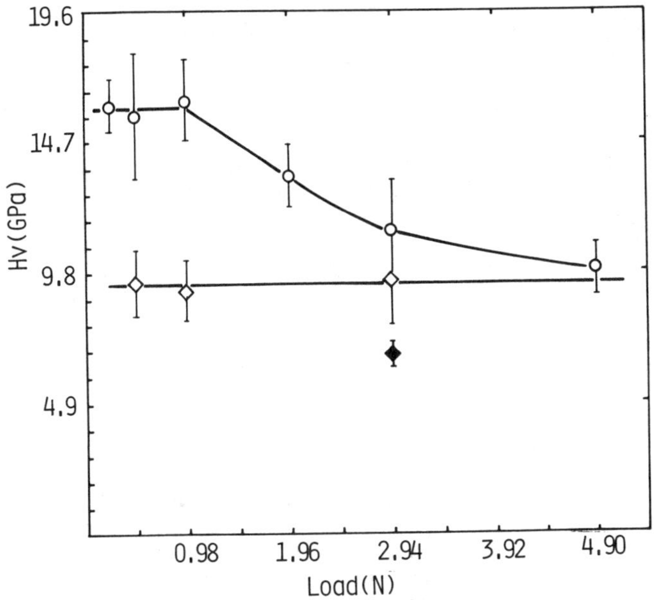

Fig. 6. Relation between Vickers hardness and applied loads for coated (open circles) and uncoated (open rhombuses) specimens. Solid rhombus represents control silica glass.

Mechanical Properties

Figure 6 shows Vickers microhardness of the uncoated and coated substrates. Uncoated substrate showed around 9.3 GPa independent of the loads. On the other hand, the coated specimen showed about 16 GPa at loads of 0.245 to 0.98 N and decreased gradually to 9.8 GPa with increase of load from 0.98 to 4.9 N. It may be caused by the influence of the substrate. Since the reported[11] Vickers microhardness of mullite is about 12 GPa, the value obtained in this work appears to be a little larger than the reported value for mullite ceramics. Increase of hardness in substrate compared with control silica glass whose value showed good agreement with the reported one[12] may be caused by the rapid cooling treatment from 1050°C to room temperature.

Four point bend strengths of uncoated and coated specimens were measured. They were from 81(7) to 89(4) MPa in coated specimens and 86(5) MPa in uncoated specimen. No significant change occurred with the coating.

Conclusions

1. Colorless, transparent and dense mullite films were prepared by the dip-coating method and adhered tightly to the silica glass substrate.
2. Formation amount of mullite in the coating films showed a maximum at the aging time of 30 days and showed orientation to (hkO). Crystalline phase of the coating films changed from mullite to spinel phase with increase of the aging time of the solution.
3. Obtained mullite films had considerable thermal stability and hardness.

Acknowledgments

The authors thank Professor M. Yamane, Tokyo Institute of Technology, for use of spectrophotometer, Professor T. Iseki for use of Vickers tester, and Professor S. Kimura for use of testing machine.

References

[1]K. Okada and N. Ōtsuka, "Characterization of the Spinel Phase from SiO_2-Al_2O_3 Xerogels and the Formation Process of Mullite," *J. Am. Ceram. Soc.*, **69** [9] 652–56 (1986).

[2]S. P. Mukherjee, "Deposition of Transparent Noncrystalline Metal Oxide Coatings by the Sol-Gel Process," Ultrastructure Processing of Ceramics, Glass, and Composites, pp. 178–88, 1984. Eds. L. L. Hench and D. R. Ulrich, John Wiley & Sons.

[3]J. Fukushima, K. Kodaira, and T. Matsushita, "Preparation of Ferroelectric PZT Films by Thermal Decomposition of Organometallic Compounds," *J. Mater. Sci.*, **19** [2] 595–98 (1984).

[4]A. Makishima, H. Kubo, K. Wada, Y. Kitami, and T. Shimohira, "Yellow Coatings Produced on Glass and Aluminum by the Sol-Gel Process," *J. Am. Ceram. Soc.*, **69** [6] C-127-C-129 (1986).

[5]F. K. Lotgering, "Topotactical Reactions with Ferrimagnetic Oxides Having Hexagonal Crystal Structures-I," *J. Inorg. Nucl. Chem.*, **9** [2] 113–23 (1959).

[6]T. Sakurai, Universal Program System for Crystallographic Computations. The Crystallographic Society of Japan, Tokyo, Japan, 1967.

[7]K. Okada, Y. Hoshi, and N. Ōtsuka, "Formation Reaction of Mullite from SiO_2-Al_2O_3 Xerogels," *J. Mater. Sci. Let.*, **5** [12] 1315–18 (1986).

[8]W. E. Cameron, "Composition and Cell Dimensions of Mullite," *Ceram. Bull.*, **56** [11] 1003–11 (1977).

[9]K. S. Mazdiyasni and L. M. Brown, "Synthesis and Mechanical Properties of Stoichiometric Aluminum Silicate (Mullite)," *J. Am. Ceram. Soc.*, **55** [11] 548–52 (1972).

[10]O. V. Mazurin, M. V. Streltsina and T. P. Shvaiko-Shvaikoskaya, Handbook of Glass Data Part A, pp. 31–44, 1983, Elsevier.

[11]G. M. Ismail, Z. Nakai and S. Somiya, "Microstructure and Mechanical Properties of Mullite Prepared by the Sol-Gel Method," *J. Am. Ceram. Soc.*, **70** [1] C-7-C-8 (1987).

[12]O. V. Mazurin, M. S. Streltsina and T. P. Shvaiko-Shvaikoskaya, Handbook of Glass Data Part A, pp. 114–17, 1983, Elsevier.

Directional Solidification in the Alumina-Silica System Microstructures and Interfaces

D. MICHEL, L. MAZEROLLES AND R. PORTIER

Lab. de Chimie Appliquée de l'Etat Solide CNRS UA302
C.E.C.M. 15 Rue Georges Urbain 94407 VITRY (France)

Abstract

Directionally-solidified samples were prepared at various compositions in the silica-alumina system. Crystals of pure mullite were formed having the 2:1 composition. Otherwise, composite materials consisting of mullite and either a silica-rich glass or alumina were obtained. Microstructural features were investigated by scanning and transmission electron microscopy.

Introduction

Mullite ceramics prepared under controlled conditions exhibit interesting optical and mechanical properties. In particular, mullite is a promising refractory material for high-temperature structural applications.[1–5] Mullite has a nominal composition of $3Al_2O_3 \cdot 2SiO_2$ with a narrow existence range in the stable alumina-silica phase diagram.[6] Stoichiometric 3:2 mullite is obtained from solid state sintering of alumina and silica powders, but the solid solution range of mullite is extended towards 2:1 and 3:1 compositions when samples are grown from the melt or quenched from high-temperature. Stable and metastable SiO_2-Al_2O_3 phase equilibria have been extensively discussed in recent paper.[7–14]

The present study is concerned with samples prepared by directional solidification at various compositions in the silica-alumina system. Depending on composition, samples were consisting of mullite crystals in a silica-rich glassy matrix, of mullite crystals or of alumina and mullite crystals. The microstructural features and the structure of the interfaces were revealed by optical and TEM observations. Relations between morphology and rupture were analyzed by SEM fractography on mullite-glass oriented composites.

Experimental Procedure

The starting materials were high-purity Al_2O_3 and SiO_2 powders. In the first step, pressed oxides were sintered at 1550°C as bars having the dimensions of 10 × 10 × 50 mm. Directional solidification was conducted using a floating zone device associated to a double ellipsoid image furnace.[15] The translation rate of the molten zone was 10 mm/hour. The grown samples were investigated by X-ray diffraction and electron microprobe analysis using the K_α lines of Al and Si with alumina and silica as standards.

Transverse and longitudinal sections were cut and diamond polished for optical observations. TEM examinations were performed on both a Jeol2000FX equipped with energy dispersive X-ray spectroscopy and a high resolution Jeol200CX microscope. Samples were oriented platelets thinned by argon bombardment.

Results

Microstructures and Crystallographic Orientation

Samples were prepared with the four following compositions: 28.0, 60.0, 66.7 and 80.0 mol% Al_2O_3. Results of micrographic, X-ray diffraction and chemical analysis are gathered in Table 1. Unit-cell parameters of the mullite phases and electron microprobe analyses are in satisfactory agreement with the values of mullite lattice constants given by Cameron.[16]

Composition 1 (28.0 mol% Al_2O_3)

As shown in Fig. 1, the microstructure is oriented along the solidification direction and crystals with a composition close to 3:2 mullite are embedded in a silica-rich glassy matrix.

The volume fraction of mullite is about 0.4. Crystals with an average size of $50 \times 5 \times 5$ μm are perfectly aligned along the c axis of mullite, as indicated by X-ray diffraction patterns on transverse sections which consisted of only the 001 and 002 lines of mullite. The micrograph given in Fig. 2 reveals the regular dispersion of crystals and their prismatic square habit.

Composition 2 (60.0 mol% Al_2O_3)

Even though the global composition is that of stoichiometric 3:2 mullite, the mullite phase composition is 65.6±/0 5 mol% Al_2O_3. A noticeable amount of a glassy phase (volume fraction about 0.1) is observed in addition to mullite (Fig. 3).

A similar microstructure was previously obtained by Yamana et al. from floating zone growth[17] and by Guse and Mateika from Czochralski pulling.[18] The compositions of mullite and silica-rich glass (11.2 ±/0.6 mol% Al_2O_3) are in agreement with the metastable SiO_2-Al_2O_3 diagram proposed by Aksay and Pask.[8]

Composition 3 (66.7 mol% Al_2O_3)

At this composition, single crystals of pure mullite were obtained. The alumina content of these samples (67.7 ±/0.5 mol% Al_2O_3) is slightly higher than that of the 2:1 starting material. This difference can be explained either by silica volatilization or by segregation during zone melting.

The X-ray and electron diffraction patterns of the mullite crystals prepared from compositions 2 and 3 displayed extra spots and diffuse scattering in addition to the fundamental spots of the orthorhombic mullite lattice (space group Pbam). The superstructure reflections obtained by X-ray De Jong patterns on a single crystal rotating along [001] are identical to that found by Tokonami et al.[19]

These complex diffraction features result from ordering in the studied mullite samples $Al_2(Al_{2+2x}Si_{2-2x})O_{10-x}$ with x~0.4 Ordering in mullite is possible between silicon and aluminum atoms (as in sillimanite) and on the anionic sublattice between oxygen atoms and vacancies. As a consequence of these order-disorder phenomena, diffuse scattering and additional spots have been observed in mullite by various authors.[13,16,18–24] Models were proposed for the "true" structure of mullites based on data from diffraction studies, high resolution electron microscopy or symmetry analysis.[18–25] As order phenomena are influenced by composition and thermal history, it is likely that variable degrees of order can be met in mullite solid solutions.

Table 1. Composition and constituent phases of directionally-solidified samples in the SiO_2-Al_2O_3 system

	Starting Composition (mol% Al_2O_3)	Constituent Phases (mol% Al_2O_3)	Mullite Parameters (Å)
1	28.0	glassy matrix + mullite crystals 10.5 62.4	a =7.5632(8) b =7.6866(8) c =2.8849(3)
2	60.0	mullite crystals + glassy phase 65.6 11.2	a =7.5831(8) b =7.6819(8) c =2.8873(3)
3	66.7	mullite crystal 67.7	a =7.5913(8) b=7.6786(3) c =2.8886(3)
4	80.0	mullite matrix + alumina 69.1 99.9	a =7.597(2) b =7.675(2) c =2.889(1)

Fig. 1. Longitudinal section of an oriented in-situ composite with the global composition of 28 mol% Al_2O_3. Elongated mullite crystals are embedded in a silica-rich glass matrix.

Fig. 2. Transverse section (TEM) of an oriented in-situ composite with global composition of 28 mol% Al_2O_3. The lateral faces of the mullite crystals are {110} crystallographic planes.

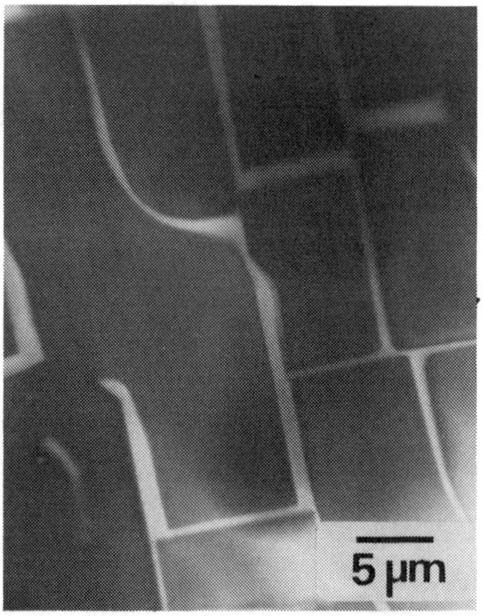

Fig. 3. Transverse section (TEM) of a sample with global composition 60 mol% Al_2O_3. The lateral faces of mullite crystals are {110} crystallographic planes.

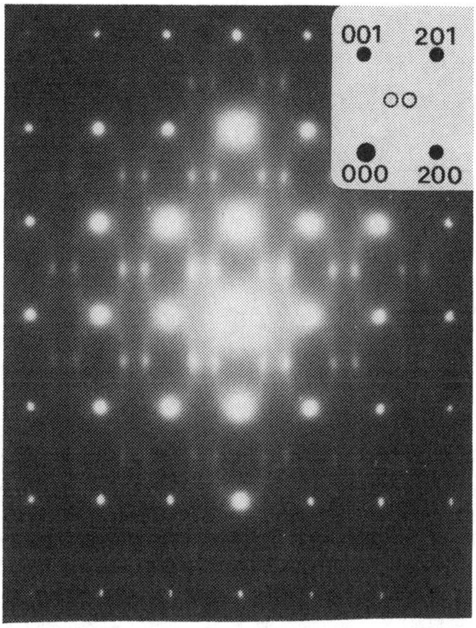

Fig. 4. Electron diffraction pattern of a (010) plane of a 2:1 mullite crystal.

Ordering in our 2:1 mullite crystals was observed on HRTEM images performed with the electron beam parallel to the [010] axis.

The corresponding diffraction pattern (Fig. 4) reveals strong extra spots and streaks typical of order in mullites. Preliminary results indicate that the studied crystals are not homogeneous at the scale of a few nanometers. Instead samples consist of a mosaic-like array of ordered domains coherent with a disordered matrix. Contrast differences are probably associated with local variations of structure or/and composition. Depending on experimental conditions, in particular on the defocus and the diffraction spots selected to form the high resolution image, the obtained image shows either the average structure of mullite (Fig. 5a) or the structure modulation only (Fig. 5b). Image simulations are under way to correlate contrast changes with structural modification, in particular oxygen vacancy ordering.

Composition 4 (80.0 mol% Al_2O_3)

Samples with a starting composition of 80.0 mol% Al_2O_3 consisted of a mullite matrix with alumina precipitates. The growth direction of the matrix was [001] mullite which was parallel to the [0001] three-fold axis of the corundum inclusions. The typical shapes of the alumina precipitates are shown in Fig. 6.

Mullite-Glass Interfaces

Interfaces between mullite and silica-rich glass in samples of composition 1 were imaged by TEM observations on (001) transverse sections. As shown on Fig. 7, the lateral faces of the prismatic crystals are well-defined {110} mullite planes.

Fracture surfaces of oriented mullite-glass materials were examined by scanning electron microscopy. Fracture occurred perpendicular to the mullite crystals when the applied stress is normal to the growth direction (Fig. 8). On the contrary, cracks

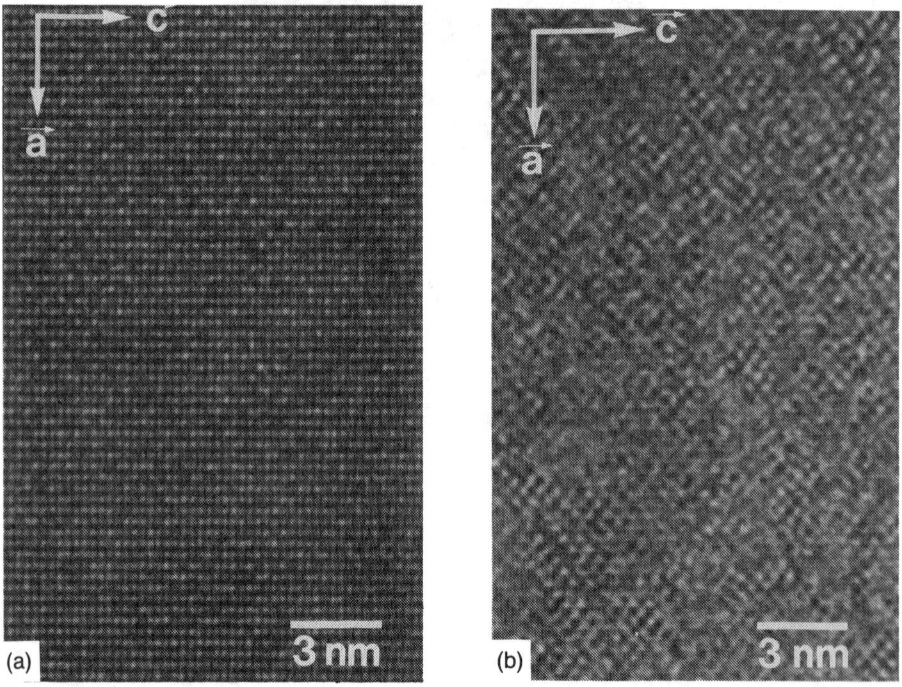

Fig. 5. High resolution images of the (010) plane of 2:1 mullite: a) objective diaphragm centered on diffuse scattering spots and including 000, 200, 001, 201 fundamental mullite spots, b) objective diaphragm centered on a central spot and including diffusion spots only.

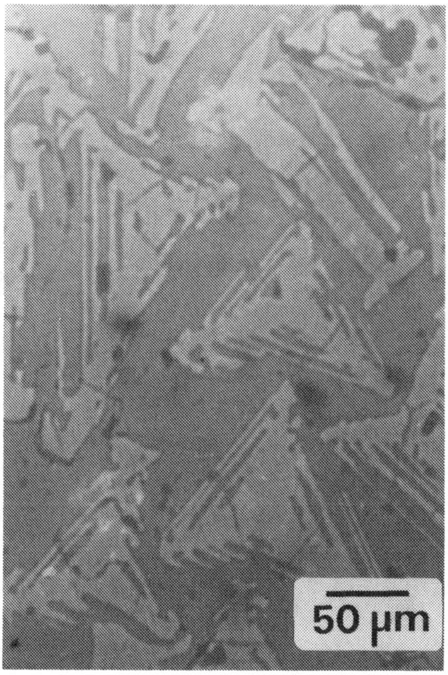

Fig. 6. Transverse section of a sample with global composition 80 mol% Al_2O_3. Alumina precipitates in a mullite matrix.

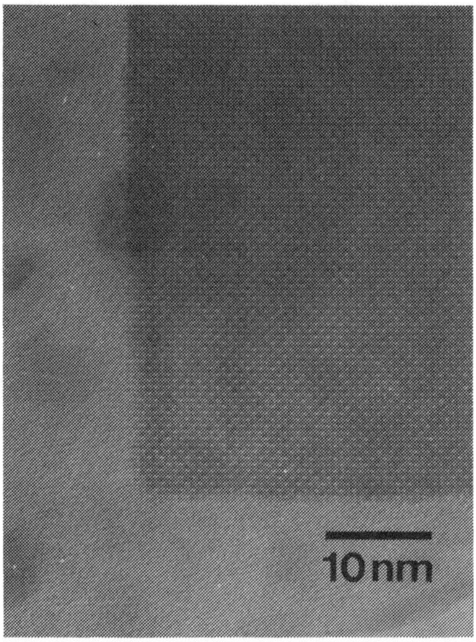

Fig. 7. High resolution image of a glass-mullite interface. Transverse section of a [001] oriented in-situ composite with global composition 28 mol% Al_2O_3. Interfaces are {110} mullite planes.

Fig. 8. Fracture surface of an oriented in-situ composite with global composition of 28 mol% Al_2O_3. Flexure stress was applied perpendicular to the growth direction.

Fig. 9. Fracture surface of an oriented in-situ composite with a global composition of 28 mol% Al_2O_3. Flexure stress was parallel to the growth direction.

propagated mainly in the glassy phase when the applied stress is parallel to the mullite crystals (Fig. 9).

Consequently, a significant difference in strength was measured in 3-point bending. A higher flexure strength, $\sigma \perp$ = 160 MPa, was obtained when the fracture of the mullite crystals was involved, in contrast of the strength value $\sigma_{//}$ = 105 MPa when fracture was primarily caused by the glassy matrix.

Interfaces between the mullite crystals and the residual glassy phase in samples with 3:2 starting composition were also studied.

Directional solidification of mullite occurred also along the [001] and lateral faces of crystals are {110} forms of planes. However, all the (110) and (110) interfaces are parallel, as shown on Fig. 3, whereas, lateral {110} faces follow any plane parallel to the [001] growth direction in the samples with composition 1 (Fig. 2). Samples with a global 3:2 composition can be therefore regarded as 2:1 mullite single crystals divided in elongated crystallites which are isolated by a glassy phase. The average thickness of the glass boundaries is approximately a few micrometers, but the glass film between mullite crystals can be extremely thin in some regions. For instance, Fig. 10 shows a high resolution image with a glass thickness of about 40 nm. On both sides of the glassy phase, the same orientation of mullite can be observed. The directions of the nearly square unit-cell (0.75 x 0.78 nm) correspond to a and b axes of mullite.

It is likely that mullite firstly grows as a single crystal and that the glassy phase is rejected subsequently during the cooling process. Another possible growth process is the simultaneous solidification of glass and 2:1 mullite crystals with a unique orientation minimizing the interfacial energy and elastic stress.

Fig. 10. High resolution image of mullite crystals separated by a glassy phase. Transverse section of a sample with global composition 60 mol% Al_2O_3. The mullite orientation is [001] and the interfaces are {110} mullite planes.

Conclusions

Principal conclusions drawn from directional solidification experiments at four particular compositions (28.0, 60.0, 66.7 and 80.0 mol% Al_2O_3) in the Al_2O_3-SiO_2 system are the following:

1. the mullite growth direction is [001] in all cases and the {110} faces are formed when mullite is associated with a glassy phase.
2. single crystals of 2:1 mullite are obtained from a 2:1 starting composition. Pure 3:2 mullite cannot be grown from the melt under our conditions (growth rate about 10 mm/hour) and 3:2 mullite crystals are always embedded in a glassy phase.
3. X-ray and electron diffraction on 2:1 mullite crystals reveal the diffuse scattering and the extra spots usually observed at this composition. High resolution images indicate a modulated structure with ordered domains of a few nanometers in size.
4. mullite-glass composites have a highly anisotropic fracture behavior related to their oriented microstructure.

Acknowledgments

We would like to thank J. Bahezre (CNRS Bellevue) for performing electron microprobe analyses on our samples.

References

[1]K. S. Mazdiyasni and L. M. Brown, "Synthesis and Mechanical Properties of Stoichiometric Aluminum Silicate," *J. Am. Ceram. Soc.*, **55**, 548–552 (1972).

[2]Tai-il Mah and K. S. Mazdiyasni, "Mechanical Properties of Mullite," *J. Am. Ceram. Soc.*, **66**, 699–703 (1983).

[3]S. Prochazka and F. J. Klug, "Infrared-Transparent Mullite Ceramics," *J. Am. Ceram. Soc.*, **66**, 874–880 (1983).

[4]S. Kanzaki, H. Tabata, T. Kumazawa and S. Ohta, "Sintering and Mechanical Properties of Mullite," *J. Am. Ceram. Soc.*, **68**, C6–C7 (1985).

[5]"Mullite," Ed. S. Somiya, Uchida Rokakuro Publishing (Tokyo), 1985.

[6]S. Aramaki and R. Roy, "Revised Equilibrium Diagram for the System Al_2O_3-SiO_2," *J. Am. Ceram. Soc.*, **45**, 229–42 (1962).

[7]R. F. Davis and J. A. Pask, "Diffusion and Reaction Studies in the System SiO_2-Al_2O_3," *J. Am. Ceram. Soc.*, **58**, 525–531 (1972).

[8]I. A. Aksay and J. A. Pask, "Stable and Metastable Equilibria in the System SiO_2-Al_2O_3," *J. Am. Ceram. Soc.*, **58**, 507–511 (1975).

[9]S. H. Risbud and J. A. Pask, "SiO_2-Al_2O_3 Metastable Phase Equilibrium Diagram without Mullite," *J. Mater. Sci.*, **13**, 2449–2454 (1978).

[10]S. H. Risbud and J. A. Pask, "Mullite Crystallization from SiO_2-Al_2O_3 Melts," *J. Am. Ceram. Soc.*, **61**, 63–67 (1978).

[11]C. M. Jantzen and H. Herman, "Phase Equilibria in the System SiO_2-Al_2O_3," *J. Am. Ceram. Soc.*, **62**, 212–214 (1979).

[12]S. H. Risbud and J. A. Pask, "On the Location of Metastable Immiscibility in the SiO_2-Al_2O_3 System," *J. Am. Ceram. Soc.*, **62**, 214–215 (1979).

[13]W. M. Kriven and J. A. Pask, "Solid Solution Range and Microstructures of Melt-Grown Mullite," *J. Am. Ceram. Soc.*, **66**, 649–654 (1983).

[14]F. J. Klug, S. Prochazka and R. H. Doremus, "Alumina-Silica Phase Diagram in the Mullite Region," *J. Am. Ceram. Soc.*, **70**, 750–59 (1987).

[15]A. Revcovleschi, "Arc Image Furnace for X-Ray Diffraction Studies to 3000°C and High Temperature Crystal Growth," *Rev. Inter. Hautes Temp. Refract.*, **7**, 73–90 (1970).

[16]W. E. Cameron, "Composition and Cell Dimensions of Mullite," *Am. Ceram Soc. Bull.*, **56**, 1003–1011 (1977).

[17]K. Yamana, M. Tokonami, K. Nobugai, N. Morimoto, M. Koizumi and I. Shindo, "Residual Stress in Siliceous Glass Between Mullite Crystals," *J. Am. Ceram. Soc.*, **64**, C43–44 (1981).

[18]W. Guse and D. Mateika, "Growth of Mullite Single Crystals ($2Al_2O_3.SiO_2$) by the Czochralski Method," *J. Cryst. Growth*, **22**, 237–240 (1974).

[19]M. Tokonami, Y. Nakajima and N. Morimoto, "The Diffraction Aspect and a Structural Model of Mullite $Al(Al_{1+2x}Si_{1-2x})O_{5-x}$," *Acta Cryst.*, **A36**, 270–276 (1980).

[20]W. Guse and H. Saalfeld, "Diffuse Scattering in Mullite $2Al_2O_3.SiO_2$," *Z. Kristallogr.*, **143**, 177–187 (1976).

[21]H. Saalfeld, "The Domain structure of 2:1 Mullite ($2Al_2O_3.SiO_2$)," *N. Jb. Miner.*, **134**, 305–316 (1979).

[22]Y. Nakajima, N. Morimoto and E. Watanabe, "Direct Observation of Oxygen Vacancy in Mullite 1.86 $Al_2O_3.SiO_2$ by High Resolution Electron Microscopy," *Proc. Jpn Acad.*, **51**, 173–178 (1975).

[23]Y. Nakajima and P. H. Ribbe, "Twinning and Superstructure of Al-rich Mullite," *Am. Mineral.*, **66**, 142–147 (1981).

[24]J. Ylä-Jääski, H. U Nissen, "Investigation of Superstructures in Mullite by High Resolution Electron Microscopy and Electron Diffraction," *Phys. Chem. Minerals*, **10**, 47–54 (1983).

[25]D. C. McConnell and V. Heine, "Incommensurate Structure and Stability of Mullite," *Phys. Rev. B*, **31**, 6140–6142 (1985).

Non-Stoichiometric Mullites From Al_2O_3-SiO_2-ZrO_2 Amorphous Materials By Rapid Quenching

MASAHIRO YOSHIMURA, YASUHIRO HANAUE AND SHIGEYUKI SŌMIYA

Research Laboratory of Engineering Materials
Tokyo Institute of Technology
4259 Nagatsuta, Midori, Yokohama 227 Japan

Abstract

In order to study the formation of zirconia dispersed mullite ceramics from homogeneous starting materials, hot-pressing and heat-treatments were carried out for rapidly quenched amorphous materials with 0 to 20 wt% ZrO_2 mullite compositions. These amorphous materials crystallized directly to mullite for 0–10 wt% ZrO_2 samples or mullite + t-ZrO_2 for 20 wt% ZrO_2 at about 970°C. An Al_2O_3-rich composition (82 wt% Al_2O_3) and also a significant solid solubility of ZrO_2 (>10 wt%) were estimated for these mullites by XRD studies. Amorphous spheres of 10 nm which were considered to be SiO_2-rich phase were produced by a phase separation in mullite grains.

Introduction

In recent years, considerable interest has been expressed in mullite-ZrO_2 composite ceramics as good candidate materials for high temperature structural applications. Many reports deal with their fabrication and properties.[1–10] Microstructure is one of the most important factors to improve the mechanical properties. In order to control the microstructures, it is necessary to understand metastable phases formed during fabrication. Almost all previous work dealt with fabrication of mullite-ZrO_2 composites from inhomogeneous starting materials, such as reaction sintering of $ZrSiO_4$-Al_2O_3,[2–4] or solid state reaction from ZrO_2-Al_2O_3-SiO_2 mixtures.[1,8,9] These fabrication processes appear to be too complicated to understand the formation of mullite-ZrO_2 composites, because they include both solid state and liquid phase reactions to form mullite. These reactions are, in turn, combined with precipitation and grain growth phenomena associated with ZrO_2. Thus, we have studied the formation of mullite-ZrO_2 composites from homogeneous amorphous materials.[10–12] In the present paper, we describe the formation and composition of metastable mullite phases from rapidly quenched amorphous materials in the system Al_2O_3-SiO_2-ZrO_2.

Experimental

High purity Al_2O_3 (99.9%), SiO_2 (guaranteed reagent grade) and ZrO_2 (99.9%) powders were used for all compositions (Table 1) and were mixed together with PVA

Table 1. The Composition and Crystallization Temperature of the Amorphous Materials.

Sample	Content (wt%)			Crystallization temperature (°C)
	Al_2O_3	SiO_2	ZrO_2	
Ms	72.0	28.1	—	974
5ZM	68.2	26.8	5.0	975
10ZM	64.6	25.3	10.0	979
20ZM	57.5	22.5	20.0	990

aqueous solutions and dried. The mixed powders were then molded into bars (about 5 mm in diameter and 150 mm in length) under an isostatic pressure of 200 MPa. The bottom part of the bar was melted using an arc-imaging furnace. The molten droplets were rapidly quenched using a steel twin-roller apparatus operated at ≈1000 rpm and which yielded amorphous films, typically 60 μm thick. The cooling rate was estimated to be $\leq 10^5$ K/s.[13] It was difficult to make completely amorphous films at the composition of Ms (stoichiometric mullite),[12] however the amorphous materials were collected by the selection of transparent regions from partially crystallized films.

The amorphous materials were heat-treated at 960°-1600°C for 0.5 h in air in order to crystallize them. Hot-pressing of ground amorphous powder of 20ZM was also performed at 1050°-1600°C under 25 MPa.[10,11] Lattice parameters of the mullite in these crystallized samples were examined by an X-ray diffractometer (XRD) with monochromatic CuKα radiation using at least 7 peaks with 2θ between 25° and 65°. The microstructures of these samples were observed by transmission electron microscopy (TEM) on ion-thinned foils.

Results and Discussion

Differential thermal analysis (DTA) of the amorphous materials showed a significant exothermic peak at 975–990°C which correlated with ZrO_2 content (Table 1). Some previous studies[14,15] mention the direct formation of mullite from a noncrystalline phase to occur at similar temperatures. The samples were quickly removed from the DTA furnace after passing through the exothermic peak, and were identified by XRD as single phase mullite for Ms, 5ZM and 10ZM, or mullite and tetragonal ZrO_2 (t-ZrO_2) for 20ZM. When heat-treated at 960°C for 0.5 h, the amorphous materials also crystallized into the same crystalline phases as above. TEM observation revealed that these samples had microstructures consisting of equiaxed 0.1–0.3 μm mullite grains, as seen in Fig. 1. No inhomogeneous concentrations of ZrO_2 were observed in the 5ZM samples, but dark contrast present at grain boundaries suggested the possibility of the presence of ZrO_2-rich noncrystalline regions in the 10ZM samples. Fine t-ZrO_2 precipitates existed in the 20ZM samples, particularly along the mullite grain boundaries. These results suggested a metastable solid solubility of ZrO_2 by ≃10 wt% into mullite.

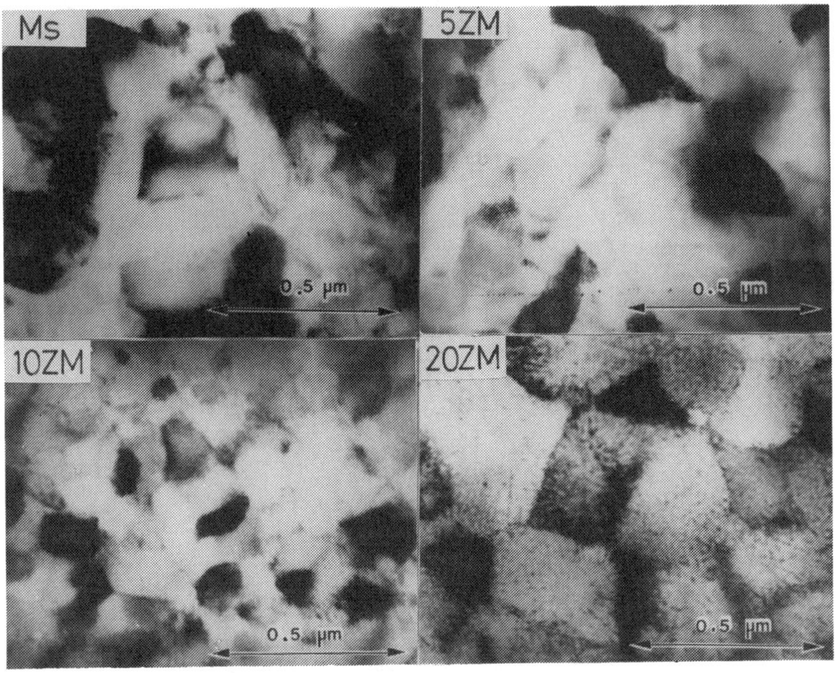

Fig. 1. TEM photographs of the samples heat-treated at 960°C.

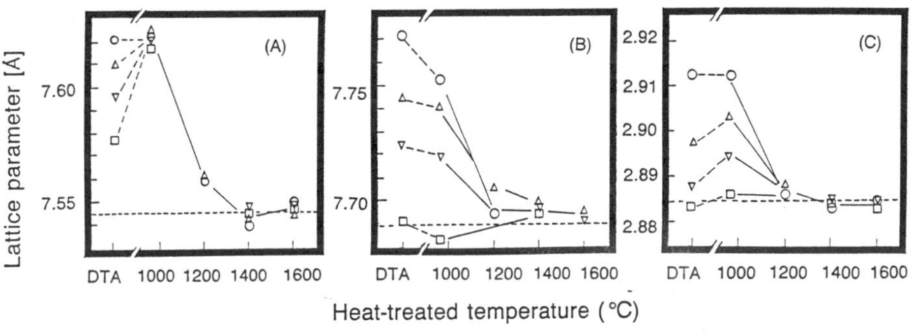

Fig. 2. The lattice parameters of mullites crystallized from rapidly quenched amorphous materials "DTA" means the samples immediately cooled after the crystallization peaks in DTA. □: Ms / ∇: 55M / Δ: 10ZM / O: 20ZM.

The mullites crystallized at lower temperatures always had larger lattice parameters than those of the stoichiometric mullite (Fig. 2). This expansion of the mullite lattice might be caused by the solid solubility of excess Al_2O_3 and/or that of ZrO_2. The relation between composition and the lattice parameters in mullite is well established[17,18] for Al_2O_3-rich mullite, as shown in Fig. 3. The mullite crystallized at low temperatures from the Ms composition had also larger lattice parameters than

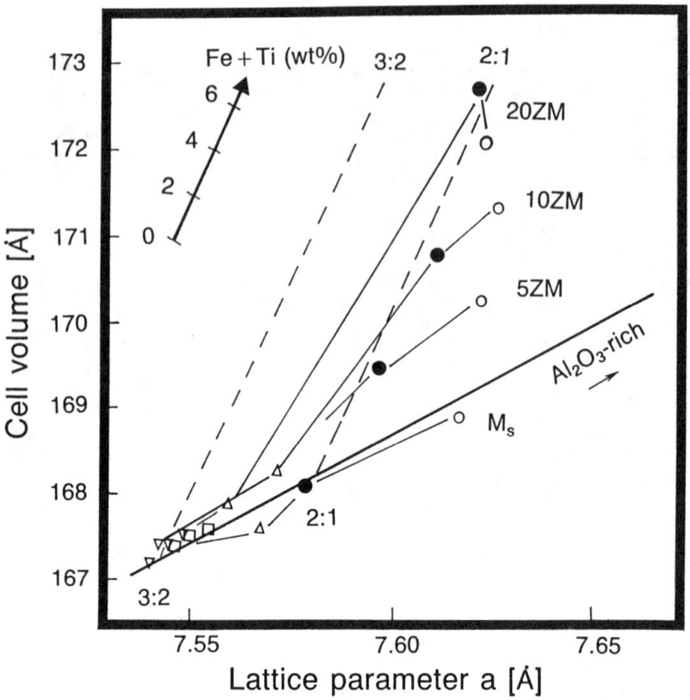

Fig. 3. The relation between lattice parameter a and cell volume of mullites. The thick solid lines are after Cameron (1977). Heat-treated samples; ○: 960°C / Δ: 1200°C / ∇: 1400°C/ □: 1600°C. ●: Immediately cooled after the crystallization peaks in DTA.

the stoichiometric mullite with the Ms composition, particularly in the a axis. According to the Cameron's relation,[17] this mullite appears to have a composition of ≈82 wt% Al_2O_3. This composition is very close to that of the metastable solubility limit of mullite in the system Al_2O_3-SiO_2.[15,16] This result is consistent with that for mullite crystallized from melts[16] having Al_2O_3-rich compositions. The deviation from the Al_2O_3-rich line in Fig. 3 could be caused by the solid solubility of ZrO_2 in the mullite. There exists no data concerning the relation between mullite lattice parameters and ZrO_2 solubility. However, we can expect similar behavior to the solubility of Fe reported previously[17,23–25] and also indicated in Cameron's data in Fig. 3.

Fe^{3+} ions can substitute for Al^{3+} ions in octahedral coordination sites, but it is difficult to substitute in tetrahedral sites because Fe^{3+} ions (0.645 Å) are much larger than Al^{3+} ions (0.535 Å). This was indicated by Mössbauer spectra.[23] This substitution of up to 12 wt% Fe_2O_3 produces the expansion of mullite lattice parameters in a, b, and c directions, but particularly along the b axis.[21–23] The Zr^{4+} ions are also expected to substitute for Al^{3+} ions in octahedral sites because the Zr^{4+} ion (0.72 Å) is even larger than the Fe^{3+} ion. This substitution also results in the expansion of the mullite lattice, particularly in the b-axis. In fact, the lattice parameters of the mullite crystallized at low temperatures had larger a, b, and c parameters, as indicated in Fig. 2. The solubility limit of ZrO_2 can be estimated to be ≈10 wt% from these lattice parameters using Cameron's a-V relation seen in Fig. 3 and the absence of ZrO_2 crystalline phases in the 10ZM sample. The scale used for Fe+Ti should be modified for the case of Zr, corresponding to the difference in their ion sizes.

The mullite crystallized at low temperatures should also be Al_2O_3-rich (≤ 82 wt% Al_2O_3-SiO_2) as indicated in Fig. 3. The a-V relations for them are shown in straight lines which related to ZrO_2 contents. The difference between the two series of the samples for the "DTA" and "annealed at 960°C" would probably be caused by the difference in the Al_2O_3 content and/or ZrO_2 content. Although the heat-treating temperatures are similar, the "DTA" samples could be heated to a higher temperature by the exothermic reaction for crystallization of mullite and additionally for crystallization of t-ZrO_2 in ZrO_2-rich samples. No visible change in lattice parameters was observed in the 10ZM samples during heat-treatment at 960°C for times between 10 min to 24 h. Thus, the heat-treatment temperature essentially determines the composition of crystallized mullites. The mullites crystallized at low temperatures have dissolved also ZrO_2. However, their exact compositions could not be determined, because the solid solubility of both Al_2O_3 and ZrO_2 influenced the lattice parameters of the mullite. In addition, the formation of an SiO_2-rich phase accompanying the crystallization of the mullite must be taken into consideration. The crystallized mullites are always Al_2O_3-rich in comparison with the starting composition ($3Al_2O_3 \cdot 2SiO_2$; 72 wt% Al_2O_3); therefore SiO_2-rich phase(s) must exist in samples crystallized at low temperatures. This should be present as an amorphous phase because no extra crystalline phases could be detected in these samples. Because TEM observations showed no such obvious amorphous phase, it is considered that this SiO_2-rich amorphous phase could exist as a very small dispersed phase within the mullite grains. High resolution electron microscopy has revealed that ≈10 nm amorphous domains are dispersed within the mullite grains having a regular lattice image with 0.29 nm spacing as shown in Fig. 4. The samples of 20ZM hot-pressed at 1050°C had a low density of 3.1_5 g/cm^3 even if it consisted of almost pore-free crystallized mullite grains containing t-ZrO_2 precipitates.[10,11] This low density also suggests the existence of an SiO_2-rich amorphous phase in the samples.

The heat treatment at temperatures higher than 1200°C yielded mullites having smaller lattice parameters, as seen in Fig. 1 and Fig. 3. Heat-treatment at 1400°C resulted in the precipitation of t-ZrO_2 within the mullite grains of samples 5ZM, 10ZM and 20ZM, as seen in the TEM photographs in Fig. 5. These results clearly indicate the following two important features;

1. The formation of Al_2O_3-rich mullites with dissolved ZrO_2 are transient upon crystallization, that is, these mullites are metastable.
2. These mullites react with the SiO_2-rich amorphous phase to form nearly stoichiometric mullite upon heating at temperatures higher than 1400°C.

The heat-treatment at 1400°C produced only slight grain growth of mullite grains in comparison with the heat-treatment at 960°C, and therefore indicates that the reaction between the metastable Al_2O_3-rich mullite and the SiO_2-rich amorphous phase occurred within each grain.

These phenomena have also been suggested from the formation of pores after post heat-treatment at 1600°C in air without any pressure for the 20ZM samples initially hot-pressed at 1050°C.[10,11] The pores seemed to be formed from the amorphous SiO_2-rich phases which reacted with the Al_2O_3-rich mullite to form the stoichiometric mullite. The samples heat-treated at 1600°C without pressure had a

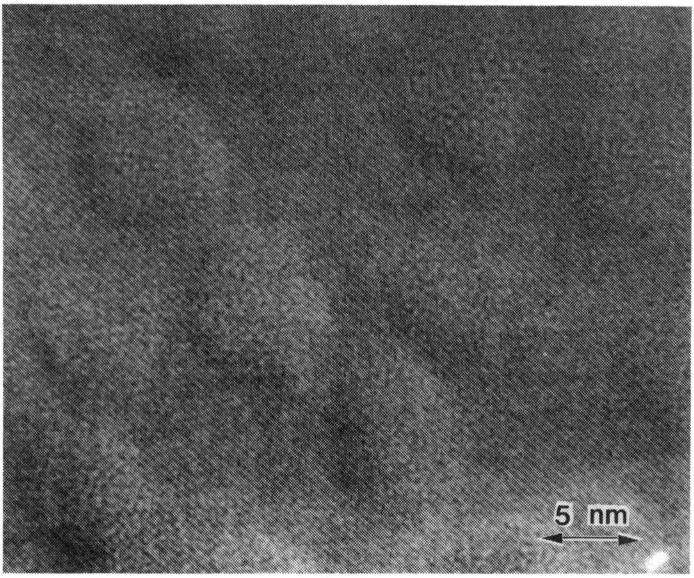

Fig. 4. High resolution TEM photograph of the 10ZM sample annealed at 960°C for 0.5 h.

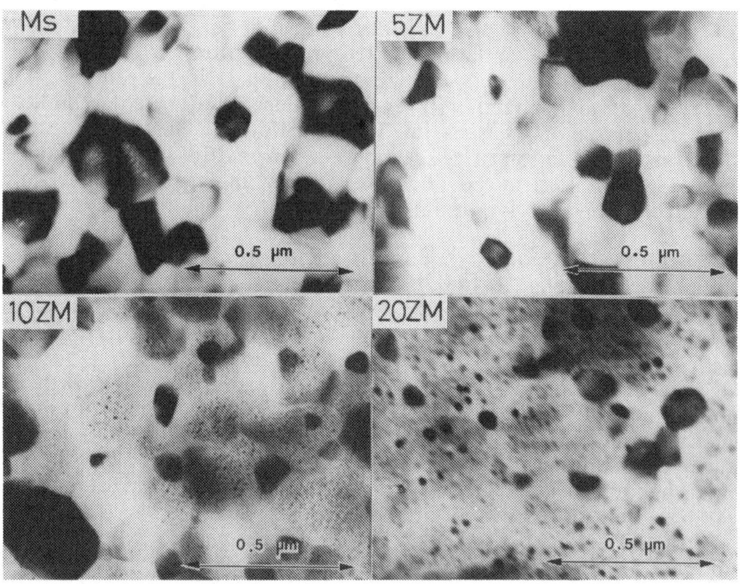

Fig. 5. TEM photographs of the samples heat-treated at 1400°C.

low density of ≈3.3 g/cm^3 because of the pores, while the samples hot-pressed at 1600°C had a higher density of ≈3.5 g/cm^3, which is nearly equal to the calculated density of 3.48–3.52 g/cm^3 for the composition of 20 wt% ZrO_2–80 wt% $3Al_2O_3$•$2SiO_2$.

The microstructural and density changes which occurred during heat-treatments, as well as the mechanical properties have been reported in our previous papers.[10,11]

After heat-treatment between 1400°-1600°C with and without applied pressures, the resulting mullite had nearly the stoichiometric composition of $3Al_2O_3$•$2SiO_2$ with little solid solubility of ZrO_2 in them. This result is consistent with the earlier reports by Moya et al.[1,7] Dinger et al.[26] previously suggested metastable solubility of ZrO_2 in mullite prepared at 1570°C by a reaction between "premullite" and 15 wt% ZrO_2, but more recent results indicate much lower solubility (few % or less).[6] Therefore the higher solubility of ZrO_2 in mullite should be metastable, where the composition would be influenced by many factors such as starting materials, processing technique, thermal history, size and location of grains. Under equilibrium conditions, however, the solid solubility of ZrO_2 as well as of Al_2O_3 should be small, as shown in the present study.

Acknowledgments

The authors wish to thank Prof. Mareo Ishgame for experimental advices, Dr. Kazuyori Urabe for high resolution TEM observation, and Dr. Tatsuo Noma for helpful discussions.

References

[1]J. S. Moya and M. I. Osendi, "Microstructural and Mechanical Properties of Mullite/ZrO_2 Composite," *J. Mater. Sci.*, **19** [9] 2909–14 (1984).

[2]J. S. Wallace, G. Petzow, and N. Claussen, "Microstructure and Property Development of in Situ-Reacted Mullite-ZrO_2 Composites," pp. 436–42, Science and Technology of Zirconia II, Advances in Ceramics vol 12, edited by N. Claussen, M. Ruhle, and A. H. Heuer, (The American Ceramic Society, Columbus, OH, 1984).

[3]P. Boch and J. P. Giry, "Preparation and Properties of Reaction-sintered Mullite-ZrO_2 Ceramics," *Mater. Sci. and Engineering*, **71**, 39–48 (1985).

[4]G. Orange, G. Fantozzi, F. Cambier, C. Leblud, M. R. Anseau, and A. Leriche, "High Temperature Mechanical Properties of Reaction-sintered Mullite/Zirconia and Mullite/Alumina/Zirconia Composite," *J. Mater. Sci.*, **20** [7] 2533–40 (1985).

[5]R. McPherson, "Preparation of Mullite-Zirconia Composites from Glass Powder," *J. Am. Ceram. Soc.*, **69** [3] 297–98 (1986).

[6]J. M. Rincon, T. R. Dinger, G. Thomas, J. S. Moya, and M. I. Osendi, "Microstructure of Mullite/Al_2O_3/ZrO_2/Tough Ceramic Composites," *Acta Metall.*, **35** [5] 1175–79 (1987).

[7]J. S. Moya and M. I. Osendi, "Effect of ZrO_2(ss) in Mullite on the Sintering and Mechanical Properties of Mullite/ZrO_2 Composites," *J. Mater. Sci. Lett.*, **2** [10] 599–601 (1983).

[8]Qi-Ming Yuan, Jia-Qi Tan, and Zheng-Guo Jin, "Preparation and Properties of Zirconia-Toughened Mullite Ceramics," *J. Am. Ceram. Soc.*, **69** [3] 265–67 (1986).

[9]Qi-Ming Yuan, Jia-Qi Tan, Ji-Yao Shen, Xuan-Hui Zhu and Zheng-Fang Yang, "Processing and Microstructure of Mullite-Zirconia Composites Prepared from Sol-Gel Powders," ibid., **69** [3] 268–69 (1986).

[10]M. Yoshimura, T. Noma, Y. Hanaue, and S. Sōmiya, "Zirconia Dispersed Mullite Ceramics through Hot-pressing of ZrO_2-SiO_2-Al_2O_2 Amorphous by Rapid Quenching," pp. 165–72, in Advanced Structural Ceramics, MRS Symp. Proc., Vol. 78, edited

by P. F. Becher, M. V. Swain, and S. Sōmiya, Materials Research Society, Pittsburgh (1987).

[11]M. Yoshimura, M. Kaneko, Y. Hanaue, and S. Sōmiya, "Mullite-ZrO_2 Ceramics from Rapidly Quenched ZrO_2-SiO_2-Al_2O_3 Amorphous Phase," Science and Technology of Zirconia III, Advances in Ceramics Vol. 24, edited by S. Sōmiya, H. Yanagida, and N. Yamamoto, accepted for publication by The American Ceramic Society.

[12]M. Yoshimura, M. Kaneko, and S. Sōmiya, "Rapid Quenching of Melts in the System ZrO_2-SiO_2-Al_2O_3," *J. de Physique, Colloque, C1*, **47** [2] 473–77 (1986).

[13]M. Yoshimura, M. Kaneko, and S. Sōmiya, "Preparation of Amorphous Materials by Rapid Quenching of Melts in the System ZrO_2-SiO_2-Al_2O_3," *J. Mater. Sci. Lett.*, **4** [9] 1082–84 (1985).

[14]K. Okada and N. Otuka, "Characterization of the Spinel Phase from SiO_2-Al_2O_3 Xerogels and the Formation Process of Mullite," *J. Am. Ceram. Soc.*, **69** [9] 652–56 (1986).

[15]T. Takamori and R. Roy, "Rapid Crystallization of SiO_2-Al_2O_3 Glasses," *J. Am. Ceram. Soc.*, **56** [12] 639–44 (1973).

[16]I. A. Aksay and J. A. Pask, "Stable and Metastable Equilibria in the System SiO_2-Al_2O_3," *J. Am. Ceram. Soc.*, **58** [11–12] 507–12 (1975).

[17]W. E. Cameron, "Composition and Cell Dimensions of Mullite," *Am. Ceram. Soc. Bull.*, **56** [11] 1003–7, 1011 (1977).

[18]A. J. Majumdar and J. H. Welch, "New Data on Synthetic Mullite," *Tr. Br. Ceram. Soc.*, **62** [8] 603–13 (1963).

[19]S. O. Agrell and J. V. Smith, "Cell Dimensions, Solid Solution, Polymorphism, and Identification of Mullite and Silimanite," *J. Am. Ceram. Soc.*, **43** [2] 69–76 (1960).

[20]M. K. Murthy and F. A. Hummel, "X-Ray Study of the Solid Solution of TiO_2, Fe_2O_3, and Cr_2O_3 in Mullite ($3Al_2O_3 \cdot 2SiO_2$)," *J. Am. Ceram. Soc.*, **43** [5] 267–273 (1960).

[21]C. M. Cardile, I. W. M. Brown, and K. J. D. Mackenzie, "Moessbauer Spectra and Lattice Parameters of Iron-substituted Mullites," *J. Mater. Sci. Lett.*, **6** [2] 357–62 (1987).

[22]H. Schneider and K. Wohllebden, "Microchemical Composition and Cell Dimension of Mullites from Refractory-Grade South American Bauxites," *Ceramics International*, 7 [4] 130–36 (1981).

[23]H. Schneider, "Iron Incorporation in Mullite," *Ceramics International*, **12** [3] 117–25 (1986).

[24]M. Tomoaki, Y. Nakajima, and N. Morimoto, "The Diffraction Aspect and a Structural Model of Mullite, $Al(Al_{1+2x}Si_{1-2x})O_{5-x}$," *Acta Cryst.*, A36, 270–76 (1980).

[25]H. Saalfeld and W. Guse, "Structure Refinement of 3:2 Mullite ($3Al_2O_3 \cdot 2SiO_2$)," *N. Jb. Miner. Mh.*, [4] 145–50 (1981).

[26]T. R. Dinger, K. M. Krishnan, G. Thomas, M. I. Osendi, and J. S. Moya, "Investigation of ZrO_2/mullite Solid Solution by Energy Dispersive X-ray Spectroscopy and Electron Diffraction," *Acta Metall.*, **32** [10] 1601–7 (1984); corrected in **33** [2] 329 (1985).

Investigation On Mullite Prepared From Refined Bauxite

LI NAN AND WANG XITANG

Department of Materials Science and Engineering
Wuhan Iron and Steel University
Wuhan, Hubei, China

Abstract

Refined bauxites as the cheap raw materials for production of mullite were studied and compared with the mixture of $Al(OH)_3$ and fine SiO_2 powder. The mullite first appeared at about 1000°C for bauxite samples. This temperature was 200°C lower than the temperature of the mixture of $Al(OH)_3$ and SiO_2 powder. A lot of mullite crystallites produced by kaoline-mullite reaction were as the seeds. Thus the densification was enhanced. The density of bauxite samples sintered at 1650°C reached 2.96g/cm^3 which was much higher than that of the sample made of the mixture of $Al(OH)_3$ and SiO_2. The mullite content in the bauxite grogs and in the sample of $Al(OH)_3$ and SiO_2 were almost the same, about 92% determined by chemical analysis. The bauxite sample containing boehmite has less expension resulted from secondary mullitization than the bauxite sample containing diaspore.

Introduction

Mullite is a very important material for industry because of its high strength at high temperature, good thermal shock resistance and chemical stability. In recent years, much research has been conducted to obtain high purity mullite from alkoxides and other precursors. However, for large scale production, such as refractories, cheap raw materials are needed. China is rich in bauxite from which mullite can be obtained. The influences of composition, particle size and firing conditions on sintering, composition and microstruction of mullite have been studied, compared with samples prepared from a mixture of $Al(OH)_3$ and a fine powder of silica. The present paper reports the important results.

Experimental

The beneficiated bauxites and the mixture of $Al(OH)_3$ and fine silica powder were used as the raw materials. Based on microscopy, x-ray diffraction and DTA studies, it was confirmed that the main mineral compositions in the bauxites were diaspore, boehmite, kaolin and small amount of quartz, rutile and anatase. The boehmite had a very fine crystal size.

The various raw materials were mixed in appropriate proportions and wet ground. The chemical composition of the batches are given in Table 1. The batch

Table 1. Chemical composition of the batches (wt%).

	Al_2O_3	SiO_2	Fe_2O_3	TiO_2	CaO	MgO	K_2O	Na_2O	Ig-loss	Al_2O_3/SiO_2
A	59.80	23.52	0.51	1.08	0.03	0.06	0.10	0.07	14.66	2.54
B	62.44	20.68	1.64	1.82	0.05	0.15	0.11	0.03	14.12	3.00
C	52.12	19.64	0.12	0.12	0.07	0.07	0.09	0.19	26.54	2.65

Fig. 1. Typical scanning electron micrograph of mullite produced from the bauxit batch A in Table 1 and sintered at 1650°C, 3 hr.

A is a mixture of two bauxites. B is a single bauxite. C is a mixture of $Al(OH)_3$ and fine powder of silica. Phase transformation during heating of the powder batch was examined by high temperature x-ray diffraction to 1350°C. The pellets were pressed with about 7% water. After drying, they were sintered at 1400°C, 1500°C, 1600°C and 1700°C for a soaking time of 3 hr. The bulk density, apparent porosity of each sintered pellet were subsequently determined. The phase composition and chemical composition were determined by x-ray quantitative analysis and chemical analysis (CA). The apparent crystal size (length and width) and their distribution were determined using an image analyzer from SEM photographs such as that shown in Fig. 1. The chemical composition of the mullite crystals was determined by electron microprobe analysis (EMP). The shrinkage curves of the pellets made of batch A, B and C were measured with a constant rate of 10°C/min.

Results and Discussions

The high temperature x-ray diffraction patterns showed that the mullite first appeared at about 1000°C for sample A and B. The mullite for sample C first appeared at about 1200°C. That meant the mullite in bauxite samples appeared earlier than that of the mixture of $Al(OH)_3$ and fine SiO_2 powder when they were heated.

The shrinkage curves of pellets A, B and C, are shown in Fig. 2. For sample B, there are three apparent turning temperatures at about 1150°C, 1350°C and 1550°C. Almost no shrinkage occurs in the temperature range from 1350°C to 1550°C for sample B. When the temperature is risen over 1550°C, all samples again shrink

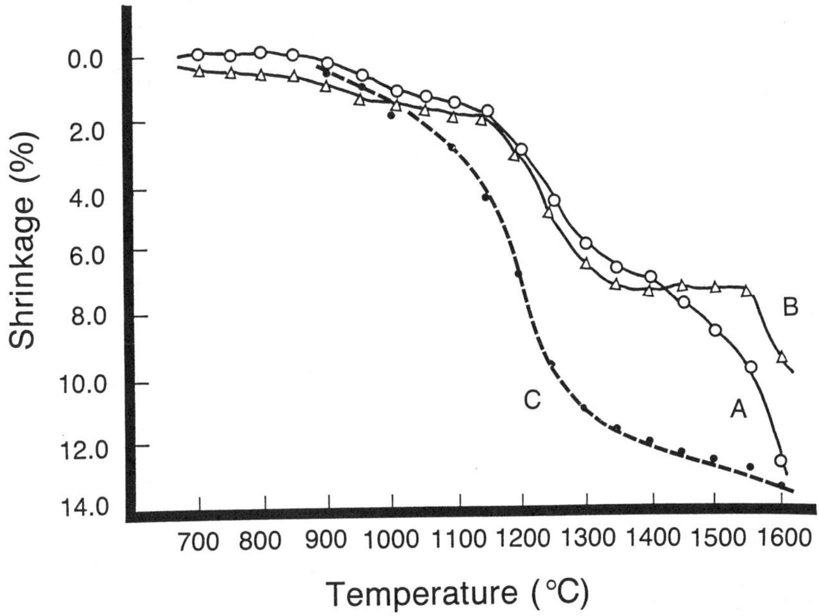

Fig. 2. Shrinkage curves for sample A, a mixture of two bauxites, B. a bauxite, C. a mixture of $Al(OH)_3$ and fine SiO_2 powder, whose compositions are shown in Table 1.

rapidly. The shrinkage curve of sample A is similar to the shrinkage curve of sample B, but does not have a level zone. The shrinkage curve of sample C shows a different feature, in that it shrinks rapidly in the temperature range from 1100°C-1300°C, when temperature rises above 1300°C, the shrinkage in this material is lowered, but rapid shrinkage does not appear again to 1600°C.

The bulk density and apparent porosity of the sintered samples as a function of sintering temperature are given in Table 2. The sample A is most sinterable, and the sample C is least sinterable among the three samples.

The content of mullite in the sintered samples determined by x-ray quantitative analysis and chemical analysis were given in Table 3. It is evident that the mullite content in three samples are almost the same. It can be seen in Table 1 that sample C has the least amount of the impurities but contains the most alkalis oxide which are much more harmful to formation of mullite than other impurity oxide. Perhaps, this is the reason why sample C has almost the same mullite content as sample A and B. Several investigations concerned with the kaolinite-mullite reaction have been conducted. Mullite is formed in this reaction via two intermediate phases, metakaolin and a cubic phase. The problem is whether the cubic phase is simple r-Al_2O_3 or a Si-Al spinel. Brindley and Nakahira noted[1] the structural continuity from kaolinite through metakaoline to the spinel-type phase, and then mullite. The mullite may appear at a temperature lower than 1050°C as a result of the atomic arrangement passing quickly through the spinel-like structure into a mullite-like arrangement. Chakraborty and Ghosh[2] reported that kaolinite was endothermically converted into metakaolin without loss of amorphous SiO_2 at a reaction temperature of $\simeq$600°C.

Table 2. Bulk density and apparent porosity of the samples as a function of sintering temperature.

Temperature	°C	1500	1600	1650	1700
A Mix.of two bauxites	Bulk density (g/cm^3)	2.54	2.89	2.95	2.92
	Apparent porosity (%)	18.9	1.8	1.5	3.2
B A single bauxite	Bulk density (g/cm^3)	2.37	2.92	2.96	2.93
	Apparent porosity (%)	26.4	1.7	1.7	2.0
C Mix.of $Al(OH)_3$ and SiO_2	Bulk density (g/cm^3)	1.65	2.10	2.19	2.54
	Apparent porosity (%)	48.1	33.3	30.8	15.8

Table 3. Mullite content in the samples sintered at high temperature for 3 hr.

Sample	Temperature (°C)	Mullite content (wt%)	
		chemical analysis	X-ray quantitative analysis
A	1650	92.28	>95
B	1650	91.99	>95
C	1700	93.00	>95

At an exothermic reaction temperature of 980°C, the metakaolinite suddenly transformed into Al-Si spinel with liberation of amorphous SiO_2, but only a small amount of weakly crystalline mullite formed. Mullite formation became rapid and the spinel phase disappeared at ≃1250°C, indicating that Si-O-Al linkages persist from kaolinite to metakaolinite to Al-Si spinel and ultimately to mullite. Comer[3] studied mullite development in kaolinites fired between 850°C and 1200°C and observed that mullite was consistently in preferred orientation on the large well-crystallized kaolinite stacks but not on poorly crystallized specimens. He postulated that well-crystallized kaolinites gave rise to the best-developed spinel phase which then provided the best condition for developing oriented mullite. The results of our investigation seem to support Brindley et al.[1] The spinel-type structure transforms to a mullite phase at about 1100°C. The mullite and cristobalite develop continuously to about 1350°C accompanied by considerable shrinkage.

The secondary mullitization occurred considerably in the temperature range 1350°C-1550°C which resulted in lower shrinkage in this temperature range (Fig. 2). Sample A shrank more rapidly than sample B, also sample A reached higher bulk density than sample B (Table 2). These might be explained by the differences of mineral composition. Sample A contained fine boehmite grains and sample B contained diaspore. Boehmite dehydrates to transform to r-Al_2O_3 which transforms to α-Al_2O_3 at temperatures above 1100°C. This temperature is near the secondary mullitization temperature, therefore, reactivity is increased. Conversely, diaspore dehydrates to transform to α-Al_2O_3 within the temperature range 450–600°C which is sintered before secondary mullitization. Thus the reactivity is decreased.

The sintered bulk density of the samples made from bauxites was much higher than that of the sample made of the mixture of $Al(OH)_3$ and SiO_2 powder (Table 2).

Table 4. The Al_2O_3/SiO_2 ratios of the various batches and mullite.

No.	Mullite: Results of CA		Mullite: Results of EMP		
1*	2.09	2.61	2.97	2.45	2.61
2*	2.41	2.65	3.39	3.02	3.07
3	1.97	2.36	2.81	2.50	2.75
4	2.45	2.68	2.93	2.51	2.77
5	2.60	2.84	3.52	2.95	3.18

*No.1, No.2 made of bauxite, the others made of $Al(OH)_2$ and fine silica powder.

Table 5. The statistic data of crystal size of mullite.

Calcination condition	Number of crystal	Parameter	Max. (u)	Min. (u)	Avge. (u)	Standard deviation
1650°C 3 hr	952	Length	12.72	0.84	3.74	3.10
		Width	5.00	0.23	1.58	0.52
		Length/Width	9.68	0.90	2.54	1.27
1650°C 7 hr	573	Length	17.58	1.49	5.14	6.33
		Width	6.35	0.39	2.10	0.61
		Length/Width	8.76	0.97	2.59	1.49

This difference could be due to the formation of mullite crystallites. For bauxite samples a lot of mullite crystallites were formed by kaolinite-mullite reaction. These mullite crystallites might be as the seeds. Thus the densification was enhanced via a liquid-phase transformation mechanism. The mullite crystallites in the mixture of $Al(OH)_3$ and fine SiO_2 powder were formed by solid reaction between Al_2O_3 and SiO_2. Thus less mullite crystallites were formed, the densification was inhibited.

The influences of the Al_2O_3/SiO_2 ratio in the raw materials on the resulting Al_2O_3/SiO_2 ratio of mullite is shown in Table 4.

It was found that with an increasing ratio in the raw materials, the Al_2O_3/SiO_2 ratio of mullite also increased as expected. The values of the ratio in the mullite as measured by EMP were higher than those results obtained by CA, especially for the samples with high Al_2O_3/SiO_2 ratios. That means the glassy phase in the sintered sample has lower Al_2O_3/SiO_2 ratio. Kriven and Pask[4] reported that the solid solution range of mullite grown from the melt was observed to 83.6 ± 0.6 wt% Al_2O_3. It was possible that the mullite with high Al_2O_3/SiO_2 ratio in our study was also mullite produced from the melt by a solution-precipitation process.

The statistical data regarding the crystal size of mullite prepared from batch A are given in Table 5.

Summary

The refined bauxite can be used as raw materials to obtain mullite grog. A lot of mullite crystallites produced by kaolinite-mullite reaction started at about 1000°C are as the seeds for the solution-precipitation process. Thus the densification is enhanced. The bulk densities of the bauxite samples sintered at 1650°C for 3 hr reaches about 2.96 g/cm^3 which are much more than that of the sample of $Al(OH)_3$ and SiO_2 mixture. The secondary mullitization occurs considerably in the temperature range 1350°C–1550°C for bauxite samples. It was found that the bauxite sample containing boehmite had less expension resulted from secondary mullitization than the bauxite sample containing diaspore. The mullite content in the grogs made from the mixture of $Al(OH)_3$ and SiO_2 and made from the bauxites were almost the same because of higher content of alkali oxide in the mixture of $Al(OH)_3$ and SiO_2 powder.

References

[1]G. W. Brindley and M. Nakahira, "Kaolinite-Mullite Reaction Series III, The High-Temperature Phase," *J. Am. Ceram. Soc.*, **42** [7] 319–42 (1959).

[2]A. K. Chakraborty and D. K. Ghosh, "Reexamination of the Kaolinit-to-Mullite Reaction Series," *J. Am. Ceram. Soc.*, **61** [3–4] 170–73 (1978).

[3]J. J. Comer, "New Electron-Optical Data on the Kaolin-Mullite Transformation," *J. Am. Ceram. Soc.*, **44** [11] 561–63 (1961).

[4]W. M. Kriven and J. A. Pask, "Solid Solution Rang and Microstructures of Melt-grown Mullite," *J. Am. Ceram. Soc.*, **66** [9] 649–54 (1983).

Microstructures and Properties of SiC Whisker-Reinforced Mullite Composites*

TERRY TIEGS, PAUL BECHER AND PETER ANGELINI

Oak Ridge National Laboratory

Abstract

Mullite is an attractive material for certain structural applications because of its excellent phase stability, low thermal conductivity, excellent creep resistance, and high temperature strength retention as compared to other engineering ceramic materials. However, mullite suffers from low strength and fracture toughness. Results at Oak Ridge National Laboratory showed significant improvement in both strength and toughness of mullite with the addition of SiC whiskers.

Typically, monolithic mullite has strength and toughness values of 200 MPa and 2.2 MPa$\sqrt{m}$, respectively. With the addition of 20 vol.% SiC whiskers the corresponding values are increased to 425 MPa and 4.7 MPa$\sqrt{m}$. The improved strength is retained up to elevated temperatures, however, at temperatures greater than 1200°C the strength is dependent on the secondary phases present at the grain boundaries.

Initially densification was by hot-pressing, but for economical production of composite material to near-net-shape, the ability to pressureless sinter is necessary. At the present time mullite-10 vol.% SiC whisker composites can be densified without pressure to greater than 94% T.D., which is nearly identical to results on our more highly-developed alumina-whisker composites.

Introduction

Reinforcement of ceramics with SiC whiskers is currently of interest because of the ability of the whiskers to increase the fracture toughness and strength of composite materials as compared to monolithics. It is anticipated that application of these improved composites will be in high temperature structural components, such as those for advanced heat engines. Fracture toughness and strength improvements have been observed in alumina,[1–5] alumina-zirconia,[5,6] zirconia,[6] and silicon nitride[7,8] matrix systems. Similar property improvements have also been reported for mullite and mullite-zirconia matrix composites.[2–5]

*Research sponsored by the U.S. Department of Energy, Assistant Secretary for Conservation and Renewable Energy, Office of Transportation Systems, as part of the Ceramic Technology for Advanced Heat Engines Project of the Advanced Materials Development Program, under contract DE-AC05-840R21400 with Martin Marietta Energy Systems, Inc.

While the other matrix composites have excellent properties, mullite matrix composites could provide advantages in particular applications because of mullite's unique combination of properties such as:

1. lower thermal conductivity,
2. better thermal shock resistance,
3. excellent creep resistance,
4. high-temperature strength retention,
5. lower density, and
6. improved high-temperature phase stability with SiC.

Development of mullite-SiC whisker composites at Oak Ridge National Laboratory has included research on composite properties, understanding of the toughening mechanisms, and development of near-net-shape fabrication techniques.

Experimental Procedure

The mullite composites were fabricated by either hot-pressing or pressureless sintering. For all of the composites, the SiC whiskers* were dispersed and mixed with the mullite-matrix powders† in a liquid medium, such as water or methanol, by using a high shear mixer‡ and ball-milling. These methods have been found to provide considerable improvement in achieving a satisfactory dispersion of the SiC whiskers and the matrix powder compared to hand mixing.[9] In addition, the mixtures for pressureless sintering contained approximately 2 wt.% polyethylene glycol as a pressing aid. After mixing, the slurries were dried and screened to –80 mesh (<180 μm). The powder-whisker mixtures for hot-pressing were consolidated in graphite dies at temperatures of 1500 to 1675°C. Densities obtained were greater than 97% of theoretical density (T.D.). Samples for pressureless sintering were uniaxially pressed into bars at 138 MPa and isopressed at 380 MPa. After binder burnout, the pressed bars were fired in closed alumina crucibles packed with excess mullite powder and SiC whiskers under argon or nitrogen atmosphere at temperatures from 1500 to 1800°C.

Fracture toughness measurements were obtained using both the applied moment double cantilever beam (AMDCB) and the multiple indent flexure strength (MIFS) techniques.[10] The latter technique employed flexure bars having nominal dimensions of 2.2 × 2.5 × >20 mm, with polished tensile surfaces that contained three Vicker's DPH indents to produce controlled flaws in the region of maximum tensile stress. The fracture strengths were determined by four-point flexure on bars having the same dimensions as the MIFS specimens, but with 180 or 220 grit diamond ground tensile surfaces.

Microstructure and Properties of Hot-Pressed Mullite-SiC Whisker Composites

The microstructure resulting from the addition of SiC whiskers into a mullite matrix is shown in Fig. 1. The whiskers are easily observed as are some of the SiC

*ARCO Chemical/Advanced Composites, Greer, SC, Grade SC-9.

†Baikowski International,Charlotte, NC, Grade CR-193.

‡Brinkmann Instrument Co., Westbury, NY, Model PT45/8.

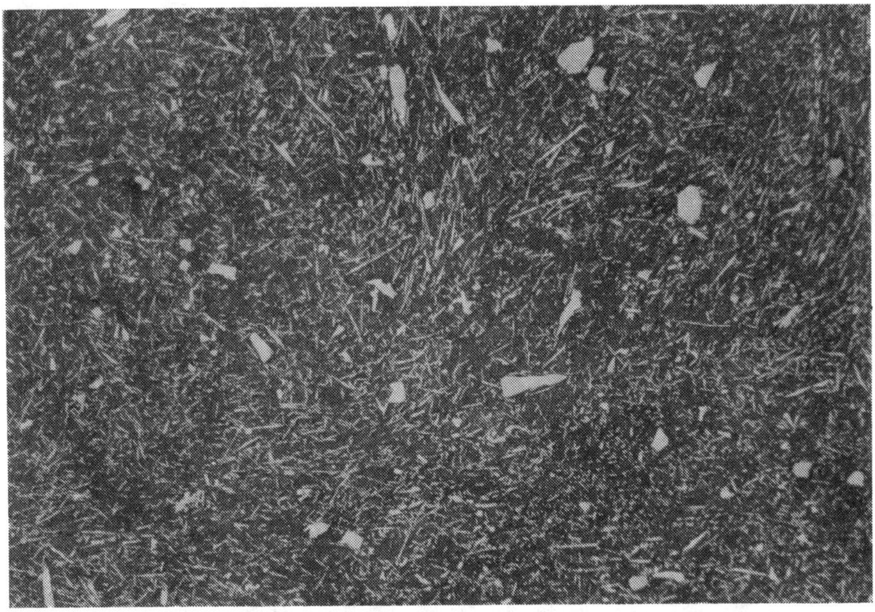

Fig. 1. Microstructure of mullite-SiC whisker composite showing whiskers and associated SiC particulate. Metallographic section is perpendicular to hot-pressed axis. Bar = 50 μm, 400 ×.

particulate (usually <25 μm in diameter) that is normally associated with the whiskers. Improvements in the fracture toughness and flexural strength as the whiskers are incorporated into the mullite matrix are given in Table 1. As indicated, significant increases are exhibited over monolithic mullite. Examination of a typical fracture surface by scanning electron microscopy (SEM) reveals considerable crack interaction with the whiskers resulting in a very rough looking surface morphology (Fig. 2). This surface roughness indicates that toughening occurs mainly by crack bridging and deflection. These mechanisms are also illustrated in Fig. 3, which shows a crack propagation path induced in a transmission election microscopy (TEM) thin foil. The crack path is clearly observed to be affected by the presence of the SiC whiskers. The mechanisms observed in the mullite matrix system are analagous to those reported in alumina-SiC whisker composites.[3,4]

Table 1. Fracture toughness and flexural strength of hot-pressed mullite-SiC whisker composites at room temperature

SiC Whisker Content (vol.%)	Fracture Toughness (K_{1c}, MPa$\sqrt{m}$)	Flexural Strength (MPa)
0*	2.2	200
10	3.6 ± 0.2	422 ± 103
20	4.7 ± 0.4	425 ± 31

*Nominal Values

Fig. 2. Fracture surface of mullite-SiC whisker composite showing rough surface morphology indicating toughening by crack bridging and deflection.

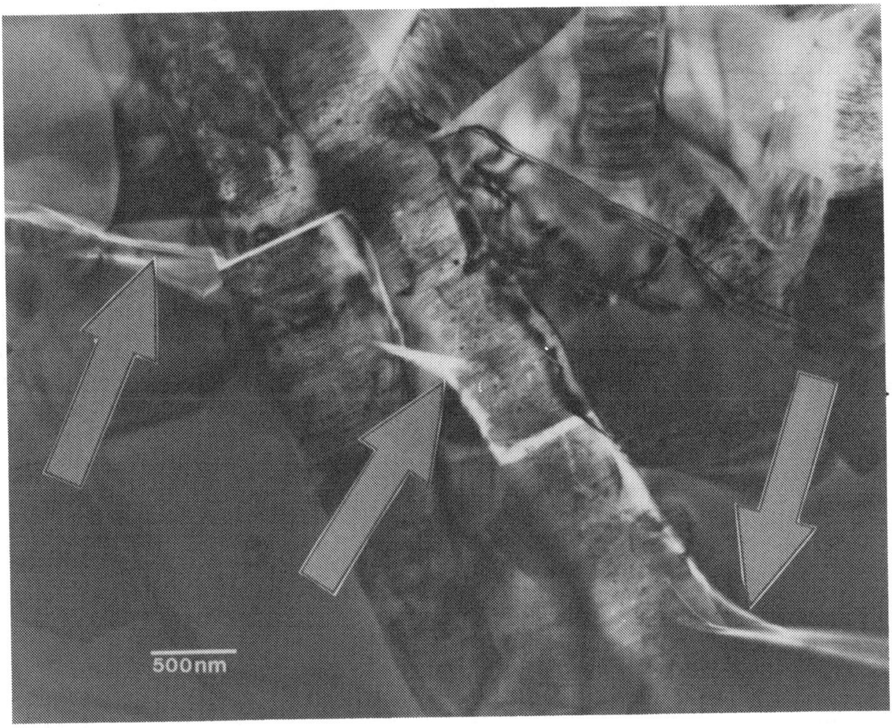

Fig. 3. TEM photomicrograph of a mullite-SiC whisker composite showing crack-whisker interaction. Arrows indicate crack path.

The flexural strength of a mullite-20 vol.% SiC whisker composite as a function of temperature is presented in Fig. 4. As shown, the significant increase in the strength of the composite is retained up to temperatures of at least 1200°C. The drop in strength at 1400°C is attributable to a degradation or softening of the intergranular glassy phase that is present in the composites. TEM examination (Fig. 5) showed pockets of silica-rich glass, high in calcium, in selected areas surrounding some of the SiC whiskers. The source of the intergranular glassy phase is believed to be due to excess silica in the mullite powder used in the present study. Similar glassy phases have been observed in other studies using the same mullite powder.[11] Previous results on mullite without glassy grain-boundary phases have shown no strength degradation up to temperatures of 1500°C.[12] It would be anticipated that elimination or crystallization of the glassy phases in the present material would improve the high temperature properties.

Pressureless Sintering

While hot-pressing of the mullite-SiC whisker composites produces dense materials with attractive mechanical properties, the ability to pressureless sinter the composites is desirable for some potential applications. Pressureless sintering is also

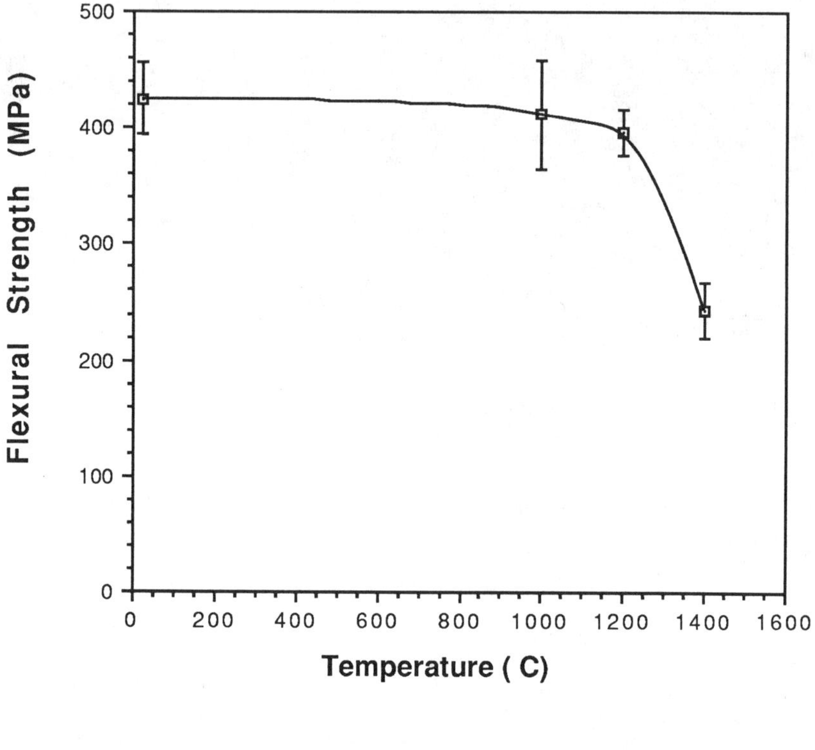

Fig. 4. Flexural strength of hot-pressed mullite-20 vol.% SiC whisker composites at elevated temperature.

Fig. 5. TEM photomicrograph of mullite-SiC whisker composite showing intergranular glass surrounding some SiC whiskers.

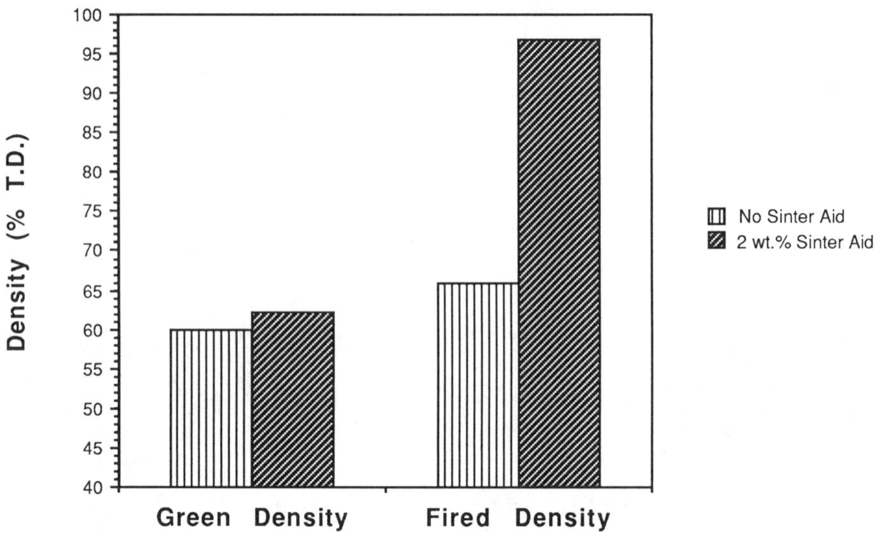

Fig. 6. Effect of sintering aid addition on densification of mullite-10 vol.% SiC whisker composites. Sintered at 1600°C.

necessary to provide complex and near-net-shape fabrication capability for these composites.

Previous results have shown that SiC whiskers inhibit sintering by interfering with particle rearrangement and shrinkage.[9] To promote densification, liquid phase sintering aids are used. Figure 6 shows the contrast in densification behavior of mullite-10 vol.% SiC whisker composites with and without sintering aids[§]. As indicated, the sintering aids are necessary to achieve acceptable densities at whisker contents of 10 vol.% or higher.

Because of the inhibiting effect of the whiskers, the ability to densify the composites is highly dependent on the whisker content. As the whisker content approaches 20 vol.%, the sintering is severely inhibited (Fig. 7). For the present time, high density mullite-SiC whisker composites (>94% T.D.) are generally limited to compositions with up to 10 vol.% whiskers.

Conclusions

SiC whisker reinforcement has been shown to significantly improve the mechanical properties of mullite ceramics. For example, typical mullite has a fracture toughness (K_{1c}) of about 2.2 MPa$\sqrt{m}$ and a flexural strength of approximately 200 MPa. We have demonstrated that by the addition of 20 vol.% SiC whiskers to hot-pressed mullite, corresponding values of 4.7 MPa$\sqrt{m}$ and 420 MPa are obtained. The improved strength is maintained up to temperatures of at least 1200°C. At higher temperatures, the strength is reduced because of intergranular glass phases associated with the mullite powder. Analysis has indicated that crack bridging and deflection are the major toughening mechanisms.

[§]**Type of sintering aids not disclosed due to patent considerations.**

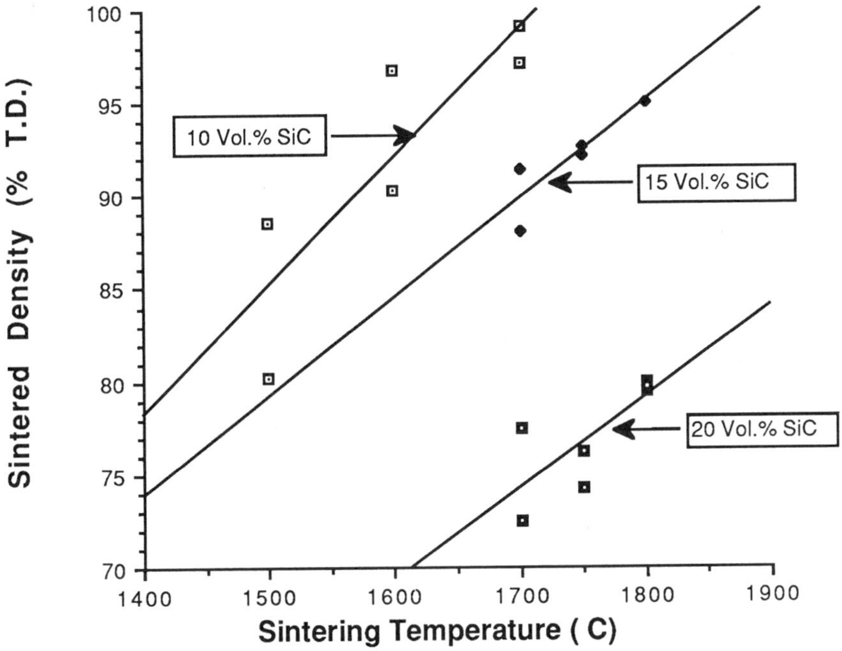

Fig. 7. Effect of whisker content and sintering temperature on density of mullite-SiC whisker composites. All composites contain 2 wt.% sintering aids.

The densities of mullite-SiC whisker composites fabricated by pressureless sintering were determined to be a function of the whisker content. Densities greater than 94% T.D. are generally limited to composites with whisker contents up to 10 vol.%.

Acknowledgments

The authors wish to acknowledge J. Geer and J. O'Niel for composite fabrication, H. Keating and W. Warwick for sample preparation, and F. Christie for manuscript preparation.

References

[1]P. F. Becher and G. C. Wei, "Toughening Behavior in SiC-Whisker-Reinforced Alumina," *J. Am. Ceram. Soc.*, **67** [12] C-267-C-269 (1984).

[2]G. C. Wei and P. F. Becher, "Development of SiC-Whisker-Reinforced Ceramics," *Am. Ceram. Soc. Bull.*, **64** [2] 298–304 (1985).

[3]T. N. Tiegs and P. F. Becher, "Whisker Reinforced Ceramic Composites"; pp. 639–47 in Tailoring Multiphase and Composite Ceramics, Materials Science Research Series, Plenum Press, New York, 1986.

[4]P. F. Becher, T. N. Tiegs, J. C. Ogle and W. H. Warwick, "Toughening of Ceramics by Whisker Reinforcement," pp. 61–73 in Fracture Mechanics of Ceramics, Vol. 7, Plenum Press, New York, 1986.

[5]P. F. Becher and T. N. Tiegs, "Toughening Behavior Involving Multiple Mechanisms: Whisker Reinforcement and Zirconia Toughening," *J. Am. Ceram.*, **70** [9] 651–54 (1987).

[6]N. Claussen and G. Petzow, "Whisker-Reinforced Zirconia-Toughened Ceramics," pp. 649–62 in Tailoring Multiphase and Composite Ceramics, Materials Science Research Series, Plenum Press, New York, 1986.

[7]R. Hayami, K. Ueno, I. Kondou, N. Tamari and Y. Toibana, "Si_3N_4-SiC Whisker Composite Material," pp. 663–74 in Tailoring Multiphase and Composite Ceramics, Materials Science Research Series, Plenum Press, New York, 1986.

[8]S. T. Buljan, J. G. Baldoni and M. L. Huckabee, "Si_3N_4-SiC Composites," *Am. Ceram. Bull.*, **66** [2] 347–52 (1987).

[9]T. N. Tiegs and P. F. Becher, "Sintered Al_2O_3-SiC-Whisker Composites," *Am. Ceram. Bull.*, **66** [2] 339–42 (1987).

[10]R. F. Cook and B. R. Lawn, "A Modified Indentation Toughness Technique," *Comm. Am. Ceram. Soc.*, **11**, C-200-1, 1983.

[11]R. D. Nixon, et al., "Creep of Hot-Pressed SiC Whisker-Reinforced Mullite," Proceedings of First International Conference on Mullite, Toyko, Japan, 1987, (to be published).

[12]T. Mah and K. S. Mazdiyasni, "Mechanical Properties of Mullite," *J. Am. Ceram. Soc.*, **66** [1] 699–703 (1983).

Zirconia-Toughened Mullite/The Role of Zircon Dissociation

P. Boch, T. Chartier and J. P. Giry

Ecole Nationale Supérieure de Céramique Industrielle
Limoges, France

Abstract

Zirconia-toughened mullite has been prepared by reaction-sintering of mixtures of alumina and zircon. The use of very fine particles and the choice of fast heating rates favor densification in comparison with mullitization. Reassociated zircon must be preferred to dissociated one. The alumina to silica ratio is a critical parameter. The 75% Al_2O_3 composition shows the lowest room-temperature mechanical strength and the 68% composition the highest. ZrO_2 inclusions do not sensibly improve strength but they increase room-temperature toughness by more than 50%. High-temperature mechanical properties are degraded by the low-viscosity of impure silica, iron oxide being particularly detrimental.

Introduction

In a related paper in these Proceedings, Boch et al.[1] have shown that reaction-sintered mullite ceramics have a room-temperature strength (σf) of about 200 MPa and a room-temperature toughness (K_{1c}) of about 2 MPa$\sqrt{m}$. Although these values may be considered as satisfactory for most applications, they are inferior to those of alumina ceramics (about 300 MPa for σf and 3.5 MPa$\sqrt{m}$ for K_{1c}). Therefore, it was of interest to determine to what extent ZrO_2 additions to the mullite ceramics improved their mechanical properties.

Claussen and Jahn[2] have determined a room-temperature strength of about 400 MPa and a room-temperature toughness of about 4.5 MPa$\sqrt{m}$ for "in-situ reacted mullite-zirconia composites."[3] They have succeeded in reaching a complete separation of the sintering and reaction steps. Subsequently, numerous studies have been devoted to ZrO_2-toughened mullite[4–16] ceramics, fabricated by various routes: reaction-sintering of mixtures of alumina and zircon, sintering of mixtures of mullite and zirconia, or chemical methods. The present paper concerns the preparation by reaction-sintering of alumina and zircon powders, and focuses on the role of the dissociation of zircon.

Reaction-Sintering of Zirconia-Mullite

Reaction-Sintering of Zirconia-Mullite Ceramics

The basic reaction is (for stoichiometric 3/2 mullite):

$$3Al_2O_3 + 2\ ZrSiO_4 \rightarrow (3\ Al_2O_3.2\ SiO_2) + 2\ ZrO_2$$

Zircon can be either in the non-dissociated ($ZrSiO_4$) or dissociated (ZrO_2 + SiO_2) state.

1. In the case of non-dissociated zircon, the original mixture consists of two crystalline compounds, $ZrSiO_4$ and α-Al_2O_3. The reaction of mullitization is expected to be concomitant with the zircon dissociation. This means that the development of crystalline mullite will be associated with the progressive consumption of zircon and alumina. However, some non-crystalline phases (amorphous silica and non-crystalline mullite) can exist during a transient period.
2. In the case of dissociated zircon, the original mixture is constituted of crystalline zirconia (generally in its monoclinic form) and amorphous silica. The question is whether the ternary mixture of ZrO_2, Al_2O_3, and amorphous SiO_2 will react to produce mullite before amorphous SiO_2 recrystallizes into cristobalite or whether amorphous silica will recrystallize first.

Both the densification kinetics and the mullitization kinetics depend on the nature of zircon (non-dissociated or dissociated) and on the recrystallization of amorphous silica. Dilatometric anomalies will also be sensitive to the nature of zircon, because dissociated zircon has a low density in comparison with its reassociated analog (3.85 vs 4.63 g cm^{-3}, respectively).[9,17]

Experimental

The alumina was RC172 Reynolds. Three grades of zircon were used:

1. RZS1 grade is a very fine (mean particle diameter: 0.5 μm) non-dissociated zircon, with a noticeable excess of free, amorphous silica. The SiO_2 content is 39 wt% whereas the SiO_2 content of stoichiometric $ZrSiO_4$ is only 32.5 wt%.
2. DZS2 grade is a dissociated zircon, produced by electric melting and fast cooling. Its ZrO_2 to SiO_2 ratio is very close to stoichiometric $ZrSiO_4$. DZS2 was attrition ground to different particle sizes, as will be indicated later on.
3. RZS2 grade is a reassociated zircon, obtained by the recombination of DZS2. The recombination was generally obtained by a thermal treatment of 2 h at 1420°C. The recombination kinetics are very sensitive to the size of zircon particles, as will be discussed later. RZS2 was attrition ground to the same mean particle sizes as DZS2: particular attention was paid to obtain not only the same mean particle size but also the same particle distribution in both materials.

All zircon grades were produced from natural zircon sand[18,19] and their purity was not very high. They contain from 0.1 to 0.2 wt% TiO_2 and from 0.05 to 0.2 wt% Fe_2O_3.

Powder mixtures were prepared and pressed using the same procedure as for mullite.[1] The sintering treatments were also similar to those used to prepare ZrO_2-free mullite ceramics.

The extent of the reaction of mullitization (alumina + zircon $\rightarrow$ mullite + zirconia) was estimated by the ratio X1, which characterizes the development of zirconia at the expense of zircon:

$$X1 = \frac{I\ mZ\ (1,1,-1) + I\ tZ\ (1,1,1)}{I\ mZ\ (1,1,-1) + I\ tZ\ (1,1,1) + I\ ZS\ (2,0,0)}$$

and the decomposition of zircon was estimated by ratio X2, which characterizes the development of mullite at the expense of zircon:

$$X2 = \frac{I\ M\ (2,1,0)}{I\ M\ (2,1,0) + I\ ZS\ (2,0,0)}$$

where I (i,j,k) are the intensities of the X-ray peaks for monoclinic and tetragonal zirconia (mZ and tZ), zircon (ZS) and mullite (M). Both X1 and X2 are equal to zero, in the non-mullitized state, and to the unity, in the fully-mullitized state (with a correction term to take into account the zircon non-stoichiometry). Both ratios will evolve in a similar way if the only reaction is mullitization. However, X2 will increase at a slower rate than X1 if SiO_2 does not react immediately or reacts to give non-crystalline mullite. The evaluation of cristobalite was conducted using its (1,0,1) X-ray peak.

Density measurements were carried out using the liquid displacement technique with distilled water. True density was derived from comparison with data from mixtures of powders of individual components. The experiments must be carried out as precisely as possible because of the large density variations. In the non-reacted state, the densities are 4.06, 3.91, and 4.33 gcm^{+3} for the mixtures with RZS1, DZS2, and RZS2, respectively. In the fully-reacted state, the corresponding densities are 3.67, 3.83, and 3.83 gcm^{-3} (if m-ZrO_2 is 30% and t-ZrO_2 is 70%). The accuracy of the density values is about 1%. It should be noted that the dissociated zircon + alumina mixture has a density close to that of the reacted material (3.91 and 3.83 gcm^{-3}, respectively). On the contrary the reassociated zircon + alumina mixture has a higher density than that of the reacted material (4.33 and 3.83 gcm^{-3}, respectively). Microstructural observations, shrinkage experiments, and mechanical tests were carried out using similar procedures as those in Ref. 1.

Results and Discussion

Dissociation and Recombination of Zircon

The SiO_2-ZrO_2 equilibrium diagram contains $ZrSiO_4$ (zircon) as the only stable binary compound. At high temperature, zircon decomposes into tetragonal zirconia and β-cristobalite. Butterman and Foster[20] found zircon to decompose between about 1675 and 1685°C, whereas Curtis and Sowman[21] reported dissociation from 1540 to 1677°C. Initial results of Anseau et al.[22] on the reaction-sintering of mixtures of zircon and alumina were in agreement with Curtis and Sowman's diagram. However, the detailed studies of Pena, de Aza, Moya, and co-workers[23–25] finally demonstrated that Butterman and Foster's diagram must be preferred, which is in agreement with other studies.[26–30] The behaviour of pure zircon cannot be inferred from data obtained on natural impure zircons. Moreover, the decomposition process must be discriminated from the first formation of liquid with partial dissolution of zircon. The studies[23–25] on the influence of additions of Al_2O_3, ZrO_2, TiO_2, CaO and MgO have shown that transitory and permanent liquids can appear at temperatures as low as 1400°C in impurity-rich zircons. However, the usual impurity content in the mixtures of zircon and alumina which were used here should correspond to the first appearance of liquid at about 1540°C.

As previously indicated, RZS2 was obtained by recombination of dissociated zircon. A study of the recombination kinetics has shown them to be very sensitive to the state of division of the material. Two mean particle sizes were chosen, namely 1 mm and 1.6 μm. The fine grade is named f-ZSD and the coarse grade c.ZSD. The recombination was followed by measuring the X3 ratio:

$$X3 = \frac{I\ ZS\ (2,0,0)}{I\ ZS\ (2,0,0) + I\ mZ\ (1,1,-1)}$$

where the different terms have the same meaning as in X1 and X2.

The recombination kinetics of f-ZSD are monotonic vs. temperature and time. Recombination is fully completed after 10 h at 1300°C, and a temperature increase always accelerates the recombination kinetics. On the contrary the recombination of c-ZSD does not obey a monotonic law. The recombination rate begins to increase when temperature increases. A maximum is achieved at about 1480°C, and then decreases again when temperature continues to increase. The recombination ratio is 92% after 16 h at 1480°C but only 46% after 16 h at 1600°C (Table 1). These data are in qualitative agreement with McPherson's studies on the recombination of plasma dissociated zircon.[31,32] They suggest that the recombination rate is controlled, not by diffusion limited crystal growth but by nucleation of zircon. Therefore, the recombination rate is increased by grinding, which provides new surfaces which act as heterogeneous nucleation sites for zircon.

Table 1. Recombination of coarse-grain zircon (The recombination ratio is expressed in %)

	t = 2 h	t = 8 h	t = 16 h
T = 1300°C	6	/	/
T = 1420°C	27	70	/
T = 1480°C	33	80	92
T = 1540°C	16	58	75
T = 1600°C	10	34	46

8 h at 1600°C + 8 h at 1480°C: recombination ratio of 93%
8 h at 1600°C + 8 h at 1540°C: recombination ratio of 42%

Sintering of Starting Materials

The understanding of the reaction-sintering behaviour of mixtures of alumina and zircon requires information on the sintering behaviour of individual starting materials (alumina and zircon). The main results of a previous study[9] are:

1. RC172 Reynolds alumina exhibits a progressive densification and reaches relative densities of more than 95% when sintered at 1550°C for 2 h;
2. Very fine, silica-rich RZS1 has a high sinterability, since almost complete densification (99%) is obtained at temperatures not exceeding 1300°C;
3. ZS2 zircon shows a behaviour which depends on its state of dissociation. Dissociated DZS2 begins to densify more rapidly than reassociated RZS2, which can be due to the presence of free, amorphous silica. However, the densification rate slows when the recombination reaction begins (at about 1300°C) and the relative density does not sensibly increase as long as the recombination develops (up to about 1420°C). The recombination is

associated with a density increase (from 3.85 to 4.63 gcm^{-3}) which should induce either a shrinkage of the whole specimen or the creation of pores. Although the shrinkage predominates, the contraction accompanying the zircon recombination is detrimental to densification, and some de-densification phenomena can even occur (Fig. 1). Therefore, the densification rate is lower for DZS2 than for RZS2.

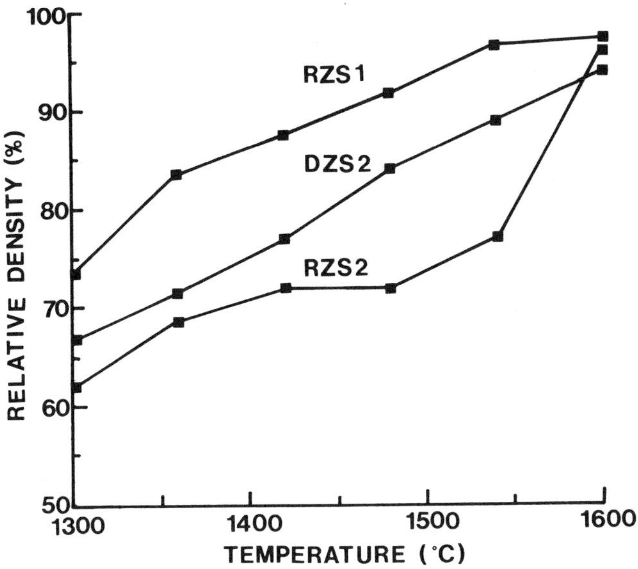

Fig. 1. Densification of alumina + zircon mixtures vs. temperature (soaking time: 2 h).

Reaction-Sintering of Mixtures of Alumina and Zircon: 3/2 Stoichiometric Mullite

Influence of the State of Dissociation of Zircon

Figures 1, 2, and 3 show the evolution of densification (d/do), zirconia development (X1 ratio), and mullite development (X2 ratio) during the sintering of alumina + zircon mixtures, for three zircon grades: RZS1 (mean particle size 0.5 μm), RZS2 and DZS2 (mean particle size: 1.1 μm). The main comments are as follows:

1. Densification kinetics are slower for the alumina + zirconia mixtures than for the individual components.[9] For instance, alumina + RZS1 mixtures densify to 91.5% after 2 h at 1480°C, whereas alumina and RZS1 densify to 96 and 99%, respectively, for the same firing conditions.
2. The recombination of dissociated DZS2 into $ZrSiO_4$ and the recrystallization of amorphous silica into β-cristobalite begin at about 1300°C for the alumina + DZS2 mixtures, whereas mullitization begins at about 1360°C only. This means that mullitization mainly develops by reaction between alumina and reassociated zircon (and/or cristobalite) and, therefore, that the role of free, amorphous silica is only limited to the early stages of sintering. A similar result has been found for the reaction-sintering of zirconia-free mullite[1] where most of the amorphous silica recrystallizes into cristobalite before

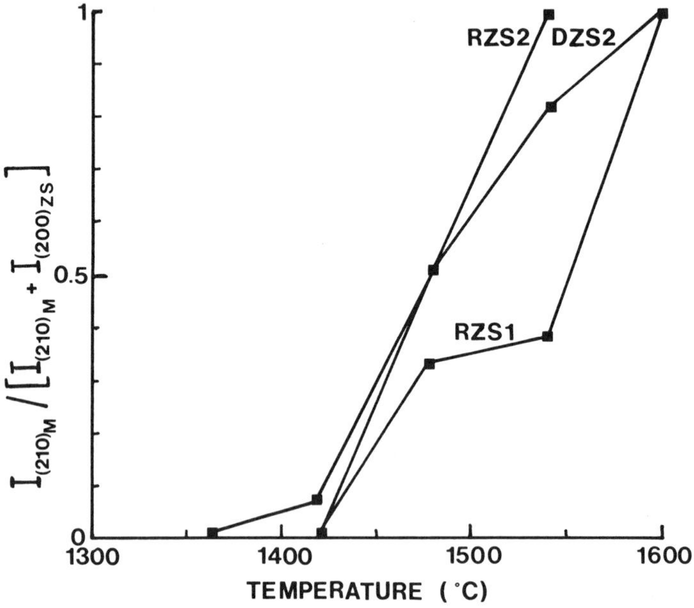

Fig. 2. Zirconia development in alumina + zircon mixtures vs. temperature (soaking time: 2 h).

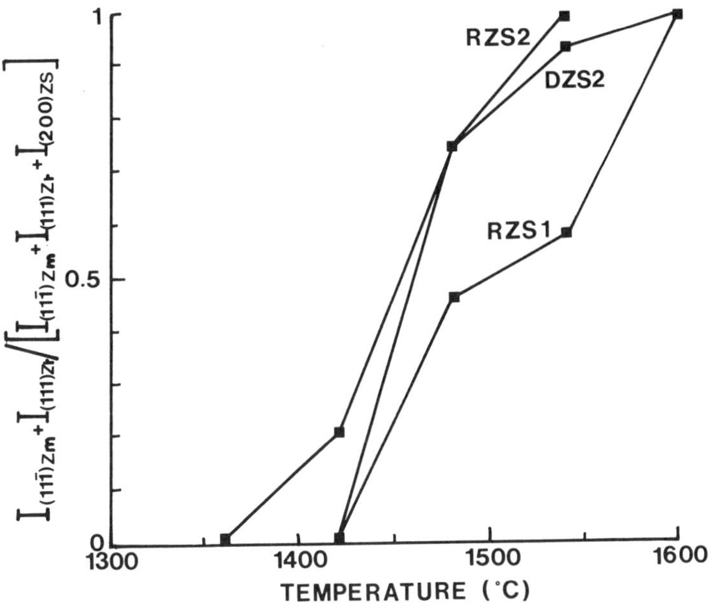

Fig. 3. Mullite development in alumina + zircon mixtures vs. temperature (soaking time: 2 h).

reacting with alumina. Mullitization begins at about 1450°C only for the mixtures with reassociated RZS1 and RZS2.

3. Zirconia (evaluated from X1 ratio) develops faster than mullite (evaluated from X1 ratio) does. This shows that the silica liberated by the zircon decomposition does not instantaneously react with alumina or reacts to form non-crystalline phases (e.g. non-crystalline mullite).
4. The mullite development slows down densification. This is particularly clear in the alumina + RZS2 mixtures, where densification remains on a plateau as long as mullitization develops (from 1450°C to 1550°C°).
5. The highest final densities (>97%) are obtained for the mixtures with the finest (RZS1) zircon.

Influence of Particle Size

Densification and mullitization kinetics were studied for mixtures of alumina (mean particle size: 0.65 μm) and zircon using two grades of RSZ2 zircon, corresponding to two mean particle sizes of 3 and 0.4 μm, respectively. As for zirconia-free mullite, the decrease in particle size increases the densification rate faster than it increases the mullitization rate, and the finest zircon leads to the best densification data. The main results are as follows (Figs. 4 to 8):

1. The mixture with the fine (0.4 μm) zircon densifies to more than 97% after 0.5 h at 1540°C, whereas the mixture with the coarse (3 μm) zircon only densifies to 88% after a longer time of 10 h at the same temperature.

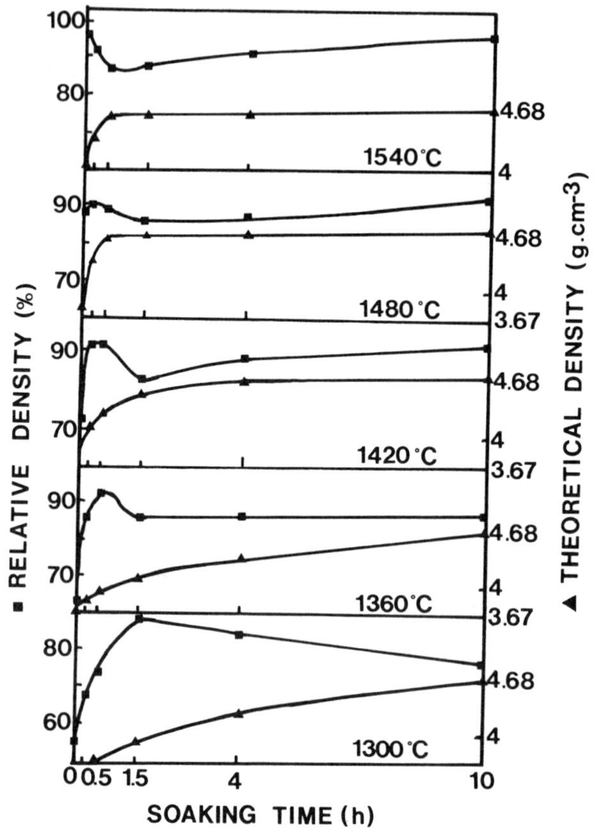

Fig. 4. Dissociated DZS2 zircon (mean grain size: 1.6 μm): variation of relative density and true density vs. time at different sintering temperatures.

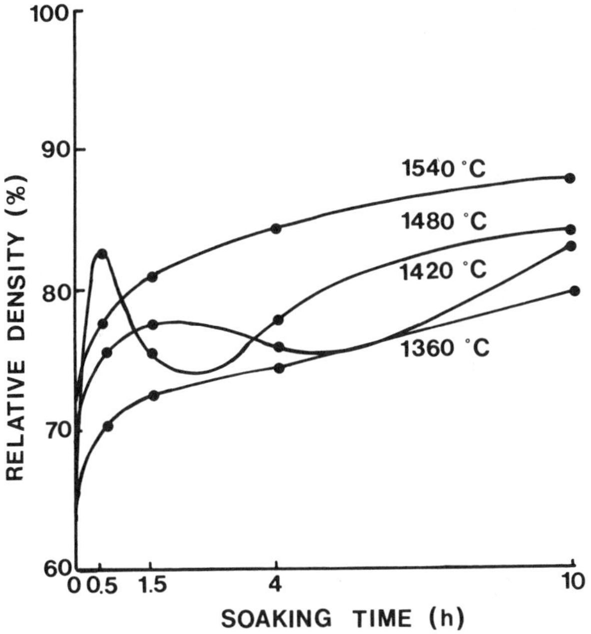

Fig. 5. Mixtures of alumina and coarse RSZ2 zircon (mean particle size: 3 μm): densification vs. time at different sintering temperatures.

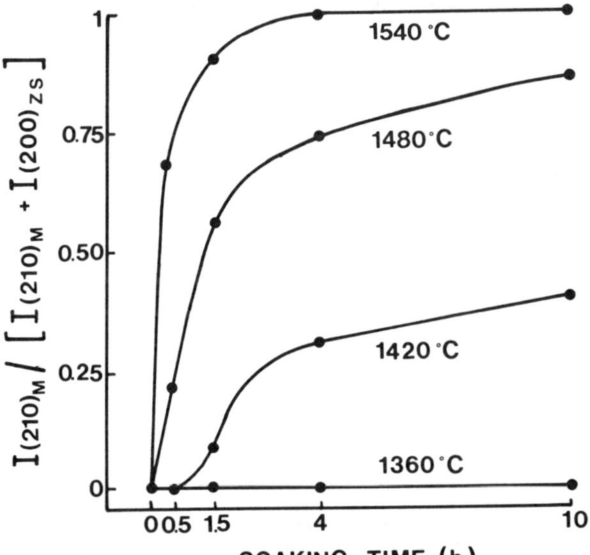

Fig. 6. Mixtures of alumina and coarse RSZ2 zircon (mean particle size: 3 μm): mullitization vs. time at different sintering temperatures.

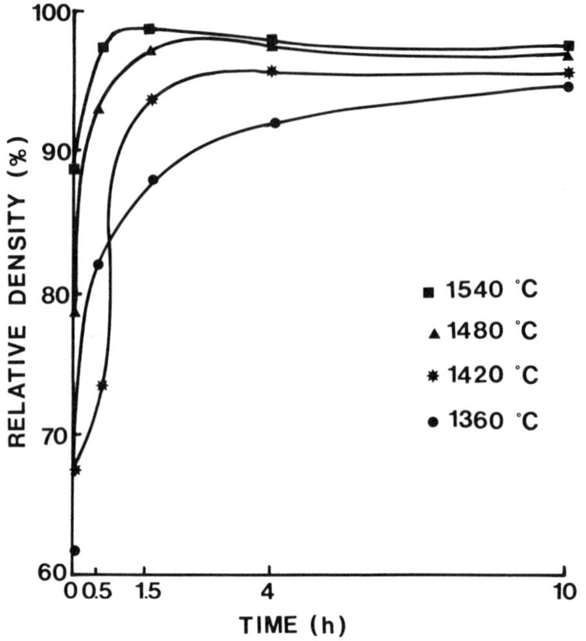

Fig. 7. Mixtures of alumina and fine RZS2 zircon (mean particle size: 0.4 μm): densification vs. time at different sintering temperatures.

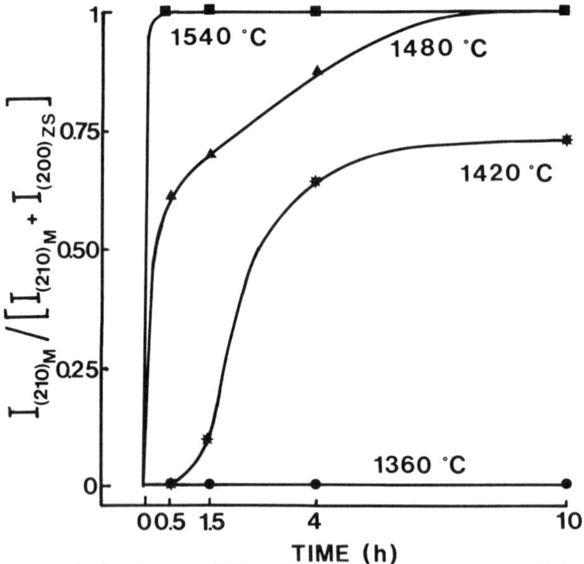

Fig. 8. Mixtures of alumina and fine RZS2 zircon (mean particle size: 0.4 μm): mullitization vs. time at different sintering temperatures.

Mullitization is completed after 0.5 h for the fine mixture and after 4 h in the coarse mixture.

2. The mixture with the fine zircon does not show a noticeable mullitization after a firing of 1.5 h at 1420°C, although it is already densified to 94%. However, the subsequent development of densification corresponds to a slow densification rate.
3. The mixture with the coarse zircon can exhibit some de-densification phenomena. This is in particular the case when it is fired at 1420 and 1480°C.

This is in qualitative agreement with what is found in zirconia-free mullite.[1] This also confirms the hypothesis of Yangyun and Brook[33] that the use of very fine powders may allow us to separate the densification stage from the mullitization one, and therefore may help to produce nearly dense zirconia-mullite ceramics.

Influence of Heating Rate

Fast heating rates were not found[34] to increase the final densification, contrary to what was found for zirconia-free mullite.[1] Slow heating rates can even slightly improve densification. However, slow heating rates result in the growth of mullite grains and the coarsening of zirconia inclusions, which yields both a decrease in the tetragonal-zirconia content and a decrease in strength (Table 2). Therefore, fast heating rates must be preferred.

Table 2. Influence of heating rate on properties of reaction-sintered zirconia-mullite ceramics

	Heating rate: 300°C h^{-1}			Heating rate: 3000°C h^{-1}		
	Density (d/do)	t-ZrO_2 (vol%)	R.T. strength (MPa)	Density (d/do)	t-ZrO_2 (vol%)	R.T. strength (MPa)
Al_2O_3 + DZS2	95	7	140	94.5	10	175
Al_2O_3 + RZS2	97	5	185	96.5	7	210

Influence of the Al_2O_3 to SiO_2 ratio. Influence of Additions of Zirconia and Ferric-Oxide

Phase Composition

Five compositions were studied, namely 68.0, 71.8, 75.0, 77.3, and 80.0 wt% alumina (in the alumina to silica ratio). Variations in this ratio are obtained by varying the alumina to zircon ratio. This means that the zirconia content is also modified: it decreases from 20 to 14.5 vol% when the alumina to silica ratio increases from 68.0 to 80.0 wt%. Only RZS1 (i.e. the finest, reassociated zircon) was studied. Table 3 gives the characteristics of the studied compositions. The firing cycle was a slow heating rate (300°C h^{-1}) to the firing temperature of 1600°C, a soaking time of 2 h and then a slow decrease (780°C h^{-1}) to room temperature. All compositions were fully reacted, and the "3/2" composition of reference was 98% dense.

The determination of the phase compositions has shown that the alumina-rich end of the mullite solid-solution in ZrO_2-mullite is 74.8 wt%, as in ZrO_2-free mullite. The silica-rich end of the mullite solid-solution was found to be 71.8 wt% instead of 70.4 wt% in ZrO_2-free mullite. As in the case of ZrO_2-free mullite, mullite begins to develop with a composition close to that of the silica-rich end of the solid-

Table 3. **Characteristics of the Al_2O_3-$ZrSiO_4$ compositions**

Composition	Starting materials Al_2O_3 (wt%)	RZS1-$ZrSiO_4$ (wt%)	Equivalent composition Al_2O_3 (wt%)	SiO_2 (wt%)	ZrO_2 (wt%)
M68.0	45.5	54.5	45.8	21.6	32.6
M71.8	49.5	50.5	49.8	19.6	30.6
M75.0	54.2	45.8	54.4	18.1	27.5
M77.3	57.3	42.7	57.5	16.9	25.6
M80.0	61.2	38.8	61.4	15.4	23.2

Table 4. Phase composition of reaction-sintered ZrO_2-mullite for various Al_2O_3-to-SiO_2 ratios (sintering: 2 h/1600°C)

Al_2O_3 to SiO_2 ratio (wt%)	Al_2O_3 content in mullite	SiO_2 content (vol%)	mullite content (vol%)	Al_2O_3 content (vol%)	ZrO_2 content (vol%)	t/m ratio	d/do (%)
68.0	71.8	5.9	74.1	/	20.0	23	>98
71.8	71.8	/	81.0	/	19.0	23	98
75.0	74.8	/	82.5	0.5	17.0	27	>97
77.3	74.8	/	77.0	7.0	16.0	31	>97
80.0	74.8	/	70.5	15.0	14.5	31	97

solution, whatever the bulk composition is. The mullite composition then reaches its final value.

The 75.0, 77.3, and 80.0% Al_2O_3 compositions contain α-corundum precipitates. The 68% Al_2O_3 composition contains an excess of about 5.9 vol% of SiO_2, and it would have been expected to contain cristobalite and/or zircon. However, the only crystalline phase which was found was mullite, which means silica did not recrystallize nor react during cooling. Subsequent annealing treatments (e.g. 1 h at 1500°C) allow the reaction between silica and zirconia to develop to give zircon. Therefore, the 68% SiO_2-materials have finally three phases: zircon, zirconia, and 71.8% SiO_2-mullite.

Table 4 gives the characteristics of reaction-sintered materials. It must be noticed that the tetragonal to monoclinic ratio of free zirconia increases from 23 to 31% when the zirconia content decreases from 20 to 14.5 vol%. This means that the content of tetragonal zirconia remains at the same low value of about 5 vol% in all compositions.

Mullitization, Densification, and Microstructures

The comparison between the kinetics of densification (d/do) and mullite development (X2 ratio) leads to the following comments:

1. For all compositions, mullitization begins between 1480 and 1540°C. However, kinetics depend on composition: they accelerate when the alumina content increases (Fig. 9). For instance, 0.5 h at 1500°C are necessary to complete mullitization in the 68% Al_2O_3 composition, whereas a simple heating (300°C min^{-1}) to 1540°C yields complete mullitization in the 80% Al_2O_3 composition. It is not clear whether this acceleration in the zircon decomposition is related to chemical reasons (modification of the diffusion

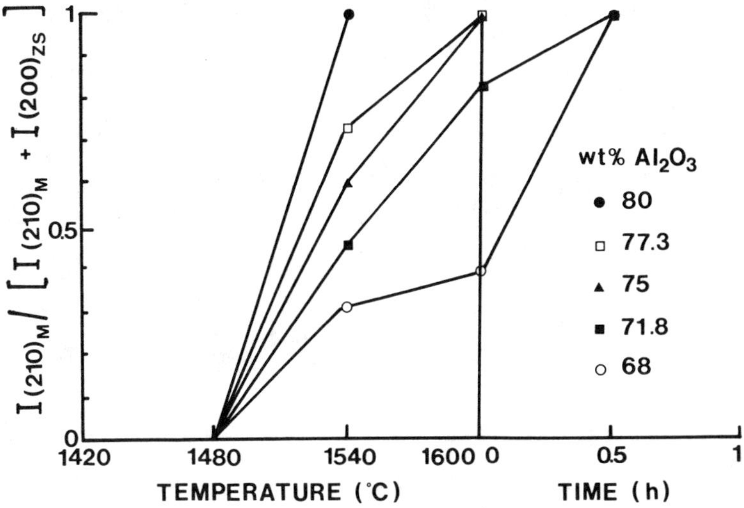

Fig. 9. Kinetics of mullitization for ZrO_2-mullite materials of different alumina to silica ratio.

coefficients) or to the increase in the number of contacts between alumina and zircon particles (the volumetric ratio of alumina to zircon increases from 1.1 to 2.1 when the alumina content increases from 68 to 80%).

2. The densification kinetics are not very sensitive to the alumina-to-silica ratio. However, low-alumina compositions reach final densities which are slightly higher than those of high-alumina compositions (see Table 4).
3. The microstructural examinations (Fig. 10) show that mullite and zirconia are homogeneoulsy distributed in all materials. However, grain size and grain shape depend on the alumina to silica ratio.

   - Mullite exhibits equiaxed grains in the 71.8, 77.3, and 80.0 Al_2O_3 materials. Corundum precipitates exist in the 77.3 and 80.0% Al_2O_3 materials. They are mainly located in intergranular positions. Zirconia inclusions show rather irregular shapes.
   - Mullite exhibits mixed (acicular and equiaxed) grains in the 68% Al_2O_3 material. This can be related to the presence of a liquid phase at the sintering temperature. Glassy residues can be seen along some grain boundaries and at 4-grain junctions. Zirconia inclusions are rounded or coarsened to a bead-like form.
   - Zirconia inclusions also show a bead-like form in the 75.0% Al_2O_3 material. Mullite is constituted of large (10–15 μm), elongated grains.

These results show that exaggerated grain growth occurs in 75.0% Al_2O_3 zirconia-containing mullite as well as in 75.0% Al_2O_3 zirconia-free mullite.[1] This demonstrates that zirconia inclusions do not exert a sufficient dragging force to hinder growth. This has also been observed in zirconia-alumina materials. Lange and Hirlinger[35] have found that ZrO_2 inclusions can sometimes exhibit sufficient self-diffusion to move with the Al_2O_3 4-grain junctions during grain growth, in which

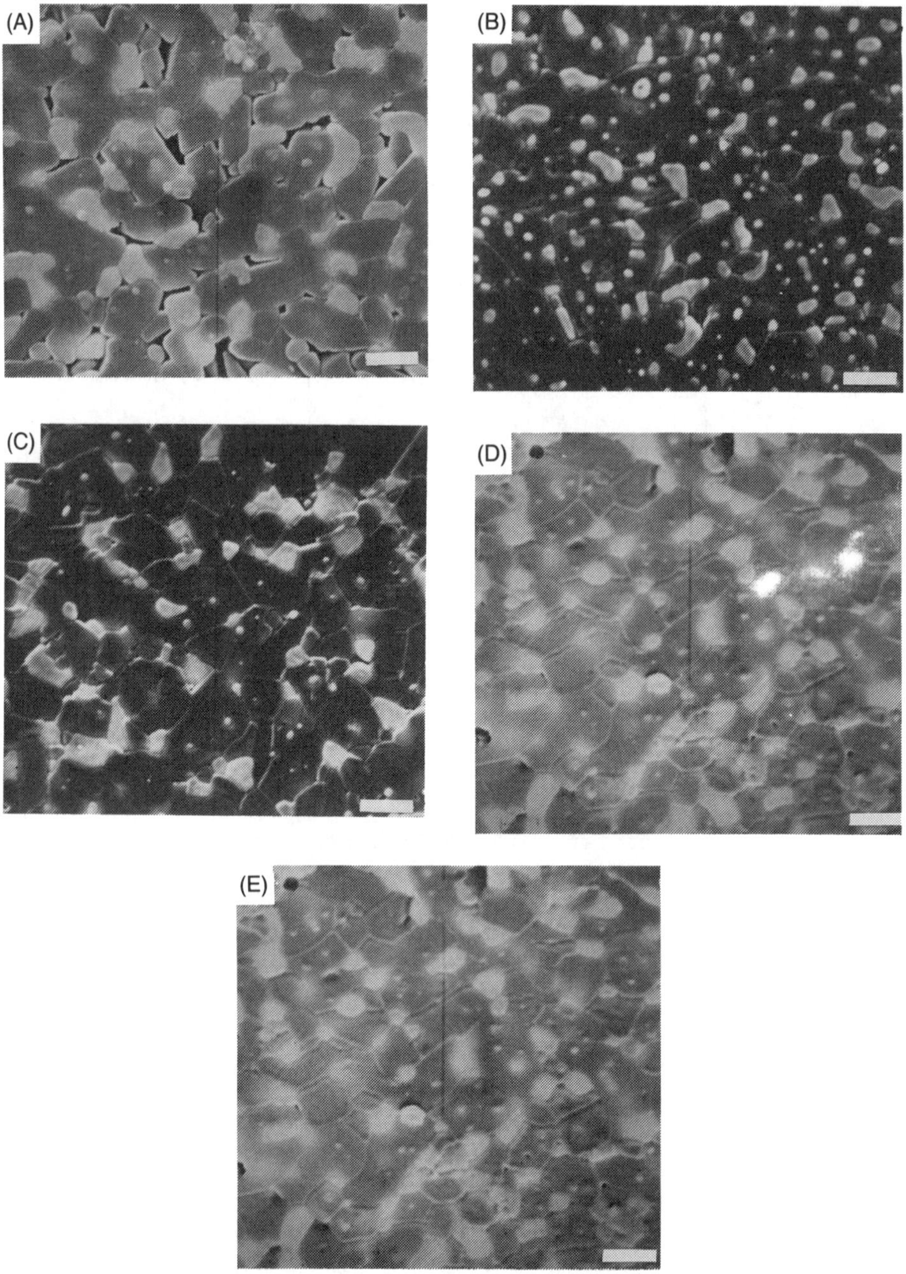

Fig. 10 Microstructures of reaction-sintered ZrO_2-mullite ceramics with different alumina to silica ratio. Bar = 3 μm.

case exaggerated grains consume both neighboring grains and ZrO_2 inclusions. Kibbel and Heuer[36] have related this behaviour to the presence of a continuous, amorphous (glassy) grain-boundary phase. This may be also valid in mullite ceramics, where silicate-containing residues are expected to remain along grain boundaries. However, this does not explain why only the 75% Al_2O_3 composition is sensitive to exaggerated grain growth. As in zirconia-free mullite, this could be related to the variation in the alumina content of mullite from a low initial value (70.4 and 71.8% for zirconia-free and zirconia-containing mullites, respectively) to a high final value (74.8% for both kinds of mullites). Further studies are required to clarify this point.

Influence of Zirconia and Iron-Oxide Additions

The additives studied were zirconium and iron oxides. ZrO_2 was selected because the comparison between mullite and zirconia-toughened mullite may be facilitated when the effects of small amounts of zirconia are known. Such small amounts cannot induce toughening mechanisms, but they can nevertheless modify the reaction-sintering behaviour and the microstructure of mullite and, thereby, the properties of the materials. And iron or iron-oxide impurities are common pollutants in natural zircon and frequent contaminants introduced during grinding operations.

ZrO_2 and Fe_2O_3 additions were introduced in the alumina + silica or alumina + zircon mixtures using soluble salts to improve homogeneity. Thus, $Fe(NO_3)_3.9H_2O$ and $Zr(NO_3)_4.5H_2O$ were added to the mixtures which were then calcined for 5 h at 800°C. For mullite materials, the final additive contents were 0.75 and 1.5 wt% of ZrO_2 or Fe_2O_3. For ZrO_2-toughened mullite materials, the final additive contents—referred to the only mullite phase because iron oxide is expected to enter the mullite solid-solution[37,38]—were 0.25, 0.50, and 1.50 wt% of Fe_2O_3.

The main results concerning the influence of the aforenoted additions are as follows:

1. Zirconia additions slightly accelerate the kinetics of densification, and they slightly improve the final densification. However, they do not have significant effects on reactions. Retardation in densification caused by both silica-to-cristobalite transformation and mullite formation occur in the same temperature intervals for mullite and ZrO_2-doped mullite.
2. Iron oxide additions strongly accelerate both reaction and densification kinetics in mullite ceramics (Fig. 11). The addition of 1.5 wt% iron oxide reduces the temperature of the silica-to-cristobalite transformation, the mullite formation and the beginning of densification of the alumina/silica mixtures by about 75°C. From 1300°C to 1500°C, the total shrinkage of the mixtures containing iron oxide is less than that of the other mixtures. This can be attributed to (1) the early formation of cristobalite causing an early retardation of densification, and (2) the lowering of the mullite formation temperature giving more mullite and hence more pronounced expansion. Despite the earlier formation of mullite which could result in a lower final densification, the mixtures containing iron oxide always densify better than the others. For example, 1.5 wt% of Fe_2O_3 results in greater than 96% of final density in 10 h at 1550°C, whereas the densifications of other mixtures subjected to the same thermal treatment are less than 86%. Furthermore, the presence of iron oxide yields a microstructure which consists of large, elongated grains. Finally, it must be pointed out that although the Fe_2O_3 additions have beneficial effects on densification they have detrimental

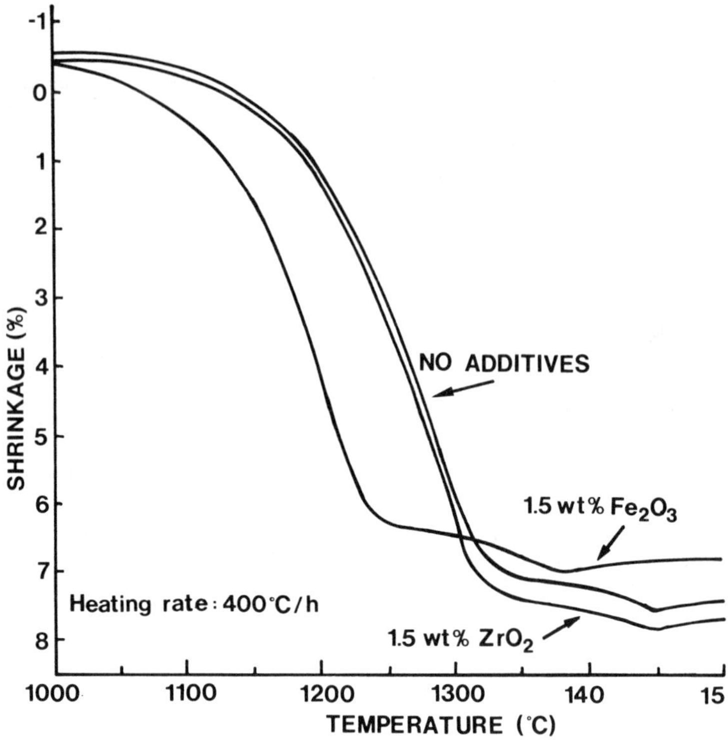

Fig. 11. Shrinkage vs. temperature for "3/2" alumina to silica mixtures containing additives.

effects on the high-temperature mechanical properties of materials, as will be seen in the last paragraph.

3. Iron oxide additions dramatically accelerate the recombination of dissociated zircon. For instance, 10 h at 1300°C are required for full recombination of "pure" DZS2 whereas 4 h and 30 min are sufficient in the 0.25% and 1.50 wt% Fe_2O_3 materials, respectively.
4. Iron oxide additions also dramatically accelerate the mullitization and densification kinetics in the alumina + zircon mixtures. For instance, heating from room temperature to 1540°C leads to final densifications of 62.5 and 96.5% for RZS2 mixtures without iron-oxide additions and with 1.5% iron oxide, respectively.
5. The microstructure of iron-rich mullite (or ZrO_2-mullite) materials is constituted of large, elongated grains. Thermally etched iron-rich materials show deep grooves along the grain boundaries, which demonstrates the presence of vitreous residues.

The existence of a liquid phase at the temperature of sintering can explain the high sinterability of iron-rich mullites in comparison with that of iron-free mullites:

1. Only 0.12 wt% of CaO is sufficient to form an eutectic liquid with 1.5 wt% Fe_2O_3 at 1203°C.[37] More than half of required amount of CaO is present as an impurity in the α-alumina used. Above 1200°C, wustite (FeO) is stable under oxygen partial pressures below about 10^{-6} atm.[38] Although the oxygen partial pressure of the firing atmosphere (air) is nearly $10^{-0.7}$ atm., strong reducing conditions can be created locally inside the specimens during burning organic binders and can remain inside pores or densely packed grains. Therefore, some zones may be enriched in Fe_3O_4 or even FeO. Two eutectic liquids are formed between FeO and SiO_2 below 1200°C: one at 1183°C containing 62 wt% FeO and the other at 1181°C containing 77 wt% FeO.[39] The Fe_3O_4-SiO_2 system shows an eutectic at 1455°C containing 20 wt% SiO_2.[40]
2. The FeO-Fe_2O_3-ZrO_2-SiO_2 equilibrium diagram[41,42] indicates that under normal oxygen partial pressures the maximum solubility of iron oxides is about 4 wt% in zircon and 5 wt% in zirconia. Consequently, no liquid phase should develop at temperatures below 1600°C in the studied mixtures where the Fe_2O_3 content remains below 1.5 wt%. However, this would be true if iron oxide was perfectly homogeneously distributed, which may not be the case here where iron salts were deposited onto the surface particles. For RZS2 mixtures (mean particle diameter of 1.5 μm) a rough assumption that the thickness of the iron-oxide layer is about 100 Å would lead to a local concentration of about 50% Fe_2O_3, in which case the liquidus temperature would be 1425°C. Taking into account the possible existence of the local reducing conditions previously mentioned, the lowest liquidus temperatures would be 1125 and 1145°C. All these data are in favour of the presence of a liquid phase at the temperature of sintering in both zirconia-free and zirconia-containing mullite materials.

A supplementary reason to permit liquid-phase sintering in iron-rich materials is that iron-free and iron-rich materials do not have the same activation energy of densification.[34] Assuming that:

$$\delta[(d/do)]/\delta t = A \exp[-E/RT]$$

where $\delta[(d/do)]\delta t$ is the densification rate, A a constant, E the activation energy, R the gas constant, and T the temperature, Arrhenius plots of Ln $[\delta(d/do)]/\delta t$ vs. 1/T lead to activation energies of about 450 kJ/mole for iron-free materials and 300 kJ/mole for iron-containing materials. The large difference between the two values confirms that the two materials do not densify by the same sintering mechanisms. The low value of E in the iron-rich materials is what could be expected in the case of a liquid-phase sintering.

Mechanical Properties

Figure 12 shows the influence of 1.5 wt% additions of ZrO_2 and Fe_2O_3 additions on the flexural strength of "3/2" mullite. At room temperature, the strength of three mullites is in the same rank that their densification. The iron-containing mullite, which has the highest densification (97.3%), shows a strength of 250 and 260 MPa at R.T. and 600°C, respectively. From 600 to 1200°C, its strength decreases from 260 to 125 MPa, with a rapid drop between 900 and 1200°C. The fracture at 1200°C was verified to be entirely intergranular, which supports the hypothesis of an intergranular-phase softening or melting around 1200°C, in agreement with what has been mentioned regarding the liquid phase in iron-rich mullites. At temperatures > 600°C, the strength of mullite ceramics containing 1.5 wt% ZrO_2 is slightly inferior

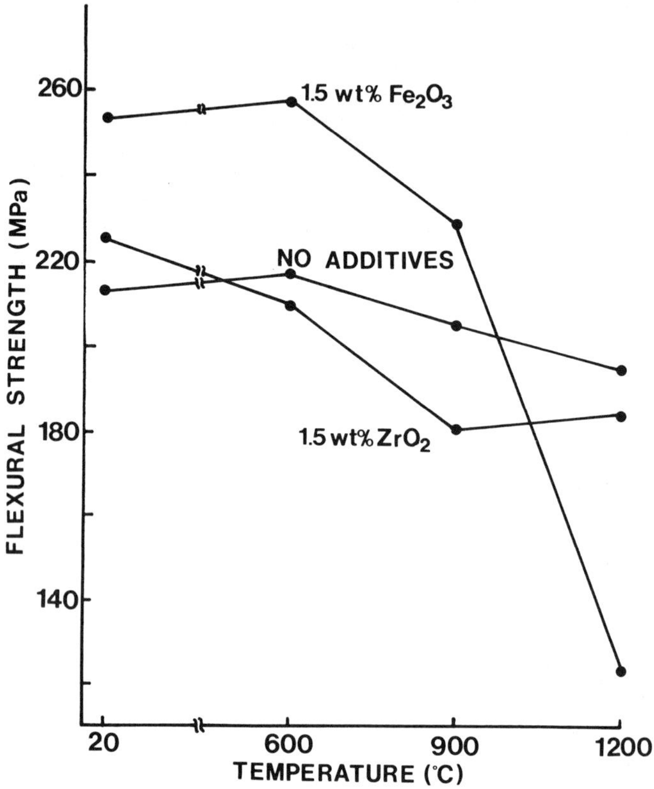

Fig. 12. Flexural strength vs. temperature for "3/2" mullite containing additives.

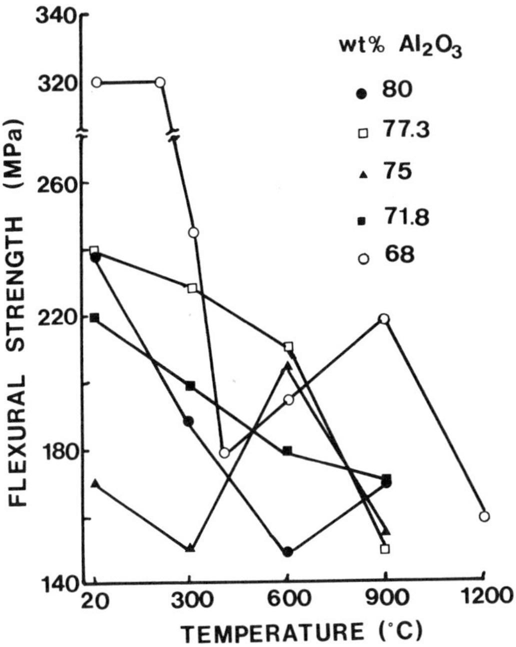

Fig. 13. Flexural strength vs. temperature for reaction-sintered ZrO_2-mullite with different alumina to silica ratio.

to that of pure mullite, although the densification of the latter is lower than that of the former. Therefore, small additions of ZrO_2 do not have a significant beneficial effect on the strength of mullite.

Figure 13 shows the influence of the alumina-to-silica ratio on the strength of ZrO_2-rich mullites. At room temperature, the highest strength (320 MPa) is found for the 68% Al_2O_3 composition, whereas the lowest (170 MPa) is found for the 75% Al_2O_3 composition. This shows that, as in ZrO_2-free mullites,[1] the silica residues in the silica-rich composition are beneficial, whereas the coarse microstructure in the 75% Al_2O_3 material is detrimental. A noticeable difference between high-purity mullites and low-purity ZrO_2-mullites is that a temperature increase can improve the mechanical properties of the former, whereas it decreases the properties of the latter. For the 68% Al_2O_3 composition, for instance, the strength decreases from 320 MPa at room temperature to 180 MPa at 400°C, then increases up to a relative maximum at 900°C, then decreases down to 160 MPa at 1200°C. This behaviour seems to be related to the viscous flow of impure siliceous phases. Additions of Fe_2O_3 were verified to be very detrimental to the high-temperature mechanical properties of ZrO_2-mullites, as they were for ZrO_2-free mullites.

At room temperature, K_{1C} values are 3.3 MPa$\sqrt{m}$ for the silica-rich ZrO_2-mullite and only 2 MPa$\sqrt{m}$ for the ZrO_2-free mullite with the same alumina content. Thus, the zirconia inclusions noticeably improve toughness. Temperature affects toughness as it affects strength: K_{1c} begins to decrease, then increases again to reach a relative maximum of 3 MPa$\sqrt{m}$ at 900°C, then falls again. At 1200°C, ZrO_2-free mullite exhibits a higher toughness than ZrO_2-rich mullite (2 MPa$\sqrt{m}$ instead of 1.85 MPa$\sqrt{m}$). Young's modulus and Poisson's ratio are 220 GPa and 0.28 (resp. 200 GPa and 0.28) for ZrO_2-mullite and ZrO_2-free mullite. The highest Young's modulus of the ZrO_2-containing materials demonstrates that no microcracks develop around the zirconia inclusions, in which case elastic moduli would be severely decreased.

Finally, it must be noticed that the mechanical strength of ZrO_2-containing mullite may be considered as rather low in comparison with certain literature data, for instance those of Claussen et al.[2,11] where the room temperature strength can exceed 400 MPa. However, it must be pointed out that most of the literature data correspond to 3 or 4-point bend tests, using small, polished bars, whereas the present data correspond to the biaxial flexure of as-sintered, 25 mm diameter discs.

Conclusions

The main conclusions of this study can be summarized as follows:

1. Nearly dense (>97%) ZrO_2-mullite ceramics can be prepared by reaction-sintering of mixtures of alumina and zircon, for firing temperatures below or equal to 1600°C, i.e. for conditions close to those which are used to sinter "reactive" aluminas.
2. Dissociated zircon is not of particular interest for the reaction-sintering of ZrO_2-toughened mullites.
3. The alumina content of the stable mullite solid-solution varies from about 71.8 to 74.8 wt% Al_2O_3 in ZrO_2-rich mullites, whereas it varies from about 70.5 to 74.8 wt% Al_2O_3 in ZrO_2-free mullites.[1]
4. ZrO_2-mullite always begin to form with a composition close to the silica-rich end (i.e. 70.5 or 71.8% Al_2O_3) whatever the bulk composition is. The composition later moves toward the final one.
5. Excessive grain growth develops in the 75% Al_2O_3 materials.
6. The alumina-to-silica ratio affects the mechanical properties. The silica-rich material (68% Al_2O_3) exhibits the best mechanical properties. Zirconia inclusions improve the room-temperature toughness more than the room temperature-strength. However, the use of low-purity zircon leads to low-viscosity silica residues which are detrimental to mechanical properties: the 68% Al_2O_3 mullite exhibits a strength which continuously increases from room temperature to more than 1200°C,[1] whereas its ZrO_2-containing counterpart does not show similar improvement.
7. Finally, iron-oxide additions accelerate the reaction kinetics as well as the densification kinetics. However, they coarsen the microstructure and give low-melting point temperature phases, which are detrimental to high-temperature properties.

References

[1]P. Boch, T. Chartier, and P. D. D. Rodrigo, "High-Purity Mullite Ceramics by Reaction-Sintering," These Proceedings.

[2]N. Claussen and J. Jahn, "Mechanical Properties of Sintered, in situ-Reacted Mullite-Zirconia Composites," *J. Am. Ceram. Soc.*, **68** [3–4] 228–29 (1980).

[3]E. Di Rupo, E. Gilbart, T. G. Carruthers, and R. J. Brook, "Reactive Hot-Pressing of Zircon-Alumina Compositions," *J. Mat. Sci.*, **14**, 3, 705–11 (1979).

[4]E. Di Rupo, M. R. Anseau, and R. J. Brook, "Reaction Sintering: Correlation Between Densification and Reaction," *J. Mat. Sc.*, **14**, 2924–2928 (1979).

[5]E. Di Rupo and M. R. Anseau, "Solid State Reactions in the $ZrO_2.SiO_2.Al_2O_3$ System," *J. Mat. Sc.*, **15**, 114–118 (1980).

[6]M. R. Anseau, E. Di Rupo, and F. Cambier, "Thermal Expansion of Zircon-Alumina Materials Prepared by Reaction Sintering," *J. Mat. Sc. Letters*, **16** (1981).

[7]M. R. Anseau, F. Cambier, and C. Leblud, "Some Comments on Ceramic Solid-State Reaction Kinetics Using Results Obtained on the ZnO-Al_2O_3 System," *J. Mat. Sc. Letters*, **16**, 1121–26 (1981).

[8]M. R. Anseau, C. Leblud, and F. Cambier, "Reaction Sintering (RS) of Mixed Zircon-Based Powders as a Route for Producing Ceramics Containing Zirconia with Enhanced Mechanical Properties," *J. Mat. Sc. Letters*, **2**, 366–370 (1983).

[9]P. Boch and J. P. Giry, "Preparation and Properties of Reaction-Sintered Mullite-ZrO_2 Ceramics," *Materials Sc. and Eng.*, **71** [1–2] 39–48 (1985).

[10]S. Prochazka, J. S. Wallace, and N. Claussen, "Microstructure of Sintered Mullite-Zirconia Composites," *J. Am. Ceram. Soc.*, **66**, 8, C 125–127 (1983).

[11]N. Claussen, "Mullite Composites," 1st International Workshop on Mullite, Tokyo, November 9–10 (1987).

[12]A. Leriche, "Mechanical Properties and Microstructures of Mullite-Based Composite Materials Prepared by Different Routes," Ibid.

[13]J. S. Moya, "Reaction-Sintered Mullite-Zirconia Ceramics," Ibid.

[14]Q. M. Yuan, J. Q. Tan, and Z. G. Jin, "Preparation and Properties of Zirconia-Toughened Mullite Ceramics," and Q. M. Yuan, J. Q. Tan, J. Y. Shen, X. H. Zhu, and Z. F. Yang, "Processing and Microstructure of Mullite-Zirconia Composites Prepared from Sol-Gel Powders," *J. Am. Ceram. Soc.*, **69**, 3, 265–69 (1986).

[15]K. Rundgreen, P. Elfving, R. Pompe, K. D. P. Lagerlf, and B. Larsson, "Comparative Study of Sol-Gel Prepared Mullite and Mullite-ZrO_2 Composites with and without Y_2O_3," to be published in the proceedings of "Zirconia 86," Am. Ceram. Soc. publ.

[16]P. L. Husum and O. T. Sorensen, "Mullite-zirconia ceramics prepared from cheap raw materials," p. 23, Ceramic Materials 87, Göteborg Sweden, June 2–3 (1987).

[17]C. Leblud, M. R. Anseau, E. Di Rupo, F. Cambier, and P. Fierens, "Reaction Sintering of ZnO-Al_2O_3 Mixtures," *J. Mat. Sc. Letters*, **16** (1981).

[18]K. N. Maiti, "Industrial Production of High Quality Zircon Opacifier," *Glass and Ceram. Bull.*, **30**, 3 (1983).

[19]R. O. Heckroodt, "South African Zircon as a Raw Material," *Interceram*, **4** [8] 395–96 (1982).

[20]W. Butterman and W. R. Foster, "Zircon Stability and the ZrO_2-SiO_2 Phase Diagram," *Am. Mineral.*, **52**, 880–85 (1967).

[21]C. E. Curtis and H. G. Sowman, "Investigation of the Thermal Dissociation, Reassociation and Synthesis of Zircon," *J. Am. Ceram. Soc.*, **36** [6] 190–96 (1953).

[22]M. R. Anseau, J. P. Biloque, and P. Fierans, "Some Studies on the Thermal Solid State Stability of Zircon," *J. Mat. Sc. Letters*, **11** (1976).

[23]P. Pena and S. De Aza, "The Zircon Thermal Behaviour: Effect of Impurities," Part 1: *J. Mat. Sc.*, **19**, 135–142 (1984); Part 2: *J. Mat. Sc.*, **19**, 143–149 (1984).

[24]J. S. Moya and M. I. Osendi, "Microstructure and Mechanical Properties of 4 Mullite/ZrO_2 Composites," *J. Mat. Sc.*, **19**, 2909–2914 (1984).

[25]P. Pena, P. Miranzo, J. S. Moya, and S. De Aza, "Multicomponent Toughened Ceramic Materials Obtained by Reaction Sintering": Part 1: "ZrO_2-Al_2O_3-SiO_2-CaO system," *J. Mat. Sc.*, **20**, 2011–2022 (1985); Part 2: "System ZrO_2-Al_2O_3-SiO_2-MgO," Ibid., 2702–2710; Part 3: "System ZrO_2-Al_2O_3-SiO_2-TiO_2," Ibid., 2711–2718.

[26]T. S. Jones, S. Kimura, and A. Muan, "Phase Relations in the System FeO-Fe_2O_3-ZrO_2-SiO_2," *J. Am. Ceram. Soc.*, **50**, 3 (1967).

[27]C. C. Sorrel, C. A. Sorel, "The Al_2O_3-SiO_2-ZrO_2 Liquidus Surface: a Discussion," *J. Am. Ceram. Soc.*, **60** [1–2] 93–94 (1977).

[28]Charles A. Sorrel and Charles C. Sorrel, "Subsolidus Equilibria and Stabilization of Tetragonal ZrO_2 in the system ZrO_2-Al_2O_3-SiO_2," *J. Am. Ceram. Soc.*, **60** [11–12] 495–499 (1977).

[29]W. D. Tuohig and T. Y. Tien, "Subsolidus Phase Equilibria in the System ZrO_2-Y_2O_3-Al_2O_3," *J. Am. Ceram. Soc.*, **63** [9–10] 595–596 (1980).

[30]P. Pena and S. de Aza, "Compatibility Relations of Al_2O_3 in the System ZrO_2-Al_2O_3-SiO_2-CaO," *J. Am. Ceram. Soc.*, **67**, 1, C3-C5 (1984).

[31]Ai Ming Wong and R. McPherson, "The Structure of Plasma Dissociated Zircon," *J. Mat. Sc. Letters*, **16**, 1732–35 (1981).

[32]R. McPherson, R. Rao, and B. V. Shafer, "The Reassociation of Plasma-Dissociated Zircon," *J. Mat. Sc.*, **20**, 2597–603 (1985).

[33]S. Yangyun and R. J. Brook, "Preparation of Zirconia-Toughened Ceramics by Reaction-Sintering," *Science of Sintering*, **17** [1] 35–47 (1985).

[34]J. P. Giry, "A study of the reaction-sintering of alumina + zircon mixtures: preparation and properties of zirconia-mullite ceramics," Ph.D. Thesis, University of Limoges (1987).

[35]F. F. Lange and M. M. Hirlinger, "Hindrance of Grain Growth in Al_2O_3 by ZrO_2 inclusions," *J. Am. Ceram. Soc.*, **67**, 3, 164–68 (1984).

[36]B. Kibbel and A. H. Heuer, "Exaggerated Grain Growth in ZrO_2-Toughened Alumina," *J. Am. Ceram. Soc.*, **69**, 3, 231–36 (1986).

[37]R. B. Sosman, H.E. Merwin, and E. J. Wash, *Acad. Sci.*, **6**, 532 (1916).

[38]N. L. Bowen and J. F. Schairer, *Am. J. Sci.*, **24**, 200 (1932).

[39]A. Muan and E. F. Osborn, in Phase Equilibria among Oxides in Steel Making," Addison-Wesley Publ.

[40]B. Phillips and A. Muan, "System FeO-Fe_2O_3 in Air: Isobaric," *J. Am. Ceram. Soc.*, **42**, 415 (1959). Reading USA (1965).

[41]K. G. Skinner, W. H. Cook, R. A. Potter, and H. Palmour, "Effect of TiO_2, Fe_2O_3, and Alkali on Mineralogical and Physical Properties of Mullite-Type and Mullite-Forming Al_2O_3-SiO_2 Mixtures," *Am. Ceram. Soc.*, **36**, 349–55 (1953).

[42]S. M. Johnson and J. A. Pask, "Role of Impurities on Formation of Mullite from Kaolinite and Al_2O_3-SiO_2 mixtures," *J. Am. Ceram. Soc.*, **65**, 1, 31–35 (1982).

Reaction Sintered Mullite-Zirconia and Mullite-Zirconia-SiC Ceramics

José S. Moya

Professor of Materials Science
Instituto de Cerámica y Vidrio, C.S.I.C.
Arganda del Rey, Madrid, Spain

Abstract

The reaction sintering of mixed, fine zircon-based powers in the presence of calcite, wollastonite or magnesia as additives which supply a transient liquid during firing has been studied. The effect of the nature of the additive, type and amount of alumina, as well as the presence of SiC on the reaction sintering process, microstructure and mechanical properties of the final sintered components are discussed.

Introduction

The reaction sintering (RS) of mixed, fine zircon-based powder is considered as an inexpensive route for producing mullite-ceramics containing dispersed zirconia grains with enhanced mechanical properties.[1–8] In the last few years an extensive research program has been conducted at the Ceramics Dept. of ICV (Madrid) on the basis of the following reaction:

$$ZrSiO_4 + Al_2O_3 + \text{additives} \rightarrow Al_6Si_2O_{13} + ZrO_2 + \text{extra phase} \quad (1)$$

CaO, MgO and TiO_2 have been used as additives in order to develop a transient liquid phase below the final temperature treatment in order to produce a tough mullite-zirconia composite with close to theoretical density.[3–5]

Two concepts are considered in this alternative sintering process: (a) two or more incompatible reactants are homogeneously mixed and subsequently heated at a suitable temperature where sintering and reaction take place; and (b) the possible appearance during heating of a transitory liquid phase which enhances sintering and/or reaction at lower temperature.

The physico-chemical nature of the additive, the presence of free alumina and finally the effect of the addition of SiC on the final microstructure and properties of the reaction-sintered compact will be analyzed and discussed in this study.

Experimental Procedure

Appropriate quantities of the raw materials listed in Table 1 required to produce mullite-zirconia, mullite-zirconia-anorthite, mullite-zirconia-anorthite-alumina, mullite-zirconia-spinel and mullite-zirconia-spinel-alumina composites, were milled

Table 1. Characteristics of the Raw Materials

Raw material	Source	Specific surface ($m^2\ g^{-1}$)	d_{50} (μm)	Chemical analysis (%)							
				Al_2O_3	Na_2O	SiO_2	Fe_2O_3	CaO	MgO	ZrO_2	TiO_2
Al_2O_3	Alcan C-71 (CAN)	0.8	3.6	99.8	0.18	0.03	0.03	—	—	—	—
	Reynolds R.C. 172 DBM (USA)	7.0	0.6	99.8	0.04	0.042	0.019	0.043	0.095	—	—
	Alcoa CT-3000 SG (USA)	8.0	0.5	99.6	0.01	0.08	0.03	0.02	0.1	—	—
$ZrSiO_4$	OPAZIR S (QUIMINSA) (SPAIN)	7.4	1.1	0.9	—	32.9	0.1	—	—	65.0	0.2
$CaCO_3$	Merck (FRG)	—	16.0	—	—	—	—	100	—	—	—
$CaSiO_3$	C6, CAMPI Y JOVE S.A.	0.7	18.0	0.25	0.04	49.49	0.59	47.86	0.15	—	—
MgO	Balkowski Int	10	3.0	—	0.02	—	—	0.02	99.6	—	—
SiC	NAVARRO, S.A.		2.0			1.98					

and homogenized by attrition milling for 1.5 h in isopropanol-2 to 1 μm average particle size.

After this operation, the mixtures were dried and sieved. The powders were compacted by isopressing at 200°C MPa. The green bars were fired at 1450°C for 1.5 h and for 3 h in the case of the MgO and CaO containing-compositions respectively. The SiC-containing composition was heat treated in a nitrogen atmosphere at 1450°C–1.5 h.

The dynamic sintering of the studied compacts was performed by using a constant heating rate (10°C/min.) to 1550°C and employed samples of 5 mm in diameter and 10 mm in length. The free additive compositions were heated at 1570°C for 2 h.

The density of the fired compacts was determined by Archimedes method, modulus of rupture, by three point bending and KIC by indentation.

The microstructure was observed on polished and thermally etched surfaces by optical microscopy, SEM and by TEM-EDX on specimens prepared by wafering, grinding and argon-ion bombardment.

Results

On the basis of the general reaction (1) a series of reaction sintering compositions were formulated considering the high temperature equilibrium relationships of the two quaternary systems: SiO_2-Al_2O_3-ZrO_2-MgO[4] and SiO_2-Al_2O_3-ZrO_2-CaO.[3] Calcite ($CaCO_3$), wollastonite ($CaSiO_3$) and magnesia (MgO) were used as additives (Table 1).

In this sense the following reactions have been studied:

$$2ZrSiO_4 + 3Al_2O_3 \rightarrow 2ZrO_2 + Al_6Si_2O_{13}$$

$$2.74ZrSiO_4 + 3.37Al_2O_3 + 0.37CaCO_3 \rightarrow 2.74ZrO_2 + Al_6Si_2O_{13} + 0.37CaAl_2Si_2O_8$$

$$2.35ZrSiO_4 + 3.35Al_3O_4 + 0.35CaSiO_3 \rightarrow 2.35ZrO_2 + Al_6Si_2O_{13} + 0.35CaAl_2Si_2O_8$$

$$2ZrSiO_4 + 6.7Al_2O_3 \rightarrow 2ZrO_2 + Al_6Si_2O_{13} + 3.7Al_2O_3$$

$$2.74ZrSiO_4 + 0.53Al_2O_3 + 0.37CaCO_3 \rightarrow 2.74ZrO_2 + Al_6Si_2O_{13} + 6.16Al_2O_3 + 0.37CaAl_2Si_2O_8$$

$$2.33ZrSiO_4 + 8.47Al_2O_3 + 0.33CaSiO_3 \rightarrow 2.33ZrO_2 + Al_6Si_2O_{13} + 5.14Al_2O_3 + 0.3CaAl_2Si_2O_8$$

$$2ZrSiO_4 + 3.3Al_2O_3 + 0.3MgO \rightarrow 2ZrO_2 + Al_6Si_2O_{13} + 0.3Al_2MgO_4$$

$$2ZrSiO_4 + 6.7Al_2O_3 + 0.3MgO \rightarrow 2ZrO_2 + 3.4Al_2O_3 + Al_6Si_2O_{13} + 0.3Al_2MgO_4$$

The dynamic sintering curves of all these reactions are plotted in Figs. 1 and 2.

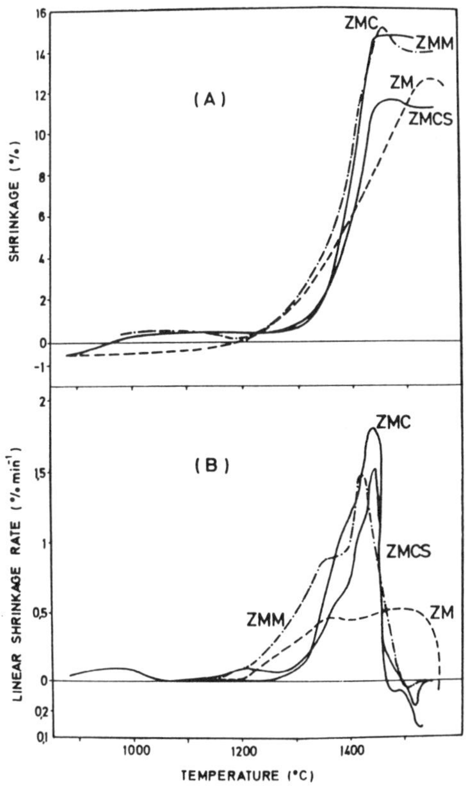

Fig. 1. (A) Linear shrinkage vs. temperature for ZM, ZMM, ZMC and CMCS compacts; (B) Linear shrinkage rate for the same compacts.

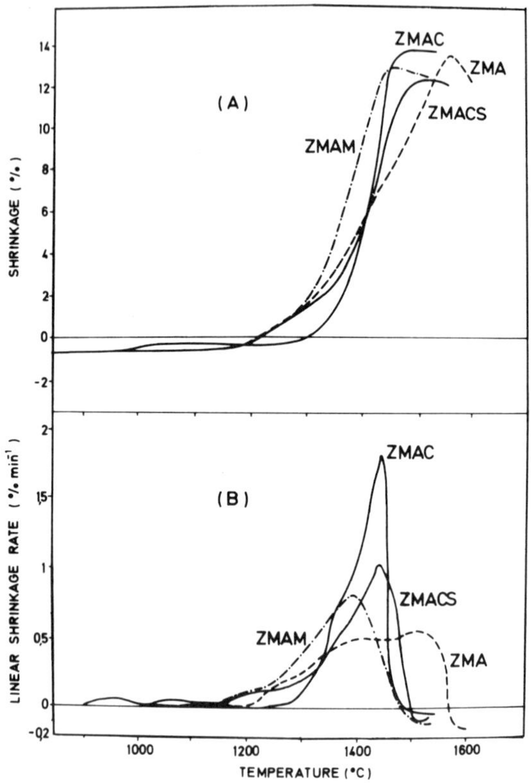

Fig. 2. (A) Linear shrinkage vs. temperature for ZMA, ZMAM, ZMAC, and ZMACS compacts. (B) Linear shrinkage rate for the same compacts.

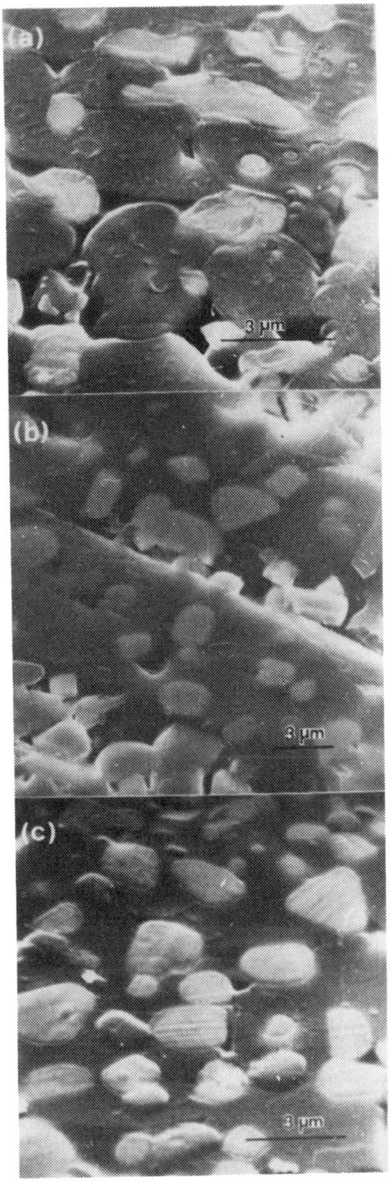

Fig. 3. SEM micrographs corresponding to ZM, ZMM and ZMC compacts.

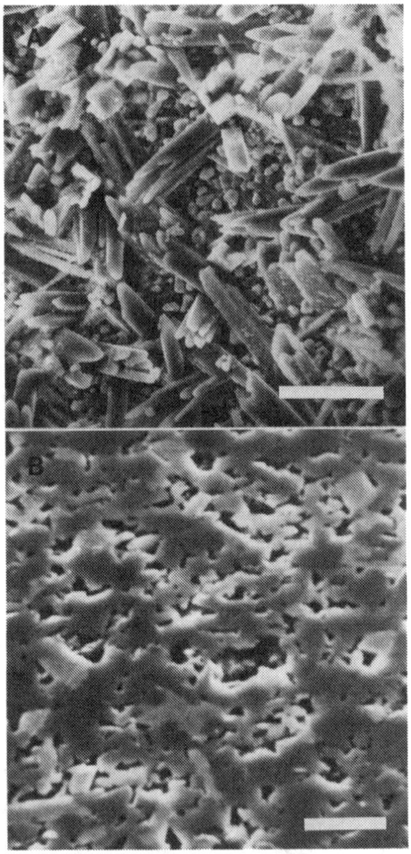

Fig. 4. SEM micrographs corresponding to ZMAM (A) and ZMAM-20% vol. SiC (bar = 10 μm) (B) compacts (bar = 2 μm).

The density of all fired compacts was always higher than 97% of theoretical. SEM micrographs of the different sintered compacts are shown in Fig. 3.

A 20 vol % SiC content composite was obtained starting from the ZMAM unreacted mixture plus the corresponding amount of 2 μm average grain size SiC powder (Table 1). The SEM micrographs corresponding to the ZMAM and the SiC containing compacts are shown in Fig. 4. The mechanical properties of the composites are shown in Table 2.

Discussion

Mullite-ZrO_2 and Mullite-Al_2O_3-ZrO_2 Based RS Composite

From Figs. 1 and 2 the following points can be stated:

Table 2. Mechanical Characterization of RS Mullite-Zirconia and Mullite-Zirconia-SiC Ceramics

Material	K_{Ic} ($MPa \cdot m^{1/2}$)	H (GPa)	σ_f (MPa)
Zm	4.4	8.5	350
ZMC	4.9	8.0	200
ZMCS	4.8	8.0	180
ZMM	4.2	9.7	164
ZMM*	5.0	8.0	300
ZMAC	5.1	10.0	300
ZMACS	5.2	10.0	330
ZMAM	5.4	12.0	341
ZMAM-SiC	4.5	14.0	400

*Composition obtained using Alcan-C-71 Alumina

1. When no additive is present (ZM, ZMA compositions) the effect of free alumina on the kinetics of the RS process can be considered negligible.
2. CaO and MgO-containing compositions show a significantly high rate of densification as compared to free-additive compositions.
3. The wollastonite-containing compositions show a smoother sintering process than those containing calcite. This effect is more significant in compositions containing free alumina.
4. In general the presence of free alumina in CaO- and MgO-containing compositions tends to decrease the linear shrinkage rate, making the overall sintering process smoother.

The ultimate explanation of all these differences has to be directly related to the nature of the transitory liquid that appears during firing.

It seems that in the case of additive-free compositions (ZM, ZMA) the parameters that control the sintering rate are the classic ones in solid state sintering, e.g. average grain size, specific surface area of the starting powder, etc.

Conversely, as has been reported elsewhere[3,4,8] in the case of additive-containing compositions, during the firing schedule, at a determined temperature, the starting powder reacts according to the prediction of the corresponding equilibrium diagram, giving a liquid with specific characteristics (composition, surface tension, viscosity) which immediately diffuses through the compact capillary network modifying the coordination number of the particulates (rearrangement effect), reacting with the different grains and consequently changing its composition and main physical properties up to its complete disappearance. This process is dynamic and much depends on the chemical nature of the starting additive. During this step the reaction products nucleate and grow forming the final microstructure and properties.

As experimental evidence of the previous statement, Fig. 3 shows SEM micrographs corresponding to ZM, ZMC, and ZMM sintered compacts, where significant differences in mullite shape and average grain size can be observed.

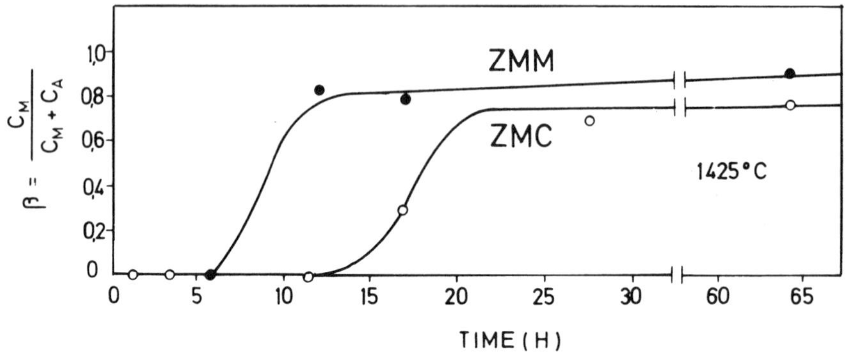

Fig. 5. Evolution of the ß parameter for ZMM and ZMC compacts as a function of time at 1425°C.

In Fig. 5 the rate of mullite formation (Beta) vs. time at a constant temperature (1425°C) for ZMM and ZMC compositions is shown. According to the equilibrium diagrams SiO_2-Al_2O_3-ZrO_2-MgO[4] and SiO_2-Al_2O_3-ZrO_2-CaO,[3] the initial transitory liquid appearance is close to this temperature in both systems; however, the nucleation period for mullite formation is very different. Consequently, the average size of the mullite crystals observed in the ZMM sintered compact is 3–4 times higher than in the ZMC sample. The reason must be related to the different nature of the transitory liquid. Both liquids were analyzed by TEM-EDX microanalysis, and the results are reported in Table 3.

As observed in Figs. 1 and 2, the presence of free alumina does not significantly effect the initial shrinkage of the compacts, probably because the nature and composition of the initial transitory liquid for the ZMAC, ZMAM and ZMACS compositions are similar to those corresponding to the ZMC, ZMM and ZMCS samples. However, the relative amount of liquid as well as the change in composition during the dynamic sintering is not the same. This fact is due to: a) The relative higher fraction of alumina, and b) the higher number of alumina grains that this liquid will wet and partially dissolve during the firing process.

It has been observed that in the case of ZMC, ZMM and ZMCS compositions, the transitory liquid partially left the bulk and appeared on the surface of the sintered compact by capillary forces. This phenomenon did not take place in the case of the free alumina containing compacts. (Fig. 6). These results clearly prove the previous statement.

Another route which would affect the chemical evolution of the initial transitory liquid is to modify the size of the capillar network of the green compact. In this

Table 3. TEM-EDX Microanalysis of the Transitory Liquid Phases

Sample	T (°C)	Time (h)	(wt%) Al_2O_3	SiO_2	ZrO_2	MgO	CaO	Localization
ZMM	1425	17	29.9	56.5	5.7	7.9	—	close to saphirine crystals
			68.9	24.2	1.1	5.9	—	at grain boundaries
ZMC	1425	17	33.4	48.4	3.4	—	14.8	close to anorthite crystals
			72.3	21.2	3.3	—	3.2	at grain boundaries

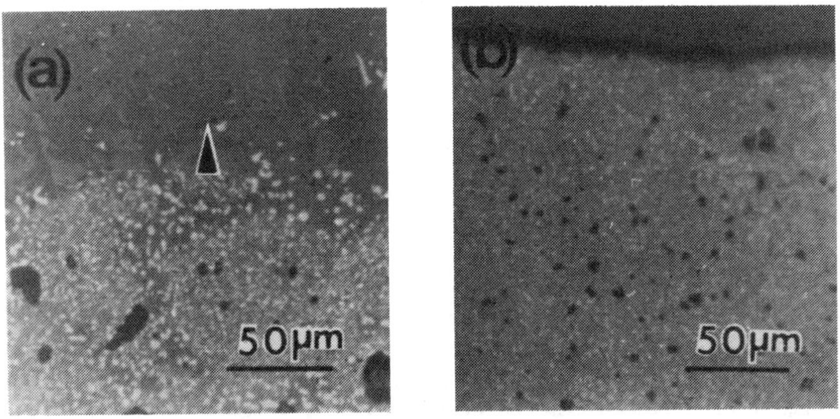

Fig. 6. Optical micrographs of a polished cross section of ZMC (a) and ZMAC (b) compacts. Arrow indicates glassy phase on the specimen surface.

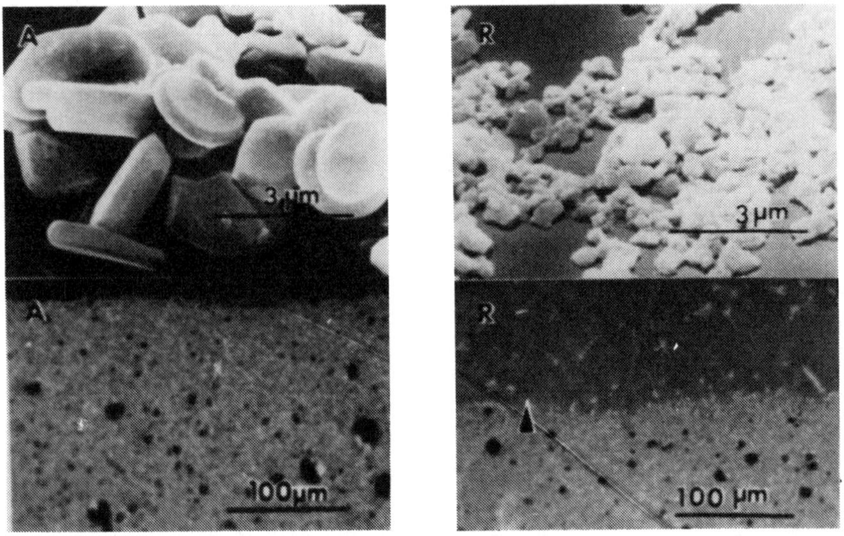

Fig. 7. SEM micrographs corresponding to Alcan C-71 (A) and Reynolds R. C. 172 DBM (R) alumina powders and the Optical micrographs corresponding to a polished cross section of ZMM compacts obtained from Alcan C-71 and Reynolds R.C. 172 DBM alumina powders.

way the following experiment was developed. The ZMM reaction was performed using two kinds of alumina powder, Reynolds RC. 172DBM with 0.6 μm average grain size and Alcan C-71 with an average grain size of 3.6 μm (Table 1). The rest of the reactant and all processing and firing conditions were kept constant.

In the case of Reynolds alumina compact, after sintering, a significant amount of liquid was observed on the sample surface while no liquid was detected in the case of the Alcan sample (Fig. 7). Important changes were also observed in the microstructure and mechanical properties of both compacts (Table 2).

Another possible way to avoid a glassy phase on the surface of the samples has been studied by Moya and Miranzo[9] by using prereacted powder. In this case the ZMM mixture was prereacted at 1450°C and attrition milled to 1 μm average grain size. This powder (ZMMP) was processed as the ZMM one.

The SEM micrographs corresponding to both samples are shown in Fig. 8. As observed, the microstructure of these samples are completely different. In the ZMM compact the mullite grains are columnar with an average size of 10 μm. The zirconia grains were mainly intergranular. Conversely, the ZMMP compact shows mainly intergranular zirconia grains and smaller (5 μm) equiaxed mullite grains. As a consequence an increase (50%) in the sigma f value was obtained in the last composites.

SiC-Containing RS Composites

Mullite-Zirconia and Mullite-Alumina-Zirconia composites have limited applications in thermal shock and wear, due to their low thermal conductivity (2–3 W/mK) and hardness (8–10 GPa). These properties cannot be dramatically increased by changing their composition in the ZrO_2-Al_2O_3-SiO_2 system. In this sense, SiC is

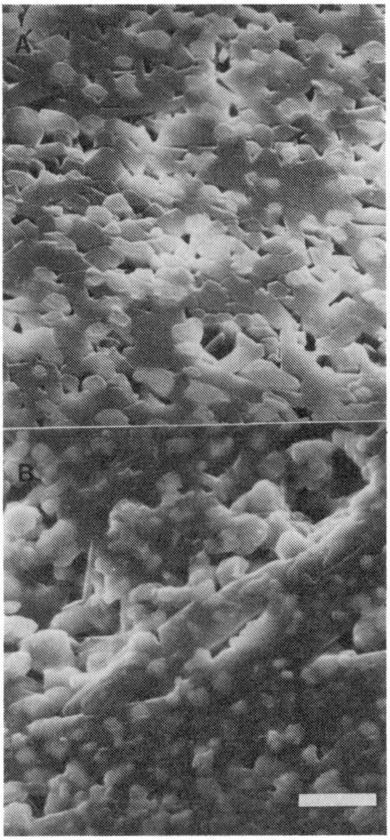

Fig. 8. SEM micrograph corresponding to (A) ZMMP and (B) ZMM compacts. Bar = 4 μm.

a potential candidate as an additive due to its high hardness (25–30 GPa) and thermal conductivity (84 W/mK) at 500°C).

When the reaction sintering of zircon and alumina mixtures occurs in the presence of SiC, two main issues have to be considered:

a. Solid-gas reactions during heat treatment; and
b. SiC-ZrO_2 incompatibility.

Baudin and Moya[10,11] have extensively studied both basic problems as a function of heating atmosphere, SiC-grain size and heating schedule. With the composition studied in the present work a dense (99% of theoretical) mullite-ZrO_2-Al_2O_3-20 vol.% SiC (ZMAM-SiC) composite was obtained in a nitrogen atmosphere at 1450°C without the presence of any nitrogen-containing compound.

As observed in Fig. 4, the mullite grain growth in this sample is drastically inhibited (from 30 μm to 1 μm) by the presence of SiC grains. This is due to the perfect solid state compatibility between mullite and SiC. This reduces the critical defect size (from 75 μm to 50 μm) improving not only the hardness (Table 2) but also the bending strength (400 MPa).

In this composite the SiC grains are surrounded by smaller mullite grains which act as a physical barrier for oxygen diffusion. This inhibits the oxidation of SiC at high temperature. Aging experiments carried out in air at 1400°C showed no microstructural and mechanical degradation in the sample.

The bending strength, hardness, toughness and relative density values obtained in the reaction sintered ZMAM-SiC composite are similar to those reported by Wei et al.[12] in mullite-20% vol. SiC-w obtained by hot pressing at 1600°C.

By adding SiC to these RS mixtures it seems plausible that one can improve their properties such that they can now be considered as wear and thermal shock resistant materials, and they may also be able to be used as high temperature structural component materials. Furthermore, these materials are much more stable than SiC, Si_3N_4 and sialon based materials in high temperature oxidizing conditions.

Conclusions

Reaction sintering has proved to be a simple and economic way for producing a wide range of mullite-matrix composites with a moderate high toughness (4.5–5.5 MPa m1/2), hardness (10–14 GPa) and bending strength (250–400 MPa).

By obtaining this new family of materials through a transitory liquid phase, we open the possibility of tailoring a multicomponent mullite-ceramic composites with different microstructures and properties, which then can be used in wear, chemical and thermal shock resistant applications as well as high temperature structural materials in oxidizing and non-oxidizing conditions up to 1400°C without the appearance of any glassy phase.

Acknowledgments

The author wishes to acknowledge the experimental assistance and helpful discussions provided by S. de Aza, P. Pena, P. Miranzo and C. Baudin.

References

[1]E. Di Rupo, R. R. Anseau, and R. J. Brook, "Reaction Sintering: Correlation Between Densification and Reaction," *J. Mater. Sci.*, **14**, 2924–28 (1979).

[2]N. Clausen and J. Jahn, "Mechanical Properties of Sintered In Situ Reacted Mullite-Zirconia Composites," *J. Am. Ceram. Soc.*, **63**, 228–229 (1980).

[3]P. Pena, P. Miranzo, J. S. Moya, and S. de Aza, "Multicomponent Toughened Ceramic Materials Obtained by Reaction. Sintering Part I: System ZrO_2-Al_2O_3-SiO_2-CaO," *J. Mater. Sci.*, **20**, 2711-2718 (1985).

[4]P. Miranzo, P. Pena, J. S. Moya, and S. de Aza. Idem. Part 2. bis, **20**, 2702–2710 (1985).

[5]M. F. Melo, P. Pena, J. S. Moya, and S. de Aza. Idem. Part 3 bis, **20**, 2711-2718 (1985).

[6]P. Pena, M. A. Rodríguez, J. S. Moya, and S. de Aza, in High Tech Ceramics, 861-870. Ed. P. Vincenzini. Elsevier. Amsterdam, 1987.

[7]J. S. Moya, P. Miranzo, P. Pena, and S. de Aza, in Processing of Advanced Ceramics, 122–133. Ed. J. S. Moya and S. de Aza, Soc. Esp. Ceram. Vidr. Madrid. Spain 1986.

[8]P. Miranzo, P. Pena, S. de Aza, J. S. Moya, J. María Rincón, and G. Thomas, "TEM Study of Reaction-Sintered Zirconia-Mullite Composites with CaO and MgO," *J. Mat. Sci.*, **22**, 2987–2992 (1987).

[9]J. S. Moya and P. Miranzo, in High Tech. Ceramics, 1317–1324. Ed. P. Vincenzini. Elsevier. Amsterdam, 1987.
[10]C. Baudin and J. S. Moya, High Temperature Reactions and Stability in Mullite-Zirconia-SiC composites. Proceedings of Int. Conference on Ceramic-Ceramic Composites. Mons. Belgium (1987) in press.
[11]C. Baudin and J. S. Moya, "Effect of the Processing Route on the Sintering of Mullite-Zirconia-Alumina-SiC Composites," *Sci. of Ceramics*, **14**, in press.
[12]G. C. Wei and P. F. Becher, Development of SiC-Whiskers-Reinforced Ceramics, *Am. Ceram. Soc. Bull.*, **64**, 298–304 (1985).

Mechanical Properties and Microstructure of Mullite-SiC-ZrO$_2$ Particulate Composite

HIDETO KAMIAKA, CHITAKE YAMAGISHI AND JUNJI ASAUMI

Central Research Laboratory
Nihon Cement Co., Ltd.
2-23, 1-chome, Kiyosumi, Koto-ku, Tokyo 135, Japan

Abstract

The present study was conducted to clarify the relation between the mechanical properties and the microstructures of the composites fabricated by the pressureless sintering method with fine and high purity Mullite, SiC and ZrO_2 powders. It was found that the fracture toughness of the pressureless-sintered body (containing SiC 20 vol% and ZrO_2 20 vol%) increased from 2.2 to 4.8 MPa·m$^{1/2}$ by improving the strength at room temperature and maintaining it at 1300°C.

Introduction

Recently a high purity Mullite ($2Al_2O_3 \cdot 2SiO_2$) ceramic has been recognized as a high-temperature structural material because of excellent mechanical properties: a high strength[1-3] and a high creep resistance[4] under high-temperature conditions.

However, the development of the structural material has been delayed[5,6] since it has low fracture toughness and relatively low strength at room temperature compared with other engineering ceramics. In order to improve the mechanical properties, Mullite composites with SiC whiskers and/or ZrO_2 particles have been developed. It is well known that the fracture toughness and the flexural strength are improved in ZrO_2 particle reinforced Mullite.[7,8] However, the flexural strength of the composite decreases under high-temperature conditions. On the other hand, the densification of SiC whisker reinforced Mullite composites is hindered with increasing whisker contents. Therefore to densify the composite containing whiskers the press-sintering method is required. Using the press-sintering method, the strength and toughness of composite were reinforced, but complex shapes could not be fabricated.

We have developed the pressureless sintering of SiC whisker reinforced Mullite composite[6] to fabricate complex shapes. It was found that the density of composite was decreased by adding more than 15 vol% SiC whisker, and even the composite with 10 vol% SiC whisker was brittle compared with other engineering ceramics.

In this paper, we have studied the influence of reinforcement for Mullite ceramics to understand the relationships between the mechanical properties and the composite materials aiming towards complex shape applications.

Table 1. Characteristics of Raw Powder

Raw Powder	Median ‡ Particle Size (μm)	Purity†	Type	Manufacture
Mullite*	0.2	99.9% Fe_2O_3 100ppm Na_2O 100ppm K_2O 60ppm TiO_2 60ppm	AM-72S	Nihon Cement
SiC	0.2	98.0% SiO_2 3000ppm C 4000ppm Fe 400ppm Al 300ppm	Ultra Fine	Ibiden
ZrO_2	0.2	99.9% Al_2O_3 260ppm SiO_2 100ppm Na_2O 12ppm	H S	Dowa Chemical

*72 Wt% Al_2O_3
†Catalogue Value
‡Measured with SEDIGRAPH

Experimental Procedure

The characteristics of the raw powders used in the composites are listed in Table 1.

The starting materials were prepared by using commercial powders: high purity Mullite, SiC and ZrO_2 powders. The powders were mixed and ground with Mullite balls and ion-exchanged water in Al_2O_3 pot mill for 3 hours. The powder mixtures were pressed under 31 MPa and formed with the cold isostatic press under 196 MPa. The pressureless sintering of the compacts were done in an electric furnace under an air atmosphere at 1700°C for 2 hours. The sintered specimens 4 mm by 20 mm by 40 mm were sliced and ground by #600 diamond wheel. Edges were chamfered with a #1000 diamond disk. The test pieces with dimensions of 3 mm by 4 mm by 40 mm were annealed in air at 1300°C for 2 hours.

Relative Density

The relative density of composites is the ratio of the bulk density to the theoretical density. The bulk densities were measured by Archimedes' method using ion-exchanged water. The theoretical densities were calculated by using the density of each powder (Mullite 3.15 g/cm^3, SiC 3.20 g/cm^3, ZrO_2 5.85 g/cm^3).

Observation of Microstructure

The samples were prepared by polishing with 0.04 μm γ-Al_2O_3 and some of the samples were etched with HF solution. Microstructures were examined by using scanning electron microscopy (SEM).

Flexural Strength

A three-point flexural strength measurement was carried out using a 30 mm outer span with a crosshead speed of 0.5 mm/min. at room temperature and 1300°C in air.

Fracture Toughness (K1c)

Fracture toughness was measured by the Indentation Microfracture (IM) method. The samples for the IM method were polished at finish with 0.04 μm γ-Al_2O_3 and indented by Vicker method at a load of 49 N for 15 sec. The samples using the IS method were indented by Vicker at the center of the test pieces with a load of 98 N for 15 sec and were bent using a 30 mm outer span with a crosshead speed of 0.5 mm/min. at room temperature.

In this paper, SC and Z indicate SiC and ZrO_2. For example, SC40Z20 indicates a sintering composite of Mullite 40 vol%, SiC 40 vol% and ZrO_2 20 vol%.

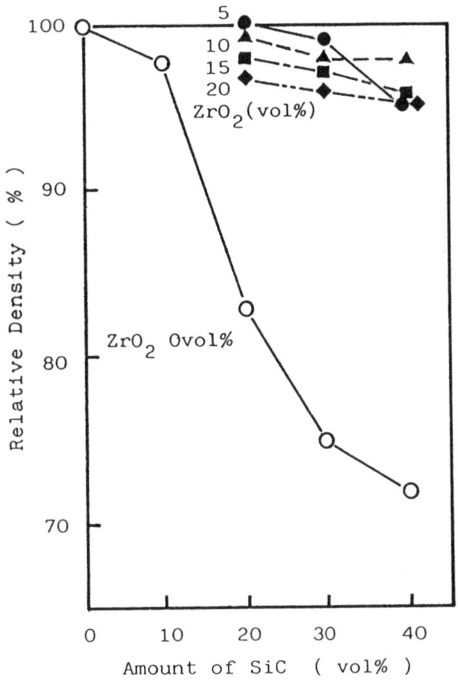

Fig. 1. Effect of SiC and ZrO_2 on Densification.

Results and Discussion

Effect of SiC and ZrO_2 On Densification

The relative densities of composites with pressureless sintering are shown in Fig. 1. Without ZrO_2 the composites having 10 vol% SiC showed 97.6% relative density. However, it was found that the relative density decreased rapidly by adding 20 vol% and more of difficult SiC. The density degradation of composites is due to the difficult sinterability with SiC additions. However, on adding 5–20 vol% ZrO_2 to the composites containing 20–40 vol% SiC these sintered bodies rapidly densified and achieved more than 95% relative density.

Composite is due to densification at temperatures higher than 1600°C is due to the Mullite solid solution region shifting to the Al_2O_3-rich side so the SiO_2 liquid exudes.[10] Also there is a eutectic point about 1675°C in the ZrO_2-SiO_2 phase diagram. In addition there is a significantly high solid solubility of mullite in ZrO_2 and ZrO_2 in mullite.[11]

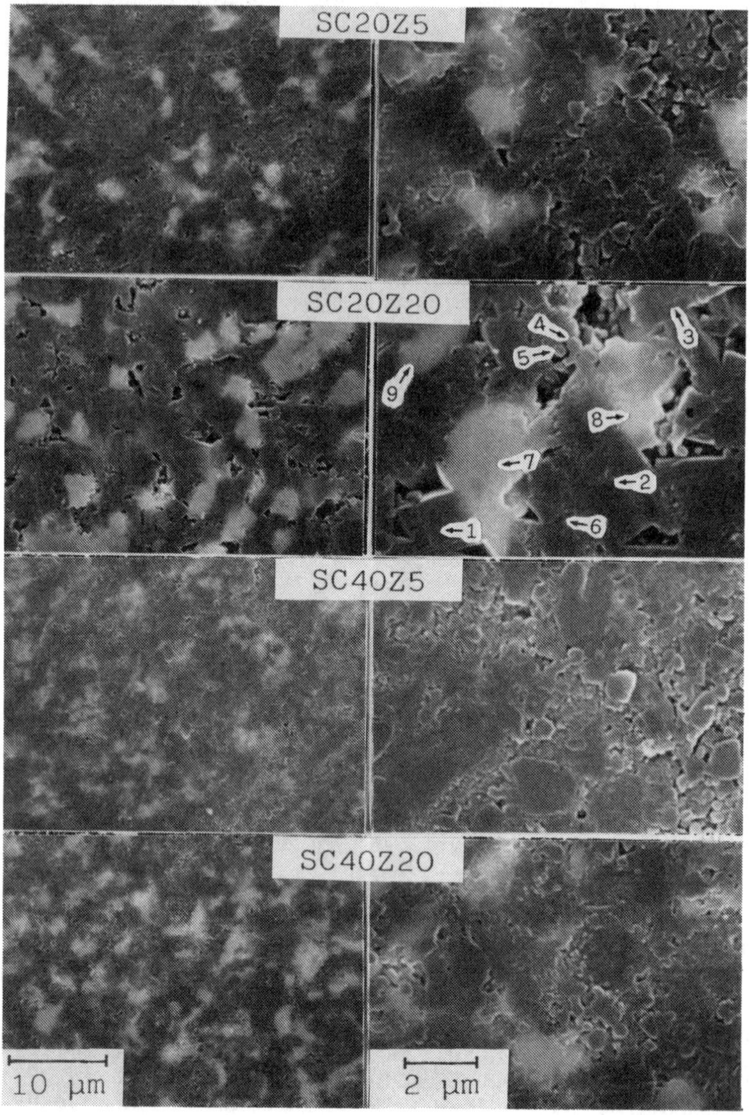

Fig. 2. Microstructure of Composites. Arrows indicate Mullite 1,2,3; SiC 4,5,6; ZrO_2 7,8,9.

Microstructure

The microstructures of the composites (SC20Z5, SC20Z20, SC40Z5, SC40Z20) are shown in Fig. 2. It is evident that only the microstructure of SC20Z20 consists of large grown grains of ZrO_2 and also grown grains of Mullite. From the observation of Fig. 2, it is found that SiC grains do not grow with increasing amounts, but they also prevent Mullite and ZrO_2 grains from growing. The ZrO_2 addition makes Mullite grains grow.

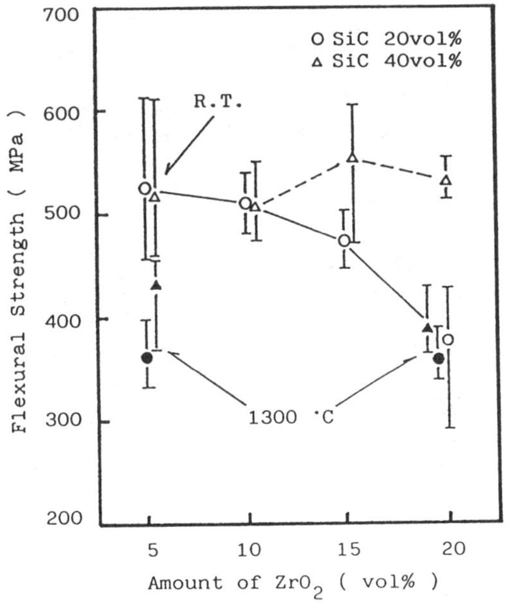

Fig. 3. Effect of SiC and ZrO_2 on Flexural Strength.

Effect of SiC and ZrO_2 Additions on the Flexural Strength

Figure 3 shows the effect of SiC and ZrO_2 additions on the flexural strength. Regardless of the amount of ZrO_2 added, the flexural strength at room temperature for the composites containing 40 vol% SiC remains almost constant at 500–600 MPa, while for composites containing 20 vol% SiC the strength decreases with increasing ZrO_2 additions above 10 vol%. Figure 4 shows the microstructure of the fracture surfaces near the fracture origin of SC20Z20 and SC40Z20. The fracture surface (A) of SC20Z20 shows a large crack and seems relatively smooth, while on the SC40Z20 surface (B) does not show any large crack and appears significantly rough. In (A) and (B) composites, since the measured amount of tetragonal ZrO_2 is about 10%, many cracks must have occurred around the ZrO_2 grains. The degradation of the strength in (A) results from the large crack that appears extensively on the fracture surface. The large crack is attributed to the expansion of ZrO_2 during the tetragonal → monoclinic transformation. The size of crack is closely related to the grain size of ZrO_2 as well as the grains in the matrix. In case of large ZrO_2 grains extremely high tensile stresses due to the expansion of ZrO_2 during the transformation develop around ZrO_2. Cracks are thus likely to grow large and with the grains

Fig. 4. Fracture Surface of Composites. Arrow indicates large crack.

growing large in the matrix, the elongated cracks occur because of cleaving the large grain or the elongated grain boundary by means of the same stresses. Increasing the ZrO_2 addition, the crack-linking grows more. The same phenomena are reported for Al_2O_3-ZrO_2 composites.[12,13] It is explained that the degradation of the strength results from the difference of thermal expansion between Al_2O_3 and ZrO_2 as well as the expansion of ZrO_2 during the transformation.

The location of the fracture origin was done on SC2OZ5, SC2OZ2O, SC4OZ5 and SC4OZ2O. The fracture origins of SC2OZ2O were mostly observed at the surface, while others were observed inside of the test pieces. It is considered for SC2OZ2O that the fracture origin forms at the surface since many large cracks grow and link together at the surface having the maximum tensile stress under flexural measurement. Figure 5 shows the fracture origin of SC4OZ2O. It was found that the inner defects were the agglomeration of SiC particles that did not densify, and the large crack attributed to the transformation of the ZrO_2 grain was caused by the agglomeration. Therefore, the flexural strength at room temperature can increase significantly by homogeneously mixing and uniformly controlling the microstructure.

The flexural strength at 1300°C was measured for SC2OZ5, SC2OZ2O, SC4OZ5 and SC4OZ20. The strength of these composites was indicated to be about 400 MPa. However, there was a tendency for the composites with the addition of 40% SiC to have a slightly higher strength than those with 20 vol% SiC.

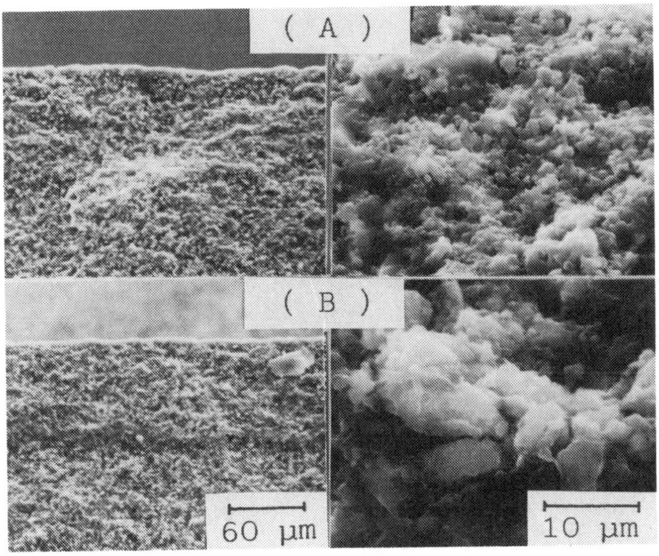

Fig. 5. Fracture Origin of Composites (SC4OZ2O).

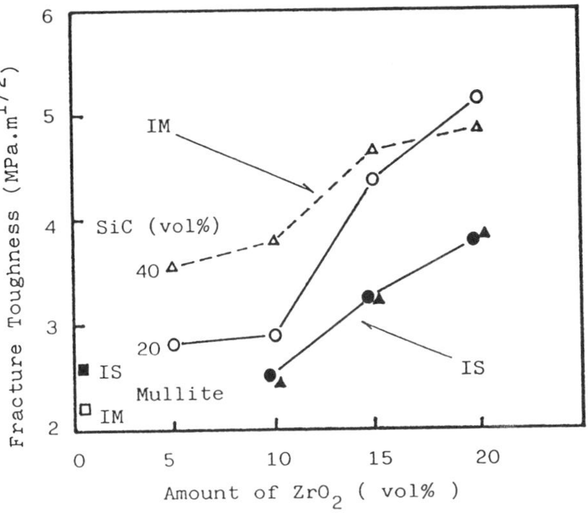

Fig. 6. Effect of SiC and ZrO_2 on Fracture Toughness.

Effect of SiC and ZrO_2 Additions on the Fracture Toughness

Figure 6 shows the effect of SiC and ZrO_2 additions on the fracture toughness (K1c) measured by IM and IS. The K1c measured by IM was 5.2 MPa•m$^{1/2}$ on SC2OZ2O and 4.8 MPa•m$^{1/2}$ on SC4OZ2O. However, the K1c measured by IS

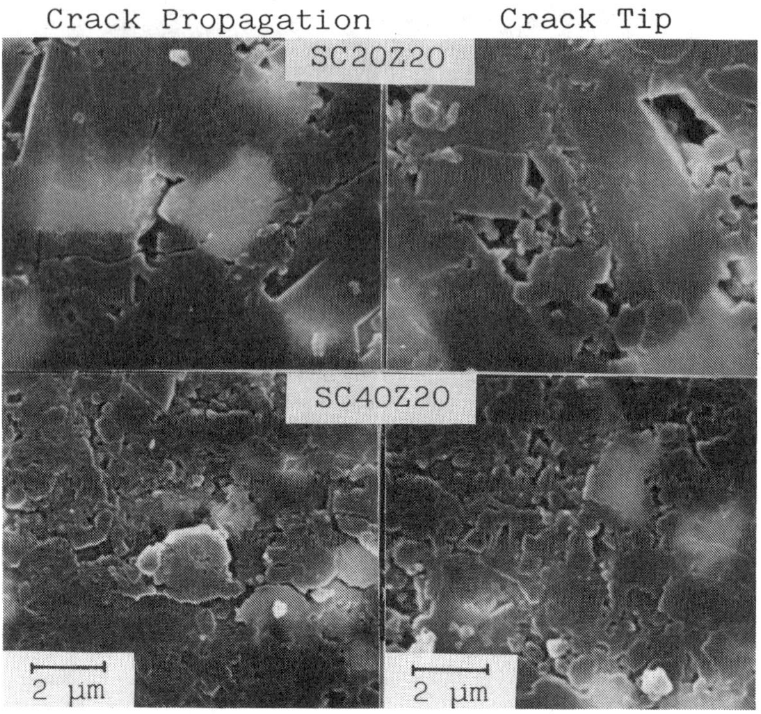

Fig. 7. Crack Propagation and Crack Tip.

indicated 3.8 MPa•m$^{1/2}$ on both composites. Thus, the K1c values measured by IM were higher than the ones measured by IS.

Figure 7 shows the relation between the microstructure and the crack induced by the indented Vicker's. The observation was done after etching the polished surface induced indentation. From Fig. 7, it was found that the mechanism of toughening differed between SC2OZ2O and SC4OZ2O. For the SC2OZ2O, the induced crack grew relatively straight, cleaved the large grains of mullite and ZrO_2 sometimes slightly deflected by the fine SiC grain and grain boundary. The front of the crack was pinned by the large grain or went within then was deflected and branched in the large grain. The deflection in the grain is attributed to the fine inclusion, since SiC grains and ZrO_2 grains are observed significantly. On the SC4OZ2O the induced crack was deflected by the fine grains of Mullite, SiC and ZrO_2 and passed along the grain boundary so that many branched cracks occurred. From the above observation, it was found that the increase of K1c on SC2OZ2O due to the pinning and deflection of the crack front by large grains, and on SC4OZ2O to the deflection and the branching of cracks by the fine grains. The K1c measured by IS was lower than by IM. Since it was difficult to search for a crack tip in the above composites, there were many deflections and branching of microcracks so that the K1c was larger.

Conclusion

The present study was conducted to clarify the relation between the mechanical properties and the microstructures of the composites fabricated with fine (0.2 μm) and highly purified Mullite, SiC and ZrO_2 powders–

1. Without ZrO_2 the relative density of composites rapidly decreased with addition of more than 10 vol% SiC. However, in case of adding 5–20 vol% ZrO_2 on the composites containing 20–40 vol% SiC these sintered bodies rapidly densified and achieved more than 95% relative density.
2. The additions of SiC prevented Mullite and ZrO_2 grains from growing, and additions of ZrO_2 made Mullite grains grow.
3. For the composites containing 40 vol% SiC, irrespective of the amount of ZrO_2 added, the flexural strength at room temperature remained unchanged and indicated 500–600 MPa, while for the composites containing 20 vol% SiC the strength decreased with the increasing of ZrO_2 above 10 vol%.
4. It was explained that, for SC20Z20, the degradation of the flexural strength at room temperature resulted from the large crack attributed to the large grains of Mullite and ZrO_2 as well as the expansion of ZrO_2 during the transformation.
5. The flexural strength at 1300°C was measured on SC20Z5, SC20Z20, SC40Z5 and SC40Z20. The strength of these composites indicated about 400 MPa.
6. The fracture toughness (K1c) measured by Indentation Microfracture (IM) method and Indentation Strength (IS) method was increased with addition of ZrO_2.
7. For the SC20Z20 and SC40Z20 the K1c measured by IS indicated the same values (3.8 MPa•$m^{1/2}$). However, the mechanism of toughening was different. The increases of K1c for SC20Z20 was attributed to the pinning of the crack front by the large grains and the crack branching, and for SC40Z20 the deflection and the branching of crack by the fine grains.

Acknowledgment

The authors acknowledge M. Sagawa for preparing test pieces and the manuscript.

References

[1]S. Kanzaki and H. Tabata et al., "Sintering and Mechanical Properties of Stoichiometric Mullite," *J. Am. Ceram. Soc.*, **68** (1) c6–c7 (1985).

[2]S. Kanzaki, T. Kumazawa, and J. Asaumi, et al., "Dependency of Mechanical Property of Sintered Mullite on Chemical Composition," *Yogyo-Kyokai-Shi.*, **93** (7) 407–408 (1985).

[3]P. C. Dokko, J. A. Pask, and K. S. Mazdiyasni, "High-Temperature Mechanical Properties of Mullite under Compression," *J. Am. Ceram. Soc.*, **60** (3–4) 150–55 (1977).

[4]P. A. Lessing, R. S. Gordon, and K. S. Mazdiyasni, "Creep of Polycrystalline Mullite," *J. Am. Ceram. Soc.*, **58** (3–4) 149 (1975).

[5]C. Yamagishi and J. Asaumi, "The Synthesis of the High Purity Mullite Powder Based on the Spray Pyrolysis Technique," *Kagaku-Souchi.*, **29** (5) 52–55 (1987).

[6]C. Yamagishi, H. Kamiaka, and J. Asaumi, "High Purity Mullite Ceramics," Mullite II. 1st ed., Uchida Rohkakuho, Inc., Tokyo (1987) pp. 81–92.

[7]P. Boch and J. P. Giry, "Preparation and Properties of Reaction-Sintered Mullite-ZrO_2 Ceramics," *Mater. Sci. Eng.*, **71** (1-2) 39-48 (1985).
[8]G. Orange and G. Fantozzi, et al., "High Temperature Mechanical Properties of Reaction-Sintered Mullite/Zirconia and Mullite/Alumina/Zirconia Composites," *J. Mater. Sci.*, **20** (7) 2533-2540 (1985).
[9]S. Prochazka and F. J. Klug, "Infrared-Transparent Mullite Ceramics," *J. Am. Ceram. Soc.*, **6** (12) 874-880 (1983).
[10]T. Kumazawa and S. Kanzaki, et al., "Sinterability of SiO_2-Al_2O_3 Powder Prepared by Spray Pyrolysis—Effect of Chemical Composition—," *Yogyo-Kyokai-Shi.*, **94** (5) 485-490 (1986).
[11]T. R. Dinger, K. H. Krishnam, and G. Thomas, "Investigation of ZrO_2/Mullite Solid Solution by Energy Dispersive X-ray Spectroscopy and Electron Diffraction," *Acta. Metall.*, **32** (10) 1601-1607 (1984).
[12]N. Claussen, "Fracture Toughness of Al_2O_3 with an Unstabilized ZrO_2 Dispersed Phase," *J. Am. Ceram. Soc.*, **59** (1-2) 49-51 (1976).
[13]N. Claussen, J. Steeb, and R. F. Pabst, "Effect of Induced Microcracking on the Fracture Toughness of Ceramics," *Am. Ceram. Soc. Bull.*, **56** (6) 559-562 (1977).

Alumina/Mullite Interfaces in Mullite/Zirconia Composites

K. SRIKRISHNA AND G. THOMAS

Materials Science and Mineral Engineering
University of California and National Center for Electron Microscopy, MCSD
Lawrence Berkeley Laboratory
Berkeley, CA 94720

J. S. MOYA

Instituto de Ceramica y Vidrio
Arganda del Rey, Madrid, Spain

Abstract

Reaction sintered mullite/zirconia composites were characterized by conventional and high resolution electron microscopy. The interphase interfaces were for most part glass free though glass was observed at triple points with which a zirconia grain was associated. Studies of the retained alumina/mullite interface revealed it to be semi-coherent and highly stepped. No simple epitaxial relationship was observed. However the hexagonal alumina facetted along its prismatic planes and in a manner to accommodate the low index planes of the growing mullite. This suggests that mullitization of alumina has preferred growth planes and that through additives this process can possibly be controlled.

Introduction

The excellent high temperature, mechanical and chemical properties of mullite have made it a contender for several advanced structural applications. Mullite/zirconia composites are the most extensively studied of all mullite based composites.[1–3] These composites exhibit toughnesses twice that of mullite (5–6 MPa$\sqrt{m}$) and strengths comparable to that of pure mullite. Primarily two processing routes have been utilized to produce mullite/zirconia composites. The first route uses mullite and zirconia as starting materials[4] (or a precursor of mullite[5]) and conventional sintering to obtain a dense composite. The second route utilizes alumina (or a precursor of alumina) and zircon and reaction sintering to obtain a dense mullite/zirconia composite.[6] The latter route has emerged as the economically more attractive route for obtaining such composites due to the availability of the starting materials and the low processing temperatures.

The reaction sintering of alumina and zircon proceeds according to the reaction

$$2ZrSiO_4 + 3Al_2O_3 \rightarrow 2ZrO_2 + 3Al_2O_3.2SiO_2 \quad (1)$$

and involves a transient liquid phase. However, in the absence of additives, the slow kinetics of reaction (1) leads to the retention of glassy phases and the reactants. Such glassy phase retention is detrimental to the final mechanical properties. Additives, such as TiO_2, enhance the solid state reaction and the sintering process. A number of researchers have studied reaction sintering in order to understand the effect of processing route,[7] the role of precursors[8] and additives.[9] This research was aimed at studying the interfaces in a mullite/zirconia composite (with 0.25 mol% TiO_2) which has been well characterized[9] by x-ray diffraction and conventional transmission electron microscopy in order to understand the nature of interphase interfaces and specifically to shed light on the mullitization of alumina in reaction sintering.

Experimental

The starting materials used in this study were Al_2O_3 R.C. 172 DBM, zircon opacifier from Quiminsa (Opazir S), USA and Merck AR grade TiO_2 (99.9%). The processing of these composites is described in detail elsewhere.[9] Post sintering annealing of the composites at the sintering temperature was carried out for various periods of time. Electron transparent specimens were prepared by mechanical grinding and ion milling in the usual way and conventional electron microscopy was carried out in a Philips EM400T analytical microscope. High resolution electron microscopy was carried out with a JEOL 200CX electron microscope.

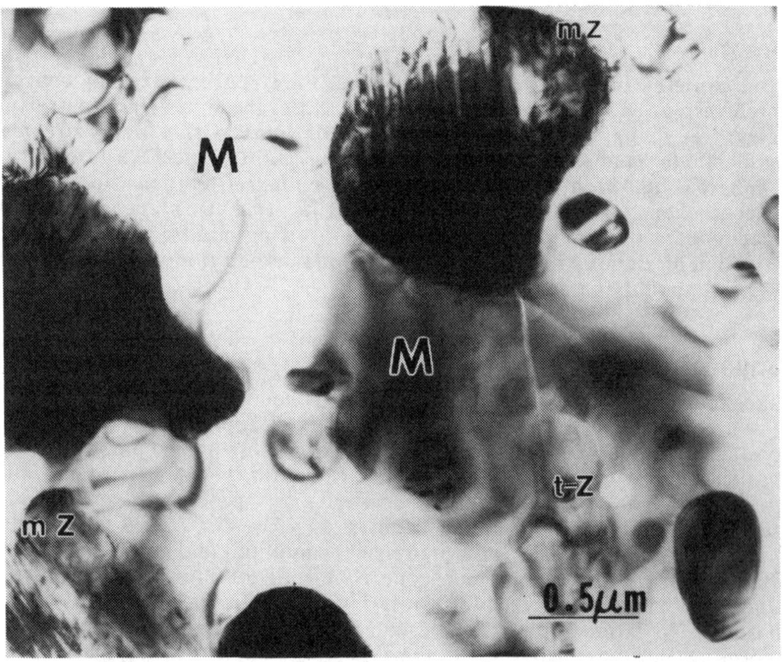

Fig. 1. Bright field micrograph depicting the general microstructure of reaction sintered mullite/zirconia composites (m = mullite, m-Z = monoclinic ZrO_2, t-Z = tetragonal ZrO_2)

Results

The general microstructure of the reaction sintered mullite/zirconia composites shown in Fig. 1, was observed to be very similar to that reported by several other researchers.[4,9] It consisted of a mullite matrix of 2–4 microns average size in which the second phase zirconia was found predominantly at intergranular sites and was monoclinic in crystal structure. Some intragranular zirconia was also observed and was mostly tetragonal. Under the electron beam the tetragonal zirconia frequently transformed to the monoclinic structure, which was easily detectable due to the resultant twinning. X-ray diffraction data indicated that no $ZrSiO_4$ or Al_2O_3 was present and that the reaction had gone to completion. However, detailed electron microscopy studies clearly identified free intragranular Al_2O_3 particles and no zircon was observed in the composites beyond one hour of annealing. These particles (Fig. 2a) were confirmed to be Al_2O_3 by ultra-thin window (capable of detecting oxygen) energy dispersive x-ray spectroscopy (EDXS) and diffraction analysis and lattice parameter measurements showed them to be hexagonal alpha alumina (Fig. 2b), the starting material. On annealing, while the intergranular zirconia underwent grain growth, the intragranular zirconia and alumina did not coarsen at all. The microstructural evolution and mechanical properties of these composites are discussed in detail elsewhere.[10]

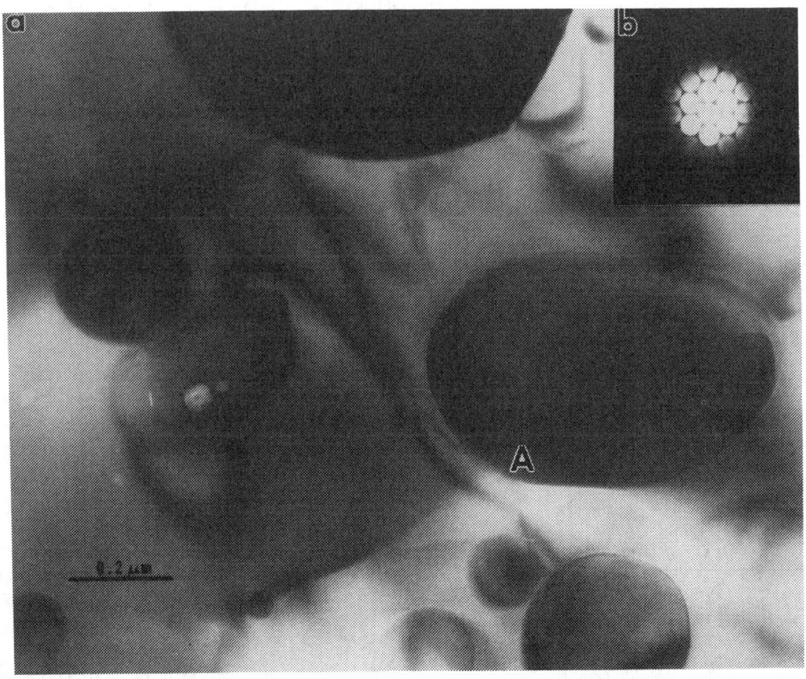

Fig. 2. (a) Bright field micrograph of an intragranular alumina particle within a mullite grain and (b) microdiffraction pattern from the alumina particle (marked A).

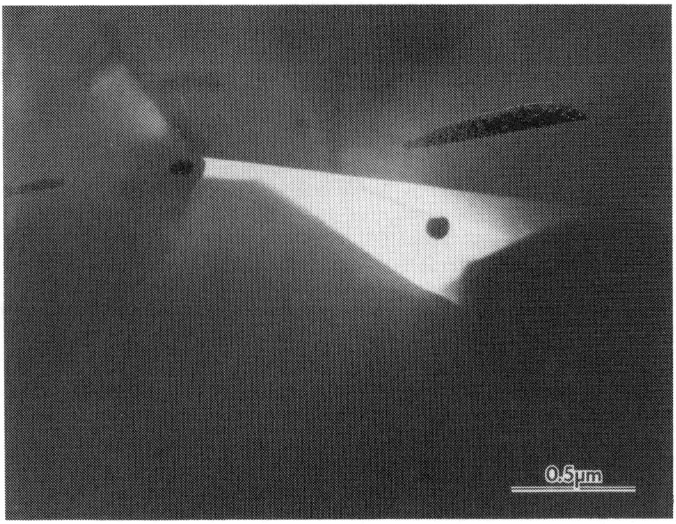

Fig. 3. Dark field micrograph showing the glassy phase at a mullite/zirconia triple point.

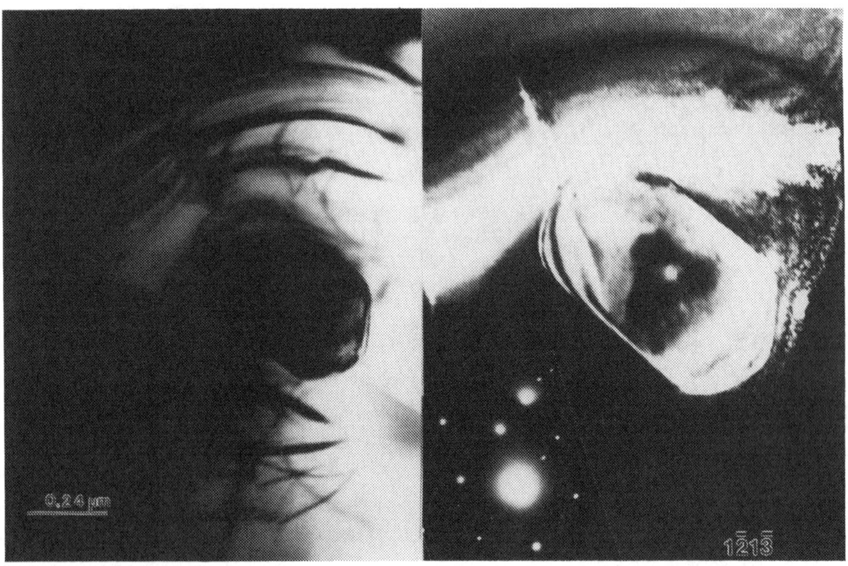

Fig. 4. (a) Bright field, (b) dark field pair of an alumina particle facetted along its {1010} planes.

Mullite/mullite and mullite/alumina interfaces were observed to be glass free. However mullite/zirconia triple grain junctions had some residual glass (Fig. 3), which acted as a sink for the impurities in the composites.[11] Melo and co-workers[12] have established that the reaction sintering process in the quaternary system Al_2O_3-SiO_2-ZrO_2-TiO_2 proceeds through the formation of a transitory liquid. This liquid corresponds to the quaternary invariant point of the $ZrSiO_4$ + $Al_6Si_2O_{13}$ + $ZrTiO_4$ + SiO_2. Overall, the liquid that was retained as a glass phase comprised less than 1 volume percent of the sintered composite. Detailed tilting experiments revealed that the alumina particles which appeared to be spherical were for most part facetted. Figures 4a and 4b show a bright field dark field pair of a facetted alumina particle. Selected area diffraction revealed that the alumina was facetted along its {1010} planes. Experiments revealed that the mullite matrix and the alumina did not have a simple orientation relationship and that the interface was semi-coherent. High resolution electron microscopy was carried out to understand the relationship between the matrix and the alumina particles. Figure 5 is a low magnification structural image of the mullite viewed down the c-axis and it is clear that the alumina is facetted in a manner to accommodate the low index mullite planes. Imaging of the individual interfaces revealed the interface to be semi-coherent and stepped. Figures 6 and 7 are HREM images of two of the alumina facets seen in Fig. 5 and it is clear that a simple coherent epitaxy of mullite on alumina does not occur. The periodicity in contrast observed in Fig. 7 suggests that the interface is semi-coherent. However,

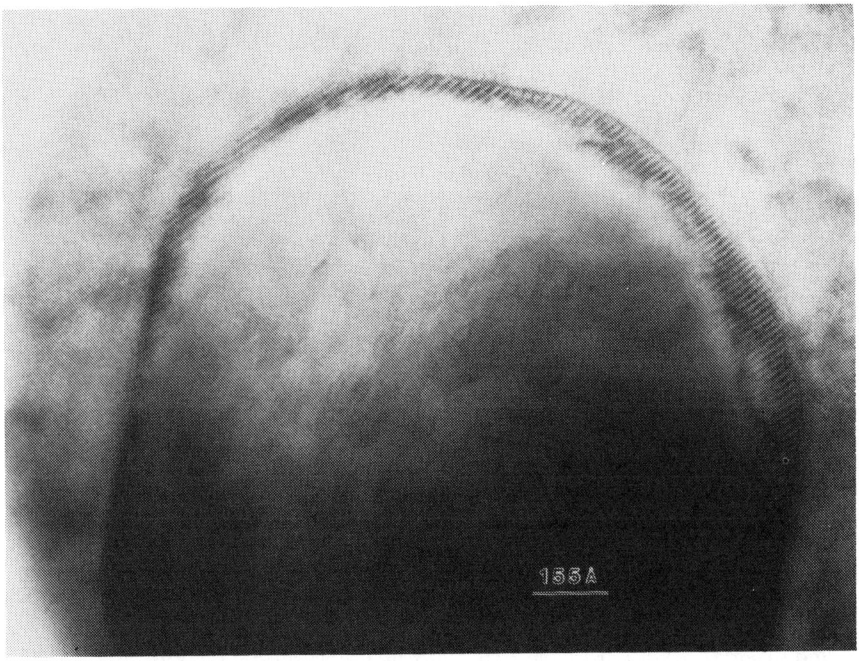

Fig. 5. High resolution image of an alumina particle and the mullite matrix, imaged along the mullite c-axis.

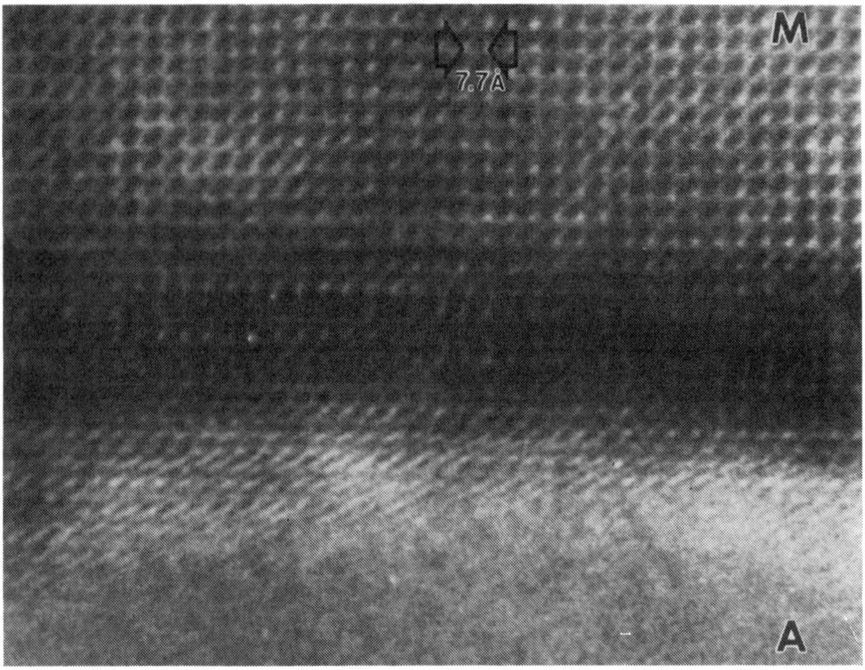

Fig. 6. Higher magnification image of the planar facet in the particle in Fig. 5 revealing the incoherent mullite/alumina interface.

Fig. 7. High resolution image of the particle in Fig. 5 depicting another facet and revealing the defect nature of the interface.

interpreting the exact nature of such an interface is a non-trivial exercise. This is so because the interfaces are not edge on at all positions and it has been impossible to find orientations upon tilting that allow complete lattice imaging simultaneously in both of the phases. This is another reason to believe the interfaces are semicoherent or incoherent.

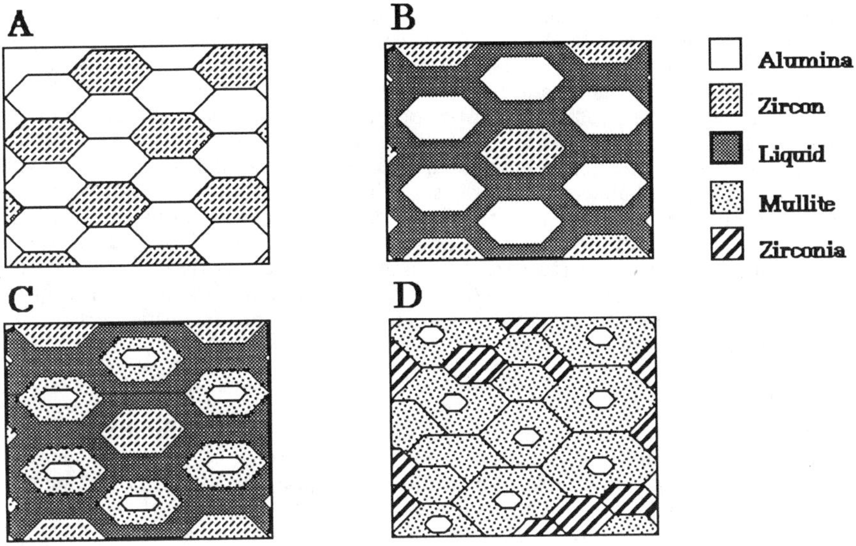

Fig. 8ad. Schematic diagram depicting the reaction sintering process in the presence of a transitory liquid phase.

Discussion

Wallace et al.[13] have observed that the mullitization of alumina in reaction sintering is a sluggish process which results in large glassy pockets and the retention of the reactants. The addition of TiO_2 enhances the reaction and sintering rate hence improving the microstructure.[9] The reaction sintering process in the presence of a liquid phase has been modeled as shown schematically in Fig. 8 a-d. It must be noted that as the densification and sintering in these composites proceed simultaneously, the scheme in Fig. 8 is a simplification primarily modeling the reaction sequence. Figure 8a shows the starting materials Al_2O_3 and $ZrSiO_4$ after compaction. On heating to the sintering temperature a silica-rich liquid (containing Ti and Zr)[11,12] phase forms (Fig. 8b). X-ray diffraction of specimens cooled from this stage show that a metastable $ZrTiO_4$ phase is present.[12] The liquid thus formed reacts with the Al_2O_3 to form mullite (Fig. 8c). The results obtained in this work suggest that the mullite which forms does so epitaxially on the Al_2O_3. On further reaction the metastable $ZrTiO_4$ decomposes to give ZrO_2. As Al_2O_3 and the liquid react, a widening layer of mullite retards further reaction. Nevertheless, the formation of a silica rich transitory liquid enhances transport in these composites in comparison to a solid state reaction between Al_2O_3 and $ZrSiO_4$ accounting for the near absence of glassy phases.

Wallace and coworkers proposed that the initial mullite that formed was non-crystalline to explain the rounded morphology of the zirconia grains. The results of this study show that the alumina/mullite interface is glass free and has a preferred orientation relationship. It is therefore more likely that mullitization of alumina occurs directly by the formation of crystalline mullite on alumina and the rounded morphology of the ZrO_2 is due to the transient liquid phase. This is also supported by the observation of the glass phase being invariably associated with ZrO_2 grains.

As mullite and alumina have different crystal structures and the difference in their unit cell dimensions is large, it is expected that mullite/alumina interfaces would be incoherent. Minimization of the surface energy would require ideally a spherical particle; however the strain energy contribution of such a morphology would be high. The facetting of the alumina along its prismatic planes indicates that these are lower energy planes and help lower the total energy of the system. The fact that these facets occur producing semi-coherent or incoherent interfaces in order to accommodate the low index planes of mullite suggests that mullitization occurs along preferred low energy planes. Rincon et al.[14] have observed the influence of additives on the overall morphology of the phases. Thus in theory it should be possible through the use of additives to affect the mullitization of alumina in a beneficial manner. Naturally further work is in progress to investigate these possibilities.

Although details of the fracture processes will not be reported here, indendation techniques yielded a toughness of 4–6 MPa$\sqrt{m}$. The toughening mechanism in these composites cannot be due to ZrO_2 transformation but appears to be mostly due to crack deflection and microcracking.[15]

Conclusions

Reaction sintered mullite/zirconia composites have for most part glass free interfaces. Some unreacted transitory liquid phase is retained as glass primarily at mullite/zirconia triple junctions. The mullite intragranular alumina interfaces were crystalline free from a second phase and semi or incoherent with large step densities. The alumina facetted along its prismatic planes in a manner to accommodate low index and hence low energy matrix planes. However no simple epitaxial relationship has yet been established.

Acknowledgments

This research was supported by the National Science Foundation under Grant No. DMR-83-1317239. The authors would also like to acknowledge the National Center for Electron Microscopy, Lawrence Berkeley Laboratory for the use of electron microscope facilities.

References

[1]E. Di Rupo, T. G. Carruthers and R. J. Brook, *J. Amer. Cer. Soc.*, **61**, 468 (1978).

[2]Claussen and J. Jahn, ibid, **63**, 228 (1980).

[3]J. S. Moya and M. I. Osendi, *J. Mater. Sci.*, **19**, 2909 (1984).

[4]S. Prochazka, J. S. Wallace and N. Claussen, *Comm. Amer. Cer. Soc.*, **66**, C-125 (1983).

[5]J. S. Moya and M. I. Osendi, *J. Mater. Sci. Lett.*, **2**, 599 (1983).

[6]E. Di Rupo, M. R. Anseau and R. J. Brook, *J. Mat. Sci.*, **14**, 2924 (1979).

[7]E. Di Rupo, E. Gilbart, T. G. Carruthers and R. J. Brook, ibid, **14**, 705 (1979).

[8]C. Zografou, D. Von Mallinckrodt and P. Schulz, *J. de Physique*, **47**, C1-429 (1986).

[9]K. Srikrishna, M.S. Thesis, University of California, May 1986.

[10]K. Srikrishna, G. Thomas and J. S. Moya, Advances in Ceramics, III International Conference on Zirconia, American Ceramic Society, Columbus OH, in press.

[11]K. Srikrishna, G. Thomas and J. S. Moya, Ceramic Microstructures 86:Role of Interfaces, ed: J. Pask, Pergamon Press, NY, 111 (1986).

[12]M. F. Melo, J. S. Moya, P. Pena and S. de Aza, *J. Mater. Sci.*, **20**, 2711 (1985).

[13]J. S. Wallace, G. Petzow and N. Claussen, Science and Technology of Zirconia, Advances in Ceramics, Vol 12, eds. N. Claussen, M. Ruhle and A. H. Heuer, 439 (1984).
[14]J. Ma. Rincon, T. R. Dinger, G. Thomas, J. S. Moya and M. I. Osendi, *Acta Metall.*, **35**, 1175 (1987).
[15]K. Srikrishna and G. Thomas, submitted to the J. Amer. Cer. Soc.

AEM and HRTEM Studies of the Eutectic System Zirconia–Mullite

M. R. Notis, V. P. Dravid and C. E. Lyman

Department of Materials Science and Engineering
Lehigh University
Bethlehem, PA 18015

Abstract

The crystallography and morphology of twins and microcracks in the eutectic system mullite-zirconia and the associated microchemistry are discussed based on preliminary analytical and high resolution transmission electron microscopy (AEM and HRTEM) observations, in view of the tetragonal to monoclinic (t-m) transformation and the associated toughening mechanism(s). Well defined twin relationships were observed in m-ZrO_2 which were consistent with previous investigations and theoretical predictions.

Highly symmetric and crystallographically well-defined microcracks were observed at the mullite-ZrO_2 interface. The microdiffraction technique was used to elucidate the crystallography of monoclinic twins and the microcracks, which were observed to be closely interrelated. HRTEM revealed an isolated glassy phase in the mullite matrix, but the majority of ZrO_2-mullite interfaces were glass free. X-ray microanalysis indicated very low solubility of mullite species in ZrO_2 and vice-versa.

Introduction

It is well known that the tetragonal to monoclinic (t-m) transformation in ZrO_2 usually involves twinning which minimizes the macroscopic shape (volume) change of the transforming grain. The twin or the variant structure has a pronounced influence on the strain energy change in the t-m transformation.[1] It is also expected that a large stress concentration at interphase interfaces would lead to microcrack formation.[2]

Although there is an extensive body of literature concerning the fracture mechanics of microcrack toughening, there is a notable paucity of information about the crystallography associated with the microcracks and its interrelationship with the twin systems in m-ZrO_2. It is important to understand this relationship in view of the recent debate in the literature regarding twinning modes in m-ZrO_2, the resultant substructure within the particles, and the possible effects of these factors upon the fracture behavior of zirconia based ceramics. The difference in twin structure associated with thermally transformed and stress induced transformed zirconia particles still appears to be an open area of investigation.[3,4] Further, although various toughening mechanisms such as transformation toughening, microcrack toughening, crack branching, etc. have been recognized, there is a considerable controversy regarding the actual contribution of each to the overall toughness.[5] This is especially so in the case of zirconia toughened ceramics such as zirconia toughened alumina (ZTA). It has recently been pointed out that the dilatational component of the residual strain field alone does not fully explain the observed toughness values of ZTA.[6,7] In addition, there are some recent reports about solid solution strengthening in mullite-ZrO_2 system.[8] Thus, the structure, morphology and the crystallography of twins and microcracks and microchemistry of interfaces deserve further attention in light of their possible contributions to the fracture toughness of zirconia toughened ceramics. A preliminary account of twinning and microcracking in the ZrO_2-Mullite eutectic system has been presented elsewhere,[9] however the principal results are included here for completeness.

Experimental

The sample used in the present study was a directionally solidified eutectic (DSE) in the system Al_2O_3-SiO_2-ZrO_2. The DSE consisted of a near stoichiometric mullite matrix with m-ZrO_2 fibers and lamellae embedded in the matrix in colonies.

Transverse sections were cut from the bulk sample and were mechanically polished to about 75 μm. These thin sections were further thinned to perforation using a Gatan twin beam ion milling system operated at 6 kV. The foils were coated with a thin layer of amorphous carbon to avoid charging under the electron beam. Transmission electron microscopy (TEM) observations were made on a Philips EM 400T analytical electron microscope (AEM) operated at 120 kV. Microdiffraction was performed using a nominal probe size of 10 nm, which was adequate to confine the electron beam within a single monoclinic domain. Microanalysis was performed with the same instrument using the scanning transmission electron microscopy (STEM) mode with a nominal probe size of 10 nm. Utmost care was taken by choosing interfaces which pointed towards the detector (specimen tilt about 5°) and by avoiding any interfacial overlap. All the analyses were carried out in foils thin enough for quantification using the Cliff-Lorimer (k-factor) approach without any absorption correction.[10] The following k-factors were measured: $k_{Al\text{-}Si} = 1.12 \pm 0.03$, $k_{Al\text{-}Zr} = 0.63 \pm 0.05$ and $k_{Zr\text{-}Si} = 1.78 \pm 0.13$. All refer to K-lines except for Zr where Zr_L was measured. HRTEM was performed using a Philips EM 430T TEM operated at 250 kV having point-to-point resolution of about 2.3 Å at Scherzer defocus. All the HRTEM micrographs in this paper were taken near Scherzer defocus and close to axial illumination conditions.

Results

Sorrell and Sorrell[11] proposed that two possible situations exist for the nature of ternary equilibrium in the ZrO_2-SiO_2-Al_2O_3 system, depending strongly upon the form of melting of the phase ($3Al_2O_3.2SiO_2$). A ternary eutectic is predicted within the ZrO_2-$3Al_2O_3$:$2SiO_2$-Al_2O_3 compatibility triangle only if mullite is a congruently melting compound. Recent work by Klug et al.[12] indicates that mullite is a congruently melting phase. However, preliminary work by Then-Xing[13] indicates no ternary eutectic present within the compatibility triangle for the conditions studied. Pask[14] suggests that in the absence of α-Al_20_3 mullite exhibits congruent melting, but when α-Al_2O_3 is present, incongruent melting is observed. In any case, as pointed out by Sorrell and Sorrell,[11] metastable behavior in both the ZrO_2-SiO_2 and Al_2O_3-SiO_2 systems, and consequently in the ternary system, is highly likely. The ternary melt composition used in the present study was determined using electron microprobe analysis to be 50.7 w/o Al_2O_3–27.3 w/o SiO_2–22 w/o ZrO_2 ($\approx$44

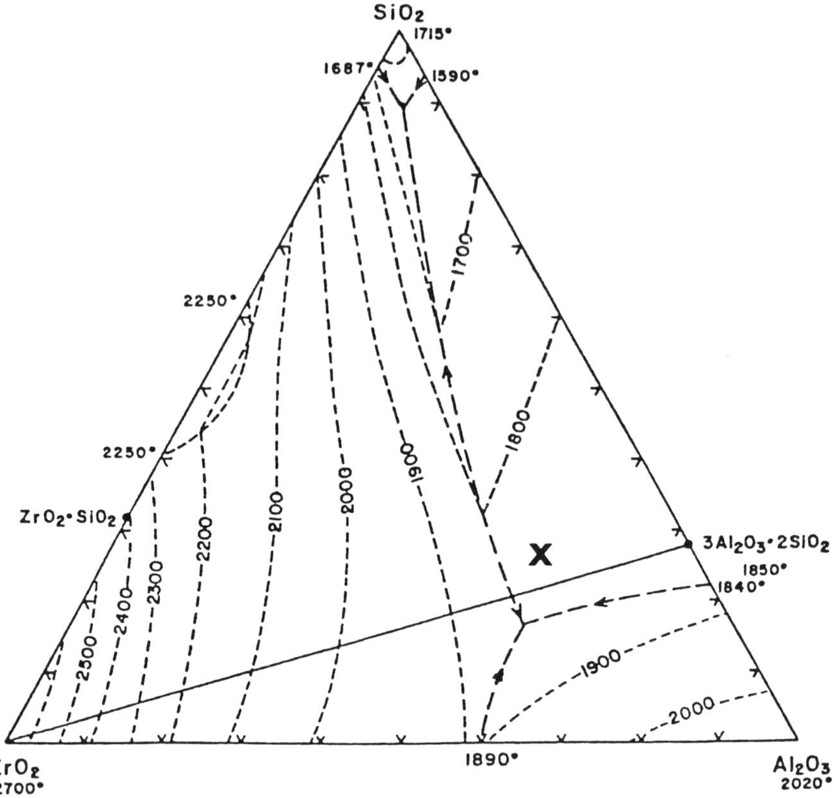

Fig. 1. Ternary phase diagram after Sorrell and Sorrell (compositions in wt.%), with starting composition indicated by "x".

m/o Al_2O_3–40 m/o SiO_2–16 m/o ZrO_2) and is shown superimposed on the phase diagram suggested by Sorrell and Sorrell[11] (Fig. 1).

Low magnification TEM observations revealed that the eutectic consisted of lamellae as well as fibers of m-ZrO_2 in a mullite matrix. The identity of the mullite matrix was determined by quantitative x-ray microanalysis and by convergent beam electron diffraction (CBED); Fig. 2 shows a [001] CBED pattern revealing b and a glide planes of mullite, (space group Pbam) indicated by the presence of orthogonal Gjonnes-Moodie (G-M) lines in alternate reflections.[15] The transverse shape of the ZrO_2 fibers was near ellipsoidal with an aspect ratio of 3–5. Although the shape of the embedded phase in a colony structure could be misleading, it does not affect the analysis presented here.

Electron diffraction indicated the general orientation relationship, within the experimental error, to be:

Growth Direction // [001] mullite // [010] m-ZrO_2

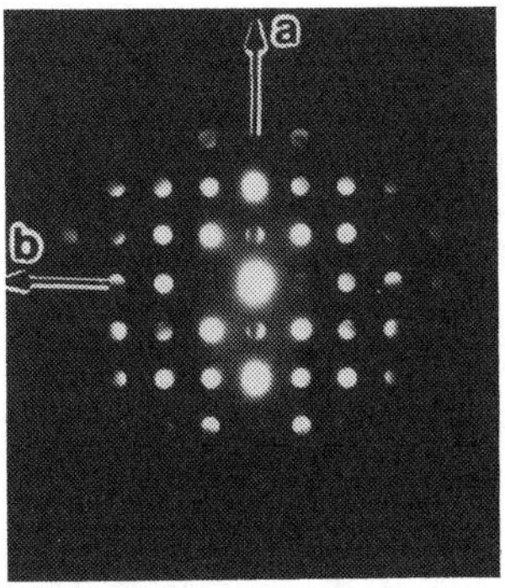

Fig. 2. CBED pattern taken along [001] zone axis of mullite. Orthogonal G-M lines indicate the presence of two glide planes in the space group.

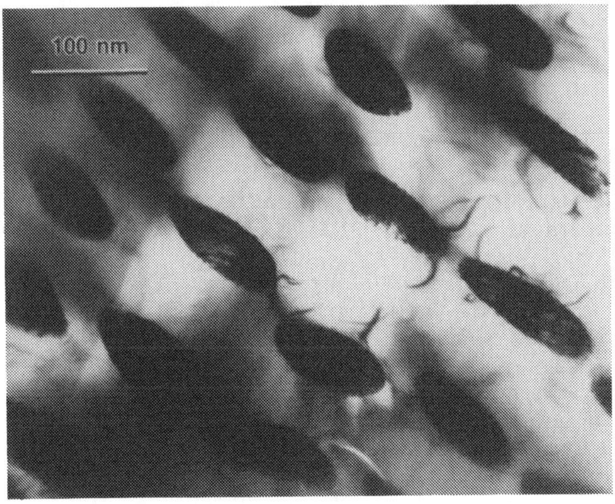

Fig. 3. Low magnification TEM image of twinned m-zirconia fibers embedded in mullite matrix.

Figure 3 shows a low magnification image of ellipsoidal m-ZrO_2 in a mullite matrix. All the m-ZrO_2 fibers are twinned and appreciable strain contrast is evident in the matrix. Very fine microcracks were observed and are revealed in the higher magnification image in Fig. 4. After examining numerous m-ZrO_2 particles of similar shape and size, it was concluded that only those particles where the twin planes run either perpendicular or at a small angle to the long dimension of the m-ZrO_2 phase show microcracks along the interphase interfaces. The occurrence of m-ZrO_2 is obviously a consequence of the displacive t-m transformation during cooling since there is no t- or c- stabilizing solute present in the system. This transformation results in high anisotropic volume expansion of m-ZrO_2 (approximate-

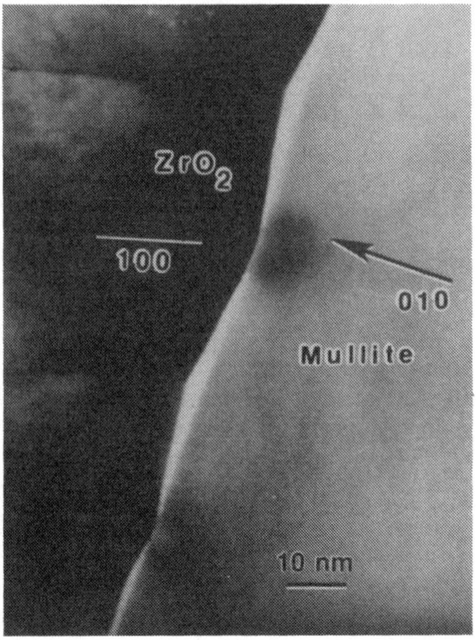

Fig. 4. A high magnification image of mullite-zirconia interface showing symmetric microcracks. The trace of the (100) planes in ZrO_2 is shown.

ly +0.4%) which usually leads to twinning and microcrack formation at the interphase interfaces depending on the crystallography of twins.

A typical example is shown in Fig. 4 where the twin planes are perpendicular to the long dimension and the morphology of the microcracks is quite clear. Note that the microcracks have similar shape, which can be approximated as isosceles triangles with large base width. The width of the microcracks is equal to the width of the two associated monoclinic variants. Note that even in the presence of microcracks, strain contrast is quite pronounced between the microcracks which implies that despite microcracking, residual strain during t-m transformation may remain at the interphase interfaces which are believed to be incoherent. Microdiffraction indicated that the twin plane in the above case is (100) m-ZrO_2, and (100) mullite being parallel to the apparent interphase interface. This situation is similar to the suggestion of Muddle et al.[3] for Mg-PSZ. Another configuration showing microcracking was observed where the twin planes were at an angle to the long dimension of the m-ZrO_2 ellipse. Although there was a pronounced effect on the microcrack shape, the width was essentially the same as the previous configuration. In this case also, we have observed (100) as the twin plane.

A third case is presented in Fig. 5 where the twin planes are almost parallel to the long dimension of m-ZrO_2. The absence of microcracks should be noted. The above three cases are summarized schematically in Fig. 6 assuming the orientation relationships between tetragonal and monoclinic variants to be those shown in each case.

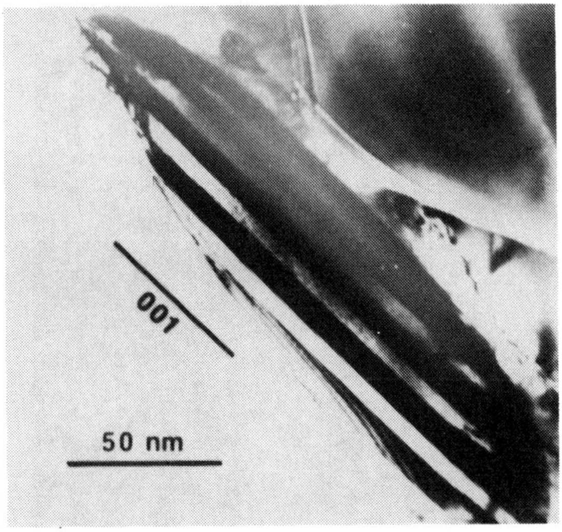

Fig. 5. Bright field TEM micrograph of m-zirconia fiber with twins parallel to the long dimension. The trace of 001 planes is shown on the left side.

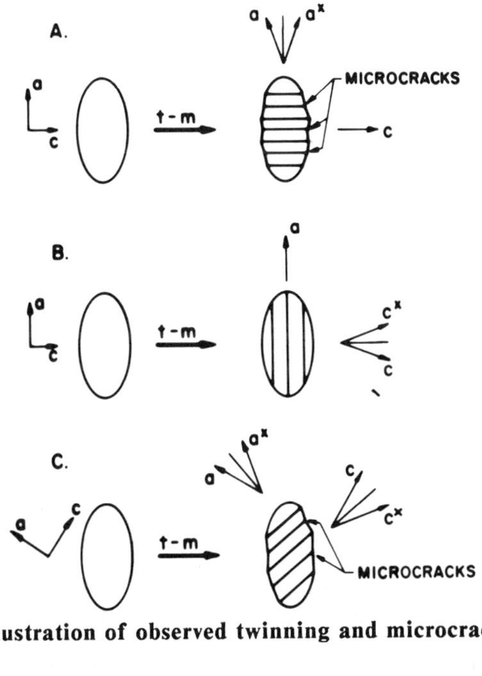

Fig. 6. Schematic illustration of observed twinning and microcracking modes in zirconia fibers.

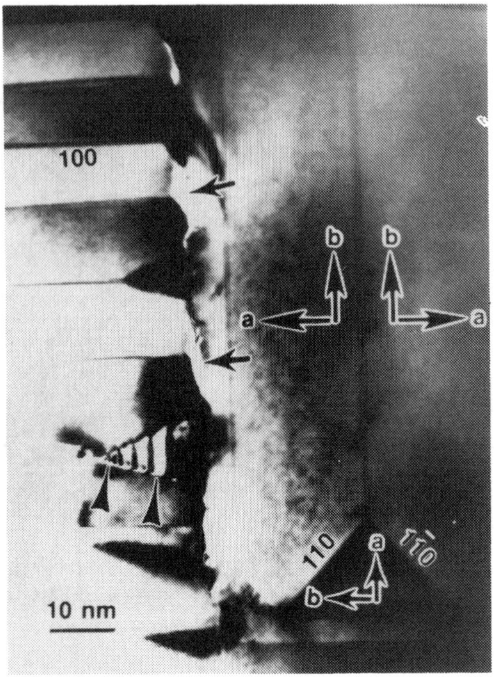

Fig. 7. Bright field TEM micrograph of complex twinning and microcracking in a large m-zirconia particle (see text).

Twinning and microcracking were observed to be more complex in lamellar m-ZrO_2. Figure 7 shows an example where there are several twin domains within single m-ZrO_2 lamellae. It is well established that the rate controlling step in t-m transformation is thee nucleation event. Therefore for larger size particles it is conceivable that many nucleation sites may be available, leading to impingement of various growing interfaces. This is likely to result in complex domain formation as in Fig. 7. The crystallography of the central twin system is similar to the findings of Van Tendeloo et al.[16] Domains of closure similar to the observations of Bischoff and Ruhle[17] are also seen in the same figure. Microcracks are seen at the intersection of perpendicular domains as well as at very fine twin variants shown by the arrows. Efforts are underway to characterize such particles more extensively.

Microanalysis was performed across a ZrO_2 mullite interface which was not associated with any microcracks. Figure 8 shows a profile depicting wt.% of Al_2O_3, SiO_2 and ZrO_2 with respect to the location from the interface. Note that there is little solubility of Al_2O_3 and SiO_2 in m-ZrO_2. Similarly, solubility of ZrO_2 in mullite is also very limited. Further, within the spatial resolution limit the profile is stepped without any transient behavior.

Figure 9 shows an HRTEM image of a flat portion of the ZrO_2-mullite interface. The light contrast at the interface possibly suggests the presence of a grain boundary amorphous phase. However, there appears to be an extra (100) plane of m-ZrO_2 after every 14 planes. A small interfaces step is also indicated on the micrograph. Such

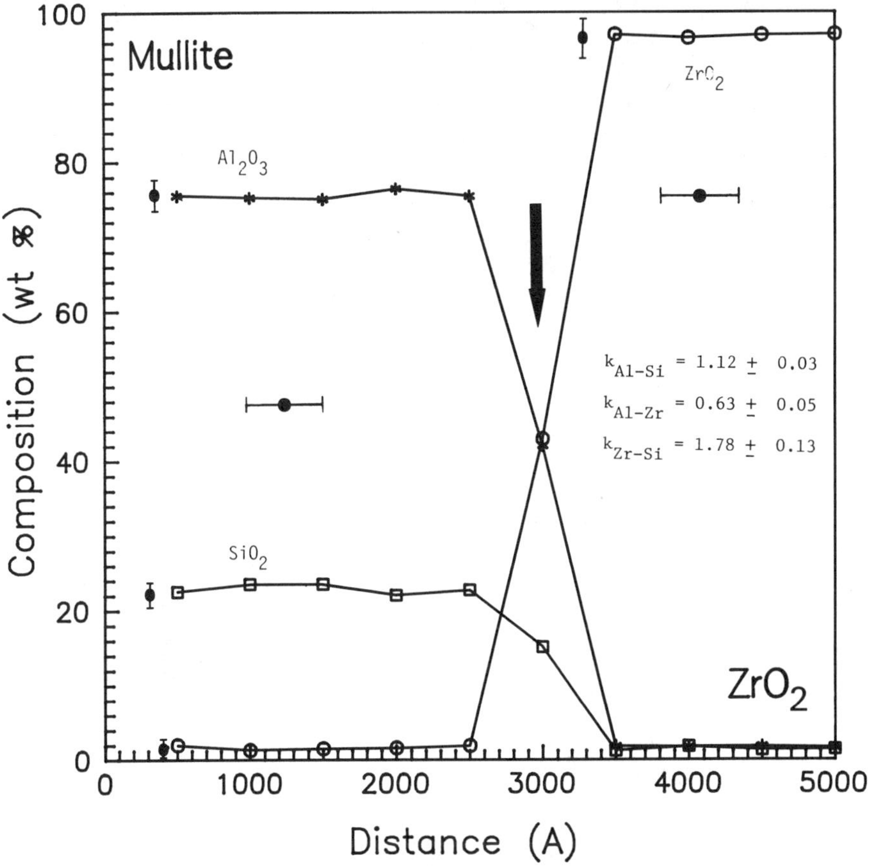

Fig. 8. X-Ray microanalysis composition profile across a mullite-ZrO_2 interface. Location of the interface is marked with an arrow. Errors due to counting statistics and beam broadening are indicated by bars.

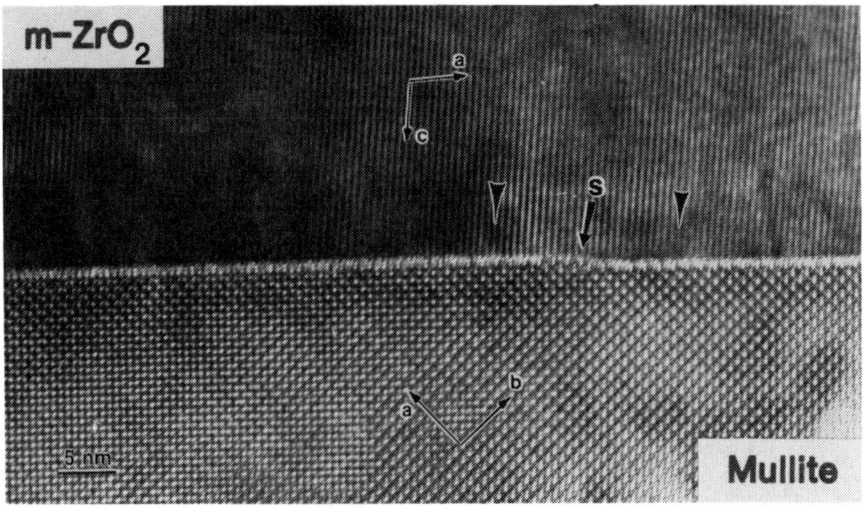

Fig. 9. HRTEM image of m-ZrO_2-mullite interface. Extra (100) planes in m-ZrO_2 are arrowed. A small interfacial step is also indicated by 's'.

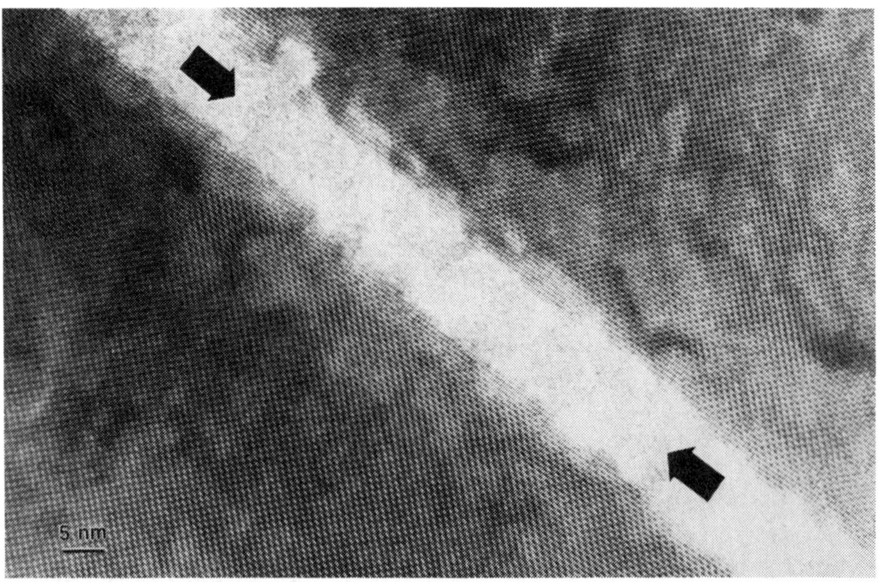

Fig. 10. HRTEM image of a colony boundary in mullite showing intergranular pocket of glassy phase which is rich in SiO_2.

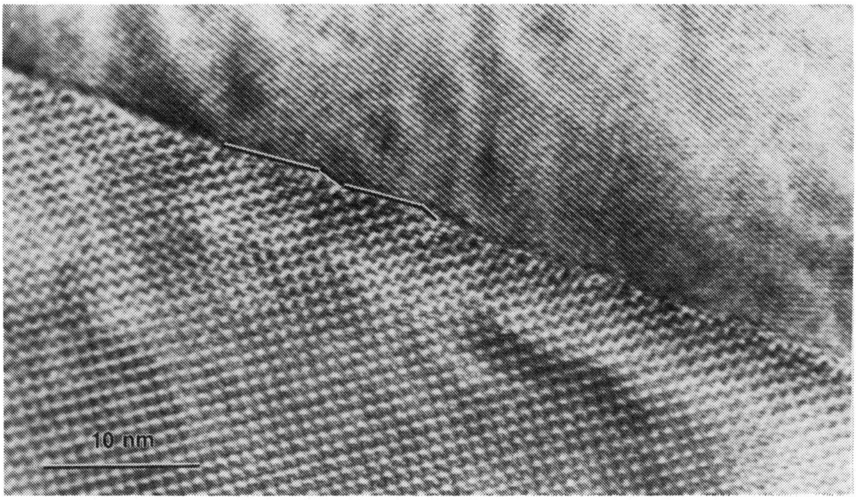

Fig. 11. HRTEM image of a curved interface. Note the presence of interfacial steps. Lines have been added to aid the eye in locating these steps.

bright contrast was rarely observed at the interphase interfaces and glassy phase was detected unambiguously only in mullite matrix colony boundaries (Fig. 10). In the case of curved ZrO_2-mullite interface, steps (ledges) were observed which may be responsible for accommodation of lattice misfit (Fig. 11).

Discussion

The (001) and (100) twin systems observed in this work are consistent with a recent analysis which shows that twinning and microcracking is entirely due to the thermal transformation of zirconia during cooling.[4] The {110} type twins are also consistent with Bischoff and Ruhle's[17] observations. The microcracking, and its correlation with the twin variants appears to have arisen due to the excess strain energy release at the interphase interface (Fig. 4) or within m-ZrO_2 (Fig. 7).

Although our microstructural analysis is consistent with previous work, the x-ray microanalysis data contradicts some recent reports in the literature.[8,18,19] Dinger et al.[8] investigated ZrO_2-mullite composites using the AEM techniques. They observed very large solubility of Al_2O_3 in ZrO_2 (≈36 wt%), SiO_2 in ZrO_2 (≈8 wt%) and ZrO_2 in mullite (≈34 wt%), using the STEM mode for microanalysis. These investigators did not give the Cliff-Lorimer k-factors used[10] for quantification of the raw data. Further, they observed an extremely unusual effect of lower values for the solubility data in the TEM mode than in the STEM mode and indicated the possible ordering of m-ZrO_2 due to large Al_2O_3 and SiO_2 solubilities.

In their subsequent technical note,[20] they attributed previously observed "extra" reflections to double diffraction effects and emphasized that the TEM mode microanalysis gave lower solubility values than the STEM mode. Due to this discrepancy they suggested that their data (ref. 8, Fig. 3) might be overestimated by a factor of about seven (Al_2O_3) to fifteen (ZrO_2). In any event it appears extremely unlikely for the TEM and STEM data to differ so appreciably. In their most recent publication,[18] they refer to their original paper for microanalytic data but report relatively small solubility of SiO_2 (≈8 wt%) and Al_2O_3 (≈1.5 wt%) in ZrO_2. Even this apparent solubility may be due to x-ray excitation of electrons scattered from one phase to the other off the lower polepiece. Further, their k-factors differ appreciably from the k-factors of Rincon et al.[19] and the ones measured in this study. Although k-factors do vary from instrument to instrument, such a large variation is not anticipated, for the same type of instrument.

Being directionally solidified from the melt, the solubility data in our system (Fig. 8) are probably the maximum solubility values. The errors associated with concentrations of the minor species in each phase are not surprising in view of the poor counting statistics afforded by AEM x-ray microanalysis. Thus mullite and ZrO_2 have very small solubility in each other.

We believe that our model DSE system has provided some clues regarding the possible toughening (and weakening) mechanism(s) in mullite-ZrO_2 composites. Very low solid solubility values for mullite in ZrO_2 and vice-versa rule out any significant contribution due to chemical activity at the interphase interfaces. The preferential segregation of the glassy phase as isolated pockets at the colony boundaries suggest those to be probable fast track growth paths. The tendency of certain m-ZrO_2 particles (Fig. 6) to twin and form microcracks indicate the possibility of microcrack toughening or weakening depending on the dilatational/shear components. Large monoclinic particles (lamellae) are likely to resist crack propagation through crack deflection. Further, crack-bridging mechanism, proposed for certain single phase ceramics may possibly play a decisive role in this system. However, a definitive answer will be given by systematic study involving fracture toughness measurements along with observations of crack propagation in this system. Such experiments are presently underway.

Significant analogous behavior is expected between the lamellar and rod type DSE morphology observed in the present study, and the variety of synthesized zirconia/mullite composites fabricated by more conventional processes. However, the DSE's represent a class of structures with well-defined crystallographic orientation and therefore represent a good model system for further basic study.

Summary

Conventional TEM observations show that the morphology, and locations of microcracks are related to the twin variants of m-ZrO_2. The twin relations and microcracking behavior are consistent with previous reports on similar systems. Microanalysis data shows limited solubility ionic species of mullite in ZrO_2 and vice versa, which contradicts previous literature. HRTEM observations revealed certain misfit accommodation features. The TEM, HRTEM and microanalysis results should supplement the ongoing experiments involving crack propagation observations. The combined results should reveal the possible toughening mechanisms in this system which are likely to be a combination of crack deflection, microcracking and crack bridging.

Acknowledgments

The authors thank Professor V. S. Stubican for supplying the sample and Mr. Jim Ciulik for some experimental assistance. Research supported by US DOE (No. DE-FG02-84ER-45150).

References

[1]A. G. Evans, N. Burlingame, M. Drory and W. M. Kriven, "Martensitic Transformation in Zirconia—Particle Size Effects and Toughening," *Acta Metall.*, **29**, 447-456 (1981).

[2]Y. Fu, A. G. Evans and W. M. Kriven, "Microcrack Nucleation in Ceramics Subject to a Phase Transformation," *J. Am. Ceram. Soc.*, **67** [9], 626-30 (1984).

[3]B. C. Muddle and R. H. J. Hannink, "Crystallography of the Tetragonal to Monoclinic Transformation in MgO-Partially-Stabilized Zirconia," *J. Am. Ceram. Soc.*, **69** [7], 547-55 (1986).

[4]P. M. Kelly and C. J. Ball, "Crystallography of Stress-Induced Martensitic Transformations in Partially Stabilized Zirconia," *J. Am. Ceram. Soc.*, **69** [3], 259-64 (1986).

[5]A. G. Evans and R. M. Cannon, "Toughening of Brittle Solids by Martensitic Transformations," *Acta Metall.*, **34** [5], 761-800 (1986).

[6]R. M. McMeeking, "Effective Transformation Strain in Binary Elastic Composites," *J. Am. Ceram. Soc.*, **69** [12], C-301-C-302 (1986).

[7]J. C. Lambropoulos, "Shear, Shape and Orientation Effects in Transformation Toughening," *Int. J. Solids Structures*, **22** [10] 1083-1106 (1986).

[8]T. R. Dinger, K. M. Krishnan, G. Thomas, M. I. Osendi and J. S. Moya," Investigation of ZrO_2/Mullite Solid Solution by Energy Dispersive X-Ray Spectroscopy and Electron Diffraction," *Acta Met.*, **32** [10], 1601-7 (1984).

[9]Vinayak P. Dravid, M. R. Notis and C. E. Lyman, "Twinning and Microcracking Associated with Monoclinic Zirconia in the System Zirconia-Mullite," Accepted for publication J. Am. Ceram. Soc., April, 1988.

[10]G. Cliff and G. W. Lorimer, *J. Micro.*, Vol. 103, 203, 1975.

[11]C. C. Sorrell and C. A. Sorrell, "The Al_2O_3-SiO_2-ZrO_2 Liquidus Surface: A Discussion," *J. Am. Ceram. Soc.*, **60** [1-2], 93-4, 1977.

[12]F. J. Klug, S. Prochazka, and R. H. Doremus, "Alumina-Silica Phase Diagram in the Mullite Region," Research Report, Corporate Research and Development, General Electric, Schenectady, NY, Aug. (1986).

[13]Then-Xin, "Phase Relations and Transformations in the System ZrO_2-$3Al_2O_3.2SiO_2$-Al_2O_3," Paper 3B-1, p. 302-3, Extended Abstracts Zirconia-III, Tokyo, (1986).

[14]J. Pask, "Current Understanding of Stable and Metastable Phase Equilibria and Reactions in the SiO_2-α-Al_2O_3 System," in Ceramic Powders, ed. P. Vincenzini, p. 21-31, (1983).

[15]J. W. Steeds and R. Vincent, "Use of High Symmetry Zone Axes in Electron Diffraction in Determining Crystal Point and Space Groups," *J. Appl. Cryst.*, **16**, 317-24, (1983).

[16]G. VanTendeloo, L. Anders and G. Thomas, "Electron Microscopy Investigation of the ZrO_2-ZrN System-II. Tetragonal and Monoclinic ZrO_2 Precipitation," *Acta Metall.*, **31** [10], 1619-25 (1983).

[17]E. Bischoff and M. Ruhle, "Twin Boundaries in Monoclinic ZrO_2 Particles Confined in a Mullite Matrix," *J. Am. Ceram. Soc.*, **66** [2], 123-7 (1983).

[18]J. M. Rincon, T. R. Dinger, G. Thomas, J. S. Moya and M. I. Osendi, "Microstructure of Mullite/ZrO_2 and Mullite/Al_2O_3/ZrO_2 Tough Ceramic Composites," *Acta Met.*, **35** [5] p. 1175-9 (1987).

[19]J. M. Rincon, J. S. Moya, and M. F. DeMelo, "Microstructural Study of Toughened ZrO_2/Mullite Ceramic Composites Obtained by Reaction Sintering with TiO_2 Additions," *Br. Ceram. Trans. J.*, **85**, 201-6 (1986).

[20]T. Dinger, K. M. Krishnan, G. Thomas, M. I. Osendi and J. S. Moya, "Technical Note," *Acta Met.*, **33** [2] p. 2 (1985).

Mechanical Properties and Microstructures of Mullite-Zirconia Composites

A. LERICHE

National Research Centre of the Belgian Ceramic Industry-C.R.I.B.C.
4, avenue Gouverneur Cornez-B-7000 Mons-Belgium

Abstract

Mullite-zirconia composites prepared by different routes are compared from the microstructural and mechanical viewpoints. The reaction-sintered materials prepared in the presence of magnesia or titania present similar mechanical properties although very different microstructures. The MgO-composites are characterized by a crosslinked mullite grain arrangement with the presence of an intergranular glassy phase, whereas the microstructure of the titania composites and the sol-gel materials are constituted of juxtaposed equiaxed grains without an intergranular glassy phase. The sol-gel composites present the highest flexural strengths and the finest and the most homogeneous microstructures. However, the toughness value is less than that of the MgO-composites because of a lower tetragonal zirconia content and the absence of a crosslinked mullite grain arrangement.

Introduction

Recently, mullite-zirconia composites have been studied because of their many interesting characteristics such as refractoriness, chemical inertness, low thermal expansion and mechanical properties, the last of which may be improved by the presence of dispersed zirconia particles.[1]

The initial preparation method was the conventional sintering of either milled mullite, alumina and silica mixtures or premullite and zirconia powder mixtures.[2–4] In this processing, the powder mixture homogenization constitutes the critical step of the processing, and the sintered materials usually have a bad zirconia grain dispersion. The zirconia grains are mainly located at the mullite grain boundaries and present little tetragonal phase retention for high zirconia contents compared to other processing methods. Table I summarizes the main characteristics of the materials produced by different processes. To enhance the microstructural homogenization and therefore the mechanical properties, two other routes, namely reactive sintering and the sol-gel process have been explored.

The reactive sintering method produces in situ mullite-zirconia composites from fine zircon and alumina powder mixtures. Dense materials can be obtained by two different techniques: i) the two-step method[5,6] which separates the densification and the reaction steps through a special firing cycle that includes successive plateaus at 1400 and 1600°C and, ii) the addition of oxides of a few percent[7–11] which combines

Table 1. Main characteristics of mullite–zirconia composites prepared by different routes.

Material		Sintering temperature (°C)	% ρ_{th}	T_r (%)	Grain size (μm) mullite	Grain size (μm) ZrO_2	Microstructure: Grain shape and location	Mechanical Properties: σ_F (20 °C) (MPa)	Mechanical Properties: K_{Ic} (20 °C) (MPa/m)
Conventional Sintering									
Fused mullite									
+ 10 v% ZrO_2	(2)	1 610	>98	65	1.2	0.7	Equiaxed mullite grains		
15				37	1.2	0.8	intergranular zirconia		
20				16	1.3	1.0	absence of amorphous phase		
25				11	—	—		200–250	3–4$^+$
Reacting $Al_2O_3 + SiO_2$ +									
10 v% ZrO_2	(3)	1 650	—	40–50	5.0	2.0	Less equiaxed mullite grains	200	2.8$^+$
17				10–20	7.0	3.0	mainly intergranular zirconia	210	2.4$^+$
							presence of amorphous phase		
Premullite +									
10 v% ZrO_2		1 570	99	—	0.8	—	Angular mullite grains	250	2.9*
15	(4)						intergranular zirconia		
20								290	3.0*
Reactive Sintering									
$2ZrSiO_4 + 3Al_2O_3 \rightarrow 2ZrO_2 +$									
$Al_6Si_2O_{13}$									
24 v% ZrO_2									
2 step method	(6)	1 400 –	98	60	4.0	1.5	Irregular shaped mullite grains	450	3.5–5.0$^+$
		1 575					inter- and intragranular zirconia	270	4.3*
additives CaO	(8)	1 500			2–5	1–3	Rectangular shaped mullite grains		
							intergranular zirconia		
							presence of amorphous phase		
MgO	(7,9)	1 500	98	50	20–25	1.0	Elongated needlelike mullite grains	320,330	3.7$^+$,4.5*
							inter- and intragranular zirconia		
							presence of amorphous phase		
TiO_2	(17,10)	1 500			3.0	2.0	Equiaxed mullite grains	290,210	3.3$^+$,4.9*
							intergranular zirconia		
							absence of amorphous phase		
Sintering of Sol-Gel Powders									
$Al(NO_3)_3$, colloidal SiO_2,									
$ZrOCl_2$	(12)								
5 v% ZrO_2		1 590	99.100	80	0.7	0.5	Equiaxed mullite grains	480	2.2$^+$
10				40			intergranular zirconia	480	3.0$^+$
15				30				520	3.2$^+$
20				10				550	3.4$^+$

Table 1. Continued.

Al_2O_3, colloidal SiO_2, $ZrOCl_2$	(14)								
5 v% ZrO_2		1 600	>98	40–45	0.8	0.4	Equiaxed mullite grains intergranular zirconia		
10								520	4.9*
15									
20									
25									
Al_2O_3, colloidal SiO_2, $ZrOCl_2$, YCl_3	(13)								
19 v/o ZrO_2		1 600	98	46		0.3	Needlelike and equiaxed mullite grains	240	3.0*
19 v/o (3m/o Y_2O_3)		1 600	98	87		0.7	Intergranular rounded zirconia particles	250	4.0*
$Al(OC_3H_7)_3$, $Si(OC_2H_5)_4$, ZrO_2, Y_2O_3	(15)								
20 v/o ZrO_2		1 600		6	2.0	1.5	Irregular shaped mullite	—	2.5+
20 v/o ZrO_2 (1→3m/o Y_2O_3)				20→35			Heterogeneous dispersion of irregular shaped zirconia particles	—	2.5-2.2+

% ρ_{th} : percent of theoretical density
T_r : relative tetragonal zirconia content
σ_F : flexural strength
K_{Ic} : critical stress intensity factor

* measured by indentation method
+ measured by Single Edge Notched Beam techinque

with silica to form a transitory liquid phase favouring the simultaneous occurrence of densification and reaction at low temperatures (1450–1500°C).

The reaction sintered materials are usually characterized by a coarser microstructure, but nevertheless with a high tetragonal zirconia content. This is probably due to homogeneous dispersion of intragranular and intergranular zirconia grains inside the mullite matrix. These materials have superior mechanical properties (Table I).

The sintering of very pure and ultra-fine mullite-zirconia powders synthesized by co-precipitation of inorganic precursors[12–14] or organo-metallic compounds[15–16] also yields dense materials usually with a very fine microstructure that is characterized by a regular dispersion of intergranular zirconia particles. The tetragonal zirconia content can be as high as that of the reaction sintered materials, and the flexural strength is usually higher, although the critical stress intensity factor is lower. The different results of Table I concerning the sintering of sol-gel powders show the difficulty of adjusting all of the synthesis parameters. Sol-gel methods also present the practical disadvantages of being very sophisticated and expensive compared to the reactive sintering process; however, the sol-gel materials are more promising to achieve high mechanical properties at high temperatures.

The purpose of this paper is to characterize mullite-zirconia composites prepared by reaction sintering in presence of magnesia[18] or titania[17] and by sintering of co-precipitated powders[12] and to compare their microstructures and mechanical properties.

Experimental Procedures

Preparation of Materials

Reaction-Sintered Materials

Commercial zircon*, alumina† and magnesia‡ or titania§ powders having a particle size of about a 1 micron were mixed in the following proportions: 53.5 wt% zircon, 44.7 wt% alumina, 1.8 wt% magnesia and 47.7 wt% zircon, 45.5 wt% alumina, 6.8 wt% titania. These compositions were chosen from previous studies[18–17] and correspond to materials which yield the best mechanical properties. Homogenization of the powder mixture was achieved by attrition milling in isopropyl alcohol for 1 hour. After drying and sieving, the powders were compacted by isostatic pressing into cylinders (length = 40 mm, diameter = 10 mm) which were reaction sintered at 1500°C for 15 minutes.

Sol-gel Materials

The sol-gel mullite-zirconia materials (alumina/silica = 71 wt%, 20 volume % zirconia) were prepared by Y. Kubota and H. Takagi using a co-precipitation method from an aqueous solution of aluminum nitrate, colloidal silica and zirconyl chloride.[12]

After ball-milling, these oxide powders were cold die-pressed into rectangular blocks followed by cold isostatic pressing. The compacts were then pressureless sintered at 1590°C for 3 hours.

---

*Opazir S–Quiminsa & Minerals (E).

†RC152DBM–Reynolds Chemicals (USA).

‡Mag Chem 10–Martin Marietta (USA).

§R1052 Kronos–NL Chemicals S.A.

Characterization of Materials

The sintered materials were analyzed by X-ray diffraction to determine the nature and the content of the crystalline phases. The relative tetragonal zirconia content was calculated from the Porter and Heuer's relation[19] modified by taking the zirconia atomic coordinates proposed by Smith and Newkirk.[20]

This new relation gives tetragonal zirconia content values very close to these obtained with the Toraya-Yoshimura-Somiya relation.[21]

The densification level (percent of theoretical density) was calculated from the bulk density value determined by the water immersion method and from the absolute density value. The microstructures of polished and chemically or thermally etched samples were observed by scanning electron microscopy and then analyzed following a previously developed[18,22] original method. This method gives a classical estimation of grain size and also an evaluation of the quality of the second phase dispersion. It consists of digitizing between 1000 and 2000 particles from SEM pictures of sintered material, calculating their mean diameters and the interparticle distances. These distances are then examined for increasing inspection distance. Grain agglomerates can be identified by computing a topological function (F) defined as the ratio between the grain set average diameter and the inspection distance.

Elastic modulus (E) was measured by the resonance frequency method (natural vibration). The flexural strength (σF) was determined from a 3 point bending test with a span of 15 mm and a loading rate of 0.1 mm/min for the reactive sintered specimens. A span of 30 mm and a load rate of 0.5 mm/min was utilized for the sol-gel specimens. The fracture toughness (K_{IC}) was calculated from single edge notched bean (SENB) test results (notch depth: 100 μm) for reactive sintered composites and from Knoop Diamond microhardness indentation results for the sol-gel composites.

Results and Discussion

Physical, Crystallographic and Thermal Characterization

The MgO and titania-reaction-sintered materials present densification levels higher than 98% and complete reaction with 24 and 21 volume % zirconia, respectively. The sol-gel mullite-zirconia composite with the 3 : 2 mullite stoichiometry (71 wt% alumina) and 20 volume % zirconia is completely dense, but presents a much lower tetragonal zirconia content than the reaction-sintered composites, as shown in Table II. The critical grain sizes of transformation (d_c) were calculated from the relative tetragonal phase contents and the volume distribution according to the hypothesis that the tetragonal zirconia grains are the smallest ones. The value of 0.5 μm for the sol-gel materials is much smaller than for the reaction sintered materials. This confirms the previous result of J. S. Wallace et al.[23] who reported a higher value of d_c in a reaction sintered matrix (1.2 μm) than for a mullite matrix obtained by conventional sintering (0.6 μm).

Microstructural Characterization

Examples of the material microstructures are presented in Figs. 1, 2 and 3. The micrograph of the MgO-reaction sintered composite (Fig. 1) was obtained after chemical etching which dissolved a large volume of the intergranular glassy phase. The microstructure is characterized by an intragranular and intergranular dispersion of zirconia grains (white grains) within a crosslinked structure of elongated mullite grains (grey grains). The micrographs of the titania-reaction sintered composites (Fig. 2) and the sol-gel composites (Fig. 3) containing very little glassy phase were obtained after thermal etching. Indeed, the composition for the titania-reaction

Table 2. Microstructural, physical and mechanical characteristics of different mullite zirconia composites.

	MgO-reactive sintered composite	TiO_2-reactive sintered composite	Sol-gel composite
Grain size (μm)			
mullite	6	2–4	0.6
zirconia	1.2	1.0	0.6
Grain number per unit area (μm^{-2})			
mullite	—	—	2.7
zirconia	0.3	0.5	1.2
Distance between zirconia grains (μm)	0.30→2.80	0.15→0.90	0.10→0.53
% theoretical density	98	98	100
Tetragonal zirconia content (%)	34	40	8
Transformation critical size (μm)	1.7	1.5	0.5
Young Modulus (GPa)	224 ± 4	205 ± 4	205
Flexural strength 20 °C (MPa)	321 ± 22	290 ± 42	560
Critical stress intensity factor (MPa/m)	3.7 ± 0.2	3.3 ± 0.3	3.1

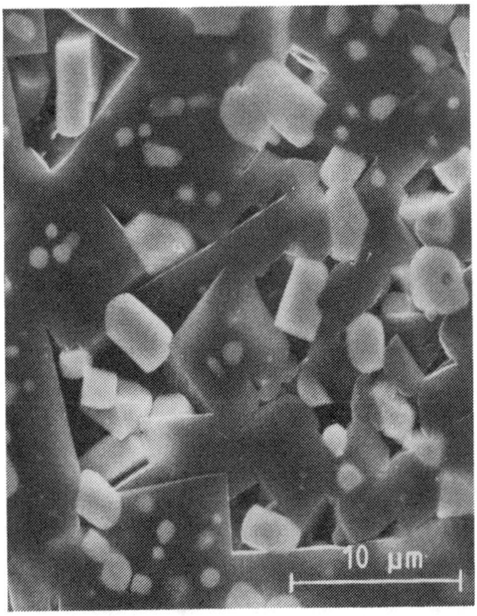

Fig. 1. Microstructure of the reaction-sintered mullite-zirconia composite containing 1.8 wt % of magnesia.

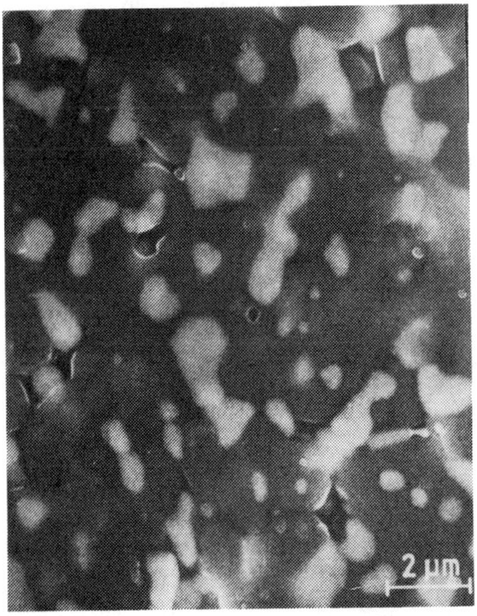

Fig. 2. Microstructure of the reaction-sintered mullite-zirconia composite containing 7 wt % of titania.

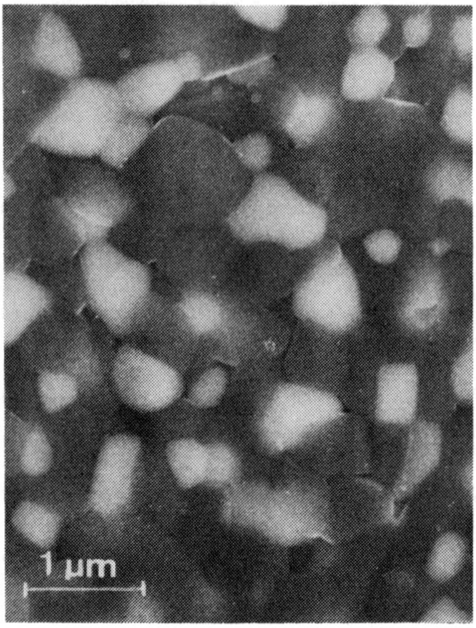

Fig. 3. Microstructure of the mullite-zirconia composite prepared from sol-gel powders.

sintered composite was chosen with the goal of preferential crystallisation of aluminum titanate within the grain boundary amorphous phase.

The material microstructural characteristics such as the average grain size, the grain number per unit area and the minimum and maximum distances between zirconia grains are presented in Table 2. It is clearly shown that the presence of an intergranular amorphous phase favours mullite and zirconia grain growth. The number of zirconia grains per unit area increases with the microstructural fineness and is therefore the largest for the sol-gel composite. Concerning the zirconia grain dispersion, the values in Table 2 and Fig. 4 of the topological function F versus the distance between particles reveals that the MgO-composite presents a heterogeneous structure with a large interparticle distance distribution, contrary to the two other materials. These three materials are, however, characterized by the presence of a few zirconia grain agglomerates consisting of joined particles. This is illustrated by the presence of a peak in the topological function for small intergrain distances. It can be noted that the maximum value of the peak increases as the grain agglomerate diameter increases. It can be concluded from these results, that the sol-gel composite produces the finest and the most homogeneous zirconia dispersion, while the MgO-composite occurs with large agglomerates of intergranular zirconia within the well dispersed intragranular zirconia.

Mechanical and Thermal Characterization

The mechanical properties of the materials are also listed in Table 2. In spite of their very different microstructures, the MgO and titania-composites have similar mechanical properties at room temperature: a Young modulus of about 210 GPa, flexural strengths of about 300 MPa and critical stress intensity factors between 3 and 4 MPa/m. The sol-gel material without porosity is characterized by a higher flexural

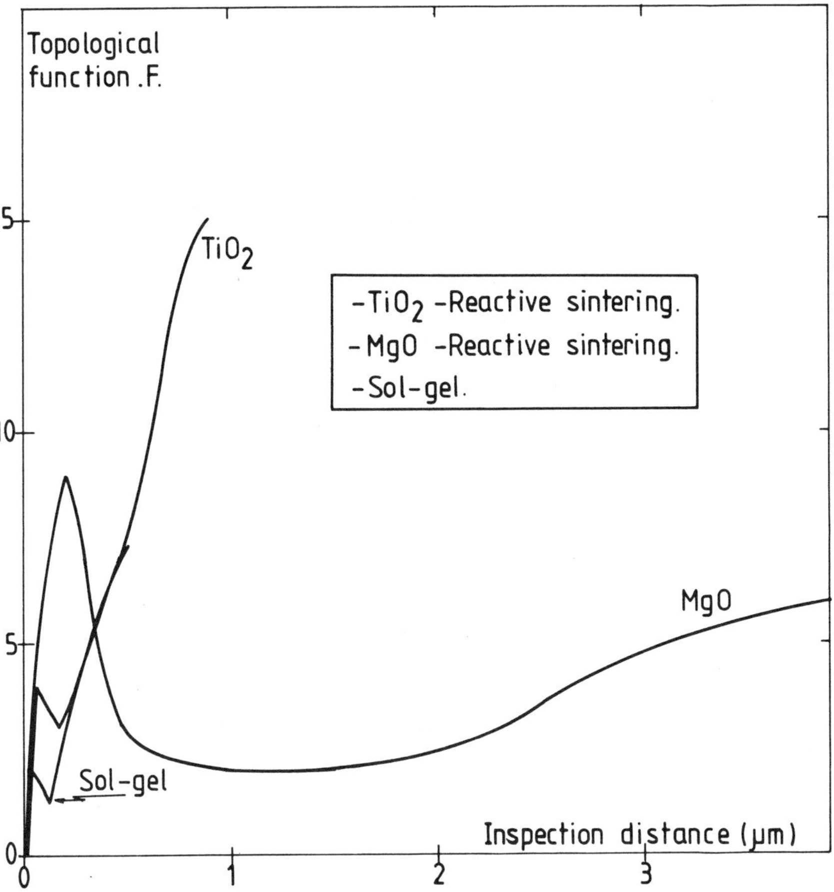

Fig. 4. Evolution of the topological function, illustrating the zirconia grain dispersion versus the inspection distance for the three materials: the MgO and titania-reaction sintered composites and the sol-gel composite.

strength. However, it also has a lower toughness value which can be related to the lower tetragonal zirconia content and the absence of a crosslinked mullite grain structure.

The high temperature changes of the flexural strengths for the three materials are shown in Fig. 5. The strength of the sol-gel material is higher than that of the reaction-sintered materials, it decreases with temperature to 900°C and then remains at a stable value of about 300 MPa to 1400°C.

The MgO-reaction sintered composite has been previously discussed by G. Orange et al.[24] who explain the increase of σ_F and of K_{IC} between 600 and 800°C by the increase of fracture energy as a result of energy dissipation through relaxation occurring within the grain boundary phase. The same authors attribute the decreases in σ_F and K_{IC} above 800°C to grain boundary sliding induced by the fluidization of

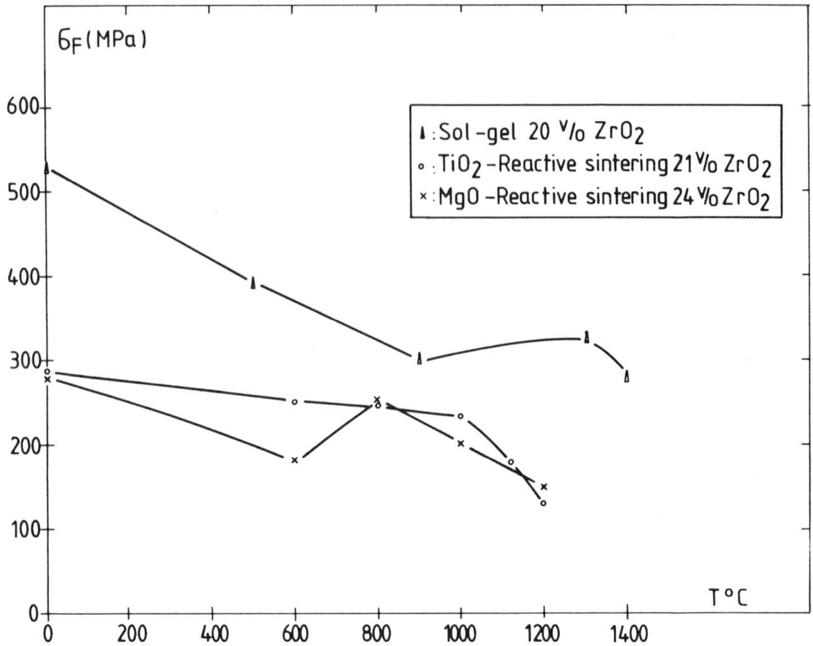

Fig. 5. Flexural strength versus temperature for the three materials: the MgO and titania-reaction sintered composites and the sol-gel composite.

the glassy phase. This phenomenon was not observed with the titania-composite which showed only a very slight decrease in σ_F to 1000°C.

Contrary to expectations, the use of titania instead of magnesia does not lead to significant enhancement of the mechanical properties of the reaction-sintered composite at high temperatures. Indeed, the titania addition does promote the crystallisation of the grain boundary glassy phase but leads to a classical mullite-zirconia microstructure without the presence of elongated mullite grain crosslinking. Therefore the reinforcement microstructural effect due to a lengthening of the fracture path and a higher proportion of transgranular fracture[24,25] does not occur.

Conclusion

In spite of their very different microstructures, the reaction-sintered mullite-zirconia composites prepared in the presence of magnesia and titania show very similar mechanical properties. In the case of the titania-composite, the boundary glassy phase crystallises as aluminum titanate, which explains the flexural strength-temperature curve. However the microstructure does not cause mullite grain crosslinking as it does for the MgO-composite. As a consequence, lower K_{IC} values do not result, in spite of the finer microstructure.

The mullite-zirconia material prepared from sol-gel powders is completely dense and homogeneous without processing defects and an intergranular glassy phase. This contributes to the high flexural strengths to 1400°C. However, intrinsic properties of the titania composite, as represented by the critical stress intensity factor, are inferior to those of the MgO-composite.

Acknowledgments

The author is especially grateful to Dr. Y. Kubota and Dr. H. Takagi who prepared the sol-gel material for characterization purpose. Thanks are also expressed to Dr. F. Cambier for suggestions regarding the manuscript.

References

[1]F. F. Lange, "Transformation Toughening," *J. Mater. Sci.*, **17**, pp. 225–54, 1982.

[2]S. Prochazka, J. S. Wallace, and N. Claussen, "Microstructure of Sintered Mullite-Zirconia Composite," *Comm. Am. Ceram. Soc.*, C-125-127, 66, 8, 1983.

[3]G. De Portu and J. W. Henney, "The Microstructure and Mechanical Properties of Mullite-Zirconia Composites," *Br. Ceram. Trans. J.*, **83**, 3, 69–72, 1984.

[4]J. S. Moya and M. I. Osendi, "Microstructure and Mechanical Properties of Mullite/Zirconia Composites," *J. Mater. Sci.*, **19**, 2909–2914, 1984.

[5]N. Claussen and J. Jahn, "Mechanical Properties of Sintered In-Situ Reacted Mullite-Zirconia Composites," *J. Am. Ceram. Soc.*, **63**, [3–4], 228–229, (1980).

[6]J. S. Wallace and N. Claussen, "Development of Phases in in-situ Reacted Mullite-Zirconia Composites" in Surfaces and Interfaces in Ceramic and Ceramic-Metal Systems. 155, 165 Edited by Joseph Pask and Anthony Evans, Plenum Publishing Corporation, 1981.

[7]A. Leriche, L. Dapra, M. R. Anseau, C. Leblud, and F. Cambier, "Microstructures and Mechanical Properties of Reaction Sintered Zirconia Toughened Mullite Ceramics," in Mechanically, Chemically and Thermally Induced Failure in Engineering Materials, pp. 341–53, Edited by J. D. Bolton and S. Hampshire, Proc. of the 2nd Irish Durability and Fracture Conference NIHE-Limerick, March 28–29, 1984.

[8]P. Pena, P. Miranzo, J. S. Moya, and S. de Aza, "Multicomponent Toughened Ceramic Materials Obtained by Reaction Sintering Part I: System Zirconia-alumina-silica-calcia," *J. Mater. Sci.*, **20**, pp. 2011–22, 1985.

[9]P. Miranzo, P. Pena, J. S. Moya, and S. de Aza, "Multicomponent Toughened Ceramic Materials Obtained by Reaction Sintering Part II: System Zirconia-alumina-silica-magnesia," *J. Mater. Sci.*, **20**, pp. 2702–10, 1985.

[10]M. F. Melo, J. S. Moya, P. Pena, and S. de Aza, "Multicomponent Toughened Ceramic Materials Obtained by Reaction Sintering Part III: System Zirconia-alumina-silica-titania," *J. Mater. Sci.*, **20**, pp. 2711–18, 1985.

[11]F. Cambier and A. Leriche, "Reactive Sintering Paths for the Synthesis of Mullite-Zirconia Composites," First International Workshop on Mullite, November 9–10, 1987, Tokyo.

[12]Y. Kubota and H. Takagi, "Preparation and Mechanical Properties of Mullites and Mullite-Zirconia Composites," 179–188, Special Ceramics 8, British Ceramic Proceedings n 37 Octobre 1986, Edited by S. P. Howlett & D. Taylor, Published by the Institude of Ceramics, UK.

[13]K. Rundgren, P. Elfving, R. Pompe, K. P. D. Lagerlof, and B. Larsson, "Comparative study of Sol-Gel Prepared Mullite and Mullite-Zirconia Composite with and without Yttria," 3rd Int. Conf. on the Science and Technology of Zirconia, Tokyo, September 9–11, 1986.

[14]M. G. M. U. Ismail, Z. Nakai, S. Somiya, "Properties of Zirconia Toughened Mullite Synthesized by Sol-Gel Method," 3rd Int Conf. on the Science and Technology of Zirconia, Tokyo, September 9–11, 1986.

[15]Q. M. Yuan, J. Q. Tan, J. Y. Shen, X. H. Zhu, and Z. F. Yang, "Processing and Microstructure of Mullite-Zirconia Composites Prepared from Sol-Gel Powders," *J. Am. Ceram. Soc.*, **69** [3] 268–69 (1986).

[16]L. Grahl-Madsen, "Preparation of Mullite-Zirconia Composites by Sol-Gel Method" (to be published).

[17]A. Leriche, M. Vivey, and F. Cambier, "Preparation of Mullite-Zirconia Ceramics by Reaction Sintering. Study of their Thermomechanical Properties" Contract number SUT-116-B. EEC Research Programme in the Materials Sector-Substitution and Materials Technologies and Ceramics 1982–1985.

[18]A. Leriche, "Influence of Processing Parameters on Mullite-Zirconia Composites Microstructure," Ph. D. Thesis, Mons, Belgium, 1982.

[19]D. L. Porter and A. H. Heuer, "Microstructural Development in MgO-Partially Stabilized Zirconia (Mg-PSZ)," *J. Am. Ceram. Soc.*, **62**, 5–6, pp. 298–305, 1979.

[20]D. K. Smith and H. W. Newkirk, "The Crystal Structure of Baddeleyite (Monoclinic Zirconia) and its Relations to the Polymorphism of Zirconia," *Acta Cryst.*, **18**, 6, pp. 983–991, 1965.

[21]H. Toraya, M. Yoshimura, and S. Somiya, "Calibration Curve for Quantitative Analysis of the Monoclinic-Tetragonal zirconia System by X-Ray Diffraction," *Comm. Am. Ceram. Soc.*, C-119-121, 1984.

[22]A. Leriche and M. Deletter, "Quantitative Estimation of Particle Dispersion within Ceramic Matrices," in Processing of Advanced Ceramics, pp. 158–167, Edited by J. S. Moya and S. de Aza, Soc. Esp. Ceram. Vidr. Arganda del Rey, Madrid, Spain, 1986.

[23]J. S. Wallace, N. Claussen, and S. Prochazka, "Properties of Mullite-Zirconia Composites Prepared by Different Processing Routes," in Symposium on Ceramic Engine Components. Hakona-Japan, 17–19 October 1983.

[24]G. Orange, G. Fantozzi, F. Cambier, C. Leblud, M. R. Anseau, and A. Leriche, "High Temperature Mechanical Properties of Reaction-Sintered Mullite/Zirconia and Mullite/Alumina/Zirconia Composites," *J. Mater. Sci.*, **20**, pp. 2533–40, 1985.

[25]F. F. Lange, "Fracture Toughness of silicon nitride as a Function of the Initial Phase Content," *J. Amer. Ceram. Soc.*, **62**, 9–10, p. 428, 1979.

Microstructures and Mechanical Properties of Mullite-Zirconia Composites Made From Inorganic Sols and Salts

K. Rundgren*, P. Elfving*, H. Tabata, and S. Kanzaki

Government Industrial Research Institute
Nagoya, Hirate-cho, Kita-ku, Nagoya 462, Japan

R. Pompe

Swedish Institute for Silicate Research
Box 5403
S-402 29 Göteborg, Sweden

Abstract

A flexural strength (four-point bending, RT) of 632 MPa was achieved for a sol-gel prepared mullite-zirconia composite (MZYB), 30 wt% ZrO_2 with 3 mol% Y_2O_3, sintered in air to 99.0% of theoretical density at 1550°C for 4 h. The retained volume fraction of tetragonal ZrO_2 on a ground surface was 94%. The presence of t-ZrO_2 particles crystallized from the intergranular phase was believed to increase the strength by a stress-induced $t \rightarrow m$ transformation in the stress field of propagating cracks. For MZYB (1600°C/4 h) with a t-ZrO_2 fraction of 52% a fracture toughness (K_{IC}) of 4.1 MPa•$m^{1/2}$ was achieved, which presumably can be explained by microcrack nucleation at transformed m-ZrO_2 particles and by a larger amount of transformable t-ZrO_2. The microstructure and the mechanical properties were closely related to the size of the starting sol particles. Mixing of alumina (particle size, 10 Å) and silica (50 Å) sols with aqueous solution of $ZrO(NO_3)_2$ and $Y(NO_3)_3$ (MZYB) resulted in a homogeneous and fine-grained microstructure with extremely low amount of residual glassy phase and improved mechanical properties.

Introduction

Today the research activities in the field of mullite ceramics is steadily increasing mainly due to the promising properties of mullite materials for high temperature applications.[1,2] However, still these materials have been reported to have lower strength and toughness compared with alumina ceramics.

***On leave from Department of Inorganic Chemistry, Chalmers University of Technology and University of Göteborg, S-412 96 Göteborg, Sweden.**

Sol-gel[1–5] and reaction sintering[6–10] methods offer two important routes for the development of mullite and mullite-zirconia composites with increased strength and toughness. Sol-gel prepared mullite and mullite-zirconia materials have been reported to have an improved microstructure and mechanical properties, due to the homogenous mixing of oxides and the low impurity level when using colloidal sols and alkoxides as starting materials.[1–3] Still, due to the drying step and a high calcination temperature used in many sol-gel methods, the amount of hard agglomerates and closed pores have been found to increase.[4,5]

In reaction sintering it is possible to improve the mechanical properties of mullite materials starting with relatively inexpensive raw materials.[6–10] However, after an in-situ reaction between zircon ($ZrSiO_4$) and alumina to form mullite-ZrO_2 bodies an essentially invariable phase composition is obtained, due to their fixed stoichiometry of the reaction. Even though mullitization and densification by this process has some shortcomings[9] Miranzo et al.[10] showed that the resulting microstructure was fine-grained and some intragranular ZrO_2 particles were found inside the mullite grains. Intragranular ZrO_2 particles were also found by Rundgren et al.[5] when inorganic sols were used as a starting material. Homogeneous mixing of oxides at a molecular level as in sol-gel methods is considered to be an important factor for the development of fine and well-dispersed ZrO_2 particles with reduced amount of residual glassy phase. When powder mixtures of mullite and ZrO_2 were used,[11] predominantly intergranular ZrO_2 particles were found and substantially less tetragonal phase was retained compared to composites prepared by reaction sintering.

The aim of the present work was to explore in more detail the powder processing step, when using a sol-gel route, and to study the effect of different sizes of the starting sol particles on the resulting microstructure and on the mechanical properties of mullite and mullite-zirconia composites.

Experimental Procedure

Starting Materials and Powder Preparation

Two series of stoichiometric mullite and mullite-zirconia composite powders were synthesized using commercial colloidal sols and salts. Powder series A (colloidal powders) was made from pure inorganic sols of silica* (particle size 50 Å), alumina† (200 Å), zirconia† (75 Å) and yttria† (100 Å) and series B (polymeric powders) was made from silica* and alumina‡ (10 Å) sols with aqueous solution of $ZrO(NO_3)_3$ and $Y(NO_3)_3$.

The silica sol was ion-exchanged with a cationic resin to remove sodium ions before mixing with the other sols. The sols in the solutions (pH ≈ 3.0 and 5 wt% in concentration), mixed for one hour, were coprecipitated with diluted aqueous ammonia (pH ≈ 9.0) and processed as illustrated in Fig. 1. The sol mixtures starting from aluminum chlorohydrate (series B) were coprecipitated with ammonia, washed with water, filtered and dissolved in nitric acid. This procedure was repeated until the resulting gel was completely free of chloride ions.

In order to remove physically bonded water and thus avoid formation of hard agglomerates, all gels were dispersed and filtered in methanol and later in acetone

*Bindzil 15-500, Eka Nobel AB, Sweden.

†Nyacol AL 20 (alumina sol) and zirconia and yttria sols, Nyacol Products Inc., USA.

‡Aluminum Chlorohydrate, Albright & Wilson, UK.

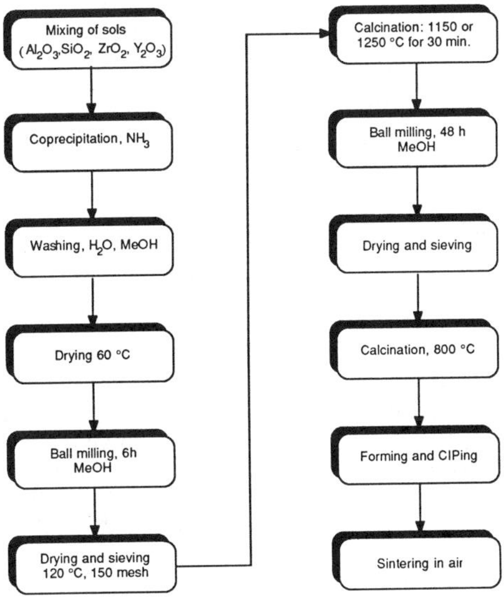

Fig. 1. Flow chart for powder processing.

and subsequently dried in acetone with a rotary evaporator. After vacuum-drying at 60°C and in oven at 120°C the powders were milled in methanol for 6 hours so as to break down hard agglomerates before calcination. The powders were calcined for 30 minutes at 1150°C for the composites and at 1250°C for the mullite powders. This procedure was carried out in order to decrease the specific surface area, from around 300 m^2/g to around 30 m^2/g, and to remove chemically bonded water and counterions (NH_4^+, NO_3^-).

The calcined powders were then milled in methanol for 48 hours in a nylon vibration mill with Si_3N_4-cylinders, dried and screened (150 mesh) for granulation. All powders were calcined at 800°C for 1 h in order to burn out organic inclusions.[12] Powder MA and MB corresponds to nearly stoichiometric mullite, with the content of 72.1 wt% Al_2O_3 for MA and 72.9 wt% Al_2O_3 for MB, determined by X-ray fluorescence analysis. MZA and MZB are mullite-zirconia composite powders with 30 wt% ZrO_2, MZYA and MZYB contains 30 wt% ZrO_2 with 3 mol% Y_2O_3. The Y_2O_3 addition was taken relatively to the amount of ZrO_2. Typical impurities determined by the ICP method are shown in Table I. The low impurity level of sodium found in powder series B is due to three consecutive ion-exchange steps of the silica sol before mixing and also washings with water.

Sintering

Green bodies, formed by die-pressing at 30 MPa and subsequent cold isostatically pressing at 200 MPa, were sintered in air between 1550 to 1650°C for 2 or 4 hours.m The sintering schedules were determined by shrinkage curves obtained in a dilatometer. The sintered plates were cut and ground with #400 diamond wheel to prepare test pieces of the size 3 by 4 by 40 mm.

Table I. Impurity level in mullite-ZrO_2 powders

Sample	Na	K	Ca	Fe	Ti	V	Zr
MA	140	100	20	60	270	20	180
MZYA	140	50	20	50	450	20	
MB	110	200	30	70	50	10	280
MZB	70	120	20	90	90	20	
MZYB	70	100	20	80	90	20	

by ICP-method, unit:ppm.

Characterization

The phase composition was determined by X-ray diffractometry (XRD). The retained volume fraction of tetragonal ZrO_2 in the composites, as-fired and ground, obtained from the ratio of intensity of the {111} tetragonal and the {111} monoclinic peaks.[13] The bulk density was measured using an immersion method in water. Theoretical densities were calculated from the density and weight percent of the pure oxides present. Young's modulus was determined by ultrasonic measurements. Flexural strength was measured by four-point bending (30/10 mm/mm span) at room temperature and at elevated temperatures up to 1300°C in air at a loading rate of 0.5 mm/min. Fracture toughness was determined using an indentation microcrack method,[14] with a load of 5 kg. Microstructural appearance of polished and etched surfaces (1% HF at 8°C for 24 hours) and fracture surfaces was examined with scanning electron microscopy (SEM). Thin specimens for transmission electron microscopy (TEM) study were prepared using standard ion thinning techniques and were subsequently investigated in a JEOL 4000 FX TEM instrument.

Results and Discussion

Powder Properties

Table II shows composition, specific surface area and mean particle size of as-processed powders. The mean particle size was found to be less than 1 μm for all powders after milling, except for MZA. The apparent particle size increased somewhat for composite powders, which could be due to the crystallization-growth of ZrO_2 and the formation of hard agglomerates during calcination.

Table II. Powder properties of calcined, 30 min., and milled, 48 h, mullite and mullite-ZrO_2 composites

Sample Composition				Specific Surface	Mean Particle
Mullite (M)	Calcination (°C)	ZrO_2 (wt%)	Y_2O_3 (mol%)	area (BET) (m^2/g)	Size (d_{50}) (μm)
MA#	1250			24.0	0.78
MZA	1150	30		67.0	1.02
MZYA	1150	30	3.0	22.0	0.99
MB#	1250			35.0	0.62
MZB	1150	30		28.0	0.81
MZYB	1150	30	3.0	21.0	0.79

#Composition of mullite. MA:72.1 wt% Al_2O_3 and MB:72.9 wt%.

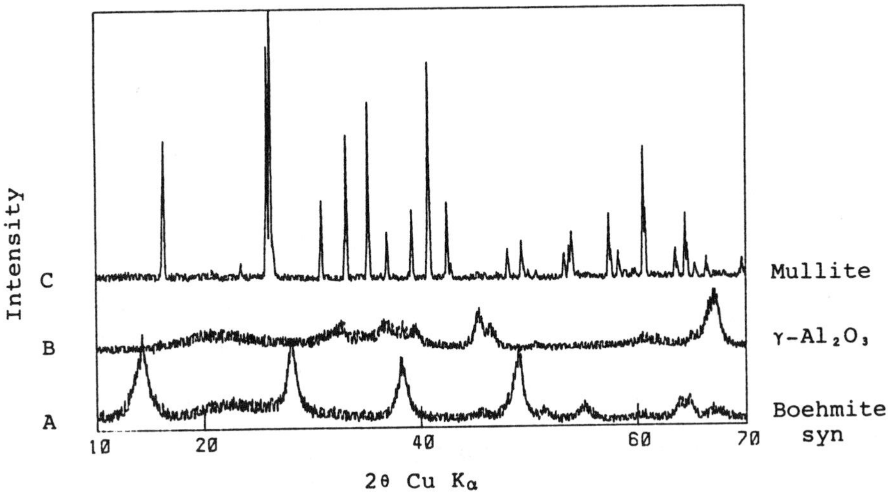

Fig. 2. X-ray diffraction patterns of; A) As-dried MA powder, B) Powder MA calcined at 1250°C for 30 min., C) Polished sample MA sintered at 1650°C for 4 h.

The colloidal powders (series A) show an exothermic peak at 1200°C in the DTA curve due to mullite formation, whereas the polymeric powders (series B) show an exothermic peak at 970°C in the DTA curve probably due to crystallization of spinel or γ-Al_2O_3. XRD patterns in Fig. 2A show that as-dried MA powder has peaks corresponding to syntethic boehmite, which originate from the starting sol particles in the Nyacol alumina sol in powder series A. As-dried powders from series B were amorphous, as shown in Fig. 3A. XRD patterns of as-processed powders, shown in Fig. 2B and 3B, suggest that a form of γ-Al_2O_3, or possibly an Al-Si spinel, along with weakly crystalline mullite and t-ZrO_2 in the composites forms at this temperature.

Sintered Bodies

Even though the green densities were only 45–50%, due to partly amorphous powders, the densification of most materials led to nearly theoretical densities. Table III summarizes some characteristics of the sintered mullite and mullite-zirconia composites. Generally, powder compacts from powder series B sintered to higher densities compared to compacts from series A. For pure mullite, MA and MB, a temperature of 1650°C was needed in order to reach nearly theoretical density.

Shrinkage curves obtained in a dilatometric study indicated an increased sintering rate for composites with only ZrO_2. As also shown by others,[11] this effect occurs at relatively low temperatures suggesting that ZrO_2 promotes densification of mullite without the formation of a liquid. This is consistent with a very low volume fraction of amorphous phase in the sintered specimens, analyzed by TEM. Similar enhancement of densification has been observed in many other binary ceramic systems and is sometimes termed as sub-eutectic sintering, its basic mechanism is not yet understood.[11]

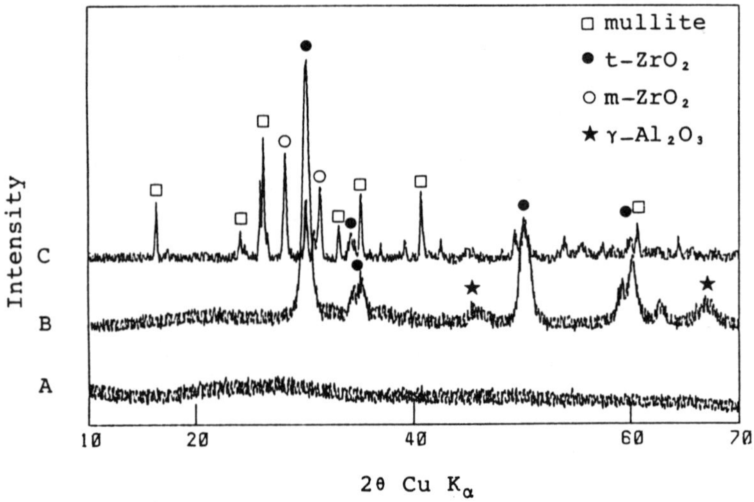

Fig. 3. X-ray diffraction patterns of; A) As-dried MZB powder, B) Powder MZB calcined at 1150°C for 30 min., C) Polished sample MZB sintered at 1600°C for 4 h.

Table III. Sintering and mechanical results of mullite and mullite-zirconia materials

Sample	Temp./time (°C/h)	Bulk Density (g/cm^3)	(%theor.)	t-ZrO_2 Fraction (%) (as-fired)	(ground)	E (GPa)	K_{IC} ($MPa \cdot m^{1/2}$)
MA	1650/4	3.11	98.5			216	1.9
	1650/2	3.04	96.2			202	2.1
MZA	1650/2	3.63	98.5	36	20	218	3.7
	1600/4	3.62	98.2	46	28	214	3.7
	1600/2	3.55	96.2	53	34	205	4.0
MZYA	1600/4	3.63	97.5	45	42	216	3.5
	1600/2	3.64	97.8	59	55	214	3.5
MB	1650/4	3.13	99.2			223	1.8
	1650/2	3.09	98.0			207	1.9
	1630/4	3.08	97.6			204	2.1
MZB	1650/2	3.65	99.0	20	14	221	3.4
	1600/4	3.60	97.7	25	20	214	3.3
	1600/2	3.51	94.2	32	20	198	3.7
MZYB	1600/4	3.62	97.4	55	52	213	4.1
	1600/2	3.64	97.9	70	67	213	4.0
	1550/4	3.68	99.0	97	94	180	3.4

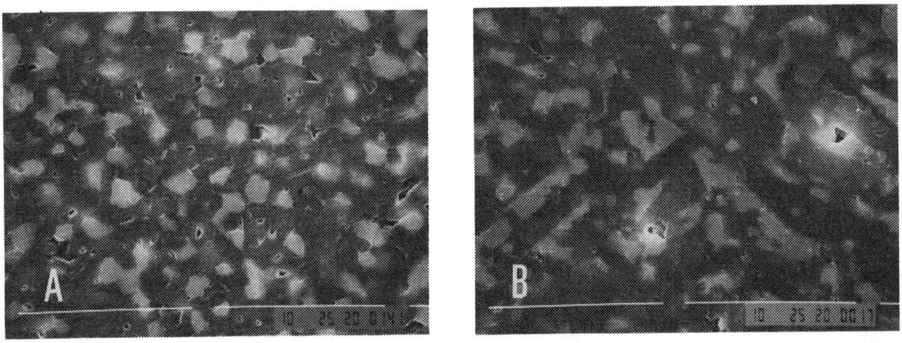

Fig. 4. Scanning electron micrograph of etched surfaces showing; A) Spherical ZrO_2 particles in sample MZYA (1600°C/2 h) and B) sample MZYB (1600°C/2 h) with sharp-edged ZrO_2 particles and whisker-like mullite grains (bar = 10 μm).

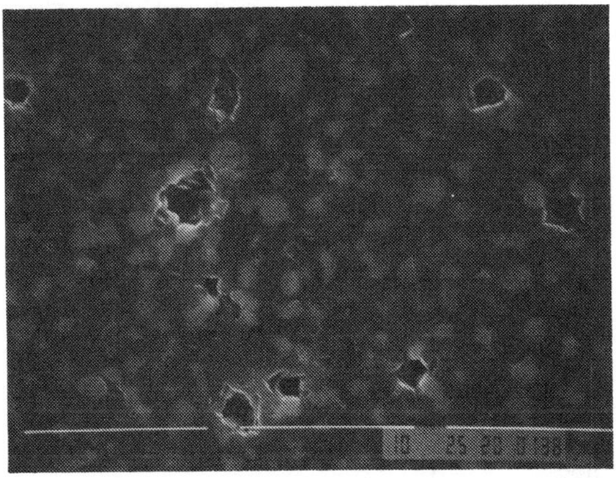

Fig. 5. Scanning electron micrograph of sample MZYA (1600°C/4 h) showing clusters of ZrO_2 with pores, probably originated from small agglomerates of ZrO_2 particles formed during drying and calcination steps (bar = 10 μm).

The bulk density increased with increasing sintering temperature and time for MZA and MZB composites. The amount of silica-rich glass was found to be larger in powder series A (colloidal powders) compared to series B (polymeric powders), as shown in Fig. 4A and B for sample MZYA and MZYB, respectively. This together with a higher impurity level and specific surface area may explain why MZA sintered to higher density than MZB at 1600°C, even though powder MZB is more fine grained than MZA.

The bulk density decreased with increasing sintering temperature and time for MZYA and MZYB composites. This indicates a liquid-phase sintering mechanism with grain and pore growth opposing the densification process at high temperatures, ≥1600°C. Intergranular ZrO_2 particles and small pores coarsened by coalescence in the presence of a liquid phase to clusters of ZrO_2 and to large pores, respectively[5,10] (see Fig. 5).

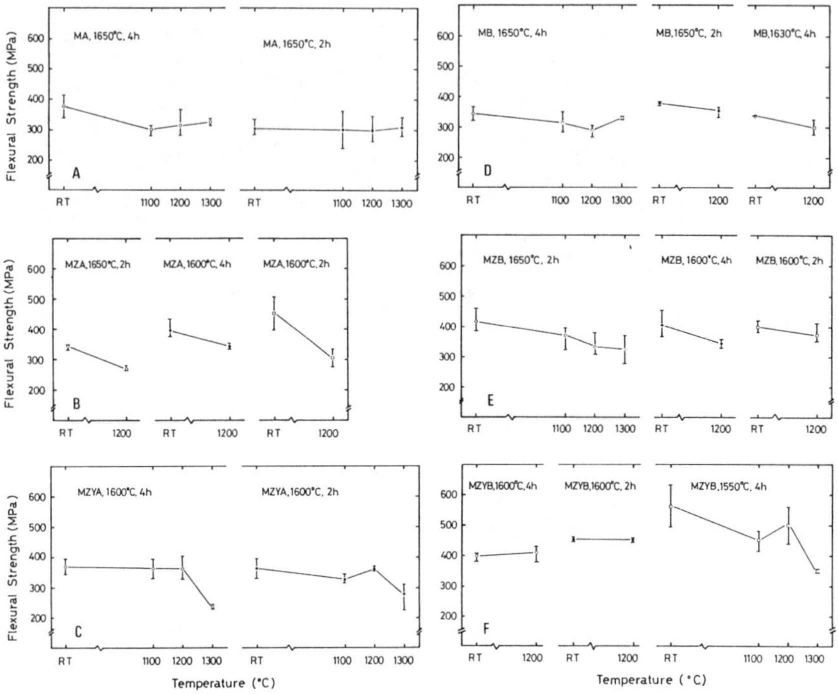

Fig. 6. Flexural strength as a function of temperature for mullite and mullite-zirconia material made from powder series A and B. The spread shown are the total spread in data for 2 to 4 specimens. A) Sample MA, B) Sample MZA, C) Sample MZYA, D) Sample MB, E) Sample MZB, and F) Sample MZYB.

Mechanical Properties

Flexural strength as a function of temperature for mullite and mullite-zirconia composites is shown in Fig. 6 A-F. The mullite specimens from series A and B, Fig. 6A and D, show retained or slightly decreased strength level from room temperature (RT) to 1300°C. This strength retention up to 1300°C, may be related to the presence of a viscous glassy phase at triple points of the mullite-mullite grain junctions. This has been described as an "apparent" strengthening of mullite at high temperatures.[1,2]

The specimens with 30 wt% ZrO_2 addition, MZA and MZB, Fig. 6B and E, show an increased strength, by approximately 25%, compared to pure mullite. The RT strength for MZA decreased with increasing sintering temperature and time, probably due to decreased amount of tetragonal ZrO_2 and grain growth. For MZB the strength at RT was slightly increased with increasing sintering temperature and increased holding time due to higher sintered densities. The strength at RT for MZYB samples decreased with increasing sintering temperature and time due to decreased amount of t-ZrO_2 and lower bulk densities as described above.

The high temperature strength of specimens with only ZrO_2 addition, MZA and MZB, decreased with increasing testing temperature, probably caused by the weakening of the grain boundary phase at elevated temperatures. For MZYA and

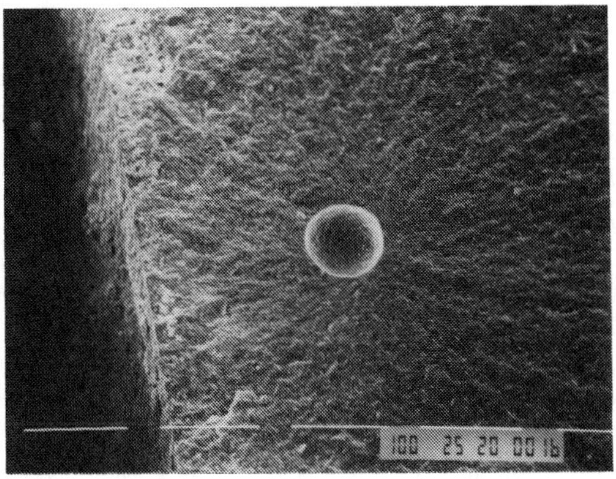

Fig. 7. Scanning electron micrograph of a fracture surface of sample MZYB (1550°C/4 h) showing a pore in the bulk. σ_b = 632 MPa (bar = 100 μm).

MZYB specimens the strength level was retained up to 1200°C. At 1300°C the strength dropped drastically, presumably due to the presence of zirconium- and yttrium-rich glassy grain boundary phase.

The fracture toughness values, K_{IC}, are shown in Table III. A K_{IC} value of 3.4 MPa•m$^{1/2}$ was achieved for sample MZYB, sintered at 1550°C for 4 h. Almost all (94%) of the ZrO_2 particles were in tetragonal form after grinding. The small size of the ZrO_2 particles makes them resistant to spontaneous transformation during cooling to RT and this suggests that no toughening by microcracking occurs. The toughness increased to 4.0 and 4.1 MPa•m$^{1/2}$ for sample MZYB with both tetragonal and monoclinic ZrO_2 sintered at 1600°C for 2 and 4 h, respectively. This can presumably be explained by a larger amount of transformable *t*-ZrO_2 particles and by a microcrack nucleation at transformed *m*-ZrO_2 particles as shown by Rühle et al.[15]

A fractographic study showed that the fracture origin was surface flaws caused by machining, and pores and agglomerates in the bulk. In pure mullite the fracture origins was mainly pores and surface defects, but for composites the relative portion of agglomerates and large closed pores increased (Figs. 5 and 7). The formation of hard agglomerates and closed pores probably occurred during the drying step of the powder and during the crystallization of ZrO_2 in the calcination step.

Microstructures

Some representative microstructures for the mullite-zirconia composites are shown in Fig. 8 for series A and in Fig. 9 for series B. The overall microstructure of samples from series B, both MZB and MZYB, was found to be much finer compared to series A, with a larger amounts of intragranular ZrO_2 particles and sharp-edged intergranular ZrO_2 particles. The average size of ZrO_2 particles are shown in Figs. 8 and 9.

SEM observation on sample MZYB (1550°C/4 h) showed that the amount of intragranular ZrO_2 particles was approximately 5 vol% with an average grain size of 0.5 μm. Samples MZYA (1600°C/2 and 4 h) show a large amount of silica-rich glass in triple junctions. This indicates that colloidal powders, i.e. powders made from relatively large boehmite sols, after mullitization have an excess of a silica-rich glassy

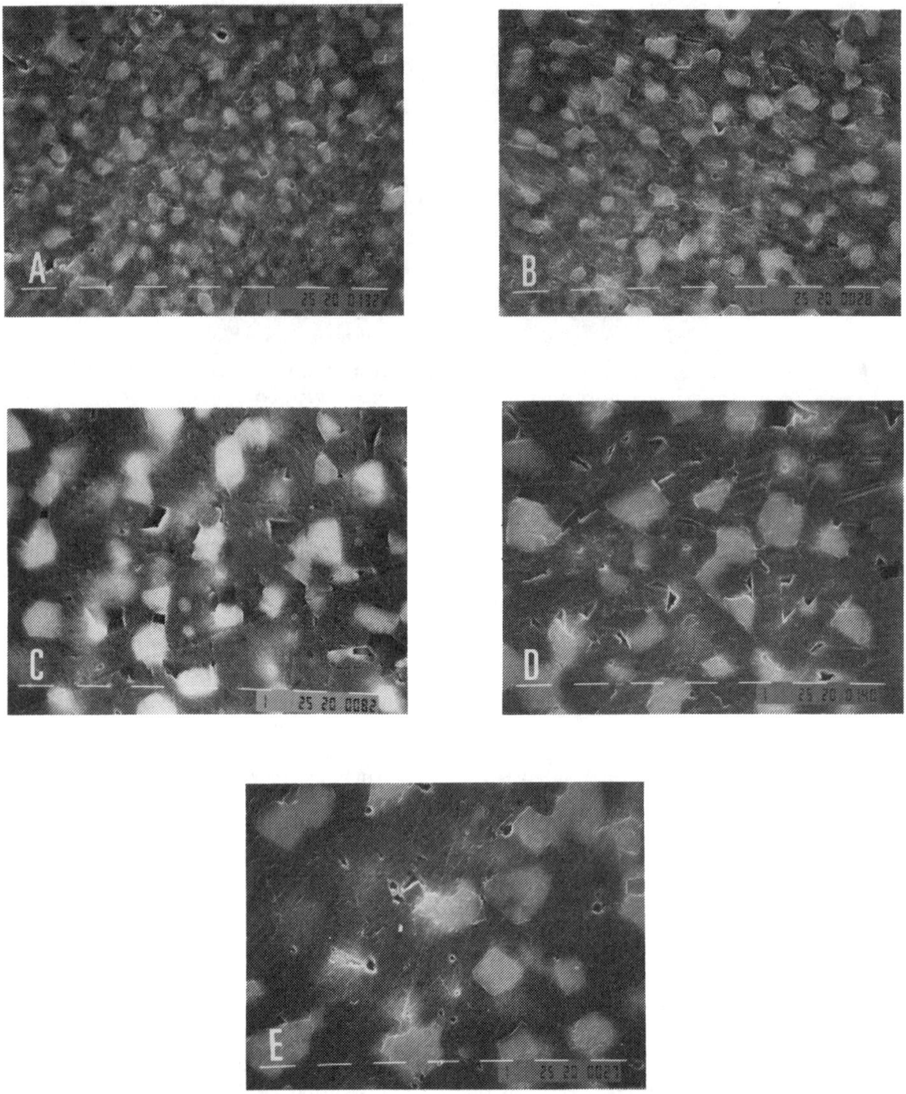

Fig. 8. Scanning electron micrographs of etched mullite-zirconia composites from series A showing average grain size (d) of dispersed ZrO_2 particles (bar = 1 μm). (A) Sample MZA (1600°C/2 h), d = 0.5 μm (B) Sample MZA (1600°C/4 h), d = 0.7 μm (C) Sample MZA (1650°C/2 h), d = 1.2 μm (D) Sample MZYA (1600°C/2 h), d = 1.0 μm (E) Sample MZYA (1600°C/4 h), d = 1.8 μm.

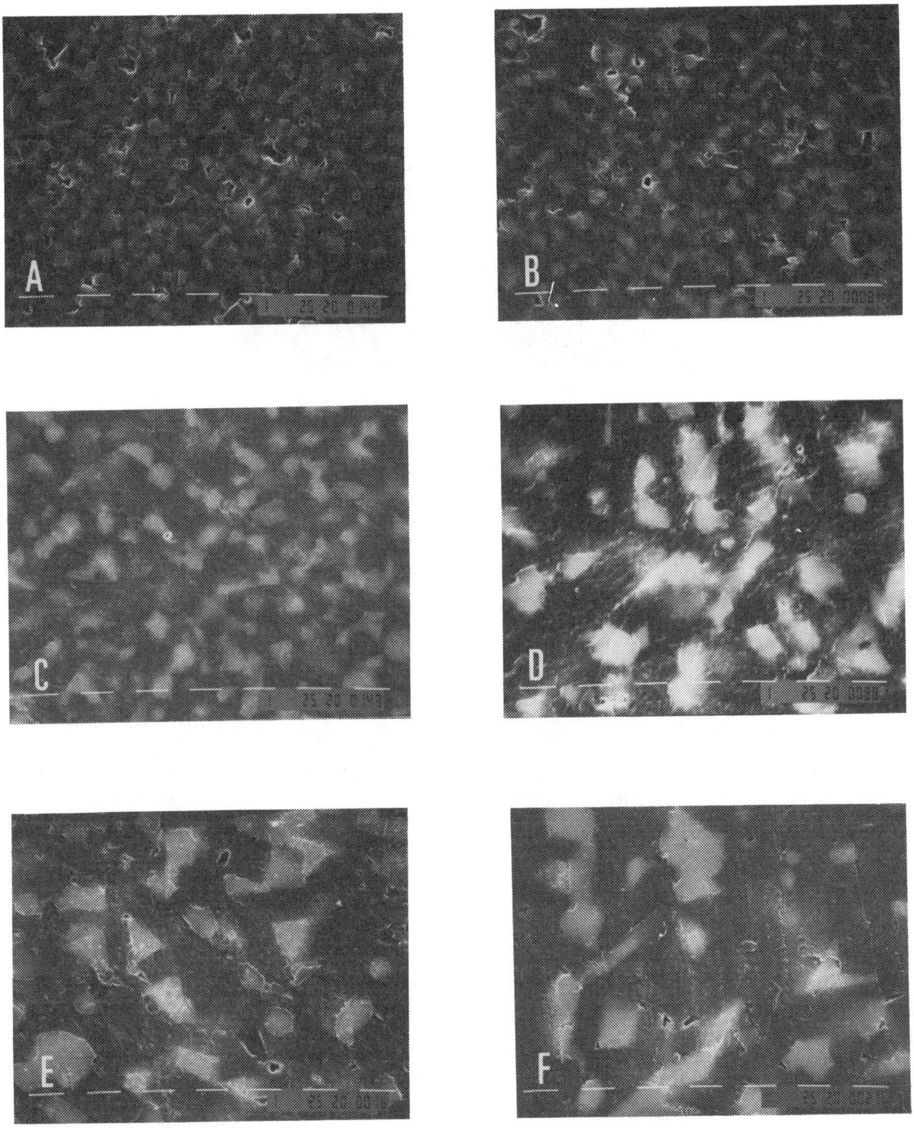

Fig. 9. Scanning electron micrographs of etched mullite-zirconia composites from series B showing average grain size (d) of dispersed ZrO_2 particles (bar = 1 μm). (A) Sample MZB (1600°C/2 h), d = 0.4 μm (B) Sample MZB (1600°C/4 h), d = 0.5 μm (C) Sample MZB (1650°C/2 h), d = 0.7 μm (D) Sample MZYB (1550°C/4 h), d = 1.0 μm (E) Sample MZYB (1600°C/2 h), d = 1.1 μm (F) Sample MZYB (1600°C/4 h), d = 1.3 μm.

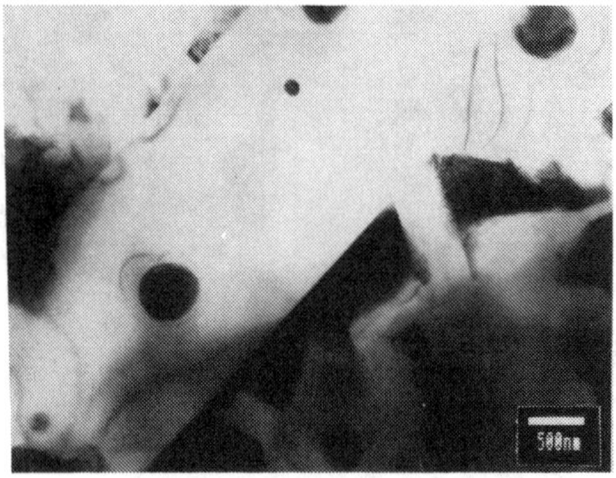

Fig. 10. Transmission electron micrograph of sample MZYB (1550°C/4 h) showing intragranular round ZrO_2 particles inside an elongated mullite grain and intergranular sharp-edged ZrO_2 particles (bar = 500 nm).

phase. Only a very small amount of residual glass was found in sample MZYB by TEM analysis, which indicates that sample series B (polymeric materials), i.e. homogeneous mixing of alumina and silica at a molecular level, results in a complete mullitization with almost ecess of silica-glass in grain boundaries.

In all samples of MZYB the ZrO_2 particles were sharp-edged and irregular in size. The ZrO_2 particles seemed often to have crystallized at the triple junctions, indicating their origin from the intergranular phase (Figs. 9 D-F, 4B and 10). The morphology of the matrix, mullite grains was found to be acicular and whisker-like, in all samples of MZYB, probably due to a large amount of glassy phase present during sintering. The average size of the mullite grains was 5 by 2 μm for sample MZYB (1600°C/2 h) shown in Fig. 4B. Elongated mullite grains were also found in silica-rich mullite materials[1,2,3] and were shown to improve the strength at elevated temperatures. The morphology of intergranular ZrO_2 particles in MZA and MZYA (Fig. 8 A-E) is equiaxed, possibly due to the addition of ZrO_2 and Y_2O_3 as large inorganic sol particles, which are thought to have easily crystallized to t-ZrO_2, and thus are believed to give only a small amount of glassy phase during sintering.

A preliminary TEM investigation of samples from series B (MB, MZB and MZYB) has been made. Thin foils of TEM samples of MZA and MZYA were easily cracked during polishing and sample preparation, probably due to large amounts of microcracks, and could therefore not be studied. Sample MB (1650°C/4 h) showed almost no residual glassy phase and the average grain size of the equiaxed mullite grains was estimated to be about 1.5 μm. Sample MZB (1600°C/4 h) in Fig. 11 shows a large amount of micro-twinned intergranular ZrO_2 particles. Twinning results from accommodative processes help to minimize strain on the particle during transformation on cooling from the firing temperature.[16] Stress lines observed in the mullite matrix around the m-ZrO_2 particles in Figs. 10 and 11 were probably induced during the t-$ZrO_2 \rightarrow m$-ZrO_2 transformation. Extremely small amount of glassy phase were observed in sample MB. For sample MZYB, with Y_2O_3 addition, the amount of glassy phase was found to increase. However, as shown in Fig. 10, the t-ZrO_2 particles formed in the triple junctions, suggest their origin from an initially large amount of amorphous grain boundary phase.

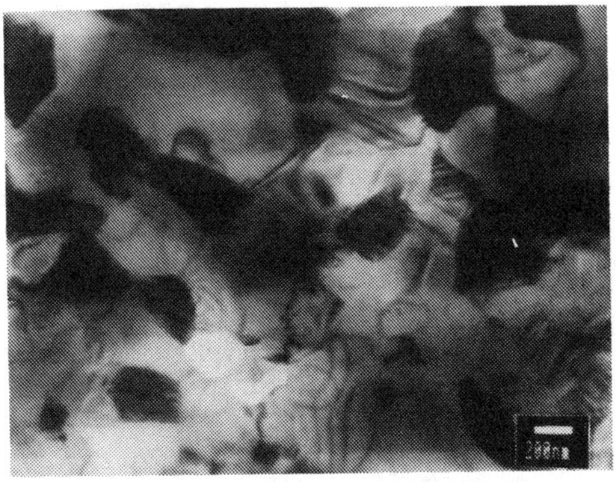

Fig. 11. Transmission electron micrograph of sample MZB (1600°C/4 h) showing stress lines around microtwinned intergranular m-ZrO_2 particles (bar = 200 nm).

Further improvement of the microstructural and mechanical properties of sol-gel prepared mullite and mullite-zirconia composites is believed to be achievable through improvement of the powder processing, i.e., circumventing the drying step from water and minimizing the amount of hard agglomerates and closed porosity formed during calcination.

Summary

1. The microstructure and the mechanical properties of mullite-zirconia composites are closely related to the size of the starting sol particles, where mixing of alumina (10 Å) and silica (50 Å) sols with aqueous solution of $ZrO(NO_3)_2$ and $Y(NO_3)_3$ (MZYB) formed a homogeneous and fine-grained microstructure with improved mechanical properties.
2. Sharp-edged intergranular ZrO_2 particles crystallized from the liquid and acicular, whisker-like mullite grains are considered to increase the strength and toughness of ZrO_2 composites.
3. Composites without Y_2O_3 densified by a presumably subeutectic sintering mechanism, i.e. an increased sintering rate with essentially no liquid phase present, whereas samples with Y_2O_3 showed a liquid-phase sintering mechanism, i.e. grain and pore growth opposing the densification process at high temperatures ≥1600°C.
4. High temperature flexural strength dropped drastically at 1300°C for composites with Y_2O_3 addition, wheras composites without Y_2O_3 showed a slowly decreasing strength at elevated temperatures.
5. An increased flexural strength with an increasing t-ZrO_2 fraction may be explained by a stress-induced $t \rightarrow m$ transformation in the stress field of propagating cracks.
6. An increased fracture toughness with an increased m-ZrO_2 fraction can presumably be explained by microcrack nucleation at transformed m-ZrO_2 particles and by an increased amount of transformable t-ZrO_2 particles.

Acknowledgments

Financial support from the Swedish Board for Technical Development (STU) is gratefully acknowledged. The authors thank Dr. T. F. Swank, Nyancol Products,

Inc., USA and Mr. B. Larsson, Eka Nobel AB, Sweden for supplying the inorganic sols. The authors also thank Mr. T. Kumazawa, Mino Yogyo Co., Ltd, Japan for his help with the TEM study.

References

[1]T. Mah and K. S. Mazdiyasni, "Mechanical Properties of Mullite," *J. Am. Ceram. Soc.*, **66** [10] 699–703 (1983).

[2]S. Kanzaki, O. Abe, M. Ohashi and H. Tabata, "Mechanical Properties of Pressureless Sintered Mullite Ceramics," in "Ceramic Materials and Components for Engines," edited by W. Bunk and H. Hausner, Verlag Deutsche Keramische Gesellschaft, Bad Honnef, FRG, pp. 625-631 (1987).

[3]B. B. Ghate, D. P. H. Hasselman and R, M. Spiggs, "Synthesis and Characterization of High Purity, Fine Grained Mullite," *J. Am. Ceram. Soc. Bull.*, **52** [9] 670–72 (1973).

[4]M. D. Sacks and J. A. Pask, "Sintering of Mullite-Containing Materials; II, Effect of Agglomeration," *J. Am. Ceram. Soc.*, **65** [2] 70–77 (1982).

[5]K. Rundgren, P. Elfving, R. Pompe, K. P. D. Lagerlöf and B. Larsson, "Comparative Study of Sol-Gel Prepared Mullite and Mullite-ZrO_2 Composites With and Without Y_2O_3," in "Advances in Ceramics," Vol. **24**, edited by S. Somiya (American Ceramic Society, 1987), to be published.

[6]N. Claussen and J. Jahn, "Mechanical Properties of Sintered, In Situ-Reacted Mullite-Zirconia Composites," *J. Am. Ceram. Soc.*, **63** [3–4] 228–29 (1980).

[7]J. S. Wallace, G. Petzow and N. Claussen, "Microstructure and Property Development of in Situ-Reacted Mullite-ZrO_2 Composites," in "Advances in Ceramics," Vol. **12**, 436–442 (1983).

[8]Baudin de la Lastra, C. Leblud, A. Leriche, F. Cambier and M. R. Anseau, "K_{IC} Calculations for Some Mullite-Zironia Composites Prepared by Reaction Sintering," *J. Mater. Sci. Letters*, **4**, 1099–1101 (1985).

[9]P. Boch, J. P. Giry and P. D. D. Rodrigo, "Reaction Sintering of Mullite and Zirconia-Mullite Ceramics," in "Ceramic Materials and Components for Engines," edited by W. Bunk and H. Hausner, Verlag Deutsche Keramische Gesellschaft, Bad Honnef, FRG, pp. 307–314 (1987).

[10]P. Miranzo, P. Pena, S. De Aza, J. S. Moya, J. Rincon and G. Thomas, "TEM Study of Reaction-Sintered Zirconia-Mullite Composites With CaO and MgO Additions," *J. Mater. Sci.*, **22**, 2987–2992 (1987).

[11]S. Prochazka, J. S. Wallace and N. Claussen, "Microstructure of Sintered Mullite-Zirconia Composites," *J. Am. Ceram. Soc.*, **66** [8] c125-c127 (1983).

[12]F. F. Lange, B. I. Davis and E. Wright, "Processing-Related Fracture Origins: IV, Elimination of Voids Produced by Organic Inclusions," *J. Am. Ceram. Soc.*, **69** [1] 66–69 (1986).

[13]P. A. Evans, R. Stevens and J. G. P. Binner, "Quantitative X-ray Diffraction Analysis of Polymorphic Mixes of Pure Zirconia," *Brit. Ceram. Trans. J.*, **83** [2] 39–43 (1984).

[14]B. Marshall and A. G. Evans, "Reply to 'Comment on "Elastic/Plastic Indentation Damage in Ceramics: The Median/Radial Crack System,"'" *J. Am. Ceram. Soc.*, **64** [12] C182-C183 (1981).

[15]M. Rühle, N. Claussen and A. H. Heuer, "Transformation and Microcracking Toughening as Complementary Processes in ZrO_2-Toughened Al_2O_3," *J. Am. Ceram. Soc.*, **69** [3] 195–97 (1986).

[16]E. Bischoff and M. Rühle, "Twin Boundaries in Monoclinic ZrO_2 Particles Confined in a Mullite Matrix," *J. Am. Ceram. Soc.*, **66** [2] 123–27 (1983).

Superplastic Flow of Mullite-Zirconia Composites*†

C. K. YOON AND I-WEI CHEN

Department of Materials Science and Engineering
The University of Michigan
Ann Arbor, Michigan 48109-2136

Abstract

Mullite-zirconia composites containing up to 50 v/o mullite have been superplastically deformed to large strains at relatively low temperatures and stresses. These composites were sintered at 1400°C and contained very fine grain sizes as low as 0.2 μm. The highest observed strain rate was 0.007 s^{-1}. At 1350°C, the flow stress of a composite containing 50 v/o mullite was 62 MPa, the strain rate was 0.001 s^{-1}. The strain rate sensitivity was higher than 0.6 in regime 2, and the grain size sensitivity was 3. The mullite grains behaved like hard inclusions in a superplastic 2Y-TZP matrix. While a threshold stress was found at lower strain rates in regime 1, regime 3 behavior was not apparent to a strain rate of 0.007 s^{-1}.

Introduction

Mullite has excellent high temperature strength. For example, single crystals of mullite stressed along the c-axis do not deform plastically at 1500°C to 900 MPa.[1] In addition, its bending strength at 1400°C is lower than that at room temperature by less than 10%.[2] Furthermore, zirconia toughened mullite has been reported to be an excellent matrix material for whisker-reinforced composites.[3] In view of these attributes, the near net shape forming technology of mullite and its composites are of potential interest.

One promising forming approach is through deformation processing in the superplastic regime. In superplastic flow, the strain rate $\dot{\epsilon}$ follows the relationship[4]

$$\dot{\epsilon} \propto \sigma^n/dP. \qquad (1)$$

where σ is the flow stress, d is the grain size, and n and p are constants. Superplastic flow differs from diffusional creep in that its strain rate is typically faster and its stress exponent, n, is typically slightly higher than that for Nabarro-Herring creep[5] or Coble creep.[6] However, the grain size dependence is similar. Due to low stress

*This research was supported by U.S. Department of Energy, Office of Basic Energy Sciences, Division of Materials Science, under grant No. DE-FG02-87ER45302.

†Presented at the First International Workshop on Mullite, Tokyo, Japan, November (1987).

sensitivity, the superplastic flow is nearly Newtonian and thus rather stable against flow localization during large deformation.[4] These attributes are thought to be highly desirable for deformation processing. Nominally stoichiometric mullite having a nominal grain size of 3 μm has been found by Dokko et al. to deform via the Nabarro-Herring mechanism at 1400 and 1500°C.[1] However, the strain rates were very low and some cavitation was observed. More recently, the feasibility of superplastic forming of fine grain ceramics below 1500°C has been convincingly demonstrated by Wakai and coworkers for zirconia[7] and for zirconia-28 v/o Al_2O_3.[8] In view of this development, we have pursued a similar approach using mullite-zirconia composites containing up to 50 v/o mullite. The procedures employed in this research, the results and a discussion of these results are presented in the following sections.

Experimental

The mullite-zirconia composites reported in this study contained a non-stoichiometric mullite (molar Al_2O_3/SiO_2 ratio = 1.77) and a yttria stabilized zirconia (with 2 m/o Y_2O_3 addition). The volume fractions of mullite in the composites were between 10 and 50 v/o. To obtain mullite powders, a solution of aluminum nitrate and tetraethoxysilane with ethanol was used, from which precipitates were formed by adding NH_4OH. The coprecipitated powders were calcined at 1000°C, then attrition-milled. Commercial 2Y-TZP powders, from TOSOH Inc., were added to the mullite powders, and the mixture was attrition-milled in water with a surfactant. In the above procedure, the alumina and silica pickup from the milling media was measured and compensated for accordingly in the starting composition. The milled slurry was pressure-cast at 95 psi into cakes, which were dried and isostatically pressed at 175 MPa. Nearly theoretical densities (less than 2% porosity) were obtained after sintering in air at 1400°C for 1 hr. Annealing at 1400°C in air for up to 20 hours was also performed to alter the grain size, which was measured by the linear intercept method. The grain size reported in this study is taken as 1.56 times the linear intercept length of zirconia grains. Mullite grain size typically scaled with zirconia grain size but was slightly smaller.

Rectangular bar specimens with dimensions of 2 x 2 x 4.6 mm were prepared for deformation experiments. Compression tests between 1250 and 1350°C were conducted in air in a platinum furnace using SiC platens and an alumina spherical seat. Specimens were deformed at constant displacement rate conditions. Since steady-state deformation could be reached with strains of the order of 0.01 to 0.02, a single specimen was used for flow stress measurements at two or three increasing displacement rates. Deformation was terminated after a height reduction of 50%.

Results

The microstructures of the as-sintered mullite-zirconia composites revealed well-dispersed mullite and zirconia (Y-TZP) grains, as shown in representative micrographs in Fig. 1. X-ray diffraction patterns showed only mullite and tetragonal ZrO_2 phases in the as-sintered, annealed, and deformed specimens. No solid solution between the constituents of mullite (Al and Si) and those of Y-TZP (Zr and Y) was detected. The grains of mullite in the present composition were essentially equiaxed. As evident from Fig. 1, at 50 v/o Y-TZP, the mullite and zirconia grains were nearly identical in size, and the zirconia grain size was the finest. At lower fractions of mullite, however, the zirconia grains were coarser and consistently larger than the mullite grains.

Deformation data is shown in Figs. 2–3 in terms of true stress σ and true strain rate $\dot{\epsilon}$, for composites containing 50 and 10 v/o mullite, respectively, and deformed

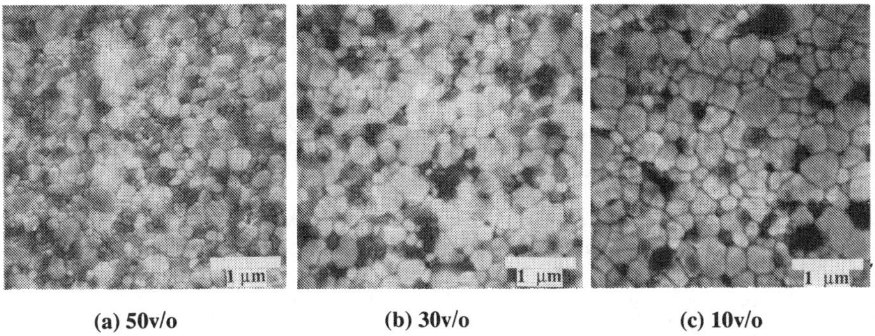

Fig. 1. Microstructures of as sintered mullite-zirconia composites. (a) 50v/o (b) 30 v/o (c) 10v/o mullite.

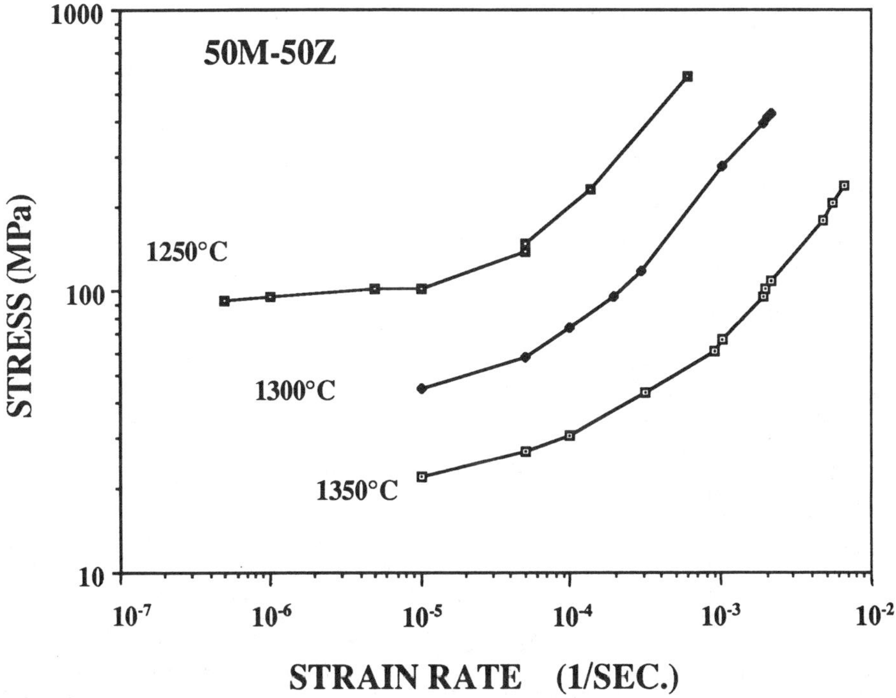

Fig. 2. Stress-strain rate curves for mullite containing 50v/o Y-TZP at different temperatures.

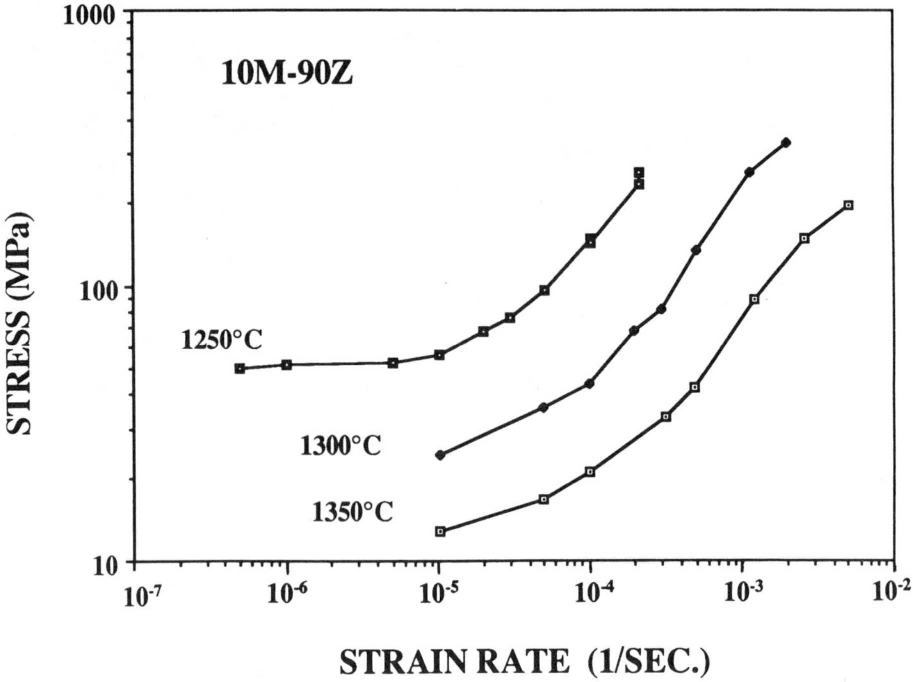

Fig. 3. Stress-strain rate curves for Y-TZP containing 10 v/o mullite at different temperatures.

at three different temperatures. The highest observed strain rate was 0.007 s^{-1}. The strain rate sensitivity, defined as

$$m = \partial \ln\sigma / \partial \ln\dot{\epsilon} = 1/n. \quad (2)$$

is the slope of the log-log curves in Figs. 2 and 3. The maximum value of m approached 0.70 in samples containing 50 v/o mullite and 0.85 in samples containing 10 v/o mullite. As mentioned before, high m values are characteristic of superplastic flow in fine grain polycrystals. An m value approaching unity is considered desirable for deformation processing, since it confers flow stability. For comparison, Wakai et al. found m = 0.5 in a Y-TZP ceramic and its alumina-containing composite.[7,8]

The high m regime (the so-called "regime 2" in the literature)[4] is of the most importance for superplastic forming in both materials. However, the results in this research at lower stresses and therefore lower m values were also of interest. In this latter regime the strain rate sensitivity increased rapidly as the strain rate increased. This phenomenon gave rise to an appearance of a threshold flow stress below which deformation was very slow (i.e. the so-called "regime 1" in the literature).[4] The value of the apparent threshold stress is dependent on both temperature and microstructure. At very high stresses, it is also expected that dislocation creep, which has an m value lower than 0.33, is operative (i.e. the so-called "regime 3" in the literature).[4] This trend was reflected only in the 10 v/o mullite samples. As mentioned before, at low

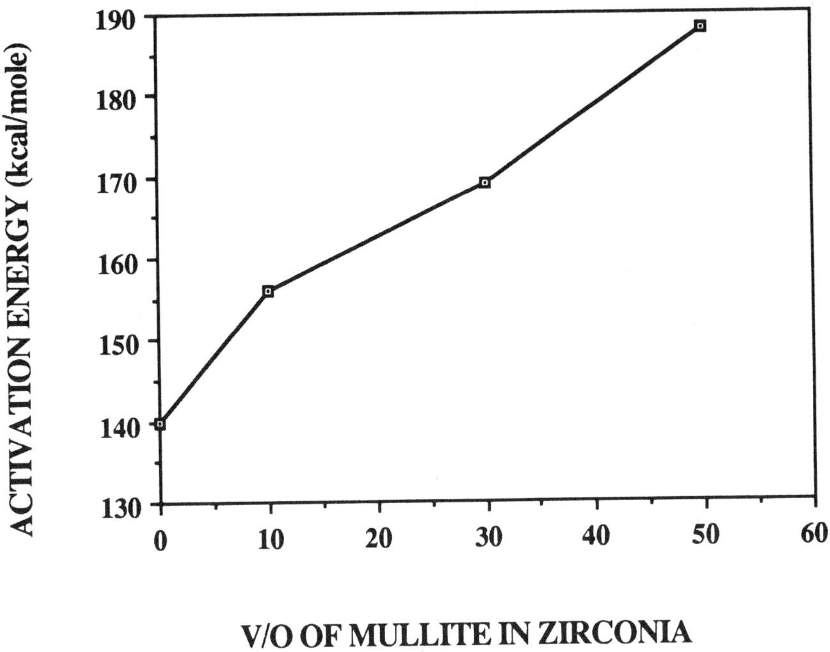

Fig. 4. Activation energy of superplastic flow versus volume fraction of mullite.

volume fractions of mullite the grain size was somewhat larger, which is less favorable for superplastic flow (see below). In all other materials, the regime 2 with high m value extended to the highest strain rate.

The dependence of superplastic flow on temperature is apparent in Figs. 2 and 3. The activation energies lie between 155 Kcal/mole for Y-TZP-10 v/o mullite and 187 Kcal/mole for Y-TZP-50 v/o mullite. The reported activation energy of superplastic flow of Y-TZP containing 3m/o Y_2O_3 is 140 Kcal/mole.[7] The dependence of the activation energy on mullite content is summarized in Fig. 4.

The dependence of superplastic flow on annealing time at 1400°C is shown in Fig. 5. The strain rates decreased with longer annealing times due to a larger grain size produced by annealing. The grain size sensitivity, defined as

$$p = -\partial \ln \epsilon / \partial \ln d. \quad (3)$$

was found to be 3, using the grain size of Y-TZP as a reference. As can be seen from Fig. 6, the grain size sensitivity is independent of the volume fraction of mullite. However, it differs from the value, p = 2, reported by Wakai et al. on Y-TZP.[7] Both p = 2 and 3 are characteristic of superplastic flow.[4]

To assess the effect of the volume fraction of mullite on superplastic flow, data on strain rates was compensated, by normalizing the zirconia grain size of each composite against the zirconia grain size of 50 v/o Y-TZP composite, according to Eqn (1) and p = 3, and replotted in Fig. 7. It is clear from Fig. 7 that the presence of mullite had a definite effect on lowering the strain rate, or, equivalently, on

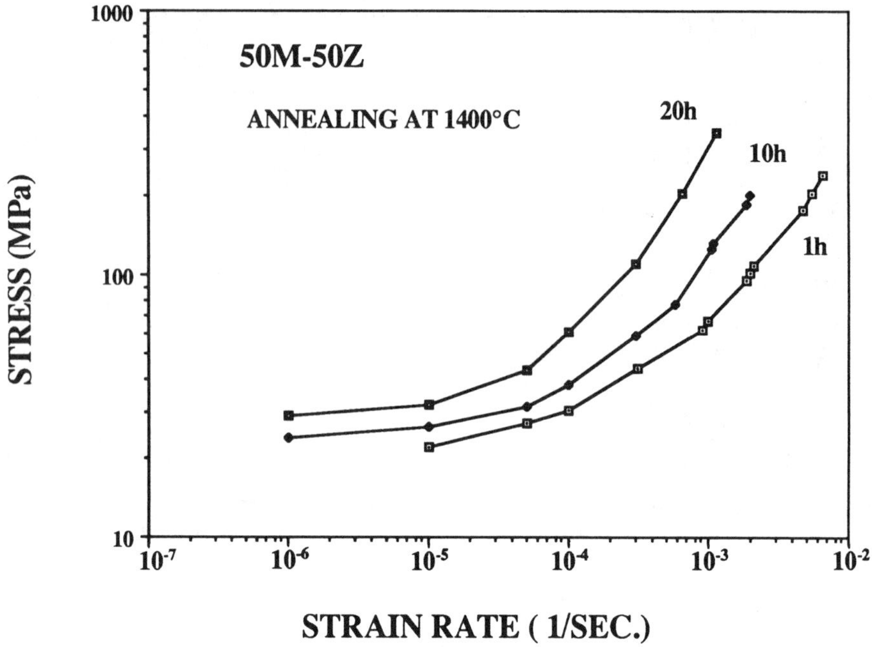

Fig. 5. Stress-strain rate curves at 1350°C for mullite containing 50 v/o Y-TZP annealed at 1400°C for different time.

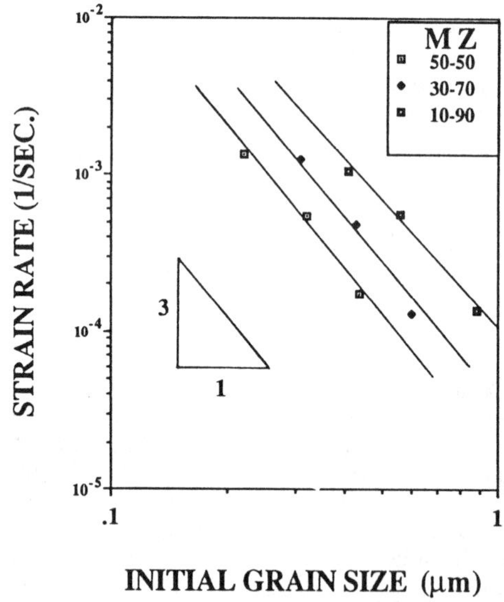

Fig. 6. Grain size dependence of strain rate at different mullite fractions.

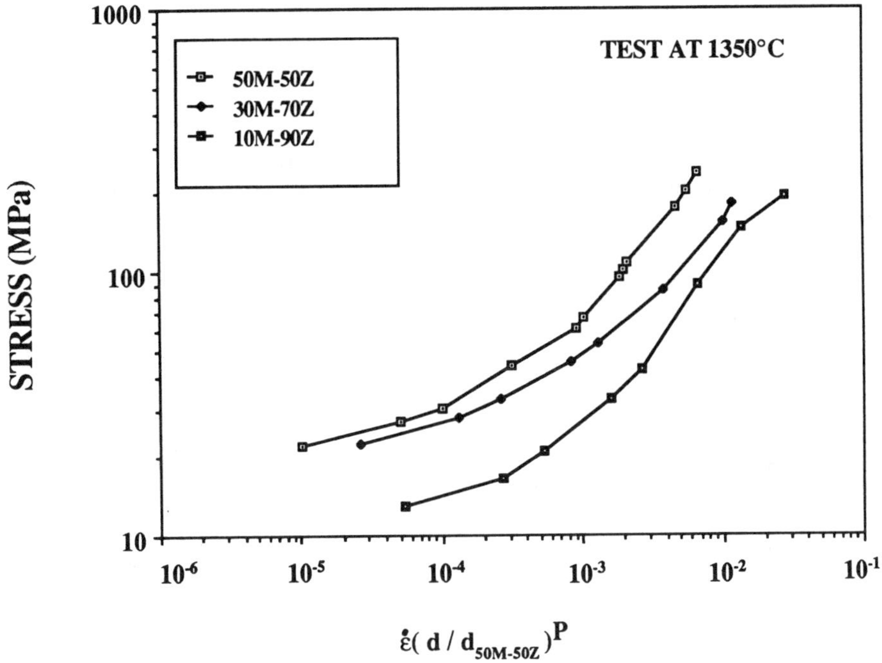

Fig. 7. Grain size compensated strain rates, showing the effect of mullite fractions.

elevating the flow stress. The threshold stress also increased with the amount of mullite. Such behavior is characteristic of superplastic flow of two-phase composites, in which the matrix flows superplastically but the included phase is essentially rigid.[9] Under such circumstances, the strain rate was predicted to depend on the volume fraction of the rigid phase, ρ, according to the following relation[9]

$$\dot{\epsilon} = \dot{\epsilon}_o (1-\rho)^{2+1/2m}. \tag{4}$$

In the above, $\dot{\epsilon}_o$ is the strain rate of the matrix phase under the same stress and temperature. The data of Fig. 7 are in reasonable agreement with the above prediction. From such observations, it is concluded that the mullite grains essentially behave like rigid inclusions in a superplastically flowing Y-TZP matrix.

Using the above information on flow characteristics, it is also possible to represent the data of different temperatures and grain sizes in terms of a normalized strain rate. Here the strain rates at different grain sizes were compensated according to Eqn (1) and p = 3, and the data at different temperatures was compensated according to Arhennius Law and the activation energies of Fig. 4. This is shown in Fig. 8 for Y-TZP containing 10 v/o mullite. As before the strain rate sensitivity reached a maximum of 0.6 for this material. Note that this value is somewhat smaller than that deduced from Fig. 2. Note particularly in Fig. 8 the varying magnitude of the threshold stress, which increases with grain size and decreases with temperature.

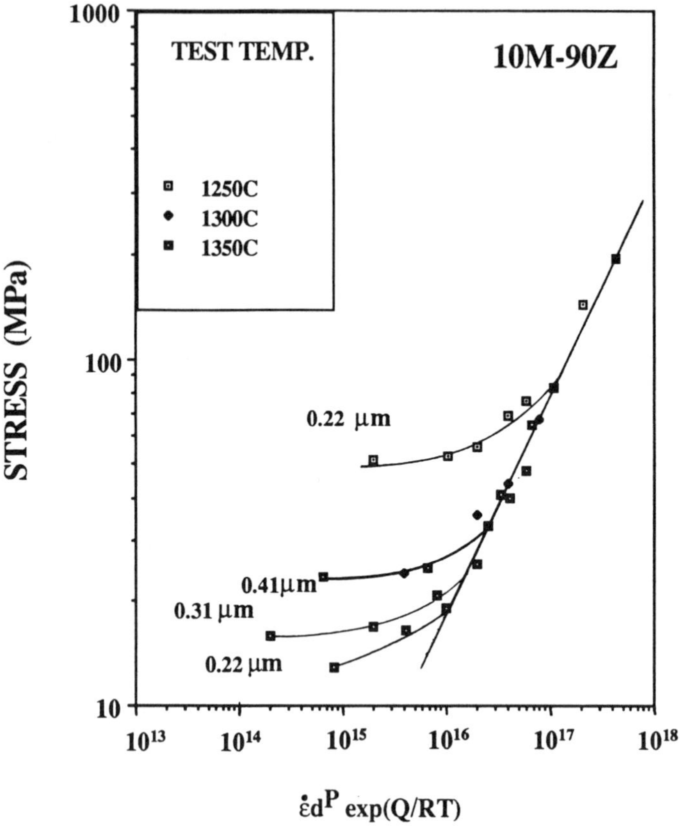

Fig. 8. Grain size and temperature compensated strain rates, for Y-TZP containing 10 v/o mullite in regime 1 and 2. Threshold stress is dependent on grain size and temperature.

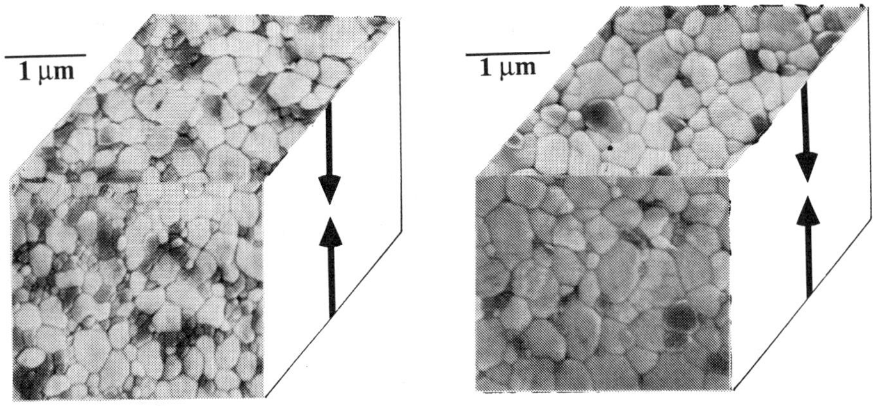

(a) 50M-50Z (b) 10M-90Z

Fig. 9. SEM microstructures of mullite-zirconia composites deformed at 1350°C; (a) 50M-50Z, $\epsilon = -0.7$ and $\epsilon = 6 \times 10^{-3}\ s^{-1}$. (b) 10M-90Z, $\epsilon = -0.7$ and $\epsilon = 5 \times 10^{-3}\ s^{-1}$. Arrows indicate loading direction.

Such behavior is contrary to the theoretical predictions in the literature of superplasticity and diffusional creep.[4]

Microstructural examination of the deformed specimens showed no apparent cavitation at a resolution of 10 nm available in a scanning electron microscope. Little grain growth was observed. Heavily compressed specimens showed very little barrelling, indicating the absence of significant friction between the specimens and the platens. A representative set of micrographs illustrating these features is shown in Fig. 9. Lastly, we found a pronounced crystallographic texture developed for both the Y-TZP matrix and the mullite phase after 50% compression. A more detailed description of the microstructural evolution during superplastic flow of mullite composites will be published elsewhere.

Discussions

The present study demonstrates that mullite-zirconia composites containing up to 50 v/o mullite can be superplastically deformed at high rates at relatively low temperatures. At a strain rate of 0.001 s^{-1}, the flow stress for the mullite-50 v/o TZP composite at 1350°C was as low as 63 MPa. In comparison, the reported superplastic flow rate of Y-TZP containing 28 v/o alumina[8] at the same temperature and flow stress was only 0.00005 s^{-1}; or, to reach a comparable strain rate at the same flow stress, the deformation temperature must be raised to 1450°C. Indeed, at 1350°C and 60 MPa, the strain rate of the mullite composite containing 50 v/o Y-TZP was three times faster than that of the Y-TZP studied by Wakai et al.[7] Similar comparisons with superplastic flow of alumina[10] and alumina-zirconia composites[11] also showed the mullite-zirconia composites reported here to possess a relatively low temperature and a relatively high strain rate at a comparable flow stress. Such attributes are favorable for superplastic forming.

The above results raise the realistic possibility of superplastic forming of mullite-zirconia composites. A strain of 100% can be achieved in 15 minutes at 1350°C at a flow stress of 60 MPa. A shorter forming time or a lower forming stress can be achieved by raising the temperature. These promising results were achieved in the present materials by attaining a finer grain size, which is possible by employing a relatively low sintering temperature (1400°C). Since the superplastic flow rate was inversely dependent on the grain size to the power of 2 ~ 3, the consequence of grain refinement is very significant. Factors controlling sinterability and grain size primarily reside with powder and green body characteristics which are beyond the scope of the present paper. It suffices to state that, with further advances in reducing the sintering temperature and refining the grain size, additional improvements in the formability of superplastic ceramics are definitely possible.

One obvious concern is the *in situ* grain growth which could occur during superplastic deformation.[10] With an increasingly larger grain size, the flow stress is expected to increase during forming. For a typical forming operation involving a strain of 100% over a time span of 15 minutes, grain growth by a factor of 2 with an attendant hardening of the flow stress by a similar factor might be expected. Fortunately, the two-phase ceramic composite studied here appears to be relatively resistant to flow-enhanced grain growth, at least to a strain of 0.5.

Although the flow mechanisms of superplasticity have been extensively discussed in the past two decades, the issue has not been resolved and there is no *ab initio* theory capable of explaining both the magnitude of the very fast strain rate and the microstructural evolution accompanying the very large deformation.[4,9] In view of this situation, we will not venture to speculate the mechanism here either. On the other hand, several micromechanical aspects of superplasticity, such as cavitation and strain localization, are becoming better understood. In this regard, the recently developed rheological model of two phase flow appears to account for the deformation behavior of several superplastic alloys (such as α-β brass),[9] and, as shown by the present study, appears applicable to the two phase ceramics as well. Despite the lack of a more basic understanding of the flow mechanisms, these advances are still significant from the viewpoint of process modelling and control.

Conclusions

1. Fine grain mullite-zirconia composites containing up to 50 v/o mullite have been superplastically deformed to large strains at 1250 to 1350°C. The highest observed strain rate was 0.007 s^{-1}.
2. In regime 2, the strain rate sensitivity was higher than 0.6, and the grain size sensitivity was 3. The mullite grains behaved like hard inclusions in a superplastically flowing Y-TZP matrix.
3. A threshold stress was found at lower strain rates in regime 1. The threshold increases with increasing grain size, increasing mullite volume fraction, and decreasing temperature. Behavior of regime 3, i.e. a low strain rate sensitivity, was not apparent to a strain rate of 0.007 s^{-1}.
4. No grain growth, grain elongation, or cavitation was evident after a strain of 0.5, even for an initial grain size of 0.2 μm.

References

[1]P. C. Dokko, J. Pask, and K. S. Mazdiyasni, "High Temperature Mechanical Properties of Mullite Under Compression," *J. Am. Ceram. Soc.*, **60**, [3–4], 150–155 (1977).

[2]S. Kanzaki and H. Tabata, "Sintering and Mechanical Properties of Stoichiometric Mullite," *J. Am. Ceram. Soc.*, **68** [1] C-6–C-7 (1985).

[3]T. N. Tiegs, P. F. Becher, and P. Angelini, "Processing and Resulting Microstructures and Properties of Whisker Reinforced Mullite Composites," Presented at the First International Workshop on Mullite, Tokyo, Japan, Nov., (1987).

[4]J. W. Edington, K. N. Melton, and C. P. Cutler, "Superplasticity," pp. 61–170 in *Progress in Materials Science*, **21** (1976).

[5]C. Herring, "Diffusional Viscosity of a Polycrystalline Solid," *J. Appl. Phys.*, **21** [5] 437–445 (1950).

[6]R. L. Coble, "A Model for Boundary Diffusion Controlled Creep in Polycrystalline Materials," *J. Appl. Phys.*, **34** [6] 1679–1682 (1963).

[7]F. Wakai, S. Sakaguchi, and Y. Matsuno, "Superplasticity of Yttria-Stabilized Tetragonal ZrO_2 Polycrystals," *Adv. Ceram. Mater.*, **1** [3], 259–263 (1986).

[8]F. Wakai, H. Kato, "Superplasticity of TZP/Al_2O_3 Composite," *Adv. Ceram. Mater.*, **3** [1], 71–76 (1988).

[9]I-W. Chen, "Superplastic Flow of Two-Phase Alloys," pp. 5.1–5.20, in *Superplasticity*, International Conference on Superplasticity–Grenoble, eds. B. Baudelet and M. Suery, Editions du C.N.R.S, Paris (1985).

[10]K. R. Venkatachari and R. Raj, "Superplastic Flow in Fine-Grained Alumina," *J. Am. Ceram. Soc.*, **69** [2] 135–138 (1986).

[11]B. J. Kellett and F. F. Lange, "Hot Forging Characteristics of Fine-Grained Al_2O_3 and Al_2O_3/ZrO_2 Ceramics," *J. Am. Ceram. Soc.*, **69** [8] C172–173 (1986).

Creep of Hot-Pressed SiC Whisker Reinforced Mullite

R. D. NIXON, S. CHEVACHAROENKUL, AND R. F. DAVIS

North Carolina State University
Department of Materials Science and Engineering
Box 7907
Raleigh, NC 27695-7907

T. N. TIEGS

Metals and Ceramics Division
Oak Ridge National Laboratory,
Oak Ridge, TN 37831

Abstract

The steady-state, constant compressive stress creep behavior of hot-pressed unreinforced and 20 vol% SiC-whisker reinforced mullite was studied in Ar within the temperature and stress ranges of 1450 to 1700K and 5–220 MPa, respectively. Both materials showed three regimes of temperature: 1450–1560K, 1560–1630K and 1630–1700K in which the activation energy of creep was different (385 kJ/mol, 1212 kJ/mol, and 866 kJ/mol, respectively, for the composite and 357 kJ/mol, 1051 kJ/mol and 754 kJ/mol, respectively, for the unreinforced material). The stress exponent for the unreinforced mullite was approximately 1 for all temperature regimes; for the composite this parameter increased from approximately 2 in the lowest temperature regime to 3 at 1700K. Transmission electron microscopy (TEM) studies as well as the low activation energy values obtained below 1560K indicated that creep in the lowest temperature regime was controlled by the viscous flow of the SiO2-rich intergranular glass phase. With increasing temperature to 1630K, extensive cavitation also occurred. Cavities were created in the glass phase at triple points and between mullite-mullite and whisker-mullite boundaries, leading to higher activation energy values as a result of the increase in free surface area. Above 1630K, no cavities were observed, indicating that the viscosity of the glass decreased sufficiently such that the cavities created at lower temperatures were filled by the intergranular phase. The decrease in activation energy in this temperature regime supports this result.

Similar research on a mullite without an amorphous boundary phase in the temperature and stress ranges of 1600–1700K and 20–300 MPa, respectively, revealed an activation energy of 748 kJ/mol and stress exponent value of 1–1.2. This data as well as Scanning Electron Microscopy (SEM) and TEM studies were

consistent with steady-state creep controlled by grain boundary sliding accommodated by lattice diffusion and cavitation.

Introduction

Mullite ($3Al_2O_3$–$2SiO_2$) possesses good chemical stability in an O_2 containing atmosphere making it a useful material for high temperature applications. Its orthorhombic crystal structure results in limited slip systems. However, when consolidated, there is usually a residual intergranular glass phase present which limits the high temperature strength and increases the deformation rate of this material. To overcome this problem, one must attain a closely controlled chemical mixture of Al_2O_3 and SiO_2 (or Al, Si and O) such that during sintering the composition occurs and remains solely within the mullite phase field.

In previous research, Dokko et al.[1] conducted constant compressive stress creep experiments on stoichiometric mullite (prepared from alkoxy-derived powders and hot-pressed to theoretical density) in air at a temperature of 1673K in the stress range of 9.8 MPa-38.2 MPa. According to the authors, this material was "presumably free of glass" on the grain boundaries; however, acid etching resulted in "severe etching on a localized scale, suggesting the possible presence of an amorphous structure along such boundaries." Under the above conditions, the stress exponent was determined to be approximately 1. A 25K cycling from 1673K resulted in an activation energy of 711 kJ/mol. In addition, the grain size dependence of creep was determined to be approximately 2. While these exponents are consistent with the Nabarro-Herring creep mechanism,[2,3] SEM studies revealed cavitation at grain boundaries and at triple points in the crept specimens. The authors concluded that the primary mechanism of creep was grain boundary sliding accommodated by lattice diffusion. Creep studies on mullite were also performed by Lessing et al.[4] They used a 2.7 μm grain size material, deformed by four-point bending under dead load conditions in air in the temperature range of 1623–1723K under the same stresses as noted above. The stress exponent and activation energy were determined to be 1.0 and 685 kJ/mol, respectively.

Creep experiments have also been performed on mullite materials known to contain a glass phase on the grain boundaries. Kumazawa et al.[5] have performed tensile creep experiments in air in the temperature range of 1623–1723K on hot-pressed mullite containing 56 to 68 mole% Al_2O_3. The stress exponents of creep changed with Al_2O_3 composition: n = 1.9 (56 mole%), n = 1.6 (60 mole%) and n = 1.8 (68 mole%). The two lowest Al_2O_3 compositions contained needle shaped (long and thin) grains and a SiO_2-rich glass on the grain boundaries. The highest Al_2O_3 composition possessed rod shaped (short and fat) grains and corundum at the grain boundaries. The activation energies of creep for all compositions in this temperature range were within the range of 830 to 960 kJ/mol. Four-point bend experiments in air have also been conducted by Penty and Hasselman on 99% dense vacuum-pressure sintered mullite[6] in the temperature and stress ranges of 1703–1785K and 7–70 MPa, respectively. Their studies resulted in an activation energy value of 698 kJ/mol and stress exponent values of 1–1.4. Furthermore, Hulse and Pask[7] have performed tensile creep experiments on mullite containing 63 mole% Al_2O_3 in the temperature range of 1589–1694K and obtained an activation energy of 740 kJ/mol. All of the kinetic data above for mullite containing a glass phase was indicative of creep controlled by a grain boundary sliding mechanism.

Since the glassy phase in mullite seriously degrades its creep resistance by allowing grain boundary sliding to occur at relatively low temperatures, it seemed important to explore whether whisker reinforcement would be effective in improving creep resistance. SiC whiskers have been shown to significantly reduce the steady-state creep rates in polycrystalline Al_2O_3.[8] Therefore, the same reinforcing agents were chosen here. In the present research, constant compressive stress creep experiments have been conducted over a wide range of stresses and temperatures on a hot-pressed unreinforced and SiC whisker-reinforced mullite containing a grain boundary amorphous phase. For comparative purposes, similar experiments have been conducted on a mullite material containing no whiskers and no boundary phase. Extensive transmission electron microscopy (TEM) and limited scanning electron microscopy (SEM) have been performed in order to elucidate the mechanisms of deformation in all materials. The procedures employed and the results obtained from this work as well as the conclusions which may be currently derived are reported in the following sections.

Experimental Procedures

Material Preparation and Characterization

Three different mullite materials were studied. Unreinforced *m*ullite (M) and SiC *w*hisker-reinforced *m*ullite (MW) was prepared at Oak Ridge National Laboratory from mullite powders* derived from reaction of a mixture of alum ($NH_4Al(SO_4)_2 \cdot 12H_2O$) and SiO_2. The material resulting from this reaction was jet milled to obtain powder having the following particle size distribution (by sedigraph): < 1 μm (30%), < 2 μm (80%), < 3 μm (95%), < 4 μm (98%) and < 6 μm (100%). The major impurities in this powder were determined via inductively coupled plasma spectrometry to be Fe (38 ppm), Na (64 ppm), K (10 ppm), and Ca (200 ppm).

To produce the starting materials for the fabrication of all bulk M and MW samples, this powder was wet (H_2O) milled in a slurry for 16 hours in a rubber lined jar mill using Al_2O_3 media. The SiC whiskers† were cleaned in H_2O to remove excess rice hulls. Twenty vol% whiskers were subsequently added to the slurry and homogenized for 2 minutes using a high shear mixer. The slurry was stirred on a hot plate to a thick consistency to keep the whiskers from settling, then dried in air (1 atm.) at 383K for approximately 12 hrs.

The typical and principal impurities in the whiskers were determined[9] by emission spectroscopy to be Al (2810 ppm), Fe (1100 ppm), Ca (2050 ppm), Mn (480 ppm) and Mg (370 ppm). Additional physical and chemical characteristics of these whiskers have been extensively studied by Nutt[10] and Sharma et al.[11] As shown by these investigators, the whiskers are nominally 0.2 to 0.5 um in diameter and up to 100 um in length. They possess a large number of planar faults lying on close packed planes perpendicular to the whisker axis which result in complex mixtures of α and β polytypes arranged in thin lamellae. Approximately 100 nm of the core region of the whiskers contains cavities 1 to 20 nm in diameter which are associated with partial dislocations lying in radial directions.

The unreinforced (M) mullite was hot-pressed in a graphite die for 0.75 hours at 1873K under a pressure of 14 MPa, resulting in a cylinder of height and diameter

---

*Baikowski International, Charlotte, NC.

†ARCO Chemical Co., Advanced Materials Division, Greer, SC. The whiskers were prepared from ground rice hulls which had been mixed with C (in the form of coke) and heated in Ar in an electrical rotary furnace at 1273K to produce a coke-like product containing C and SiO_2. This precursor material was subsequently converted to SiC whiskers in an rf heated graphitizing furnace at 2073K in Ar [7].

19 mm and 63.5 mm, respectively. The composite (MW) was hot pressed for 0.75 hours at 1948K under a pressure of 14 MPa, resulting in a cylinder of height and diameter 28.5 mm and 38 mm, respectively.

In addition to mullite materials containing glass, one billet of *m*ullite containing *n*o *g*lass (MNG) was prepared from powder produced by a sol-gel technique.[‡] In this case, a mixture of AlOOH and tetraethoxysilane was gelled by hydrolysis and subsequently crushed to form powder of submicron size. This powder was pressed at room temperature and then pressureless sintered in air at 1523K for 2 hrs resulting in a disk of height and diameter 4 mm and 17 mm respectively.

This disk was finally annealed at 1873K for 48 hrs in order to increase the grain size.

Creep Sample Preparation and Experiments

Sections 8 mm in thickness were cut from the original pressed disks of M and MW. Cylindrical specimens were subsequently ultrasonically trepanned perpendicular to the hot-pressing direction, such that the long axis of the whiskers were contained in the planes parallel to the stressing direction during creep. They were prepared as described in Refs. 12, 13 such that they possessed an average height and diameter of 7.6 mm and 4.8 mm, respectively. Constant compressive stress creep experiments were conducted in a continuously purified and circulating Ar atmosphere at 0.11 MPa pressure using pushrods and secondary platens of Poco graphite and primary platens of chemically vapor deposited SiC nominally oriented along the (111) perpendicular to the applied stress. In this apparatus, uniaxial compressive stresses, constant to within 1% for strains up to 10%, can be applied to the sample. The resulting strains can be read with an accuracy of 0.5 μm. A complete description of this instrument is presented in Ref. 13. In these materials, the creep runs were conducted 1) at constant temperatures within the range 1530K to 1700K and as a function of stress within the range 5 MPa to 220 MPa and 2) at constant stresses within the range 20 MPa to 100 MPa and as a function of temperature within the range 1450K to 1700K.

In order to prepare creep specimens from the MNG billet, the procedures described above were followed except that the specimens possessed a height and diameter of approximately 7.5 mm and 3 mm, respectively. Creep experiments were conducted 1) at constant temperature within the range 1600K to 1700K and as a function of stress within the range 20 MPa to 300 MPa and 2) at a constant stress of 100 MPa and as a function of temperature within the range 1580K to 1700K. The various conditions for each experiment are given in Table 1.

Sample Preparation For TEM and SEM Analysis

Following each creep experiment, sections were cut for TEM studies from the center of each crept sample perpendicular to the direction of the applied stress. In addition, sections from some MW specimens were cut parallel to the stress axis. All TEM sections were lapped flat, dimpled with 1 μm diamond paste on both sides to a center thickness of 25 μm and thinned to electron transparency using an ion beam gun with an accelerating voltage of 6 KV, 1 amp at 12 degrees gun tilt until perforation followed by 3 KV, 1 amp at 5 degrees for 45 seconds to remove damage created at the higher voltage. Finally the samples were C-coated to prevent charging during TEM observations.

[‡]**University of Washington, Dept. of Materials Science and Engineering, Roberts Hall, FB-10, Seattle, Wash. 98195.**

Table 1. Experimental creep data for M (hot-pressed unreinforced mullite), MW (hot-pressed whisker-reinforced mullite) and MNG (mullite containing no amorphous phase on the grain boundaries)

Sample #	Temp. (K)	Stress (MPa)	Time (ks)	Total Time (ks)	True Strain	Total True Strain	Steady State Creep Rate (s^{-1})	Kinectics n and/or Q
M1	1520	100	44.4	44.4	0.0019	0.0019	2.94E-8	Q=357±44
	1540	100	40.2	84.6	0.0017	0.0036	3.92E-8	kJ/mol
	1560	100	42.0	126.6	0.0026	0.0062	6.08E-8	
	1580	100	12.0	138.6	0.0021	0.0083	1.67E-7	Q+1051±7
	1600	100	3.9	142.5	0.0019	0.0102	4.51E-7	kJ/mol
	1620	100	1.2	143.7	0.0015	0.0117	1.26E-6	
	1640	100	0.6	144.3	0.0016	0.0133	2.79E-6	Q=754±19
	1660	100	0.3	144.6	0.0018	0.0151	5.84E-6	kJ/mol
	1680	100	0.1	144.8	0.0023	0.0174	1.09E-5	
	1700	100	0.1	144.9	0.0020	0.0194	2.01E-5	
M2	1520	25	66.6	66.6	0.0006	0.0006	4.47E-9	n=
	1520	50	57.6	124.2	0.0007	0.0013	1.04E-8	1.21±0.02
	1520	75	48.6	172.8	0.0010	0.0023	1.77E-8	
	1520	100	42.0	214.8	0.0011	0.0034	2.59E-8	
	1520	125	39.6	254.4	0.0015	0.0049	3.26E-8	
	1520	150	44.4	298.8	0.0018	0.0067	3.94E-8	
	1520	175	42.6	341.4	0.0021	0.0088	4.60E-8	
	1520	200	40.2	381.6	0.0023	0.0111	5.54E-8	
	1540	200	18.0	399.6	0.0017	0.0128	9.02E-8	
	1560	200	15.0	414.6	0.0026	0.0154	1.61E-7	
	1580	200	12.3	426.9	0.0060	0.0214	4.53E-7	
M3	1450	100	75.0	75.0	0.0008	0.0008	5.33E-9	Q=409±7
	1470	100	41.4	116.4	0.0003	0.0011	8.15E-9	kJ/mol
	1490	100	40.2	156.6	0.0005	0.0016	1.35E-8	
	1510	100	37.2	193.8	0.0007	0.0023	2.06E-8	
	1530	100	52.8	246.6	0.0016	0.0039	3.09E-8	n=
	1530	120	22.2	268.8	0.0008	0.0047	3.97E-8	1.09±0.05
	1530	140	21.0	289.8	0.0010	0.0057	4.87E-8	
	1530	160	29.4	319.2	0.0016	0.0073	5.32E-8	
	1530	180	23.4	342.6	0.0015	0.0088	6.24E-8	
	1530	200	34.8	377.4	0.0024	0.0112	6.79E-8	
	1530	220	51.6	429.0	0.0039	0.0151	7.36E-8	
M4	1510	20	84.6	84.6	0.0005	0.0005	2.72E-9	Q=392±20
	1520	20	48.6	133.2	0.0001	0.0006	3.16E-9	kJ/mol
	1530	20	37.8	171.0	0.0001	0.0007	4.06E-9	
	1540	20	63.0	234.0	0.0003	0.0010	4.74E-9	
	1550	20	40.2	274.2	0.0002	0.0012	6.08E-9	
	1560	20	57.0	331.2	0.0004	0.0016	8.08E-9	
	1570	20	36.0	367.2	0.0004	0.0020	1.22E-8	Q=1053±34
	1580	20	35.4	402.6	0.0007	0.0027	2.12E-8	kJ/mol
	1590	20	18.0	420.6	0.0005	0.0032	3.48E-8	
	1600	20	21.0	441.6	0.0012	0.0044	6.03E-8	
	1600	30	33.6	475.2	0.0032	0.0076	9.04E-8	n=
	1600	40	16.2	491.4	0.0021	0.0097	1.24E-7	1.19±0.03
	1600	50	8.4	499.8	0.0014	0.0111	1.62E-7	
	1600	60	6.6	506.4	0.0014	0.0125	2.05E-7	
	1600	70	7.2	513.6	0.0017	0.0142	2.39E-7	
	1600	80	3.6	517.2	0.0011	0.0153	2.89E-7	
	1600	90	2.4	519.6	0.0008	0.0161	3.54E-7	
	1600	100	1.2	520.8	0.0005	0.0166	4.16E-7	
M5	1640	25	6.3	6.3	0.0034	0.0034	4.65E-7	n=
	1640	35	4.2	10.5	0.0028	0.0062	6.52E-7	1.01±0.01
	1640	45	3.6	14.1	0.0030	0.0092	8.28E-7	
	1640	55	3.6	17.7	0.0038	0.0130	1.01E-6	
	1640	65	2.1	19.8	0.0025	0.0155	1.22E-6	
	1640	75	2.4	22.2	0.0035	0.0190	1.42E-6	

Sample #	Temp. (K)	Stress (MPa)	Time (ks)	Total Time (ks)	True Strain	Total True Strain	Steady State Creep Rate (s^{-1})	Kinectics n and/or Q
M6	1600	7	33.0	33.0	0.0009	0.0009	2.29E-8	Q=697±27
	1620	7	38.4	71.4	0.0019	0.0028	4.94E-8	kJ/mol
	1640	7	18.6	90.0	0.0020	0.0048	1.04E-7	
	1660	7	12.0	102.0	0.0025	0.0073	2.06E-7	
	1680	7	7.2	109.2	0.0027	0.0100	3.81E-7	
	1700	7	4.8	115.0	0.0031	0.0131	6.32E-7	n=
	1700	16	0.8	115.8	0.0024	0.0155	1.56E-6	1.10±0.02
	1700	25	0.8	116.6	0.0021	0.0176	2.60E-6	
	1700	35	0.8	117.4	0.0024	0.0200	3.83E-6	
	1700	44	0.8	118.2	0.0031	0.0231	4.93E-6	
MW1	1520	100	87.6	87.6	0.0001	0.0001	1.85E-9	Q=385±48
	1540	100	87.0	174.6	0.0002	0.0003	2.98E-9	kJ/mol
	1560	100	86.4	261.0	0.0004	0.0007	4.05E-9	
	1580	100	63.0	324.0	0.0008	0.0015	1.24E-8	Q=1212±15
	1600	100	13.2	337.2	0.0005	0.0020	4.12E-8	kJ/mol
	1620	100	8.5	345.7	0.0011	0.0031	1.27E-7	
	1640	100	14.3	360.0	0.0043	0.0074	3.05E-7	Q=866±47
	1660	100	3.8	363.8	0.0029	0.0103	7.45E-7	kJ/mol
	1680	100	1.7	365.5	0.0028	0.0131	1.59E-6	
	1700	100	1.2	366.7	0.0035	0.0166	2.88E-6	
MW2	1520	102	90.2	90.2	0.0004	0.0004	1.83E-9	n=
	1520	136	160.2	250.4	0.0004	0.0008	3.23E-9	1.95±0.07
	1520	170	88.8	339.2	0.0004	0.0012	4.73E-9	
	1520	204	111.0	450.2	0.0008	0.0020	7.26E-9	
	1550	204	82.8	533.0	0.0011	0.0031	1.42E-8	
	1580	204	45.6	578.6	0.0028	0.0059	6.40E-8	
MW3	1530	100	51.0	51.0	0.0000	0.0000		n=
	1530	120	84.6	135.6	0.0001	0.0001	2.29E-9	2.22±0.06
	1530	140	81.0	216.6	0.0001	0.0002	3.28E-9	
	1530	160	172.2	388.8	0.0005	0.0007	4.14E-9	
	1530	180	141.6	530.4	0.0007	0.0014	5.81E-9	
	1530	200	87.0	617.4	0.0004	0.0018	7.00E-9	
	1530	220	161.4	778.8	0.0012	0.0030	8.84E-9	
MW4	1570	102	70.8	70.8	0.0009	0.0009	1.17E-8	n=
	1570	136	33.0	103.8	0.0007	0.0016	2.35E-8	2.27±0.05
	1570	170	43.2	147.0	0.0016	0.0032	3.80E-8	
	1570	204	34.2	181.2	0.0019	0.0051	5.69E-8	
MW5	1600	60	90.0	90.0	0.0011	0.0011	1.18E-8	n=
	1600	70	46.8	136.8	0.0009	0.0020	1.90E-8	2.87±0.05
	1600	80	42.6	179.4	0.0012	0.0032	2.81E-8	
	1600	90	78.6	258.0	0.0030	0.0062	3.81E-8	
	1600	100	84.0	342.0	0.0044	0.0106	5.16E-8	
MW6	1640	22	86.4	86.4	0.0006	0.0006	6.65E-9	n=
	1640	27	48.0	134.4	0.0006	0.0012	1.24E-8	3.20±0.08
	1640	32	43.2	177.6	0.0009	0.0021	2.17E-8	
	1640	37	43.2	220.8	0.0015	0.0036	3.61E-8	
	1640	42	28.8	249.6	0.0016	0.0052	5.50E-8	
	1640	47	9.6	259.2	0.0008	0.0060	8.77E-8	
	1700	6	30.0	30.0	0.0002	0.0002	2.71E-9	n=
	1700	10	44.4	74.4	0.0005	0.0007	1.05E-8	3.02±0.12
	1700	16	46.2	120.6	0.0015	0.0022	3.40E-8	
	1700	22	41.4	162.0	0.0040	0.0062	9.34E-8	
	1700	27	6.6	168.6	0.0016	0.0078	2.30E-7	
	1700	32	3.9	172.5	0.0016	0.0094	4.06E-7	
	1700	37	2.6	175.1	0.0016	0.0110	6.30E-7	
	1700	42	1.8	176.9	0.0017	0.0127	9.53E-7	
	1700	47	1.2	178.1	0.0016	0.0143	1.32E-6	

Sample #	Temp. (K)	Stress (MPa)	Time (ks)	Total Time (ks)	True Strain	Total True Strain	Steady State Creep Rate (s^{-1})	Kinectics n and/or Q
MNG1	1580	100	231.0	231.0	0.0015	0.0015	6.39E-9	Q=748±35 kJ/mol
	1600	100	173.4	404.4	0.0027	0.0042	1.61E-8	
	1620	100	121.8	526.2	0.0047	0.0089	3.77E-8	
	1640	100	46.2	572.4	0.0034	0.0123	7.07E-8	
	1660	100	13.8	586.2	0.0019	0.0142	1.37E-7	
	1680	100	9.3	595.5	0.0022	0.0164	2.33E-7	
	1700	100	7.4	602.9	0.0027	0.0191	3.61E-7	
MNG2	1600	20	249.0	249.0	0.0012	0.0012	3.00E-9	n= 1.10±0.07
	1600	40	94.8	343.8	0.0008	0.0020	7.05E-9	
	1600	140	65.4	409.2	0.0026	0.0046	2.82E-8	
	1700	140	6.6	415.8	0.0043	0.0089	6.29E-7	n= 1.24±0.07
	1700	180	5.7	421.5	0.0045	0.0134	7.38E-7	
	1700	220	3.3	424.8	0.0034	0.0168	9.44E-7	
	1700	260	1.4	426.2	0.0018	0.0186	1.18E-6	
	1700	300	1.2	427.4	0.0019	0.0205	1.50E-6	

Sections for the SEM studies from the MNG material were cut, lapped flat, polished to a one micron diamond finish, etched in a heated solution of 15 vol% HF in distilled water and subsequently Au-coated to prevent charging.

Results and Discussion

Experimental Creep Results (M and MW Materials)

Figure 1 shows the complete (1(A)) sequential creep curves and, for clarity, the higher strain rate regions (1(B) and (C)) of these curves obtained for both materials crept under a constant stress of 100 MPa and in the temperature range of 1520K to 1700K. Each curve is characterized by a period of primary creep in which the strain rate is followed by a period of steady-state creep, as shown by plots of the true strain rate as a function of true strain in Figs. 1(D) and (E). Table 1 lists the steady-state creep rates ($\dot{\epsilon}_{ss}$) for each condition of temperature for these materials. It is clearly evident from Fig. 1(A) that whisker reinforcement increased the creep resistance of this material.

The steady-state creep rates derived from Fig. 1 and plotted for both materials as a function of 1/T in Fig. 2 were grouped in three distinct regimes of temperature: 1520K-1560K, 1560K-1630K, and 1630K-1700K. As a consequence, the activation energy (Q) of creep from the equation $\dot{\epsilon}_{ss}$ α exp(-Q/RT) had different values. For the unreinforced material, the values of Q in these respective ranges were 357 kJ/mol, 1051 kJ/mol and 754 kJ/mol. In the composite material, the corresponding values were 385 kJ/mol, 1212 kJ/mol and 866 kJ/mol.

Activation energy values were also obtained for the unreinforced material under the conditions of 1450K-1530K at 100 MPa, 1510K-1600K at 20 MPa and 1640K-1700K at 7 MPa. As shown by the associated data in Table 1, the low activation energy regime extended down to at least 1450K. Moreover, the activation energy was independent of stress within these temperature and stress ranges.

The stress dependence of creep from the equation $\dot{\epsilon}_{ss}$ α σ^n was obtained for each material within the ranges of stress and temperature of 7-220 MPa and 1520K-1700K, respectively.

Plots of the true strain rate as a function of true strain for the conditions of 1600K in the stress ranges of 20 to 90 MPa (unreinforced) and 60 to 100 MPa (reinforced) in Figs. 3(A) and (B) show that primary strain occurred for each change in condition prior to steady-state creep. Figure 4 shows representative plots of the steady-state creep rates as a function of stress at the temperatures of 1530K, 1600K

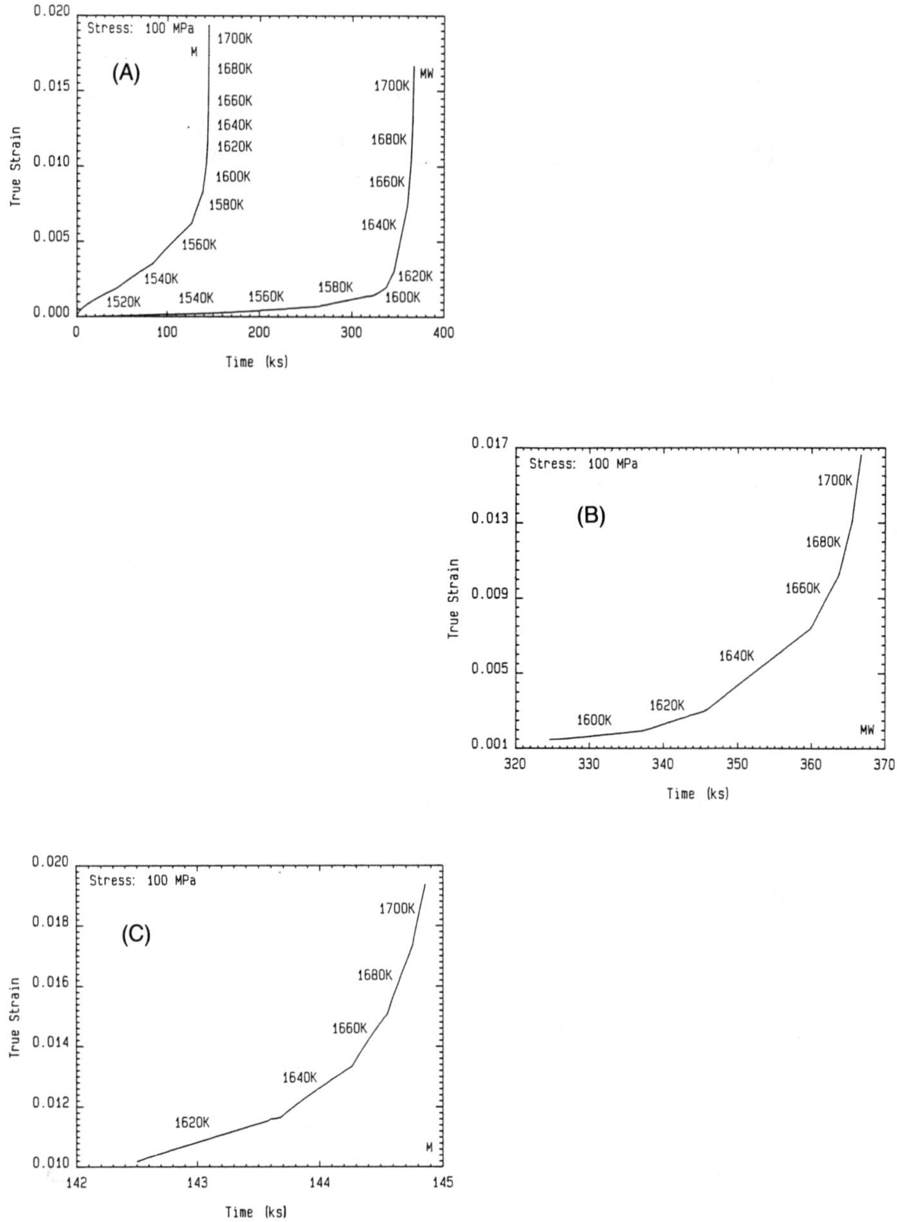

Fig. 1. (A) Creep curves sequentially obtained for both the M and MW materials using a constant stress of 100 MPa and the ten temperatures shown in the graph. For the M and MW materials, respectively, (B) and (C) are sequential creep curves

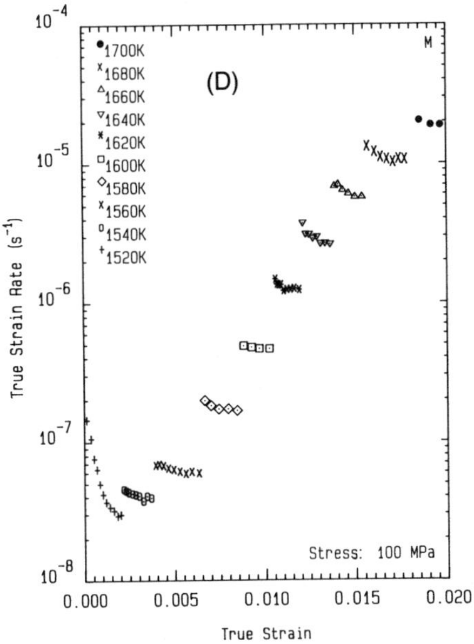

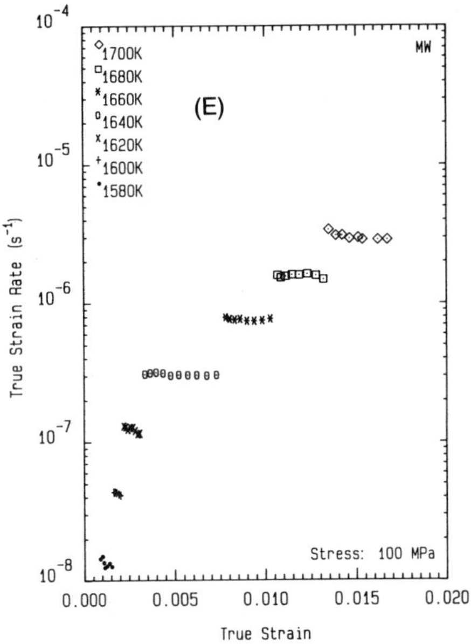

from the higher strain rate regions in (A) and (D) and (E) are the values of the true strain rate as a function of true strain from (A)-(C).

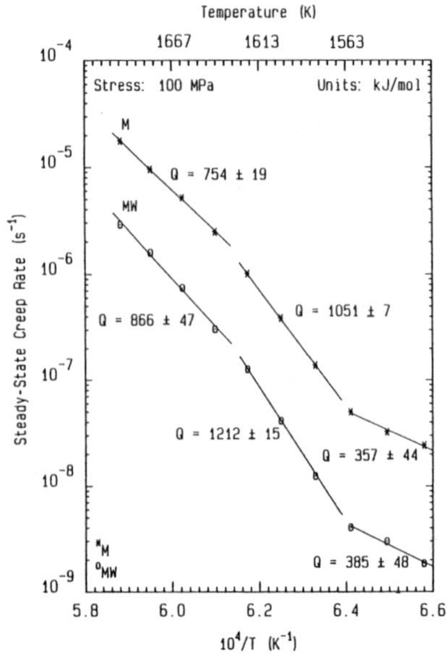

Fig. 2. Steady-state creep rate values derived from Figs. 1(D) and (E) as a function of reciprocal temperature obtained for the M and MW materials under a constant stress of 100 MPa.

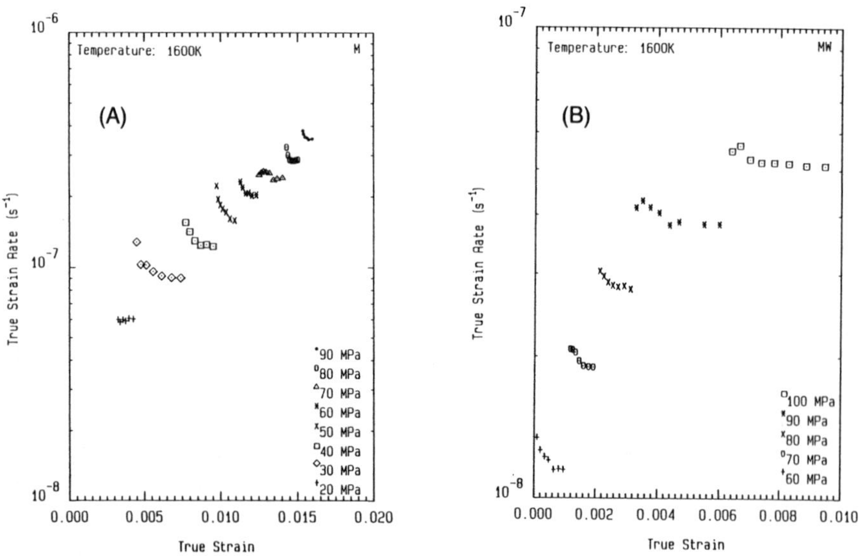

Fig. 3. Values of true strain rate as a function of true strain at a constant temperature of 1600K and stresses of (A) 20–90 MPa for the M material and (B) 60–100 MPa for the MW material.

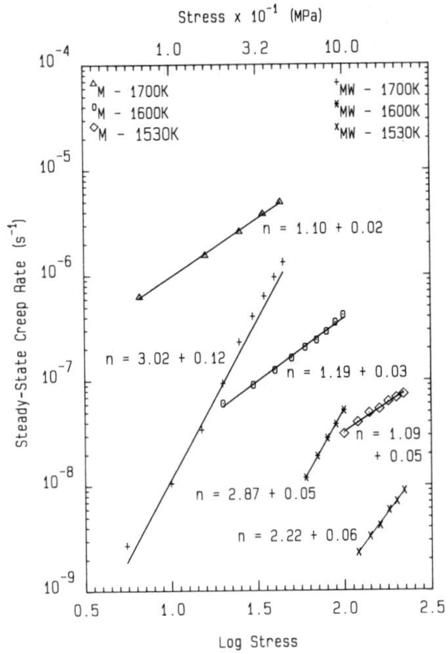

Fig. 4. Steady-state creep rate values as a function of log stress obtained for the M and MW materials at the temperatures of 1530K, 1600K and 1700K.

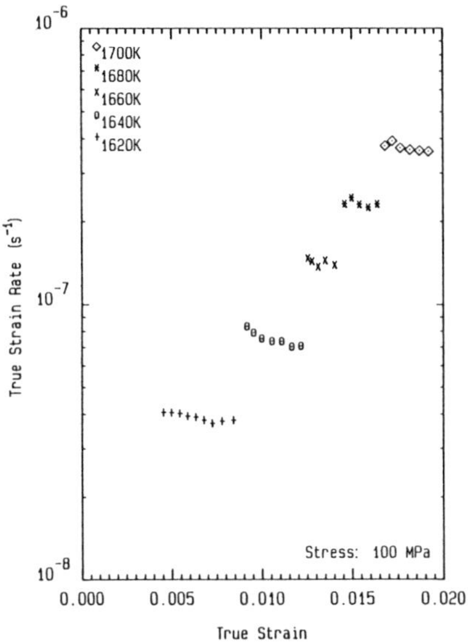

Fig. 5. True strain rate values as a function of true strain for the MNG material crept in the temperature range of 1620K–1700K and under a constant stress of 100 MPa.

and 1700K for both materials. Other steady-state creep rate vs. log stress curves were omitted from this figure for clarity; however, the data is given in Table 1. The values of the stress exponents for the unreinforced mullite were approximately one at all temperatures; whereas, the stress exponent values for the composite increased from approximately two at 1530K to three at 1700K.

Experimental Creep Results (MNG Materials)

The MNG material was crept under two sets of conditions used for the M material (see Table 1) in order to compare the kinetics of creep in mullite with and without an amorphous grain boundary phase. Figure 5 shows the resulting true strain rate versus true strain curves for creep in the temperature range of 1620K to 1700K. As before, primary creep occurred at each condition. The steady-state creep rate as a function of 1/T in Fig. 6 yielded a single activation energy value of 748 kJ/mol. The exponent values shown in Fig. 7 for 1600K and 1700K were approximately one. These values are comparable to the unreinforced material in the same temperature range; however, the creep rate is lower by two orders of magnitude. Also shown in this figure is the data obtained by Dokko et al.[1] for mullite materials apparently free of glass and with average grain sizes of 2.7 μm (sample (A) in[1]) and 4.2 μm (sample (C) in[1]). Their results are consistent with the new data presented here.

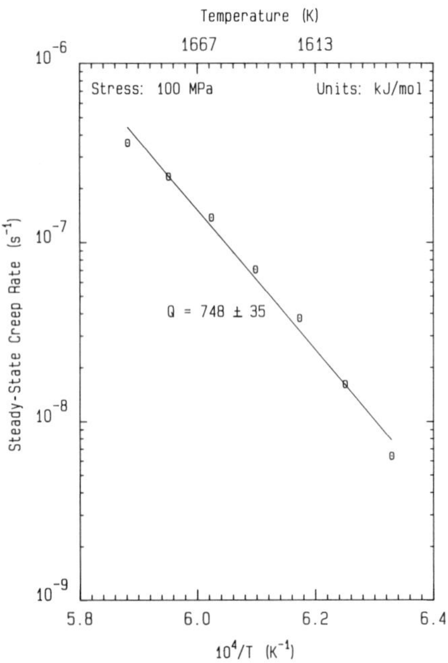

Fig. 6. Steady-state creep rate values as a function of reciprocal temperature obtained for the MNG material at a constant stress of 100 MPa for the MNG material.

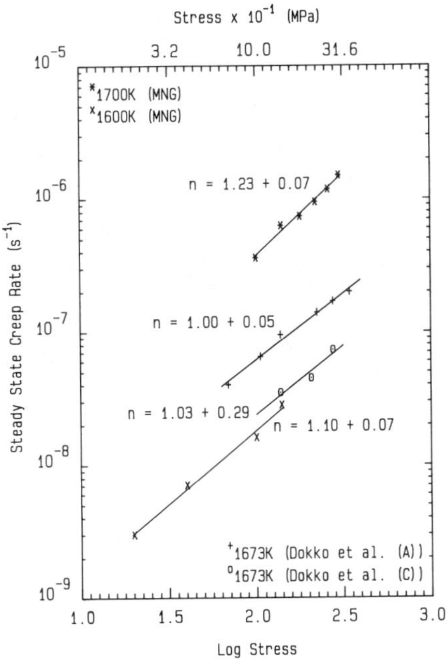

Fig. 7. Steady-state creep rate values as a function of log stress for the MNG material and for a mullite material studied by Dokko et al.[1] which contained virtually no glass on the grain boundaries.

Microstructural Examination of the M and MW Materials

In order to interpret the kinetic data and to determine the mechanisms of creep in each temperature regime, the microstructural changes occurring in each material must be characterized relative to the as-received material. The primary technique employed in this research was transmission electron microscopy.

Microstructures of the as-received M and MW samples are shown in Figs. 8 (A)-(E) and 9 (A)-(C), respectively. The mullite grains in both materials were equiaxed and quite dense except for a few which contained occasional oval shaped voids (see arrows, Figs. 8(A-C) and (B) and 9(A)). The mullite grain size in the M material, as determined from mean intercept lengths, was in the range of 0.8–1.0 um; the mullite grain size in the MW material was in the range 0.5–1.0 um. A thin, SiO_2-rich, amorphous layer occurred between the mullite grains (Fig. 8(C)) and also between the mullite and SiC whiskers (Fig. 9(B)). This phase was also observed at many triple points (Figs. 8(C) and 9(B)) and other multi-grain junctions (Fig. 8(D)). The unreinforced material also contained approximately 5 percent porosity (see Fig. 8(A)) which occurred in localized regions throughout the microstructure (compare Figs. 8(A) and 8(B)). More specifically, the porosity occurred at four or more grain junctions (Figs. 8(B) and 8(E)). All of the pores in the M material possessed sharp corners and were surrounded by the SiO_2-rich amorphous layer (Fig. 8(E)). By contrast, the SiC-whisker reinforced mullite was fully dense (Fig. 9(A)) except for the very occasional intragranular porosity noted above. The whiskers in the

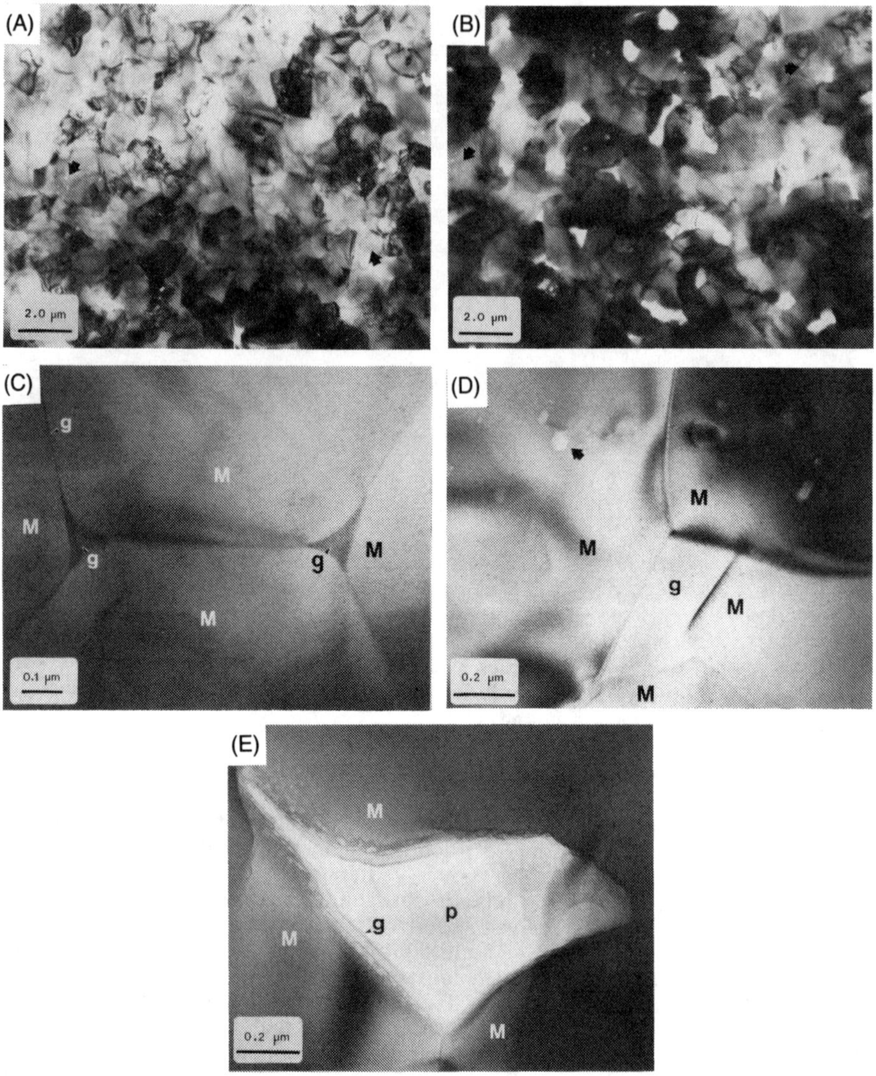

Fig. 8. TEM micrographs of the as-hot pressed (1873K, 14 MPa, 0.75 hr.) M material showing (A) equiaxed grains and intragranular oval-shaped pores (see arrows) (B) intergranular porosity and SiO_2-rich glass (C) on the grain boundaries and at triple points, (D) at a four-grain junction and (E) surrounding an intergranular pore.

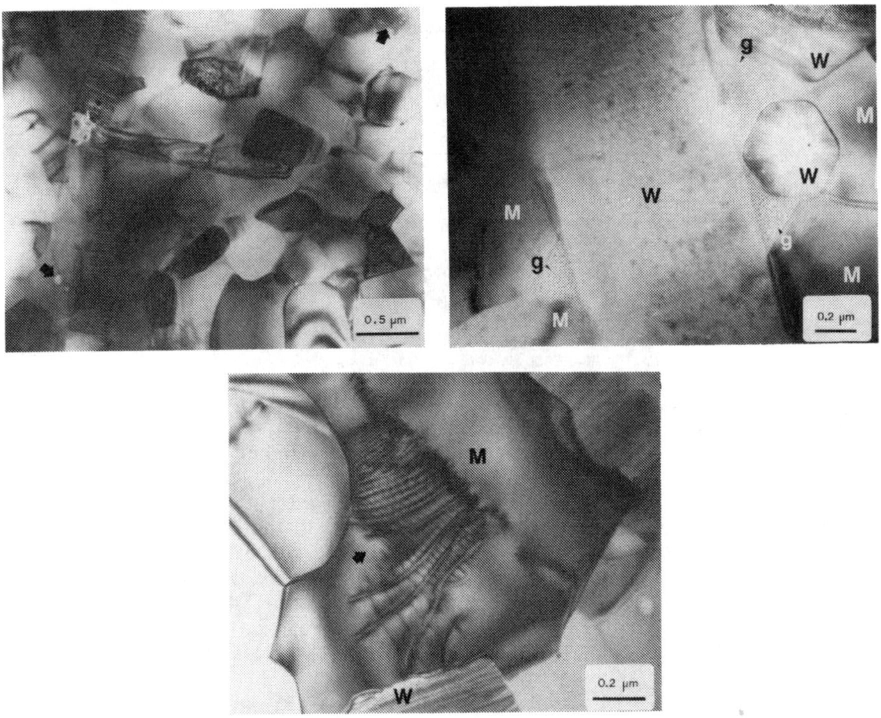

Fig. 9. TEM micrographs of the as-hot pressed (1948K, 14 MPa, 0.75 hr.) MW material showing (A) equiaxed mullite grains, long SiC whiskers and intra-granular oval-shaped pores (see arrows), (B) SiO_2-rich glass at the triple points and (C) a dislocation network in a mullite grain.

composite were hexagonal in cross section and were equal to or smaller in diameter than the mullite grains of the matrix (Figs. 9 (A) and (B)). However, the aspect ratio of these whiskers was in the range 30:1 to 100:1, and thus their length is much greater than the mullite grain size. The whiskers contained defects, particularly stacking faults, as observed in the earlier research by Nutt[9,10] and Sharma et al.,[11] and shown in Fig. 9(B).

In addition to the features described above, a number of mullite grains in both materials contained dislocations. Figure 9(C) shows a dislocation network which developed in a mullite grain during hot-pressing. It appears that these dislocations were generated by the whisker pressing against the grain as shown by the arrow. Dislocation processes, however, are not expected to contribute significantly to creep in this material, because of the limited number of slip systems.

The separate activation energy curves in Fig. 2 indicate that different creep mechanisms may be operative. To discern these mechanisms, it is imperative that the microstructures of the crept M and MW materials be analyzed. To effect this study, creep runs were purposely stopped within each of the three temperature regimes at the final temperatures of 1530K, 1600K and 1700K.

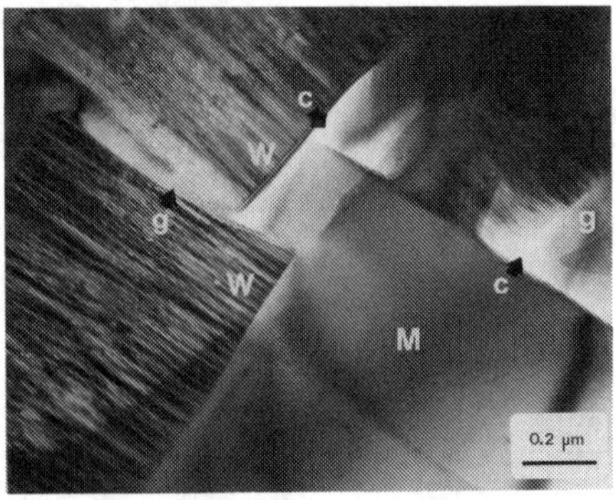

Fig. 10. TEM micrograph of the MW material crept under the final conditions of 1530K and 220 MPa (0.30 percent true strain). Cavities (see arrows) which developed during hot pressing are present at mullite-whisker interfaces.

The microstructure of the M material crept under the final conditions of 1530K and 220 MPa (1.51 percent true strain) was identical to that of the as-received M material shown in Fig. 8. Similarly, microstructures for the SiC-whisker reinforced mullite crept under the same final conditions (total true strain of 0.3 percent) showed no change except a small amount of cavitation which was observed at occasional whisker-mullite interfaces, as shown in Fig. 10. Most areas, however, remained quite dense. This is to be expected, for as shown in Table 1, the creep rate and final true strain were very low.

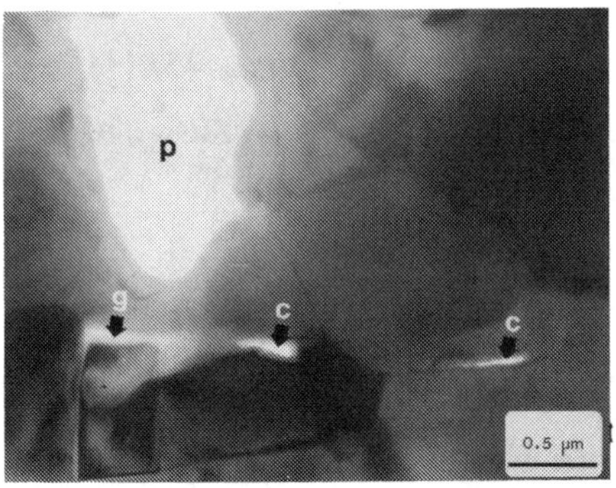

Fig. 11. TEM micrograph of the M material crept under the final conditions of 1600K and 100MPa (1.66 percent true strain). Cavities (see arrows) which developed during creep are present in the grain boundary amorphous phase.

When the temperature was increased into the middle regime, a significant change in the microstructure of the M and MW materials was observed. At low magnifications, changes were not apparent, but at high magnifications cavities were clearly discerned along grain boundaries. Figure 11 shows the microstructure which developed in the unreinforced material crept under the final conditions of 1600K and

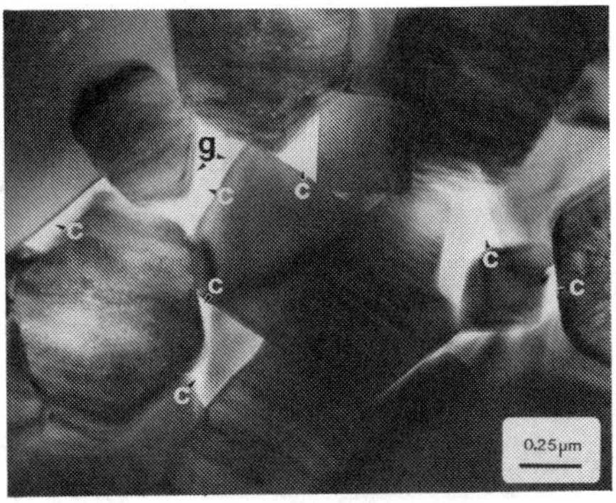

Fig. 12. TEM micrograph of the MW material crept under the final conditions of 1600K and 100 MPa (1.06 percent true strain). Cavities (see arrows) which developed during creep occur only along the whisker-whisker and whisker-mullite interfaces.

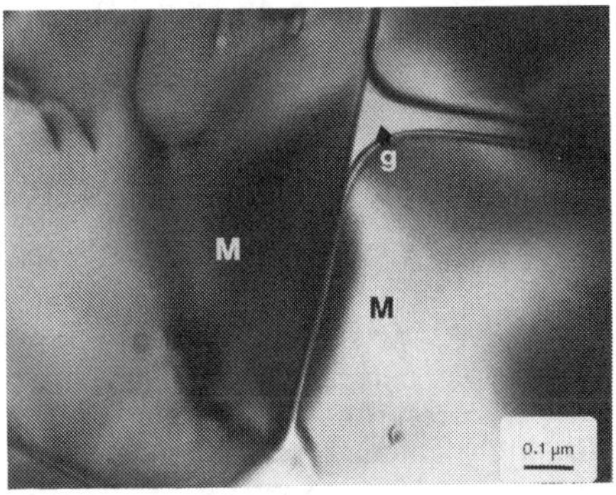

Fig. 13. TEM micrograph of the M material crept under the final conditions of 100 MPa and 1700K (1.94 percent true strain). The grain boundary cavity which was created at 1600K was filled by the amorphous phase.

100 MPa (total true strain of 1.66%). These cavities are different than the pores described above. As shown by the arrows, part of a single grain boundary contained excess glass while another portion contained a cavity. The cavity was surrounded by the amorphous grain boundary phase. It was not an artifact from the preparation method but a consequence of the creep process.

Cavities produced as a result of deformation were also observed for the SiC-whisker reinforced mullite crept under the final conditions of 1600K and 100 MPa (total true strain of 1.06%), as shown in Fig. 12. In the case of the MW material, however, cavities occurred almost solely at whisker-whisker and whisker-mullite interfaces. Cavities were rarely observed between mullite grains.

Figure 13 represents the microstructure of the unreinforced mullite crept under the final conditions of 1700K and 100 MPa (total true strain of 1.94%). The grain boundary is separated as shown by the arrows, but no cavitation occurred in this area.

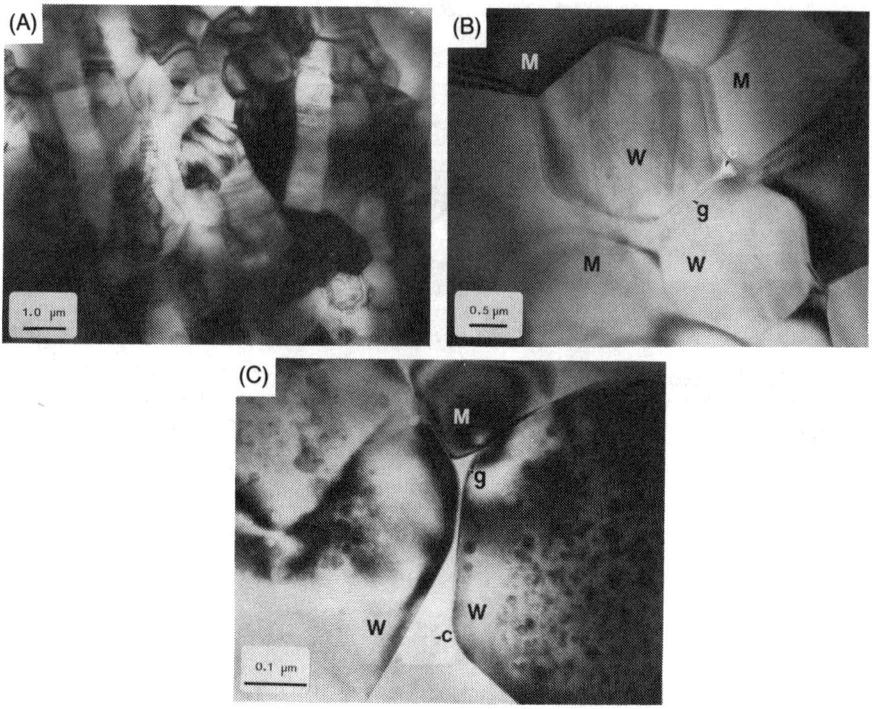

Fig. 14. TEM micrographs of the MW material crept under the final conditions of 1700K and 100 MPa (1.66 percent true strain) showing (A) a dense microstructure which developed as a result of the cavities created at 1600K being filled by the glass phase at the higher temperature, (B) evidence for grain separation due to grain boundary sliding and (C) an unfilled cavity in the center of the glass phase at a triple point.

Similar results were observed for the composite. Figures 14(A)-(C) show micrographs of the SiC-whisker reinforced material crept under the final conditions of 1700K and 100 MPa (total true strain of 1.66%). Note that although more strain has been incurred in this sample than those crept at 1600K, less cavitation is observed (compare Figs. 14(A) and (B) to Fig. 12). Figure 14(B) and (C) confirm that boundaries between whiskers and between whiskers and mullite grains that would show separation following creep at lower temperatures are now filled by the glass phase during creep at 1700K.

Microstructural Examination of the MNG Material

Microstructures of the as-received MNG material are shown in Fig. 15. The mullite grains were equiaxed, having an average grain size of approximately 4 μm. This material contained approximately 1.3 percent residual porosity with the pores located at multigrain junctions, as shown in Fig. 15(A). Arrays of low angle boundary dislocations were also commonly observed within the grains. In addition,

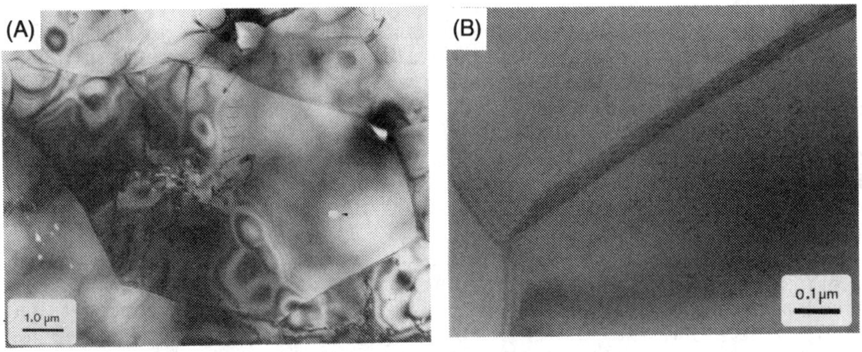

Fig. 15. TEM micrographs of the as-sintered (1523K, 2 hrs.) MNG material showing (A) equiaxed grains containing SiO_2-filled pores and (B) grain boundary dislocations indicating the absence of the amorphous grain boundary phase.

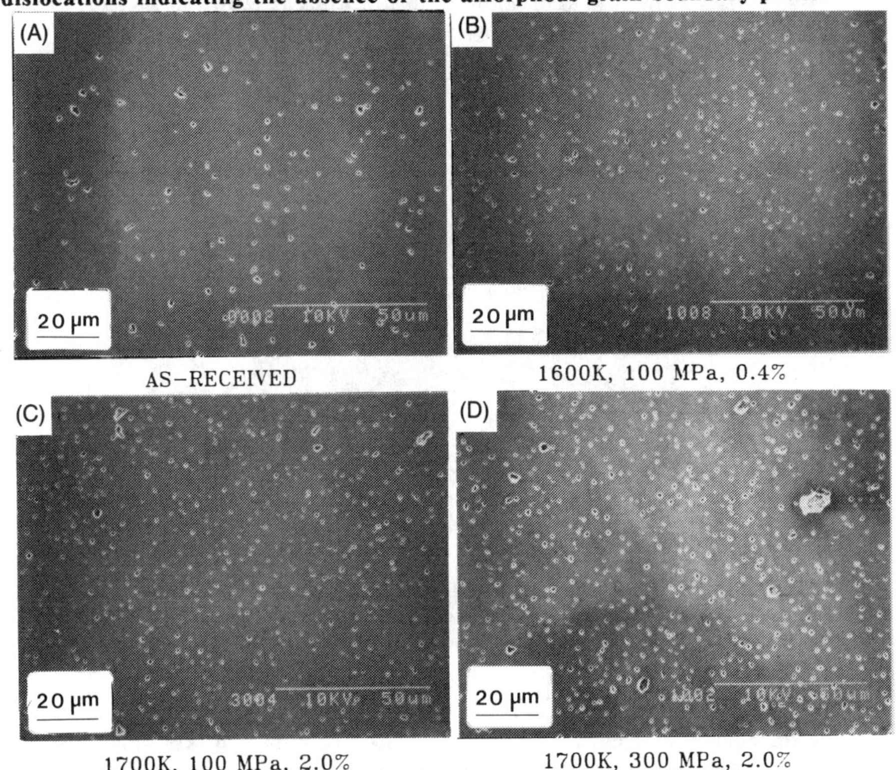

Fig. 16. SEM micrographs of polished sections of the MNG material (A) as-sintered and crept under the final conditions of (B) 1600K and 100 MPa (0.4 percent true strain), (C) 1700K and 100 MPa (1.9 percent true strain) and (D) 1700K and 300 MPa (2.0 percent true strain). Percent porosities for (A)-(D) are 1.3, 3.0, 4.8 and 5.5, respectively.

small oval shaped SiO_2-filled pores were present inside the grains (see arrow in Fig. 15(A)). A unique feature of this material was the absence of an amorphous phase between grains; indeed, some boundaries were constituted of grain boundary dislocations as shown in Fig. 15(B).

Following creep, the microstructure of the MNG material showed a significant increase in porosity. Figures 16(A)-(D) are SEM micrographs of polished sections of the uncrept material and of this material crept under the final conditions of 1600K and 100 MPa (0.4% true strain), 1700K and 100 MPa (1.9% true strain) and 1700K and 300 MPa (2% true strain), respectively. The volume fraction of porosity in each of these specimens was obtained by placing a 37 × 29 grid of points over each micrograph, counting the number of points which hit a pore, and dividing this number by the total number of points on the grid (=1073).[14] This procedure resulted in the respective porosities of 1.3%, 3.0%, 4.8% and 5.5%, corresponding to the conditions noted in Figs. 16(A)-(D). It is therefore apparent that cavitation processes occurred during creep and that the amount of cavitation increased with increasing true strain and applied stress.

These cavities were located on multigrain junctions as shown in Fig. 17. It was observed that surfaces surrounding the cavities were smooth, indicating that mass transport had occurred inside these cavities during sintering and creep. Dislocation sources were activated to some degree during creep; however, there were no obvious signs of dislocation generation or annihilation at the cavities. Instead, the majority of dislocations interacted to form subboundaries as shown in Fig. 17.

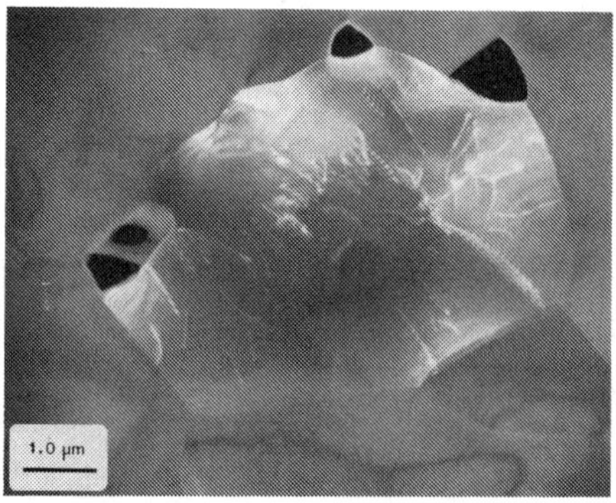

Fig. 17. Dark field TEM micrograph of the MNG material crept under the final conditions of 1600K and 100 MPa. Typical features are cavities at the triple points and dislocation networks which have formed stable subboundaries.

Integrated Analysis of Results

Effect of Porosity and Grain Size on Creep Rate in the M and MW Materials

As noted in an earlier section the M unreinforced material contained approximately five percent porosity; whereas, the MW whisker reinforced material was almost fully dense. The presence of porosity is known to increase the creep rate of ceramics (e.g., SiC[12]); so its effect must be taken into account for mullite before comparing the creep resistance of the M and MW materials. The creep rate of a porous material has been expressed by Langdon[15] as

$$\dot{\epsilon}_{ss}\ (\text{porous}) = \dot{\epsilon}_{ss}\ (100\%\ \text{dense}) \times [(1+(\beta P/(1-(\beta+1)P))^{1-n}/(1-P^{2/3})^n] \quad (1)$$

where P is the percent porosity, β is a constant (= -4 for Al_2O_3) and n is the stress exponent (=1 for unreinforced mullite). Using these factors the creep rates of the M material crept in the temperature range of 1520K–1700K under a constant stress of 100 MPa were estimated for a material of 100 percent density. The adjusted values are listed in Table 2 and plotted in Fig. 2. The difference in the uncorrected and density compensated values was approximately 18%. Note that the estimated steady-state creep rate for the 100% dense M material would still be higher than that of the MW material under identical conditions. Since the average grain sizes in both the M and MW material are approximately equal, it can be concluded that the whiskers do indeed increase the creep resistance of this material.

Table 2. Comparison of the steady-state creep rates ($\dot{\epsilon}_{ss}$) of the density compensated M (hot-pressed unreinforced mullite), MW (hot-pressed whisker-reinforced mullite) and MNG (mullite containing no amorphous phase on the grain boundaries) materials. These values are denoted as A, B and C, respectively. $\dot{\epsilon}_{ss,gs}$, denoted by D, is the steady-state creep rate of the MNG material corrected for a 1 μm grain size. T is the absolute temperature. All of these values were obtained under a constant at 100 MPa.

T(K)	$\dot{\epsilon}_{ss}$ (M) A	$\dot{\epsilon}_s$ (MW) B	$\dot{\epsilon}_{ss}$ (MNG) C	$\dot{\epsilon}_{ss,gs}$ (MNG) D	A/B	A/C	A/D	B/C	B/D
1580	1.37E-7	1.24E-8	6.39E-9	1.02E-7	11.0	21.4	1.34	1.94	0.12
1600	3.70E-7	4.12E-8	1.61E-8	2.58E-7	8.98	23.0	1.43	2.56	0.16
1620	1.03E-6	1.27E-7	3.77E-8	6.03E-7	8.11	27.3	1.71	3.37	0.21
1640	2.29E-6	3.05E-7	7.07E-8	1.13E-6	7.51	32.4	2.03	4.31	0.27
1660	4.79E-6	7.45E-7	1.37E-7	2.19E-6	6.42	35.0	2.19	5.44	0.34
1680	8.94E-6	1.59E-6	2.33E-7	3.73E-6	5.62	38.4	2.40	6.82	0.43
1700	1.65E-5	2.88E-6	3.61E-7	5.78E-6	5.73	45.7	2.85	7.98	0.50

Mechanisms of Creep in the M and MW Material

The salient features to be described by the mechanism of creep in these mullite materials are as follows. First, the presence of SiC whiskers clearly increases the creep resistance of mullite by approximately one order of magnitude as shown in Table 2 and Fig. 2. Second, there are three separate temperature regimes for the M and MW materials, each with their characteristic activation energies. Values for the activation energies for the M and MW materials in each regime are comparable. Third, TEM studies corresponding to each of the three regimes show, for both materials, that grain boundary sliding is the dominant creep mechanism. In the low temperature regime the grain boundary phase is sufficiently viscous that relatively little displacement along the grain boundaries is observed. In the medium temperature regime a lower viscosity results in intergranular sliding and cavitation. In the high temperature regime extensive intergranular sliding appears without cavitation. Fourth, and the major distinction between the unreinforced and whisker reinforced material, values for the stress dependence of $\dot{\epsilon}_{ss}$ for the M and MW materials are different. The creep rate varies linearly with the applied stress, i.e., n = 1, for the unreinforced material, consistent with viscous flow, and varies with n = 2 to 3 for the reinforced material, indicating a stronger stress dependent process.

The behavior and observations on unreinforced (M) mullite are entirely consistent with the mechanisms of grain boundary sliding in the presence of a viscous phase analyzed by Raj et al.[16] and Tsai and Raj.[17] That is to say at low temperatures, grain boundary sliding occurs by the sequential transport of matter by its dissolution and diffusion through the glassy phase and subsequent reprecipitation on another site. At medium temperatures the viscosity is lower and cavitation occurs particularly at triple points or grain boundary ledges due to negative pressures created there.[17] At high temperatures the liquid phase is fluid enough to fill the cavities as they are formed. Lange[18] rationalizes the same situation by proposing that the process is simply a matter of grains sliding relative to one another with a rate inversely proportional to the viscosity of the amorphous phase.

The question then is how do the silicon carbide whiskers in the MW material affect the viscous flow mechanism? It has been proposed in previous work[8] where higher n values were found for whisker-reinforced alumina compared with unreinforced alumina that the increase in creep resistance is caused by creep in the whisker itself and that dislocation processes are the rate controlling mechanism. We believe neither our temperatures nor stresses are high enough to initiate this mechanism. Nevertheless the notion that the load may be transferred from the mullite phase to the silicon carbide phase during creep may be significant. The critical sliding interface then becomes the silicon carbide whisker—glassy phase rather than the mullite grain boundary—glassy phase. Cavitation in the middle temperature range in the reinforced material is observed at the whisker-glass interface rather than the mullite-mullite interface suggesting this interface is critically involved.

The main difference between intergranular sliding and cavitation in the unreinforced mullite and whisker sliding in the reinforced mullite is geometrical. The relatively equiaxed granular structure of the mullite always presents triple lines normal to the stress where cavitation can take place and intergranular sliding is relatively uninhibited. A whisker on the other hand has a high aspect ratio. Sliding over its surface produces small cavities only at the ends of the whiskers. Thus the rate controlling step is the interfacial sliding between the silicon carbide whisker and the glass along the cylindrical tube which represents the surface of the whisker. We are not aware of any analysis of the creep rate of rigid rods placed in a viscous medium which this model requires, to know whether a higher stress exponent follows. It is interesting to note, however, that earlier work by Kumazawa et al.[5] on needle shaped mullite grains also showed n = 2. A higher value of n accompanying grain boundary sliding may be characteristic of an anisotropic microstructure.

Finally, as shown in Table 2, the difference between the steady-state creep rates of the M and MW material becomes less significant with increasing temperature. This result, as well as the stress dependence values shown in Fig. 4, show that under sufficiently high temperature and stress, the steady-state creep rates of both the M and MW materials become the same.

Mechanisms of Creep in the MNG Material

As shown in Table 1, the steady-state creep rate of the MNG material is between one to two orders of magnitude less than the density compensated values of the M material and one order of magnitude less than that of the MW material under identical conditions. The increase in creep resistance is attributed to two factors: first, a larger average grain size and second, the absence of an amorphous grain boundary phase.

Dokko et al.[1] determined the grain size dependence of creep in polycrystalline mullite containing no boundary phase to be 2 (i.e. $\dot{\epsilon}_{ss} \alpha d^{-2}$, where d is the average grain size). Therefore, in order to compare the MNG mullite with the M and MW materials, the steady-state creep rate values of MNG must be increased by a factor of 16 since its average grain size (4 μm) is four times greater than that of the M and MW materials (1 μm). The adjusted $\dot{\epsilon}_{ss}$ for the MNG material (see "D" in Table 2) is still significantly less than that for the M material (see "A" in Table 2) under comparable stresses and temperatures (see ratio A/D in Table 2), but it approaches or exceeds $\dot{\epsilon}_{ss}$ for the MW material (see "B" and ratio B/D in Table 2). Therefore, the grain size accounts for a major part, but not all, of the increased creep resistance of the MNG mullite.

It is evident that the second factor above accounts for not only the additional difference between the steady-state creep rates of the MNG and M and MW materials, but also for the increasing differential between these values as the temperature is increased. This latter result is demonstrated by the increasing values of the ratios A/C, A/D, B/C and B/D in Table 2, which correspond to the rapid (exponential) decrease in viscosity of the glass phase with an increase in temperature in the M and MW materials. Thus the absence of the glass phase also contributes to the superior creep resistance of MNG as compared with M and MW.

With these two factors in mind, the mechanism of creep in the MNG material must first be evaluated in terms of the kinetic data. The kinetics of this work ($n = 1$, $Q = 748$ kJ/mol) as well as the work by Dokko et al.[1] ($n = 1$, $m = 2$ and $Q = 711$ kJ/mol) and Lessing et al.[4] ($n = 1$ and $Q = 685$ kJ/mol) point to grain boundary sliding as the controlling mechanism. At low temperatures, grain boundary sliding is accommodated by diffusion and transfer of atoms from boundaries under compression to boundaries under tension (Nabarro-Herring mechanism[2,3]). If the stress is sufficiently low, the diffusion process should fully accommodate grain boundary sliding without cavity formation. At higher temperatures and stress levels the diffusion process is not fully responsive to grain boundary sliding and cavitation at multigrain junctions occurs (Fig. 16). The extent of cavitation increases with the applied stress and total true strain. We therefore conclude that the creep mechanism in the glass phase free mullite (MNG) is grain boundary sliding in the entire temperature range with the strain discontinuities accommodated by lattice diffusion and cavitation.

The difference in creep rate for the MNG material reported here and the lower rate reported by Dokko et al.[1] (see Fig. 7) is believed to be due to the difference in microstructure and not creep mechanism. Although Dokko's work was performed in air and the present work in Ar, the n values and activation energies are similar indicating that the same creep mechanisms are operative. As shown by the SEM micrographs in Fig. 18, the microstructures of the crept materials from both the work of Dokko et al.[1] (Fig. 18(A)) and the present work (Fig. 18(B)) are very similar. The main difference is the presence of the lath shaped grains in Fig. 18(A). These lath-shaped grains increase the creep resistance by effectively increasing the grain size.

Finally it is interesting to note that some creep occurred by the movement of dislocations; however, the amount was limited and dislocation processes did not make a significant contribution to steady-state creep. This reflects the complex crystallographic structure of this material and the difficulty experienced by dislocations to move even at high levels of stress and temperature. The operative slip systems in mullite were observed to have a Burgers vector of a [001]. These dislocations were observed to react to form stable nodes as illustrated in Fig. 17 and subgrain boundaries which restrict further dislocation glide.

Conclusions

1. Constant stress creep experiments have been conducted within the ranges of temperature and stress of 1450–1700K and 5–220 MPa, respectively, on polycrystalline mullite with and without 20 vol% SiC whiskers and containing a SiO_2-rich amorphous grain boundary phase. The steady-state creep rate of the composite was decreased by a factor of approximately 10 at all

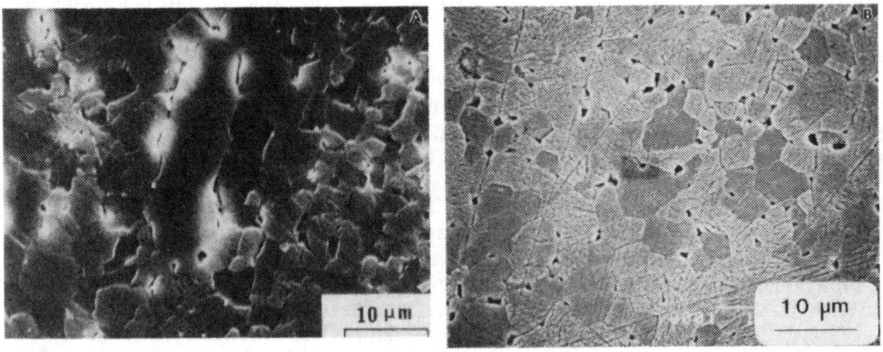

Fig. 18. Optical micrographs of polished and etched sections of (A) the mullite material prepared and studied by Dokko et al.[1] and (B) the as-sintered MNG material. Microstructures of both materials are similar; however, the mullite material in (A) contains a number of lath-shaped grains which effectively increase the grain size.

temperatures, relative to the density compensated values of the unreinforced material.

2. The activation energies of the unreinforced mullite and the composite were essentially the same at all temperatures. However, the values of this parameter were distributed within three temperature ranges: 1450–1560K (Q = 385 kJ/mol), 1560–1630K (Q = 1212 kJ/mol) and 1630–1700K (Q = 866 kJ/mol). The stress exponent for the unreinforced mullite was approximately 1.0 at all temperatures; that for the composite increased from approximately 2.0 at 1530K to 3.0 at 1700K.
3. Kinetic data and TEM results indicate similar controlling mechanisms for both the M and MW materials. The primary factor operating was the changing viscosity of the amorphous grain boundary phase as a function of temperature. Within the lowest temperature regime, the creep rates were most likely controlled by grain boundary sliding accommodated by diffusion in the glass phase. Some viscous flow may also be present. In the intermediate temperature range, these mechanisms were joined by cavitation primarily along grain boundaries in the unreinforced mullite and almost solely at whisker-whisker and whisker-mullite interfaces in the composite. Within the highest temperature range, cavitation was not observed, indicating that creep was controlled by a grain boundary sliding by viscous flow and diffusion through the glass phase.
4. Similar creep experiments on pressureless sintered mullite containing no amorphous grain boundary phase were performed in the temperature and stress ranges of 1600–1700K and 20–300 MPa, respectively. The steady-state creep rate of this material was approximately an order of magnitude less than that of the hot-pressed whisker-reinforced mullite. The activation energy was 748 kJ/mol. The stress exponent value was 1–1.2. The kinetic data and TEM and SEM studies indicate that creep was controlled by grain boundary sliding which was accommodated by lattice diffusion and the formation of cavities on the grain boundary.

References

[1]P. C. Dokko, J. A. Pask and K. S. Mazdiyasni, "High Temperature Mechanical Properties of Mullite Under Compression," *J. Am. Ceram. Soc.*, **60** [3–4] 150–5 (1977).

[2]F. R. N. Nabarro, "Steady-State Diffusional Creep," *Phil. Mag.*, **16**, 231–7 (1967).

[3]C. Herring, "Diffusional Viscosity of a Polycrystalline Solid," *J. Appl. Phys.*, **21** [5], 437–45 (1950).

[4]P. A. Lessing, R. S. Gordon and K. S. Mazdiyasni, "Creep of Polycrystalline Mullite," *J. Am. Ceram. Soc.*, **58** [3–4] 149 (1975).

[5]T. Kumazawa, S. Ohta, S. Zanzaki and H. Tabata, "Influence of Powder Characteristics on Microstructure and Mechanical Properties of Mullite Ceramics," 1st International Workshop on Mullite, Abstract #MP-14, Nov. 9–10, 1987, Tokyo, Japan.

[6]R. A. Penty and D. P. H. Hasselman, "Creep Kinetics of High Purity, Ultra-fine Grain Polycrystalline Mullite," *Mat. Res. Bull.*, 7 [10] 1117–23 (1972).

[7]C. O. Hulse and J. A. Pask, "Analysis of Deformation of a Fireclay Refractory," *J. Am. Ceram. Soc.*, **49** [6] 312–8 (1966).

[8]A. H. Chokski and J. R. Porter, "Creep Deformation of an Alumina Matrix Composite Reinforced with Silicon Carbide Whiskers," *J. Am. Ceram. Soc.*, **68** [6] C-144-6 (1985).

[9]S. R. Nutt, "Microstructure and Growth Model for Rice-Hull-Derived SiC Whiskers," *J. Am. Ceram. Soc.*, **71** [3] 149–56 (1988).

[10]S. R. Nutt, "Defects in Silicon Carbide Whiskers," *J. Am. Ceram. Soc.*, **67** [6] 428–31 (1984).

[11]N. K. Sharma, W. S. Williams and A. Zangvil, "Formation and Structure of Silicon Carbide Whiskers From Rice Hulls," *J. Am. Ceram. Soc.*, **67** [11] 715–20 (1984).

[12]C. H. Carter, Jr., R. F. Davis and J. Bentley, "Kinetics and Mechanisms of High Temperature Creep in Silicon Carbide: I, Reaction-Bonded," *J. Am. Ceram. Soc.*, **67** [6] 409–17 (1984).

[13]C. H. Carter, Jr., C. A. Stone, R. F. Davis and D. R. Schaub, "High Temperature, Multi-Atmosphere, Constant Stress Compression Creep Apparatus," *Rev. Sci. Instrum.*, **51** [10] 1352–7 (1980).

[14]E. Thomson, "Quantitative Microscopic Analysis," *J. Geol.*, **38** [3] 193–222 (1930).

[15]T. G. Langdon, "Dependence of Creep Rate on Porosity," *J. Am. Ceram. Soc.*, **55** [12] 630–1 (1972).

[16]R. Raj, R. L. Tsai, J. G. Wang and C. K. Chyung, "Superplastic Flow in Ceramics Enhanced by a Liquid Phase," p. 353–78 in Deformation of Ceramic Materials II, edited by R. E. Tressler and R. C. Bradt, Plenum Press, New York (1984).

[17]R. L. Tsai and R. Raj, "Creep Fracture in Ceramics Containing Small Amounts of a Liquid Phase," *Acta Metall.*, **30** [6] 1043–58 (1982).

[18]F. F. Lange, "Non-elastic Deformation of Polycrystals With A Boundary Phase," pp. 361–81 in Deformation of Ceramic Materials, edited by R. C. Bradt and R. E. Tressler, Plenum Press, New York, 1975.

High Temperature Mechanical Properties of Mullite Ceramics

H. Ohnishi, K. Maeda, T. Nakamura, and T. Kawanami

Nippon Kagaku Togyo Co., Ltd.
3-2-24 Oriono-cho, Sakai, 590 Japan

Abstract

The microstructure and mechanical properties of high-purity, near theoretically dense mullite, sintered using powder prepared by the Sol-Gel method, have been investigated. The formation of a glassy phase or α-Al_2O_3 particles in the mullite grain boundary influenced both grain growth and grain shape. The values of the SiO_2-rich compositions were higher at high temperature than at room temperature; those of the Al_2O_3-rich compositions were similar in both regions of temperature. The creep strain-rate was influenced by the grain size and the presence of a glassy phase or α-Al_2O_3 particles. The strain-rates of the 74 wt% Al_2O_3 composition were lower than in the compositions with more or less Al_2O_3. In this study, the creep mechanisms of mullite was considered to be the Coble mechanism of grain boundary sliding accommodated by grain boundary diffusion.

Introduction

Mullite is one of the technological attractive materials[1] due to its excellent strength and creep resistance[2] at high temperature, low thermal expansion and conductivity, good thermal stability and chemical inertness. Mullite is not a stoichiometric composition, as shown in the phase equilibrium diagrams for SiO_2-Al_2O_3 system reported by Aksay et al.[3] and Prochazka et al.[4] (Fig. 1). Therefore, the microstructure and mechanical properties of mullite should change as a function of Al_2O_3/SiO_2 ratio. In this study, the influence of the composition and microstructure on the mechanical properties of mullite were investigated.

Experimental Procedure

Mullite powder was prepared using the Sol-Gel method.[5] An Al_2O_3 sol and a SiO_2 sol were mixed in several ratios which ranged from 68 wt% Al_2O_3/32 wt% SiO_2 to 76 wt% Al_2O_3/24 wt% SiO_2 and subsequently calcined. The powders were ground by ball-milling, uniaxially pressed in to specimens under 98 MPa and sintered at 1600°C and 1700°C for 2 or 4 hours in air. Table 1 shows the chemical compositions of each specimen. Phase identification was carried out by X-ray powder diffraction analysis using Ni-filtered CuKα radiation. The flexural strength was measured at R. T. and 1400°C in air in 3-point bending on a 30mm span using 3×4×40mm test bars at a

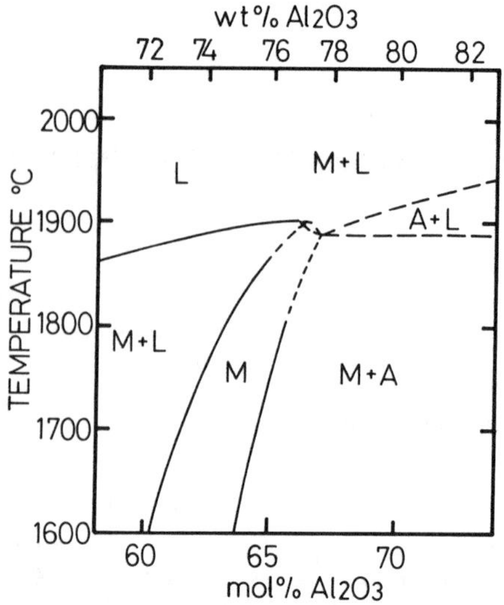

Fig. 1. The phase equilibrium of SiO_2-Al_2O_3 (from Ref. 2).

Table 1. Chemical compositions of each specimen.

Al_2O_3/SiO_2	68/32	70/30	71.8/28.2	73/27	74/26	75/25	76/24
Al_2O_3(wt%)	67.92	69.87	71.27	73.12	74.65	75.36	76.10
SiO_2(wt%)	32.03	30.00	28.62	26.81	25.07	24.57	23.80
Alkali(ppm)	110	140	200	130	60	180	270

cross head speed of 0.5mm/min. Creep tests were conducted at 1500°C in air using 3-point bending on a 30mm span using 3x10x40mm test bars.

The strain-rate and stress-exponent were calculated using Eqs. (1) and (2).[6]

$$\sigma = \frac{3PL}{2BD^2} \cdot \frac{2N+1}{3N} \quad ---- \qquad (1)$$

$$\epsilon = \frac{2D(N+2)}{L^2} \, \dot{y} \quad ---- \qquad (2)$$

where σ is the maximum stress, P is the applied load, L is the span length, N is the stress-exponent, B and D are the width and the thickness of specimen, respectively, and $\dot{y}$ is the deflection rate at the load point.

The microstructure was observed by SEM after thermal etching, and the average grain size of each specimen was determined by the intercept method.[7]

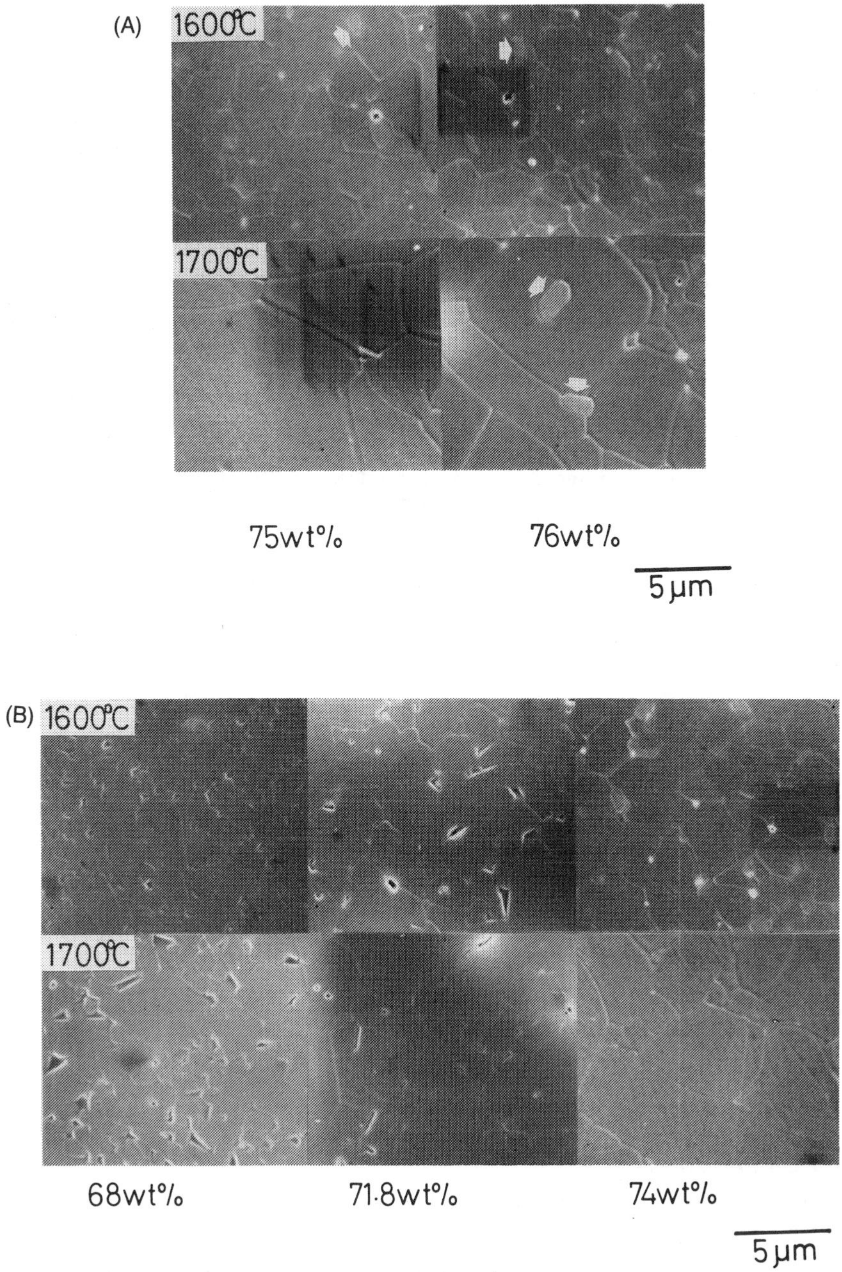

Fig. 2. SEM of 68, 71.8, 74, 75 and 76 wt% Al_2O_3 composition sintered at 1600°C and 1700°C.

Results and Discussion

Microstructure

SEM of 68, 71.8, 74, 75 and 76 wt% Al_2O_3 compositions sintered at 1600°C and 1700°C are shown in Fig. 2. A perusal of these micrographs reveals that the microstructure was influenced by both the Al_2O_3/SiO_2 ratio and the sintering temperature.

The compositions with an Al_2O_3 content less than 73 wt% contained a glassy phase at the triple points of the mullite grain boundaries. Alpha-Al_2O_3 particles, smaller than the mullite grains, were observed (See arrows in Fig. 2) as an inter- and intragranular phase in compositions with 75 wt% Al_2O_3 or greater and sintered at 1600°C and in these compositions with 76 wt% Al_2O_3 or greater and sintered at 1700°C. The results were in good agreement with the identification by XRD analysis and with the phase equilibrium for the SiO_2-Al_2O_3 system reported by Prochazka et al.[4] (Fig. 1) rather than by Aksay et al.[3]

The SiO_2-rich compositions with less Al_2O_3 content than 73 wt% contained mainly enlonged grains. By contrast, the Al_2O_3 rich compositions (>73 wt%) contained primarily large granular particles in the microstructure.

Figure 3 shows the average grain size dependence on the chemical composition and the sintering temperature. The average grain size in the compositions containing no glassy phase or α-Al_2O_3 particles was larger than that in the SiO_2-rich or the Al_2O_3-rich compositions. It is considered that the formation of liquid phase during sintering hinders the grain growth and causes the formation of enlonged mullite grains in the SiO_2-rich compositions. The presence of α-Al_2O_3 particles also hinder the grain growth.

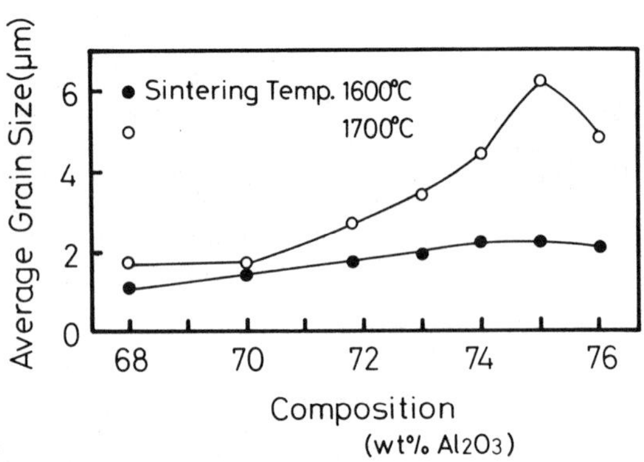

Fig. 3. Grain size dependence of chemical composition of samples sintered at 1600°C and 1700°C.

Mechanical Properties

Figure 4 shows the relationship between the chemical composition and density. The density of the compositions increased with increasing Al_2O_3 content. More over the density of each composition sintered at 1600°C was higher than those sintered at 1700°C. All compositions had 98% of theoretical density or greater. The flexural strength of each composition sintered at 1600°C is shown in Fig. 5. The strength of the SiO_2-rich compositions increased at 1400°C. By contrast the Al_2O_3-rich composi-

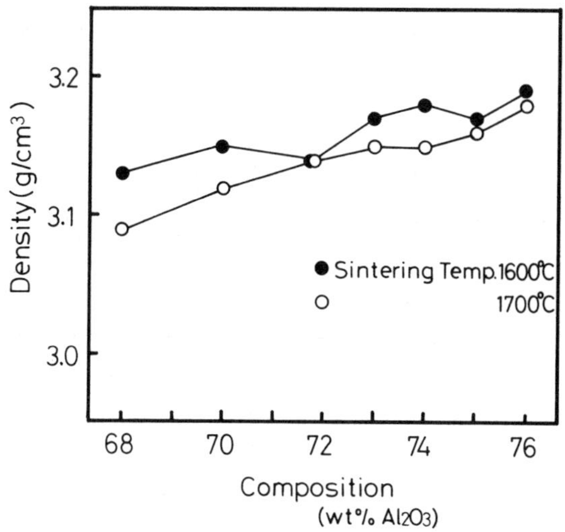

Fig. 4. Relationship between the chemical composition and density of samples sintered at 1600°C and 1700°C.

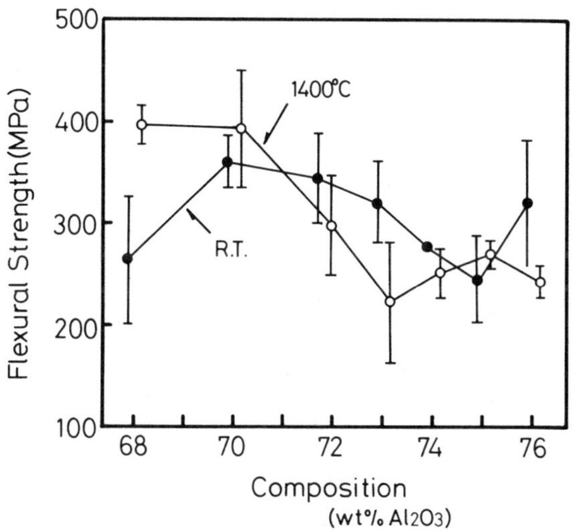

Fig. 5. Flexural strength at R.T. and 1400°C vs. chemical composition of samples sintered at 1600°C.

tions showed little significant difference between R.T. and 1400°C. The above results suggest that the significant increase in the strength of the SiO_2-rich compositions was caused by plastic deformation at the crack front in the glassy phase.

Log-log plots of steady-state creep rates vs. stress from 10 to 70 MPa at 1500°C for each composition are shown in Fig. 6 (A) and (B). The strain-rate of the 74 wt% Al_2O_3 composition was lower than the compositions with more or less Al_2O_3 and sintered at the same temperature. The stress-exponent, N, of the 73–75 wt% Al_2O_3 compositions showed the same value. The value of N increased with increasing SiO_2 or Al_2O_3 content outside these compositions.

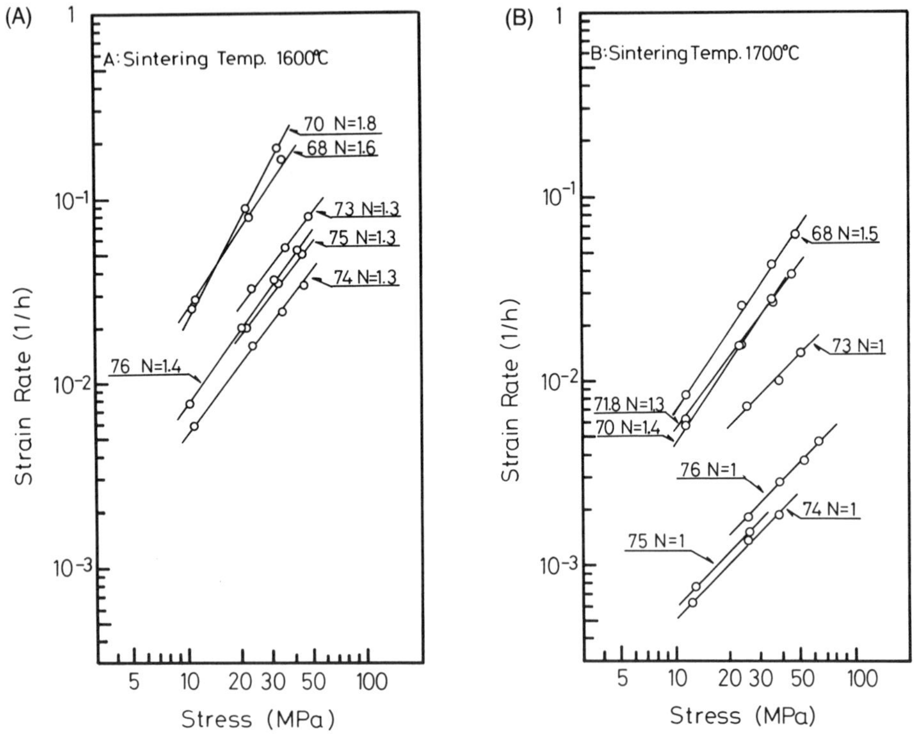

Fig. 6. Steady-strain creep vs. stress at 1500°C of each composition sintered at (A); 1600°C and (B); 1700°C.

In samples with the same grain size, the strain-rate was influenced by the amount of the glassy phase. For example, the amount of glassy phase in the 70 wt% Al_2O_3 composition is estimated to be less than that in the 68 wt% Al_2O_3 composition and the strain-rate of 70 wt% Al_2O_3 composition sintered at the same temperature. The strain-rate was also influenced by the presence of α-Al_2O_3 particles. For example, the strain-rate of the 74 wt% Al_2O_3 composition was lower than the 76 wt% Al_2O_3 composition sintered at the same temperature.

The grain-size dependence of the strain-rate under 25 MPa for 68, 74 and 76 wt% Al_2O_3 compositions is shown in Fig. 7. The grain-size dependence, m, for all compositions was about 3. The value of the stress-exponent for 74 and 76 wt% Al_2O_3

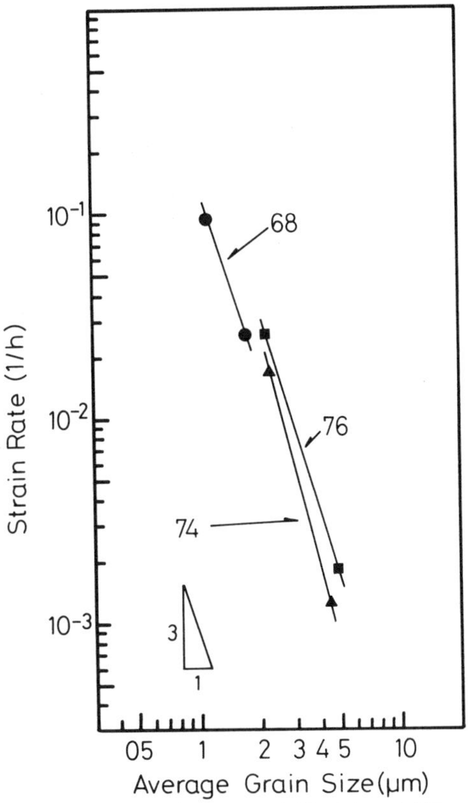

Fig. 7. Effect of grain size on strain-rate of 68, 74 and 76 wt% Al_2O_3 compositions.

compositions was nearly 1 and that of 68 wt% Al_2O_3 composition was 1.5 or 1.6 under this applied stress. Dokko et al.[2] has suggested that the creep mechanism of mullite is the Nabarro-Herring mechanism. However, the present results suggest that the Coble mechanism is operative.

Conclusions

The microstructure and high temperature mechanical properties of high-purity, near theoretically dense mullite, sintered using powder produced by the Sol-Gel method, have been investigated. The Al_2O_3/SiO_2 ratios ranged from 68 wt% Al_2O_3/32 wt% SiO_2 to 76 wt% Al_2O_3/24 wt% SiO_2.

1. The compositions with less Al_2O_3 than 73 wt% contained a small amount of glassy phase at the triple points of the mullite grains. In the compositions with 75 wt% Al_2O_3, α-Al_2O_3 particles were observed as an inter- and intragranular phase, the latter were found in the large mullite granular particles.

2. The formation of a liquid phase during sintering hindered the grain growth and allowed enlonged mullite grains to form. The α-Al_2O_3 particles also hindered grain growth.
3. The flexural strengths of the SiO_2-rich compositions at high temperatures exceeded those measured at R.T. because of the plastic deformation by sliding of the grain boundary and the stress relaxation at the crack fronts by the glassy phase. Compositions with no glassy phase had approximately the same flexural strengths at high temperature as those measured at R.T.
4. The strain-rate of the 74 wt% Al_2O_3 composition was lower than that of the compositions with more or less Al_2O_3. The strain-rate increased with increasing SiO_2 or Al_2O_3 content beyond the 74 wt% Al_2O_3 composition. The strain-rate was also influenced by the grain size, the glassy phase and the α-Al_2O_3 particles.
5. In this study, the creep mechanism of mullite was considered to be the Coble mechanism, which is controlled by grain boundary diffusion.

References

[1]H. Ohnishi, T. Kawanami, T. Hiraiwa, and T. Miyazaki, "Mechanical Properties of Mullite," p. 633-641 in processing of 2nd International Symposium of Ceramic Components for Engines, edited by W. Bunk and H. Hausner, Deutsche Keramische Gesellschaft Bad Honnef, 1986.

[2]P. C. Dokko, J. A. Pask, and K. S. Mazdiyasni, "High-Temperature Mechanical Properties of Mullite Under Compression," *J. Am. Ceram. Soc.*, **60** [3–4] 150–55 (1977).

[3]I. A. Aksay and J. A. Pask, "Stable and Metastable Equilibrium in the System SiO_2-Al_2O_3," *J. Am. Ceram. Soc.*, **58** [11–12] 507–12 (1975).

[4]S. Prochazka and F. J. Klug, "Infrared-Transparent Mullite Ceramics," *J. Am. Ceram. Soc.*, **66** [12] 874–80 (1983).

[5]T. Hiraiwa, K. Miyazaki, Y. Kimura, T. Kawanami, and H. Ohnishi, "Synthesis and Sintering of High Purity, Fine Grained Mullite Powders," Presented at the 169th Meeting of The Electrochemical Soc. (1986).

[6]G. W. Hollenberg, G. R. Terwilliger, and R. S. Gordon, "Calculation of Stresses and Strains in Four-Point Bending Creep Tests," *J. Am. Ceram. Soc.*, **54** [4] 196–99 (1971).

[7]R. L. Fullman, "Measurement of Particles Sizes in Opaque Bodies," *J. Metal.*, [3] 447–52 (1953).

Andalusite Derived Mullite-Matrix Refractories

ISRAEL PERETZ

North American Refractories Company (NARCO)
Research and Technology Center
3127 Research Drive
CATO Industrial Park
State College, PA, 16801, U.S.A.

RICHARD C. BRADT

University of Washington
Dept. of Materials Science and Engineering
Seattle, WA, 98195, U.S.A.

Abstract

Development of a mullite matrix in commercial refractory systems in combination with various mineralogically based aggregates is a highly desirable objective for aluminosilicate refractories. This paper describes an alumina-andalusite-ball clay mixture as the primary matrix components to develop mullite on firing. It also reports the related investigation of the properties of mullite matrix refractory systems incorporating structurally related aggregate grains, including: a calcined kaolin, an andalusite and a tabular alumina. Elastic moduli, modulus of rupture, fracture toughnesses and the work-of-fracture values were measured and are reported. Properties are discussed relative to the three different aggregates and the role of the thermal expansion mismatch between the mullite matrix and the aggregate grain and its effect on structural microcracking and related mechanical properties.

Introduction

Advantages of the presence of mullite in commercial refractories include: (i) its relatively high melting point,[1–8] (ii) the low coefficient of thermal expansion,[9,10] which leads to excellent thermal shock resistance, (iii) a high refractoriness under load, or a low creep rate[11–16] and (iv) its quite good resistance to slag attack.[17] These highly desirable characteristics are among the many reasons that underlie the increasing demand for a high mullite content in many aluminosilicate refractories.[18]

In the simplest sense, a refractory body may be considered as a two-component material system: the matrix and the aggregate grain. Most forms of chemical or mechanical degradation and failure of materials will naturally proceed through the weakest portion of the structure that is usually assumed to be the matrix of a refractory body. Consequently, when considering aluminosilicate refractory systems, a matrix consisting of a high content of mullite is an excellent basis on which to design the microstructure and to engineer specific physical properties into the refractory. Along with the matrix, the aggregate grain is an equally important refractory component, as it can also profoundly influence the properties of a refractory. There are several important characteristics to be considered for the aggregate grain, including the particle size and its distribution, the volume fraction of grain in the mix, and the refractoriness of the aggregate, etc. Each of these features of the aggregate grain is within the control of the refractory engineer during processing. When these considerations are coupled with the choice of the matrix, it enables the refractory engineer to microstructurally design the individual properties of the refractory body to meet the specific demands of the particular technical application of interest.

In many applications, refractories require a balance or a compromise of a number of individual properties, although frequently accentuating one or another. However, many of the properties of refractories are completely interrelated through their structural dependence. These interrelationships are such that maximizing any single property often results in the unfortunate consequence of a drastic reduction of another property or properties. For this reason the best refractory product for a particular application is often one that develops the most desirable compromise of properties. It is within that spirit of engineering compromise that this paper reports a study of the effects of microstructural design of a refractory mullite matrix with different aggregates on the physical and mechanical properties, as well as the thermal shock damage resistance of a refractory system designed about an initially optimized mullite matrix in combination with calcined kaolin, andalusite and tabular alumina aggregate grains.

Experimental

Materials: The Mullite Matrix and Aggregates

In principle, nearly all natural and synthetic raw materials containing Al_2O_3 and/or SiO_2 are potential candidates for inclusion in the list of possible constituents for development of mullite. However, in the practical sense, the raw materials for the synthesis of a high mullite content refractory matrix must meet a number of technical criteria which include a high level of reactivity and a limit to the concentration levels of specific impurities. Furthermore, for potential commercial utilization, availability and cost must also be seriously considered. For these combined reasons, the sources of "pure" Al_2O_3 which were selected for study included both a tabular alumina and a more reactive calcined alumina. Silica fume and potter's flint were chosen as the sources for "pure" SiO_2. Commercial grades of andalusite, domestic calcined kaolins of different alumina contents, and a ball clay were used in various combinations as the initial batch constituents to survey for the development of a workable mullite matrix refractory system.

A preliminary mullite matrix synthesis survey was completed for ten mixes of the above using only two-component raw material systems. A subsequent derivative mullite matrix synthesis study addressed the influence of additions of 5, 10, and 15 wt./o ball clay to the most promising two-component matrices to develop a workable three-component system. The three-component matrix systems which were investigated included combinations of calcined and tabular aluminas, fumed silica, andalusite, calcined kaolin and a calcined bauxitic kaolin.

Three types of aggregate grains were chosen to develop refractory systems in combination with the mullite matrices: a calcined kaolin, an andalusite, and a tabular alumina. These yielded three distinct refractory systems: one with an aggregate grain that is considerably more refractory than the mullite matrix, the tabular alumina, one with an aggregate grain that is considerably less refractory than the mullite matrix, the calcined kaolin, and also one with a similar, but somewhat less refractory aggregate grain, the andalusite.

The aggregate grain size distribution for each refractory system was determined on the basis of the Furnas[19] continuous grading equation for maximum density. An initial -3+150 aggregate size distribution was further modified to yield two additional aggregate grain size distributions, one a slightly coarser grain size distribution (-3+65) in which the finest aggregate size was 65 mesh and the other a slightly finer grain size distribution (-6+150).[20] For each aggregate type, three matrix contents were investigated: low, medium and high, containing 25%, 35% and 45% matrix, respectively. Combining the three aggregate size distributions and the three matrix contents results in a test matrix of nine mullite matrix refractory bodies. The designations of the refractory systems are L, M and H for the low, medium and high matrix contents and C, I and F for the coarse, intermediate and fine aggregate grain size distributions, respectively.

Results and Discussion

The Mullite Matrix

Figures 1 and 2 illustrate the effects of the firing temperatures and soaking times on the reaction rates of the two types of alumina with the fumed silica. The extent of mullite formation as measured by the X-ray method originally described by Tuttle and Cook[21] and some of the physical properties of these two systems are indicated. It is evident that the calcined alumina/fumed silica system yields the higher mullite content as compared to the tabular alumina based system. However, this system also possesses the lowest density and the highest volume changes, two features which are not particularly desirable for most commercial refractory manufacturing processes. A similar dilemma exists when the other raw material systems are compared. For example, the porosity of the fumed silica/tabular alumina system is the lowest through 1600°C as shown in Figure 3, however, in contrast, the mullite content development of this system is also one of the lowest, see Figure 4.

The addition of the ball clay to develop a commercial system further complicates the problem, although it significantly improves the refractory body pressability and reduces the porosity as shown in Figure 5. It also increases the impurity content, and thus may be expected to affect the high temperature mechanical properties and, subsequently, the refractory performance at elevated temperatures. However, the room temperature physical properties for each of the systems, except the fumed silica-based ones, are improved when the ball clay is incorporated into the matrix, provided that the firing temperature does not exceed 1600°C.

The apparent contradiction of good physical properties and low mullite content exists for many two and/or three-component mullite matrix systems. For firing temperatures in the conventional aluminosilicate refractory firing range, that is between 1400° and 1500°, the optimal compromise mullite matrix appears to be that of the tabular alumina-andalusite -15% ball clay system. It is highly ranked relative to all of the others as to its physical properties as well as its high mullite content. For those reasons, that component combination system was utilized for the subsequent mullite matrix development in all of the refractory formulations.

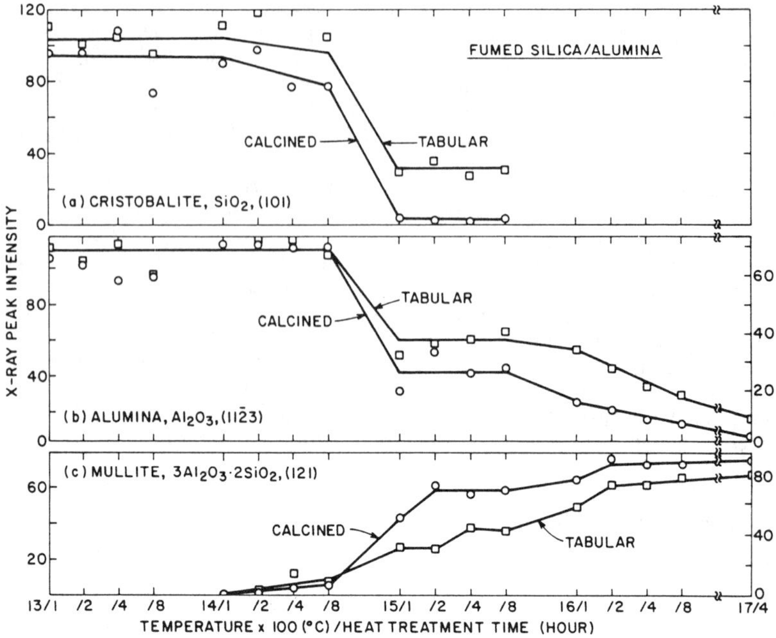

Fig. 1. X-ray diffraction intensities of the two component matrix systems based on fumed silica and alumina as a function of firing temperature and soaking time.

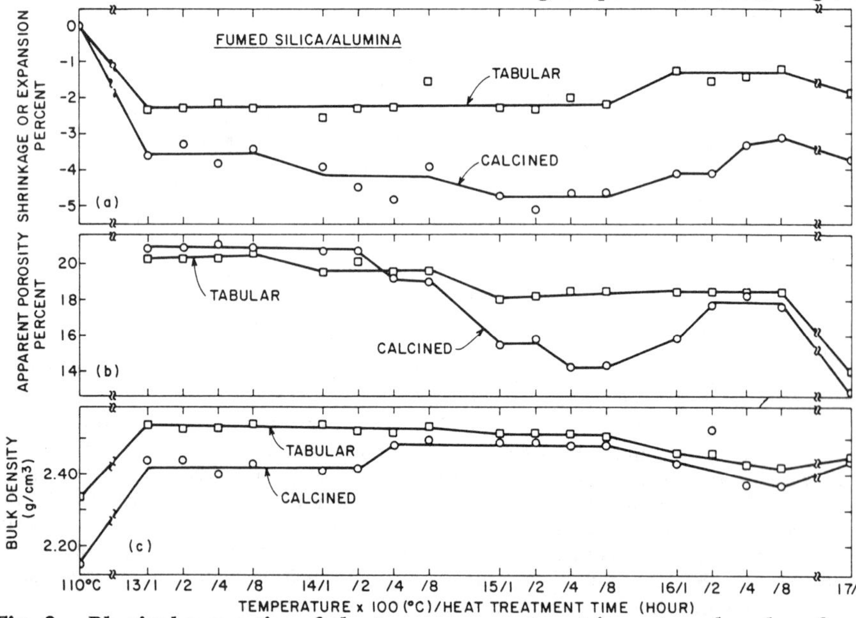

Fig. 2. Physical properties of the two component matrix systems based on fumed silica and alumina as a function of firing temperatures and soaking time.

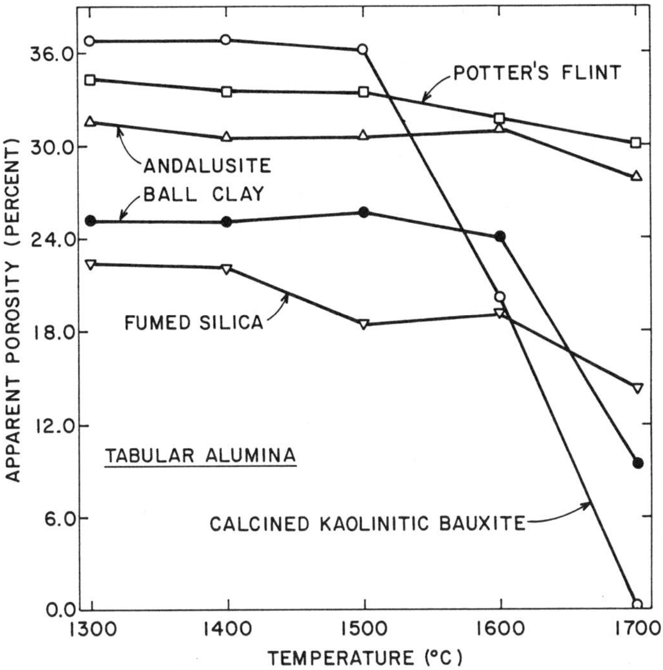

Fig. 3. Apparent porosity variation with firing temperature for the tabular alumina based systems (4 hours at temperature).

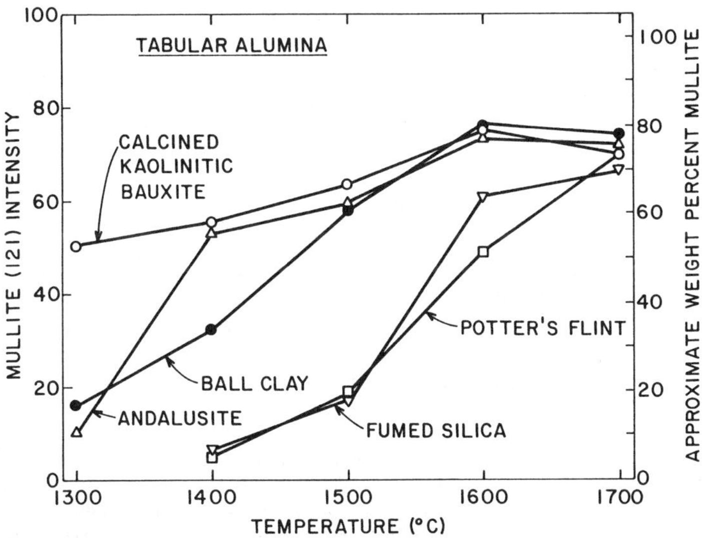

Fig. 4. Mullite (121) peak intensity as a function of firing temperature for the tabular alumina based systems (4 hours at temperature).

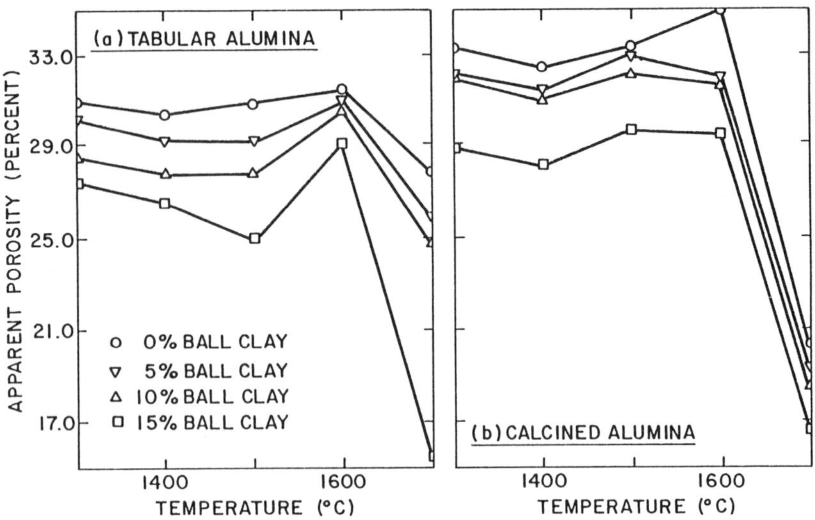

Fig. 5. Influence of ball clay additions on the apparent porosity in the andalusite-alumina systems.

Refractory Body Physical Properties

The green density of an aluminosilicate refractory body is usually directly related to the densities of the individual constituents, the aggregate and the matrix size distributions, the forming or pressing pressure and the moisture content of the mix. If the last two processing parameters are kept nearly constant during processing, then the density becomes a function solely of the densities of the individual components and the size distributions of the aggregates and of the matrix. Table 1 illustrates this effect on the green densities of the andalusite aggregate refractories. If the densities of these andalusite aggregate bodies are examined, where the aggregate and the matrix have approximately the same density, then it is evident that as the matrix content increases the bulk density decreases. The same trend is also apparent within the tabular alumina aggregate refractory systems, where both the microstructure and these are much higher densities for the lower matrix content systems as compared to the higher matrix content systems.

The density after firing is also shown in Table 1. It is a complicated combination of the green density, the sintering process and the expansion during mullitization of the bodies during firing, however, the density trends summarized in Table 1 are quite similar to those of the previously discussed green densities. It is appropriate to also note that the fired densities are consistently lower than the green densities, a somewhat unusual event. The primary contributions to this effect are water loss and the decrease of density during mullitization.

Mechanical Properties

Table 2 summarizes the effects of the refractory aggregate grain, its size distribution and the refractory body matrix contents on the Young's modulus of the three refractory systems. Generally as the matrix content increases and the aggregate size distribution becomes finer, the elastic moduli will be expected to increase.

Table 1. Green and fired densities of the andalusite grain refractory systems.[+]

	C	I	F
L	2.80/2.61	2.80/2.62	2.75/2.59
M	2.80/2.59	2.77/2.60	2.74/2.56
H	2.71/2.53	2.69/2.50	2.67/2.46

[+](green/fired) (g/cc)

Table 2. Young's moduli of the fired refractory bodies (GPa).

	C	I	F
Calcined Kaolin			
L	15.65	18.08	17.70
M	13.79	15.20	27.69
H	21.73	27.31	22.17
Andalusite			
L	24.84	27.25	31.42
M	33.32	35.10	42.75
H	42.17	31.18	36.60
Tabular Alumina			
L	15.29	16.10	18.01
M	22.06	14.69	18.65
H	21.84	21.45	25.06

Similar trends exist for the room temperature modulus of rupture. As the matrix contents increase and a finer aggregate is incorporated into the mix, the flexural strengths generally increase as shown in Table 3. From these results, it is apparent that both the elastic moduli and the flexural strengths of the andalusite aggregate grain refractories are usually much higher than the equivalent calcined kaolin and tabular alumina aggregate grain refractory bodies.

Figure 6 is a comprehensive plot of the room temperature flexural strengths or cold modulus of rupture values for all of the refractory bodies versus their Young's moduli. As normally expected, these two physical properties are directly related within each type of aggregate grain refractory body. The slopes of the individual lines for each of the individual aggregate grain types are of further interest. The fact that the plots of room temperature flexural strengths versus Young's moduli yields straight lines for each of the aggregate types suggests that these results may be

Table 3. Flexural strengths (MOR) of the refractory bodies (MPa).

	C	I	F
Calcined Kaolin			
L	6.60±0.30	7.47±0.27	8.80±0.44
M	10.04±0.45	9.56±0.53	11.85±0.49
H	10.56±0.33	10.91±0.44	11.55±0.47
Andalusite			
L	10.82±0.33	12.05±0.48	12.40±0.32
M	12.41±0.44	12.52±0.53	13.55±0.49
H	13.93±0.57	11.91±0.64	14.06±0.72
Tabular Alumina			
L	5.68±0.27	6.35±0.24	7.49±0.37
M	8.84±0.41	9.38±0.36	10.04±0.22
H	10.04±0.22	12.06±0.15	12.96±0.36

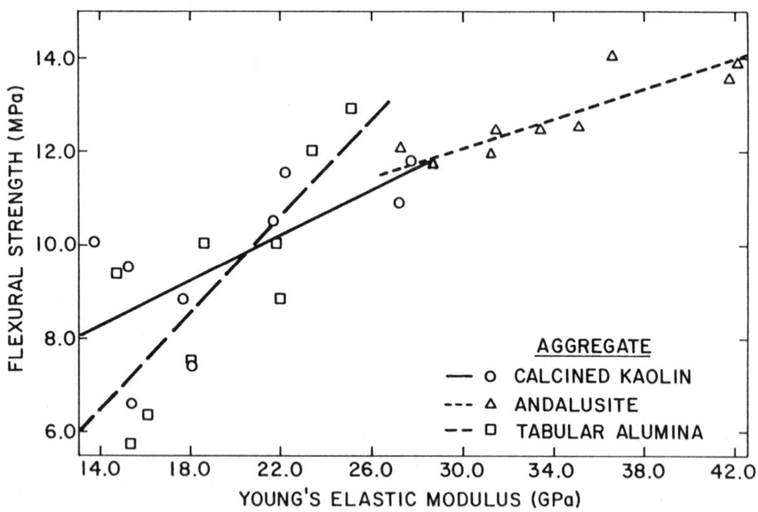

Fig. 6. Flexural strength variation as function of the Young's elastic modulus for the refractory systems.

further considered in terms of the strengths of the individual aggregates.[22] The slope of the line for the andalusite aggregate system is 1.60×10^{-4} and that of the calcined kaolin is similar, but just slightly greater at $2.35 \cdot 10^{-4}$, while the tabular alumina is more than twice as large, 5.23×10^{-4}.

The general coordinates of the various aggregate grain structures in the flexural strength-Young's modulus diagram reflects some aspects of the reactions which these materials experience during firing. Although both the calcined kaolin aggregate grain and the andalusite aggregate grain systems are located in different regions of the strength/modulus diagram, their slopes are similar. Were it not for the separate refractory system containing the tabular alumina aggregate grain, one might even attempt to apply a single straight line to describe these two aggregate systems. The reason for the similarity of these two refractory systems is the emergence of the mullite component from the aggregate decomposition process. Although the amounts and character of the mullite from the calcined kaolin and that from the andalusite are certain to be quite different, Figure 6 very clearly suggests that the general effects of that mullite component on the Young's modulus/flexural strength relationship are quite similar.

If the fracture toughness[23] trends in Table 4 are compared with the flexural strength trends, it is evident that they are very nearly the same, a trend also reported by Schumacher.[24] As the refractory matrix contents increase and the aggregate grain size distributions become finer, the fracture toughness values increase. Moreover, the fracture toughness of the andalusite refractory system is the highest, as might have been expected. Figure 7 is a comprehensive plot of the room temperature

Table 4. Fracture toughnesses of the refractory bodies ($MPa \cdot m^{1/2}$).

	C	I	F
Calcined Kaolin			
L	0.46±0.05	0.47±0.04	0.47±0.02
M	0.57±0.02	0.71±0.06	0.76±0.08
H	0.63±0.05	0.72±0.07	0.65±0.05
Andalusite			
L	0.78±0.03	0.80±0.05	0.82±0.04
M	0.84±0.03	0.89±0.04	0.90±0.03
H	0.86±0.05	0.84±0.03	0.93±0.02
Tabular Alumina			
L	0.38±0.05	0.45±0.05	0.49±0.04
M	0.62±0.06	0.60±0.04	0.62±0.05
H	0.65±0.04	0.70±0.05	0.77±0.06

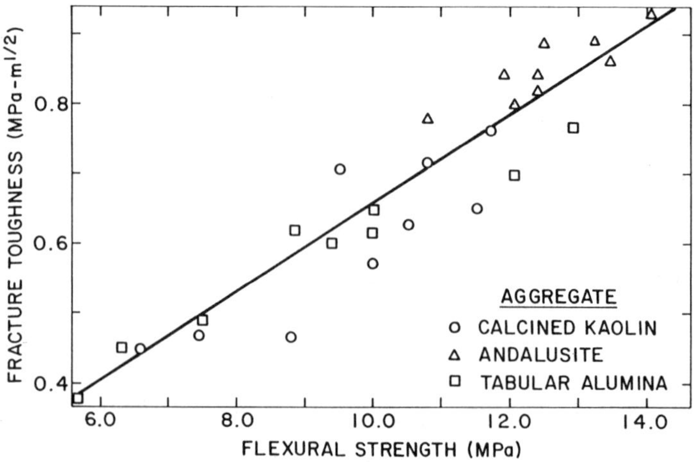

Fig. 7. Flexural strength variation as function of fracture toughness for the refractory systems.

flexural strengths versus the fracture toughness values for all 27 of the mullite matrix refractory bodies in the three different aggregate grain systems. All of the refractories lie on the same trend line. An important consequence of this result is that for all of these mullite refractory systems, the critical flaw size is relatively constant. This confirms that the flexural strengths and fracture toughness of these refractories are directly related and that the structural modification of one of these parameters may be expected to result in a corresponding variation of the other.

The effects of temperature on the flexural strengths of the calcined kaolin aggregate grain refractories are presented in Figures 8 and 9. It is evident that the finest aggregate distributions and the highest matrix contents impart the highest flexural strengths to this refractory system over the temperature interval investigated here, room temperature to 1400°C. Figure 10 summarizes the general effects of the aggregate grain type on the flexural strengths. The calcined kaolin aggregate grain refractory system has the lowest flexural strengths over the complete temperature interval. The andalusite aggregate grain refractory system, which has a lower liquid silicate content, has the highest strengths at low temperatures, but only moderate strengths at 1400°C. The tabular alumina aggregate refractory system possesses only moderate strengths at lower temperatures. However, the glassey silicate content of this refractory system is precisely the correct one to permit crack healing, or blunting, thus enabling the strengths to increase at moderately elevated temperatures.

The same effects of temperature which were observed for the flexural strengths also persist for the fracture toughnesses as might be expected from the previous correlation of the two properties. Higher matrix contents as well as finer aggregate size distributions, consistently yield higher fracture toughnesses. Figure 11 illustrates the matrix content and temperature effects on the fracture toughnesses of the tabular alumina aggregate grain refractory systems, while Figure 12 illustrates the temperature dependence of the fracture toughnesses for all three aggregate grain systems.

Distinct differences exists in the resistance to stable crack propagation between the different aggregate refractory systems at the elevated temperatures near to their

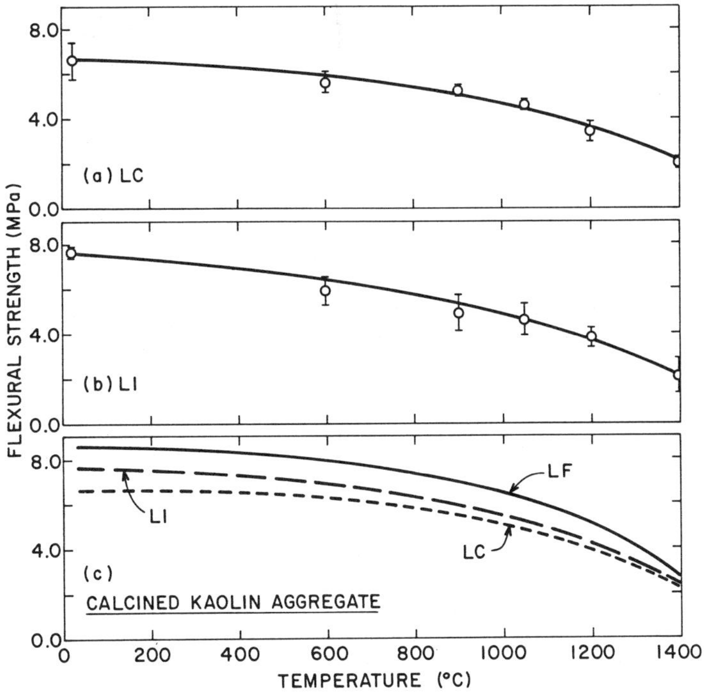

Fig. 8. **Temperature effects on the flexural strengths of the low matrix content calcined kaolin aggregate refractory systems.**

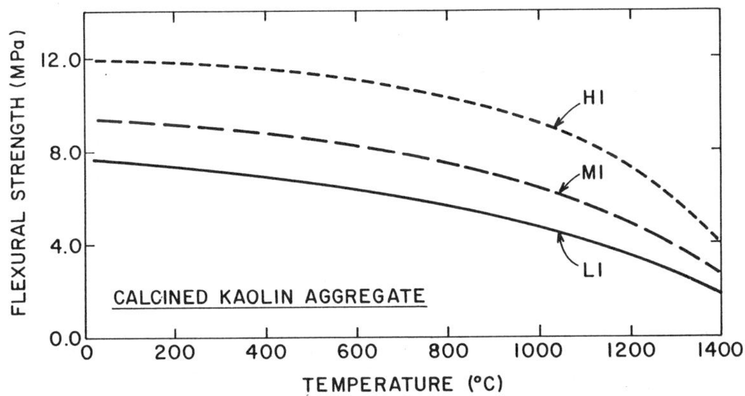

Fig. 9. **Matrix content and temperature effects on the flexural strengths of the calcined kaolin aggregate refractory systems.**

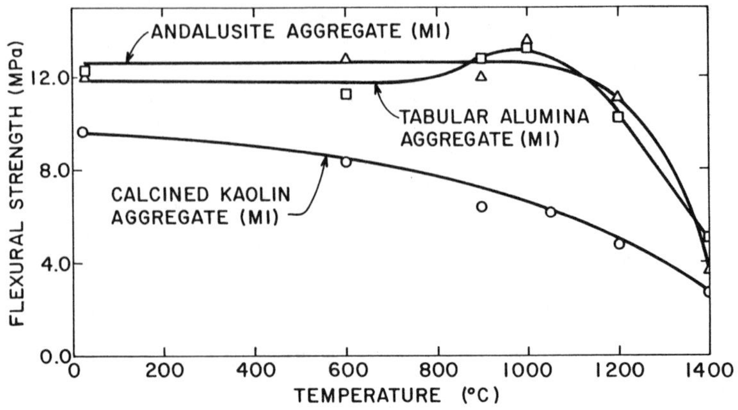

Fig. 10. Aggregate type and temperature effects on the flexural strengths of the refractory systems.

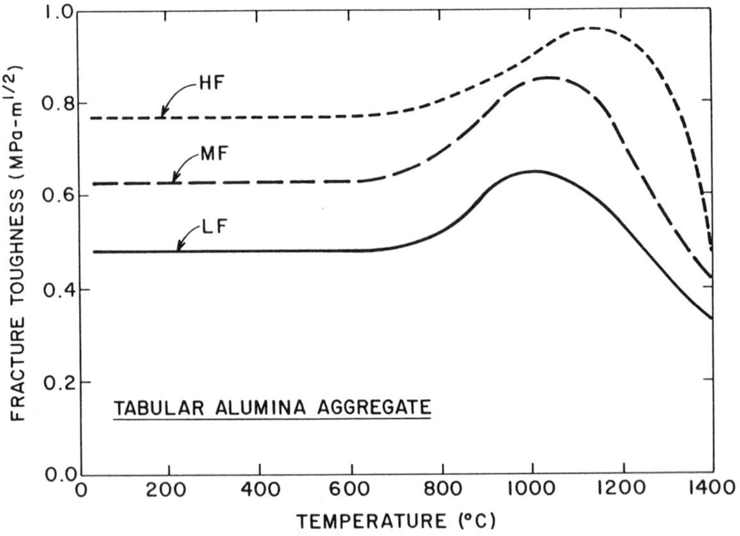

Fig. 11. Matrix content and temperature effects on the fracture toughnesses of the tabular alumina aggregate refractory systems.

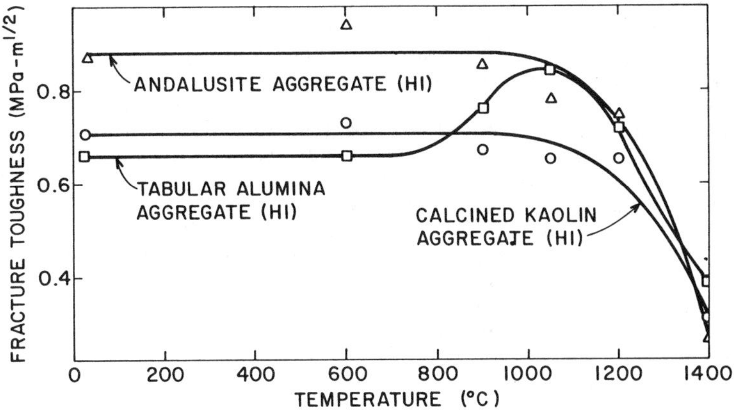

Fig. 12. Aggregate type and temperature effects on the fracture toughnesses of the refractory systems.

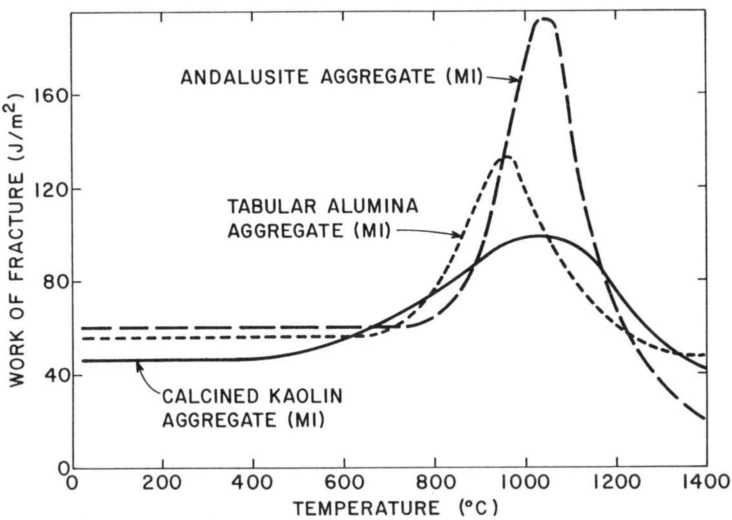

Fig. 13. Typical aggregate type and temperature effects on the work-of-fracture values of the refractory systems.

respective high temperature maxima. These differences derive primarily from the aggregate grain purity levels which affect the viscosity of the liquid silicates in these refractory systems. However, the internal flaw population in these refractories may also assume a secondary role. These differences are clearly illustrated with the work-of-fracture results measured by the notched-beam technique[25–26] and shown in Figure 13. For the calcined kaolin aggregate grain refractory system, which is the system with the highest impurity levels, a softening process resulting in an increase work-of-fracture begins as low as about 500°C. A maximum in the work-of-fracture occurs between 1000°C and 1100°C. However, because of the high liquid silicate contents,[27] the structural bonding is also affected, and the maximum work-of-fracture values are only about twice the room temperature level.

The softening and flow of silicates in vicinity of the crack tip process zone does not become very significant until about 800°C in the andalusite aggregate refractory grain system for which very high work-of-fracture values are obtained between 950°C and 1150°C. Because of the lower total structural content of the liquid silicate phases in these refractories, the cohesive strength is much greater than that of the calcined kaolin aggregate refractory bodies and thus the work-of-fracture values are substantially higher. Viscous silicate crack tip blunting processes do not to become very effective within the tabular alumina aggregate grain refractory system until about 700°C and are less effective by 1050°C. Nevertheless the tabular alumina aggregate refractory system retains relatively high work-of-fracture values to 1400°C, mainly as a consequence of the low, perhaps nearly optimum, liquid silicate content.

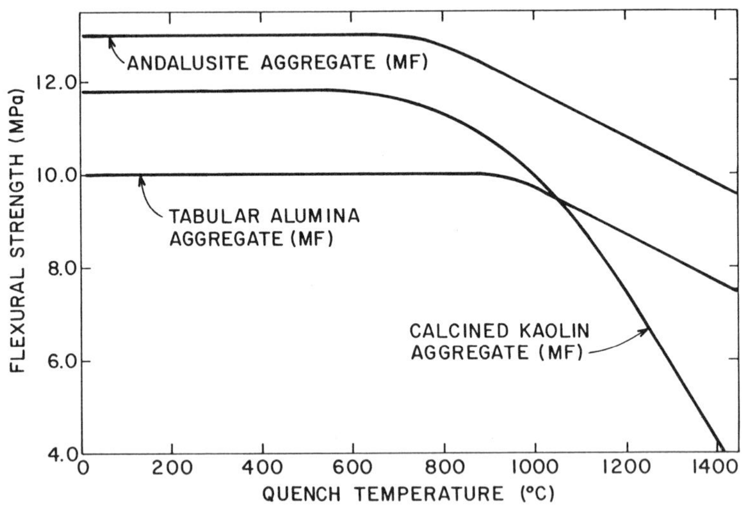

Fig. 14. Strength variations with severity of thermal shock quench temperature for the three aggregate refractory systems.

Thermal Shock Damage Resistance

From an experimental thermal shock damage initiation viewpoint the most susceptible of these mullite-matrix refractory systems are the calcined kaolin aggregate bodies. The most thermal shock damage resistant bodies appear to be those with the tabular alumina aggregate grains, while the andalusite aggregate grain bodies are intermediate. This behavior is illustrated in Figure 14. The fact that the tabular alumina aggregate grain refractories are the least susceptible to thermal shock crack initiation might be attributed to the difficulty of crack propagation through the dense tabular alumina grains. However, it might also be directly related to the mullite (low thermal expansion) and corundum (high thermal conductivity) contents for they should reduce the thermally generated stresses during the thermal shock tests.[28–31] On the same basis, the calcined kaolin aggregate refractories which are characterized by

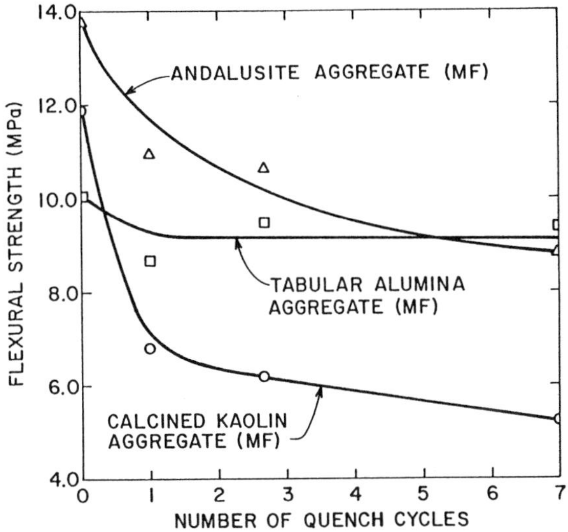

Fig. 15. Strength variations for multiple quench cycles from 1200°C for the three aggregate refractory systems.

the lowest mullite and corundum contents may be expected to have the poorest resistance to thermal shock damage.

Figure 15 compares the medium matrix content, finest aggregate distribution bodies of the three different aggregate refractory systems with regard to the effects of multiple thermal shock quench cycles on the strength degradation. Unlike the calcined kaolin refractory systems which lost a major portion of their strengths after the initial thermal shock quench cycle, or the andalusite refractories which also lost a substantial portion of their strengths during the initial thermal shock and continued to degrade during the next few cycles, the tabular alumina aggregate grain refractories incurred very little damage in terms of strength losses. The tabular alumina aggregate refractories retained nearly all of their strength, even after seven thermal shock cycles.

The stresses which are generated during these thermal shock quenches are related to the thermal expansion, the thermal conductivity, and the glass and the crystalline phase contents of these refractories. Thus, the tabular alumina aggregate refractories, which incorporate the strongest aggregates, retained most of their strength after the first thermal shock quench cycle, while the calcined kaolin aggregate refractories lost the highest portion of their strength during the initial thermal shock.

Thermal Shock Damage Resistance Parameters

The thermal shock damage resistance parameter, R'''', was specifically derived by Hasselman[32–33] for application to high strength materials with relatively short crack lengths which are believed to be susceptible to kinetic crack growth. The thermal shock damage resistance parameter R_{st} was theoretically derived to apply to quasi-static crack extension and is the most logical one to be applied to most refractory materials with relatively large initial crack sizes. The values of R'''' and R_{st} were calculated for the 27 refractory systems based on the following equations:

$$R'''' = \frac{E\gamma_{WOF}}{\sigma_f^2 (1-\nu)} \quad (1)$$

$$R_{st} = \frac{\gamma_{WOF}}{\alpha^2 E} \quad (2)$$

where γ_{WOF} is the work of fracture surface energy, E is the elastic modulus, σ_f is the strength, ν is Poisson's ratio and α is the thermal expansion coefficient. Considering the defining equation for R'''', it can further be shown that:

$$R'''' \alpha \frac{\gamma_{WOF}}{\gamma_{NBT}} = \frac{2E\gamma_{WOF}}{K_{Ic}^2} \quad (3)$$

Tables 5–7 summarize the effects of matrix content and aggregate size distribution for the three aggregate system on the thermal shock damage resistance parameters R'''', and R_{st} and the ratio $\gamma_{WOF}/\gamma_{NBT}$. These tables clearly illustrate that as the matrix content increases and, generally, as the aggregate grain size distribution becomes finer, the thermal shock damage resistance is negatively affected. However, very poor correlation exists between the thermal shock damage resistance parameters and the strength loss, which was observed during the prism quench test.

Table 5. Calculated thermal shock damage resistance parameters for the calcined kaolin aggregate refractories.

	C	I	F
$R''''(m\times10^2)$			
L	2.09	1.85	1.38
M	1.06	0.87	1.29
F	1.22	1.40	0.92
$R_{st}(m^{1/2\circ}C)$			
L	11.19	9.79	9.44
M	12.94	10.27	7.99
F	9.70	8.08	8.86
$(\gamma_{WOF}/\gamma_{NBT})$			
L	7.76	8.36	8.26
M	5.92	5.81	5.52
F	5.30	5.47	5.06

Table 6. Calculated thermal shock damage resistance parameters of the andalusite aggregate refractories.

	C	I	F
$R''''(m \times 10^2)$			
L	1.41	1.33	1.29
M	1.56	1.71	1.53
F	1.33	1.10	1.05
$R_{st}(m^{1/2}{}^\circ C)$			
L	9.01	8.68	8.01
M	7.75	7.56	6.99
F	6.53	7.13	6.77
$(\gamma_{WOF}/\gamma_{NBT})$			
L	5.01	5.44	5.64
M	5.15	5.42	6.36
F	6.88	4.28	4.20

Table 7. Calculated thermal shock damage resistance parameters of the tabular alumina aggregate refractories.

	C	I	F
$R''''(m \times 10^2)$			
L	2.47	2.53	1.75
M	2.19	1.14	1.16
F	1.23	0.99	1.05
$R_{st}(m^{1/2}{}^\circ C)$			
L	7.24	7.36	6.34
M	7.18	8.95	7.59
F	6.98	7.82	6.72
$(\gamma_{WOF}/\gamma_{NBT})$			
L	10.14	9.06	6.98
M	7.44	4.46	5.50
F	5.58	5.34	5.03

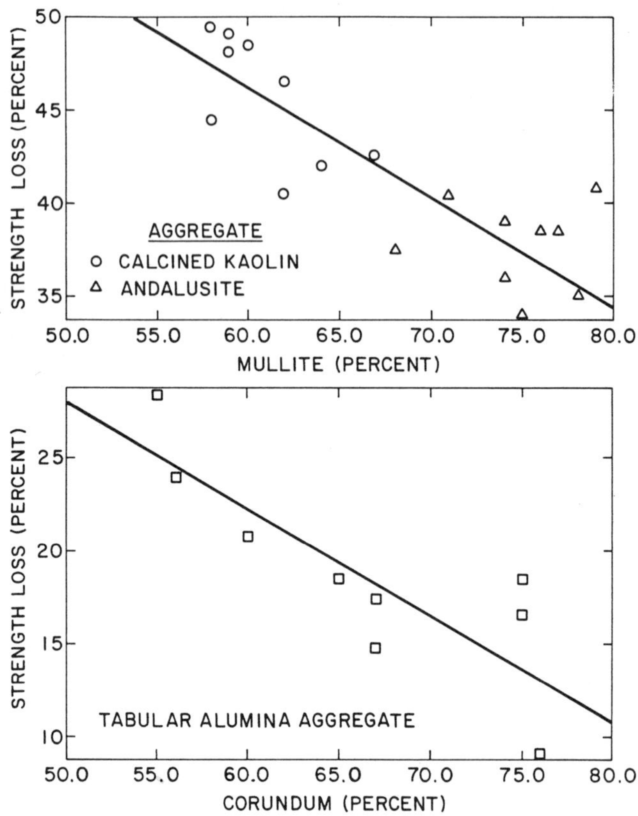

Fig. 16. Percentage strength loss variation with: (a) the mullite percentage of the calcined kaolin and andalusite aggregate refractories, (b) the corundum percentage of the tubular alumina aggregate refractories.

In contrast, Figure 16 shows that a good correlation exists between the strength loss after the thermal shock test to the mullite content in the calcined kaolin and andalusite systems and to the corundum content in the tabular alumina system, independently. In Figure 16, the percentage strength loss versus mullite and corundum content diagrams illustrate and further confirm the concepts advanced relative to the strength versus elastic modulus diagram in Figure 6. The calcined kaolin and andalusite aggregate grain refractories both ascribe to the same line, while results for the tabular alumina follow a separate line in an entirely different coordinate region of the general diagram. It might be speculated that the reason for the much lower percentage strength loss, or reduced thermal shock degradation of the tabular alumina aggregate grain bodies is the higher strengths of the individual tabular aggregate grains and the interaction of the thermal stress generated cracks with those tabular alumina aggregate grains.

Summary and Conclusions

The concept of a mullite-matrix refractory body was investigated for several different aggregate grain types. After exploring the use of different constituents, an optimal matrix composition based on tabular alumina, andalusite and ball clay was developed. It was observed during this matrix development that even though many different raw material combinations are possible to achieve mullite formation, the minor impurity levels and the specific details of the component reactions assume major roles in determining the final mullite content and its properties.

Combination of a mullite matrix with various refractory aggregate grains resulted in satisfactory refractory bodies. Calcined kaolin and andalusite aggregate grains exhibit many similarities in trends of physical properties, a not too surprising observation as both of these aggregates yield a substantial percentage of additional mullite from decomposition during firing. Although both of these types of aggregate grains fall on the same lines for strength and toughness variation, the andalusite aggregate bodies are generally superior in most instances. This point further illustrates the importance of the character of the mullite source on the resulting physical properties of the refractory body.

The refractory body utilizing a tabular alumina aggregate grain had distinctly different physical properties, usually not quite as outstanding as the andalusite based system. However, the high strength tabular alumina grains did impart characteristic strength features to these refractory systems and these exhibited improved characteristics for applications where extensive crack propagation and direct crack-aggregate grain interactions are of importance.

References

[1]N. L. Bowen and J. W. Greig, "The System Al_2O_3-SiO_2," *Journal of the American Ceramic Society*, **1**, pp. 238–254 (1924).

[2]E. S. Shepherd, G. A. Rankin, and W. Wright, "The Binary Systems of Alumina with Silica, Lime with Magnesia," *American Journal of Science*, **28**, pp. 301–307 (1909).

[3]N. A. Toropov and F. Y. Galakhov, "New Data on the Alumina-Silica System," *Dokladay Akadenii Nauk, S.S.S.R.*, **78**, (2), pp. 299–302 (1951).

[4]N. A. Toropov and F. Y. Galakhov, "Solid Solutions in the System Al_2O_3-SiO_2," Izvestiya Akademii Nauk S.S.S.R., Otdetenie Khimicheskikj Nauk, No. 1, pp. 8–11 (1958).

[5]K. Konopicky, "The Equilibrium Diagram of the System SiO_2-Al_2O_3," Bulletin de la Societe Francaise de Deramique, No. 33, pp. 33–36 (1956).

[6]G. Troenel, K. H. Olst, K. Konopicky, H. Baver, and I. Patzah, "Investigation in the SiO_2-Al_2O_3 System," *Berichte de Deutche Keramische Gesellschafti*, **34**, (12), pp. 397–402 (1957).

[7]A. Aramaki and R. Roy, "Revised Phase Diagram for the System Al_2O_3-SiO_2," *Journal of the American Ceramic Society*, **45**, (5) pp. 229–242 (1962).

[8]I. A. Aksay and J. A. Pask, "The Silica-Alumina System: Stable and Metastable Equilibria at 1.0 Atmosphere," *Science*, **183** (4210), pp. 69–71 (1974).

[9]W. D. Kingery, H. K. Bowen, and D. R. Uhlmann, Introduction to Ceramics 2nd Edition, John Wiley and Sons, Inc. pp. 594–595 (1976).

[10]K. S. Mazdiyasni and L. H. Brown, "Synthesis and Mechanical Properties of Stoichiometric Alumina Silicate (Mullite)," *Journal of the American Ceramic Society*, **55**, (11), pp. 548–552 (1972).

[11]V. A. Burdick and D. E. Day, "Creep of High-Alumina Refractories," *American Ceramic Society Bulletin*, **48** (12), pp. 1109–1113 (1969).

[12]J. R. Kreglo and W. J. Smothers, "Creep of Selected High-Alumina Brick," *Journal of Metals*, **19** (7), pp. 20–22 (1967).
[13]C. O. Hulse and J. A. Pask, "Analysis of Deformation of Fireclay Refractories," *Journal of the American Ceramic Society*, **49**, (6), pp. 312–317 (1966).
[14]R. E. Beyer, R. D. Ek, J. L. Scott, G. A. Zevgrer and O. J. Whittemore, "Creep of Superduty Fireclay Brick under Compression," *American Ceramic Society Bulletin*, **55**, (12), pp. 1045–1051 (1976).
[15]J. F. Clements and J. Vyse, "Creep Measurements of Some High-Alumina Refractories," *Transaction of the British Ceramic Society*, **65**, (2), pp. 59–85 (1966).
[16]P. A. Lessing, R. S. Gordon, and K. S. Mazdiyasni, "Creep of Polycrystalline Mullite," *Journal of the American Ceramic Society*, **58**, (3–4), pp. 149 (1975).
[17]J. H. Chesters, Refractories: Production and Properties, The Iron and Steel Institute, London, pp. 314–315 (1973).
[18]J. C. Hicks, "Are Refractory Materials Becoming a Problem?" *American Ceramic Society Bulletin*, **54**, (7), pp. 644–646 (1975).
[19]C. C. Furnas, "Grading Aggregates, Mathematical Relations for Beds of Broken Solids of Maximum Density," *Industrial and Engineering Chemistry*, **23**, (9), pp. 1052–1058 (1931).
[20]I. Peretz, Mullite Matrix Refractories: Mechanical Properties and Thermal Shock Damage, Ph.D. Thesis, Ceramic Science, The Pennsylvania State University, May, (1983).
[21]M. A. Tuttle and R. L. Cook, "Fundamental Study of Crystalline and Glassy Phases in Whiteware Bodies," *Journal of the American Ceramic Society*, **32**, (9), pp. 279–294 (1949).
[22]I. Peretz, S. Perrella and R. C. Bradt, "Aggregate Particle Strengths and the Strength of Aluminosilicate Refractories," 549–559 in Proc. 1st Int. Conf. Ref., Tokyo Japan, Nov. (1983).
[23]W. F. Brown and S. E. Snawley, "Plane Strain Crack Toughness Testing of High Stress Metallic Materials," American Society for Testing and Materials, ASTM Special publication, 40, No. 410, pp. 13–15 (1966).
[24]C. A. Schumacher, Fracture of Alumina Refractories, M.S. Thesis, Ceramic Science, The Pennsylvania State University, August, (1980).
[25]J. Nakayama, "Direct Measurement of Fracture Energies of Brittle Hetrogeneous Materials," *Journal of the American Ceramic Society*, **48**, (1), pp. 583–587 (1965).
[26]J. H. Ainsworth and R. G. Herron, "High Temperature Fracture Energy of Refractories," *American Ceramic Society Bulletin*, **55**, (7), pp. 655–665 (1976).
[27]F. Nadachowski, Assessment of the Phase Composition of Fireclay Refractories Based on Ternary Diagrams," *Refractory Journal*, (4), pp. 126–131, April (1965).
[28]W. D. Kingery, "Factors Affecting Thermal Stress Resistance of Ceramic Materials," *Journal of the American Ceramic Society*, **38**, (1), pp. 3–15 (1955).
[29]K. T. Farber, M. D. Huang and A. G. Evans, "Quantitative Studies of Thermal Shock in Ceramic Based on a Novel Test Technique," *Journal of the American Ceramic Society*, **64**, (5), pp. 291–301 (1981).
[30]Y. T. Chen, F. T. Lee, H. Y. Chang, and Y. C. Ko, "Thermal Expansion of Aluminosilicate Refractory Brick," *American Ceramic Society Bulletin*, **61**, (8), pp. 866–871 (1982).
[31]Y. T. Chen, T. F. Lee, and Y. C. Ko, "Thermal Conductivity of Fireclay and High-Alumina Refractory Brick," *American Ceramic Society Bulletin*, **61**, (7), pp. 718–724 (1982).
[32]D. P. H. Hasselman, "Elastic Energy of Fracture and Surface Energy as Design Criteria for Thermal Shock," *Journal of the American Ceramic Society*, **46**, (11), pp. 535–540 (1963).

[33]D. P. H. Hasselman, "Unified Theory of Thermal Shock Fracture Initiation and Crack Propagation in Brittle Ceramics," *Journal of the American Ceramic Society*, **52** (11), pp. 600–6–5 (1969).

The Behavior of Mullite Particles Contained in Cermet Friction Materials

MASAAKI MISUMI, TAKASHI HANAZAWA, AND MASAFUMI OHKUBO

Department of Mechanical Engineering
Faculty of Engineering
Seikei University
3-3-1 Kichijoji-kita-machi, Musashino-shi, Tokyo 180

Abstract

In this study, three kinds of cermet were tested with a dynamometer test machine under several conditions. The observation of the sliding surface showed that the superiority of the cermet depends on the crushing behavior of mullite particles. The crushing behavior of mullite, silica, alumina and sillimanite was investigated by a slow compression test of a thin powder layer using a mortar. The load-displacement relation of powdery layer fitted Matsui's Equation, but the true energy for crushing of mullite and silica was larger than that calculated by Matsui's Equation. Their fracture mode was crumbling and different from that of glassy alumina. The crushing rate after compression was proportional to the compressive stress, and the gradient could be a simple measure of crushing easiness.

Introduction

The cermet friction material is a composite which has the three components of (1) metals, (2) ceramics and (3) solid lubricants. This material is used for severe sliding conditions such as that of aircraft braking. It's sliding surface, in that case, is exposed, partially, to high temperatures (800~1000°C) and high pressures (several hundreds of kgf/cm^2). High velocity shearing forces, in addition, act on the surface of the cermet. The metal component of the cermet is plastically deformed under such conditions. The ceramic component, on the other hand, hardly deforms plastically because it is supported by metal component, or ploughs the mating plates, or is crushed and distributed over a sliding surface. Such mechanical behavior of the ceramic component produces profound effects upon (1) the mating-plate-wear, (2) the friction coefficient and (3) the wear resistance of the cermet.

This work emphasizes the importance of the crushing behavior of ceramic particles in the cermet. It clarifies the relationship between the friction properties of cermet and the crushing behavior of ceramic particles.

The superiority of mullite containing cermet for aircraft brake linings probably relies on the crushing behavior of mullite particles. Compared with the crushing behavior of the other ceramics (alumina, sillimanite, etc.), the mullite particles are characteristic. The origin of the above difference is considered from the view point of the crushing rate and the microstructure of the ceramic particles in cermet.

Table 1. Composition of samples (Wt.%).

Table.1 Composition of samples (Wt.%)

Sample	Metal Matrix			Wear Resisting Component				Lubricant
	Cu	Sn	Zn	Fe	Mo	SiO_2	Mullite	Graphite
A	47	3	5	.8	5	24		8
B	85	4	-	-	-	-	6	5
C	85	4	-	-	-	6	-	5

Testing Material and Experimental Procedure

Sliding Test

Table 1 shows the composition of three test samples. Sample A (Fig. 1a) which includes 24% ceramic particles of silica and mullite is the brake lining for aircraft, which was slid against Cr-Mo steel plate under the condition of 65 times of absorbed energy 290 $Kgfm/cm^2$ (common use) and one time of 475 $Kgfm/cm^2$ (emergency use), as determined with a dynamometer test machine. Sample B (Fig. 1b) which includes 6% mullite particles and C (Fig. 1c) which includes 6% silica particles are the facings for automobile clutches which were slid against FC25 cast iron plate 5000~10000 times under the condition of absorbed energy 5~26 $Kgfm/cm^2$. The maximum temperatures were 800°C for brake lining and 300°C for clutch facing. The sliding surface and cross section near the surface were metallographically observed by a microscope under a magnification of ×400.

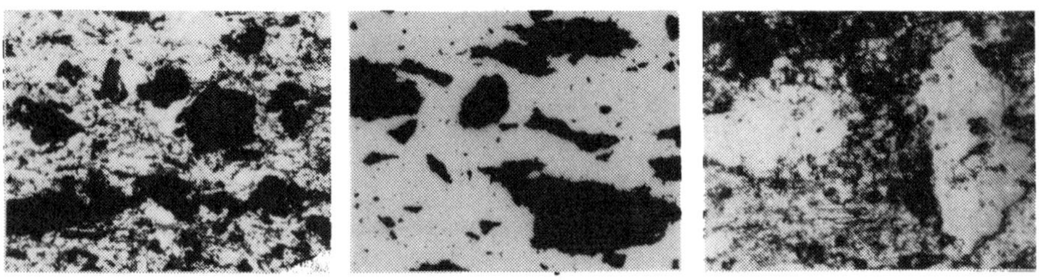

(a) Sample A (b) Sample B. (c) Sample C.

Fig. 1. Samples for sliding test.

Compression Crush Test of Ceramics Powder

A static compression test was carried out to study the crushing characteristics of ceramics particles, under a loading speed of 10 ton/min for several kinds of powder layers (1 mm thickness, 2.7~4.7gr weight), using a quenched high speed steel mortar (HRC=60) (Fig. 2). Any influence of upper and lower wall friction on the load-displacement curve of a powder layer had not been observed for thicknesses less than 5 mm.[1] The dimensions of ceramic particles varied from 74 to 149 μm (through #100 but not through #200 JIS standard sieve). SEM photographs of a particle from each ceramic powder are shown in Fig. 3. The surface of mullite was rough and irregularly porous, that of alumina and silica were glassy with sharp edges, and that

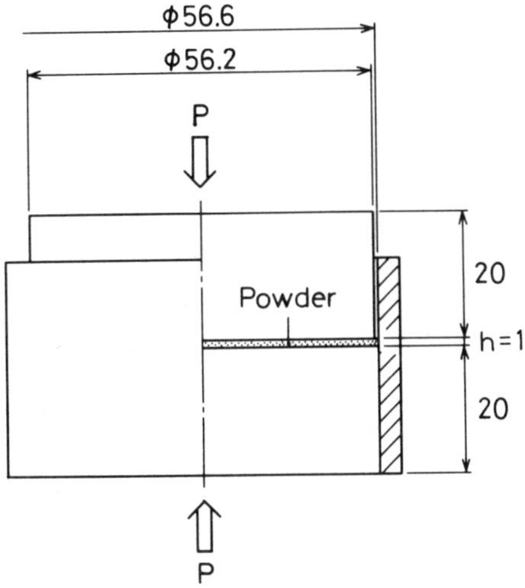

Fig. 2. Mortar for slow compression test.

(a) Alumina (×700)

(b) Mullite (×700)

(c) Silica (×1000)

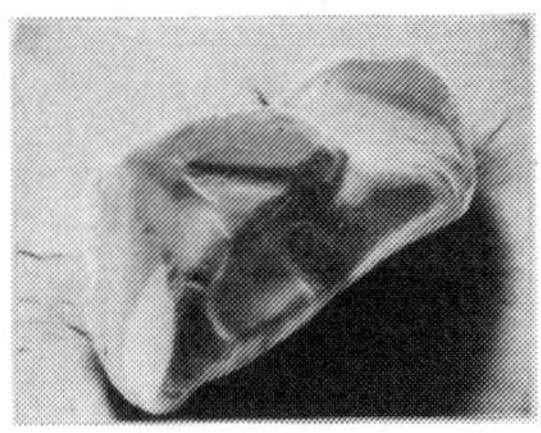

(d) Sillimanite (×700)

Fig. 3. A particle in powder for compression test.

of sillimanite was round and smooth. This surface feature might relate to the crushing behavior under compression.

The load-displacement relation of the mortar was recorded, and the true load-displacement curves of a powder layer were calculated by subtracting the displacement of the mortar itself. After compression, the crushing ratio of amount through #200 to the initial mass was determined.

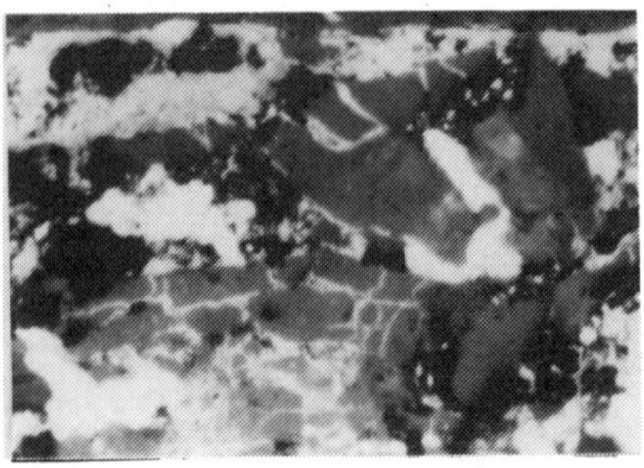

(a) Netty Fracture of mullite.

(b) Crushed mullite under surface oxide layer.

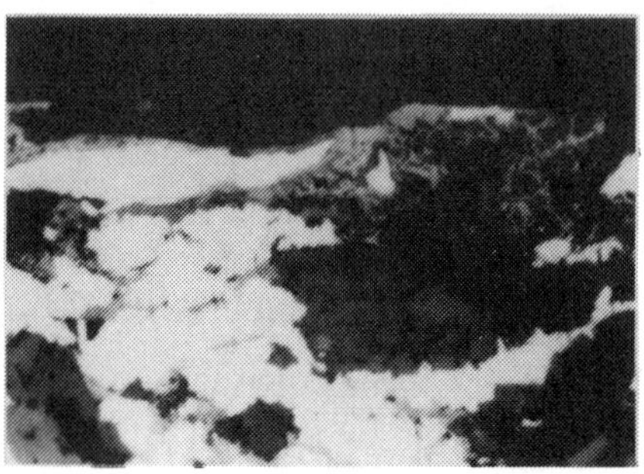

(c) Convex of a fractured mullite

Fig. 4. Sample A after sliding; cross section near surface.

Sliding Test Results

Sample A (Brake Lining For Aircraft)

Virgin Material

It had the distinctive feature of the cermet: large mullite particles comprising about 50% of cross-section and a great deal of netty graphite (Fig. 1a). The surface could not be distinguished from the inner part. The surface of sample A was statically compressed in order to examine the pressure dependence of mullite fracture, but mullite fining was not observed.

After Sliding

The sliding surface was covered by a thick oxide layer because of high temperature and pressure. Thus no crushed mullite was observable. However, observation of the cross-section showed that large mullite grains were fractured into small grains separated by another phase (Fig. 4a), which finally became the oxide at the surface after being dispersed and mixed into surface copper oxide as shown in Fig. 4b. Mullite is harder than metal, so mullite projects to be a convex from the surrounding metal matrix (Fig. 4c). Figure 5a shows a large crack in the mullite at the surface, and Fig. 5b shows some small cracks in a mullite grain. Mullite behaved as an impalpable powder near the sliding surface.

(a) Large cracks in mullite.

(b) Fine cracks in mullite.

Fig. 5. Sample A after sliding; cracks in mullite at the surface.

Fig. 6. Sample B after sliding; cracks in mullite at the surface.

Sample B (Clutch Facing For Automobile)

Virgin Material

It had a composite structure of dispersed mullite and large graphite particles in a Cu-Sn alloy matrix (Fig. 1b).

After Sliding

Fined mullite particles were easily observed at the sliding surface without an oxide layer because of low absorbed energy. Figure 6 shows many fine cracks in a mullite grain.

Sample C

Virgin Material

Sample C is B with silica substituted for mullite. Figure 1c shows two silica grains at the surface.

After Sliding

It was easy to observe the silica particles which were partially fractured near the boundary but mainly worn (Fig. 7).

Fig. 7. Sample C after sliding; weared silica at the surface.

Compression Test Results

Figure 8 shows the gradually increasing cyclic load-displacement curves of powder layer which suggests differences in particle strength and crushing behavior.

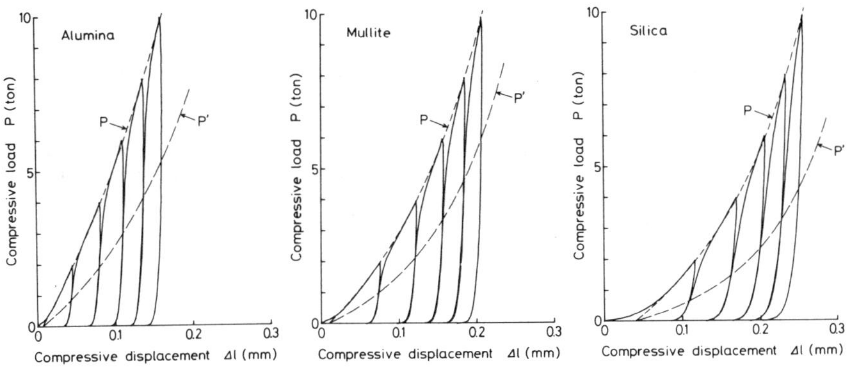

Fig. 8. Load deformation curves.

The relation of load P to displacement Δl had been recognized as being expressed as follows by Matsui[2]:

$$P = \alpha A \phi \tag{1}$$

with the newly defined compressive strain

$$\phi = \Delta l/(\Delta l_{\infty} - \Delta l) \tag{2}$$

where α is the modulus of compressibility of powder layer depending on the crushing characteristics, A is the cross sectional area of plunger, Δl is the compressive displacement of powder layer after eliminating the displacement for the initial rearrangement, and $\Delta l\infty$ is that in the case of $P=\infty$. With the initial vacant space ratio of powder layer $\epsilon_o = \Delta l\infty/(\Delta l\infty + l\infty)$, $l\infty$ being the final thickness of a powder layer at $P=\infty$, equation (1) becomes

$$P = \alpha A/\{\epsilon_o/(1-\epsilon_o) \cdot (l\infty/\Delta l) - 1\} \tag{3}$$

The shortly dashed curves of Eq. (3) shown in Fig. 8, which is the load-displacement diagram for the no-load-history material, show good agreement with the experimental data. A connecting link between the intersections of hysteresis curves in Fig. 8, the area under it means the compaction energy, might be expressed as follows[3]:

$$P' = (1-\epsilon)P = \alpha A(1-\epsilon)\phi \tag{4}$$

with the vacant space ratio $\epsilon = (\Delta l\infty - \Delta l)/(\Delta l\infty - \Delta l + l\infty)$ (5)

Equation (4) showed good agreement for alumina but not for mullite and silica, probably because the fracture mode may be different. The Authors introduced a correction factor β, that is

$$P' = \alpha A(1-\epsilon)\phi \cdot \beta \tag{6}$$

β was 0.8 and 0.6 for mullite and silica. The area between the curves of P and P′ means the true energy for crushing, i.e., the crushing of silica was easier than mullite under slow compression. The data of crushing ratio supported such result. It was approximately proportional to the compressive stress over the testing range (Fig. 9). Solid points show the data for 0.5 mm thick powder layer, almost similar with the 1 mm thickness. The line gradient is a simple measure of crushing easiness, so it is easy to compare the crushing characteristic of a ceramic powder. The crushing ratio of mullite is between silica and alumina.

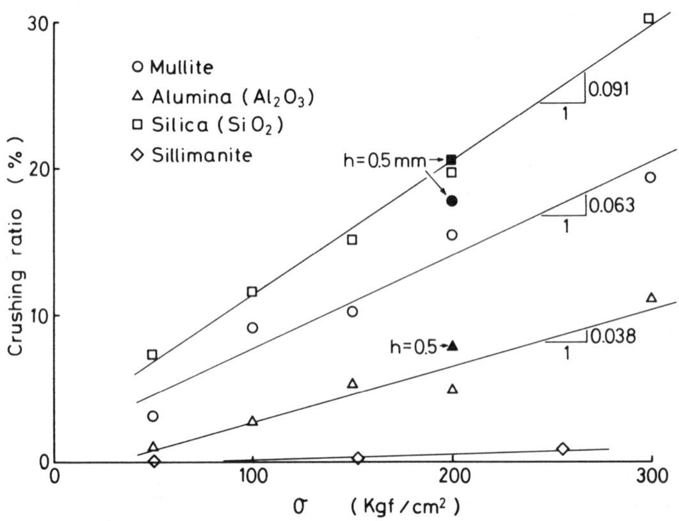

Fig. 9. Relation between Crushing ratio and compressive stress.

Discussion

Hard ceramic particles in the metallic alloy increases the deformation resistance of metal matrix by the formation of true contact point, ploughing, cohesion, and so on. Error of ceramic powder selection, however, leads to increase of wear and damage of mating material. In order to avoid these, finer powder may be used, but the strength of sintering material decreases with addition of fine ceramic powder. From this point of view, mullite has a very desirable crushing character as an additive for brake lining. It does not damage the mating material by self-crushing, disperses into the flowing matrix at true contact points and finally merges with oxide layer formed at sliding surface.

The degree of crushing easiness is an important problem in practice. Moderate crushing behavior corresponding to a sliding condition is desired for the ceramic powder. The crushing characteristics of mullite particles were too complicated to express only by a slow compression test in room temperature. The compression test showed that some crumbling fracture of mullite, which seemed to be very different from the sharp brittle fracture of alumina, might provide the key effect in a severe sliding condition.

Conclusions

The sliding test at a severe sliding condition showed that the crushing characteristics of mullite in metal matrix played an important role on the wear resistance superiority of mullite containing cermet. The slow compression test showed that (1) the crushing rate was proportional to the compressive stress, (2) the load-displacement relation of no-load-history powder layer was expressed by Matsui's Eq. (3) the load at the intersections of hystereses was observed to be 40% and 20% lower for silica and mullite than that indicated by Matsui's Eq. The differences might be due to a different mode of fracture, some form of crumbling rather than clear brittle fracturing.

References

[1]M. Hirota, T. Ohshima, T. Ishihara, and A. Kanazawa, "Compaction Behavior of Powder at High Temperatures," *J. Soci. Mat. Sci. Japan*, Vol. 33, No. 372,1125 (1984).

[2]K. Matsui, "Compression Equation for Powdery Layers being Crushed in a Steel Mortar," *J. Soci. Mat. Sci. Japan*, Vol. 32, No. 356,478 (1983).

[3]K. Matsui, T. Sekiguchi, H. Ikahata, K. Miyata, H. Honma, and M. Tsukagoshi, "Compaction Energy of Powdery Layers and Tensile Splitting Strength of the Compacted Layers," *J. Soci. Mat. Sci. Japan*, Vol. 20, No. 213,762 (1971).

INDEX